S0-BGS-327

PROTOSTARS & PLANETS

PROTOSTARS & PLANETS

STUDIES OF STAR FORMATION AND OF THE ORIGIN OF THE SOLAR SYSTEM

Edited by

TOM GEHRELS

With the assistance of

MILDRED SHAPLEY MATTHEWS

With 51 collaborating authors

THE UNIVERSITY OF ARIZONA PRESS
TUCSON, ARIZONA

COLLABORATING AUTHORS

H. A. Abt, *323*
L. Alaerts, *439*
H. Alfvén, *533*
G. E. Assousa, *368*
C. Bertout, *648*
D. C. Black, *288*
L. Blitz, *341*
P. Bodenheimer, *288*
K. H. Böhm, *632*
W. V. Boynton, *427*
D. E. Brownlee, *134*
A. G. W. Cameron, *453*
E. J. Chaisson, *189*
C. R. Chapman, *599*
D. D. Clayton, *13*
B. Donn, *100*
B. G. Elmegreen, *341*
N. J. Evans II, *153*
G. B. Field, *243*
T. Gehrels, *3*
R. Greenberg, *599*
W. K. Hartmann, *58, 599*
E. Herbst, *88*
W. Herbst, *368*
J. M. Herndon, *502*
C. L. Imhoff, *699*
M. Jura, *165*
R. F. Knacke, *112*
L. V. Kuhi, *708*
C. J. Lada, *341*
R. B. Larson, *43*
E. H. Levy, *516*
R. S. Lewis, *439*
T. Ch. Mouschovias, *209*
H. Reeves, *399*
T. V. Ruzmaikina, *545*
A. E. Rydgren, *690*
V. S. Safronov, *545*
J. M. Scalo, *265*
D. N. Schramm, *384*
J. Silk, *172*
C. P. Sonett, *516*
R. K. Ulrich, *718*
F. J. Vrba, *189*
J. F. Wacker, *599*
J. T. Wasson, *488*
W. D. Watson, *77*
G. Welin, *625*
G. W. Wetherill, *565*
L. L. Wilkening, *502*
H. W. Yorke, *648*

THE UNIVERSITY OF ARIZONA PRESS

Manufactured in the U.S.A.

Library of Congress Cataloging in Publication Data
Main entry under title:

Protostars and planets.

Includes bibliographies and index.
1. Stars—Evolution. 2. Solar system—Origin.
I. Gehrels, Tom, 1925- II. Matthews, Mildred Shapley.
QB806.P77 521'.58 78-10269
ISBN 0-8165-0674-4
ISBN 0-8165-0657-4 (pbk.)

CONTENTS

Part III – CLOUDS AND FRAGMENTATION

Part IV – ASSOCIATIONS AND ISOTOPES

GLOSSARY, ACKNOWLEDGMENTS, AND INDEX

PART I

Introduction and Overviews

INTRODUCTION

TOM GEHRELS
University of Arizona

The purpose of this book is to develop the interface between studies of star formation and those of the origin of the solar system. The production of this source book is described and an overview is given of the contents. The last section mentions missing topics, particularly the detection of planets of other stars.

With this book we hope to stimulate a new discipline (Sec. I). Next, in Sec. II, I shall give some details of how we are developing a new method for the production of source books that are usable as graduate texts. After some remarks regarding the cover and the contents of this book (Sec. III), I will describe the plans for the next volume in Sec. IV.

I. A NEW DISCIPLINE

Until the advent of space programs in the late 1950's, all observations of planets were a part of astronomy because the remote sensing could be done only with astronomical telescopes. *Planetary astronomy* since that time has been broadened into *planetary sciences* to include aspects of meteorology, geology, meteoritics, cosmochemistry and plasma physics. Because of this diversification of planetary sciences on the one hand and exciting discoveries of new objects in astronomy on the other, there is a growing separation between astronomers and planetary scientists. There is an increasing abundance of topics to be studied and fields to be mastered, and the need for a new interdisciplinary science is indicated.

The space programs have brought great change to planetary astronomy, so much so that some colleagues begin to see an end to the remote sensing with groundbased telescopes. The study of the lunar surface, for example, was done solely with groundbased telescopes until the RANGER program, while now relatively few optical observations of the lunar surface are made with earth-based telescopes. By the end of this century there may be a similar situation for Venus, Mars, and Jupiter such that most if not all of the observational data may then be obtained with instruments on spacecraft. Planetary scientists are forced to accept new types of data and a broader view of the solar system as a stellar configuration.

In astronomy too there is a steep increase in data and understanding that require a broadening of fields of study. With the Space Telescope and the Shuttle opportunities the space programs will offer more to stellar, galactic, and extragalactic astronomy.

The ultimate goal of planetary sciences is the understanding of the origin and evolution of the solar system as a component of the universe. That has obviously much to gain from studies of star formation, in which a range of stellar types is considered.

The ultimate goal in astronomy is the understanding of the origin of the universe. The formation of the components of the universe – that is, of the stars and interstellar clouds – is of the greatest interest here. The studies are enriched by interfacing with planetary scientists who have a large amount of detailed information for one system. The origin of that system probably is similar to that of an early stage in the formation of many, if not all, stars.

Cross-fertilization of information and understanding is bound to occur when investigators who are familiar with the stellar and interstellar phases meet with those who study the early phases of solar system formation. Some problems to be discussed are grouped together as follows. What is the association of large molecules with regions of star formation, and what is the origin of pre-planetary matter? What is the extent of the role of plasma and hydromagnetic processes in the interstellar medium, in star formation and in the early evolution of the solar system? What is the chemical composition of interstellar matter, when is the presence of pre-condensed matter important, and was the circumsolar nebula hot or cold? Which clouds and stars form planets? How does the fragmentation of a dense cloud begin, and can we follow its development and its final products in terms of protoplanets? And particularly at this time, there should be an interesting discussion between astronomers and planetary scientists regarding the role of supernovae in the formation of the solar system and of stars in general.

A new discipline will come about naturally as the expansion of science develops into the future. Names considered thus far are "Originary Sciences," "Astrogony" and "Cosmogony."

II. THE MAKING OF A SOURCE BOOK

The need for a source book in cosmogony fits in with our development of a new method for producing technical books. The purpose is to make a "Space Science Series" for the University of Arizona Press. There has been previous experience namely with the books edited by Lynds (1971), Gehrels (1971, 1974, 1976), Burns (1977), while Gehrels (1979) is in preparation.

Our goal is to produce a book that covers a discipline with fair completeness and that is used by workers in the field as a reference volume and by graduate students as a textbook. Because of the variety of topic areas on protostars and protoplanets and the rapid development in these fields it does not seem feasible for a single author to write such a book. A group of authors would offer better expertise in the various topic areas, and also greater objectivity which is so important when describing the frontier of science. One could therefore next consider the traditional compendium-type book having a dozen or so major chapters, with the authors selected by a diversified committee. The writing of such a compendium, however, is slow for lack of stimulation by a meeting of authors and other interested participants. The authors should get together and their conference should become an essential part of the book's production.

The disadvantage, compared with a single author writing a text, is the lack of unity and some repetition among chapters. For an advanced text, however, where one explores the frontier, it is instructive to see the difference in methods and results of various authors. The editors must provide a detailed Glossary, an Index and cross-referencing among chapters in order to unify and complete the coverage, while volunteers who are familiar with certain areas should write introductory overview chapters.

We held a meeting which was widely advertised and financially supported in order to make it possible for participants to come from afar. But this book is not a mere Conference Proceedings. Many conference reports are inferior volumes; I believe they should no longer be published without special care. I will now describe in some detail the special procedures we followed which might be useful for future books of this kind.

The idea of "Protostars and Planets" came up in 1976 during discussion with colleagues at the Lowell Observatory regarding the future of planetary astronomy. A first circular letter was distributed to see if the idea of a book and a meeting would receive sufficient support from authors and referees. An Organizing Committee was formed from those who responded, critically as well as positively, providing a broad representation of expertise (members of the Organizing Committee and the Chairpersons are indicated with an asterisk in the acknowledgments at the end of this book). We debated the size of an organizing committee; while the International Astronomical Union accepts not more than 10 members – who must provide wide international representation – we found it productive to have as many as 20 or 30, depending on the range of topic. Their duties were to check and advise on ~4 circulars sent out before the meeting and particularly on the topics of the chapters: who should cover them and how, length of presentation at the meeting, and the question of review chapter versus contributed paper. In practice we found that certain members, say about 6, became quite involved, even in organizing some of the sessions and urging experts to participate, while again about 6, not necessarily the same 6, were most helpful in the refereeing stages. We had no need for a Local Organizing Committee because the arrangements were expertly made by S.J. Marinus and M.S. Matthews.

About one year before the meeting we mailed a printed circular with the widest possible distribution. One hundred and eighty-three people came, of which 38 were from abroad. The meeting was endorsed by the International Astronomical Union as its Colloquium No. 52. It started – Tuesday evening, January 3, 1978 – with A.B. Weaver, Executive Vice-President of the University of Arizona, introducing the theme. The meeting ended – Saturday noon, January 7 – after synopses on the cloudy, stellar and planetary states were presented by D.D. Clayton, R.B. Larson and W.K. Hartmann (see their chapters following this one). The meeting had the aspect of a working conference, with long sessions in the morning and afternoon, and a discussion session in the evening. After each presentation, about 50% of its length was spent in discussion (this percentage seemed a bit low for theoreticians!). There were no social functions such as organized luncheons, but we did make a pilgrimage to Kitt Peak and the customary hike to Seven Falls.

The principle of volunteering has become important. There is a practical reason, namely that such volunteering promises prompt production of manuscripts. With the first printed announcement a set of review topics was proposed – and it could be augmented by suggestions – so that potential authors could volunteer. In cases of more than one volunteer for the same topic, the Organizing Committee made a choice, or we suggested a "shotgun wedding" where two authors, preferably from different backgrounds, combined their views into a single chapter.

In addition to the review chapters, we invited contributed papers describing mostly personal research; these are published in a Feature Issue for "Protostars and Planets" in a 1978 issue of *The Moon and the Planets.* There also was the opportunity of giving short contributions at the meeting. Any of these could have resulted in a paper in the Feature Issue, or possibly a review chapter in this book. In some cases, especially when the topic could not be incorporated in a paper or chapter, the short contribution was published as a COMMENT.

All of this material has been carefully refereed and edited. It was sent to an expert and to an outsider but interested person with the request to inspect it thoroughly. When merely a remark was returned endorsing publication, we asked an additional referee to review it in order to obtain a detailed reading with suggestions for improvement. We tried to select referees who would be interested in the material and involved in the "Protostars and Planets" project. Regarding refereeing, we had been warned to look out for the opposite case where an author may rely too much on getting some of his work done by the referee; we found no evidence of abuse, but it would have been hard to distinguish from genuine refereeing.

Preliminary manuscripts were due at the meeting; a few new ones were volunteered or solicited afterwards. On the other hand, the refereeing resulted in the rejection of a few manuscripts. The effective date for the material in this book and Feature Issue is March 1978. The contributed papers were forwarded for editing and publication to the office of Z. Kopal, editor of *The Moon and the Planets,* for the Feature Issue on protostars and planets.

A firm restriction was applied against duplicate publication. The material in the Feature Issue had to be ~80% new. The authors agreed that the material in this book and in the Feature Issue would not be taken from or submitted to other publications.

We received at least one criticism that Conference Proceedings have the advantages of speed in delivery and of opportunity in trying out new ideas, and that it is expensive in time and money to apply our special procedures (i.e., open organization and attendance; participation independent of conference attendance; the principle of volunteering; "shotgun weddings;" thorough refereeing and editing; avoidance of duplicate publication, detailed Glossary and Index; cross-referencing among chapters). Our answer is that the few months' delay is worth the quality, while proliferation of scientific literature should be minimized. Not to kill a new idea is, however, an editor's responsibility. The expenditure involved in careful processing is not large, ~20% of the total effort. This is seen in Table I that contains rough estimates of the time and money for the writing of the manuscripts (but not for the original research being reviewed); for the meeting attendance by authors and colleagues (I estimate that ~2/3 of the effort went into this book); for the time spent by referees, authors and editors in the processing; and for the publishing which includes typing and photo-offset.

TABLE I

Estimates of Time and Costs for this Book

Activity	Time in 1000 hr	Cost in $10,000
Writing	4	7
Meeting	4	8
Processing	3	4
Publishing	1	1
TOTAL	12	20

The greatest improvement since our previous book (Gehrels, 1976) is to have this one set on an IBM Composer, each chapter as soon as it was ready, and the book produced by photo-offset. G. McLaughlin produced many of the figures in the proper size for photo-offset. The quality of appearance is as good as that of type setting at the printers. The improvement is in speed (~9 months versus 14 1/2 for *Jupiter*) and price; to judge the latter fairly, however, one must take into account that this time the composing was largely supported by grants and that without such support this book might have cost twice its present price.

We are still learning to make better source books, and suggestions for improvement are welcome.

III. REMARKS ON THE CONTENTS OF THIS BOOK

Barnard's Loop is on the front cover as a symbol of the Great Debate, contained in this book, regarding the relative role of supernovae and density waves in the formation of stars and the solar system. I quote from a personal letter by W. Herbst (see his chapter with G.E. Assousa).

> "Barnard's Loop is not generally recognized as a supernova remnant. It may simply be a product of the stellar winds and radiation pressure of the numerous early-type Orion stars. But it is an important and often overlooked feature of the whole Orion region, which is certainly a place where much star formation is occurring. Since it could well be a supernova remnant, it highlights that aspect of the presentations in this book, but in a subtle way. This subtlety is important to maintain. The "supernova connection" is new and exciting but may be out of fashion in a few years. Hopefully the book will be useful much longer. Therefore it is better to use a feature like Barnard's Loop, which may or may not be a supernova remnant, but is certainly connected in some way to star formation, than to use a well-established supernova remnant, as the Cygnus Loop, which might turn out to be unrelated to star formation."

The subtlety already occurs in the discovery of Barnard's Loop. Part of it was first observed by Herschel as a nebulosity, and the entire loop by W.H. Pickering as a spiral nebula! Lada, Blitz and Elmegreen associate spirals with density waves, while Reeves relates Barnard's Loop to an example for a glorious "bing bang" supernovae creation of the solar system. In extreme contrast, Kobrick and Kaula (1978) revive a version of Jeans' theory, with the origin of the solar system caused by tidal effects in a stellar association.

Barnard (1894 *a,b,* 1903, 1927) acknowledged the previous discoveries but, because of his greater interest, his name became attached to the Loop. Sketches of Barnard's Loop are found in Fig. 1 of Reeves' and Fig. 6 of Field's chapters. The feature needs special photography because of its great size (~10°) and its faintness, but it is said to be a naked eye object. The picture on the cover, extended on the back cover to show the λ Ori nebula, is from an Hα negative kindly supplied by Isobe (1973); it is printed again as the Frontispiece of the Feature Issue (Gehrels, 1978).

I recommend that one starts the reading of this book by looking through the Glossary. Because of the diversity of backgrounds of authors and readers it contains terms that may be fully familiar to some, but essentially needed by others. For readers new to the field, most of the basic concepts are also found in this rather extensive Glossary.

The overviews by Clayton, Larson and Hartmann are written to provide synopses for and cross references among the chapters in this book. The order of chapters is more or less in a time sequence for the formation process, but some chapters discuss different topics and stages and the Index may be useful for cross reference.

The leading question was posed to the participating astronomers: "What clouds and stars form planets?" We concentrated on the formation of solar-type stars, but the chapters are not restricted to these stars. For the planetary scientists the confinement was held to the early stages of formation and evolution of the solar system. We came to the remarkable conclusion that *all* clouds and stars form planet building material, and the book is subdivided by considering our protoplanets and planetesimals as a common circumstellar disk.

I believe the collection of ideas in this book will cause a discontinuity, a leap forward, in understanding the formation of stars and planets. These formations are so complex that parts of many theories will be needed for an integrated description. Much of the essential description and theory is found in this book. A synthesis may lead to a new unified, although complicated, theory of formation of stars and planets.

IV. FUTURE WORK

A new discipline cannot be launched by definition or declaration, but the active participation of such a large number of the best scientists is a sign of success. The intense interchanges in discussions and refereeing among astronomers and planetary scientists seems to confirm the need for a new combined discipline.

The interchange should be continued through local seminars and national gatherings, say once a year, like those of the Penrose Conferences in Geology or the Gordon Conferences, that is, with informal presentations and not necessarily followed by publication. In addition we propose to assemble again as authors of another source book; we state this now in order to solicit support and advance planning because certain valuable scientists could not participate this time for lack of sufficient advance notice. We propose to produce the next volume on cosmogony of stars and planets with a meeting the first week of 1984 in Tucson.

We can see improvements to be made next time. Some of the questions posed in Sec. I still have not been answered. Large areas of investigation are not represented in this book. In the remainder of this section I mention a few examples, realizing that this listing also will be incomplete.

The lifetime work of Bok (see Goldberg, 1971) and of other investigators on gas and dust, have inspired detailed investigations regarding the connection between dust and star formation. Bok (1978) himself, however, has pointed out that in the Magellanic Clouds there appears to be star formation going on without evidence of any dust.

Alfvén would agree with me, and there is evidence on following pages, that the roles of plasmas and magnetohydrodynamics are insufficiently explored.

We have not heard much this time directly from observers at infrared wavelengths. Possibly, they could tell us more regarding the early stages of star formation, before most of the outer nebula has coalesced and cleared. The remarkable system of ϵ Aur, that may have such an outer cloud, is not even mentioned in this book.

We failed to obtain the participation of solar physicists. It seems impossible to accept the conclusion that solar physics and the detailed studies of our nearest star would not have close connections with star formation and the origin of the solar system. At what stage of solar system formation did the sun become a star? Regarding the solar neutrino problem, I quote from a personal letter by Demarque.

> "It may seem strange, but there is no obvious connection between the origin of the solar system and the present solar neutrino problem. Of course, the neutrino emission represents one of the constraints that any detailed theory of solar evolution of the future, which should also include the origin of the solar system, will have to satisfy. Some of the suggestions for a resolution of the neutrino problem involve the influence of rapid internal rotation on the structure of the solar interior and its possible effects on a neutrino flux either through mixing by circulation currents, or through the cooling effects of centrifugal acceleration. But a concern for axial rotation is hardly a new development in considerations of the origin of the solar system. Another perhaps far-fetched possibility has to do with solar models with an accretion core. Such models have been suggested in the context of the solar neutrino discrepancy by Hoyle and Prentice and perhaps others. If this latter picture of solar formation turns out to be correct, it could have a definite effect on our views of the early phases of the solar system."

Finally, there is the topic of detection of planets of other stars. It may well become one of the most important topics, and in the near future at that. The discovery of (the possibility for) life elsewhere in the universe will be one of the most startling experiences in the history of mankind. Such an experience would be comparable to the invention of gunpowder with its effect on the feudal structure of society, to Columbus' discovery of the Americas with its stimulation of the Golden Age in Western Europe, to man finally gaining his third free dimension in the public balloon launches of 1783 in Paris with their influence on the French revolution, or to the impact of the space program upon the questioning in our society in the 1960's. Factually and technically, however, there is not as yet much to report. The results from proper motions have been contradictory and a gain to a precision of $\pm 10^{-2}$ arcsec is needed. Black (1978) believes in a possible improvement to $\pm 10^{-4}$ arcsec with special care using earthbased telescopes, and to $\pm 10^{-5}$ arcsec with telescopes in space. For detection of radial velocities, 12 m sec^{-1} is needed to detect a Jupiter/Sun type of system, and 9 m sec^{-1} for Earth/Sun, while presently a typical precision in astronomy is on the order of 100 m sec^{-1}. The instrument of Serkowski (1977) is close to a precision of 30 m sec^{-1} and still being improved. Black (1978) also describes how the detection of planets about other stars is being developed in direct imaging, with special techniques such as those of interferometry and apodization.

The discovery of planets of other stars would also cause a discontinuity in cosmogony. The goal may then become re-oriented towards the understanding of man's origin which is a topic beyond, but using the results of, the studies of formation of stars and the solar system, reaching towards a global view on the future of mankind.

Acknowledgments. We received essential financial support from the International Astronomical Union, the National Science Foundation, the National Aeronautics and Space Administration, the University of Arizona Foundation, and the University of Arizona Press. I cannot begin to make acknowledgments to individuals as so much is owed to so many and I would not know where to stop. Towards the end of the book we have given a list of members of the Organizing Committee, referees and authors, and others who supported this venture.

REFERENCES

Barnard, E.E. 1894*a*. *Astron. Astrophys.* 13: 811.

____. 1894*b*. *Popular Astronomy* 2: 151.

____. 1903. *Astrophys. J.* 17: 77.

____. 1927. In *A Photographic Atlas of Selected Regions of the Milky Way* (E.B. Frost and M.R. Calvert, eds.). Washington, D.C.: Carnegie Institution of Washington.

Black, D.C. 1978. *Space Sci. Rev.* In press.

Bok, B.J. 1978. *Moon and Planets* (special Protostars and Planets issue). In press.

Burns, J.A., ed. 1977. *Planetary Satellites.* Tucson: Univ. Ariz. Press.

Gehrels, T., ed. 1971. *Physical Studies of Minor Planets.* NASA SP-267, Washington, D.C.: U.S. Government Printing Office.

Univ. Ariz. Press.

Gehrels, T., ed. 1974. *Planets, Stars and Nebulae, Studied With Photopolarimetry.* Tucson: Univ. Ariz. Press.
_____., ed. 1976. *Jupiter.* Tucson: Univ. Ariz. Press. Transl. into Russian; Moscow: Publ. House "Mir".
_____., ed. 1978. *Moon and Planets* Feature Issue "Protostars and Planets".
_____., ed. 1979. *Asteroids and Planets X.* Tucson: Univ. Ariz. Press.
Goldberg, L. 1971. In *Dark Nebulae, Globules and Protostars* (B.T. Lynds, ed.), p. 147. Tucson: Univ. Ariz. Press.
Isobe, S. 1973. In *Interstellar Dust and Related Topics* (J.M. Greenberg and H.C. van de Hulst, eds.), Dordrecht: Reidel.
Kobrick, M., and Kaula, W.M. 1978. *Moon and Planets* (special Protostars and Planets issue). In press.
Lynds, B.T., ed. 1971. *Dark Nebulae, Globules and Protostars.* Tucson: Univ. Ariz. Press.
Serkowski, K. 1977. *Astron. Quarterly* 1: 5.

THE CLOUDY STATE OF INTERSTELLAR MATTER

DONALD D. CLAYTON
Rice University

An introduction to the morphology, dynamics and chemistry of the interstellar medium is presented. The purpose is in part to summarize the kinds of information known about the interstellar medium and in part to highlight those aspects that relate directly to the formation of protostars and planets. The material surveys also the relevance of the contents of this book to the cloudy state that is the ancestor of stars and planetary systems.

By the cloudy state we mean the state of the diffuse interstellar matter, with emphasis on its denser and more opaque regions. An entire book would be required merely to introduce the several types of astronomical observations and the many astrophysical arguments that are relevant to it. From an impressive and explosively growing body of such knowledge we seek those features that seem germain to the formation of protostars and planets. In a strict sense, such a distillation of relevant knowledge is probably not possible; more probably, everything known or knowable about the galaxy and its interstellar medium will play some role in the understanding of the origins of our solar system and others similar to it.

Nonetheless, a beginning can be made by identifying obvious relationships of the interstellar medium to the problem at hand. One relationship is morphological, consisting of a descriptive map of cloudy structures and their evident relationship to regions of new star formation. Another is dynamic, focusing on both the galactic mechanics that could lead to the observed distribution of interstellar matter and also the collapse dynamics that could describe the conversion of clouds into stars. Another evident concern is chemical; we should attempt not only to understand the chemical composition of interstellar structures as a problem in the evolution of the galaxy but also to unveil chemical clues to the processes by which planetary bodies have accumulated. Although it cannot be stated with certainty, it appears probable that we are ourselves a part of the interstellar medium (as opposed to a spin-off from the sun or some other passing star), in which case we may hope to find relatively direct connections between the planetary state and the cloudy state. That hope gives rise to controversy. Some workers argue that traces of the cloudy presolar state can be found in certain (perhaps all) solar system bodies, whereas others present evidence that memory of prior history has been largely eradicated.

Many papers in this book address the cloudy state and its connection to protostars and planets. To put them into a larger perspective is the purpose of this overview of the cloudy state.

MORPHOLOGY

Our host galaxy has a mass of about $1.8 \times 10^{11}\ M_\odot$, distributed similarly to other spiral galaxies in a relatively flat differentially rotating disk with a central mass point of $0.07 \times 10^{11}\ M_\odot$ and a spheroidal distribution of $0.82 \times 10^{11}\ M_\odot$ interior to the sun and about $0.93 \times 10^{11}\ M_\odot$ outside the sun (Schmidt, 1965). Roughly 90% of this mass is contained in $\sim 2 \times 10^{11}$ stars comprising the galaxy. The sun lies $\sim$ 10 kpc from the center, where the volume density of matter, ρ, is $0.15\ M_\odot/\mathrm{pc}^3$ and the surface density is 114 $M_\odot/\mathrm{pc}^2$. A total diameter for the resolvable disk is $\sim$ 30 kpc, much greater than the thickness of the disk. The A type stars, for example, have a roughly gaussian distribution $\exp(-z^2/a^2)$ in height z above and below the midplane, with $a \simeq 150$ pc (Wooley, 1965). This distribution is fairly characteristic of the matter in our galaxy, although some classes of objects (e.g., globular clusters) are much less confined (if at all) to a disk, thereas other classes of objects (*e.g.*, H II regions) are even more strongly confined there. Because this stellar distribution comprises most of the galactic mass, it dominates gravitational effects. The diffuse interstellar medium is, accordingly, distributed in similar fashion. The motions of such a star system are reviewed, for example, by Oort (1965).

The diffuse matter between the stars is called the interstellar medium (ISM). In bulk its chemical composition is believed to be identical to that of young stars forming from it today, although certain details of that picture admit scientific doubts. By total mass the most common ingredients of the ISM are H atoms (in H I regions), H_2 molecules predominantly in cool clouds, He gas, CNO ions, dust particles, and molecules composed primarily, like life, of H, C, N, and O. The spatial distributions of these different phases are not uniform, but instead cluster into differing collective features. Also important in the ISM are certain energy densities capable of effecting ion chemistry: The ultraviolet starlight at energies less than that of Ly α is responsible for maintaining many common species (especially CNO) in ionized form; cosmic radiation penetrates optically opaque regions to produce ionization there; X-radiation permeates much of the ISM; and the universal 3 K background radiation provides a physically significant lower limit to the ambient photon background. It is in the mutual interactions of these several phases that much of the interesting astrophysics of the interstellar medium is to be found. The associated morphology of the galactic disk is clearly fundamental to appreciation of the problem of the formation of new systems.

Neutral Hydrogen

Until rather recently, the 21-cm line radiation resulting from the hyperfine transition of the ground level of atomic hydrogen was thought to provide a representative map of the interstellar medium. With no compelling evidence to the contrary, it was natural to assume that neutral hydrogen, H I, was a constant fraction of the ISM and therefore could trace its total morphology. This belief required serious revisions as a hierarchy of temperature-density phases were discovered in the ISM. In the past decade, a totally different morphology from that of H I has been found to be appropriate to the problem of star formation. Atomic hydrogen is unique in its distribution (Burton, 1976). The 21-cm galactic disk is fully twice as large as that defined by ionized (H II) and molecular (H_2) states of hydrogen, by other molecules (especially CO), by supernova remnants, by pulsars, by γ radiation, and by synchrotron radiation.

The 21-cm hyperfine line of H I is a beautiful astronomical signal because the natural width of the line is negligibly small and the receiver frequency resolution of modern radio telescopes is very accurately determinable. As a result H I emissions are fully resolvable in velocity according to the Doppler shift from the emitting region and are finely resolvable in angle. The overall width of the H I line along the galactic disk is therefore set by the differential rotation of the galaxy. Burton (1976) demonstrates how the observations allow construction of a rotation curve for the interstellar medium with the aid of a differentially rotating model. Simply, one assumes that the rotational velocity depends only on the distance from the galactic center, which yields a radial mass distribution in the galaxy similar to that obtained from much more limited optical observations (Schmidt, 1965). The perturbations to the simplest rotation curve are largely attributed to gravitational torques produced by density fluctuations in the overall galactic mass distribution. These irregularities provide evidence for the validity of density-wave phenomena in our galaxy, although a grand design of H I spiral structure has not been established. The basic procedure is to locate an H I complex with the aid of the rotation model by finding the distance, in the direction of observation, where the differential velocity due to galactic rotation matches the observed Doppler shift. This procedure no doubt has some validity, but it also contains pitfalls and limitations appreciated by the workers in this area. Systematic streaming motions of H I gas produce velocity irregularities amounting to several percent of the rotational velocity, and these irregularities probably dominate density manifestations and cause substantial errors in distance estimates. Study of the kinematic patterns caused by systematic H I motions may nonetheless remain one of the best ways to study the perturbations due to density waves (Lin *et al.*, 1969). Organized 5% variations in the density of underlying stars cannot be measured, but the gravitational perturbation caused by that irregularity can noticeably influence the kinematic patterns of the H I gas. Such organized motions have been detected and will provide ultimately, when the total system is much better understood, the best quantitative evidence of a spiral density wave.

The total column density of H I is usually not measurably directly. Only if the line profile is optically thin at all velocities is the column density $N_{\mathrm{H\ I}}$ (cm^{-2}) measured by the so-called brightness temperature. In that case the volume density over a path length Δr is obtainable by division. Surveys are reviewed in Burton (1976). Many aspects of the projection on the plane of the sky of H I emission observed at low latitudes are approximately reproduced by profiles from a model of a smooth distribution of gas with temperature 120 K and average density, $n_{\mathrm{HI}} = 0.33\ \mathrm{cm}^{-3}$ rotating with a simple differential rotation and a gaussian distribution normal to the plane approximated by $\rho(z) \simeq \rho(0) \exp - (z/0.12\ \mathrm{kpc})^2$. This average density n_{HI} is roughly constant over the galactic disk, quite unlike the total mass density which increases strongly toward the galactic center. The fractional decrease in H I in the inner parts is a characteristic displayed by other external spiral galaxies for which H I surface densities have been measured. Clearly H I is a less significant fraction of the mass toward the galactic center. The basic working hypothesis of a smoothly distributed one-state gas has been adopted by most surveys as a point of reference, even though evidence for a non-uniform distribution of both temperature and density is well established. The detailed distributions suggest a two-phase model, causing Clark (1965) to suggest a "raisin-pudding" model in which cool opaque clouds are immersed in a hot transparent medium. Such a two-phase medium is also suggested by considerations of thermal stability, a subject to which we will return shortly.

The specific parameters of the cold gas are best given by absorption measurements since the absorption coefficient varies inversely with temperature. The average H I cloud could have a column density 3×10^{20} cm^{-2}, an internal velocity dispersion of $1.3\ \mathrm{km\ s}^{-1}$, a temperature of 60 K, and would be 333 pc from the next cloud, according to model interpretations reported by Burton (1976). The clouds should then be about 5 pc in diameter, contain about $10^2\ M_\odot$ and occupy $\sim 1\%$ of the disk volume. The cloud density of H I is roughly the quotient of column density and cloud diameter, amounting to $n_{\mathrm{H\ I}}$ (clouds) = 20 cm^{-3}. By contrast, the neutral intercloud medium has $n_{\mathrm{H\ I}}$ (intercloud) = 0.17 cm^{-3}. However, the mass in the two phases is comparable, being a total of $1.2 \times 10^9\ M_\odot$ of H I in opaque clouds and $1.4 \times 10^9\ M_\odot$ of H I in the intercloud medium. Bear in mind that these are only characteristic properties of those diffuse clouds that are responsible for most of the 21-cm absorption. Wide variations no doubt exist and, more importantly, it is now known that many more massive and dense clouds exist that are not well monitored by 21-cm absorption – the so-called dense molecular clouds.

The large-scale morphology of the neutral hydrogen is interesting, and perhaps unique. The average number density $n_{\mathrm{H\ I}}(R)$ peaks between $8 < R <$ 13 kpc from the galactic center at about 0.3 cm^{-3} (Burton, 1976). Density greater than 0.1 cm^{-3} exists outward to 15 kpc, a galactic extent that proves, surprisingly, to be much larger than indicated by other key signposts of galactic activity. Neutral H I gas is both more diffuse and occupies a much larger volume of the galaxy than does the region where protostars appear to form in abundance.

Molecular Regions

Of the more than forty known interstellar molecules several, especially OH, CO and H_2CO, are abundant enough and widely enough distributed to also provide morphological radio maps. They tell a quite different story from that of H I. The radial distribution of CO centers at ~ 5.8 kpc, and the emission has fallen to half its maximum level by 8 kpc, beyond which it drops sharply. These dimensions are well within the solar orbit near 10 kpc, whereas the H I extends out to 15 kpc. The densest molecular regions do not show enhancement of 21-cm radiation, undoubtedly because hydrogen is primarily H_2 molecules in these regions. A minor astronomical revolution occurred with the growing certainty that it was molecular radiation, not 21-cm radiation, that indicates the dense complexes wherein star formation occurs. By the same token, 21-cm radiation is no complete measure of the mass of interstellar hydrogen because it omits the huge masses of H_2 in the molecular clouds. Unfortunately, H_2 itself has no observable radio transitions that would allow surveys of it. Its existence is however well documented by ultraviolet transitions (Spitzer and Jenkins, 1975) in the solar neighborhood. The abundance of H_2 on a galactic scale, as well as its location, is inferred from that of CO. The ^{13}CO line must be used because ^{12}CO is saturated by optical thickness in the clouds. If one takes normal $^{13}C/^{12}C$ ratios and assumes reasonably that most C is in the CO molecules, a plausible estimate of H_2 abundance follows. Scoville and Solomon (1975) showed in this way a column density of 1 to 5×10^{22} H_2 cm^{-2} toward the galactic center. Conversion to a volume density requires a model of the number and sizes of the cold molecular clouds in which the transitions predominantly arise. Several lines of evidence suggest roughly ~ 0.5 clouds per kpc along lines of sight, each of about 5 pc diameter, containing a mass of $4 \times 10^4\ M_\odot$ with number density $n_{H_2} \approx 10^4\ cm^{-3}$. Such masses are typical of globular clusters, rendering the larger molecular clouds as being among the most massive objects in the galaxy. From the average path density $\langle n_{H_2} \rangle \simeq 2\ cm^{-3}$ one finds, in comparison with 21-cm observations, that about 93% of the hydrogen in the inner part of the galaxy is H_2, as opposed to outside the solar orbit where hydrogen becomes mostly H I. Clearly some dynamic processes conspire in the inner galaxy to maintain the interstellar medium in large molecular clouds.

H II Regions

An H II region is a sphere of ionized hydrogen surrounding very luminous main-sequence O stars. Our galaxy probably contains about 700 such objects. The ionization is caused by the intense ultraviolet flux emerging from the very hot and very luminous stars. Their importance to galactic morphology lies in their providing a signpost of recent star formation. H II regions in our galaxy are almost invariably found in association with large molecular cloud complexes that are usually considerably more massive than the ionized and stellar material in the same region. These bright objects provide the best tracers of spiral arms in other galaxies, as shown for example by the H α photograph in Fig. 1. From our vantage point in the dusty disk of our own galaxy, however, most H II regions are heavily obscured optically. They are detected by radio telescopes tuned to the radio frequency transitions between very high lying Rydberg states of the hydrogen atom. These states are continuously replenished by radiative capture of free electrons by protons, and the numerous radio transitions identify compact sources of not only hydrogen recombination but also helium recombinations. These H II regions cluster most abundantly between $4 < R < 8$ kpc from the galactic center, closely comparable to the distribution of CO brightness (Burton, 1976). Lockman (1976) surveyed the disk in H 166 α and found a radial distribution of ionized hydrogen that again matches that of CO rather than the much larger H I distribution. It is difficult to determine whether this unbiased survey at regular longitude intervals is measuring a superposition of H II regions or a diffuse component of hot gas, but the radial distribution clearly associates with that of intense star formation.

The brightest compact H II regions are on many lines of evidence known to be less than 10^6 years old. Strom *et al.* (1975) present this scenario for their evolution:

1. An O star arrives on the main sequence; dust grains near the star are heated and evaporated so that a small H II region forms near the star. The propagation of the ionization front is controlled by the absorption by the grains of Lyman continuum photons.

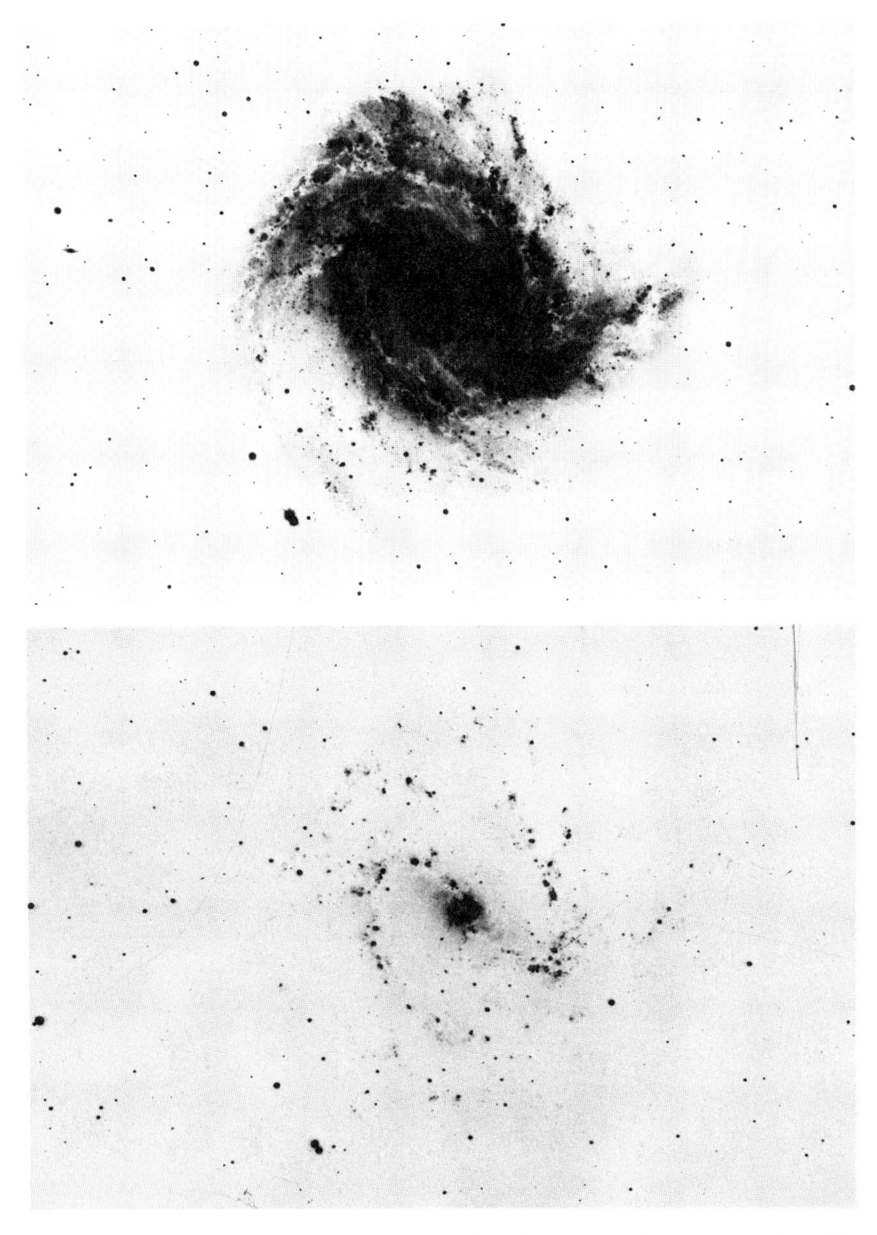

Fig. 1. (Top) A visual (yellow light) photograph of the giant spiral galaxy M 83 (NGC 5236) taken with the Cerro Tololo 4-meter telescope. (Bottom) Another photograph taken through a narrow band H α interference filter. The top photograph recorded the distribution of stars in M 83, while the bottom photograph recorded primarily the H II regions. Note how the H II regions show a clearer spiral structure than apparent from the stellar component. (Photographs of the Cerro Tololo observatory by R. J. Dufour and R. J. Talbot, Jr.)

2. Heating of the grains sufficiently heats and expands the dark cloud material surrounding the star, so that absorption of Lyman continuum photons by the gas dominates over absorption by dust grains.
3. The ionization front propagates preferentially toward regions of low density. Higher density regions act as walls and as the H II region tries to penetrate the walls, material flows from the walls toward the lower density cloud material.

One of the puzzling aspects is the observational evidence that these luminous H II regions form on the periphery of the dark molecular clouds. Mounting evidence indicates that high-mass stars form preferentially on the skin, rather than throughout the volume, of the dark clouds. This suggests, but does not require, that the cloud was compressed by an external pressure which initiated star formation. Theoretical calculations of the formation and evolution of these regions are needed. Some evidence suggests that the overpressure resulting from the H II region attempting to eat its way into the cold walls causes star formation to propagate. An inconclusively developed suggestion (Elmegreen and Lada, 1977; see the chapter of Lada, Blitz and Elmegreen) is that low-mass stars, and perhaps planetary systems like our own, form throughout the cloud volume. Possibly the formation of massive stars is dynamically different from the formation of solar-type stars.

Infrared Sources and Masers

Also of importance are the H_2O and OH masers found near bright infrared sources in dark molecular clouds. In some cases these maser sources are coincident with peaks in the intensity of the radio continuum that can be identified as compact H II regions, whereas in other cases no radio continuum source is seen (Wynn-Williams and Becklin, 1974). Perhaps some masers are driven in dense condensations interacting with H II regions, whereas others may be associated with protostellar collapse. The conclusive identification of protostar formation seems in either case imminent. Such considerations lead directly to discussion of protostellar collapse. For our present purposes we list in Table I the observed properties of four well-studied regions of star formation (Werner, Becklin and Neugebauer, 1977) – W3, Orion Molecular Clouds 1 and 2, and Sagittarius B2 near the galactic center. For each region the tabulated properties refer to a dense central core which is observed at radio wavelengths to be enveloped by a much larger and more massive cloud of lower density molecular gas. Infrared observations confirm the existence of compact luminous infrared sources with little associated visible radiation or ionized material embedded within dense clouds. Some star formation is occurring in the central portions of large clouds. Very interesting studies of the compact infrared source in Sharpless 140 have been presented by Harvey *et al.* (1978) and Blair *et al.* (1978).

Properties of these bright infrared sources are: (1) compact unresolved size; (2) deep absorption or lack of emission near λ=10 μm, believed to be

TABLE I

Properties of Star-Forming Molecular Clouds[a]

Region	Distance Light Year	Size Light Year	Luminosity ($L_\odot$)	T_{gas} (K)	n_{gas} (cm^{-3})	M_{gas} ($M_\odot$)	T_{dust} (K)	M_{dust} ($M_\odot$)
W3	9000	7	1.1×10^6	30	10^4	2×10^3	80	70
Orion 1	1500	2	4×10^5	70	10^5	500	85	10
Orion 2	1500	0.5	10^3	55	10^4	5	40	0.1
Sgr B2	3×10^4	30	10^7	20	5×10^4	5×10^5	35	5×10^3

[a]From Werner, Becklin and Neugebauer (1977), who give operational definitions of each quantity.

due to silicate dust or polysaccharides (Hoyle and Wickramasinghe, 1977) and near 3 μm due to ice or polysaccharides; (3) an otherwise approximately blackbody spectrum in the 2-20 μm wavelength range with color temperature of 300-700 K; (4) weak or non-existent radio continuum emission; and (5) probable association with H_2O and OH masers. The H II regions are often found near the infrared source, making the relationship of these to each other and to the masers a thought-provoking one. Harvey *et al.* (1978) listed three equally plausible explanations for the high ratio of infrared luminosity to radio luminosity: (a) the central luminous object is not a main-sequence star, but is still considerably cooler for its luminosity; (b) the object is still at the earliest phase of the evolution of an H II region, with a large amount of dust still near the star absorbing its ultraviolet photons before these can ionize the gas; or (c) the gas density is so high that the H II region is very compact and radiatively thick at radio wavelengths. Each interpretation supports a very young age for the central object.

Coronal Gas in the Galaxy

The advent of ultraviolet spectroscopy in space has led to the discovery of another new component of the interstellar medium. Rogerson *et al.* (1973) discovered absorption features of five-times-ionized oxygen (O VI) in the ultraviolet spectra of blue stars as recorded by the *Copernicus* satellite. Jenkins (1978) has presented a new survey of the O VI column densities in the direction of 40 stars of type O and B. The inferred spatial densities are variable but of a magnitude suggesting significant amounts of mass and pressure. Cox and Smith (1974) inferred a widely distributed network of coronal-type hot gas exists between other phases of the ISM and is produced and maintained by supernova explosions. If smoothly distributed, a density $n_{\mathrm{O\ VI}} \simeq 10^{-7}$ to 10^{-8} cm^{-3} would be characteristic of the solar neighborhood. By comparison, an average intercloud density with $n_{\mathrm{H\ I}}$ (intercloud) = 0.17 cm^{-3} corresponds with solar abundances to an average intercloud oxygen density, n_{O} (intercloud) $\simeq 10^{-4}$ cm^{-3}. But since the temperature of the O VI regions lies between 10^5 and 10^6 K, and since O VI is not likely to constitute more than 10% of O in this coronal phase, the distributed pressure of the coronal phase may be as much as that of the distributed intercloud medium. In the coronal regions themselves, which may be fingerlike tunnels through the ISM, the pressure must balance roughly the ISM pressure. This suggests that the coronal gas fills about half of the intercloud volume. One dramatic association with supernovae is found in absorption from three stars behind the Vela supernova remnant which show O VI densities an order of magnitude higher than average. The interpretation of this phase remains controversial, however. Castor *et al.* (1975) have argued that O VI reflects a hot circumstellar zone that has resulted from mass loss from the star.

Major Components of the Interstellar Medium

The morphological description of the ISM is summarized in Table II. It lists the five major dynamic phases in order of increasing average temperature (or decreasing density). Characteristic ranges are indicated. The percentages by mass and volume are intended to provide a physical feeling for the

TABLE II

Major Dynamic Phases of the Interstellar Medium

Phase	Indicator	Temperature (K)	Density n(cm^{-3})	Fraction (%) of ISM by mass	Fraction (%) of ISM by volume
Molecular Clouds	CO, OH	10-60	$10^2 - 10^7$	40	½
H I Clouds	21-cm	50-100	1-50	40	5
Intercloud	21-cm	7000	0.2	20	40
H II regions	Hα, O III	10^4	$10^2 - 10^3$	small	small
Coronal gas	O VI	$10^5 - 10^6$	$10^{-3} - 10^{-4}$	0.1	~50

fractions of the total ISM occupied by these phases, and should not be regarded as very accurate numbers, especially for the coronal gas indicated by O VI.

DYNAMICS

It has been in the observational tradition of astronomy to regard the universe and the galaxy within it as being a collection of objects, each living a more or less independent existence. One could so regard the clouds and phases of Table II. It is more of a theoretical impulse to consider how things got to be the way they are – *e.g.*, is a certain object or phase stable, how did it come into existence, and how long will it persist? The science of theoretical astrophysics has come to be dominated by such questions as applied to all the objects of the sky. When such considerations are extended to the entire system, one arrives at a kind of "galactic ecology," wherein the birth, survival, and death of one phase or object is related to those of all others. Material passes from one phase to another, undergoing slow chemical and nuclear transmutations as it does so. The science of galactic evolution is concerned with such a collective description, in which the galaxy becomes more of an "organism" and it is with our own ecology, in the astrophysical sense, that the present book is concerned.

Two-Phase Interstellar Medium

An important step in understanding concerns the naturalness of having the observed division of the ISM into a hot, rare intercloud medium and cold, dense clouds, the so-called two-phase medium. If one contemplates (unrealistically) an isothermal ISM having density fluctuations, one finds that denser regions cool much faster than low-density regions. In a short while (compared to galactic ages) the denser portions have become cooler and they may even attempt to become more dense owing to the associated loss of internal pressure. Such an imagined initial condition is therefore quite far from a steady state. To avoid violent hydrodynamic motions in the ISM one is led instead to consider that its several fluctuations should be in a steady condition and in hydrostatic equilibrium with one another. One then envisions a constant pressure throughout the ISM (as a first approximation) and requires a temperature difference between high- and low-density regions in just such a way that the pressure balances. Since $P = n_T kT$ is an appropriate gas equation of state, it then requires a temperature inversely proportional to the density of free particles, n_T. The next problem is that the number of free particles per gram of matter depends strongly on the local temperature and density. Protons and free electrons dominate very high-T phases, such as in H II regions and coronal gas, giving two free particles per atomic mass unit, whereas H_2 molecules dominate the cold molecular clouds, giving only one-half free particle per atomic mass unit. The heavier species seek even more complicated configurations, ranging from highly ionized atoms such as O VI to large molecules and dust grains. The many complicated chemical questions can be simplified, at least for purposes of computing the number of free particles, by assuming that the matter is in equilibrium with the physical processes of the thermal environment. This state is describable as a steady-state rather than a thermal equilibrium. Steady-state requires only that the rates of creation and destruction of individual particles and of their states of excitation balance, so that their number densities are constant in time. This is commonly called microscopic steady-state or equilibrium.

Even with this description, one still faces the prospect of differential cooling of all regions as they radiate away their thermal content. To achieve a *macroscopic* steady state requires a source of heat input to balance the rate of heat loss. The heating of the interstellar medium is a complicated subject (Dalgarno and McCray, 1972), but its present understanding emphasizes heat input by energetic events (ionizations by cosmic rays or X rays) and cooling by a large number of low-energy events (infrared photons, *etc.*). The most popular picture has been to assume that atoms and molecules have a certain fixed lifetime against violent ionization by interactions with an ubiquitous cosmic ray flux of moderate energy protons. In transparent regions the ionization of atomic carbon by starlight plays a major role, but this diminishes rapidly within opaque clouds. These processes maintain a steady density of free electrons n_e. The cooling, on the other hand, is dominated by fine-structure excitations of atoms, for example, $C\,(^3P_0) + H \rightarrow H + C\,(^3P_{1,2})$ followed by de-excitation, or in colder denser regions by excitation of rotational states of molecules followed by de-excitation, or by collisional heating of dust grains followed by infrared radiation. Whatever the density, one can compute the temperature and degree of excitation consistent with the assumed ionizing device by assuming heat balance.

Figure 2 shows the results of such a calculation of Shu *et al.* (1972; also see Field *et al.*, 1969). The lifetime against cosmic ray ionization was taken to be 0.8×10^{15} sec, and it was assumed to be the major heat input. It was chosen to give a steady-state density n_e of free electrons in rough agreement with the densities implied by pulsar dispersion measures. The gas was taken to have solar composition. The ordinates show both the pressure $P/k = (n + n_e)T$ and n_e as a function of the number density, n, of all atoms (neutral or

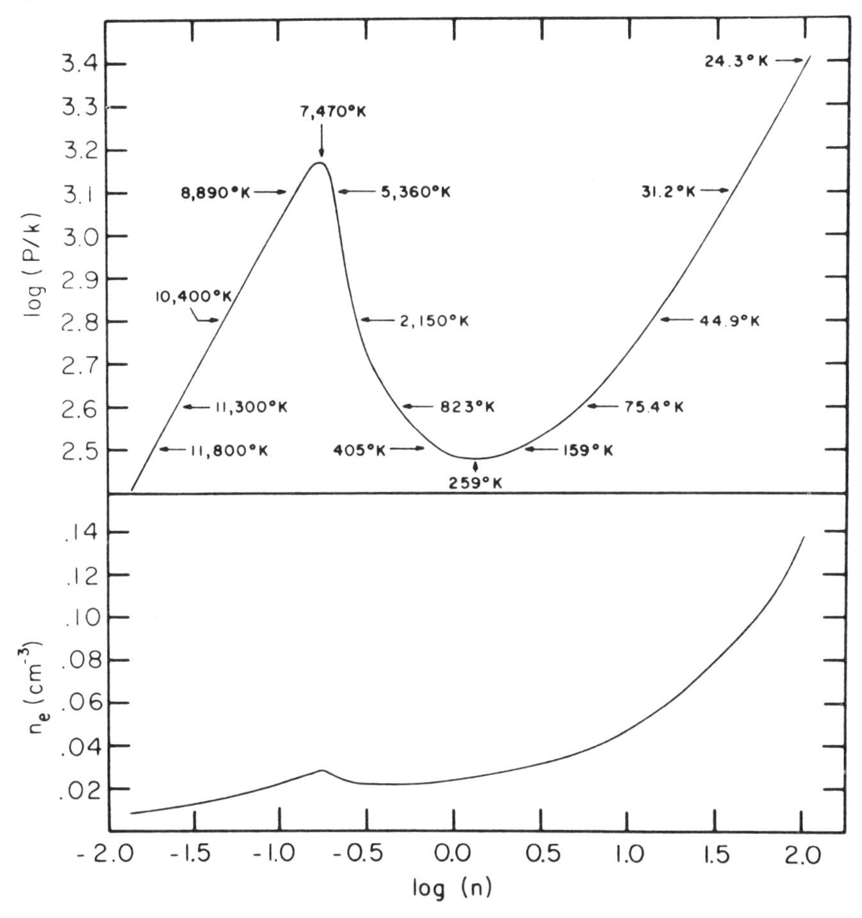

Fig. 2. Steady-state thermal balance in a gaseous medium, showing the pressure (upper) and free-electron density (lower) as a function of the total density of atoms and ions. Steady-state temperature is a balance between constant energy input from cosmic rays and cooling by radiation, and is shown as labels on the pressure curve. The middle range of n_e is unstable because the pressure decreases as the density increases. (After Shu *et al.*, 1972.)

ionized). The most striking result is the variation of the pressure with density, as well as the behavior of the steady-state temperatures which label the pressure curve. These results show that, when cooling processes are included in detail, the pressure of a gas with constant heat input is not monotonically increasing with increasing density, n. Quite to the contrary, at density values just in excess of the first peak the cooling becomes so much more efficient with increasing density that the temperature drops rapidly, so that the total pressure must do so also.

It is clear from Fig. 2 that the steady-state phases in an isobaric medium occur at discreet temperatures and densities. Imagine, for example, a medium with log (P/k) = 3.0, not unlike the interstellar medium. The steady-state solutions in Fig. 2 intersect that isobaric line at three points: (1) near $T = 9000$ K, $n = 0.1$ on the ascending curve; (2) near $T = 5000$ K, $n \simeq 0.3$ on the descending curve; and (3) near $T = 35$ K, $n \simeq 30$ on the second ascending portion. A preliminary conclusion therefore is that hydrostatic forces generated by heating and cooling imbalances would tend to cluster the medium into these three stationary states. Furthermore, the middle solution (2) is not stable. For points in the plane to the right of the descending curve, the medium cools much faster than the assumed rate of heating, so that if a perturbation moves it to higher density, it will continue. An equivalent conclusion follows from the sequence of stationary states described by the descending curve itself. Those steady states have the property that increased density is accompanied by reduced pressure. Such a situation is hydrodynamically unstable, even though an actual sequence of dynamic states would not trace out the descending curve shown unless the dynamic compression were slow enough to allow the heating and cooling rates to balance at each step. Although such balance would not normally be maintained in a dynamic situation, the identification of two stable phases has been achieved – one near 9000 K and one near 35 K. It should take only a few million years (Goldsmith, 1970) for some other distribution of phases to relax to these two. It is widely believed that these arguments allow proper understanding of the tendency of the ISM to separate into hot phases and cool phases. It is hard to ascertain whether galactic evolution mixes the material of these two phases efficiently or if each phase has a relatively long-lived independence of the other.

Three-Phase Interstellar Medium

The simple foregoing picture is probably deficient in its neglect of the very hot coronal-type gas of the ISM. Only in the past few years has evidence for its existence mounted. Most important are the soft X ray background (Burstein *et al.*, 1976) which requires a large volume of low-density high-temperature gas for its production, and the O VI absorption lines that seem to exist everywhere (Jenkins *et al.*, 1978). A wide ranging theoretical study of this problem has been carried out by McKee and Ostriker (1977). They follow the creation of volumes of hot gas by supernova expansion, along the lines suggested by Cox and Smith (1974), but with more detail and with attention to the development of an integrated picture of the ISM. They find that hot supernova remnants must fill a large fraction of the galactic volume. Its average characteristics are $n = 10^{-2.5}$ cm^{-3} and $T = 10^{5.7}$ K, and it fills 70% of the volume, producing X rays and O VI. One then has a three-phase rather than a two-phase interstellar medium in approximate pressure balance. The warm ($\sim$ 8000 K) intercloud medium discussed previously is confined to large shells surrounding the cold clouds, which are being evaporated by the X ray and ultraviolet heating. This analysis seems logical so for the time being we shall think of a three-phase ISM governed, however, by principles similar to those previously described. It is the mechanical heating of supernova shock waves that makes the big difference.

Density-Wave Compression of Clouds

Continuing toward densities higher than n = 30 in Fig. 2, the pressure continues to increase in this model. It therefore gives no indication why the diffuse clouds should become, if indeed they do, dense molecular clouds. Self gravity will not do this, because it is too weak in typical diffuse clouds to surmount the resistence of the internal pressure. The comparison of the two leads to the often-discussed Jeans criterion for self-gravitation. Spitzer (1968) provides a discussion of this and other problems of the interstellar medium. If self-gravitation, rather than external pressure, is to hold a cloud together against the disruptive tendency of its internal pressure, the density of the cloud must be sufficiently high and/or its internal temperature sufficiently low. This Jeans criterion requires the number density to be greater than a critical value $\langle n \rangle_{crit}$ for which gravity and internal pressure just balance. Numerical evaluation yields for a cloud of mass M and temperature T

$$\langle n \rangle_{crit} = 10^3 \frac{T^3}{(M/M_\odot)^2} \text{ cm}^{-3}$$

A typical large diffuse cloud may have $M = 10^3 M_\odot$ and T = 100 K, in which case its observed density $n \simeq 10$ cm^{-3} is only 10^{-2} of $\langle n \rangle_{crit} \simeq 10^3$ cm^{-3}. In the spectrum of diffuse clouds one finds little prospect for gravitational collapse, suggesting that an increase in external pressure is needed to cause a transition from hydrostatic equilibrium to gravitational collapse.

The majority view is that the transition is brought about by the overpressure of a shock wave traversing the interstellar medium. The picture most in favor today was described lucidly in the formulation of Roberts (1969), whose purpose was to develop the density-wave theory of spiral structure. The common existence of spiral arms, especially two dominant arms, is an old and perplexing problem. Re-inspection of Fig. 1 reveals the clear structure of the two major spiral arms of M 83 as recorded by the H α photograph emphasizing the H II regions of new star formation. The appearance of such spiral arms has been interpreted in two different types of theories. The first type assumes that the spiral arms are *material* arms that maintain their integrity for long times, so that the active galactic mass rotates like a curved bar about its center. The second type of theories, which is now the only one seriously maintained, is that the spiral arms represent persistent *patterns* of especially observable matter (bright young stars and dust lanes) within a relatively uniform disk of visually unimpressive matter. Figure 1 confirms that the underlying disk of stars is indeed relatively uniform, and that it is the new star formation which defines the observed spiral arms. Theories of the first type were unable to solve the problem of differential rotation about the center of mass, which would quickly wind the arms into a tight spiral rather than allow them quasi-rigid rotation. Theories of the second type assume a relatively uniform disk with increasing orbital period at increasing distance from the galactic centers. In these latter theories, the spiral arm persists as a wave pattern despite the differential rotation of matter through that spiral pattern. In a simplistic sense this can be compared to the standing wave maintained as water flows past a fixed disturbance in a rapid stream.

The way in which the H II regions align themselves like beads on a curved string emphasizes a major requirement of the process for forming massive stars – it must be very well synchronized on a galactic scale. These massive H II regions are less than 10^7 years old, showing that some mechanism orchestrates massive-star formation over distances of more than 10^4 pc. This orchestrating mechanism is thought to be a standing shock-wave enhancement of the density of matter that persists while the matter is flowing through it. These regions of new star formation (H II regions) occupy an even narrower band than does the density enhancement that is assumed to cause them. In the linear theory of small perturbations, the density enhancement extends over a broad region and could not be expected to provide a sufficiently rapid triggering mechanism to produce the narrow spiral strips like that in Fig. 1. In radio studies of our galaxy, Westerhout (1968) found that the young stellar associations and H II regions actually lie along the inner sides of the observed gaseous spiral arms. Roberts (1969) added the necessary ingredients to the Lin-Shu density wave theory to account for these basic facts. As the gas in orbital motion approaches the spiral pattern it is supersonic with respect to it. In the leading edge of the pattern it passes through a standing shock wave, wherein matter is compressed by almost a factor of 10. This compression, which occurs along a galactic wave front, causes some of the more massive clouds to collapse, leading to fragmentation and an association of new stars. In a typical calculation (Fig. 3 of Roberts, 1969) gas enters the standing shock at a supersonic velocity of 31 km sec^{-1} and leaves the shock at a subsonic velocity of 3.2 km sec^{-1}. The density contrast across that shock is a factor of 9.6. The spiral pattern itself rotates, but at a speed considerably lower than the rotational speed of the orbiting gas. And the non-axisymmetric gravitational potential, generated in a self-consistent way by the density enhancement of the spiral arms, is only about 5% of the gravitational potential of the relatively uniform disk of old stars. Finally, one notes that it is massive clouds rather than individual stars that are triggered into collapse; since $\langle n \rangle_{crit} \propto M^{-2}$, one expects the largest masses to collapse first, leading to fragmentation and associations of stars rather than to individual stars.

Shu *et al.* (1972) considered the effects of shocks on clouds with considerably more detail. They began with clouds in hydrostatic equilibrium with the interstellar medium rather than with isolated clouds. The more realistic boundary condition (external pressure) greatly modifies the Jeans criterion. They showed that a factor 5 increase in external pressure reduces the critical mass of a gravitationally bound cloud by a factor 5^2. Figure 2 shows how co-existence of the 8000 K intercloud medium with cold clouds requires the ISM pressure to be between the limits $\log (P_{min}/k) = 2.48$ and $\log (P_{max}/k) = 3.18$. For pressures lower than P_{min} no cold cloud exists (it would evaporate) and for pressure higher than P_{max} the intercloud medium is compressed into the cold-cloud state. Shu *et al.* (1972) used Fig. 2 as an equation of state for cold clouds and constructed polytropic models of cold clouds in hydrostatic equilibrium with an external pressure. They found a critical mass M_{crit} above which the clouds would be unstable to gravitational collapse; specifically they found $M_{crit} = 3000\ M_\odot$ for $P = P_{min}$, $n = 1$, $T = 260$ K and $M_{crit} = 120\ M_\odot$ for $P = P_{max}$, $n = 30$, $T = 30$ K, where the density and temperature are those near the surface of the cold cloud as determined from Fig. 2 for that steady-state intercloud pressure P. The average density within the cloud is only about a factor of two greater than that at the surface of their polytropic models. Even so, the masses of marginally stable clouds are considerably less than would have been inferred from the Jeans criterion. Interpreted literally, the results of Shu *et al.* imply that no clouds in excess of $3000\ M_\odot$ exist anywhere near $R = 10$ kpc.

When the orbiting ISM enters the standing shock wave, the compression will convert the intercloud phase into cold clouds, as would happen if P exceeds P_{max} in Fig. 2. Such a phase transition reduces P, so that one may expect $P \simeq P_{max}$ to characterize the intercloud-cloud equilibrium after the shock. In that case, clouds greater than $M_{crit} = 120\ M_\odot$ would collapse when they pass through the spiral-wave shock. Since the shock wave is self-consistent with the spiral arm perturbation causing it, this seems a natural explanation for the obvious occurrence of new H II regions on the inner or trailing edges of the spiral arms.

It seems worthy of emphasis that these considerations, coupled with the large-scale linearity of the H II regions in Fig. 1, offer convincing proof that galactic density waves trigger star formation, at least of massive stars (see the chapter by Lada, Blitz and Elmegreen). One of the leading contemporary issues is whether our solar system, the only one we can study in detail, also had such an origin. On the basis of known isotopic anomalies, Cameron and Truran (1977) have suggested that a much less well-documented process, the pressure wave from a neighboring supernova, has caused the collapse leading to the solar system. Schramm discusses this process in this book, along with observational evidence gathered by W. Herbst and Assousa (1977; see their chapter) that supernova-triggered star formation seems to have occurred elsewhere. It will be of continuing importance to study whether low-mass stars are commonly formed in this way, in which case such an origin for the solar system would not have to be considered exceptional.

Collapsing Clouds?

One difficulty wherein current understanding is not yet adequate is the relationship of the massive molecular clouds to the diffuse clouds. Somehow the less massive diffuse clouds perhaps agglomerate into large clouds, but the connection is not yet clear. The point is that star formation in our own galaxy is observed to be occurring in these massive molecular clouds rather than in small clouds. The Jeans criterion for collapse suggests that these massive dense clouds are in fact gravitationally bound. Field discusses associated problems in this book. If the H_2 clouds were in fact collapsing, the rate of star formation in the galaxy would greatly exceed its observed rate. Some physical causes give these clouds long life against collapse, but the age of a molecular cloud as an independent entity is much disputed. The other important dynamic conditions would appear to be turbulence, magnetic fields, and rotation. These are discussed in this book by Field, Silk, Mouschovias, Chaisson and Vrba, and others. Thermal pressure and turbulence seem unable to halt collapse, and a magnetic field cannot do so along the direction of the field. It would appear that rotation is left as the major dynamic inhibition, and it is also a major factor in the collapse calculations described in this book. Mouschovias describes the importance of magnetic braking in slowing down the rotation of a cloud, stressing that a cloud maintains a constant *angular velocity* during compression rather than a constant angular momentum. He argues, following Mestel, that the cloud first flattens along the field lines and then fragments, eventually expelling the magnetic field. The roles of magnetic fields can be seen to be important, but full hydrodynamic models are very difficult to construct.

It is commonly assumed that when a massive cloud does collapse it will naturally fragment into smaller gravitationally bound units. Since the Jeans critical mass for self-gravitation is proportional to $\rho^{-\frac{1}{2}}$, increasing density makes increasingly smaller masses bound. Fluctuations that were previously confined by external pressure become confined by their own self-attraction when the density is increased. It is commonly felt that this process is the cause of the simultaneous births of stars in an association. The complexity of this problem has made it resistive to definitive solution, however. From a computational standpoint one has a three-dimensional hydrodynamic problem dependent upon initial conditions. In this book Silk reviews the attempts that have been made to clarify it – attempts that have been rather qualitative owing to the numerical complexity.

Chaisson and Vrba review the evidence of magnetic fields in the dark clouds. The assumption that linear polarization is caused by magnetic field alignment of grains continues to indicate very large fields of almost 10^{-3}G in the clouds. However, the radio measurements of the Zeeman splitting in the OH molecule do not reveal such strong fields in the interior of the clouds. These fundamental puzzles remain to be resolved, at which time new understanding of the dynamical conditions of the dark clouds will follow. Magnetic instabilities wherein matter sinks along field lines owing to its weight (Parker, 1966) seem also to play a role in star formation. Appenzeller, (1971, 1974) presents evidence for such sagging pockets of matter around some young stars. One objective for the star formation problem must be to ascertain the frequency of this mode of star formation.

Each different physical mechanism for collapsing the interstellar medium leads to its own dynamic models. Any theoretical study of star formation must begin by defining the basic lines of the calculation. To define the dynamically relevant initial conditions of such calculations is a major objective of studies of the ISM – at least among those scientists concerned with protostars. One must know the thermal environment, the mass distribution, the chemical constitution, the states of turbulence and rotation, and the magnetic field strength and morphology in order to estimate the response of a cloud to any of several interesting physical stimuli. No wonder that the formation of stars and planets remains a difficult and controversial topic.

Discussions of equations of state for the interstellar medium stumble a bit over the question of heating and ionization. Cosmic rays are commonly regarded as the most effective mechanism, but X rays and impact ionization behind shock waves also have merited attention. These heating mechanisms are mostly treated as given. What attention their origins has received has for the most part suggested supernova explosions as a prime cause. The supernova origin of cosmic rays is an old debate, still unresolved. A similar cause for the X ray emitting coronal gas and for strong interstellar shock waves is more recent (e.g., Cox and Smith, 1974). But even these mechanisms cannot be taken for granted within massive molecular clouds. Their self-shielding is substantial to all external energizers. Perhaps only the mechanical conversion of internal mechanical energy remains in the centers of these massive clouds. It is a subject worthy of more attention. The problem should ultimately be related to the resistance of molecular clouds to gravitational collapse.

CHEMISTRY

Astronomy has come a long way from the day when the interstellar medium was thought to be composed of single atoms, or singly charged ions in those cases where the electron binding is weak enough to allow ionization by interstellar starlight. Well over forty interstellar molecules are now observed experimentally (see Table I of E. Herbst's chapter in this book), containing up to eleven atoms in known cases; and the stage of interstellar dust has grown into a fantasia of astrophysical theory. Chemistry has become one of the most vital and relevant areas of interstellar research, especially where the problems of our own origins are concerned. The chemical composition of our solar system is important in its own right and also serves as a point of references to compositions found elsewhere (or coming from elsewhere!). Cameron (1973*a*) has provided a critical evaluation of solar abundances.

One point must be stated at once. Even the abundances of the interstellar medium cannot for purposes of chemistry be regarded as given, but rather viewed as a changeable and evolving quantity reflecting the interchange of mass among phases. Burbidge *et al.* (1957) have described at length an outline of nuclear evolution of the ISM through nucleosynthesis in stellar fusion. Arnett and D. Clayton (1970) focused that picture on the explosive processing that accompanies violent ejection from supernovae. The ISM is a reservoir that receives newly synthesized matter, mixes it to varying degrees with the average ISM, and provides the chemical mixture for new protostars and planets. In the simplest picture the reservoir at a given galactocentric distance can be regarded as a homogeneous function of time, leading to investigations of such quantities as the Fe/H ratio, or the N/O ratio, as examples, versus galactic age. Talbot and Arnett (1971, 1973) constructed a mathematical formulation of this homogeneous approximation, using astrophysical advances to parameterize both the rates of star formation and the rates of nucleosynthesis. The astrophysical viewpoint has in this way switched from that of the given composition to that of the evolution of composition. This evolution concerns not only the total abundances but also their distribution among physical phases (dense clouds, coronal gas, etc.) and among chemical phases (molecules, dust, etc.). One can imagine a dark sea of half-glimpsed forms, slowly permuting their identities, a vastly different world from that of the bright and twinkling stars. It is the evolution of the galactic organism.

Radioactive Chronologies

One of the proofs that nucleosynthesis has been distributed over galactic times derives from the persistence of natural radioactivity, which would long ago have vanished if the atomic nuclei were arbitrarily old. Conversely, their present ratios indicate that they could not have been created just before the birth of the solar system. The compromise seems to be to distribute the production, probably in galactic supernova explosions, over the galactic age. The most useful natural radioactivities are $^{235,\,238}U$, ^{232}Th, and ^{187}Re. Their abundances and those of their daughters, $^{207,206}Pb$, ^{208}Pb, and ^{187}Os, indicate the following:

1. Much U nucleosynthesis happened within 2 Gyr before the solar formation in order to account for the high abundance of short-lived ^{235}U (about 1/3 of ^{238}U) at that time (Fowler and Hoyle, 1960);
2. Much U nucleosynthesis happened as long ago as 5 Gyr before solar formation or it would not be possible to simultaneously explain the large $^{232}Th/^{238}U$ abundance ratio (Fowler and Hoyle, 1960).
3. Much nucleosynthesis occurred earlier than 5 Gyr before solar formation or it would not be possible to understand the large cosmoradiogenic contributions to the abundances of the daughter isotopes (D. Clayton, 1964), especially of ^{187}Os (Browne and Berman, 1976). One concludes that at the time the solar nebula fragmented from the mother molecular cloud, the gas and dust contained nuclei whose age distribution ranged from roughly 10^8 yr to 10^{10} yr. A recent revaluation is provided by Hainebach and Schramm (1977).

An equally exciting story is told by shorter-lived radioactivities that no longer exist in easily detectable amounts, but whose prior existence can be documented by abundance excesses in their daughter nuclei. These are the so-called *extinct radioactivities*, and their relevance to the early days of the solar system is very great. One supposes that these radioactive nuclei also existed in the gas and dust in the molecular cloud (presumably) from which the solar system developed, and that their daughter abundances measure the radioactivity remaining at the time when today's samples (largely meteorites) formed. The most useful of these are the ^{129}I and ^{244}Pu detected in Reynolds' laboratory (e.g., Reynolds, 1967) as excess Xe isotopes, ^{26}Al detected as excess ^{26}Mg (Lee *et al.*, 1977), and ^{146}Sm detected as ^{142}Nd excess (Scheinin *et al.*, 1976). They are used to establish a relative chronology for the formation of meteorites in the solar system. Reviews related to his topic can be found in this book by Reeves, and by Schramm. Meteorite formation is spread out over $\sim 20 \times 10^6$ yr time span according to these classical interpretations of the data.

The detectable extinct radioactivities are sensitive not only to the relative formation times of solar system objects but also to the history of the production of fresh nucleosynthesis in the few hundred-million year interval just proceeding the formation of these objects. One must think of the presolar molecular cloud as containing the fallout of previous galactic nuclear explosions. Contamination levels within the cloud reflect that history. No doubt the most important puzzle is that excess ^{129}Xe concentrations from ^{129}I decay suggest that the last nucleosynthesis occurred about 10^8 yr before formation of meteorites, whereas ^{26}Mg concentrations suggest fresh contamination within 10^6 yr of that formation epoch. This discrepancy is not easily resolved, but Reeves and Schramm discuss the following approach. Fresh nuclei in the presolar cloud are created every 10^8 yr when the solar neighborhood passes through the standing spiral shock wave, resulting in massive star formation leading to supernovae after about 10^7 yr; then the t = 0 passage results in solar collapse and the injection of ^{26}Al from a nearby supernova, whereas the prior ($t \simeq - 10^8$ yr) passage had produced the presolar ^{129}I concentration ($t_{\frac{1}{2}} = 17 \times 10^6$ yr). The general treatment of radioactive chronology under such periodic production was presented by Trivedi (1977), following an idea by Reeves. The supernova associated with the formation of the solar system could then be regarded either as a trigger for solar collapse (Cameron and Truran, 1977) or as an impregnation of an early solar nebula or disk by a chance nearby supernova. Certain cogency is lost by the latter alternative, but it may yet prove the more realistic of the two. Reeves argues that up to 10 supernovae in a large association may contaminate its low-mass nebulae.

The differences in meteorite ages on this picture provide a severe constraint to the length of the epoch of meteorite accumulation. Some alternate interpretations have been set forth in order to relax that constraint. Cameron (1973*b*) suggested an incompletely mixed presolar cloud, an approach that Reeves develops to an advanced stage in this book. D. Clayton (1975*a*,1977*b*) has called attention to an interstellar chemistry interpretation of the data. He appeals to mechanical fractionation of dust from gas in the presolar nebula, accompanied by the expectation that radioactive parent/daughter abundance ratios will differ in dust and gas. In particular, extinct radioactivities have been interpreted by him as having decayed in interstellar dust particles rather than in the meteorite itself. As a test of this mechanism, ^{22}Na, with $t_{\frac{1}{2}} = 2.6$ yr, was interpreted as an extinct radioactivity (D. Clayton, 1975*b*). And excess ^{26}Mg would be expected from extinct ^{26}Al in the interstellar aluminum oxides. Although these new ideas usefully challenge the conventional interpretations, they are largely regarded as being incorrect. Most workers feel that chemical evidence suggests convincingly enough that the decays in question actually occurred *in situ* in the meteorite. Wherever the decays occurred, very significant information about the early solar system and its relation to the constitution of the parent cloud is carried by these extinct radioactivities.

The problem of age differences inferred from radioactive chronologies has been complicated by the discovery of isotopic anomalies in other elements. These discoveries have made it necessary to argue independently that inferred age differences are real. The answer will vastly influence our views of the origin of the planetary system, because the relative ages of the primitive meteorites will probably remain our best indicators of the time scale of accumulation processes in the early solar cloud. Although the conventional chronological interpretations remain the majority view, they do have uncomfortable inconsistencies; namely, age differences from initial $^{87}Sr/^{86}Sr$ are about seven times greater than age differences from $^{129}Xe/^{127}I$ (Gray *et al.*, 1973), whereas age differences of inclusions of C3 meteorites from $^{26}Mg/^{27}Al$ are about ten times less than the $^{129}Xe/^{127}I$ time span for chondrites. Because different objects with different histories are involved, these are not concrete contradictions, but they do force a contemporary re-evaluation of the dynamic plausibility of the accumulation models.

Isotopic Anomalies

For many years evidence indicated that the isotopic composition of the elements was identical in different solar system samples. Reynolds (1967) judged this the most important fact about isotopes. Exceptions were restricted to effects of cosmic-ray irradiation, radioactive decay, and fractionation patterns resulting from the differing masses of the isotopes of a given element. That situation has now changed dramatically. The key experiments were measurements of the isotopic composition of oxygen (R. Clayton *et al.*, 1973, 1977). Terrestrial variations had long been utilized for geochemical studies, but the effect employed was that of mass fractionation. When measured ratios $^{17}O/^{16}O$ and $^{18}O/^{16}O$ were compared with standard ocean water, the terrestrial variation $\delta\ ^{18}O$ was always twice the variation $\delta\ ^{17}O$ simply because ^{18}O is twice as many mass units from ^{16}O as is ^{17}O. One describes this by saying that the fractionation line has a slope of 1/2. Measurements on extraterrestrial samples have, however, revealed at least two classes of anomalies (and perhaps more):

1. Carbonaceous-chondrite minerals plot along a line of slope 1, suggesting dilution of normal oxygen by extra ^{16}O;
2. Fractionation lines in less primitive objects, although having a slope of 1/2, do not lie on top of the earth's fractionation line. The mean O of L chondrites, for example, is not chemically derivable from the mean O of Earth, whereas that of the moon is.

The first type of O anomaly was interpreted by its discoverers as a mixing line between unfractionated normal oxygen and a microscopic mineral carrier of ^{16}O. They concluded that the carrier minerals were presolar – that they had condensed in an ^{16}O-rich region near a star which had synthesized new ^{16}O and that these minerals later escaped vaporization in the early solar system. Both conclusions raised wholly new issues. A similar interpretation of Ne-E, a very ^{22}Ne-rich gas, had been made earlier by Black (1972), and his interpretation became generally accepted following the relatively unambiguous oxygen discovery. D. Clayton (1975*a,b*) presented a formation scenario for the carrier grains, namely that they naturally condensed within the interiors of supernovae as they were cooled by adiabatic expansion. The ^{16}O-rich grains are to be expected there because the major refractory synthesized elements (Mg, Al, Si, Ca), which are bathed in pure ^{16}O gas, cannot escape in totally gaseous form. In a similar manner the Ne-E carriers can be formed by chemically condensing 2.6 yr ^{22}Na in minerals during the expansion of the interior.

Other scenarios of the origin of the carrier grains involve growth in a dense nebula around a highly evolved star that has lost freshly nucleosynthesized matter, perhaps as a supernova or perhaps by something less dramatic. These circumstellar regions could even be in the same dense molecular cloud in which the sun formed. A convincing experimental proof of the site of origin has not yet been possible, so theoretical arguments have been relied upon. These experiments and theoretical arguments have opened a most exciting new frontier in the studies of solar system formation, suggesting unsuspected ties between the formation of the solar system, the chemical state of the interstellar medium, and the nucleosynthesis of the chemical elements. It may not be too much to predict that the isotopic anomalies will even provide the long sought experimental proof that nucleosynthesis occurs in exploding stars. D. Clayton (1973) had previously argued that the confirmation would first come from the γ ray astronomy of fresh radioactivity.

The issues raised by isotopic anomalies can perhaps be clarified at this point by describing the models used to account for them. These models are of basically three types.

1. A supernova explosion associated with the formation of the solar system injected gas of special composition into the solar nebula, where minerals condensed trapping these anomalies before the injecta could be homogeneously mixed (Cameron and Truran, 1977). The isotopic merits and liabilities of this model are discussed by D. Clayton (1977*a*) and by Schramm and by Reeves in this book. The model needs more dynamic study.
2. The solar nebula was initially uniformly mixed, but isotopic differences between dust and gas and between different types of dust generate isotopic anomalies because of various fractionations of these phases in the accumulation processes. The model was initially suggested by the Ne-E and O anomalies described above. No special event is called for. This theory has been developed by D. Clayton, prompted by the discovery that supernova condensates are expected to profusely populate the interstellar medium with dust of anomalous chemical and isotopic composition. That aspect of the theory has been strengthened by detailed computations of the equilibrium condensation sequences anticipated in the supernova expansion (Lattimer *et al.*, 1978). This approach is amplified by D. Clayton (1978). It needs more study of the processes that fractionate the chemical phases initially present in the solar system.
3. Different macroscopic solar-system samples ("marbles") formed in heterogeneous parts of the parent molecular cloud and drifted, like small

distant comets, from the gravitational influence of one star upon another in a stellar association. Isotopic inhomogeneities in the parent cloud were established by incomplete mixing of ten or so supernovae within the association. These drifting samples were captured by the otherwise homogeneous solar nebula, which fragmented also from the parent cloud, and were ultimately incorporated into meteorites. Reeves advances this picture in this book.

Other hybrids of these basic approaches are possible. Schramm describes in his chapter the advantages of injecting special dust from the supernova trigger, as opposed to gaseous injection from it. Approach (3) could envision its own variation of approach (2) by utilizing young dust from the ten or so supernova explosions, whereas approach (2) contemplated a collection of supernova condensates (SUNOCONS) of all ages from the entire previous history of galactic nucleosynthesis. Unquestionably the key point is the excess ^{26}Mg in Al-rich minerals (Wasserburg, personal communication, 1978). The chemistry of the samples seems to indicate that the ^{26}Al was live when that sample formed, so that if it was formed in the solar system, theories (1) or (3) are needed. D. Clayton (1977*a,b*) devised a chemical-kinetic interpretation by which already extinct interstellar ^{26}Al, or a specific portion thereof, could maintain its observed correlation with Al abundance, enabling theory (2) to account for all anomalies. Perhaps no single cause will explain all anomalies, but one does have the scientifically appropriate situation wherein advocates of each of these three approaches are attempting to utilize only that approach in the solution of all isotopic anomalies. As the weaknesses of each approach are exposed by continuing criticism, it will become clearer what mixtures of theories will best do. This is a very exciting issue, because these anomalies are perhaps the clearest remaining fingerprints of our presolar condition. Whatever the correct explanation, new diagnostic information concerning the way in which cloud matter has accumulated into planets will surely be established; but the correct path may be a long and thorny one. Podosek (1978) usefully reviews the field of isotopic anomalies, which now extends to more than fifteen elements.

A central chemical issue is that of vaporization of the presolar dust. Presolar-grain explanations of isotopic anomalies clearly require that dust survives in part until it has been accumulated into larger objects. The Ne-E and ^{16}O anomalies came as a shock to the cosmochemical community partly because it had been commonly supposed that the solar nebula began as a hot gas. This belief was based on chemical evidence of the type reviewed by Wasson in this book. The Ca Al-rich inclusions in C3 meteorites were even regarded as the first condensates as the nebula cooled – the pristine examples of a condensation sequence that fractionated elements according to their chemical volatility rather than to some other chemical property. Grossman (1972) describes the chemical thermodynamics of that scenario. Now that evidence exists that *some* dust survived, the compromise view has become that only a small fraction (~ 5%) was not vaporized in a nebula which was sufficiently hot to vaporize the rest of the solids. Wasson defends this position in this book. It has, however, been widely overlooked that the interstellar medium itself should be characterized by similar fractionation, so that the chemical possibility of a slightly warm (e.g., 300-600 K) solar nebula bearing prefractionated matter requires investigation. That possibility has been advocated by D. Clayton (1978, and references therein). Fig. 3 illustrates how, in that view, the fraction f_A of condensed element A that had

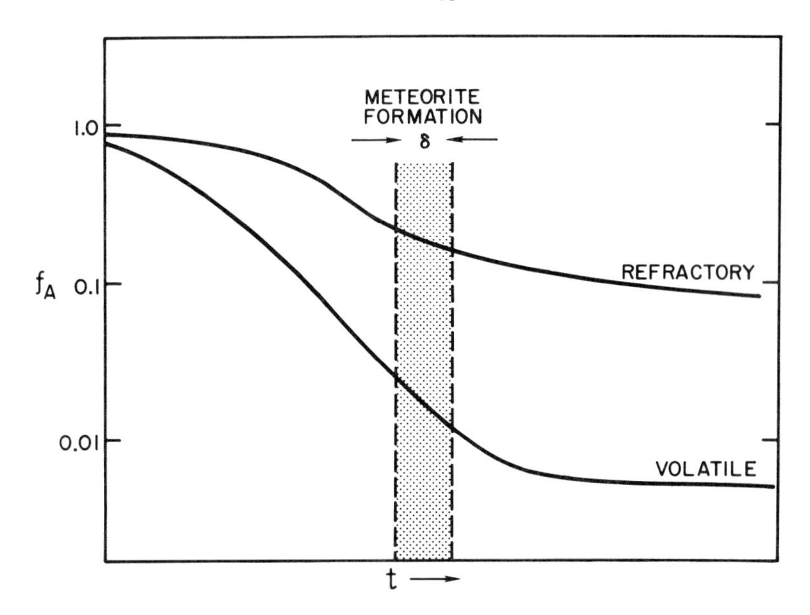

Fig. 3. In a cold solar system accumulation, the fraction of solid element A that was already solid in the intercloud medium falls with time as gaseous element A is condensed on the cold grain mantles. For a volatile element, say Pb, f_{Pb} falls to a very low value since it began largely gaseous. A moderately refractory element falls to a value only somewhat less than unity, whereas a very refractory element (not shown) remains near unity since it is almost all solid in the interstellar medium. Meteorite accumulation is idealized here as happening during this process, in which case early accumulates will have less of the volatile component than will late accumulates. (After D. Clayton, 1975*a*.)

already been condensed in refractory carriers in the ISM, falls while still gaseous species of it are added from the cold molecular cloud. A volatile element, such as Pb, falls to a low fraction by the condensation of cold gaseous matter, whereas a moderately refractory element, such as Si, cannot fall to such a low value because much of it was precondensed in refractory minerals (SUNOCONS and STARDUST). Very refractory species (Ca, Al and Ti) are not shown but would remain as high as f_A = 0.9 since more than 90% of these elements were already condensed as refractory ISM dust particles.

Early accumulation could perhaps then lead to a new interpretation of the Ca Al-rich inclusions (D. Clayton, 1977*b*). The fractionation occurs during a hierarchy of accumulation process, as described by Morfill *et al.* (1978). Fig. 3 suggests that the meteorites accumulated during the transition from precondensed to totally condensed.

One sees from this discussion that the state of chemical fractionation of the ISM cloud be crucial in the deciphering of the early accumulation processes. Field (1975) reviewed his arguments that condensation in stellar atmospheres leads to ubiquitous interstellar depletion from the gas phase of the more refractory elements. These interstellar depletions have been demonstrated by the relative strengths of far-ultraviolet absorption lines in the interstellar medium, as obtained with the *Copernicus* satellite (Spitzer and Jenkins, 1975 and references therein). On the other hand, that theory is not quite selective enough to account for the high average depletions of Ca, Al and Ti. Snow (1975) argued that the depletions were related to ionization potentials and derivative effects, a point of view developed quantitatively by Duley and Millar (1978). D. Clayton (1978) has summarized the idea that Ca, Al and Ti are selectively depleted immediately following nucleosynthesis as supernova condensates in the expansion of the supernova interior where they were synthesized. This variation of Field's thermodynamic arguments, since it depends on special initial abundances, condenses much more of Ca, Al and Ti than of Mg or Si in the expansion of the interior nuclear shells. Although the answer is cloudy at present, one can expect these studies of chemical depletion in the interstellar medium to strongly impact the related issue of refractory SUNOCONS, which would be themselves carriers of isotopic anomalies.

When one turns to astrophysical arguments whether or not a solar nebula was hot enough to vaporize dust, one finds a confusing and uncertain history. The main problem has been that such arguments depend on unknown dynamic initial conditions. For example, studies of the pre-main-sequence evolution of the sun have usually shown that it is consistent with the assumption of a very luminous early evolution ($L > 10^3 L_\odot$) called the *Hayashi track* in the Hertzsprung-Russell diagram (e.g., Larson, 1967). Luminosities in excess of $10^3 L_\odot$ would vaporize much of the inner solar system and, if surrounded by an optically thick solar nebula (another assumption), could vaporize most of the relevant regions. The problems with this picture lie in the initial conditions. If the protosolar nebula rotates, it will most likely collapse into a disk rather than into a central sun. Cameron has argued that the sun must then be "grown" by transport of low-angular-momentum matter inwards, accompanied by the transport of angular momentum outwards. Viscosity transports the angular momentum and, because it is dissipative, also heats the nebula (Lynden-Bell and Pringle, 1974). This mode of heating remains the last viable heat source for disk models. If the mass of the forming star is large, the temperature and luminosity of the disk may also be adequate to vaporize grains in the opaque parts of the disk. For a model of a low-mass star like the sun, Cameron shows in a figure in his chapter that the temperature never exceeds ~ 400 K outside the orbit of Mars. If that model is relevant, most of the primitive samples in our museums must be accumulations of dust that have been modified only by internal temperatures in the accumulated body. This fundamental controversy remains at the center of all discussions of chemical evidence concerning the origin of the planetary system.

Interstellar Chemistry

Both ion-molecule gas phase chemistry and solid-state chemistry play significant roles in the interstellar medium. The subjects are voluminously studied. Watson (1976) provides a recommendable review of molecular chemistry, as does Wickramasinghe (1967) of interstellar grains (see also Aannestad and Purcell, 1973). The chapters of Watson, E. Herbst and Donn relate these problems to protostars and planets.

The molecular chemistry provides both a diagnostic of molecular clouds and some control over quantities such as n_e that effect the dynamic evolution of the cloud. E. Herbst in this book reviews the 47 molecular interstellar species now known. He addresses especially the question of whether they are formed by ion-molecule chemistry or on the surfaces of grains. The answer is not known but the optimistic view is that the solution of very large chemical reaction networks will soon be possible and that our understanding of the conditions in the cloud will be deepened by understanding the origin of these molecules. Watson presents here a rather broader and therefore less detailed summary of his view of the major chemical problems, which are (1) poor knowledge of the physical environment, (2) major uncertainty over the meaning of the CH^+ abundance and the OH/H_2O abundance ratio, (3) the significance of big isotopic variations in H/D and $^{13}C/^{12}C$, (4) the role of dust in the chemistry, and (5) the formation of the complex molecules.

One hope not discussed is that the molecular constitution could eventually yield information on the ages of the molecular clouds, which are not known. They are commonly assumed to live only a few million years, but Townes (1978) has expressed the opinion that the differing $^{13}C/^{12}C$ ratios in different clouds may indicate that the clouds have quite long independent lifetimes, perhaps greater than 10^9 yr. If any observable chemical properties, such as the growth of a specific population or the alteration of an isotopic ratio, evolve slowly, they could provide an age indicator.

Complex hydrocarbon chemistry is a topic of growing importance. Hoyle and Wickramasinghe (1977 and references therein) argue that polysaccharides dominate the opacity of molecular clouds, provide interstellar parents for the organic polymer in carbonaceous meteorites, and perhaps provide a parent even of terrestrial life. J. M. Greenberg (1976, 1978) describes how the chemical explosions of heated grains may produce many complex interstellar molecules and polymerized forms. The meteoritic polymers carry isotopically anomalous gas that may reflect nucleosynthetic components, providing an unbroken line between the origin of the nuclei, the origin of the solar system, and perhaps even the origin of ourselves. Our planet and ourselves may still carry ancient fingerprints of the cloudy state. Others disagree, thinking that past memory was erased by a volatilization in a solar nebula. This fundamental controversy particularly enlivens the subject today.

Acknowledgments. The preparation of this report was supported by grants from the National Science Foundation and the National Aeronautics and Space Administration.

REFERENCES

Aannestad, P., and Purcell, E. 1973. *Ann. Rev. Astron. Astrophys.* 11: 309.
Appenzeller, I. 1971. *Astron. Astrophys.* 12: 313.
____. 1974. *Astron. Astrophys.* 36: 99.
Arnett, W.D., and Clayton, D.D. 1970. *Nature* 227: 780.
Black, D.C. 1972. *Geochim. Cosmochim. Acta* 36: 377.
Blair, G.; Evans, N.; Vanden Bout, P.; and Peters, W. 1978. *Astrophys. J.* 219: 896.
Browne, J.C., and Berman, B.L. 1976. *Nature* 262: 197.
Burbidge, E.M.; Burbidge, G.R.; Fowler, W.A.; and Hoyle, F. 1957. *Revs. Mod. Phys.* 29: 547.
Burstein, P.; Borken, R.J.; Kraushar, W.L.; and Sanders, W.T. 1976. *Astrophys. J.* 213: 405.
Burton, W.B. 1976. *Ann. Rev. Astron. Astrophys.* 14: 275.
Cameron, A.G.W. 1973*a*. *Space Sci. Rev.* 15: 121.
____. 1973*b*. *Nature* 246: 30.
Cameron, A.G.W., and Truran, J.W. 1977. *Icarus* 30: 447.
Caston, J.; McCray, R.; and Weaver, R. 1975. *Astrophys. J.* 200: L107.
Clark, B.G. 1965. *Astrophys. J.* 142: 1398.
Clayton, D.D. 1964. *Astrophys. J.* 139: 637.
____. 1973. In *Gamma-Ray Astrophysics* (F. Stecker and J. Trombka, eds.), Washington, D.C.: U.S. Government Printing Office.
____. 1975*a*. *Astrophys. J.* 199: 765.
____. 1975*b*. *Nature* 257: 255.
____. 1977*a*. *Icarus* 32: 255.
____. 1977*b*. *Earth Planet. Sci. Letters* 35: 398.
____. 1978. *Moon and Planets* (special Protostars and Planets issue). In press.
Clayton, R.N.; Grossman, L.; and Mayeda, T.K. 1973. *Science* 182: 485.
Clayton, R.N.; Onuma, N.; Grossman, L; and Mayeda, T.K. 1977. *Earth Planet. Sci. Letters* 34: 209.
Cox, D., and Smith, B. 1974. *Astrophys. J.* 189: L105.
Dalgarno, A., and McCray, R. 1972. *Ann. Rev. Astron. Astrophys.* 10: 375.
Duley, W.W., and Millar, T.J. 1978. *Astrophys. J.* 220: 124.
Elmegreen, B., and Lada, C. 1977. *Astrophys. J.* 214: 725.
Field, G. 1975. In *The Dusty Universe* (G. Field and A.G.W. Cameron, eds,), p. 89. New York: Neale Watson.
Field, G.; Goldsmith, D.; and Habing, H. 1969. *Astrophys. J.* 155: L149.
Fowler, W.A., and Hoyle, F. 1960. *Annals of Physics* 10: 280.
Goldsmith, D.W. 1970. *Astrophys. J.* 161: 41.
Gray, C.; Papanastassiou, D.; and Wasserburg, G.J. 1973. *Icarus* 20: 213.
Greenberg, J.M. 1976. *Astrophys. Space Sci.* 39: 9.
____. 1978. *Moon and Planets* (special Protostars and Planets issue). In press.

Grossman, L. 1972. *Geochim. Cosmochim. Acta* 36: 597.
Hainebach, K.L.; and Schramm, D.N. 1977. *Astrophys. J.* 212: 347.
Harvey, P.; Campbell, M.; and Hoffman, N. 1978. *Astrophys. J.* 219: 891.
Herbst, W. and Assousa, G. 1977. *Astrophys. J.* 217: 473.
Hoyle, F., and Wickramasinghe, N.C. 1977. *Nature* 268: 610.
Jenkins, E.B. 1978. *Astrophys. J.* 220: 107.
Lattimer, J.; Schramm, D.; and Grossman, L. 1978. *Astrophys. J.* 219: 230.
Larson, R.B. 1967. *Mon. Not. Roy. Astr. Soc.* 145: 271.
Lee, T.; Papanastassiou, D.; and Wasserburg, G.J. 1977. *Astrophys. J.* 211: L107.
Lin, C.C.; Yuan, C.; and Shu, F. 1969. *Astrophys. J.* 155: 721.
Lockman, F. 1976. *Astrophys. J.* 209: 429.
Lynden-Bell, D., and Pringle, J.E. 1974. *Mon. Not. Roy. Astr. Soc.* 168: 603.
Morfill, G.; Röser, S.; Tscharnuter, W.; and Völk, H. 1978. *Moon and Planets* (special Protostars and Planets issue). In press.
McKee, C.F., and Ostriker, J.P. 1977. *Astrophys. J.* 218: 148.
Oort, J.H. 1965. In *Galactic Structure* (A. Blaauw and M. Schmidt, eds.), Chicago: University of Chicago.
Parker, E.N. 1966. *Astrophys. J.* 145: 811.
Podosek, F. 1978. *Ann. Rev. Astron. Astrophys.* 16. In press.
Reynolds, J.H. 1967. *Ann. Rev. Nuclear Sci.* 17: 253.
Roberts, W. 1969. *Astrophys. J.* 158: 123.
Rogerson, J.; York, D.; Drake, J.; Jenkins, E.; Morton, D.; and Spitzer, L. 1973. *Astrophys. J.* 181: L110.
Scheinin, N.; Lugamir, G.; and Marti, K. 1976. *Meteoritics* 11: 357.
Schmidt, M. 1965. In *Galactic Structure* (A. Blaauw and M. Schmidt, eds.), Chicago: University of Chicago.
Scoville, N., and Solomon, P. 1976. *Astrophys. J.* 199: L105.
Shu, F.; Milione, V.; Gebel, W.; Yuan, C.; Goldsmith, D.; and Roberts, W. 1972. *Astrophys. J.* 173: 557.
Snow, T.P. 1975. *Astrophys. J.* 202: L87.
Spitzer, L. 1968. *Diffuse Matter in Space.* New York: Interscience.
Spitzer, L., and Jenkins, E. 1975. *Ann. Rev. Astron. Astrophys.* 13: 133.
Strom, S.; Strom, K.; and Grasdalen, G. 1975. *Ann. Rev. Astron. Astrophys.* 13: 187.
Talbot, R.J., Jr. and Arnett, W.D. 1971. *Astrophys. J.* 170: 409.
____. 1973. *Astrophys. J.* 186: 51.
Townes, C. 1978. In *Cosmochemistry,* Houston: Robert A. Welch Foundation. In press.
Trivedi, B. 1977. *Astrophys. J.* 215: 877.
Watson, W.D. 1976. *Revs. Mod. Phys.* 48: 513.
Werner, M.; Becklin, E.; and Neugebauer, G. 1977. *Science* 197: 723.
Westerhout, G. 1968. *Bull. Astron. Inst. Neth.* 14: 215.
Wickramasinghe, N.C. 1967. *Interstellar Dust.* London: Chapman and Hall.
Wooley, R. 1965. In *Galactic Structure* (A. Blaauw and M. Schmidt, eds.), p. 85. Chicago: University of Chicago.
Wynn-Williams, C.G., and Becklin, E. 1974. *Publ. Astron. Soc. Pacific* 86: 5.

THE STELLAR STATE: FORMATION OF SOLAR-TYPE STARS

RICHARD B. LARSON
Yale University

A brief overview of research on the formation of solar-type stars is given. Topics mentioned include some observed properties of stars and regions of star formation that set constraints on possible theoretical models and some problems of current interest in the dynamical modeling of collapsing clouds, protostars and solar nebulae. In particular, accretion processes and accretion disks have received considerable attention and seem likely to play an important role in the formation of stars and solar systems. Some unsolved problems that still need further study are also pointed out.

Most current theories of the formation of the solar system postulate that the sun was once surrounded by a dusty, disk-like solar nebula in which the planets and other solid bodies of the solar system condensed. The only time when the sun could plausibly have possessed or acquired such a pre-planetary disk was at a very early stage of evolution when it was still embedded in a "placental cloud," from which the solar nebula could have formed. Thus the origin of the solar nebula was probably intimately connected with the formation of the sun, and to understand the origin of the solar nebula it is necessary to understand the formation of the sun. It is also important to understand something about the early evolution of the sun, since the sun must have had a strong influence on the conditions in the solar nebula at the time when the planets were forming.

Astronomical observations now provide considerable information about the conditions under which star formation occurs, as well as about the properties of young stars, and this information provides boundary conditions for theories or models of the star-formation process. The properties of the molecular clouds in which stars form are reviewed in this book by Evans. In the densest parts of these clouds the Jeans mass estimated from the observed densities and temperatures is only a few solar masses, so it appears possible for solar-type stars to form directly in these conditions. As discussed by Mouschovias in his chapter, this conclusion is probably not much altered when the effects of magnetic fields are taken into account, although somewhat higher densities may be required before magnetic effects become unimportant. The fact that molecular clouds often show marked density variations and condensations on the smallest resolvable scales, sometimes coinciding with embedded infrared sources that appear to be protostars, is a strong indication that parts of these clouds have already begun to collapse and form stars.

In addition to these large molecular clouds and cloud complexes, there are also a number of smaller and more isolated dark "globules", reviewed by Bok (1977, 1978); their role in star formation is not yet so well understood. Perhaps they represent fragments left over from the disruption of large cloud complexes, but it is also possible that some of them will accrete together with other dark clouds to form cloud complexes. Some of them may collapse to form stars or small groups of stars in relative isolation, but it seems unlikely that all of them can do so, and the possibility that they are stabilized against collapse by rotation is discussed in Field's chapter. In any case, the globules are particularly well suited for individual study, and most of them show centrally condensed structures suggesting that self-gravity is important.

Some information about the way in which star formation proceeds can be gained from observations of associations of young stars, as discussed by Elmegreen and Lada (1977) and in the chapter of Lada *et al.* The massive stars often occur in subassociations showing a progression of age with spatial position, and the youngest stars appear to have recently formed near the boundaries of dark clouds, as if their formation had been triggered by external compression of these clouds. In contrast, the T Tauri stars, which are believed to be less massive (solar-type) newly formed stars, are found scattered throughout the dark clouds, which suggests that they can form directly by gravitational instability and collapse without the need for any external triggering mechanism.

At present we have no very direct observational information about the collapse processes that produced the T Tauri stars seen embedded in or closely associated with the dark clouds. However, recent work on T Tauri stars discussed by Kuhi in this book (see also Cohen and Kuhi, 1978) shows that these stars typically have hydrostatic contraction time scales between 10^5 and 10^6 yr, as inferred from their positions in the Hertzsprung-Russell (HR) diagram. According to current collapse models, hydrostatic contraction of the core of a collapsing protostar takes place concurrently with the buildup in mass of the core by accretion of matter from an extended envelope. Thus the total time required to form these stars by infall and accretion of interstellar matter cannot exceed $\sim 10^5 - 10^6$ yr, which is comparable to the range of free-fall times estimated from the densities of molecular clouds. The small distances that the T Tauri stars have moved from their parental dark clouds, combined with their estimated velocities (Herbig, 1977), also suggest ages no greater than $\sim 10^6$ yr. Thus we can infer that all of the action associated with the formation of the T Tauri stars takes place within a time interval not significantly exceeding the initial free-fall time. The fact that T Tauri stars do not as a class show evidence for dense circumstellar dust shells (see the chapter by Rydgren) indicates that in most cases their initial protostellar envelopes have already been dispersed or have collapsed into disks; thus the formation time for any flattened solar nebula must also not significantly exceed the free-fall time.

Two general properties of stars that are directly determined by the process of star formation are the spectrum of stellar masses or initial mass function (IMF), and the statistics of binary and multiple stars. Observational and theoretical aspects of the initial mass function are reviewed in this book by Scalo and Silk. According to Scalo, a lognormal function provides a better fit to the observations than the power laws conventionally used. The form of the IMF depends on the way in which collapsing clouds fragment into protostars of various masses, and therefore the observed IMF sets constraints on models for the collapse and fragmentation process. For example, the observations might in principle help to decide between a picture involving successive fragmentation into smaller and smaller units, and a coagulation model involving the accretion of small fragments into larger ones (see Silk's chapter). Eventually 3-dimensional collapse calculations (e.g., Larson, 1978) will have to reproduce the observed IMF. At present, however, a variety of models claim to provide satisfactory agreement with the observed IMF (see chapters by Scalo and Silk).

The frequent occurrence of binary and multiple stellar systems, as reviewed in the chapter by Abt, must also find explanation in a complete theory of star formation. Particularly important in the present context is the relation between binary systems and planetary systems, and the question of whether there is a continuity between them, as is often supposed. The observations are at least consistent with this possibility, and extrapolation to low-mass companions suggests that planetary systems may not be rare (Abt's chapter). A promising feature of many of the collapse calculations discussed in this book is that they suggest the frequent formation of binary stars, and some of them also suggest that planetary systems can form in a similar way.

COLLAPSE MODELS

Because the collapse of the densest condensations in molecular clouds to stellar densities occurs rapidly in small regions that are heavily obscured and/or unresolvable with most observing techniques, direct information about protostellar collapse is difficult to obtain, and a clear observational picture must probably await the further development of techniques. Meanwhile we must rely on theoretical collapse calculations to provide information about the types of processes that might occur during this "dark age" of stellar evolution. A thorough review of this work is given in this book by Bodenheimer and Black; other recent reviews have been given by Larson (1977) and Woodward (1978). In this introductory chapter we present a brief overview of the current state of the art.

In considering the results of these collapse calculations, it is important to bear in mind that they all represent rather highly idealized models and are based on uncertain assumptions concerning the initial conditions and the relevant physical processes. Since details of the results may be strongly dependent on these assumptions, the greatest attention should be given to the general features that seem to be most reproducible in different calculations.

Spherical Collapse

Although spherical collapse is the most idealized case, calculations with spherical symmetry are still important because only in this case can the calculations be followed all the way to stellar densities; also, this is the only case in which we have some assurance of the numerical accuracy of the results and a reasonably complete understanding of them. The general features of spherical collapse are by now well established: a runaway increase in central condensation, formation of a small stellar core, and infall of the remaining envelope into the core through an accretion shock. All of these qualitative features are found with sufficient generality, despite considerable differences in the assumptions made, that it seems likely that they would appear even when the assumption of spherical symmetry is relaxed. Indeed this expectation is supported by the available results for non-spherical collapse; thus the spherical calculations appear to provide at least a qualitatively correct guide to many features to be expected more generally.

The most recent and detailed calculations of spherical collapse have been reported by Kondo (1978) and Winkler (1978). Kondo uses a computational method that is different from the finite grid methods used by all other investigators, and thus his work provides an independent confirmation of results such as the formation of a hydrostatic core bounded by an accretion shock. Winkler (1978) has repeated, in more detail and with more numerical precision, the same calculation of the collapse of a 1 $M_{\odot}$ protostar that was previously done by Larson, Appenzeller and Tscharnuter, and Westbrook and Tarter; these authors had obtained discrepant values for the radius of the resulting star. It is clear that an important factor determining the final radius of the star after accretion of the envelope is the amount of energy radiated away from the accretion shock during the collapse process

(Woodward, 1978); this energy loss is important in the models of Larson and of Appenzeller and Tscharnuter, which yielded final radii of 2 $R_{\odot}$ and 4 $R_{\odot}$ respectively, but unimportant in the calculation of Westbrook and Tarter, which yielded a final radius of about 90 $R_{\odot}$. An accurate treatment of the shock region is necessary to resolve this discrepancy. Winkler has paid particular attention to this problem and has used a much finer grid than was used in any of the other studies; he finds a final stellar radius of 3 $R_{\odot}$, basically confirming the results of Larson and of Appenzeller and Tscharnuter. The discrepant result of Westbrook and Tarter is probably attributable to an excessively coarse grid in the critical region around the accretion shock (Woodward, 1978).

Although the more accurate calculations are thus probably quantitatively reliable, at least within factors of ~ 2, the numerical results of course still depend on the input assumptions; for example, a higher initial density leads to a larger final radius (Larson, 1969). A higher initial density than that determined by the Jeans criterion might result, in some circumstances, from the effects of magnetic fields (see the chapter by Mouschovias), or from dynamical effects such as coalescence of protostars (see below).

Axisymmetric Collapse

Much attention has recently been devoted to calculating the collapse of axisymmetric rotating clouds (for reviews see the chapter by Bodenheimer and Black; and Woodward, 1978). All calculations predict that the collapse is strongly nonhomologous, just as in the spherical case, and most calculations show the formation of a ring-like structure near the center of the cloud. However, some calculations do not show rings. The results are clearly sensitive to the numerical scheme used, apparently because the formation of a ring depends very sensitively on the exact conservation or nonconservation of angular momentum during the collapse. The continuing disagreement between different calculations suggests that the problem may not even be well-behaved mathematically; possibly there is an instability that leads either to a ring or to a single central condensation, depending on small differences in the initial conditions, the amount of viscous transfer of angular momentum, etc. Thus, the formation of a ring in an axisymmetric rotating cloud (if such clouds exist in nature) may be something of an accident, depending on small perturbations or details of the dynamics that are difficult to predict.

A new development in the study of axisymmetric collapse, as reviewed by Bodenheimer and Black (this book), is the calculation of the collapse of a cloud with a frozen-in magnetic field. In this case the collapse of the outer part is impeded by the magnetic field, while the core collapses to a high density and becomes strongly flattened. The fact that the core tends to collapse along the field lines means that the rise in density is great enough compared with the rise in magnetic field to allow fragmentation to occur, at least in the plane of flattening (Mouschovias' chapter). However, much more theoretical (and observational) work will be required before the role of magnetic fields in protostellar collapse is fully understood.

The axisymmetric calculations say nothing about what would happen if the constraint of axial symmetry were relaxed. It has been conjectured that rings, if they form at all, are unstable and will rapidly break up into two or more condensations orbiting around each other, possibly resulting in the formation of a binary or multiple system of stars. This could provide an attractive explanation for the observed prevalence of binary systems, and would allow most of the original cloud angular momentum to go into stellar orbital motions. (However, Mestel has emphasized that the angular momentum problem for individual stars is still present because the orbiting condensations would have significant *spin* angular momentum which must somehow be redistributed in order for stars to form.) The desire to test the conjecture of ring fragmentation has motivated much of the recent work on 3-dimensional collapse dynamics.

3-Dimensional Collapse

The results of Norman and Wilson (1978) for the fragmentation of isothermal rings make it appear almost certain that rings will generally fragment into several blobs orbiting around each other; the most common number of objects formed in these calculations is 2 or sometimes 3. Often a larger number of condensations forms first, and then they merge into a smaller number of objects. This result confirms to some extent earlier expectations that merging of protostellar fragments might be important in collapsing clouds (Layzer, 1963; Arny and Weissman, 1973). These calculations have not been followed further, so they do not indicate what the final state of the system will be, but binary formation is certainly strongly suggested as a frequent outcome.

Tohline (1978) has made 3-dimensional collapse calculations starting with a nearly uniform rotating cloud, and finds ring formation even when axial symmetry is not imposed; somewhat surprisingly, non-axisymmetric initial perturbations decay at first and do not start to grow until after the ring has formed. Apparently in Tohline's calculations the growth of density perturbations is inhibited by the tendency for condensations to be stretched out into spiral filaments by differential rotation. More calculations are evidently needed to explore such effects.

Other 3-dimensional collapse calculations reviewed by Bodenheimer and Black also frequently show fragmentation into two orbiting condensations. However, all of these results refer only to the early stages of collapse, and do not necessarily indicate the final outcome. For example, if two condensations form near the center of a collapsing cloud, gravitational or viscous drag forces exerted by the surrounding gas could cause them to spiral together and merge into a single object. On the other hand, it is also possible that even if only a single central object forms at first, a second object could form later from leftover gas orbiting around it, if the Jeans mass in this gas is small enough. Thus, it will be important to pursue the 3-dimensional collapse calculations as far as possible, and to explore a wide range of possible collapse geometries; the existing results involving ring formation and fragmentation may represent only rather special cases, and may not indicate the complete range of possibilities for 3-dimensional collapse.

A different computational approach using finite-size "fluid particles" to simulate 3-dimensional gas dynamics has been used by Larson (1978) to study the fragmentation of collapsing clouds. Although quite crude due to the limited number of particles used, these results suggest how collapse and fragmentation might proceed in clouds with various ratios of total mass to Jeans mass, and how the formation of single stars with pre-planetary disks might be related to the formation of binary systems. The scheme makes use of an "artificial viscosity" to represent the dissipative effects of shock fronts in a collapsing cloud; such dissipation is necessary for the formation of tightly bound condensations, and the viscosity also produces significant outward transfer of angular momentum. Therefore these results differ in detail from the results of the other calculations mentioned above, in which no viscosity was assumed to be present.

It is found that when the total mass is nearly equal to the Jeans mass, collapse proceeds much as in the spherical case with the formation of a single central dense core, even when the cloud has substantial rotation. No ring structure and no fragmentation are found, presumably because of the viscosity which is present in this case and transports angular momentum outward, allowing low-angular-momentum material to accumulate at the center. The leftover material forms a flattened rotating envelope, whose inner part is essentially an accretion disk in which matter spirals inward to accrete on the central core. This result suggests that, in the presence of sufficient viscosity, the collapse of a Jeans-mass cloud or fragment can lead to a single object surrounded by a flattened accretion envelope or disk, a configuration of interest for the origin of the solar system (Larson, 1972; Lynden-Bell and Pringle, 1974; Cameron, 1978). The hypothesis that the solar nebula originated as an accretion disk has been developed in detail in the solar nebula model of Cameron (see his chapter in this book).

When the Jeans mass is reduced to about half of the total mass, the particle simulations typically yield a binary system of two condensations orbiting around each other. Usually one object forms first, and then a second "satellite" condensation forms from material orbiting around it. This suggests that there may be a continuity between the formation of binary stars and the formation of single stars with planetary systems, the outcome depending on the amount of leftover material orbiting around the primary object and on the Jeans mass in this leftover gas. An interesting question to explore more fully with other 3-dimensional hydrodynamic codes will be whether or not there is a critical disk mass such that for smaller masses a stable "solar nebula" forms, whereas for larger masses a binary or multiple system of stars is formed. The particle simulations suggest the possibility that if the disk is as massive as the central object (as in the solar nebula models of Cameron), the result may be the formation of a binary system.

When the Jeans mass is much smaller than the total mass, many condensations are formed, including many binary and hierarchial multiple systems, in qualitative agreement with the observations. A wide range of masses is also found, and it may even be possible to predict a mass spectrum similar to the observed IMF (Larson, 1978). These resemblances with the observed properties of stars may be a fairly general consequence of the basic physical processes represented by these simulations, especially the viscous effects and accretion processes which determine the distribution of matter among the various condensations that form. The fact that such processes also lead naturally to the formation of accretion disks supports the idea that the solar nebula originated as an accretion disk.

A somewhat different picture for the formation of stars and planetary systems is suggested by Alfvén (see his chapter in this book). In this view dust particles collect by gravity at the center of a cloud and form a nucleus around which additional material condenses to form a star; several such condensation nuclei may eventually coalesce into a single star. Similar processes may also act to form planets. It is noteworthy that viscous effects and accretion processes play an important role in this picture, just as in the gas-dynamical collapse calculations described above; the essential difference is the special importance assigned to dust grains in the Alfvén picture. More detailed calculations will be of interest to clarify the role of dust accumulation in star formation (e.g., Flannery and Krook, 1978).

Alfvén has also emphasized that most theorizing about astrophysical plasmas has been based on simple and probably unrealistic assumptions of homogeneity, which are not valid in laboratory nor solar system plasmas, where extreme inhomogeneities are observed. Even apart from plasma effects, it is true that the great majority of fluid flows in nature are not smooth and homogeneous but are turbulent. There seems to be no reason to believe that astrophysical flows such as those in collapsing clouds would not also be turbulent, and this should be taken into account in any realistic model. Turbulence can dissipate large-scale motions by converting them into smaller-scale motions whose energy is ultimately dissipated by viscosity, and this may well be an important process in view of the central role of dissipative effects in star formation.

PROPERTIES OF PROTOSTARS AND YOUNG STARS

During most of its evolution a protostar is expected to be observable only as an infared object, and detailed radiative transfer calculations have been made to predict the infrared spectra of protostars with both spherical and axial symmetry; this work is extensively reviewed in this book by Bertout and Yorke. Infrared observations of regions of star formation (Werner, Becklin, and Neugebauer, 1977; Wynn-Williams, 1977) have revealed many infrared sources with properties similar to those expected for protostars, and it even appears possible to distinguish different stages of evolution of such objects. By adjustment of model parameters, it is possible to achieve good fits between calculated and observed spectra, including absorption features such as the silicate feature near 10 μm, which in the models is produced by cool dust in the outer part of an extended protostellar envelope.

The best studied "protostellar" sources are mostly considerably more luminous and hence presumably more massive than solar-mass protostars. Less massive protostars are expected to be similar in appearance, except for their smaller luminosities, and this expectation is consistent with the available data for the fainter protostellar sources. For a solar-mass protostar most of the infrared flux is predicted to come from the part of the envelope with radii between $\sim 1/2$ AU and ~ 100 AU and temperatures between ~ 1000 K and ~ 50 K; this is just the region in which one might expect a flattened "solar nebula" to form and condense into planets. The presently attainable resolution of infrared observations is not adequate to resolve regions of this size and show whether disk-like structures exist in them, but infrared interferometers now under consideration (Cudaback, personal communication) would potentially have the capability of probing the structure of protostellar envelopes and revealing the formation of such flattened "solar nebulae" in them.

The earliest stage of "normal" hydrostatic stellar evolution is believed to be represented by the T Tauri stars, reviewed in this book by Kuhi, by Ulrich, by Imhoff and by Rydgren. There has been much interest in comparing the predicted properties of protostellar cores after envelope accretion is completed with the observed properties of T Tauri stars. As noted above, most of the spherical collapse models yield final core radii in the range $2 - 4$ $R_\odot$. The majority of the T Tauri stars whose positions in the HR diagram have been plotted by Cohen and Kuhi (1978) fall in the predicted region near the lower end of the Hayashi track, and in fact have radii between 2 and 4 $R_\odot$. The T Tauri stars studied by Imhoff and Mendoza (1974) and by Rydgren, Strom and Strom (1976) mostly also have radii in the same range, although the most luminous ones (which are probably more massive than 1 $M_\odot$) have radii up to ~ 8 $R_\odot$. The predicted radius of a newly formed star depends mainly on the total core accretion time, which is comparable to the initial free-fall time and for spherical collapse is expected to be in the range $\sim 10^5 - 10^6$ yr; this also gives approximately the predicted apparent contraction age of the star. Larger radii and smaller ages are obtained if the initial density is higher than estimated from the Jeans criterion. Thus, the spread in radius and apparent contraction age inferred from the observations (e.g., Rydgren *et al.*,

1976; chapter by Kuhi; Cohen and Kuhi, 1978) is consistent with theoretical expectations for collapse from densities comparable to or somewhat greater than those given by the Jeans criterion (or those observed in molecular clouds). This rough agreement shows that the time scale for the collapse or dispersal of protostellar envelopes does not greatly exceed the free-fall time, and thus is not much prolonged by effects such as rotation, magnetic fields, etc.

Questions of particular interest for the formation of the solar system are whether or not residual dusty envelopes are still present around some of the very youngest stars, and if the infall of the last remnants of these envelopes, representing the final stages of the accretion process, can be detected spectroscopically. Unfortunately it has proved difficult to relate the peculiarities of the T Tauri stars unambiguously to the final stages of the star-formation process. Dusty circumstellar envelopes are clearly present around some T Tauri stars, as shown by the presence of strong excess infrared emission, but this does not seem to be a general characteristic of T Tauri stars (Rydgren's chapter), and it is not yet clear whether it indicates a particularly early stage of evolution. The spectroscopic properties of T Tauri stars are even more difficult to interpret and show a bewildering complexity (chapters by Kuhi and by Ulrich). Indications of gas outflow, infall, and rotation (perhaps in a circumstellar disk) are all found in different stars, and sometimes all of these phenomena seem to occur simultaneously in the same star. Much attention has recently been given to the subclass of T Tauri stars called YY Ori stars, which sometimes show redshifted absorption lines indicative of infall (chapter by Bertout and Yorke; Appenzeller, 1978). According to Appenzeller, observations with high spectral resolution show that the YY Ori stars are more common than previously thought, possibly constituting up to 40 – 50% of all T Tauri stars. As reviewed by Bertout and Yorke, radiative transfer calculations for collapse models predict "inverse P Cygni" line profiles resembling those seen in the YY Ori stars; thus, these objects may be showing evidence of the last stages of the protostellar accretion process. However, some of these stars also show evidence of mass outflows, often occurring simultaneously with infall; a possible explanation of this is that matter near the star continues to fall into it, while matter farther out in the envelope is driven outward by radiation pressure.

Part of the difficulty in linking the T Tauri stars directly to the final stages of protostellar collapse may arise from the fact that the T Tauri stars apparently often generate strong outflowing stellar winds, the origin of which is not yet well understood. Interactions between outflowing winds and residual infalling gas might account for many of the complex observed phenomena. For example, any interface between hot outflowing wind and cool ambient or infalling gas would probably develop Rayleigh-Taylor instabilities, causing the cool gas to be concentrated into clumps or streams which continue to fall into the star while the wind blows outward in other directions, producing simultaneous inflow and outflow in different directions.

The compact emission nebulae known as Herbig-Haro objects, reviewed by Böhm (see his chapter in this book), are believed to result from interactions between very young stars and the surrounding molecular cloud gas. In particular, it appears that the line spectra of the Herbig-Haro objects can be explained by emission from a layer of hot gas heated by a shock front having a velocity of several tens of km sec^{-1}; such a shock could be produced by an outflowing wind or an expanding blast wave from a hidden T Tauri star impinging on nearby dark cloud material (Böhm's chapter). If an outflowing wind becomes important at a very early stage of evolution when the star is still hidden by the dark cloud in which it formed, this might explain why circumstellar shells and remnant infalling material are not more often seen in T Tauri stars, since the wind may have blown away most of the close circumstellar matter by the time the star becomes optically visible.

The total amount of mass that can be lost in a T Tauri wind is limited by the total amount of energy available from gravitational contraction, as discussed by Weidenschilling (1978). In most cases it seems unlikely that more than $\sim 0.1\ M_{\odot}$ can be lost during the entire pre-main-sequence evolution of a star; this is not enough to seriously alter the internal evolution of the star. Nevertheless, the wind may play an important role in slowing down the rotation of solar-type stars during post-T-Tauri stages of evolution. New data on the rotational velocities of pre-main-sequence stars by Kuhi (1978) show rapid rotation at nearly all stages of pre-main-sequence evolution; since solar-type stars are observed to rotate slowly on the main sequence, a large loss of angular momentum must occur during late pre-main-sequence or early main-sequence evolution.

THE SOLAR NEBULA

The collapse calculations that have been made illustrate the essential importance of dissipative phenomena such as shock fronts for the formation of stars, and suggest that dissipative processes would also result in the formation of accretion disks from the leftover material which has too much angular momentum to fall into stars. The accretion disks forming around stars in close multiple systems may soon be disrupted, but for the more isolated stars these disks might survive and form planetary systems. Because such a pre-planetary disk forms in a collapsing cloud whose motions are probably very chaotic and turbulent, the disk would initially be quite turbulent, with large noncircular and noncoplanar motions. Turbulent viscosity (or other sources of viscosity, if present) would tend to transfer angular momentum outward in the disk, allowing matter to spiral inward and accrete on the central star. The evolution of such viscous disks has been studied by Lynden-Bell and Pringle (1974), and this forms the basis of the solar nebula models described by Cameron in his chapter.

The initial turbulence in the solar nebula should decay quickly, so the viscosity will soon become unimportant unless the turbulence is somehow regenerated. Accretion-disk flows are not intrinsically unstable, but Cameron suggests that at least three processes might continue to generate turbulence in a solar nebula: meridional circulation currents, gravitational instabilities, and continuing infall of matter into the disk. Collapse calculations suggest that infall into the disk should continue for some time, and that gravitational instabilities can also sometimes occur, but whether or not these processes can generate the amount of viscosity assumed in Cameron's models remains speculative. An important feature of these models is heating of the disk material by viscous dissipation as the gas spirals inward; this is the dominant heating process controlling the temperature of the gas, which in turn governs whether or not local gravitational instabilities can occur.

In Cameron's disk models, the temperature never becomes high enough to completely vaporize interstellar grains. This is also true in protostellar envelopes in general, since their predicted (and observed) temperatures are typically only a few hundred degrees. However, it is possible that the dissipation of supersonic infall motions by shock fronts during the formation of the solar nebula would produce strong transient heating sufficient to partly or completely vaporize the grains. Thermalization of the infall kinetic energy would heat the gas to temperatures of many thousands of degrees, which is amply sufficient to vaporize the grains if the high temperature is maintained for a long enough time before the gas cools again. For example, if the solar nebula forms by steady infall of gas from large distances at nearly the free-fall velocity, an accretion shock will form at the surface of the disk and all of the material entering the disk will undergo transient heating on passing through this shock front. A related possibility is that if the disk is initially highly chaotic and turbulent (as is suggested by the quite noncircular and noncoplanar orbits of the asteroids), the dissipation of supersonic turbulent motions by shock fronts will produce significant transient heating of the gas during an early stage of solar nebula formation. It will be interesting to investigate quantitatively whether or not such shock heating effects can account for the meteoritic evidence, reviewed by Wasson in this book, indicating high temperatures during the early evolution of the solar nebula.

After an initial chaotic period, the solar nebula presumably settled into a relatively quiescent equilibrium state, and the most important external factor influencing its evolution was the radiation from the early sun, including the solar wind. It has been suggested that the final dispersal of the solar nebula after solid objects formed in it was caused by the early solar wind, and this process has been analyzed by Elmegreen (1978). The material in the midplane of the nebula is shielded from the wind, and in any case the wind cannot initially eject the nebula bodily. According to Elmegreen, the dominant process of gas loss is actually viscous spiraling inward and accretion by the sun due to the effects of turbulence generated at the surface of the nebula by interaction with the outflowing wind. This process could remove the gas in less than 10^6 years, in which case the solar nebula would not last very long after the end of the accretion process by which it forms; the "solar nebulae" around some forming stars might even be largely dissipated by the time the stars become optically visible. If some gas remains in a disk close around some of the youngest T Tauri stars, interactions between this residual disk gas and the star could help to account for some of the spectroscopic complexities of T Tauri stars (see Kuhi's and Ulrich's chapters). It will be important to try to obtain more spectroscopic or photometric evidence for material orbiting around T Tauri stars, possibly in a disk, since this would greatly enhance our presently very meager understanding of the incidence and properties of "solar nebulae" around young stars.

SOME UNSOLVED PROBLEMS

Although we can perhaps begin to glimpse in outline how the sun and the solar nebula were formed, major gaps remain in our understanding of many of the processes involved, and much work remains to be done before a complete and convincing picture is possible. Present models of protostellar collapse may need revision not only in details such as the treatment of opacities and radiative transfer but also in their basic assumptions, such as the adopted initial and boundary conditions. Real protostars do not collapse in isolation but as parts of more extended clouds that are themselves collapsing or are being acted upon by external perturbing forces; therefore, the most appropriate initial and boundary conditions could be rather different from the "standard" ones. Further calculations with different assumptions, such as a higher initial density or a contracting boundary, are needed to obtain a better theoretical fit to the distributions of T Tauri stars in the HR diagram.

The present models may be even more fundamentally deficient in the omission of important physical processes. For example, fluid flow instabilities and turbulence, which are nearly ubiquitous in fluid dynamics, must almost certainly play a role in collapsing clouds as well, but such effects have so far barely been considered. Turbulence may control the way in which a collapsing cloud fragments into protostars of various sizes, and even the initial conditions with which protostars are formed; possibly the stellar mass spectrum results from a turbulent hierarchy of motions on different scales. It is noteworthy that imposed on the large-scale regularity of the solar system are many smaller-scale irregularities, which may have had their origin in initially chaotic motions occurring during the collapse of the solar nebula.

Other processes that are not well understood include the dissipative processes and the sources of viscosity in collapsing clouds. In addition to subsonic turbulence, supersonic motions and shock fronts may be expected to play a role, not only in dissipating energy during the collapse but also in compressing the gas to much higher densities. The detailed effects of shock fronts in collapsing or colliding clouds have not been extensively studied, but will almost certainly turn out to be of great importance for star formation. Also, the possible role of magnetic fields in producing an effective viscosity and transferring angular momentum has often been noted, but is still not adequately understood. Dissipative and viscous effects are very important in determining the rate of accretion of material by protostellar cores and the final distribution of matter among stars and their residual "solar nebulae". Thus a better understanding of such effects is needed to decide, for example, to what extent the sun may have formed by direct infall of matter, by slow accretion of gas from a disk, or even by the spiraling together and coalescense of several prestellar objects or "giant gaseous protoplanets". Multi-dimensional hydrodynamical calculations in which viscosity is included, even if only as an adjustable parameter, could cast some light on these and other questions. Also, further numerical studies of the instabilities of rotating, centrifugally supported configurations such as disks, bars, or rings can help to clarify the conditions under which a solar system rather than a binary or multiple star system will form.

It seems clear that many of the observed properties of very young stars result from interactions with surrounding material, but an understanding of the nature of these interactions is just beginning. In particular it will be important to study the possible instabilities that can arise when radiation pressure or outflowing winds from newly formed stars act on the ambient residual gas of the protostellar cloud, since such effects may not only help to disperse the cloud but may also cause parts of it to condense into lumps or filaments that continue to fall inward or orbit around the new star; one can speculate that comets possibly form this way.

Finally, an under-exploited field of research is the study of models for the structure and evolution of the solar nebula; the work of Cameron stands almost alone in this area. Many physical processes (gas infall, turbulence and turbulent viscosity, shock heating, radiative cooling, gravitational instabilities, dust accumulation, etc.) almost certainly play a role, so a systematic exploration of many effects and many types of models will be required. Viscous and dissipative effects will determine the distribution of matter among the sun and planets, while heating and cooling processes will determine the thermal conditions under which solids condense from the nebula. Detailed calculations of all of these effects can profitably be undertaken to illuminate the way in which various bodies of the solar system formed, and ultimately to build up a coherent picture of the relations between protostars and planets.

REFERENCES

Appenzeller, I. 1978. In *The Interaction of Variable Stars with their Environment*, Bamberg. (R. Kippenhahn, J. Rahe and W. Strohmeier, eds.) In Press.
Arny, T., and Weissman, P. 1973. *Astron. J.* 78: 309.
Bok, B. J. 1977. *Publ. Astron. Soc. Pacific* 89: 597.
____. 1978. *Moon and Planets* (special Protostars and Planets issue). In press.
Cameron, A. G. W. 1978. In *The Origin of the Solar System* (S. J. Dermott, ed.). New York: Wiley. In press.
Cohen, M., and Kuhi, L. V. 1978. *Moon and Planets* (special Protostars and Planets issue). In press.
Elmegreen, B. G. 1978. *Moon and Planets.* "On the Interaction Between a Strong Stellar Wind and a Surrounding Disk Nebula." (special Protostars and Planets issue). In press.
Elmegreen, B. G., and Lada, C. J. 1977. *Astrophys. J.* 214: 725.
Flannery, B. P., and Krook, M. 1978. *Astrophys. J.* 223: In press.
Herbig, G. H. 1977. *Astrophys. J.* 214: 747.
Imhoff, C., and Mendoza, V. E. 1974. *Revista Mex. Astron. Astrofis.* 1: 25.
Kondo, M. 1978. *Moon and Planets* (special Protostars and Planets issue). In press.
Kuhi, L. V. 1978. *Moon and Planets* (special Protostars and Planets issue). In press.
Larson, R. B. 1969. *Mon. Not. Roy. Astr. Soc.* 145: 271.
____. 1972. In *On the Origin of the Solar System* (H. Reeves, ed.), p. 142. Paris: C.N.R.S.
____. 1977. In *Star Formation* (T. de Jong and A. Maeder, eds.), p. 249. Boston: Reidel.
____. 1978. *Mon. Not. Roy. Astr. Soc.* 184: In press.
Layzer, D. 1963. *Astrophys. J.* 137: 351.
Lynden-Bell, D., and Pringle, J. E. 1974. *Mon. Not. Roy. Astr. Soc.* 168: 603.
Norman, M. L., and Wilson, J. R. 1978. *Bull. Am. Astron. Soc.* 9: 567.
Rydgren, A. E.; Strom, S. E.; and Strom, K. M. 1976. *Astrophys. J. Suppl.* 30: 307.
Tohline, J.E. 1978. *Bull. Am. Astron. Soc.* 9: 566.
Weidenschilling, S. J. 1978. *Moon and Planets* (special Protostars and Planets issue). In press.
Werner, M.W.; Becklin, E.E.; and Neugebauer, G. 1977. *Science* 197: 723.
Winkler, K. -H. 1978. *Moon and Planets* (special Protostars and Planets issue). In press.
Woodward, P. R. 1978. *Ann Rev. Astron. Astrophys.* 16: In press.
Wynn-William, C.G. 1977. In *Star Formation* (T. de Jong and A. Maeder, eds.), p. 105. Boston: Reidel.

THE PLANET-FORMING STATE: TOWARD A MODERN THEORY

WILLIAM K. HARTMANN
Planetary Science Institute

Some of the chapters in this book are briefly reviewed in the larger context of current work on the formation of planetary systems. Emphasis is on problems associated with the exact mechanism by which the initially dispersed nebular gas and dust aggregated into planetary masses. Most of the evidence suggests that terrestrial planetary material grew step by step through a hierarchy of planetesimal sizes from dust to planets. This evidence includes detection of silicate dust around young stars, chemistry and mineral aggregates of carbonaceous chondrites, numerical modeling of collisional evolution, estimates of asteroidal sizes for meteorite parent bodies, and crater populations of all explored planet and satellite surfaces. However, several authors find that the giant planets are better explained by single-step gravitational collapse directly from gas-dust form to planetary or super-planetary masses. Thus, the relative roles of simultaneous competing processes, such as condensation, gravitational collapse, and collisional accretion, remain as a current research problem. Similarly, the details of the sticking processes by which dust grains or planetesimals united, once they struck each other, remain unclear. The relation of binary- or multiple-star formation to planet formation also remains unclear. Until these problems are resolved, it is difficult to estimate what fraction of star systems may contain planets. Binary and multiple star systems, which are in the majority, may not contain planets.

IS THERE A MODERN THEORY OF PLANET FORMATION?

Solar system researchers are familiar with the several-century debate between those who attributed planet formation to an unusual catastrophe, such as stellar collision, and those who attributed it to a relatively common stellar evolutionary process. Around 1800, as a result of the work of Laplace and Kant, many naturalists thought the planets evolved from a solar nebula that broke into rings. The plurality of worlds – planets in other star systems – was widely accepted. By about 1900, lacking an explanation for the low angular momentum of the sun relative to the planets, many workers turned toward the catastrophic theories, accepting the idea that planetary systems might be rare. By the middle of our century, researchers felt they could account for solar angular momentum loss through magnetic fields and solar winds, and the pendulum swung back again to evolutionary theories. Today, there seems to be widespread satisfaction with the idea that planetary material forms as a by-product of star formation.

But, an obvious cautionary note is that we still have no proof of the existence of planetary masses near other stars, let alone systems of coplanar, circularly-orbiting planetary masses. At first glance, the best evidence is the astronomic companion of Barnard's star (van de Kamp, 1963, 1969, 1975). But this has been seriously questioned after additional work (Gatewood and Eichhorn, 1973). Published analyses of the system now include models with 0, 1, 2, and 4 planetary companions (see also Jensen and Ulrych, 1973). Thus, to be true to the scientific method in the face of this lack of evidence, we have to retain as a basic question: Did the solar system form by:

a) a rare accident (classical catastrophic theory that predicts planetary systems are very rare);
b) normal stellar evolution (classical evolutionary theory that predicts planetary systems are common); or
c) stellar evolution with moderately specialized circumstances such as (1) angular momentum conditions that prevent binary star formation, or (2) interaction with other protostars, possibly exploding.

Chapters in this book trend strongly away from theory (a), since silicate and possibly icy raw materials have been observed as dust around many presumed newly-forming stars, and since theoretical modeling suggests a certain efficiency for evolutionary aggregation of this material. Of theories (b) and (c), several chapters in this book suggest a possible new trend away from (b), toward (c), the idea that planet formation may require moderately specialized, favorable conditions. Some chapters implicitly or explicitly suggest that one such condition may be the absence of binary or multiple-star formation (perhaps a consequence of initial angular momentum conditions). For instance, Cameron (see his chapter) describes a model solar nebula originally containing 1 – 2 solar masses of material well beyond the present planetary orbits. If so much material was present we must explain why (in the case of our particular system) it was shed to produce planets rather than a multi-star system. Tscharnuter (1978), Bodenheimer and Black (this book) and Tohline (personal communication, 1978), and Boss and Peale (personal communication, 1978) describe numerical models of disk evolution that produce rings likely to break up into massive separate bodies. Boss and Peale suggest these may produce binary and multiple-star systems in many cases. We need to understand better when stars would result and not planets. Greenberg *et al.* (see their chapter in this book) find that formation of one or more additional stars might produce such gravitational perturbations that dust/planetesimal collisions would occur at too high a speed to permit accretionary growth of planets. By analysis and extrapolation of binary secondary star frequencies among solar-type F3 - G2 primary stars, Abt (this book) infers that about 54% have stellar companions and about 13% black dwarf companions, while only about 11% may have companions within the normal planetary mass range.

To answer the question raised in the title of this section, there is not yet a unified, well-accepted modern theory of planet formation, but a body of ongoing work moving sluggishly in a certain direction. It suggests that dust and gas universally form circumstellar nebulae around new stars (see chapter of Bodenheimer and Black in this book; Thompson and Strittmatter, 1978). Some of the dust may survive intact from earlier origin in supernovae or other stellar ejection processes (Clayton, 1978; Greenberg, *et al.* in this book). In any case, the material is likely to be isotopically affected by nearby supernovae (see chapters by Reeves and by Schramm in this book). New dust is almost certain to condense in the cooling circumstellar nebula (Donn, this book; Stephens and Kothari, 1978; Larimer, 1967; Lewis, 1972). The problem of planetary development is the problem of further evolution of the dust and gas components of the nebula, to be discussed in more detail below. Some papers (see Cameron, this book; Bodenheimer, 1976) suggest that gravitational instability of the whole gas/dust complex was the governing process, causing collapse into objects of mass greater than or equal to the final masses of the planets to be explained. (Commonly these giant objects are called *protoplanets*.) Other papers and, I believe, a variety of evidence suggest that at least the terrestrial planets formed by aggregation of a hierarchy of different-sized meteoritic and planetary bodies that began with widely dispersed microscopic material (commonly called *dust grains*), which aggregated into macroscopic material (*planetesimals* – a term used for dimensions of millimeters up to about 1000 km), which in turn finally aggregated into the bodies we see today (*planets*). Cameron and Pollack (1976) even give a planetesimal-type theory of Jupiter's origin, contrary to the protoplanet-type theories mentioned above. To complicate matters, the dust component may have had its own history independent of the larger gas component. The problem of precisely how the planetary material aggregated into planets is a fundamental problem with different chemical consequences. I will organize much of the rest of this review around this problem.

IS THERE STELLAR EVIDENCE RELEVANT TO PLANET FORMATION?

Because we lack good evidence of planets near other stars, as mentioned above, any evidence about planetary matter near other stars would be very important. It could come in several ways. A first way would be actual astrometric or other detection of planet-size masses near other stars. However, the present evidence on this subject is confused. As mentioned above, the well-known proposed companion(s) to Barnard's star have been unconfirmed in recent searches, and Heintz (1978) reviews data on 11 suspected unseen companions, finding no evidence for *any* companion of substellar mass. This includes cases such as Barnard's star, 61 Cygni, and Epsilon Eridani.

Even if a Jupiter-size companion were detected, what does that tell us? An important concept is that mere detection of planet-sized masses near other stars is not sufficient to claim discovery of "other solar systems" generically like ours. To claim other systems truly like our own, we need to confirm roughly circular coplanar orbits, which are indicative that the planetary objects grew in a disk-shaped nebular environment which restricted the major bodies to circular orbits, as apparently happened in our system. The need to confirm this arises because it is conceivable that planet-sized masses could also originate in some entirely different way, perhaps related to the origin of certain types of binary stars that have nothing to do with planetary systems. Is it possible, for instance, that gravitational collapse of a cloud around a new star could produce a star-star or star-planet system with eccentric orbits? Could Jupiter-scale "black dwarf" stars be captured during close encounters between objects in a crowded young star cluster containing temporary resistive interstellar or circumstellar media? Why do multi-star systems generally have non-circular, non-coplanar orbits, while our planetary system has circular coplanar orbits? Answers are unclear, but bear on the question of whether planet formation is a normal by-product of star formation.

A second way to acquire knowledge of extrasolar planetary material would be detection of preplanetary matter around young stars. This has apparently been accomplished. Rydgren, Strom, and Strom (1976) review evidence that many T Tauri stars have silicate emission features, interpreted as coming from optically thin circumstellar silicate dust. According to their review, the dust is typically finally divided, with temperatures around 200 K and distances of several astronomical units from the stars. For a sample of seven stars, Rydgren *et al.* find 10^{-8} to 2×10^{-7} solar masses of silicate particles. They believe the stars are generally less than 10^6 yr old and have masses from 0.5 to 3 $M_{\odot}$.

The varying ratios of thermal infrared to visual luminosity among similar types of proposed young stars is well known, with the infrared excesses being

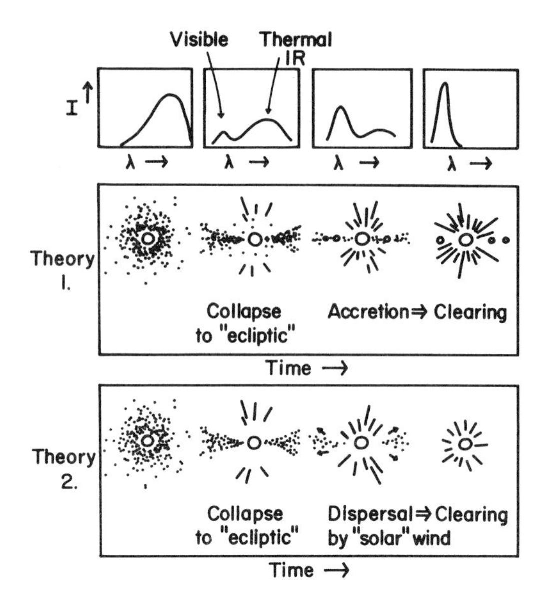

Fig. 1. Schematic outline of problems in interpreting T Tauri and infrared stars. Can the various types of presumed young stars be arranged in a sequence of infrared to visible light ratios, as shown at the top, that represents an evolutionary clearing of circumstellar nebulae? If so, is this clearing due to an aggregation of the dust into planetary bodies, thus reducing cross-section and allowing the light to escape (Theory 1)? Or, is the clearing due to physical removal of the dust from the system (Theory 2)?

attributed to energy re-radiated by the dust after absorption of light from the central star. Some probable young stars are entirely infrared; some are mostly infrared with a little visual light sneaking through; some are mostly visual with slight infrared excess; some are entirely visual. An important question is whether this series represents an evolutionary sequence, as sketched in Fig. 1. If so, which of two possible theories is correct: is the infrared excess and obscuration disappearing because the dust is aggregating into a few large planets (Theory 1 in Fig. 1)? Or is it disappearing because the dust is all being blown away by a strong "solar" wind (Theory 2)? T Tauri stars do show evidence of solar wind mass loss, but no one knows if all the dust will be carried away, if an asteroid-like swarm of bodies is left behind, or if a few planets may be left. Further modeling of young-star mass loss and gas/dust interaction, and more observational knowledge of T Tauri and other young stars might be useful in attacking this problem. Few models of gas blow-off and associated dust behavior in young circumstellar nebulae, as the star "lights up," exist.

In summary, to answer the question posed in the title to this section, yes, there is direct stellar evidence that planetary matter commonly forms around stars, but we have poor evidence on how commonly it aggregates into true planets.

WERE THERE MULTIPLE PROCESSES?

A review of the contents of this book and of earlier years of work highlight an aspect of the question about the basic planet-aggregation process. We, who have theorized about planet formation, have tended to study the problem by considering one process at a time without really comparing the relative importance of likely competing processes. One worker or group of workers may select one process, such as collisional accretion, and demonstrate by calculations that this process might have led to aggregation of particles over certain time scales and size ranges. Another group may select a process such as gravitational collapse or electrostatic attraction, and make a similar demonstration. Another group may show that inclusion of a resisting gaseous medium with certain properties may negate the effectiveness of one process, or perhaps negate its effectiveness as applied to growth over a certain size interval. This is perhaps not an unreasonable state of affairs at the current early stage of this scientific process, particularly since we cannot be sure that all stars produce pre-planetary nebulae with identical properties. Nonetheless, if we are prepared to assert that the early solar system (and perhaps some other star systems) had a surrounding nebular medium with at least roughly known properties, then we must be prepared to come to grips with the question of which processes were most important on what scale.

Too often in the past, the question of aggregation mechanisms degenerated into a question of competing alternative theories. Instead, what should concern us is competing simultaneous processes. Many processes may have occurred in the solar nebula. Perhaps one way to clarify this is to make a careful study of the time scales on which each proposed process would operate at a given stage characterized by a given size distribution of particles. At any given state the aggregation or disruption process with the shortest time scale (fastest growth rate or destruction rate) seems likely to be the most important process. This leaves open the possibility that one process might have dominated at an early stage, say when particles were small, only to be replaced by another process at a later stage.

WHAT WERE THE MAJOR PROCESSES OF AGGREGATION?

Among the processes that have been suggested to aggregate the diffuse material into planets are:

1. gravitational collapse of the nebula, producing large-scale gas and dust protoplanets;
2. condensation, producing growth by chemical combination one molecule at a time;
3. gravitational collapse of the dust components alone, producing small

planetesimals;

4. collisional accretion at low velocities, in which incoming planetesimals rebound or create ejecta moving too slowly to escape;
5. accretion by electrostatic forces;
6. accretion invoking sticky coatings to unite planetesimals;
7. accretion by vacuum welding;
8. accretion by coalescence of semi-molten droplets.

And competing with these is collisional fragmentation, occurring when particles hit each other so fast that the mass of debris exceeds the mass of the incoming body, and flies off so fast that it escapes.

One could arrange the hypothetical growth processes in several ways for further study. However, Process (1) stands out from the rest in a fundamental way. Adoption of this process in a theory leads to the prediction of planets formed from material gathered in a single stage of aggregation with all of the material gravitationally bound, and with special additional mechanisms needed to remove any unwanted mass or chemical constituents. For example, the heavy noble gases would be gravitationally bound to the planet in their cosmic abundances, unless driven off by additional mechanisms. The depletion of such gases in the earth has driven some researchers away from such theories for the terrestrial planets, while others have proposed ways to strip off the massive protoplanetary atmospheres of the terrestrial protoplanets, for example, by a violent solar wind process or tidal forces. Wetherill (see his chapter in this book) also notes that production of planets by gravitational collapse requires a more massive nebula than some researchers are willing to accept. Cameron (see his chapter) discusses such a theory, with the giant planets formed from the original protoplanets and the terrestrial planets having their protoplanetary atmospheres stripped away through their inner Lagrangian points by tidal forces between the sun and protoplanet, reminiscent of Kuiper's (1951) discussion in which protoplanets were formed and limited in size by tidal forces. Wetherill (this book) further discusses the "special series of events" required to form terrestrial planets by gravitational instability. Williams (1975) has also given a good review of protoplanet and related theories. Smoluchowski (1976, p. 5) and Williams (1978) emphasize the difference between Process (1) and processes (2) - (8) in their review papers on planet origin.

In this context, the rest of the processes have the virtue of building planetary material out of many stages of accumulation, starting with small objects that gradually evolve to larger and larger sizes, with different mean sizes and size distributions at different times. These processes, which may be lumped as involving planetesimals, agree with considerable evidence derived from terrestrial planetary material (lunar, terrestrial, and especially meteoritic material). This evidence is summarized in the cartoons of Fig. 2 showing the wide range of sizes extant at different times, indicated by various kinds of evidence. The planetesimal theory accounts for the depletion of noble gases,

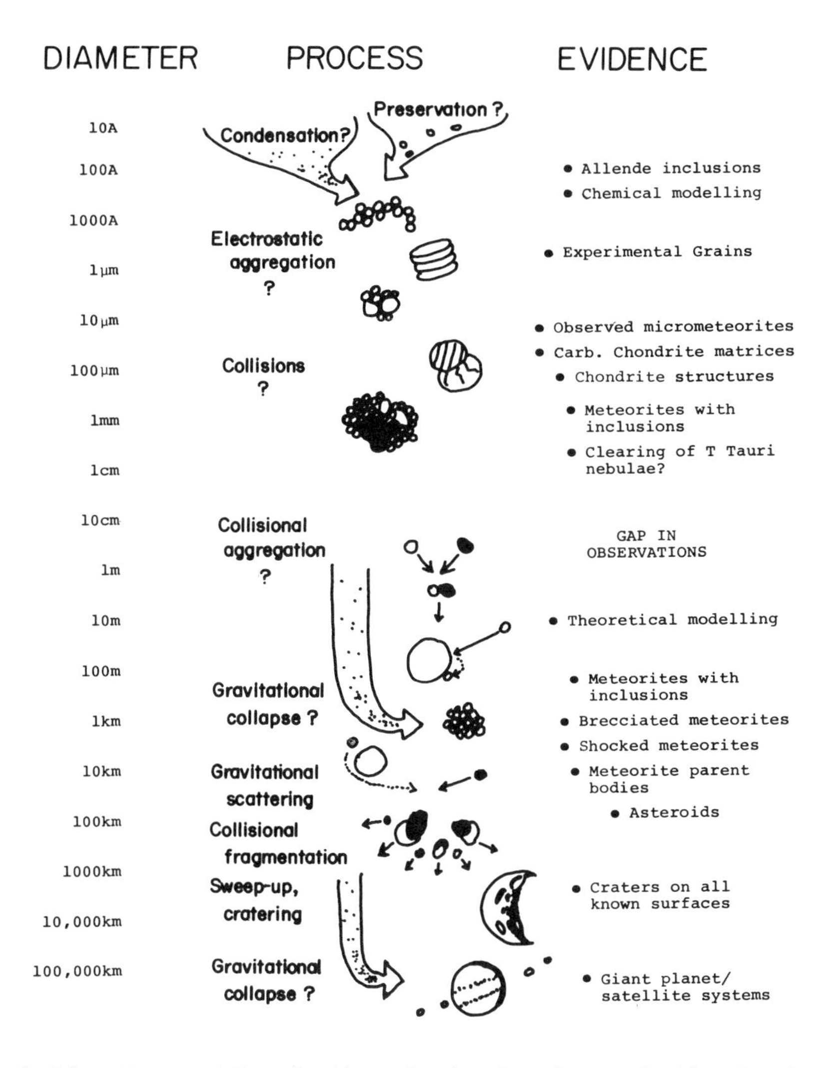

Fig. 2. Schematic presentation of evidence favoring the existence of a hierarchy of planetesimals of many different sizes during the evolution of planets. The cartoons and notations, on the right, summarize direct observational and inferred evidence for structures of various sizes beginning with dust grains which are created in the solar nebula, either by condensation or preservation of interstellar grains, or both. The evidence for objects of these many different sizes at different times in the history of terrestrial material may be inconsistent with single-step formation of terrestrial planets by gravitational collapse into planet-scale masses.

since these would not be retained by low-gravity small objects and would be lost when the remaining gaseous nebula dissipated. Boynton, Herndon and Wilkening, Wasson (see their respective chapters in this book), Trivedi (personal communication) and Margolis (1978) describe theoretical and empirical evidence that grains had different chemical histories in different environments before uniting in our present meteoritic samples; this is hard to explain in a protoplanet theory, but planetesimal theory predicts that small bodies would be gravitationally scattered into other regions by encounters with larger bodies. Physical microstructures such as grain clumps and impact-united chondrules (Lange and Larimer, 1976; Herndon and Wilkening, this book) are explained by collisional accretion of small bodies, some of which survived and preserved the original aggregate structure (though Kerridge [personal communication, 1978] cautions that some of the microstructures in question may result from post-accretion alteration, not aggregation). Derived sizes of meteorite parent bodies – a few hundred km – are also consistent with planetesimal theory and not with protoplanet theory. Safronov and Ruzmaikina (see their chapter in this book), and Greenberg *et al.* (this book) describe physical details and numerical modeling of collisions and near-collisions among planetesimals. Greenberg *et al.* show that simple mechanical collisions can lead to growth. Hartmann (1977) has summarized additional direct evidence that a wide range of planetesimal sizes existed as the planets finished forming; evidence includes the craters on surfaces of various planets, made by the sweep-up of planetesimals ranging from microscopic-size to 100 km or more in diameter. Whipple (1978) describes cometary properties indicating that comets are survivors from the ancient swarm of interplanetary material.

Thus, while dynamical analyses of the giant planets may favor the large-scale gravitational collapse of Process (1) – thus conveniently explaining the regular satellite systems of the giant planets as miniature solar systems – detailed geophysical data seem to favor a hierarchy of interparticle collisions, as in processes (2) - (8) for the terrestrial planets. This raises the scientifically unpalatable possibility of two fundamentally different processes in the solar nebula, leading to planets of different scales. Further progress in determining if one or both types occurred might come from better understanding of the processes and consequences of dissipation of the nebular medium. How much gas was needed to maintain circularized orbits of the growing planetary particles? What was its effect on interparticle velocities, and what became of it (Donnison and Williams, 1977; Weidenschilling, 1978)? Greenberg (1978) suggests that the gaseous dissipative medium may explain some properties of resonant structure in the Trojan L_4 and L_5 clouds of Jupiter's orbit and in the asteroid belt; these properties might thus also shed light on the history and dissipation time scale of the gas. A different clue to the problem may come from the understanding of possible resonances among the planetary bodies (Greenberg, 1977, 1978; see also Greenberg *et al.* in this book). For example, if it could be shown that Jupiter's presence is required to explain the resonant structure or spacing of orbits among other bodies, particularly the terrestrial planets, then this would imply that Jupiter had formed first. Such a formation sequence might again imply formation of giant planets by a gravitational collapse process operating on a faster time scale than the slower interparticle accretion that probably occurred in the inner solar system.

But if interparticle interactions such as processes (2) - (8) did occur in the inner solar system, which of them was most efficient? Which dominated growth? Process (2) – condensation – was almost certainly responsible for the initial microscopic dust component of our solar nebula and of circumstellar material observed around other young stars. Relevant chemical theory has been developed by Larimer (1967) and Lewis (1972). Their explanation of planetary composition as a function of distance from the sun was judged a major advance; it involved the isolation of a condensed dust component from a blown-away gas at a certain stage of nebular cooling. However, controversy remains about the degree to which this sort of model can explain details of planetary chemistry. Other processes such as gravitational stirring (Kaula and Bigeleisen, 1975, Wetherill, 1977) may have complicated the final chemistry. The final few percent of material, added to each growing planet including material that originated in parts of the solar system different from the zone that produced the bulk material of that planet, thus causing a "last-minute" change in the chemistry of the outer layers (Hartmann, 1976).

According to Safronov (1972), Process (3) – the gravitational collapse of the dust layer alone into planetesimals – was first discussed by Edgeworth (1949) and Gurevich and Lebedinskii (1950). It was later developed by Safronov (1972) and Goldreich and Ward (1973), who found that the newly condensed dust would settle toward the ecliptic and collapse into discrete clumps. Safronov found that the clumps in the terrestrial zone would have a mass equivalent to dimensions of a few kilometers, at density 2 gm cm^{-3}, and in the giant planet region about 100 km. He found these would later grow by collisional accretion. Goldreich and Ward similarly found for the terrestrial zone a sequence of dust condensation stages, starting with gravitational collapse into a first generation of objects whose radii range up to 100 m, which aggregate gravitationally into a second generation with radii on the order of 5 km, subsequently to grow by collisional accretion.

Some support for these types of models is found by Greenberg *et al.* (this book) who study Process (4). By using experimental results on detailed mechanics of collision, they show that bodies larger than about one kilometer radius, with a wide variety of compositions and structures, will grow by collisional accretion, but that bodies smaller than dimensions of a few meters (or perhaps a few hundred meters, depending on the nebular model) will not grow by collisional accretion. The growth occurs basically because relative velocities of co-orbiting bodies in a swarm relax to values less than the escape velocities of the larger bodies in the swarm. Thus, the impacts on the larger bodies commonly are so slow that ejecta or rebounding debris rarely escape. For smaller bodies, velocities are maintained above critical minimal values by gas drag effects or by thermal Brownian motions. Thus, Greenberg *et al.* conclude that some other process such as gravitational collapse of the dust causes the initial growth from dust grains to multi-meter planetesimals, whereas collisional accretion takes over after that. Wetherill (1977; see his chapter in this book) also finds that collisional and near-collisional interactions during later growth accounts for solar system properties such as the known decline rate of early intense bombardment on the moon.

Greenberg *et al.* (this book) emphasize that no additional sticking mechanisms are required to produce growth once the planetesimals reach kilometer-scale (or perhaps less), since the mechanics of ordinary collisions will do the job. This does not rule out the possibility of increased efficiency due to "stickiness," however. Alfvén and Arrhenius (1976, pp. 116-119) have emphasized that Process (5) – electrostatic sticking – might be quite important in bonding microscopic dust grains. They demonstrate the process with lunar dust. Even sticky organic coatings (Process 6) on carbonaceous-chondrite-like early particles have been invoked to account for initial aggregation. This idea has been little developed. Similarly, a form of vacuum welding (Process 7) might be important in bonding early bodies that came into contact. This process has been discussed very little and may deserve further study. If newly condensed dusty material is maintained by any mechanism at near-molten temperatures, coalescence of near-molten droplets (Process 8) may also be possible. To illustrate this process, I include two hitherto-unpublished photographs (Fig. 3) of chondrule-like spherules contained in soil samples from ejecta thrown out of the Dial Pack Crater (Roddy *et al.*, 1977), an 82 m diameter crater formed by explosion of 500 tons of TNT. These spherules appear to be glassy droplets of fused material, among which smaller spherules commonly fused to larger neighbors upon collision during free flight. (The soil sample contains numerous smaller examples of single and multiple spherules as well, ranging from about 0.4 to 4 mm diameter.) I am unaware of further documentation of the chemistry or structure of these spherules, but further study of such materials might be productive.

The significant point is that if any of processes 5 through 8 were effective, they could strongly promote collisional growth (any growth resulting from sticking during collisions) through the diameter range of microns to meters, in which classical collisional accretion (Process 4 – growth resulting from collisions, but with bonding by gravity only) is ineffective according to Greenberg *et al.* In other words, if processes 5 through 8 were effective, planets might grow neither by gravitational collapse of the dust component (Process 3), nor by gravitational instability in the larger gas/dust cloud (Process 1), but simply by collisions of neighboring particles. Without processes 4 through 8, gravitational instabilities would presumably be required to obtain planets. The fundamental differences between the competing large-scale gravitational Process (1), the small-scale gravitational Process (3),

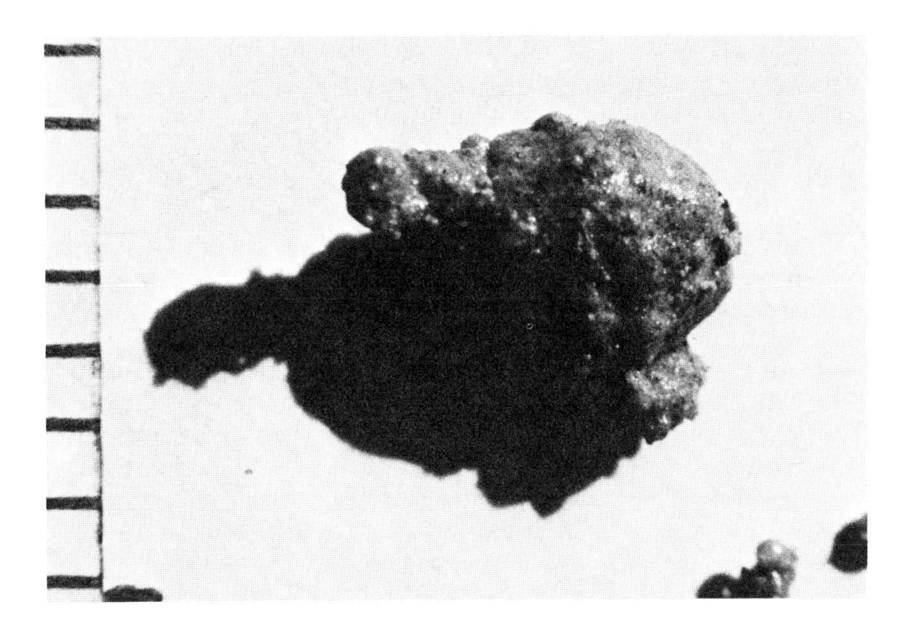

Fig. 3. Photographs of two examples of particles formed by partial coalescence of chondrule-like spherules, apparently in free flight, found in ejecta from Dial Pack Crater. Scales are in millimeters.

the chemical Process (2), and the purely collisional ones (4-8) suggest that much further study of nebular physics is required before we can truly say that we understand planetary formation.

HISTORICAL EPILOG

The previous discussion has been arranged to highlight areas of uncertainty. But to be more positive, we may prefer to take satisfaction in thinking that a new modern view of the evolution of dust into planets has emerged in the last decade or so after detailed analysis of lunar rocks and meteorites, new statistics of asteroids and comets, close-up observations of other planets, and astrophysical modeling of stellar systems.

In this view, dust is ejected from certain stars, and concentrated or re-created by condensation in opaque nebulae around other new stars. In the nebular medium the dust aggregates, and the medium itself guarantees a disk-shaped, planet-growing environment with relatively circular orbits. Some of the newly formed bodies are broken into fragments; some are ejected by near-encounters with large planets, perhaps to form the reservoir of comets; some of the last ones crash into planets causing craters on lunar and planetary surfaces. The largest ones affect the obliquities of the planets. Finally, the planets are left and the planetesimals have nearly disappeared. And this evolutionary view suggests that growth of similar systems around some other stars may occur.

If these ideas seem contemporary and productive, we should pause for a moment and turn with some interest back to the writings of the peculiar turn-of-the-century astronomer. T.J.J. See, whose "unparalleled discoveries" were described by Webb (1913) in a book that included several papers by See himself. In a tabulation of his results (directly quoted, but slightly re-organized in sequence from pages 182, 226-237), See wrote:

> "If there be an incessant expulsion of dust from the stars to form nebulae, with the condensation of the nebulae into stars and stellar systems, . . . certainly the building of these magnificent systems might well engage the attention of the natural philosopher.
>
> "When it was discovered that the roundness of the planetary orbits had arisen from the secular action of a nebular resisting medium, a new ground became available for concluding that the nucleae of the planets were formed by accretion at a great distance from the sun . . .
>
> ". . .disintegration [of comets gave] a resisting medium of cosmical dust for building up the planets and decreasing the semi-major axes and eccentricities of their orbits.
>
> "What, therefore, can the comets be but survivals out of the ancient nebula out of which the solar system has been built up?

"By the preponderance of direct revolving satellites, and the collision of such masses with the planets, their globes are given direct rotation on their axes; and their obliquities thus tend to disappear. . .

"The problem of the obliquities of the planets long presented great difficulty to the astronomers and was not solved until 1908, when it was shown . . . that it is determined by the capture and absorption of small bodies evolving about the sun in planes nearly coinciding with the planets. The countless collisions of these small bodies with a planet like the earth are illustrated by the meteors vaporized in our atmosphere . . . The same downpour of cosmical dust occurs on the other planets . . . as we see by the imbedded satellites which produced the lunar craters.

"The resisting medium implies collision of the larger bodies with the smaller ones, and thus the craters and maria on the moon bear witness to the terrible impacts involved in the creative processes by which the present beautiful order of the solar system was developed.

"Similar impacts once indented the terrestrial globe . . . but since the atmosphere and ocean developed, geological changes have modified the earth by destroying the original craters and building up the mountains . . . running as great walls along the margins of the oceans.

"The roundness of the orbits of the satellites shows that they have suffered innumerable collisions with smaller masses; and naturally they, too, are covered with craters and maria which produce the variability in their light established by the photometric researches of Guthnick.

"Having established how the planets were formed about our sun, we are enabled to affirm that as all stars developed from nebulae, they also have developed about them corresponding planetary systems, or become spectroscopic or visual binary stars . . .

"As these distant planets have the same chemical elements which we are familiar with, and in many cases are habitable, it follows that they also are inhabited by intelligent beings; otherwise, it is necessary to hold that life upon the Earth in the solar system is contrary to the general order of the Universe."

I have used the word "peculiar" in describing T.J.J. See, because he was apparently regarded as an ornery individual who went beyond the bounds of acceptable egotism about his own work, wrote with a very argumentative style and criticized colleagues by name in print for not reading papers he had sent them. He developed enemies as well as friends, and was regarded as having become mentally unbalanced, if not megalomaniacal, in his later life (Ashbrook, 1962). Might not this impressive list of direct quotes make us re-examine our attitudes toward our modern work? What have we really accomplished that is substantially new if such a turn-of-the-century individual could produce such a group of conclusions? The answer to this question should hold some lessons. Presumably, it is not just the list of conclusions that is most important in research, but rather the marshaling of a reasonably significant body of supporting data, such as direct observation, responsible calculations, or chemical analysis.

Therein lies a paradox of summary-chapter-writing. Merely to review conclusions of the chapters in this book, while it may be useful to indicate the content, establishes no more veracity than the listing of See's "unparalleled discoveries." The reader must investigate for himself or herself, the content of the supporting arguments as presented in the rest of this book.

Acknowledgments. I thank P. Schultz for assistance in locating the quotations from T.J.J. See, F. Herbert, and my P.S.I. colleagues, and a referee for helpful comments, A. Olson for assistance with the illustrations, and M.S. Matthews for patient editorial assistance.

REFERENCES

Alfvén, H., and Arrhenius, G. 1976. *Evolution of the Solar System,* NASA SP-345. Washington, D.C.: U.S. Government Printing Office.
Ashbrook, J. 1962. *Sky and Telescope* 24: 193.
Bodenheimer, P. 1976, *Icarus* 29: 165.
Cameron. A.G.W., and Pollack, J.B. 1976. In *Jupiter* (T. Gehrels ed.), P. 61. Tucson: Univ. Ariz. Press.
Clayton, D.D. 1978. *Moon and Planets* (Special Protostars and Planets issue). In press.
Donnison, J.R., and Williams, I.P. 1977. *Mon. Not. Roy. Astr. Soc.* 180: 289.
Edgeworth, K.E. 1949. *Mon. Not. Roy. Astr. Soc.* 109: 600.
Gatewood, G., and Eichhorn, H. 1973. *Astron. J.* 78: 769.
Goldreich, L.E., and Ward, W.R. 1973. *Astrophys. J.* 183: 1051.
Greenberg, R. 1977. *Vistas in Astronomy* 21: 209.
____. 1978. *Icarus* 33: 62.
Gurevich, L.E., and Lebedinskii, A.I. 1950. *Izv. AN SSSR, seriya fizich* 14(6): 765.
Hartmann, W.K. 1976. *Icarus* 27: 553.
____. 1977. in *Comets, Asteroids, Meteorites* (A.H. Delsemme, ed.), p. 277. Toledo, Ohio: Univ. of Toledo Press.
Heintz, W.D. 1978. *Astrophys. J.* 220: 931.
Jenson, O.G., and Ulrich, T. 1973. *Astron. J.* 78: 1104.
Kaula, W.M., and Bigeleisen, P.E. 1975. *Icarus* 25: 18.
Kuiper, G.P. 1951. In *Astrophysics* (J.A. Hynek, ed.), p. 351. New York: McGraw-Hill.
Lange, D.E., and Larimer, J.W. 1976. *Science* 182: 920.
Larimer, J.W. 1967. *Geochim. Cosmochim. Acta* 31: 1215.
Lewis, J.S. 1972. *Icarus* 16: 241.
Margolis, S.H. 1978. *Moon and Planets* (special Protostars and Planets issue). In Press.
Roddy, D.J.; Pepin, R.O.; and Merrill, R.B., eds. 1977 *Impact and Explosion Cratering.* New York: Pergamon.
Rydgren, A.E.; Strom, S.E.; and Strom, K.M. 1976. *Astrophys. J. Suppl.* 30: 307.
Safronov, V.S. 1972. *Evolution of the Protoplanetary Cloud and the Formation of the Earth and the Planets.* Jerusalem: Israel Program for Scientific Translation.
Smoluchowski, R. 1976. In *Jupiter* (T. Gehrels, ed.), p. 3. Tucson: Univ. Ariz. Press.
Stephens, J.R., and Kothari, B.K. 1978. *Moon and Planets* (special Protostars and Planets issue). In press.
Thompson, R.I., and Strittmatter, P.A. 1978. *Moon and Planets* (special Protostars and Planets issue). In press.

Tscharnuter, W.M. 1978. *Moon and Planets* (special Protostars and Planets issue). In press.
van de Kamp, P. 1963. *Astron. J.* 68: 515.
____. 1969. *Astron. J.* 74: 757.
____. 1975. *Astron. J.* 80: 658.
Webb, W.L. 1913. *Brief Biography and Popular Account of the Unparallelled Discoveries of T.J.J. See.* Lynn, Mass.: T.P. Nichols and Sons.
Weidenschilling, S.J. 1978. *Moon and Planets* (special Protostars and Planets issue). In press.
Wetherill, G.W. 1977. *Proc. Lunar Sci. Conf 8th.* p. 1. New York: Pergamon.
Whipple, F.L. 1978. *Moon and Planets* (special Protostars and Planets issue). In press.
Williams, I.P. 1975. *The Origin of the Planets.* London: Adam Hilger, Ltd.
____. 1978. *Moon and Planets* (special Protostars and Planets issue). In press.

PART II

Grains and Chemistry

CURRENT PROBLEMS IN INTERSTELLAR CHEMISTRY

WILLIAM D. WATSON
University of Illinois

Selected, current problems related to the basic ideas in interstellar chemistry are summarized. These include: (1) poor knowledge of the physical conditions, (2) the CH^+ abundance and OH/H_2O ratio in diffuse clouds, (3) isotope fractionation, (4) the role of dust grains and (5) formation of complex molecules.

A knowledge of interstellar chemistry is relevant to "Protostars and Planets" in at least two general ways. In the study of the collapsing interstellar cloud, molecular abundances can serve as a probe for certain quantities such as the fractional ionization and the fractional CO abundance for which alternative diagnostic techniques are insensitive or inaccurate. The fractional ionization is a major factor in determining the importance of magnetic fields on the collapse of the cloud, and the observed CO abundance, along with an assumed CO-to-hydrogen ratio, is frequently utilized as an indicator for the mass of the gas cloud. More indirectly, a knowledge of the chemistry in dense clouds will allow a continuous transition in understanding the molecular constituents of gas clouds from the interstellar to the protostar phase. The "initial conditions" of the solar nebula – especially of the dust grain component – which are determined by the collapsing interstellar cloud may have influenced, for example, certain isotope ratios in meteorites.

The field has not yet reached the stage that a quantitative chemical description of the initial conditions for the protostar can be given. We are still attempting to establish the chemistry of the interstellar cloud phase where we have observational checks. Thus an explicit connection between the interstellar research and the protostar phase does not normally appear. Nevertheless, as a result of research in the mid 1970s, especially into the influence of ion-molecule reactions, it is reasonable to begin serious efforts to consider this relationship in a quantitative manner. Because individual investigators tend to be much more familiar with either the interstellar or solar nebula phase, opportunities for cross fertilization are particularly valuable.

A number of reviews of interstellar chemistry have become available (Dalgarno and Black, 1976; Herbst and Klemperer, 1976; Watson, 1976,

1977*a*), as well as that by E. Herbst in this book. I will thus attempt to minimize duplication by providing a discussion of certain key problems in interstellar chemistry. Except for the treatment of isotope fractionation, the solutions of these problems are directed more toward establishing the basic ideas of interstellar chemistry which must first be understood than toward a direct application to protostellar conditions (i.e., attention is limited to normal interstellar conditions – ambient cloud kinetic temperatures $T \lesssim 100$ K and total gas densities $\lesssim 10^6$ cm^{-3}). Element abundances are assumed to be approximately equal to the "cosmic" values. I will first outline uncertainties about the physical conditions in interstellar clouds which cause difficulty in evaluating the chemical reactions (Sec. I). I will then discuss reactions in diffuse interstellar clouds (Sec. II). These conditions are furthest away from those of protostars, but provide the best opportunity for clear, quantitative tests of molecule reaction processes in the interstellar medium. Here the physical conditions are known best and the number of reaction processes which must be sorted out is most limited. I will continue in Sec. III with topics that can be formulated in a quantitative manner by summarizing work on isotope fractionation – mainly D/H and $^{13}C/^{12}C$ in molecules. Difficulties in assessing the role of reactions at the surface of dust grains will be outlined in Sec. IV. No quantitative proposals to understand the abundances of complex molecules ($\gtrsim$ 4 atoms) in the interstellar medium have been made, but efforts in this direction will be mentioned in Sec. V.

I. PHYSICAL CONDITIONS IN INTERSTELLAR CLOUDS

Quantitative studies of interstellar chemistry coupled with observational verification are severely hampered by poor knowledge of total gas densities, element abundances, grain composition, cosmic ray and starlight fluxes, grain and, to some degree, gas temperatures, as well as the rate of change with time of the conditions. Except in diffuse clouds, total gas densities are known to an order of magnitude at best and there is no information on element abundances other than the constituents of the observed molecules. Uncertainty about element abundances in the gas phase in dense clouds is mainly due to lack of knowledge about the importance of depletion by freezing onto the grains. In dense clouds, the influence of magnetic fields in shielding the interiors from cosmic ray particles is unknown. Although the observed optical extinction across the dense clouds is large and will screen out the starlight incident upon the exterior, the radiation field inside the cloud may be affected significantly by stars which are immersed in the cloud and whose presence cannot be detected from the earth. The importance of most gas-phase reaction schemes that have been proposed is not sensitive to temperature within the range ($T \gtrsim 100$ K) likely to occur. Exceptions are the isotope exchange reactions. In contrast the importance of reactions on surfaces of grains is likely to be extremely sensitive to the grain temperature. Infrared observations now indicate that warmer-than-average ($T \gtrsim 30$ K) grains may be present in a significant fraction of the volume of dense interstellar clouds. Grain temperatures in the general interstellar medium are expected to be less than about 20 K. Although limited information concerning the bulk composition of dust grains is available, surface reactions are governed by the outer few layers about which there is no direct, observational information.

Comparisons of predictions about interstellar clouds are of course hampered by the limitation that column densities are the measured quantities. Thus the observations often reflect a sum over regions of different physical conditions and can thereby introduce further ambiguity.

Most computations of molecular abundances for application to interstellar clouds have, for simplicity, proceeded on the assumption that the abundances are time-independent. Time scales for reaching equilibrium are frequently (1-10) x 10^6 yr even for simple molecules, depending upon the particular species and physical conditions. Since this time scale can be comparable with that for free-fall collapse of the cloud, as well as that for other possible disturbing events, the steady-state assumption may not always be of sufficient accuracy. Investigations of time-dependent effects include those of Aannestad and Field (1973), Oppenheimer and Dalgarno (1975), Langer and Glassgold (1976) and Gerola and Glassgold (1977). In Sec. II, a specific case will be presented in which non-steady-state effects drastically alter predicted abundances in diffuse clouds.

II. DIFFUSE INTERSTELLAR CLOUDS

The diffuse interstellar clouds are the classical ones, observed for many years through their optical absorption in the spectra of bright stars. Gas densities are less than about 10^3 cm^{-3}, and typically are probably 10-100 cm^{-3}. The essential difference in the physical conditions between these and the dense clouds observed mainly at radio wavelengths is the presence in the former of relatively unattenuated galactic starlight at ultraviolet wavelengths. Except for hydrogen, the gas is kept chiefly in atomic form by this starlight. It also ionizes elements with ionization energies below 13.6 eV. Because the abundances, total density, temperature, radiation field and cosmic ray flux are reasonably well known, these clouds serve as laboratories in which to test basic ideas about interstellar chemistry. There seem to be two key problems here – the CH^+ abundance and the H_2O/OH ratio.

The CH^+ Abundances

Since the earliest theoretical studies of interstellar molecule abundances beginning in 1951, the CH^+ abundance has received more attention than that of perhaps any species. The abundance of CH^+ is presumably related closely to the basic reaction processes for carbon in the interstellar medium. It has been established that the observed abundance cannot reasonably be reproduced by proposed mechanisms if the rate for

$$CH^+ + e \rightarrow C + H \tag{1}$$

is near that which is typical for molecular ions (i.e., $\langle \sigma v \rangle \simeq 10^{-7}\ cm^3 s^{-1}$) where σ is the cross-section and v the velocity (e.g., see the review by Dalgarno, 1976 for an extensive survey of the CH^+ problem). Subsequently, Mitchell and McGowan (1978) have measured the rate down to 0.0044 eV and find $\langle \sigma v \rangle = 1.4 \times 10^{-7} cm^3 s^{-1}$ at 0.01 eV.

This desperate situation has led to the proposal (Elitzur and Watson, 1978) that CH^+ is not in fact produced under the steady-state conditions in the interstellar clouds. Instead it may be generated in the hot, neutral gas behind a shock propagating outward from the H II region around the bright star toward which the observations are made. Temperatures near 4000 K may be attained in such a shock so that the endothermic reaction ($\Delta E = 0.4$ eV),

$$C^+ + H_2 \leftrightarrows CH^+ + H \tag{2}$$

can be driven to produce CH^+. The essential ingredients in this proposal are the presence of a small H_2 fraction in the pre-shock gas, a shock with adequate power, and an accurate knowledge of the cooling time behind the shock. The cooling time is uncertain due to lack of knowledge about the cross sections for

$$H + H_2(J) \rightarrow H + H_2(J') \tag{3}$$

where J is the rotational quantum number. Although the presence of the shock is almost required by the hot star, there is no observed information about its velocity. The required pressure behind the shock is, however, indicated by that of the H II region.

If this extremely time-dependent phenomenon is responsible for CH^+, it is likely that molecule formation in general will be significantly affected by such processes. The calculations for CH^+ also examine CH, OH and H_2O in diffuse clouds. Predicted column densities of CH are unaffected. Whether the predicted OH and H_2O column densities will be affected depends upon the exact shock conditions.

The H_2O/OH Ratio

A key reaction scheme in recent investigations of interstellar chemistry in the gas phase is

$$\begin{aligned} H^+ + O &\rightarrow H + O^+ \\ O^+ + H_2 &\rightarrow OH^+ + H \\ OH^+ + H_2 &\rightarrow H_2O^+ + H \\ H_2O^+ + H_2 &\rightarrow H_3O^+ + H \end{aligned} \tag{4}$$

followed by

$$
\begin{aligned}
H_3O^+ + e &\rightarrow OH + H_2 \\
&\rightarrow H_2O + H.
\end{aligned}
\tag{5}
$$

Although the branching ratio in the above electron recombination is not known, OH and H_2O would each normally be expected to result in a non-negligible fraction of the events. In, for example, the cloud toward Zeta Ophiuchi, gas densities are such that the rate for destruction is not greatly different for the two species. It is probably due mainly to

$$C^+ + OH \rightarrow CO + H^+, \text{ or } CO^+ + H \tag{6}$$

and

$$C^+ + H_2O \rightarrow HCO^+ + H. \tag{7}$$

It is thus quite surprising that an upper limit,

$$OH/H_2O \gtrsim 50 \tag{8}$$

has been deduced (Smith and Zweibel, 1976). Better information on the oscillator strength at ultraviolet wavelengths through which the upper limit for H_2O is derived would be most desirable. Herbst (1978) has recently studied the electron recombination of H_3O^+ and concludes that OH may be favored by a factor of ten. If so, miscellaneous other uncertainties can probably account for the remaining factor of five.

III. ISOTOPE FRACTIONATION

The Deuterium-Hydrogen Ratio

Fractionation of deuterium in interstellar molecules is severe and for molecules in dense clouds it cannot be predicted without a knowledge of quantities (e.g., electron density) whose values ordinarily are unavailable. Thus deuterium from molecule observations does not provide an indication of its galactic distribution. Since the D/H ratio (all forms) probably does not vary by more than a factor of two to three, except possibly in the galactic center, the D/H ratio in molecules can be utilized as a probe of the physical conditions.

Because of their direct formation process, HCO^+ and N_2H^+ seem to be the best species to examine. The result is that the upper limits to the electron density, as well as to the CO and N_2 abundances, can be deduced (Watson, 1977*b*),

$$
\begin{aligned}
[e]/[H_2] &\lesssim 3 \times 10^{-8} \sqrt{T/205\ K}\, B^{-1} \\
[CO]/[H_2],\ [N_2]/[H_2] &\lesssim 2 \times 10^{-6} B^{-1}
\end{aligned}
\tag{9}
$$

where B is the ratio of the abundances of the deuterated species to that with hydrogen. Uncertainties arise in deducing the above limits, due both to interpretation of the astronomical data and to the reaction cross-sections. Radio astronomers observe the emission lines from the $J = 1\text{-}10$ transitions of HCO^+ and N_2H^+. Conversion of these astronomical data to the column densities needed for the chemical analysis involves uncertainty due to line saturation effects. The key processes in the chemical reaction scheme are then

$$\begin{aligned} H_3^+ + HD &\rightleftarrows H_2D^+ + H_2 \\ H_3^+ + e &\rightarrow 3H, H_2 + H \\ H_3^+ + CO &\rightarrow HCO^+ + H_2 \\ H_3^+ + N_2 &\rightarrow N_2H^+ + H_2 \end{aligned} \tag{10}$$

Rate coefficients for the last two reactions seem well established, though the direct relevance of the laboratory data on the first two has been challenged. A laboratory measurement of the cross-section for the electron recombination has been made down to 0.01 eV whereas the energy in the astrophysical application is near 0.002 eV. Although the laboratory data show no evidence for a deviation from the $\langle \sigma v \rangle \propto T^{-1/2}$ dependence, theoretical arguments have been presented in favor of a substantially smaller rate coefficient at energies below those of the measurements (Porter, 1977*a*). The inability of straightforward calculations to reproduce the measured rate coefficient for the first reactions suggests that a subtle mechanism may be involved and thus casts some doubt on the extrapolation of the measured rate to interstellar gas temperatures (Porter, 1977*b*). Further, the foregoing analysis assumes that the rate coefficients for dissociative electron recombination of H_3^+ and H_2D^+, HCO^+ and DCO^+, N_2H^+ and N_2D^+ are approximately equal. Laboratory studies have now found the rate coefficient for H_3^+ to be no more than a factor of two different from that for HD_2^+ and D_3^+ (McGowan, 1978, personal communication).

The fractional electron densities ($\lesssim 10^{-8}$) derived from the relations given above when B is near one, as suggested by observation, are sufficiently low that separation of the magnetic field and the bulk gas is likely. More laboratory studies of the reactions are clearly needed before confidence can be placed in these electron densities.

Other than hydrogen, carbon is the only element for which specific mechanisms involving isotope fractionation have been proposed. There is reason to believe that fractionation is unlikely for other elements (Watson, 1977*b*). Possible scenarios for carbon isotope fractionation rest upon the exchange reaction,

$$^{13}C^+ + {}^{12}CO \leftrightarrows {}^{12}C^+ + {}^{13}CO + \Delta E \tag{11}$$

for which the rate coefficient at 300 K has been measured to be 5×10^{-10} $cm^3 sec^{-1}$ (Watson, Anicich and Huntress, 1976) and for which the exothermicity is $\Delta E/k = 35$ K. Carbon monoxide is normally thought to contain most of the gaseous carbon in the dense interstellar clouds. Hence the above reaction cannot alter significantly the $^{13}C/^{12}C$ ratio in carbon monoxide as long as the carbon in the gas is conserved.

To understand the apparent enhancement of the ratio in CO, Watson *et al.* (1976) proposed that the high vapor-pressure of CO in comparison with that of other carbon-bearing molecules prevents CO from freezing out of the gas onto grains. When other carbon-bearing molecules do freeze out, carbon-12 is then preferentially lost from the gas since they contain relatively less of the carbon-13 as a result of the above exchange reaction. The remaining carbon in the gas, which is recirculated among the various molecules including CO through the formation and destruction processes, is thus enhanced in carbon-13. Factors of two-to-three enhancement can be attained if the conditions are favorable. Detailed numerical analyses of the proposed effect under various interstellar cloud conditions have been performed by Langer (1976) and by Liszt (1978).

Enhancement of the $^{13}C/^{12}C$ ratio in interstellar molecules over the solar system value (see Rahe and Vanýsek, 1978) may of course be due in part, or solely, to nuclear processing of material by stars. Observations do suggest molecules-to-molecule variations, as well as variations across sources and from source-to-source. In one case (HC_3N), a variation of $^{13}C/^{12}C$ in the various locations for carbon atoms may have been detected (Churchwell, Walmsley and Winnewisser, 1977). If these interpretations of the observational data are correct, at least some chemical fractionation is indicated.

IV. ROLE OF DUST GRAINS IN INTERSTELLAR CHEMISTRY

The status of this topic has recently been summarized (Watson, 1977*c*). Unfortunately, little progress has been made in the past five years when considerable activity based on gas-phase reactions occurred. It remains clear that molecular hydrogen is formed primarily through recombination at the surfaces of dust grains. Interpretation of observations by the *Copernicus* satellite strongly support this conclusion reached earlier from a theoretical viewpoint. The key difference between H_2 and other molecules with regard to surface reactions is the low adsorption energy for H_2 on likely interstellar grain surfaces. It is sufficiently low that the product H_2 can be returned to the gas for grain temperatures expected for the general interstellar medium ($\simeq$10-20 K). In contrast, even the physical adsorption energy for other species is large enough that they will freeze onto the grains below 20 K. Infrared observations which suggest that grain temperatures greater than about 40 K may be present in a large fraction of the volume of molecular clouds are especially relevant to this question.

Because accurate predictions about the products of surface reactions cannot be made, it is normally difficult to consider the presence or absence of any species other than H_2 to be evidence that such reactions are contributing to the observed molecular abundances. The nitrogen hydrides in diffuse interstellar clouds are an exception to this statement. If hydrogen attachment to carbon and oxygen on grains is a major process in forming CH, CH^+ and OH, NH should also be present at a detectable level. The observational upper limit is significantly below the predicted abundance.

V. FORMATION OF "COMPLEX" MOLECULES

Quantitative proposals for the formation of complex (greater than 4-atom) molecules are not available. It is an open question whether surface or gas-phase processes dominate in the interstellar medium. Because gas-phase synthesis is normally associated with the requirement that a number of successive reactions must occur for the conversion of atoms to large molecules, surface processes have seemed more likely to many investigators. The difficulty is again the low temperature of the grains. However, the larger molecules are found in the relatively dense gas clouds from which infrared emission by the grains is frequently observed. Higher grain temperatures may thus be correlated with the presence of complex molecules. More extensive studies to establish whether the correlation exists would be valuable.

On the other hand, detection of HC_7N in a dark cloud may be highly significant. Dust clouds are not normally thought to contain the bright stars needed to heat the grains to temperatures much above that due to the background galactic starlight. Further studies of complex molecules in dark clouds along with infrared observations to determine whether warm grains are present would be most valuable in delineating the formation processes for complex molecules.

Formaldehyde H_2CO is the most complex molecule for which quantitative proposals to understand its abundance have been made. Formaldehyde occurs widely in the galaxy. Understanding its abundance through gas-phase reactions may be suggestive that gas-phase reaction schemes produce the more complex molecules. Although the formaldehyde problem has received considerable attention, it is unclear at present whether the proposed mechanisms are adequate to produce the observed abundances within the constraints of the astrophysical environment (e.g., Watson, 1977*a*).

VI. SUMMARY

I have emphasized in this chapter the deficiencies in our understanding of interstellar molecule reactions. The successes have also been considerable, especially in view of the poor knowledge of physical conditions in the gas clouds and of the lack of much-needed laboratory data. The accomplishments are however summarized adequately in the reviews cited at the beginning of this chapter.

Acknowledgment. The author's research is supported by the National Science Foundation and by an A. P. Sloan Foundation Fellowship for Basic Research.

COMMENTS

H. Alfvén: It is not very likely that cosmic plasma is due to cosmic rays. The presence of magnetic fields shows that there are electric currents. If one uses homogeneous models one can show that, in order to produce the galactic magnetic field, the drift velocity of electrons need not be very large. Homogeneous models are, however, often misleading. In all regions of space where *in situ* measurements have been made, the currents flow predominately in filaments which surface, and the electrons have large velocities. [Ed. Note: also see Comments following the chapter by Mouschovias.]

W. D. Watson: We have obtained information on the ionization rate of hydrogen in interstellar clouds by various means. In diffuse interstellar clouds, the abundance of the HD molecule provides such data. In dense interstellar clouds, the probes are less precise. However, order-of-magnitude information is provided by the abundances of the probably dominant molecular ions (HCO^+, N_2H^+) and by the electron density inferred from the (D/H) ratio in molecules. All the data are compatible with the ionization expected from high-energy cosmic rays ($\gtrsim$ 100 MeV/nucleon).

B. Donn: Data used on all kinds of effects on small grains (e.g., cosmic ray sputtering, ionization, ultraviolet radiation) are based on experiments on bulk samples. Results may be very different on submicron grains. Experiments on such systems are needed and, although difficult, they are, I believe, possible.

W. D. Watson: Our information of such effects on small grains is indeed incomplete. Some such effects have been examined theoretically (e.g., the thermal spikes that occur due to the low heat capacity).

L. Spitzer: According to Evans (this book) the ratio of molecules to hydrogen is greater in the low-temperature clouds (his Type A) than in those of somewhat higher temperature (Type B). Can this result be explained on the basis of gas-phase reactions?

W. D. Watson: I see no clear explanation for this phenomenon based on chemical reaction. It seems more likely that, if the effect is real, it is due to other causes such as, for example, the more rapid the rate at which heavy elements condense onto grains at higher gas densities of Type B, the greater the shielding of cosmic rays by larger mass of Type B clouds. In the case of ions (e.g., HCO^+), exchange reactions are likely to reduce their relative abundances at the higher gas densities. Also, some investigators (F. Arnold; D. Smith and N. Adams) have advocated radiative association rates that increase rapidly as the temperature decreases.

H. J. Völk: Regarding the formation of complex molecules and the significance of the $n[HCO^+]/n[H_2]$ vs $n[H_2]$ relation, Arnold (1977) suggested that the radiative cluster formation of, say, HCO^+ with H_2 may be proceeding rapidly in dense, cold molecular clouds. This could happen because of a possibly strong decrease of the molecular decomposition rate with decreasing temperature. Under these circumstances n$[HCO^+]$/n$[H_2]$ would in fact become rather proportional to 1/n$[H_2]$ instead of being proportional to $1/\sqrt{n[H_2]}$ (as would be true if HCO^+ was primarily lost by electron recombination).

A. Wootten: Are the limits on the $[N_2D^+]/[N_2H^+]$ ratio in L 134 consistent with H_2D^+ + CO producing all of the DCO^+ (i.e., can a contribution from CH_2D^+ + O be ruled out)?

W. D. Watson: The most reliable information on $[DCO^+]/[HCO^+]$ comes from the $H^{13}CO^+$ observations at L 134N, a different location in L 134 from where the $[N_2D^+]/[N_2H^+]$ ratio is measured. In any case, optical depth effects in N_2H^+ and possibly DCO^+ as well as uncertainty about the exact $^{13}C/^{12}C$ ratio prevent comparisons that are closer than factor-of-three accuracy. The observations are quite compatible with all the DCO^+ being produced from H_2D^+ + CO, but do not exclude comparable contributions from CH_2D^+ + O. From the viewpoint of models for cloud chemistry, I believe the latter is unlikely.

B. G. Elmegreen: What is the H_2 density in the region where DCO^+ and HCO^+ are observed?

W. D. Watson: The current accepted value is about $10^4 cm^{-3}$.

REFERENCES

Aannestad, P., and Field, G. 1973. *Astrophys. J. Suppl.* 25: 205.

Arnold, F. 1977. In *Proc. 21st Liège Astrophys. Symp.* In press.

Churchwell, E.; Walmsley, C. M.; and Winnewisser, G. 1977. *Astron. Astrophys.* 54: 925.

Dalgarno, A. 1976. In *Atomic Processes and Applications* (P. G. Burke and B. L. Moisewitsch, eds.) p. 109. Amsterdam: North-Holland.

Dalgarno, A., and Black, J. 1976. *Rep. Prog. Phys.* 39: 573.

Elitzur, M., and Watson, W. D. 1978. "Formation of Molecular CH^+ in Interstellar Shocks." Submitted to *Astrophys. J.* (Lett.).

Gerola, H., and Glassgold, A. 1977. *Astrophys J. Suppl.* In press.

Herbst, E. 1978. *Astrophys. J.* In press.

Herbst, E., and Klemperer, W. 1976. *Phys. Today.* 29: 32.

Langer, W. D. 1976. *Astrophys. J.* 212: L39.

Langer, W. D., and Glassgold, A. 1976. *Astron. Astrophys.* 48: 395.

Liszt, H. 1978. "Time Dependent CO Formation and Fractionation." In press.

Mitchell, J. B., and McGowan, J. W. 1978. "The Dissociative Recombination of CH^+." Submitted to *Astrophys. J.* (Lett.).

Oppenheimer, M., and Dalgarno, A. 1975. *Astrophys. J.* 200: 419.

Porter, R. 1977*a*. *J. Chem. Phys.* 66: 2756.

______. 1977*b*. *Adv. Chem.* In press.

Rahe, J., and Vanýsek, V. 1978. *Moon and Planets* (Special Protostars and Planets issue). In press.

Smith, W., and Zweibel, E. 1976. *Astrophys. J.* 207: 758.

Watson, W. D. 1976. *Rev. Mod. Phys.* 48: 513.

______. 1977*a*. *Accounts Chem. Res.* 10: 221.

______. 1977*b*. In *CNO Isotopes in Astrophysics* (J. Andouze, ed.) p. 105. Dordrecht: Reidel.

______. 1977*c*. In *Proc. 21st Liège Astrophys. Symp.* In press.

Watson, W. D.; Anicich, V.; and Huntress, W. T. 1976. *Astrophys. J.* 205: L165.

THE CURRENT STATE OF INTERSTELLAR CHEMISTRY OF DENSE CLOUDS

ERIC HERBST
College of William and Mary

During the 1970s, the field of interstellar chemistry has exanded rapidly. Much attention has been focused on dense clouds, the environment in which all complex interstellar molecules have been located. In this article, attention is paid to the two principal theories of molecule formation – gas-phase reactions and reactions on the surfaces of interstellar dust grains – as they pertain to dense clouds. Progress in each of these theories is discussed, with the emphasis on gas-phase reactions because these processes are better understood. It is concluded that, although our understanding of dense cloud chemistry has improved markedly, much remains to be learned.

As of 1977, forty-seven different molecular species have been detected in the gas phase of interstellar clouds. These molecules range in complexity from diatomics such as H_2 and CO to "complex" nine-atom species such as C_2H_5OH (ethanol) and $(CH_3)_2O$ (dimethyl ether). Table I contains a listing of the observed interstellar molecular species. All of the polyatomic molecules in this table have been detected by radio and microwave techniques, whereas some of the diatomic species have been observed in the visible and/or ultraviolet regions of the spectrum. In addition to the forty-seven species listed in Table I, numerous isotopic variants have been observed. Most molecules have been seen only in our galaxy although several have been sighted in extragalactic sources.

Interstellar clouds, the environment of the observed molecules, contain roughly 10% of the mass of our galaxy. Individual clouds range in extent from 1 to 30 pc and in gaseous density from 10 to 10^6 cm^{-3}. The gas is mainly hydrogen, in its H I, H II, and H_2 forms. Regions where hydrogen is chiefly H II are within ~10 pc of a hot star and are not hospitable to molecular development. In addition to a gaseous phase, interstellar clouds typically contain dust grains of size ~0.1 μm, uncertain composition, and total mass 1% of the cloud mass. Clouds where the hydrogen is chiefly neutral possess translational temperatures in the range 5 to 100 K.

Neutral interstellar clouds can be labeled conveniently, if simplistically, as diffuse or dense. Diffuse clouds have gaseous densities less than 10^3 cm^{-3} and visual extinctions less than one magnitude. The gas in these regions is

TABLE I

Observed Interstellar Molecules

		DIATOMIC	
H_2	hydrogen	C_2	diatomic carbon
CH	methylidyne	OH	hydroxyl
CH^+	methylidyne ion	SiO	silicon monoxide
CN	cyanogen	SiS	silicon sulfide
CO	carbon monoxide	NS	nitrogen sulfide
CS	carbon monosulfide	SO	sulfur monoxide
		TRIATOMIC	
H_2O	water	HCO^+	formyl ion
H_2S	hydrogen sulfide	HN_2^+	diimine radical ion
C_2H	ethynyl	HNO	nitroxyl
HCN	hydrogen cyanide	OCS	carbonyl sulfide
HNC	hydrogen isocyanide	SO_2	sulfur dioxide
HCO	formyl		
		4-ATOMIC	
NH_3	ammonia	HNCO	isocyanic acid
H_2CO	formaldehyde	C_3N	cyanoethynyl
H_2CS	thioformaldehyde		
		5-ATOMIC	
H_2CNH	methanimine	HCOOH	formic acid
CH_2CO	ketene	HC_3N	cyanoacetylene
H_2NCN	cyanamide		
		6-ATOMIC	
CH_3OH	methanol	$HCONH_2$	formamide
CH_3CN	cyanomethane		
		7-ATOMIC	
CH_3NH_2	methylamine	C_2H_3CN	cyanoethylene
CH_3C_2H	methylacetylene	HC_5N	cyanodiacetylene
CH_3CHO	acetaldehyde		
		8-ATOMIC	
$HCOOCH_3$	methyl formate	CH_3C_3N	methyl cyanoacetylene
		9-ATOMIC	
CH_3OCH_3	dimethyl ether	C_2H_5CN	cyanoethane
C_2H_5OH	ethanol	HC_7N	cyanotriacetylene

principally atomic except, in some cases, where hydrogen is divided equally between H I and H_2. No polyatomic molecules have been observed in very diffuse clouds. Dense clouds possess densities greater than $10^3 cm^{-3}$ and high visual extinctions ($\geqslant$7 mag). The high extinction protects molecules from interstellar radiation which would, if not extinguished, result in photodestruction within ~100 years. There is convincing although indirect evidence that H_2 is the overwhelmingly dominant molecular constituent of the gas in dense clouds. Microwave emission spectra from dense clouds are signatures for specific molecular species and can be analyzed to yield order-of-magnitude estimates of molecular column densities. Unfortunately, only molecules with a permanent electric dipole moment possess strong microwave (rotational) spectra so that molecules such as N_2, CO_2, CH_4 and C_2H_2 cannot be observed directly. Table II contains a list of selected molecular fractional abundances in dense clouds. The large assortment of unusual and "unstable" chemical species indicates that dense clouds are not in thermodynamic equilibrium. From the observational viewpoint, it is unclear whether or not there are significant residual heavy element atomic abundances in the interiors of dense clouds. The normal cosmic abundance ratios (Allen, 1973) of H(1), He(0.09), C(3×10^{-4}), N(9×10^{-5}), and O(7×10^{-4}) are not necessarily indicative of C,N,O atomic abundances since heavy elements such as C,N,O, can be depleted significantly from the gas phase onto the grains.

TABLE II

Selected Molecular Fractional Abundances[a]

$>10^{-6}$	10^{-6}	10^{-7}	10^{-8}	10^{-9}	10^{-10}
CO	HCN	OH	CH	HNCO	H_2CS
N_2 [b]	HNC	CS	CN	NH_2CN	HCOOH
H_2O(?)	NH_3	SO	OCS	CH_3C_2H	CH_2NH
		HCO^+	H_2S		$HCONH_2$
		N_2H^+	HCO		CH_3NH_2
		C_2H	H_2CO		C_2H_3CN
		SO_2	HC_3N		CH_3CHO
		CH_3OH			HC_5N
					$HCOOCH_3$
					CH_3C_3N
					CH_3OCH_3
					C_2H_5OH

[a]Watson, W. D. 1976. *Rev. Mod. Phys.* 48: 513 and selected references.
[b]Indirect evidence only.

The field of interstellar chemistry has arisen in an attempt to understand how interstellar molecules are formed. Much fruitful work has gone into explaining the chemistry of diffuse clouds (Solomon and Klemperer, 1972; Watson, 1973*a;* Black and Dalgarno, 1973*a;* Glassgold and Langer, 1976*a;* Mitchell, Ginsburg and Kuntz, 1977; Black and Dalgarno, 1977), but it is our primary goal here to review attempts at explaining the molecular complexity of dense clouds. In this chapter, the well-developed explanations – local gas-phase reactions and grain surface reactions – will be discussed, and the more speculative approaches – external molecular sources, shock waves, grain spallation – will be neglected. A recommended review of the important physical processes in interstellar clouds has been written by Watson (1976; also see Watson's chapter in this book).

"EARLY" CALCULATIONS 1972-75

The formation of polyatomic interstellar molecules on grain surfaces was first considered in a detailed manner by Watson and Salpeter (1972*a,b*) although previous investigators (e.g., Hollenbach, Werner and Salpeter, 1971) had studied the formation of H_2 on grains. At the low grain temperatures of neutral interstellar clouds (5-20 K), the sticking of gaseous species onto grain surfaces by physical (van der Waals) adsorption is likely. It is thought that adsorbed species, if sufficiently mobile, can associate to form larger molecules because the grain can take up the chemical bonding energy. In a binary gas-phase collision, on the other hand, the two colliding species can associate only if the chemical bonding energy is radiated away.

Watson and Salpeter (1972*a,b*) found that in regions where H I (a mobile species) existed on grain surfaces, reactive atoms such as C,N,O hitting a grain would react to form some hydrogenated molecule, either a radical such as CH,NH,OH or a saturated species such as CH_4,NH_3,H_2O, by successive association with hydrogen atoms. Similar conclusions had been reached by Duley (1970). Association with H_2 would be more difficult because of probable "activation energy barriers" (potential barriers) to reaction. (Such barriers are quite common in chemical reactions between two stable molecules.) Molecules could leave the grain surface by ejection during the chemical reaction forming them or by photoejection (Greenberg, 1973). Complex molecules could be formed by reaction of molecules remaining on the grain surface if the reactions occurred without activation energy. Destruction of molecules released into the gas phase would occur by gas-phase reactions (producing different chemical species) and photodestruction. Many of these aspects of surface reactions were independently discussed by Aannestad (1973*a*) for diffuse clouds.

Based on the above analysis, Watson and Salpeter (1972*b*) estimated abundances for various molecules. However, two specific difficulties with their arguments arise in the context of dense cloud chemistry; the abundance of atomic hydrogen is probably quite low, and sufficient radiation to promote photoejection may not be present. This latter difficulty has been possibly overcome by Allen and Robinson (1975, 1976, 1977) who suggested in 1975 that molecules could be formed on interstellar grains so small ($\leqslant 0.04\mu m$) that the energy liberated when a chemical bond is formed could heat up a grain sufficiently to release adsorbed species. Earlier, Aannestad (1973*b*) had considered shock waves as a mechanism for driving adsorbed molecules off grain surfaces. Allen and Robinson (1977) also suggested a resolution of the atomic hydrogen dilemma (see below). However, all grain models suffer because of extreme uncertainties in our knowledge of the relevant physical processes. Gas-phase processes are better understood and, with the exception of H_2, explain diffuse cloud chemistry very well (Black and Dalgarno, 1977). But can gas-phase processes produce complex interstellar molecules?

Herbst and Klemperer (1973) proposed a model of polyatomic molecule formation in which binary gas-phase reactions were utilized exclusively. (Three-body reactions are exceedingly unlikely under typical cloud conditions.) We noted that exothermic ion-molecule reactions of the type $A^+ + B \rightarrow C^+ + D$ often proceeded without activation energy in the laboratory and, consequently, could occur rapidly even at interstellar cloud temperatures. Positive ion production could occur in dense clouds via cosmic ray-induced ionization processes. Polyatomic ions would then be synthesized by ion-molecule reactions and polyatomic neutrals produced by ion-electron recombination reactions of the type $A^+ + e \rightarrow B + C$. Utilizing the initial conditions that H_2 and CO were present at their maximum cosmic abundance limited values, we calculated steady-state (*not* equilibrium) abundances of 35 molecular species containing the elements H,C,N,O of up to five atoms as a function of gas intensity in the range 10^4-10^6 cm^{-3}. The role of ion-molecule reactions in interstellar chemistry was also explored thoroughly by Watson (1973*b*, 1974). The subsequent confirmation of the presence of the molecular ions HCO^+ (Woods *et al.*, 1975) and N_2H^+ (Saykally *et al.*, 1976) in dense clouds gave strong support to the idea that ion-molecule reactions were an important ingredient of dense cloud chemistry.

Although the detailed model of Herbst-Klemperer (1973) provided qualitative and even semi-quantitative agreement with observation, it was flawed in at least several respects:

(a) The assumption that H_2 and CO were present initially ignored the question of how these species had been formed and if they were anywhere near steady-state abundances.
(b) Elements other than H,He,C,N,O were not considered.
(c) Molecules of more than five atoms were excluded.
(d) Some important reactions were neglected.

Improvements on our model were not long in coming. Black and Dalgarno (1973*b*) showed how C^+, an abundant ion in the Herbst-Klemperer scheme, could associate radiatively with H_2 to produce CH_2^+, an important precursor to organic molecule formation. Dalgarno, Oppenheimer and Black (1973) found a more efficient method of producing formaldehyde (H_2CO) than the Herbst-Klemperer mechanism. Oppenheimer and Dalgarno (1974*a,b*) extended the ion-molecule scheme to include the elements sulfur, silicon, and various metals. One effect of these additional elements in a steady-state calculation was to increase the dense cloud fractional ionization from 10^{-7} -10^{-8} (according to Herbst and Klemperer, 1973) to $\sim 10^{-6}$. Glassgold and Langer (1975) started their investigations on how C^+, the dominant carbon species in diffuse clouds, is transformed by gas-phase reactions into CO in dense clouds. Thus, even by the end of 1975, the gas-phase reaction model of dense cloud chemistry was becoming more complex and comprehensive.

RECENT MODELS

Langer and Glassgold (1976) investigated the time scales for molecule formation by ion-molecule reactions. (Time dependence in diffuse clouds had been discussed previously by Oppenheimer and Dalgarno [1975].) They found that the necessary times for CO, O_2 and H_2O to reach steady-state abundances were at least comparable to evolutionary (e.g., gravitational collapse) time scales in dense clouds. Herbst and Klemperer (1973) had shown that once large amounts of CO were formed, most polyatomic molecules could reach steady state rapidly. But, if the time scale for CO formation is long compared with the evolutionary time scale, then the chemistry and evolution of a dense cloud are coupled. This is an unfortunate situation because the evolution of dense clouds is poorly understood. Time evolution can be incorporated into a detailed cloud model in several ways as follows:

(a) We can ignore it.
(b) We can assume fixed physical conditions and known initial chemical conditions (e.g., pure atoms) and then solve the time-dependent chemical kinetic differential equations (or approximate them as Langer and Glassgold [1976] did).
(c) We can assume free-fall (or slightly more complicated) gravitational collapse to derive a time-dependent gas density and proceed as in (b):
(d) We can attempt to solve a hydrodynamic model incorporating gravitational, thermal and chemical evolution.

Is there any justification for approach (a)? In our view, there are two possible reasons for ignoring time dependence. Firstly, the physical evolution of a dense cloud may take considerably longer than a free-fall collapse time. Secondly, the additional mathematical complexities introduced by abandoning the steady-state concept make a complete gas-phase chemical model very difficult, especially if one is interested in extending our analysis to larger species. Mitchell, Ginsburg and Kuntz (1978) have utilized a steady-state model with

100 species, 454 gas-phase reactions and one grain reaction (to form H_2). These authors abandoned the Herbst-Klemperer initial conditions of considerable H_2 and CO abundances and they explicitly considered H_2 and CO formation. Their calculated column densities are in quantitative to semi-quantitative agreement with observation for a variety of molecules. There are still some departures from observed abundances. Yet their calculated values are, on the whole, sufficiently good to make one wonder about the need to consider time-dependent effects. Mitchell, Ginsburg and Kuntz (1978) did not include any "large" (>4 atoms) neutral molecules in their calculations. It is interesting to note that both the Herbst-Klemperer calculation and this more recent one are very dependent on the choice of cosmic abundance ratios.

Prasad and Huntress (1978) are completing an even more ambitious calculation than that of Mitchell *et al.* These authors have included time dependence by Scheme (b) in a gas-phase model (H_2 is, however, formed on grains as it is in all realistic models) that includes ~2000 reactions and ~250 gas-phase species of up to six atoms in size (CH_3OH). The calculated abundances should prove especially interesting. Suzuki *et al.* (1976) have utilized ~135 gas-phase reactions in a time-dependent calculation based on Scheme (c). They found that as a cloud's density reaches $\sim 10^4 cm^{-3}$, the fractional abundances of molecular species become "frozen" at values differing from steady-state values. However, most polyatomic molecules were not considered in their calculation. Langer (1976) has shown how time-dependent effects increase the abundance of H_2CO. Iglesias (1977) has recently completed a time-dependent calculation using Scheme (b).

As for Scheme (d), the most physically realistic time-dependent model is that of Gerola and Glassgold (1977) who have developed a one-dimensional hydrodynamic picture of self-gravitating interstellar clouds. (See also the earlier study of Glassgold and Langer [1976*b*].) Further details of the chemical aspects of the model are to be given in subsequent publications. The model indicates that the standard dense cloud ion-molecule chemistry does not become effective until high visual extinction and gas densities in excess of $3 \times 10^3 cm^{-3}$ have developed. Large CO abundances are reached in 3-4 x 10^6 years. The model of Gerola and Glassgold (1977) includes the effect of H_2 formation heating on the evolution of a cloud. The efficiency of such heating can speed up or delay the onset of the typical dense cloud chemistry. A fuller understanding of the effects of chemical energy release on cloud evolution is a necessity if we are to understand how protostars and stars evolve from dense clouds. A paper by Cravens and Dalgarno (1978) contains a discussion of cosmic-ray heating in H_2 regions and lists references on other chemical heating sources.

All time-dependent models discussed so far have been gas-phase theories. Allen and Robinson (1977) have published a time-dependent grain model using Scheme (b) that contains 598 possible association reactions of the type $A + B \xrightarrow{grain} AB$ where A and B are free radicals (species with unpaired electrons); these reactions presumably occur without activation energy. Molecules are both formed and destroyed by grain reactions only. Large molecules (up to 11 atoms) are considered in this model; atomic constituents such as H, present in abundance in the early stages of cloud evolution, play key roles in the synthesis of larger molecules although the atomic abundances decline as the cloud reaches maturity. The obvious difficulty with a model such as this is that our understanding of the relevant physical and chemical processes is not great. In addition, it is difficult to justify the neglect of well-understood and rapid gas-phase reactions.

The question of whether polyatomic neutral molecules are formed in the gas phase, on grain surfaces, or via both mechanisms is, of course, still a puzzle that continues to intrigue astrochemists. Huntress *et al.* (1977) maintain that the H_2O/NH_3 abundance ratio in Orion *A* agrees well with recent ion-molecule predictions but poorly with predictions based on grain models. On the other hand, Kroto *et al.* (1977) feel that the formation of the cyano polyacetylenes (HC_3N, HC_5N, HC_7N, and, just recently, HC_9N) cannot be accounted for by gas-phase theories. In our view, the following points can be made concerning this issue:

(a) H_2 is formed on interstellar grains.
(b) Gas-phase theories can explain much of the interstellar chemistry involving species up to intermediate size (~4 atoms).
(c) The formation of larger molecules is not very well understood.

Common to all dense cloud models discussed here is the assumption that photodestructive processes are sufficiently slow in dense clouds that they are not important. However, studies by Sandell (1978) and Sandell and Mattila (1975) indicate that photodestruction might be a competing and even dominant destruction process for many molecules in homogeneous clouds of density $\leqslant 10^4$ cm^{-3} and $A_V \leqslant 8$ mag. If these studies are correct, dense cloud chemistry, at least for the *lower* gas densities, may be even more complex than assumed.

RECENT WORK ON COMPLEX MOLECULE FORMATION

No published gas-phase model has included neutral molecules of more than four atoms. (The forthcoming calculation of Prasad and Huntress [1978] will contain molecules as complex as CH_3OH, a six-atom species.) Yet various investigators have considered how larger molecules can be produced by gas-phase reactions. One necessity for considering the production of more complex molecules is additional laboratory data on ion-molecule reactions. Different laboratory groups have been studying ion-molecule reactions of possible astrophysical interest; of particular note are the contributions of Huntress and coworkers at the Jet Propulsion Laboratory. (See, e.g., Huntress [1977].) Other active research groups include Bohme and Schiff at York, Ontario, Fehsenfeld and Ferguson of the National Oceanic and Atmospheric Administration, and Smith and Adams at Birmingham. Herbst and Klemperer (1976) have discussed some of the classes of reactions likely to be important in large molecule synthesis. In particular, reactions involving C^+ and simple carbonium ions (CH_3^+, $C_2H_2^+$, $C_2H_3^+$, etc.) may be critical.

In addition to ordinary ion-molecule reactions, it was recognized by Herbst and Klemperer (1973), Millar and Williams (1975) and Herbst (1976) that radiative association reactions ($A + B \rightarrow AB + h\nu$) could play a role in polyatomic molecule synthesis. These reactions become increasingly efficient as the size of the collision partners increases. Herbst (1976) has emphasized the special importance of radiative association reactions between abundant ions such as CH_3^+ and molecular hydrogen. His calculations showed that at a temperature of 50 K the CH_3^+-H_2 radiative association occurred in one out of every 10^6 collisions. Radiative association of polyatomics has not been studied in the laboratory. However, Smith and Adams (1977, 1978) have examined three-body associations ($A + B + C \rightarrow AB + C$) of various ions and neutrals and used their results to estimate the corresponding radiative association rates at low temperatures. (See also the earlier study of Fehsenfeld *et al.* [1974].) Smith and Adams (1978) have estimated that at 50 K the CH_3^+-H_2 radiative association to produce CH_5^+ occurs in more than one out of every 10^4 collisions. Smith and Adams (1978) have also found that the radiative association of CH_3^+ with interstellar neutrals such as CO, H_2O, NH_3 are even more likely; the radiative association with H_2O at 50 K occurs on *every* collision! If these rate coefficients are as rapid as suggested for a variety of ions, then radiative association reactions play an important role in the synthesis of large species. One result of the rapid CH_3^+-H_2 association is to promote the efficient production of methane via the reaction $CH_5^+ + e \rightarrow CH_4 + H$.

The reaction between CH_5^+ and electrons is an example of a polyatomic dissociative recombination reaction. These reactions have been studied extensively in the laboratory but product species have never been measured. Herbst (1978) has devised an approximate procedure to calculate the branching ratios for the many sets of possible products that can be formed. The method applied to the $NH_4^+ + e$ reaction shows that NH_3, NH_2, and NH are all produced at roughly equal rates.

To summarize, different investigators have been exploring both normal ion-molecule routes and radiative association processes leading to complex molecular ions. Ion-electron reactions have also been investigated theoretically since these reactions produce polyatomic neutrals and model builders need to know the branching ratios for production of the various neutral channels.

INTERPRETATION OF OBSERVATIONS

The bulk of this review chapter has contained a discussion of dense cloud models and related studies. There is another aspect of the chemistry of dense clouds that should not go unmentioned. Various investigators have utilized a limited number of gas-phase reactions to relate specific observational results to important pieces of information about dense clouds. Studies of this type depend on the ability to decouple a small subset of ion-molecule reactions from all the others needed to model the cloud. One example of this approach was the suggestion of Herbst *et al.* (1977) that the abundances of neutral non-polar molecules such as N_2, CO_2, C_2H_2, CH_4 could be determined by radio observation of the corresponding protonated ions HN_2^+, HCO_2^+, $C_2H_3^+$, CH_5^+. As of 1977 only HN_2^+ has been observed and the analysis of Herbst *et al.* (1977) leads to a crude fractional abundance $N_2/H_2 \sim 10^{-5}$-10^{-6}. Another approach of this type has been the use of measured isotopic abundance ratios in molecules, especially deuterium to hydrogen. In particular, the high DCO^+/HCO^+ (Watson, 1977; Guélin *et al.*, 1977) and N_2D^+/N_2H^+ (Snyder *et al.*, 1977) abundance ratios in cool dark clouds have been used to derive upper limits to the fractional electron abundance of $\sim 10^{-8}$ and to the CO and N_2 abundances of $\sim 10^{-5}$. It is somewhat ironic to note that the low electron abundance obtained for cool dark clouds is closer to the value of Herbst-Klemperer (1973) than to more recent calculated values obtained when metals were included in the calculation. In fact, Guélin *et al.* (1977) have estimated that metals must be depleted by more than two orders of magnitude from their cosmic abundance values.

CONCLUSIONS

What then is the current state of dense cloud interstellar chemistry? Detailed gas-phase models of dense clouds including many of the smaller molecules detected have been and are being proposed. These models differ in their approach to the physical and chemical time dependence of dense clouds. On the question of calculated molecular abundances, the models are certainly in semi-quantitative agreement with observation for many species. Gas-phase reactions have also been used to help interpret observational results.

The most intriguing and still unanswered question is whether gas-phase reactions can produce the larger interstellar molecules observed or whether we must fall back on the far less well understood surface chemistry of grains to explain their presence. Simple order-of-magnitude estimates of the efficacy of various ion-molecule pathways are necessary before inserting these reactions into a general model. Otherwise, the bewildering number of reactions to be considered will eventually overwhelm the investigators. It is my belief that estimates of complex molecule abundances based on ion-molecule reactions will be made in the near future.

Radio observations of dense clouds are improving continually and our knowledge of these regions will continue to grow. In the mid 1960s interstellar chemistry hardly existed; today it exists but has yet to reach maturity. We have progressed a great deal but we are still far from a complete understanding of dense interstellar clouds.

Acknowledgments. I wish to thank the donors of The Petroleum Research Fund, administered by the American Chemical Society, for partial support of my research program, and the Alfred P. Sloan Foundation for their support by awarding me their Fellowship.

REFERENCES

Aannestad, P. 1973*a*. *Astrophys. J. Suppl.* 25: 205.
______. 1973*b*. *Astrophys. J. Suppl.* 25: 223.
Allen, C. W. 1973. *Astrophysical Quantities.* London: Athlone Press.
Allen, M., and Robinson, G. W. 1975. *Astrophys. J.* 195: 81.
______. 1976. *Astrophys. J.* 207: 745.
______. 1977. *Astrophys. J.* 212: 396.
Black, J. H., and Dalgarno, A. 1973*a*. *Astrophys. J.* 184: L101.
______. 1973*b*. *Astrophys. Lett.* 15: 79.
______. 1977. *Astrophys. J. Suppl.* 34: 505.
Cravens, T. E., and Dalgarno, A. 1978. *Astrophys. J.* 219: 750.
Dalgarno, A.; Oppenheimer, M.; and Black, J. H. 1973. *Nature Phys. Science* 245:100.
Duley, W. W. 1970. *Roy. Astron. Soc. Canada J.* 64: 331.
Fehsenfeld, F. C.; Dunkin, D. B.; and Ferguson, E. E. 1974. *Astrophys. J.* 188: 43.
Gerola, H., and Glassgold, A. E. 1977. Preprint; submitted to *Astrophys. J.*
Glassgold, A. E., and Langer, W. D. 1975. *Astrophys. J.* 197: 347.
______. 1976*a*. *Astrophys. J.* 206: 85.
______. 1976*b*. *Astrophys. J.* 204: 403.
Greenberg, L. 1973. In *Interstellar Dust and Related Topics* (J. M. Greenberg and H. C. van de Hulst, eds.) Dordrecht: D. Reidel.
Guélin, M.; Langer, W. D.; Snell, R. L.; and Wootten, H. A. 1977. *Astrophys. J.* 217: L165.
Herbst, E. 1976. *Astrophys. J.* 205: 94.
______. 1978. *Astrophys. J.* In press.
Herbst, E.; Green, S.; Thaddeus, P.; and Klemperer, W. 1977. *Astrophys. J.* 215: 503.
Herbst, E., and Klemperer, W. 1973. *Astrophys. J.* 185: 505.
______. 1976. *Phys. Today* 29: 32.
Hollenbach, D.; Werner, M. W.; and Salpeter, E. E. 1971. *Astrophys. J.* 163: 165.
Huntress, Jr., W. T. 1977. *Astrophys. J. Suppl.* 33: 495.
Iglesias, E. 1977. *Astrophys. J.* 218: 697.
Kroto, H. W.; Kirby, C.; Walton, D. R. M.; Avery, L. W.; Broten, N. W.; Macleod, J. M.; and Oka, T. 1977. *Bull. Amer. Astron. Soc.* 9: 303.
Langer, W. D. 1976. *Astrophys. J.* 210: 328.
Langer, W. D., and Glassgold, A. E. 1976. *Astron. Astrophys.* 48: 395.
Millar, T. J., and Williams, D. A. 1975. *Mon. Nat. Roy. Astron. Soc.* 173: 527.
Mitchell, G.F.; Ginsburg, J.L.; and Kuntz, P.J. 1977. *Astrophys. J.* 212: 71.
______. 1978. *Astrophys. J.* In press.
Oppenheimer, M., and Dalgarno, A. 1974*a*. *Astrophys J.* 187: 231.
______. 1974*b*. *Astrophys. J.* 192: 29.
______. 1975. *Astrophys. J.* 200: 419.
Prasad, S. S., and Huntress, Jr., W. T. 1978. In preparation.
Sandell, G. 1978. Preprint; submitted to *Astron. Astrophys.*

Sandell, G., and Matilla, K. 1975. *Astron. Astrophys.* 42: 357.
Saykally, R. J.; Dixon, T. A.; Anderson, T. G.; Szanto, P. G.; and Woods, R. C. 1976. *Astrophys. J.* 205: L101.
Smith, D., and Adams, N. G. 1977. *Astrophys. J.* 217: 741.
———. 1978. *Astrophys. J.* In press.
Snyder, L. E.; Hollis, J. M.; Buhl, D,; and Watson, W. D. 1977. *Astrophys. J.* 218: L61.
Suzuki, H.; Miki, S.; Sato, K.; Kiguchi, M.; and Nakagawa, Y. 1976. *Progr. Theor. Phys.* 56: 1111.
Solomon, P., and Klemperer, W. 1972. *Astrophys. J.* 178: 389.
Watson, W. D. 1973*a*. *Astrophys. J.* 182: L73.
———. 1973*b*. *Astrophys. J.* 183: L17.
———. 1974. *Astrophys. J.* 188: 35.
———. 1976. *Rev. Mod. Phys.* 48: 513.
———. 1977. In *CNO Processes in Astrophysics* (J. Audouze, ed.) Dordrecht: D. Reidel.
Watson, W. D., and Salpeter, E. E. 1972*a*. *Astrophys. J.* 174: 321.
———. 1972*b*. *Astrophys. J.* 175: 659.
Woods, R. C.; Dixon, T. A.; Saykally, R. J.; and Szanto, P. G. 1975. *Phys. Rev. Letters* 35: 1269.

CONDENSATION PROCESSES AND THE FORMATION OF COSMIC GRAINS

BERTRAM DONN
NASA Goddard Space Flight Center

An analysis is presented of the assumptions and the applicability of the three theoretical methods for calculating condensations in cosmic clouds where no pre-existing nuclei exist. The three procedures are: thermodynamic equilibrium calculations, nucleation theory and a kinetic treatment which would take into account the characteristics of each individual collision. Thermodynamics provides detailed results on the condensation temperature and composition of the condensate provided the system attains equilibrium. Because of the cosmic abundance mixture of elements, large supersaturations in some cases, and low pressures, equilibrium is not expected in astronomical clouds. Nucleation theory, a combination of thermodynamics and kinetics, has the limitations of each scheme. Kinetics, not requiring equilibrium, avoids nearly all the thermodynamics difficulties but requires detailed knowledge of many reactions which thermodynamics avoids; it appears to be the only valid way to treat grain formation in space. To do this will require experiments on condensation that will provide an understanding of the process as it occurs in astronomical systems and also will provide the data on cross-sections, rates and related characteristics of the separate steps. A review of experimental studies concludes the chapter.

Analyses of the cloud collapse phase of star formation, accumulation of solid grains into larger solar system objects, trapping of elements later incorporated into larger objects, thermal balance and temperature of clouds, are among the processes in which cosmic grains play an important role. The formation mechanism of grains largely determines their composition, concentration as a function of time and position, radiative characteristics and accumulation probability. Consequently, a review of the status of formation mechanism of grains is very pertinent to the subject of this book.

In this chapter I treat the case of condensation from a gas in the absence of any pre-existing grains or nuclei upon which condensation can occur. Some of the analysis is expected to apply to condensation on grains already present. The chapter is not a summary of the state-of-the-art in describing the formation of grains. It is, rather, an attempt to examine the basic assumptions of the various procedures used to describe condensation and to examine their applicability to astronomical systems.

The case of heterogeneous nucleation on ions or of growth on pre-existing grains is not discussed here. Although ions may be important for stellar or circumstellar condensation, conclusions about the nature of the grain and the mechanism are not expected to change appreciably. Nucleation would occur at higher temperatures or lower concentrations than with a neutral vapor. This problem will be taken up at a later time. Very small grains may be present in some of the dense clouds considered here which formed from low-density clouds.

If interstellar grains survive the collapse of a diffuse cloud, grains less than 100 Å are probably present. Some aspects of the growth of grains are given by Greenberg (1978). The process, in outline, is described by Greenberg (1973) and by Field (1975) but detailed effects of gas-grain interaction (Hirth and Pound, 1963) are not included. As part of the study of cosmic grains at Goddard Space Flight Center, an analysis of this problem is intended.

Three mechanisms have been used to describe condensation. In historical order they are: (1) thermodynamics (Wildt, 1933), (2) kinetics (Lindblad, 1935), and (3) nucleation theory (Kamijo, 1966, 1969; Donn *et al.*, 1968) which is a combination of thermodynamics and kinetics. These analyses have subsequently undergone considerable development. A comprehensive review of the thermodynamic procedure has been given by Grossman and Larimer (1974), and a brief review of thermochemistry and nucleation theory has been published by Salpeter (1977).

Kinetic analyses require the most detailed input data and few general treatments have been attempted (ter Haar, 1944; Kramers and ter Haar, 1946; van de Hulst, 1949; Hoyle and Wickramasinghe, 1962; Donn, 1976). These papers indicate the complexity and consequent limitations of a kinetic analysis.

I. THERMODYNAMIC EQUILIBRATION CALCULATIONS

The simplicity of thermodynamics, in which the equilibrium composition and phases in a system depend only on the final state, allows detailed results to be calculated. A determination of the chemical composition of the gas and any condensates require knowledge only of the free energies of all species which may occur. Procedures for obtaining the data and carrying through the

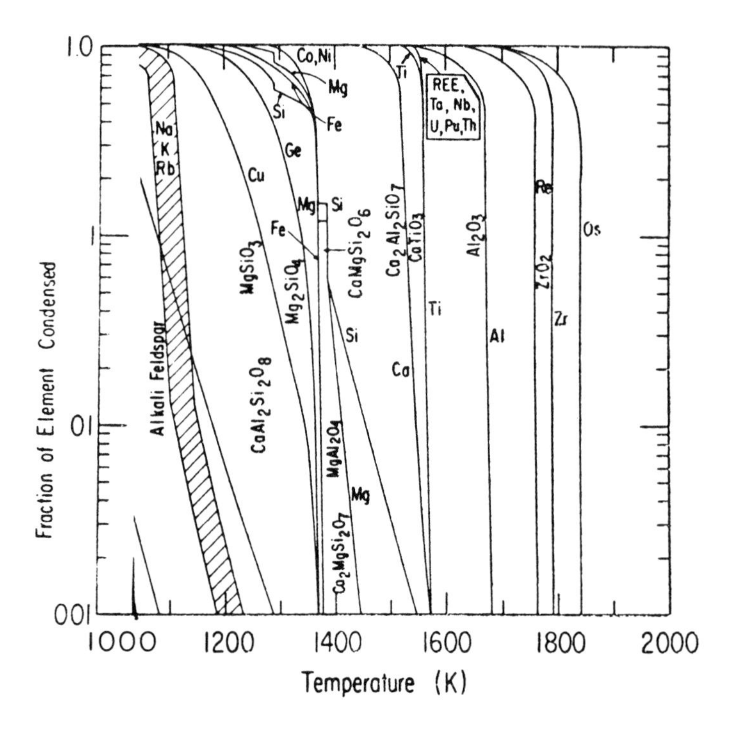

Fig. 1. Summary of thermodynamic equilibrium calculations from a nebula of solar composition, $P = 10^{-4}$ atm (from Grossman and Larimer, 1974).

calculations are given in the review of Grossman and Larimer (1974). They adopted a cloud of "solar" composition at a pressure of 10^{-4} atm. Figure 1, taken from that paper, summarizes their condensation computations. The chart shows the fraction of each element that has condensed and the mineral in which it occurs, as a function of temperature. Of the three methods for treating condensation, only thermodynamics permits one to obtain such detailed and comprehensive results. Indeed, for most elements, it is the only procedure that yields any results at the present time.

Unfortunately, there is a price that must be paid for this information; the system must be in thermodynamic equilibrium. To the degree that the system is out of equilibrium, calculated results will deviate from the actual values. Several factors exist which may lead to disequilibrium. The most important among these appear to be the low pressures in clouds, and formation of metastable species which may persist indefinitely. The effect of each of these factors will be accentuated by the non-existence in the gas of the nominal molecule of the condensed mineral.

Condensation is the tendency of the system to attain equilibrium. As with many chemical reactions an activation barrier exists which may cause a non-equilibrium state to be "frozen in." With vapor phase transitions, the requirement for supersaturation makes condensation an inherently non-equilibrium process which can have significant effects (Blander and Katz, 1968; Blander and Abdul-Gawad, 1969; Blander, 1971). The experimentally determined supersaturation for iron (Frurip and Bauer, 1977) is 20 at 2000 K and increases to 500 at 1650 K. The theoretically and experimentally well established need for supersaturation thus introduces a major deviation from equilibrium. For condensation of iron at 1600 K, the iron vapor has to be undercooled by 300°. The kinetic effects of low pressure and possible metastable species will be considered in Sec. III where an analysis of the kinetics of condensation is treated. We should emphasize that classical thermodynamics predicts the equilibrium composition and phase distribution and not the results of a phase transition which, for the reasons indicated, may yield non-equilibrium, metastable products.

II. NUCLEATION THEORY

The essence of nucleation theory is the transition from the gaseous to the condensed phase, liquid or solid. In its usual formulation (Frenkel, 1955; Hirth and Pound, 1963; Abraham, 1974) it is a combination of thermodynamics and kinetics. The process can be represented by the set

$$A_1 + A_j \leftrightarrows A_{j+1} \quad J = 1,2 \ldots . j^* \tag{1}$$

where A_1 is the monomer, A_j a cluster of j molecules or atoms, and A_{j^*} the critical cluster or condensation nucleus. The distribution $n(A_j)$, for j not too near j^*, is given by thermodynamics. Monomeric addition of an existing gaseous molecule is generally assumed, although dimers and trimers may be involved in some cases, e.g., carbon. Only one type of condensible molecule occurs and a cluster changes size only by collisional capture or evaporative loss. Several treatments of binary systems have been made (Kiang *et al.*, 1973; Shugard *et al.*, 1974; Wilemski, 1975). Kiang and Cadle (1975) treated ternary systems. In these analyses, multicomponent particles form without reaction. Condensation with reactions to form grains for which the molecular composition is different from that of the vapor has been considered explicitly by Courtney (1964) and in a more general but formal sense by Hirschfelder (1974). Characteristics of clusters which occur in these two investigations are not known, therefore preventing application to actual systems.

Salpeter (1974) proposed that polymers incident on a growing cluster may fragment through release of the binding energy of the particle. Part is captured and the escaping fragment carries away the excess energy, stabilizing the now larger cluster. This may be an important suggestion as we shall see later.

For the formation of the most stable major minerals e.g., Al_2O_3, $CaTiO_3$ (Grossman and Larimer, 1974), nucleation is expected to be seriously inhibited. The building blocks are small molecules e.g. Al_2O, TiO, CaO, and not the larger, nominal molecule of the crystal. The nominal molecules yielding very stable macroscopic crystals cannot exist until clusters of many molecules occur to form the bulk material, possibly in a very poor crystalline array.

At the low pressures in astronomical systems two additional factors enter. At a total pressure of 10^{-4} atm, the partial pressure of condensible species will be below 10^{-9} atm. The interval between successive collisions of condensible species with a small cluster will be greater than one second, whereas the interval for all molecular collisions will be about 10^{-5} sec. Therefore, the rate of cluster growth is slow on molecular time scales and the stability of small clusters becomes an important factor. The binding energy per atom for metal clusters has been studied theoretically by several investigators and one experimental measurement for iron exists (Freund and Bauer, 1977). The results are summarized in Fig. 2 where it is seen that the binding energies of very small clusters are about 20% of that for the bulk metal. No data exist for oxides. The stability and hence concentration of these clusters will be much less than would be found using the bulk binding energies.

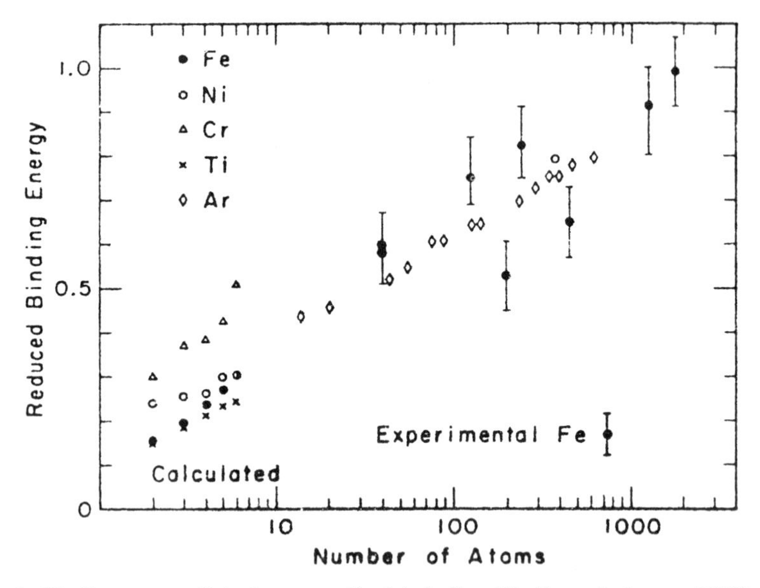

Fig. 2. Binding energy of clusters normalized to bulk solids (from Anderson, 1976).

A second important effect of low pressure is a marked deviation from thermal equilibrium. The consequences for chemical reactions in the interstellar medium have been pointed out by Donn (1973). An analysis of non-equilibrium molecular vibrational energy in late-type giant and supergiant stars were discussed by Thompson (1973). His criterion for a non-equilibrium distribution, radiative relaxation rate equal to collisional rate, is qualitative and very conservative. However, we shall use the same simple measure for deviation from equilibrium here. At $p = 10^{-4}$ atm, the molecular collision frequency is 10^5 sec^{-1}. The efficiency of translational to vibrational energy exchange is very poor (Cottrell and McCoubrey, 1961; Kondrat'ev, 1964*a*); for simple molecules at ~ 1000 K an efficiency of 10^{-3} may be adopted; i.e., 10^3 collisions are required for energy transfer to occur. This yields a transition probability for collisional excitation of about 100 sec^{-1}. Representative probabilities of radiative vibrational transition are in the range 100-1000/sec mol (Penner, 1959). Therefore, significant deviations from a Boltzmann equilibrium distribution in the sense of a deficiency in excited states will appear at 10^{-4} atm.

In summary, nucleation theory makes the following assumptions:

1. Pressures are high enough for thermodynamics to exist and for the time scale of nucleation to be short (~ 10^{-5} sec).
2. There are unreactive, condensible species.
3. The condensed molecule exists in the vapor.
4. No reactive species exist.

None of these assumptions apply to astronomical systems.

Even with the simplified approximations, nucleation theory has only been extensively used for graphite (Kamijo, 1966; Donn *et al.*, 1968; Walker, 1975; Tabak *et al.*, 1975). Mineral condensation sequences based on undercooling required by the large supersaturations have been derived by Blander and his associates in the references given above.

III. KINETIC MECHANISMS

The well-known distinction between thermodynamics and kinetic theory shows up in the astronomical condensation problem. Whereas thermodynamics can yield detailed calculations without needing detailed processes and data as long as equilibrium exists, kinetics requires no special conditions for its calculations. We must, however, have a complete knowledge of all important processes, energies, cross-sections and other relevant data. The consequence is that calculations for specific condensates have been attempted only for ice mixtures (van de Hulst, 1949) and graphite (Hoyle and Wickramsinghe, 1962). An analysis that examined for the first time the formation kinetics of diatomic molecules as the initial step in grain formation was given by ter Haar (1942) and Kramers and ter Haar (1946). An amorphous particle consisting of essentially all elements was proposed. Cernuschi (1947) reconsidered some of the basic steps in grain formation after the initial clusters formed. Hoyle and Wickramsinghe (1962) discussed graphite formation in the atmosphere of cool carbon stars. They made use of experimental values of the heat of reaction of small carbon polymers and theoretical value for the equilibrium vapor pressure of carbon.

With regard to these investigations we emphasize the difficulty, initially, of correctly writing down the series of processes that occur. Following this, the serious problem of obtaining energies, cross-sections and rates required for the calculation must be resolved. In the papers mentioned above, mechanisms were highly simplified. In spite of the large concentration of reactive molecules, only those steps were included which added to the cluster, e.g., for graphite formation, only reactions of the form

$$C + C_n \leftrightarrows C_{n+1} \quad (2)$$

were included. A much more frequent step would be

$$H + C_n \xrightarrow{k} \text{products} \quad (3)$$

where the products, CH, CH_2 etc. (Balooch and Olander, 1975) are produced at a rate k. Another expected reaction series is

$$CH_m + C_n \rightarrow C_{n+1} + H_m \quad (4)$$

where H_m may be an unstable fragment which yields $H + H_2$. This is the type of process by which pyrolytic graphite is produced. The possible significant effects of hydrocarbon reactions on carbon condensation has been mentioned by Donn *et al*., (1968). They neglected these mechanisms because they lacked the knowledge of how to treat them. Similar considerations apply to the analysis of ter Haar (1942) and the other kinetic studies.

Several other factors having important consequences on the kinetics of grain formation are briefly noted here. At low pressures, three-body collisions which are needed to stabilize the association of small molecules are very infrequent. The ratio of two-to-three body collisions is approximately 10^{-22} n (Kondrat'ev, 1964*b*) where n is the total particle density per cm^3. In the pressure interval 10^{-4} to 10^{-8} atm, this ratio drops from 10^{-7} to 10^{-11}. Salpeter's suggestion of fragmentation of the incident molecule can be effective; the escaping fragment carries away the excess binding energy and allows part of the incident molecule to add to a cluster with less than about eight atoms.

A second factor that may enhance condensation is the low vibrational temperature at low pressures which depletes the higher vibrational states of the molecules. The primary energy mode that affects condensation appears to be the internal energy of the molecule, cluster or surface rather than the molecular kinetic energy (McCarroll and Ehrlich, 1964; Shade, 1964). Thus, at high kinetic temperatures, the non-Boltzmann distribution means that the system is effectively much colder. This may be the dominant factor which allows grains to form in "hot" clouds. A similar effect applicable to small aggregates which behave as bulk solids was pointed out by Arrhenius and De

(1973).

Finally, a major contributor to condensation may be ionization which is well known to be very effective in causing nucleation in many systems (Hirth and Pound, 1963). A more general possible consequence of ionization on condensation in clouds has been considered by Arrhenius and Alfvén (1970; see also Lehnert, 1970). These papers consider grain formation where plasma effects dominate cloud processes whereas this chapter is concerned with regions where ionization has limited, specific consequences.

IV. THE THEORETICAL MECHANISMS

The analysis of the three mechanisms of condensation is now summarized. Thermodynamics yields a precise, detailed evolution of condensation in clouds if the system were always at equilibrium. However, because of the fact that condensation is a transition process, not an equilibrium state, various factors may interfere with the equilibrium state being attained. Of particular significance are the low pressures characteristic of astronomical systems, and the large supersaturation required to drive the condensation process. Nucleation theory also suffers from the requirement for thermodynamic equilibrium. In addition, it has been developed only for very simple laboratory systems. The basic assumptions are not valid for astronomical cases. A generalization of nucleation theory, more valid for present circumstances may be made (Donn, 1976). This procedure will require much more detailed knowledge than can readily be obtained and leads us to the third scheme – a kinetic treatment. Kinetics is valid under any circumstances provided all the steps and quantities are known. For a cosmic abundance system at low pressures we are very far from being able to carry out a detailed calculation. A simple, extreme calculation in which all refractory atoms and molecules add to the cluster yields a boundary condition grain. This has been done previously (Donn, 1976) with the resultant composition $Fe_7Mg_6CaAlNi(SiO)_5$. This procedure and result is somewhat analogous to that of ter Haar (1942) who was dealing with low-temperature diffuse clouds where the total concentration was about unity. The composition for the grain just given may be compared with the thermodynamic results of Fig. 1. Because of the considerable deviations from thermodynamic equilibrium, those calculations will not be accurate descriptions of the grains.

The general character of the thermodynamic results is expected to provide guidelines for the grain composition and condensation temperatures. Inhibition of nucleation will cause calculated temperatures to be upper limits. Elements, e.g., aluminum, titanium and calcium, are not expected to have distinct condensation temperatures or condense in unique crystalline minerals. The tendency may be for groups of elements to condense together over a range of temperatures in an amorphous aggregate. Solid state diffusion at high temperatures could cause annealing, leading to a multicrystalline grain (Stephens, personal communication, 1978). The actual grain composition can be expected to fall between the two extremes. As pointed out by Donn (1976), a unique composition is unlikely. We should emphasize that the nature of the system and the factors favoring condensation indicate that non-equilibrium solids will form.

V. CONDENSATION EXPERIMENTS

It should be clear that condensation experiments appropriate to the conditions in astronomical systems are needed. Experiments on condensation of smokes have been carried out by Lefèvre (1967, 1970) who produced grains of iron, iron oxide, silicon carbide and silica by an arc discharge in argon. Kamijo *et al.* (1975) used heaters, arcs, and plasma jets to vaporize a variety of similar materials, again in argon. Steyer *et al.* (1974) used an arc to vaporize and recondense silica. In these experiments, condensation takes place under uncontrolled conditions with unknown temperatures and pressures. Except for iron oxide and silicon carbide, there is only a single condensible species. The analysis of condensation conditions in clouds in the previous sections has shown that these experiments are valuable for producing small grains of various materials and studying their properties, but they do not contribute much to our understanding of astronomical condensation. Meyer (1971) has experimented with condensation on surfaces initiated by sputtering of selected targets by argon atoms. His procedure and results are difficult to relate to smoke formation in clouds. Of significance to the present analysis is Meyer's conclusion that some films formed by sputtering magnesium silicate targets could not be identified with any known magnesium silicate compound. Perhaps the experiment most pertinent to grain formation in astronomical clouds was the attempt by Dorfeld and Hudson (1973) to nucleate graphite. They simulated a cool stellar atmosphere with an expanding beam of a hydrogen-methane mixture equilibrated at temperatures on the order of 2000 K, but they obtained a negative result with only an upper limit of 10% of carbon condensing. Further work along this line could yield data on the kinetics and products of carbon condensation in hydrogen-rich atmospheres.

There are two series of experiments on nucleating smokes from vaporized refractory materials which extend the work already referred to in the direction of being more realistic for cosmic clouds. The work of Arnold and his colleagues is reported by Stephens and Kothari (1978). To date their experiments have been done at pressures above one atmosphere. At Goddard Space Flight Center, Day and I have vaporized magnesium and silicon monoxide in separate, adjacent crucibles (Day and Donn, 1978). These were independently heated, allowing different Mg/Si ratios to be obtained in the vapor. An ambient argon atmosphere at about 2 torr pressure could be heated to control the condensation temperature. Temperatures from 0 to 500°C were employed. The strong dependence of condensate yield on temperature, described below, indicates that the ambient temperature was the controling parameter. Calculated partial pressure of the reactant Mg and SiO vapors were between 0.1 and 1 torr. The smoke yield decreased as the ambient temperature was raised. At 500°C no smoke was collected unless the crucible temperatures were increased substantially, raising the reactant partial pressures. At 500°C the supersaturation for experiments in which no grains were collected is estimated to be $>10^5$. At all temperatures the grain size was a few hundred angstroms. The particles were amorphous, with variable Mg/Si ratios. Their infrared spectra showed structureless 10 and 20 μm features very similar to those found in astronomical sources. When the grains were annealed in vacuum at 1000°C for one hour, a well-developed forsterite (Mg_2SiO_4) spectrum developed. Figure 3 shows the infrared spectrum of condensed and annealed grains. The dependence of smoke yield on temperature, the high supersaturation and the amorphous, non-stoichiometric (unsaturated chemical bonds) character of the grains, all support the kinetic mechanism for grain formation. These experiments are being extended to include other metals and more complex mixtures as well as ionization and other phenomena. The experiments will also be made more quantitative.

The concept of organic grains proposed by Hoyle and Wickramasinghe

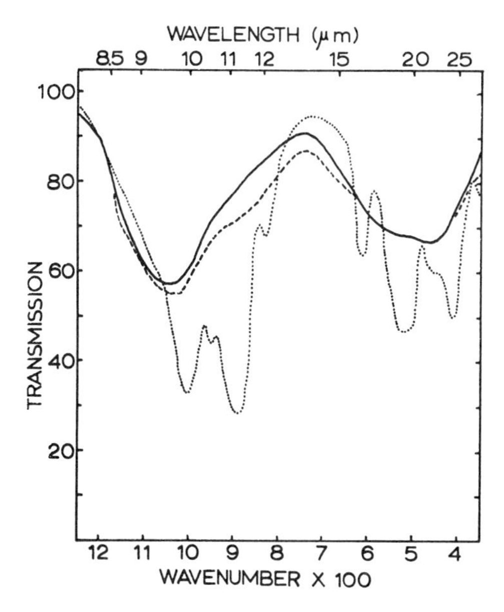

Fig. 3. Infrared spectrum of smoke condensed from SiO + Mg mixture at 300 K. Dashed curve represents initial condensate; solid curve the mixture heated to 800 K in vacuum; and dotted curve the mixture heated to 1300 K in vacuum.

(1976) has not been considered here. The kinetic treatment is expected to apply to any composition.

The ambitious, ultimate goal of our investigation of grain formation at Goddard Space Flight Center is to develop a method to permit approximate but realistic calculations of condensation in astronomical systems.

Acknowledgment. I thank J.R. Stephens for numerous comments on this chapter.

REFERENCES

Abraham, F.F. 1974. *Homogeneous Nucleation Theory.* New York: Academic Press.
Anderson, A.B. 1976. *J. Chem. Phys.* 64: 4046.
Arrhenius, G., and Alfvén, H. 1970, *Earth Planet. Sci. Lett.* 10:253.
Arrhenius, G., and De, B. 1973. *Meteoritics* 8: 296.
Balooch, M., and Olander, D. R. 1975. *J. Chem. Phys.* 63: 4772.
Blander, M. 1971. *Geochim. Cosmochim. Acta* 35: 61.
Blander, M., and Abdul-Gawed, M. 1969. *Geochim. Cosmochim. Acta* 33: 701.
Blander, M., and Katz, T.L. 1968. *Geochim. Cosmochim. Acta* 31: 1025.
Cernuschi, F. 1947. *Astrophys. J.* 105: 241.
Cottrell, T.L., and McCoubrey, J.C. 1961. *Molecular Energy Transfer in Gases.* London: Butterworths.
Courtney, W.G. 1964. *Prog. Astronaut. Aeronaut.* 15: 677.
Day, K.L., and Donn, B. 1978. *Astrophys. J.* 222: L45.
Donn, B. 1973. In *Molecules in the Galactic Environment* (M.A. Gordon, and L.E. Snyder, eds.), p. 305. New York: Wiley.
____. 1976. *Mem. Soc. Roy. Sci. Liege, Ser. 6,* 9:199.
Donn, B.; Wickramasinghe, N.C.; Hudson, J.B.; and Stecher, T.P. 1968. *Astrophys. J.* 153: 451.
Dorfeld, W.G., and Hudson, J.B. 1973. *Astrophys. J.* 186: 715.
Field, G.B. 1975. *The Dusty Universe* (G.B. Field and A.G.W. Cameron, eds.), p. 89. New York: Neale Watson.
Frenkel, J. 1955, *Kinetic Theory of Liquids,* Ch. 7. New York: Dover.
Freund, H.J., and Bauer, S.H. 1977. *J. Phys. Chem.* 81: 1001.
Greenberg, J.M. 1973. In *Molecules in the Galactic Environment* (M.A. Gordon and L.E. Snyder, eds.), p. 93. New York: Wiley.
____. 1978. *Moon and Planets* (special Protostars and Planets issue). In press.
Grossman, L., and Larimer, J.W. 1974. *Rev. Geophys. Sp. Phys.* 12: 71.
Hirschfelder, J.O. 1974. *J. Chem. Phys.* 61: 2690.
Hirth, J.P., and Pound, G.M. 1963. *Condensation and Evaporation.* New York: McMillan.
Hoyle, F., and Wickramsinghe, N.C. 1962. *Mon. Not. Roy. Astr. Soc.* 124: 417.
____. 1976. *Nature* 264: 45.
Kamijo, F. 1966. In *Trieste Colloquium on Late Type Stars,* (M. Hack, ed.), p. 252. Trieste: Trieste Observatory.
____. 1969. *Physika* 41: 163.
Kamijo, F.; Nakada, Y.; Iguchi, T.; Fujimoto, M.-K.; and Takada, M. 1975. *Icarus* 26: 102.
Kiang, C.S., and Cadle, R.D. 1975. *Geophys. Res. Lett.* 2: 41.
Kiang, C.S.; Stouffer, D.; Mohnen, V.A.; Bricard, J.; and Viglas, D. 1973. *Atmos. Environ.* 7: 1279.
Kondrat'ev, V.N. 1964*a*, *Chemical Kinetics of Gas Reactions,* Ch. 6. (translated from Russian by J.M. Crabtree and S.N. Carruthers). Reading, Mass.: Addison-Wesley.
____. 1964*b*. ibid, p. 322.
Kramers, H., and ter Haar, D. 1946. *Bull. Ast. Inst. Neth.* 10: 137.

Lefevre, J. 1967. *Ann. d'Astrophys.* 30: 731.
———. 1970. *Astron. Astrophys.* 5: 37.
Lehnart, B. 1970. *Cosmic Electrodyn.* 1: 219.
Lindblad, B. 1935. *Nature* 135: 133.
McCarroll, B., and Ehrlich, G. 1964. In *Condensation and Evaporation of Solids* (E. Rutner, P. Goldfinger, and J.P. Hirth, eds.) p. 521. New York: Gordon Breach.
Meyer, C. 1971. *Geochim. Cosmochim. Acta.* 35: 551.
Penner, S.S. 1959, *Quantitive Molecular Spectroscopy and Gas Emissivities,* Ch. 2. Reading, Mass.: Addison-Wesley.
Salpeter, E.E. 1974. *Astrophys. J.* 193: 579.
———. 1977. *Ann. Rev. Astron. Astrophys.* 15: 267.
Shade, R.W. 1964. *J. Chem. Phys.* 40: 915.
Shugard, W.J.; Heiss, R.H.; and Reiss, H. 1974. *J. Chem. Phys.* 61: 5298.
Stephens, J.R. and Kothari, B.L. 1978. *Moon and Planets* (special Protostars and Planets issue). In press.
Steyer, T.R.; Day, K.L.; and Huffman, D.R. 1974. *Appl. Optics* 13: 1589.
Tabak, R.G.; Hirth, J.P.; Meyrich, G.; and Roark, T.P. 1975. *Astrophys J.* 196: 457.
ter Haar, D. 1942. *Astrophys. J.* 100: 288.
Thompson, R.I. 1973. *Astrophys J.* 181: 1039.
van de Hulst, H.C. 1949. *Rech. Astron. Obs. Utrecht* 11, Pt. 2.
Walker, G.H. 1975. *Astrophys. Lett.* 16: 155.
Wildt, R. 1933, *Z. Astrophys.* 6: 345.
Wilemski, G. 1975. *J. Chem. Phys.* 62: 3763.

MINERALOGICAL SIMILARITIES BETWEEN INTERSTELLAR DUST AND PRIMITIVE SOLAR SYSTEM MATERIAL

R. F. KNACKE
State University of New York at Stony Brook

The mineralogy of the interstellar dust is compared with that of primitive material in the solar system. The available data indicate that the silicates in the dust resemble the silicate composition of carbonaceous chondrites, comets, and interplanetary dust. There is also some evidence that interstellar grains contain magnetite and carbonaceous compounds similar to the circumsolar material. Thus, the mineralogy appears to support the suggestion, based on current ideas of star formation and the isotopic abundance anomolies, that interstellar dust and solar nebula condensates are related.

The most ancient minerals in planets, satellites, and meteorites contain traces of events which took place in the early solar system. The chronologies of lunar rocks, cratering events on solid surfaces, and isotopic abundances in meteorites (Clayton, 1975; Cameron and Truran, 1977) record such information. Another type of record may be found in the mineralogical composition and structure of materials in the solar system. Interesting and perhaps profound mineralogical similarities among certain solar system solids such as meteorites, asteroids, comets, and interplanetary dust have emerged in recent years. There is now considerable evidence that the mineralogy that is common to these objects also characterizes the interstellar dust. If these mineralogical properties imply a close relationship between the interstellar dust and nearby circumsolar material, we would have direct access to a component of the interstellar medium. In this chapter we will compare the mineralogy of solar system material that is believed to be "primitive" with the observationally inferred mineralogy of interstellar dust.

The observational evidence for the participation of dust in star formation is extensive. Stars form in dusty regions, and most protostar precursers may be dusty globules (Bok, 1977, 1978; see also Herbig, 1974 and the chapter by

Field in this book). Circumstellar shells are observed around many pre-main-sequence objects (Strom, Strom and Grasdalen, 1975) and may be an integral part of the process of star formation. The grains in shells may be part of the pre-existing interstellar grain population, or may be newly formed in the high-density surroundings of young stars, or both. In the latter case, young grains may subsequently be ejected to replenish the interstellar grain population (Herbig, 1970). Some of the grains are presumably incorporated into ensuing planetary systems.

The association of dust with star and planet formation may also be revealed in present solar system solids. One approach to the investigation of this topic is through the composition of the solids since they can retain much of their history in their mineralogy. This is, of course, familiar in the study of meteorites, but has only recently become possible in interstellar dust investigations. To survive in relatively unaltered form, minerals must remain in a sufficiently inert environment. Planets, as well as many of the satellites where geological processes have been active, are unlikely to be good sources of such material. The asteroids and meteorites, in particular the carbonaceous chondrites in which fairly limited alteration has taken place since formation, are better candidates. These bodies formed about 4.5×10^9 yrs ago (Wood, 1968), very close to the origin of the solar system. Another possibility is found in comets which, because of their usually remote location, are protected from destructive processes in the inner solar system. The interplanetary dust is a third possibility, although it may, in fact, contain mainly meteoritic and cometary material. These probable reservoirs of primitive dust are discussed below.

Field and Cameron (1975) give an introduction to the subject. In this chapter we shall emphasize research completed since then. Reviews of interstellar dust are given by Huffman (1977), Salpeter (1977), Aannestad and Purcell (1973), and Wickramasinghe and Nandy (1972). Reviews of meteoritics include the books by McCall (1973) and Wood (1968). A recent discussion of comet composition is given by Ney (1976), and interplanetary dust is reviewed by Elsasser and Fechtig (1976).

I. COMPARISON OF CARBONACEOUS CHONDRITE AND INTERSTELLAR DUST MINERALOGY

The carbonaceous chondrites are believed to be the most primitive meteorites. Some of them contain minerals and trapped gases that indicate that they have not been heated appreciably above 500 K since formation (Wood, 1968). Elemental abundances closely follow solar abundances. Consequently, the carbonaceous chondrites appear to be relics of early, or possibly even presolar system material. Most of the other meteorites such as irons and stones, show evidence of processing through high temperatures and pressures which presumably occurred in parent meteorite bodies. Therefore, the minerals in these meteorties are unlikely to retain much evidence of early mineralogy.

The carbonaceous chondrites are divided into three classes labeled Type C1 to Type C3, according to increasing content of melted material in the form of chondrules, and to decreasing content of volatile compounds. The C1 meteorites contain few or no chondrules. The chondrules in C2 and C3's are embedded in a matrix somewhat similar to C1 material. This matrix consists of fine-grained mixtures of minerals, some of which appear likely to have condensed from the vapor phase. Therefore, C1 meteorites are often suggested to be cemented accumulations of dust grains (Wood, 1975). While the carbonaceous chondrites are quite rare as finds, this seems entirely a result of their fragile structure. Spectral analysis of asteroid reflectivities suggests that carbonaceous material is far more common in the asteroid belt, with 50 to 90% of the objects having this composition (Chapman, Morrison and Zellner, 1975).

A summary of the major mineral species in carbonaceous chondrites is shown in Fig. 1 from Anders (1964). Several of the minerals (olivine, hydrated silicates, magnetite, carbonates) have been suggested as possible interstellar grain constituents independently of meteorite analogies. We shall consider the observational evidence that the mineral composition of the interstellar grains resembles that shown in Fig. 1.

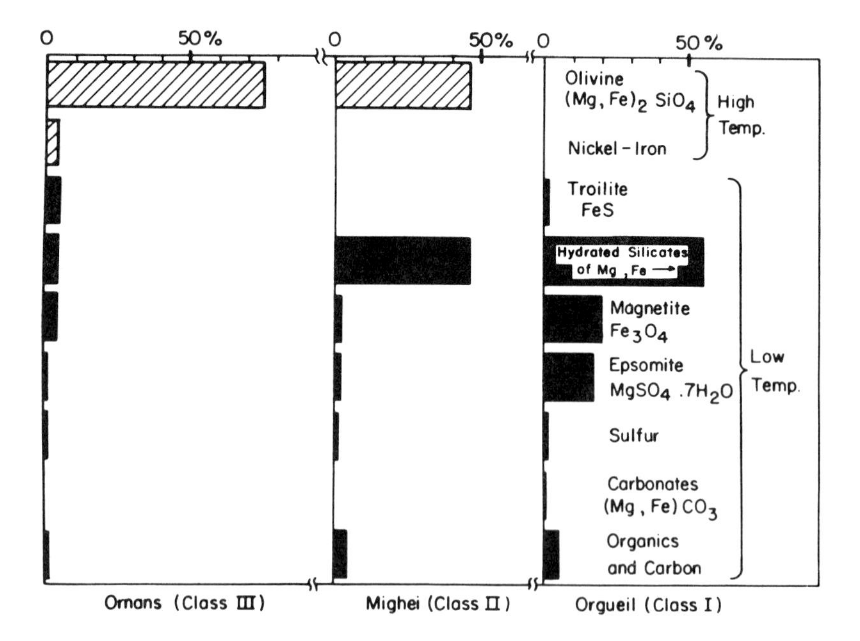

Fig. 1. Mineralogy of carbonaceous chondrites. The ratio of high-temperature to low-temperature minerals increases from Class I (C1) to Class III (C3) (Anders, 1964).

Silicates

The most abundant minerals in the carbonaceous chondrites are silicates. The C1 silicates are mainly the hydrated forms, although traces of olivines have been found in meteorites of this classification. Hydrated silicates are characterized by bonded hydroxyls or free water molecules and have a layer or sheet structure. While no terrestrial silicate may exactly duplicate the meteorite mineralogy, the C1 silicates resemble chlorites, serpentines, and montmorillonites (DuFresne and Anders, 1962; Bass, 1971). The C2 meteorites contain olivine and enstatite, usually in chondrules embedded in the hydrated silicate matrix. The type C3 contain little or no hydrated silicate.

In the equilibrium condensation model (Barshay and Lewis, 1976; Grossman, 1972), a gas of cosmic composition at pressure typical of the solar nebula cools and forms condensates. Olivine and enstatite minerals condense at temperatures near 1300 K while hydrated silicates form as alteration products near 300 K. Consequently, olivines and pyroxenes are characterized as "high-temperature" silicates, and the hydrated material as "low-temperature" silicates. This is a convenient division, although the actual history of the meteorites, especially with regard to the question of gas-solid equilibrium (Arrhenius and De, 1973), is probably more complicated.

Analysis of the interstellar grain mineralogy depends largely on remote sensing through spectroscopy. For the silicates most of the progress has come in the infrared spectral region with the observation of two "silicate" spectral features. The observations through 1972 have been reviewed by Woolf (1973). Since that time, the features have been observed in many more sources without, however, changing the interpretation significantly. The more extensive observations are of a band assigned to the stretch of the Si-O bond. When seen in absorption in dark clouds, it is characterized by a well-defined maximum at 9.7 ± 0.1 μm. It is also observed in optically thin emission in circumstellar shells where the emission peaks very near 9.7 μm. Gillett *et al.* (1975) modeled the absorption band in a number of H II regions with the wavelength dependence derived from the circumstellar emission. They found it possible to fit the spectra accurately by this procedure as shown in Fig. 2. The extensive observations of Merrill and Stein (1976) can also be modeled this way. These and other observations show that the 9.7 μm feature is remarkably similar in emission or absorption in a large number of objects with widely different physical conditions. A notable exception among ~ 30 sources is OH 231.8 + 4.2 (Gillett and Soifer, 1976) in which the feature is broader and shifted to longer wavelengths than in other sources. Since the exact position and shape of the Si-O stretch band is dependent on the mineralogy of the silicates, the interstellar grain mineralogy is fairly uniform in the galaxy, with perhaps some variations in localized sources. This may indicate that most interstellar grains have had a similar history and makes it possible to make comparisons with mineralogies in the solar system.

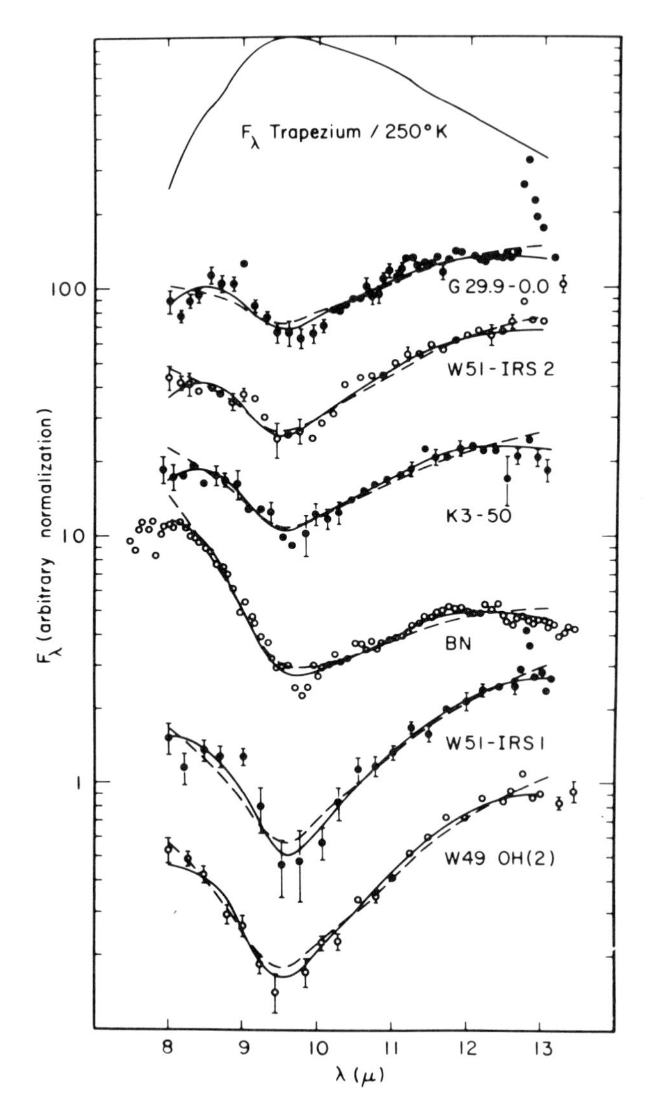

Fig. 2. Spectra of compact H II regions. The model fits to the absorption bands are based on the band shape derived from the emission spectrum due to hot grains in the Orion nebula (shown at the top). Two models with different temperature distributions are shown (Gillett *et al.*, 1975).

Silicates also have a bending mode near 20μm (±3μm). Such a feature (Fig. 3) has been observed in absorption near 18 μm in several sources (Simon and Dyck, 1975, Forrest *et al.*, 1978). The corresponding emission feature in hot dust has also been observed (Treffers and Cohen, 1974). The detection of the second strong silicate feature corroborates the silicate identification. As observations in this wavelength region become more routine, this band should also become a useful mineralogical probe.

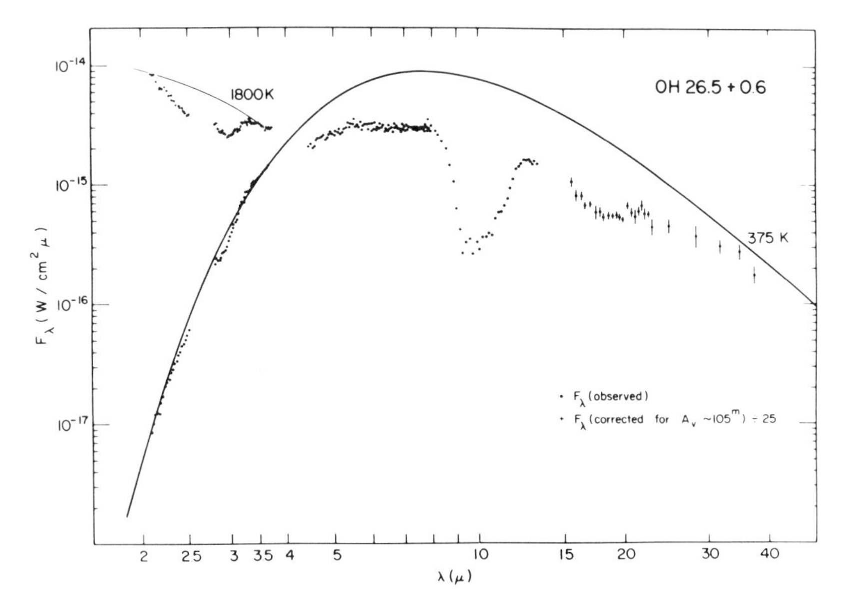

Fig. 3. Spectrum of OH 26.5 + 0.6 showing the 9.7 μm and 18 μm silicate features and the 3.1 μm ice feature. The solid line is a 375 K blackbody fit to the 2-4 μm data. The plus signs in the upper part of the figure show a reddening correction of A_V = 105 mag (Forrest *et al.*, 1978).

The silicate identification is not generally accepted. In particular Wickramasinghe *et al.* (1977), in a series of papers, have suggested several organic compounds including polysacharides and polymers of various types to account for the observed bands. We believe that the band coincidences with these compounds are accidental because the infrared features occur in the spectra of oxygen-rich stars where organic molecules are not observed. The 9.7 μm band does *not* occur in the spectra of carbon stars where such compounds would seem likely, while a feature near 11 μm has a ready explanation as silicon carbide (Treffers and Cohen, 1974). Finally, the hypothesis requires that the interstellar 3.07 μm band (Gillett and Forrest, 1973) is due to the same absorber as the 9.7 μm band. This also seems unlikely because the relative strengths of the bands in cold material vary greatly (Merrill, Russell and Soifer, 1976) suggesting that two compounds (ice and silicates) are responsible.

Infrared absorption bands have long been used in mineralogical studies of silicates (Hunt, Wisherd and Bonham, 1950; Lazarev, 1972; Farmer, 1974) and the techniques of infrared mineralogical analysis are now beginning to be applied to the interstellar particles. Hackwell (1972) pointed out the qualitative similarity of the position and strength of the interstellar 9.7 μm band with that of a sample of Cold Bokkeveld, a Type C2 carbonaceous chondrite. This comparison is also discussed by Woolf (1973).

A comparison of the interstellar band with the absorption band of several terrestrial silicates, including both high- and low-temperature types, was made by Zaikowski, Knacke and Porco (1975). Typical mineral spectra are shown in Figs. 4 and 5 and can be compared with the interstellar spectra in Fig. 2. It was pointed out above that chlorite, serpentine, and montmorillonite minerals, whose spectra are shown in Fig. 4, are found in carbonaceous

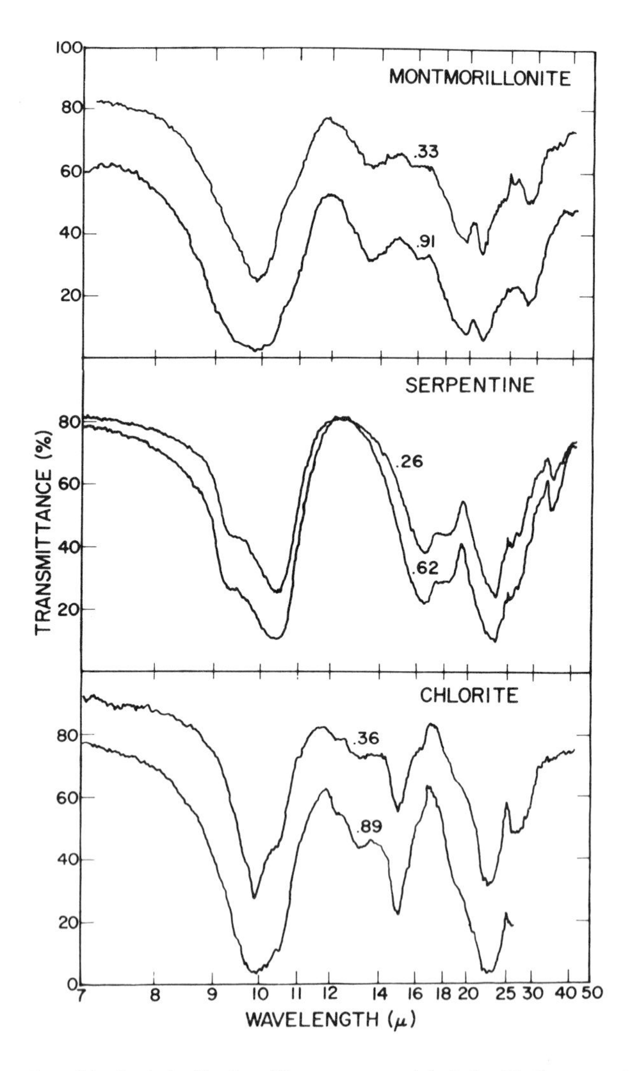

Fig. 4. Spectra of hydrated silicates. The curves are labeled with the mass density in mg cm^{-2} (based on Zaikowski *et al.*, 1975).

chondrites. However, all the hydrated silicate spectra are very similar in the 10 μm region. A strong band occurs near this wavelength, and subsidiary band structure at neighboring wavelengths is usually weak or absent. This is not characteristic of high-temperature silicates like olivines or enstatites which often have multiple absorptions occurring at wavelengths in disagreement with the position of the interstellar band (Fig. 5). This property of the olivines and enstatites is also seen in the optical constants of enstatites and olivines measured by Steyer, Day and Huffman (1974).

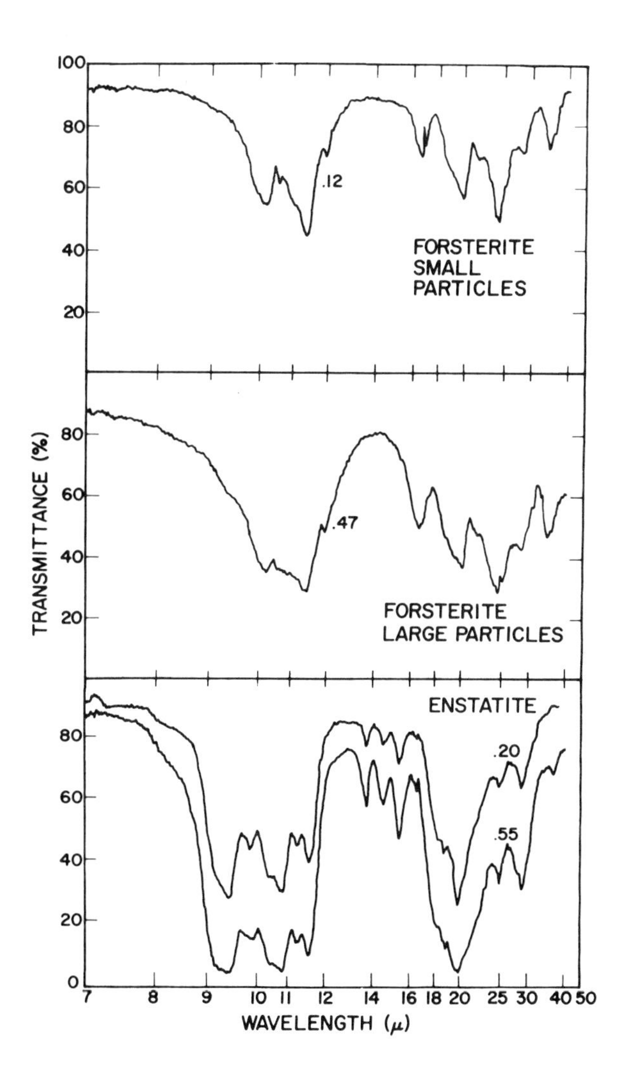

Fig. 5. Spectra of high temperature silicates (based on Zaikowski *et al.*, 1975).

Zaikowski *et al.*, concluded that a major fraction of the interstellar silicates are hydrated. The general similarity of the hydrated silicate spectra near 10 μm seems to be due to the similar layer structure of these minerals. Their common tendency to a single, strong band near 9.7 μm is the property by which they most resemble the interstellar bands. Thus, no particular hydrated silicate was proposed for the interstellar absorber, but rather a class of silicates has been identified with the interstellar bands. The identification does not require the interstellar dust to be homogenous, nor is it based on comparison with just a single compound which is more likely to give accidental agreement. On the basis of this identification, Zaikowski *et al.* pointed out that the interstellar silicate should resemble the carbonaceous chondrite matrix.

Direct comparisons between carbonaceous chondrites and interstellar dust were obtained by Zaikowski and Knacke (1975). As expected on the basis of the hydrated silicate identification, the meteorite bands resemble the interstellar bands quite closely with the Type C1 giving better agreement with the interstellar absorber than the Type C2 that contain olivine (Fig. 6).

It would be desirable to specify the mineralogy of the interstellar silicate even more closely, perhaps to determine which of the hydrated silicates are present. This is difficult to do on the basis of the 9.7 μm band alone, but the band near 20 μm appears to be more specific to different hydrated minerals and may prove more useful for the purpose (Figs. 4 and 5). However, it presents observational difficulties as well as ambiguities in interpretation because of radiation transfer effects (Kwan and Scoville, 1976). The observations are not yet extensive enough to allow one to infer much about the mineralogy from this band.

Meyer (1971) and Day (1974, 1976) conducted experimental investigations of likely condensation processes which may form interstellar grains and they synthesized several silicates. Meyer condensed crystal silicates in conditions resembling those in circumstellar clouds. Some of the condensates closely resemble the matrix material in Type C1 carbonaceous chondrites and provide experimental support for the idea that these minerals can condense from the vapor.

Infrared spectra of Day's condensates fit carbonaceous chondrite spectra quite well (Fig. 7). These silicates resemble the hydrated minerals although they tend to have less well-defined crystallinity. In this property they may be more like meteoritic silicates which are found in widely different structural order from amorphous to crystalline (Kerridge, 1969, 1971). Comparison of the condensates, the interstellar absorber, and a Type C2 carbonaceous chondrite, Murchison, showed that all three have similar infrared spectra near 10 μm. Day (1976) later found that his synthetic silicate has a weaker 9.7 μm absorption than most crystalline silicates. Both spectral (Gillette *et al.*, 1975) and polarization data (Martin, 1975; Capps and Knacke, 1976) indicate that the interstellar band has an absorption coefficient near 2-3000 cm^2 g^{-1}

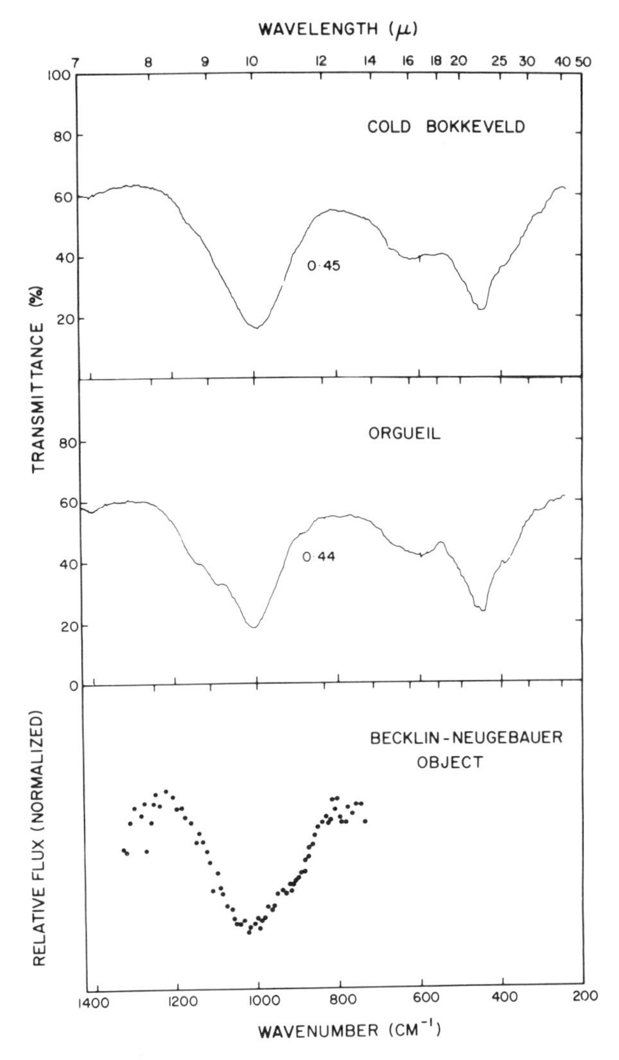

Fig. 6. Spectra of Orgueil (Type C1 carbonaceous chondrite), Cold Bokkeveld (Type C2 carbonaceous chondrite), and the BN object in Orion (based on Zaikowski and Knacke, 1975).

rather than the 6-10,000 cm^2 g^{-1} typical of olivines and enstatites. This is further evidence that the interstellar grains are not made of pure high-temperature silicates. However, it is not clear whether this implies a silicate similar to Day's or just that other minerals are present in addition to silicates. The latter would be expected from the carbonaceous chondrite analogy.

The infrared mineralogical analyses have been partially corroborated by Penman's (1976) measurements of optical constants. He also concluded that hydrated silicates, especially carbonaceous chondrite-like materials, best fit the interstellar dust absorptions. These data have the advantage that they are based on reflection measurements and may be less subject to distortions caused by particle size effects in KBr disks (Martin, 1971). The optical constants show that most terrestrial hydrated silicates have a 9.7 μm absorption band which is too narrow to give a good fit to the interstellar band. The meteorite absorptions are broader, giving a much better fit, probably because of the amorphous and heterogeneous mineralogy. The presence of olivine or other high-temperature minerals in a chlorite-type matrix, as in some Type C2 chondrites, can also broaden the band.

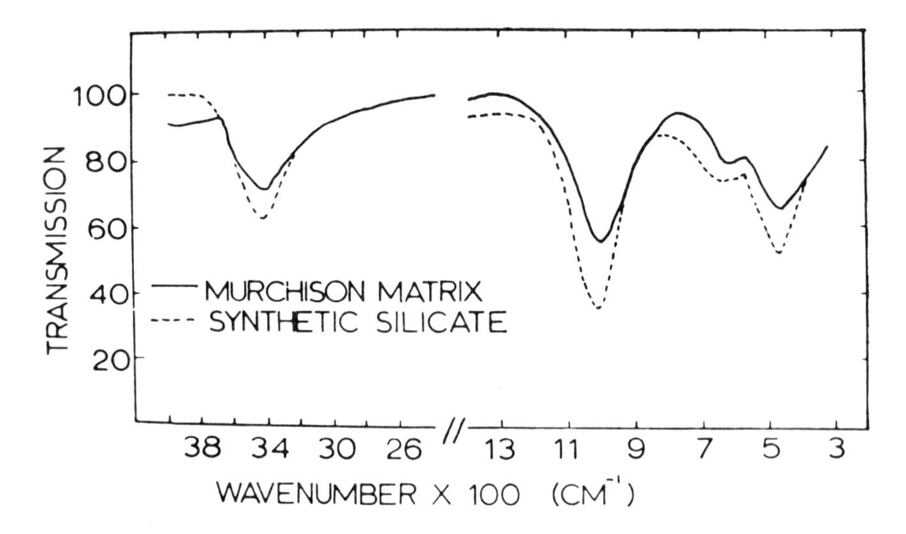

Fig. 7. A synthetic silicate and the Murcheson meteorite (Type C2 carbonaceous chondrite) spectra (Day, 1976).

The results of these several investigations are in general agreement. High-temperature crystalline silicates like olivines and enstatites do not give good fits to the astronomical data, but could be present in small quantities. The mineral composition that best fits the data appears to include one or more hydrated silicates to give the band position at 9.7 μm and other silicate components either structurally or chemically different to give the broadening of the band. The latter could be high-temperature silicates. This model closely resembles the Type C1 or Type C2 carbonaceous chondrite mineralogy.

This seemingly happy situation may be deceiving, however. Two groups have recently synthesized amorphous silicates related to olivines whose spectra also give good fits to the interstellar bands. Day and Donn (1978; see also Donn's chapter in this book) condensed a magnesium silicate from magnesium and silicon monoxide vapor. Amorphous grains formed, with a spectrum near 10 μm which appears to give as good a fit to the interstellar absorptions as the spectra discussed above. On heating of this material characteristic forsterite absorptions appeared. Huffman and Kraetschmer (1978) irradiated olivines with heavy particles and found that an amorphous surface layer formed on the material. Reflection spectra of the amorphous layer indicate that it too gives a good fit to the interstellar bands. It is not clear what, if any, relationship exists between these minerals and meteoritic silicates. Perhaps the critical parameter is the amorphous nature of these silicates, a property which may, in fact, have some bearing on silicates found in interplanetary particles discussed in Sec. III below. Evidently much more laboratory work remains to be done.

Evidence for the presence of the other minerals of Fig. 1 in interstellar grains is considerably more problematical, because their spectral features are weaker, if present at all, in interstellar spectra. This is not surprising if the grains are similar to carbonaceous chondrites. The meteorite spectra are dominated by the silicate features in the middle infrared.

Magnetite

A ferromagnetic or superparamagnetic mineral in the interstellar grains has long been suspected because the observed polarization requires a degree of alignment which is greater than predicted from paramagnetic processes (Davis and Greenstein, 1951; Jones and Spitzer, 1967). The argument is difficult to make quantitative because the alignment of grains is not understood completely; Purcell (1975) has even suggested a mechanism that might make paramagnetic alignment possible after all. In any event, the grains are magnetic and if this does imply cooperative magnetic behavior in them, abundance arguments suggest that iron is the most likely magnetic ion, quite possibly in the form of magnetite, Fe_3O_4. Magnetite grains are present in C1 carbonaceous chondrites in peculiar structures suggesting that the magnetite condensed from a gas phase (Jedwab, 1971). Shapiro (1975) has shown that magnetite particles could give the observed polarization in the visible region of the spectrum.

Huffman (1977) compared apparent broad structure in the interstellar extinction curve (Hayes and Rex, 1978) with the absorption of magnetite and with the absorption of the Orgueil meteorite. As shown in Fig. 8, there are broad features at roughly the same wavelengths in all three curves. The Murchison meteorite, which contains little magnetite, does not show this structure. Thus, there is now some direct spectral evidence that magnetite is an interstellar constituent. Note also that since the silicate feature appears in the polarization-wavelength curve (Dyck *et al.,* 1973), the silicates must be aligned. Magnetite particles located in the silicate matrix, as in Orgueil, in a "raisin pudding" structure (Jones and Spitzer, 1967) could be the aligning agent in the grains.

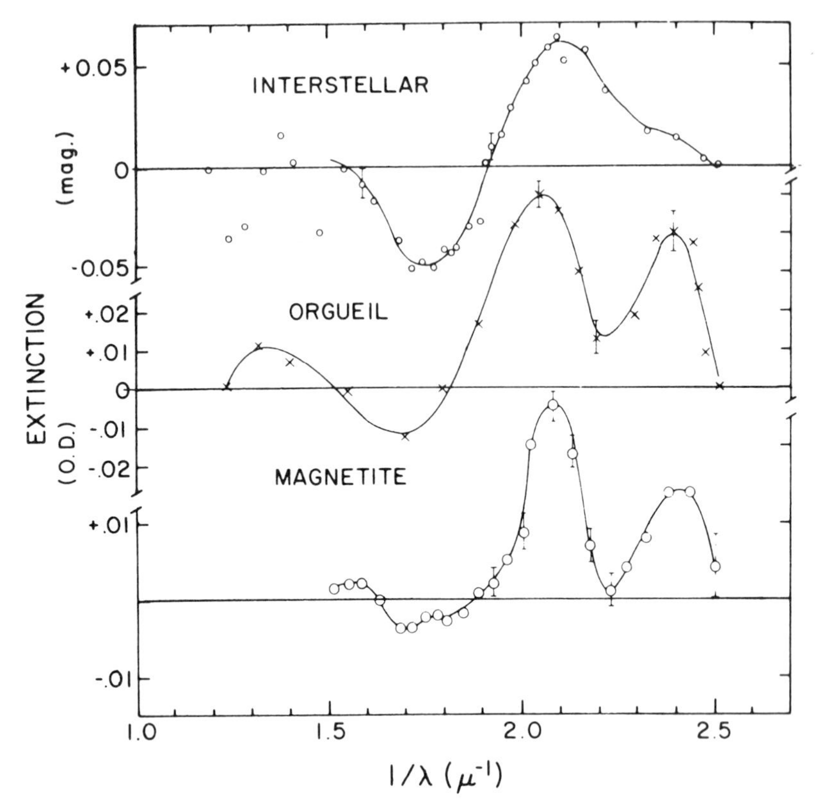

Fig. 8. Spectra of Orgueil, magnetite, and the broad structure in the interstellar extinction. Residuals obtained by subtracting the continuum are shown (Huffman, 1977).

Carbonates

That the somewhat rare carbonates in carbonaceous chondrites might have their interstellar analogue is indicated by a band near 11.3 μm which has been tentatively identified as emission of $MgCO_3$ or $FeCO_3$ (Gillett, Forrest and Merrill, 1973; Bregman and Rank, 1975). This band is observed in emission in planetary nebulae and H II regions. The wavelength coincidence and band shape agreement are good and mass estimates of the required amount of carbonates appear to be reasonable. A serious difficulty is that stronger carbonate bands at 7.7 μm and 33 μm have not been found (McCarthy, Forrest and Houck, 1977). It may be possible that solid state or excitation effects could mask bands in parts of the spectrum. Establishing the presence of carbonates in the grains would be interesting because carbonates are found only in carbonaceous chondrites, although probably as secondary products of reactions after the parent meteorites formed (McCall, 1973). The carbonates in carbonaceous chondrites are apparently not abundant enough to stand out in laboratory spectra of the meteorites.

Sulphates

The hydrated sulphate minerals gypsum ($CaSO_4 \cdot 2H_2O$), epsomite ($MgSO_4 \cdot 7H_2O$), and bloedite ($Na_2Mg(SO_4)_2 \cdot 4H_2O$) are found in C1 carbonaceous chondrites (McCall, 1973). The presence of magnesium sulphate shows that the meteorites were exposed to liquid water. Sulphate compounds have not yet been observed in interstellar grain spectra. It may be less likely that secondary minerals like carbonates and sulphates have an interstellar analogue than is the case for the condensation products.

Organics and Carbon

The organic compounds in meteorites have been the subject of controversy regarding possible connections with terrestrial life. It seems generally accepted now that the compounds and structures in the meteorites need not be the product of living things (Nagy, 1975).

There are indications that carbon is in the form of graphite in the interstellar grains. The observation of a feature near 2200 Å in reddened stars (Stecher, 1969; Bless and Savage, 1972) is strong observational evidence for graphite which has a band at this wavelength (Gilra, 1972). Huffman (personal communication, 1977) pointed out that this is a problem for an interstellar grain analogue in the carbonaceous chondrites. The identification, although based on only one band, is fairly convincing because well-defined absorption bands (rather than edges) are unusual in solids except for graphite (Huffman, 1975). Graphite is not abundant in carbonaceous chondrites and this could be an argument against an interstellar dust-meteorite similarity, at least for all of the dust.

It might be useful in this regard to investigate the ultraviolet spectra of the carbonaceous materials in meteorites. Much of this material consists of insoluble polymer-like compounds which are relatively hydrogen-poor. The organic compounds such as amino acids which have received more attention are part of the much smaller soluble fraction. The polymer-like compounds may deserve more consideration as possible interstellar grain substances. Sakata *et al.* (1977) found a feature near 2200 Å in an extract of Murchison material; they suggested that this may be the source of the 2200 Å interstellar band. The procedure for obtaining this substance or its quantity in the meteorite were not specified very closely, however.

Certain emission bands observed in nebular objects (Russell, Soifer and Willner, 1977) might be resonances of solid carbonaceous material like that found in meteorites (Knacke, 1977). This suggestion is based partly on the coincidence of an emission band at 3.3 μm with the resonance of the C-H bond near this wavelength. An interstellar band at 6.2 μm could be due to C-C vibrations. Other coincidences with nebular emission bands (7.7 μm, 11.3 μm) and the spectrum of the carbonaceous material (Meinschein *et al.*, 1963) are not striking; they may be absent.

In the context of solids which may be present in interstellar dust but not in carbonaceous chondrites, we should also mention silicon carbide. Treffers and Cohen (1974) found evidence for SiC in the emission of carbon-rich stars; this had been predicted by Gilman (1969). SiC may not be a major grain component since it would be formed only in carbon-rich stars; the bands are not observed in the interstellar clouds having the silicate feature. On the other hand, the 2200 Å feature is very ubiquitous and clearly represents a major mineral component of the grains.

II. COMETARY SOLIDS

Maas *et al.* (1970) observed a 9.7 μm emission feature in the spectrum of Comet Bennett. Subsequently, the feature has been observed in comets Kohoutek, Bradfield and West. The most extensive spectroscopy is that of Comet Kohoutek observed by Merrill (1974) with a filter wheel spectrometer of intermediate resolution. The spectrum (Fig. 9) closely resembles the silicate spectra of circumstellar dust, emission in the Orion nebula, and other cometary spectra obtained at lower resolution. This observation of silicates in comets, together with the evidence for water or ice in their nuclei, supports the "dirty snowball" model proposed by Whipple (1950).

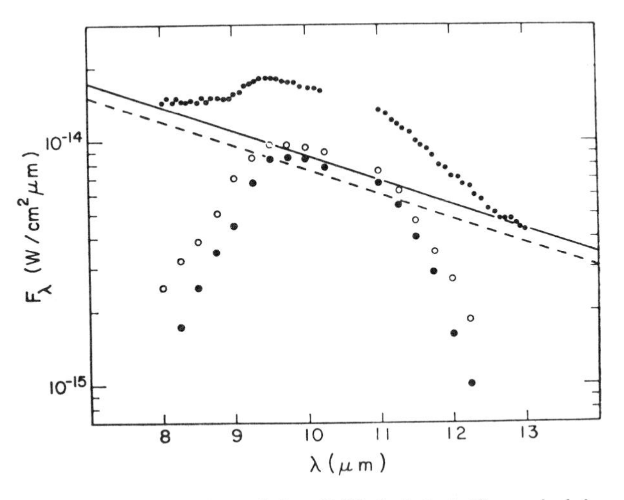

Fig. 9. Spectrum of Comet Kohoutek (small filled circles). The underlying continuum is approximated by a 600 K blackbody. Subtraction of blackbody curves, shown as solid and broken lines, yields the excesses shown by the larger circles (filled and open circles, respectively). Figure is from Merrill (1975).

The marked resemblance of the 9.7 μm cometary feature and the feature in interstellar grains strongly suggests that there are structural and compositional similarities between these objects. To pursue this aspect of the mineralogy, Rose (1977) carried out laboratory measurements of the infrared emission spectra of silicates and compared them with cometary spectra. Some of his spectra of silicate minerals and meteorites are shown in Fig. 10. Rose concludes that three mixtures of silicates could match the cometary spectra: (1) a hydrated layer silicate in combination with a high-temperature silicate like olivine or anorthite to broaden the feature, (2) an amorphous magnesium

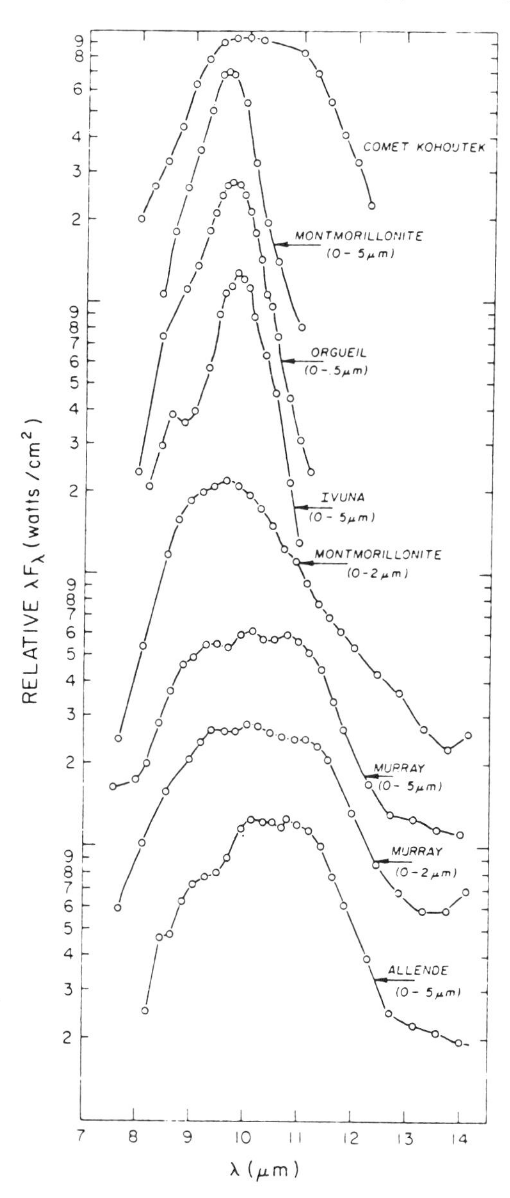

Fig. 10. Emission spectra of hydrated minerals and meteorites compared to cometary emission (Rose, 1977).

silicate in combination with a high-temperature condensate, and (3) glassy olivine and glassy anorthite in approximately equal proportions. The first possibility resembles a Type C2 carbonaceous chondrite while the second recalls Day's (1974) suggestion of amorphous material.

It seems difficult to relate the ices in comets and interstellar grains. Ice is, of course, much more volatile than silicates and has only three low-temperature and pressure phases (Bertie and Whalley, 1967). Thus, it is less likely to survive nebular processes in unaltered form, and it lacks the mineralogical complexity that makes it possible to trace silicate histories. The 3.1 μm ice band would also be difficult to observe in comets. Nevertheless, the association of ice and silicates in comets is significant and the spectral similarities with interstellar grains provides further support for solar system mineral analogues.

The cometary silicate measurements are important to the present discussion. The accumulating astronomical and laboratory emission and absorption spectroscopy of comets and meteorites gives one confidence in the infrared techniques for determining mineralogy in astronomical remote-sensing applications. Carbonaceous chondrites and comets should be mineralogically similar if both are remnants of similar condensation processes in the solar nebula. The infrared data suggest that they may be. The observations then tie the primitive solar system material to the interstellar dust as well. Within observational accuracy, the interstellar silicates cannot be distinguished from the circumsolar silicates in comets and carbonaceous chondrites.

III. INTERPLANETARY DUST

The interplanetary dust is probably composed largely of cometary and meteoritic debris (Millman, 1975). Recently, collections of this material with NASA U-2 aircraft have begun to provide fairly detailed information on the composition of this dust (Brownlee, Tomandl and Hodge, 1975; Brownlee *et al.*, 1976; and Brownlee's chapter in this book). In their elemental abundances, mineralogy, and inclusion content, the identified cosmic dust particles resemble C1 and C2 carbonaceous chondrites (these properties are used to separate the extra-terrestrial particles from terrestrial ones in the stratosphere). This type of material makes up a fairly large fraction of collected particles. Of over six hundred particles, more than half have abundances which are similar to C1 and C2 carbonaceous chondrites, and over 150 resemble them in all the properties listed above. The results generally corroborate ealier conclusions from meteor spectroscopy (Millman, 1974).

The particle collections provide important supporting evidence for the interpretations of the asteroid and cometary spectroscopy discussed above, namely, that both contain carbonaceous chondrite-like material. However, there are also significant differences between the dust and carbonaceous chondrites. Brownlee *et al.* make comparisons of the grain structure of the interplanetary particles and meteorites. Grain sizes in the particles vary from less than 100 Å to micron dimensions (Fig. 11), while grains in meteorites are larger and tend to a more sheet-like structure. The interplanetary dust appears to be somewhat more amorphous than carbonaceous chondrites and may contain less hydrated layer-lathic silicate.

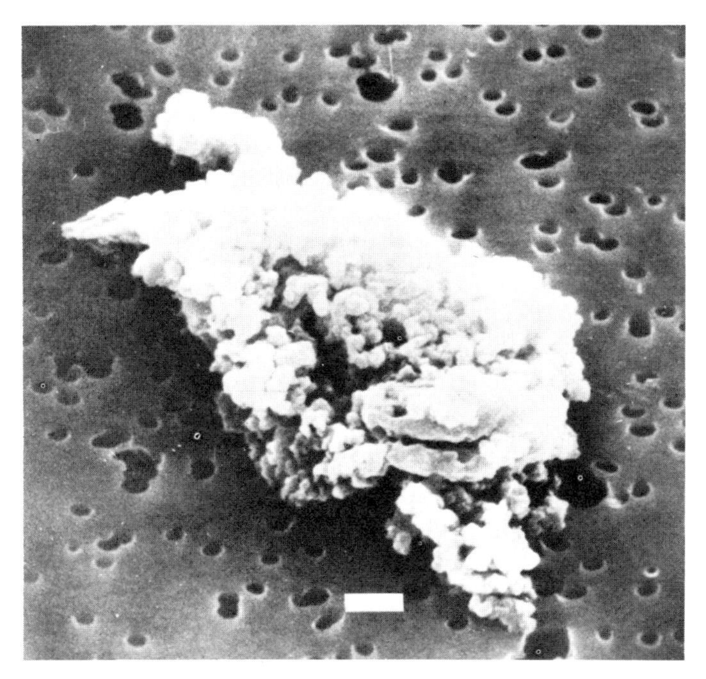

Fig. 11. Carbonaceous interplanetary particle. The line corresponds to 1 μm (Brownlee *et al.*, 1977).

Brownlee in his chapter discusses possible reasons for the structural and compositional differences between the collected dust and metorites. If, as Brownlee suggests, this is a new form of extra-terrestrial material, it could be significant for interstellar grain mineralogy. Thus, it is noteworthy that abundances of Ca, S, and Na indicate the dust may be less processed or "weathered" than C1 meteorites. A cometary origin for the dust could mean that it is composed of very ancient material, even interstellar grains, which have remained in an inert environment until passage of the comet through the inner solar system.

Laboratory spectra of the interplanetary particles are not yet available. Based on the similarity to carbonaceous chondrites one could expect infrared spectra similar to those of interstellar grains. Their more amorphous structure might not affect the infrared spectrum significantly. As described in Sec. I, spectra of some amorphous silicates closely resemble the hydrated silicate spectra.

If the collected dust particles do indeed originate in comets, they are giving us laboratory access to minerals which are not otherwise available and quite probably have a different history than that of the meteorites in our collections. The structural differences may be significant for this reason and continued collections of this material will be valuable.

IV. DISCUSSION

The evidence for some mineralogical similarities among carbonaceous chondrites, comets, interplanetary dust, and the interstellar grains seems fairly persuasive. At this time it is based largely on the silicates and the weaker evidence that magnetite may also be present in the dust. Still, the final word, even on the silicates, is certainly not in, and this chapter is really more a progress report on a picture which seems to be emerging, but which may yet change drastically.

The mineralogy of the meterorites and interplanetary particles is better established than that of the material that we cannot get into the laboratory. Further progress will depend on refinement of the remote sensing techniques to better understand the interstellar silicate and, especially, to identify and describe the non-silicate minerals in the interstellar dust.

Although the focus of this chapter has been observational, we shall add a few comments on theoretical and laboratory condensation investigations. The mineralogical consequences of theoretical condensation schemes in solar nebulae seem somewhat ambiguous. The most successful models are based on equilibrium between the gases and condensates (Barshay and Lewis, 1976). In this process hydrated silicates do not form as condensates, but appear at low temperatures as an alteration product of magnesium silicates. The temperatures reached at the asteroid belt in the solar nebula would give silicates and water, that is, carbonaceous chondrite-like solids. This model has been criticized by Arrhenius and Alfvén (1971) and Arrhenius and De (1973) who point out that dust, because of its high infrared emissivity, will cool to temperatures of 200-400 K while immersed in a hot gas ($>$1000 K) of pressure 10^{-3}-10^{-5} atm, as is usually attributed to the solar nebula. In this case, condensation is not well understood; Arrhenius and Alfvén suggest that condensation of hydrated silicates on the cool substrates might be expected. Perhaps it is significant that both equilibrium and non-equilibrium processes could lead to the same silicate types. In any event, condensation processes are very difficult to treat theoretically if there are departures from equilibrium (Arrhenius and De, 1973; Donn, 1976; Salpeter, 1977).

Ultimately, one must confront the question of how deep the solar system-interstellar grain similarities really are. Are the alleged mineralogical similarities accidental; are they a result of similar formation processes of the solids in interstellar space and the solar system; or does the circumsolar material contain interstellar dust left over from the formation of the solar system? Cameron (1975) has pointed out that interstellar grains cannot be completely evaporated until temperatures exceed 1800 K. These temperatures are achieved only at distances less than 2 AU in the solar nebula (see Wasson's chapter). Thus, the comets and the interplanetary dust might contain the remnants of interstellar grains. Even meteorites could possibly contain interstellar grain cores from which volatile mantles evaporated. Perhaps more likely in view of recent results, is the possibility of interstellar grains in interplanetary particle collections. Interstellar grains may be structurally and mineralogically similar to the grains shown in Fig. 11.

A direct connection between interstellar dust and solar system condensates has not yet been established, but their mineralogy, as well as the isotopic anomolies in meteorites, indicate that it is a distinct possibility. This remains perhaps the most intriguing question posed by the results of dust mineralogy.

REFERENCES

Aannestad, P.A., and Purcell, E.M. 1973. *Ann. Rev. Astron. Astrophys.* 11: 309.

Anders, E. 1964. *Space Sci. Rev.* 3: 583.

Arrhenius, G., and Alfvén, H. 1971. *Earth Planet. Sci. Lett.* 10: 253.

Arrhenius, G., and De, B.R. 1973. *Meteorites* 8: 297.

Barshay, S.S., and Lewis, J.S. 1976. *Ann. Rev. Astron. Astrophys.* 14: 81.

Bass, M.N. 1971. *Geochim. Cosmochim. Acta* 35: 139.

Bertie, J.E., and Whalley, E. 1967. *J. Chem. Phys.* 46: 1271.

Bless, R.C., and Savage, B.D. 1972. In *The Scientific Results from the Orbiting Astronomical Observatory (OAO-2)* (A.D. Code, ed.), p. 175. Washington, D.C.: U.S. Government Printing Office.

Bok, B.J. 1977. *Publ. Astron. Soc. Pacific* 89: 597.

____. 1978. *Moon and Planets* (special Protostars and Planets issue). In press.

Bregman, J., and Rank, D.M. 1975. *Astrophys. J.* 195: L125.

Brownlee, D.E.; Tomandl, D.; Blanchard, M.B.; Ferry, G.V.; and Kyle, F. 1976. *An Atlas of Extraterrestrial Particles Collected with NASA U-2 Aircraft – 1974-1976.* NASA: TM X – 73, 152.

Brownlee, D.E., Tomandl, D., and Hodge, P.W. 1975. In *Interplanetary Dust and Zodiacal Light,* (H. Elsässer and H. Fechtig, eds.), p. 279. Berlin: Springer-Verlag.

Cameron, A.G.W. 1975. In *The Dusty Universe* (G.B. Field and A.G.W. Cameron, eds.), p. 1 New York: Watson.

Cameron, A.G.W., and Truran, J.W. 1977. *Icarus* 30: 447.

Capps, R.W., and Knacke, R.F. 1976. *Astrophys. J.* 210: 76.

Chapman, C.R.; Morrison, D.; and Zellner, B. 1975. *Icarus* 25: 104.

Clayton, D.D. 1975. *Astrophys. J.* 199: 765.

Davis, L., and Greenstein, J.L. 1951. *Astrophys. J.* 114: 206.

Day, K.L. 1974. *Astrophys. J.* 192: L15.

____. 1976. *Icarus* 27: 561.

Day, K.L., and Donn, B. 1978. Preprint.

Donn, B. 1976. *Mém. Roy Soc. Sci. Liège, Ser. 6,* 11: 499.

DuFresne, E.R., and Anders, E. 1962. *Geochim. Cosmochim.* Acta 26: 1085.

Dyck, H.M.; Capps, R.W.; Forrest, W.J.; and Gillett, F.C. 1973. *Astrophys. J.* 183: L99.

Elsässer, H., and Fechtig, H. 1976. *Interplanetary Dust and Zodiacal Light.* Berlin: Springer-Verlag.

Farmer, V.C., ed. 1974. *The Infrared Spectra of Minerals.* London: Mineralogical Society.
Field, G.B., and Cameron, A.G.W., eds. 1975. *The Dusty Universe* New York: Watson.
Forrest, W.J.; Gillett, F.C.; Houck, J.R.; McCarthy, J.; Merrill, K.M.; Pipher, J.L.; Puetter, R.C.; Russell, R.W.; Soifer, B.T.; and Willner, S.P. 1978. *Astrophys J.* 219: 114.
Gillett, F.C., and Forrest, W.J. 1973 *Astrophys. J.* 179: 483.
Gillett, F.C.; Forrest, W.J.; and Merrill, K.M. 1973. *Astrophys. J.* 183: 87.
Gillett, F.C., Forrest, W.J.; Merrill, K.M.; Capps, R.W.; and Soifer, B.T. 1975. *Astrophys J.* 200: 609.
Gillett, F.C., and Soifer, B.T. 1976. *Astrophys. J.* 207: 780.
Gilman, R.C. 1969. *Astrophys. J.* 155: L185.
Grossman, L. 1972. *Geochim. Cosmochim. Acta* 36: 597.
Hackwell, J.A. 1972. *Astr. Astrophys.* 21: 239.
Hayes, D.S., and Rex, K.H. 1978. Preprint.
Herbig, G.H. 1970. *Mém. Roy. Soc. Sci. Liège, Ser. 5,* 19: 13.
____. 1974. *Publ. Astron. Soc. Pacific* 86: 604.
Huffman, D.R. 1975. *Astrophys. Space Sci.* 34: 175.
____. 1977. *Adv. Physics* 26: 129.
Hunt, J.M.; Wisherd, M.P.; and Bonham, L.C. 1950. *Anal. Chem.* 22: 1478.
Jedwab, T. 1971. *Icarus* 15: 319.
Jones, R.V., and Spitzer, L. 1967. *Astrophys. J.* 147: 943.
Kerridge, J.F. 1969. In *Meteorite Research* (P.M. Millman, ed.), p. 500. Dordrecht: Reidel.
____. 1971. *Nature Phys. Sci.* 230: 66.
Knacke, R.F. 1977. *Nature* 269: 132.
Kwan, J., and Scoville, N. 1976. *Astrophys. J.* 209: 102.
Lazarev, A.N. 1972. *Vibrational Spectra and Structure of Silicates* New York: Consultants Bureau.
Maas, R.W.; Ney, E.P.; and Woolf, N.J. 1970. *Astrophys. J.* 160: L101.
Martin, P.G. 1971. *Astrophys. Lett.* 7: 193.
____. 1975. *Astrophys. J.* 202: 393.
McCall, G.J.H. 1973. *Meteorites and their Origins.* New York: Wiley.
McCarthy, J.; Forrest, W.J.; and Houck, J.R. 1977. In *Symposium on Recent Results in Infrared Astrophysics.* (P. Dyal, ed.), p. 190. Washington, D.C.: U.S. Government Printing Office.
Meinschein, W.G.; Nagy, B.; and Henessy, D.J. 1963. *Ann. New York Acad. Sci.* 108: 553.
Merrill, K.M. 1974. *Icarus* 23: 566.
Merrill, K.M.; Russell, R.W.; and Soifer, B.T. 1976. *Astrophys. J.* 207: 763.
Merrill, K.M., and Stein, W.A. 1976. *Publ. Astron. Soc. Pacific* 88: 285.
Meyer, C. 1971. *Geochim. Cosmochim. Acta* 35: 551.
Millman, P.M. 1974. *J. Roy. Astr. Soc. Canada* 68: 13.
____. 1975. In *The Dusty Universe* (G.B. Field and A.G.W. Cameron, eds.), p. 185. New York: Watson.
Nagy, D. 1975. *Carbonaceous Meteorites.* Amsterdam: Elsevier.
Ney, E.P. 1976. In *The Study of Comets* (B. Donn, M. Mumma, W. Jackson, M. A'Hern and R. Harrington, eds.), p. 334. Washington, D.C.: U.S. Government Printing Office.
Penman, J.M. 1976. *Mon. Not. Roy. Astr. Soc.* 175: 149.
Purcell, E.M. 1975. In *The Dusty Universe* (G.B. Field and A.G.W. Cameron, eds.), p. 155. New York: Watson.
Rose, L.A. 1977. Ph.D. Dissertation, University of Minnesota, Minneapolis, Minn.
Russell, R.W.; Soifer, B.T.; and Willner, S.P. 1977. *Astrophys. J.* 217: L149.
Sakata, A.; Nakagawa, N.; Iguchi, T.; Isobe, S.; Morimoto, M.; Hoyle, F.; and Wickramasinghe, N.C. 1977. *Nature* 266: 241.
Salpeter, E.E. 1977. *Ann. Rev. Astron. Astrophys.* 15: 267.
Shapiro, P. 1975. *Astrophys. J.* 201: 151.
Simon, T., and Dyck, W.M. 1975. *Nature* 253: 101.
Stecher, T.P. 1969. *Astrophys. J.* 157: L125.

Steyer, T.R.; Day, K.L.; and Huffman, D.R. 1974. *Appl. Opt.* 13: 1586.
Strom, S.E.; Strom, K.M.; and Grasdalen, G. 1975. *Ann. Rev. Astron. Astrophys.* 13: 187.
Treffers, R., and Cohen, M. 1974. *Astrophys. J.* 188: 545.
Whipple, F.L. 1950. *Astrophys. J.* 111: 375.
Wickramasinghe, N.C.; Hoyle, F.; Brooks, J.; and Shaw, G. 1977. *Nature* 269: 675.
Wickramasinghe, N.C., and Nandy, K. 1972. *Rep. Prog. Phys.* 35: 159.
Wood, J.A. 1968. *Meteorites and the Origin of Planets.* New York: McGraw-Hill.
____. 1975. In *The Dusty Universe* (G.B. Field and A.G.W. Cameron, eds.), p. 245. New York: Watson.
Woolf, N.J. 1973. In *Interstellar Dust and Related Topics* (J. Greenberg and H.C. van de Hulst, eds.), p. 485. Dordrecht: Reidel.
Zaikowski, A., and Knacke, R.F. 1975. *Astrophys. Space Sci.* 37: 3.
Zaikowski, A.; Knacke, R.F.; and Poroco, C.C. 1975. *Astrophys. Space Sci.* 35: 97.

INTERPLANETARY DUST: POSSIBLE IMPLICATIONS FOR COMETS AND PRESOLAR INTERSTELLAR GRAINS

D. E. BROWNLEE
California Institute of Technology

Laboratory studies of micrometeorites show that typical interplanetary dust particles are fine grained, black aggregates which contain approximately chondritic elemental abundances for the eleven most abundant elements in chondrites. Abundances of C, S, Na and Mn indicate that the particles are at least as volatile-rich as the most primitive meteorites. A small fraction (<10%) of the particles have hydrated silicate microstructures somewhat similar to C1 and C2 meteorites. These particles may be samples of materials related to meteorite parent bodies. Most of the particles, however, have a different, finer grained and more porous microstructure and they appear to be a new type of extraterrestrial material not seen in meteorites. The basic construction of these particles is a porous aggregate of ~1000 Å sized grains. Some of the grains are single crystals, others are themselves aggregates of smaller grains ranging in size down to less than 100 Å. Forsterite, enstatite, pyrrhotite, pentlandite and magnetite occur in the particles as minor phases. A significant fraction of the particles may be amorphous or poorly crystalline. While the origin of interplanetary dust is not firmly established it is very likely that at least some of the more than 300 particles studied are debris from contemporary comets. The volatile contents of comets indicate formation far from the sun, possibly by accretion of presolar interstellar grains. The 1000 Å grains in interplanetary dust may be ancient interstellar grains.

Individual interplanetary dust particles are short-lived solar system bodies. They are destroyed on short time scales by interparticle collisions (Kresák, 1976), solar heating during close perihelion passage, rotational bursting resulting from photon pressure (Paddack and Rhee, 1976), or by ejection from the solar system by the radiation pressure process of the β meteoroids (Zook and Berg, 1975). The integral effect of these destruction processes is complex and depends on physical and orbital properties of the particles. For typical nanogram particles at 1 AU, lifetimes are determined by catastrophic collisions and are on the order of 10^4 yrs (Kresák, 1976; Whipple, 1976). Although individual particles are short lived, the existence of microscopic impact craters on lunar and meteorite samples, with ancient exposure histories, indicate that dust has existed in the interplanetary medium for most of the age of the solar system (Horz *et al.*, 1975; Brownlee and Rajan, 1973; Goswami *et al.*, 1976).

The dust cloud of the solar system is maintained in quasi-equilibrium by fairly continuous injection of new particulate material. When the solar system is inside an interstellar cloud, the major dust source may be the interstellar medium (Newman and Talbot, 1976). Usually, and at the present time, the major suppliers are sources within the solar system. It has been suggested that the sun itself may produce submicron refractory particles (Hemenway *et al.*, 1972). Mercury and satellites without atmospheres expel dust liberated by impact but the major producers of micron-sized dust are asteroids and comets. The asteroids have a large cumulative surface area and generate large amounts of dust due to collisions with particles ranging in size from microns to kilometers (Dohanyi, 1976). Comets generate dust when they penetrate the inner solar system. Sublimation of cometary ice releases and expels dust grains contained in the dust ice matrix which forms comet nuclei. Radiation pressure complicates the release process and places surprisingly large lower limits on the sizes of particles which can be put into bound orbits by comets with highly eccentric orbits (Harwit, 1963; Jambor, 1976).

Interplanetary dust is important to studies of the origin of the solar system because it is material from comets and asteroids – the smallest surviving bodies from the early solar system. The parent materials which replenish the dust cloud of the solar system may, in some cases, be relict planetesimals or fragments of planetesimals from the solar nebula. Because comets and asteroids escaped capture by planets and because they are essentially gravitationless bodies, they may have suffered only minimal heating and alteration during the accretional and post-accretional phases of planetesimal formation.

PARENT BODIES OF INTERPLANETARY DUST

The relative importance of asteroids versus comets as sources of interplanetary dust is unfortunately not known. The fact that comets are the major suppliers of the millimeter meteoroids which produce optical meteors suggests that a significant fraction of the submillimeter portion of the meteoroid complex is also cometary material (Millman, 1976). While comets are probably the major source, it is important to remember that the dust presently impacting the earth was probably generated by a number of different parent bodies. Because of radiation pressure effects, dust particles can have orbital parameters significantly different from those of their parent materials. This produces diffusion and a tendency for dust to become somewhat homogeneously distributed within the inner solar system. While it has been estimated that meteorites may have come from less than a dozen parent bodies (Anders, 1971), it is likely that dust now hitting the earth has come from a larger number of bodies which have been in the inner solar system over the past 10^5 yr.

A complicating factor in interpretation of interplanetary dust is that after release from a parent body the dust can still be altered in the interplanetary medium. Besides solar flare and solar wind effects, particles can be altered by collisions and close approaches to the sun.

Asteroids

Main-belt asteroids are believed to be fragments of planetestimals which did not accumulate to form planets. They appear to be products of the inner solar system and presumably are accumulations of original condensates and reworked materials which existed in the Mars-Jupiter region of the solar nebula. Both studies of meteorites (Wasson, 1974) and of asteroid spectral reflectivity (Gaffey and McCord, 1977) indicate that some asteroids experienced heating which produced effects ranging from mild metamorphism to melting and igneous differentiation. Because nearly all meteorites are breccias (compacted fragments of previously existing rocks) it is probable that asteroids are covered with a layer of impact generated debris (regolith) somewhat analogous to the lunar regolith. Dust from main-belt asteroids should have compositions compatible with reflectivity studies (Gaffey and McCord, 1977) and some should be similar to known meteorite types. The reflectivity measurements indicate that most should be similar to carbonaceous chondrites. Some asteroidal particles should show effects of thermal metamorphism and evidence of residence in an impact generated regolith. Regolith effects could be in the form of implanted solar wind, tracks of solar flare particles, shock induced transformations, micrometeorite craters and surface accreta from nearby microcraters.

Comets

Dust particles from comets may be the original dust particles which accreted along with ices to form kilometer-sized "dirty snowball" planetesimals. Because of their volatile content and remoteness from the sun, comets are believed to be the most pristine bodies in the solar system (Delsemme, 1977). If comets are products of the solar system, then they are definitely products of its outer regions. They must have originally formed either in the Saturn-Neptune region of the nebula or possibly much farther out in satellite nebulae (Cameron, 1972). The condensation temperature for the ices in comets has been estimated to be only on the order of 100 K (Delsemme and Rud, 1977). Most of the dust from comets is probably relatively unmodified material because it has been stored at cryogenic temperatures for the age of the solar system. Many of these particles may be pristine samples of materials from the outer regions of the solar nebula. Most of the particles probably have not resided in impact generated regoliths because the time scale for sublimation of active, dust-producing comets is shorter than the time scale for regolith formation.

The comets which are major dust producers are probably active ones that do not alter the properties of constituent grains. Some dust particles however will come from less benign environments which exist at the surfaces of old volatile depleted comets. These comets may build crusts of non-volatile materials which in some cases will be heated to temperatures of 500 C or more at perihelion. Hydrous alteration as well as other reactions could seriously alter grains in such comets.

Asteroids apparently formed by accumulation of nebular condensates. Comets, on the other hand, formed in distant regions of the nebula, probably in some cases beyond the region where presolar interstellar grains had been evaporated (Cameron, 1975). For these comets, formation would occur by accumulation of presolar interstellar grains, icy condensates from the nebula and possibly some higher temperature fractions transported outward from the inner regions of the nebula. It is a distinct possibility that the dust grains in some comets may be pure presolar interstellar grains with no nebular condensate fraction.

COLLECTION OF INTERPLANETARY DUST

Because of high relative velocity and low spatial density, non-destructive collection of interplanetary dust in space is exceedingly difficult. Fortunately small particles can non-destructively enter the earth's atmosphere and at appropriate altitudes they can be collected from an unpolluted environment where the fall velocities are low and spatial densities are correspondingly high. Although all interplanetary material that enters the atmosphere is heated, micron-sized particles with moderately low entry velocities can survive without melting. Small particles decelerate at high altitudes where the rate of generated frictional energy is low enough to be thermally radiated without melting of the particles. The maximum size of particles that enter without melting depends on the composition of the particles and their entry parameters. For 15 km sec^{-1} entry velocity the maximum size for a refractory silicate like enstatite ($MgSiO_3$) is ~100 μm; for FeS it is only 10 μm. The heating occurs as a short pulse usually lasting less than 10 sec. Interplanetary dust particles that survive entry without melting are called "micrometeorites" (Whipple, 1951). Although micrometeorites escape melting, they are heated and their maximum temperatures can vary anywhere from ~100 C to >1000 C depending on entry parameters and particle size, shape and density (Brownlee *et al.*, 1977).

Modern attempts to collect interplanetary particles in the atmosphere began with the historic Venus Flytrap mesospheric rocket experiment (Hemenway and Soberman, 1962). A large number of collections by rocket, balloon, aircraft and surface means were conducted but most of the results were inconclusive. Reviews of these collections can be found by Rosen (1969), Parkin and Tilles (1968) and Brownlee (1978). The results discussed in this chapter are those of Brownlee *et al.* (1977) who collected particles in the stratosphere first with balloons at 34 km and then with NASA U-2 aircraft at 20 km. The extraterrestrial origins of the particles were proven on the basis of their distinctive elemental abundances and mineralogy and because they contain large amounts [0.1 cm^3 (STP) g^{-1}] of implanted solar wind helium (Rajan *et al.*, 1977).

The stratospheric micrometeorites were collected by inertial deposition from a high velocity windstream onto a collection plate covered with a thick film of silicone oil of high viscosity. For analysis, particles were individually picked off, washed and put into electron microscopes for evaluation. In all, more than 300 extraterrestrial particles in the size range 2 μm to 50 μm were subjected to detailed scrutiny. For the $\geqslant$2 μm size range studied in the U-2 collections, contamination was not a serious problem. The major contaminant was aluminum oxide, injected into the stratosphere by solid fuel rockets which use aluminum powder as a fuel additive (Brownlee *et al.*, 1976). Disregarding aluminum particles, the extraterrestrial particles constitute over half of the particles collected during five years of collection. The density of 5 μm interplanetary dust in the stratosphere was $10^{-3} m^{-3}$ and did not appear to vary significantly with time. Some of the particles were spheres and had melted during entry; most however were true micrometeorites and had not experienced even partial melting. The helium measurements indicate that more than half of the particles were small bodies in space and were not fragments of larger bodies which broke up in the atmosphere.

The earth's atmosphere provides a finite window for micrometeorite collection. Particles larger than ~25 μm are strongly heated during entry and those smaller than 1 μm are essentially uncollectable because of the high concentrations of submicron stratospheric sulfate aerosols. The most valuable, least heated particles are those of low density and ~5 μm size.

THE PHYSICAL PROPERTIES OF INTERPLANETARY DUST

Of the ~300 interplanetary dust particles analyzed, more than half have cosmic (solar) elemental abundance patterns. This is not a selection effect but it is an expected result for primitive, undifferentiated solar system material. Most of the particles which do not have cosmic abundances are single mineral grains which appear to have been at one time imbedded in a matrix material of cosmic elemental composition. The non-cosmic abundance particles are usually either the transparent mafic silicates enstatite [Mg,SiO_3] and olivine [$(Mg,Fe)_2SiO_4$] or opaque particles composed of magnetite [Fe_3O_4] and pyrrhotite [$Fe_{1-x}S$]. The mafic silicate particles and the magnetite-sulfide particles are often found with fine-grained material of cosmic abundance adhering to their surfaces. The study of micrometeorites has made it clear that most interplanetary dust particles that survive atmospheric entry are opaque aggregates of small grains whose cumulative composition is close to that of solar abundances. Optically, even particles only a few microns in size are exceedingly black. Careful examination in the optical microscope usually reveals the existence of several micron-sized reflective sulfide grains and transparent silicates imbedded in opaque material.

Structure

With few exceptions the cosmic abundance micrometeorites are aggregates of small grains (Figs. 1-5). The particles are called chondritic aggregate micrometeorites where the term "chondritic" merely implies solar composition. The grains range in size from <100 Å to several microns. In the scanning electron microscope (SEM) most of the particles appear to have a characteristic grain size of 2000 Å – 3000 Å. This may be somewhat of an illusion because higher resolution observations in transmission electron microscopes (TEM) reveal the existence of many grains that are very much smaller. Some of the typical 1000 Å sized grains are themselves aggregates of smaller grains or are coated with smaller grains. Each dust particle contains millions of constituent grains. The shapes of the grains tend to be irregular and relatively equidimensional although rods, platelets and euhedral crystals do occur. As observed in high-voltage electron microscopy some of the grains are rounded and are coated with 500 Å amorphous coatings (Bibring *et al.*, 1978). The coatings and rounded edges are similar to solar wind irradiation effects observed on lunar soil grains.

The external shapes of the particles are strongly influenced by their basic 1000 Å size grain structure. Particles only a few microns in size tend to have complex highly convoluted shapes with dramatic cavities and protrusions. Particles larger than 10 μm are large in comparison to their typical 1000 Å size constituent grains and they are comparatively smooth on size scales on the order of 10 μm.

A property which shows significant range from particle to particle is porosity. The particles range from compact with no apparent porosity to highly porous, open aggregate structures. The most porous particles (rare) have densities probably on the order of unity or possibly somewhat less. Typical particles have probable densities in the range 1-3 g cm^{-3}. Actually, the effective density of the smaller particles (<10 μm) is difficult to define. If the volume of a particle is defined by a smoothly varying surface which

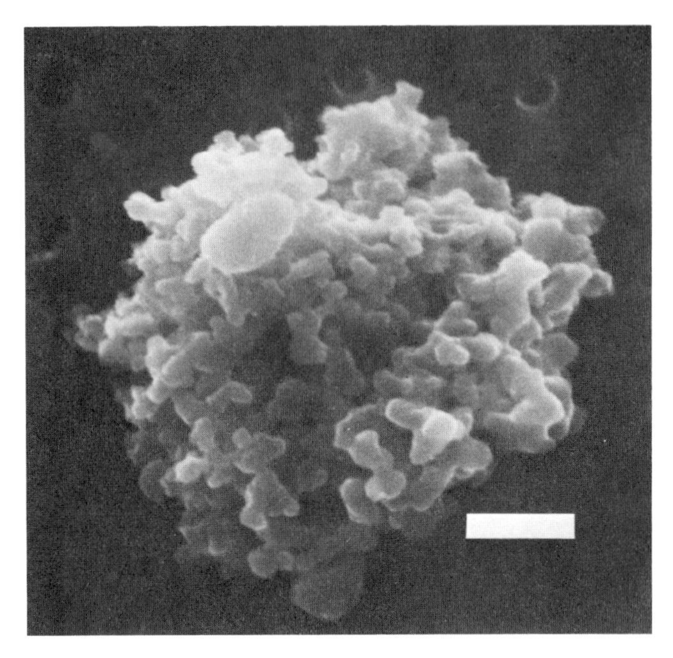

Fig. 1. Chondritic aggregate interplanetary dust particle. Scale bar = 1 μm.

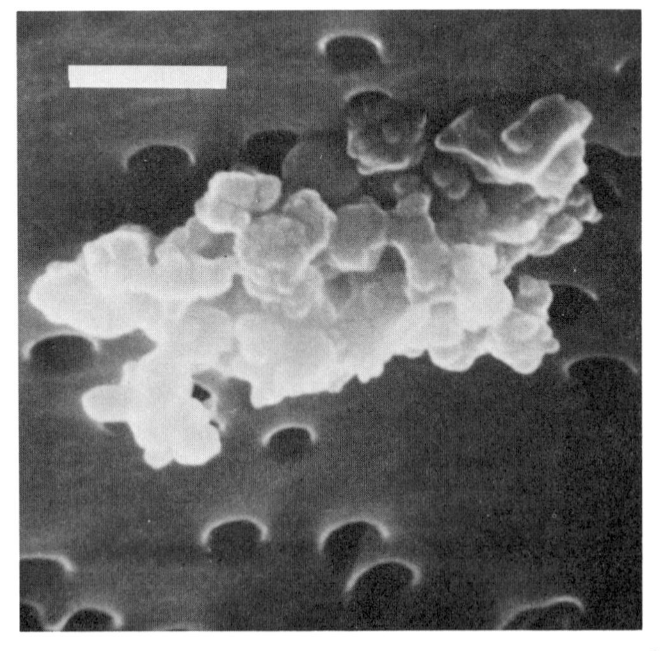

Fig. 2. Chondritic aggregate interplanetary dust particle. Some of the 3000 Å grains are aggregates of smaller grains. Scale bar = 1 μm.

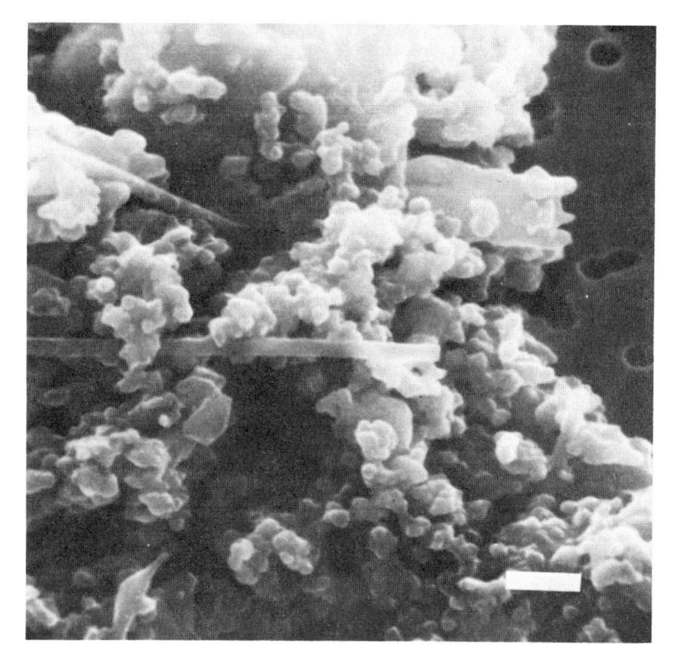

Fig. 3. Portion of a chondritic aggregate dust particle which contains rare whisker-shaped grains. Scale bar = 1 μm.

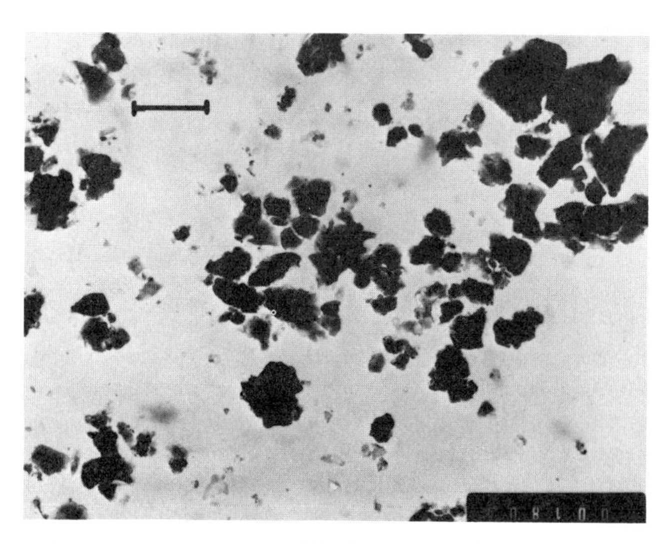

Fig. 4. Transmission electron micrograph (TEM) of an interplanetary dust particle which has been disaggregated and mounted on a 200 Å carbon film. Note the equidimensional grain shapes. Scale bar = 1 μm.

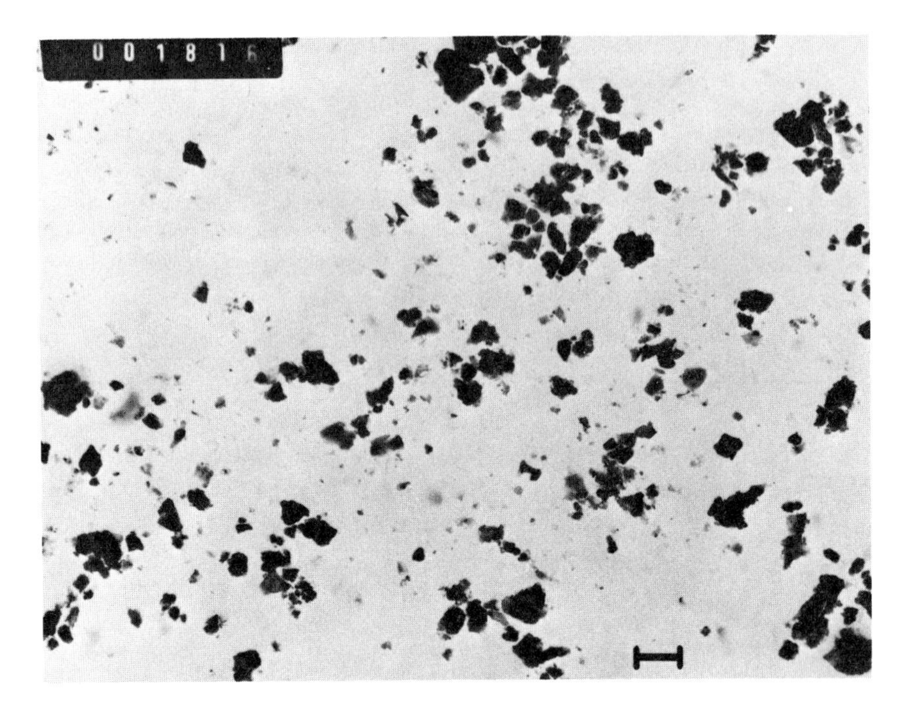

Fig. 5. Transmission electron micrograph (TEM) of a disaggregated interplanetary particle showing submicron grain shapes. Scale bar = 1 μm.

contacts the particle's high points, then all of the smaller particles have low densities because of protruding grain structures. It is not known whether the range in porosities in the particles is intrinsic or was caused by alteration in the atmosphere, in space or on the parent body.

Elemental Composition

The elemental compositions of chondritic aggregate particles are close to those of chondritic meteorites. In Table I the average composition of thirteen porous chondritic aggregate interplanetary dust particles are compared with chondritic meteorites (Brownlee *et al.*, 1977). The masses of the analyzed particles range from 5×10^{-11}g to 10^{-8}g and it is remarkable that such small particles agree so well with "cosmic" abundances. From Table I it is clear that the particles have "chondritic" abundances and agree reasonably well with abundances from all chondrite groups. The high-sulfur abundance (and also carbon) indicate that the particles are volatile-rich, similar to C1 and C2 carbonaceous chondrites. The S/Ca and Na/Ca ratios agree with the accurately determined solar photospheric values of Holweger (1977) and are higher than those of all meteorite types except C1 chondrites.

TABLE I

Comparisions of Average Elemental Compositions (Atomic Fractions) and RMS Deviations of thirteen 3 μm to 25 μm Chrondritic Aggregate Micrometeorites with Meteorites and Fine Grained Material.

	Interplanetary Dust	C1[a]	C1[b] Matrix	C2[a]	H[a]	L[a]
Mg/Si	0.85 ± 0.15	1.06	0.92	1.04	0.97	0.94
Fe/Si	0.63 ± 0.26	0.90	0.54	0.84	0.81	0.58
S/Si	0.35 ± 0.13	0.46	0.13	0.23	0.10	0.10
Al/Si	0.063 ± 0.023	0.085	0.093	0.084	0.061	0.061
Ca/Si	0.048 ± 0.017	0.071	0.011	0.072	0.050	0.048
Na/Si	0.049 ± 0.048	0.060	0.016	0.035	0.042	0.043
Ni/Si	0.037 ± 0.020	0.051	0.047	0.046	0.049	0.030
Cr/Si	0.012 ± 0.0047	0.013	0.012	0.012	0.011	0.011
Mn/Si	0.015 ± 0.0017	0.0092	0.0046	0.0062	0.0068	0.0068
Ti/Si	0.0022 ± 0.0006	0.0024	0.0012	0.0024	0.0021	0.0021

[a] Bulk analyses of carbonaceous (C1 and C2) and ordinary (H and L) chondrite meteorites (Mason, 1971).
[b] Analyses of fine grained material (matrix) in Type 1 carbonaceous chondrites (McSween and Richardson, 1977).

Mineralogy

The mineralogy of recovered interplanetary dust samples is determined by X-ray diffraction, electron diffraction and microprobe analysis. Unfortunately the mineral contents of typical chondritic interplanetary dust particles are very poorly known. X-ray diffraction patterns on pristine-looking samples often reveal nothing. Most patterns do show the existence of magnetite (Fe_3O_4) and the magnetic sulfide, pyrrhotite ($Fe_{1-x}S$). Because of compositional considerations neither of these minerals can be major phases. The magnetite may have been produced by heating during atmospheric entry. Analysis of individual submicron grains in crushed aggregates shows that grains of olivine $(Fe,Mg)_2SiO_4$, enstatite $MgSiO_3$, Ni bearing pyrrhotite and pentlandite $(Fe,Ni)_9S_8$ are numerous but again they are not the major mineral phases. The constitutent olivine grains in a given particle are unequilibrated; they have various Fe/Si ratios. The vast majority of 1000 Å grains are iron magnesium silicates with variable compositions but yet unidentified crystal structures. While a significant portion of the grains are crystalline they may be poorly ordered and thus produce weak X-ray diffraction patterns. Amorphous material is fairly common in crushed particles (Flynn *et al.*, 1978; Bibring *et al.*, 1978).

Rare micrometeorites (<10%) contain identifiable hydrated silicates as the major mineral phase. Table II shows the X-ray diffraction results for one such particle (XP-36) compared with two terrestrial hydrated silicates and the Murchison C2 carbonaceous chondrite. Although the chondritic dust particles containing hydrated silicates are not well characterized, it appears that they are a different class of material than the normal chondritic aggregate interplanetary dust particles.

TABLE II

X-ray Diffraction Results on a Rare 15 μm Chondritic Interplanetary Dust Particle Compared with the Murchison C2 meteorite and the Hydrated Silicates Serpentine and Chamosite.

Interplanetary Dust Particle XP-36[a]		Serpentine 9-444		Chamosite 7-339		Murchison[d]	
d[b]	*I*[c]	*d*	*I*	*d*	*I*	*d*	*I*
7.21	100	7.33	100	7.12	100	7.2	100
4.63	40	4.60	60	4.68	50	4.3-4.7	(Band)
				4.30	20		
				3.93	30		
3.60	70	3.66	100	3.55	100	3.58	50
2.93 (Band)	20			3.07	20		
2.65	40	2.62	30	2.71	50	2.70	20
2.53	60	2.50	100	2.53	100	2.53	80
		2.34	70			2.43	30
2.06	50	2.15	60	2.15	70	2.16	40
1.80	20	1.96	70	1.78	60	1.79	20

[a] This particle contained a hydrated silicate as the major mineral phase. It is unusual and not like typical interplanetary particles collected in the stratosphere.
[b] The *d* values are lattice spacings in angstroms.
[c] The *I* values are diffraction line intensities (maximum intensity ≡ 100, lines less than 20 not listed).
[d] Results are by Fuchs *et al.* (1973).

Although more than half of the analyzed micrometeorites have chondritic compositions and are aggregates of large numbers of small grains, numerous micrometeorites are single mineral grains. Usually these are opaque grains of sulfide and magnetite or are transparent grains of olivine or pyroxene (Fig. 6). Often these particles contain clumps of chondritic aggregate material on their surfaces indicating that the particles were at one time in a matrix of chondritic aggregate material. X-ray diffraction of the sulfide particles indicates the mineral is $Fe_{1-x}S$ with variable Ni content up to several percent. The iron deficiency varies from zero to over 10%. The sulfides are found in a wide variety of morphologies including hexagonal crystals, octahedrons, spheres, stacks of platelets, irregular masses and aggregates. The micron-sized transparent silicates are usually olivine or pyroxene. As shown in Fig. 7 the iron contents of the pyroxenes are usually close to zero while in the olivines the iron content ranges from zero to Fe/Mg = 1. Other particles such as NiFe metal are found but they are very rare.

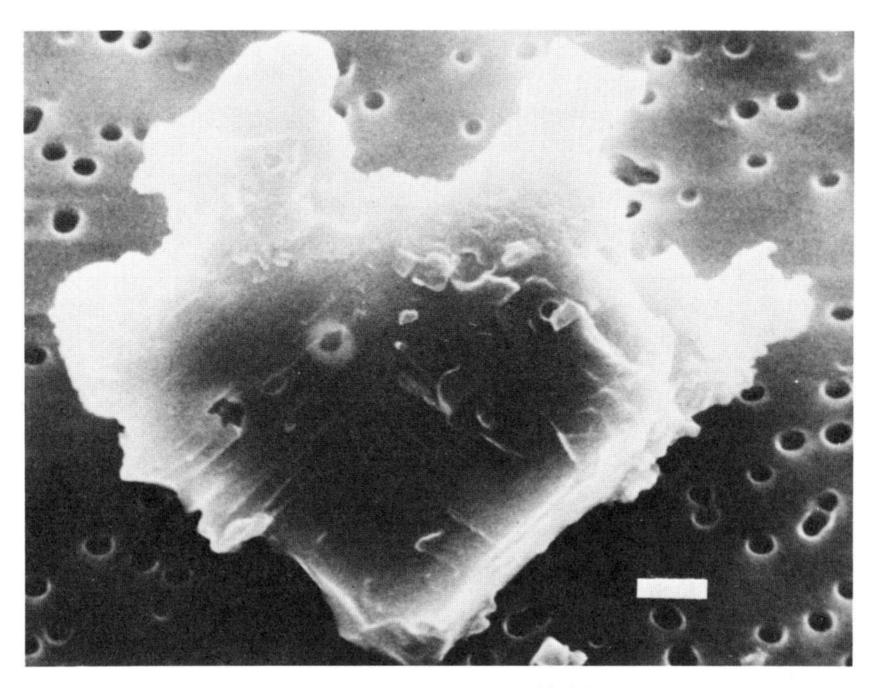

Fig. 6. A micrometeorite composed of pure enstatite ($MgSiO_3$). Scale bar = 1 μm.

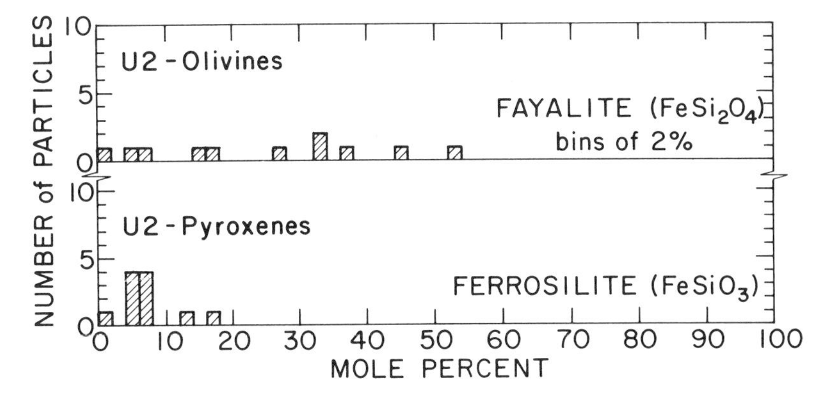

Fig. 7. Iron contents of interplanetary olivine and pyroxene particles collected as single crystals ranging in size from 5-50 μm.

COMPARISON OF INTERPLANETARY DUST AND METEORITES

The small grain sizes, presence of magnetite and pyrrhotite, low reflectivity, high C and S content, presence of pure Mg olivine, rare presence of hydrated silicates, and the presence of Ni-rich sulfides, distinguish interplanetary dust from most meteorites and indicate that typical dust particles are more similar to C1 meteorites and the fine-grained matrix of C2 meteorites (see Wasson, 1974, for classification) than any other known material. However, in most cases interplanetary dust particles are distinctly different from C1 and C2 meteorites. Most of the fine-grained material in the meteorites is a hydrated layer lattice silicate which forms tangled masses of crumpled sheets and fibers. This material is not really very porous and although the grains may be small, they are small only in one or sometimes two dimensions. In the scanning electron microscope (SEM) the meteorite particles appear smooth (Fig. 8) in comparison with interplanetary dust. In the transmission electron microscope (TEM) the meteorite particles almost always show distinctive fibrous, platey textures, features almost never seen in the dust particles (Bibring *et al.*, 1978; Flynn *et al.*, 1978; Brownlee *et al.*, 1977). The rare hydrated silicate bearing micro-meteorite in Table II does have a smooth exterior similar to meteorite particles, and TEM analyses show that many internal grains are platey. This particle is very different from typical chondritic aggregate interplanetary dust particles and is similar,

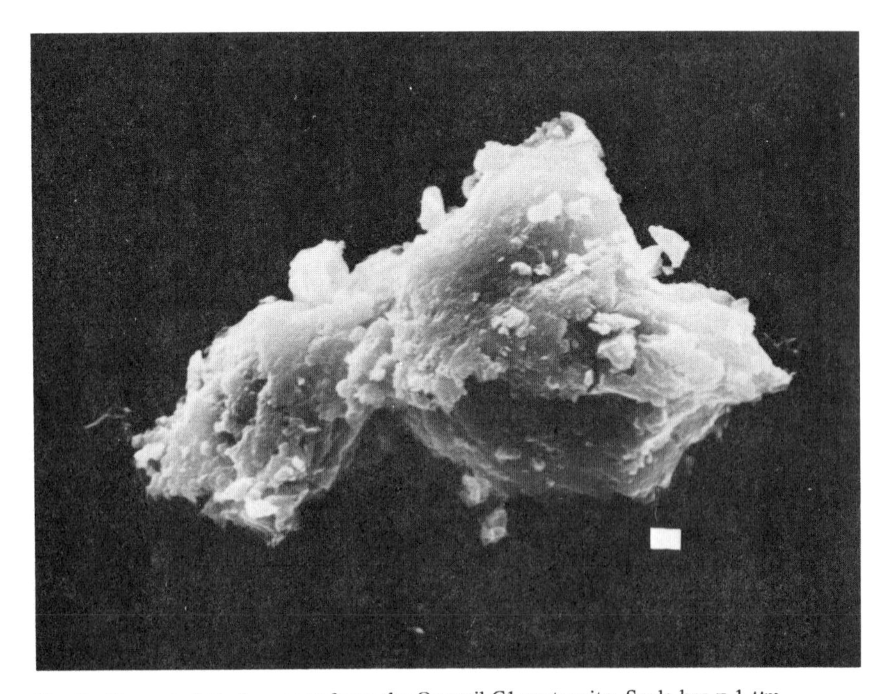

Fig. 8. Characteristic fragment from the Orgueil C1 meteorite. Scale bar = 1 μm.

although not identical, to common C1 meteorite fragments. In most cases, however, SEM and TEM observations indicate that interplanetary dust is a material distinctly different from meteorites. Most of the dust particles are more porous, finer grained and in some ways more "complex" than meteorites. It does not appear possible that the dust material could have been produced from known meteorite materials.

The high C and S contents and other properties of the chondritic interplanetary dust particles indicate that they fall into a region of meteorite classification occupied only by the primitive C1 and C2 meteorites. However, the data in Table I indicate that for most of the other elements the average particle is more similar to the ordinary L group chondrites. If the data in Table I are representative for interplanetary dust compositions then it indicates that the dust has small but significant deviations from C1 meteorites. Although C1 meteorites are often used as a definition of "solar" abundances, there are only five such meteorites and they may well have come from the same parent body.

It is instructive to compare the interplanetary dust data in Table I with the fine grained material in C1's (C1 matrix). Here one sees that the fine grained interplanetary dust has much better agreement with "cosmic" abundances. Sulfur, calcium and sodium are strongly depleted in the fine grained fraction of the meteorites. These elements have been mobilized and redistributed in the C1 parent bodies, possibly due to leaching by water. The C1's contain veins of water soluble minerals. Bunch and Chang (1978) and others have presented evidence for extensive hydrous alteration (weathering) in the carbonaceous chondrite parent body or bodies. The dust particles in Table I have not been depleted in Na, S and Ca and it does not seem likely that their parent bodies could have experienced appreciable hydrous alteration.

POSSIBLE EXPLANATIONS FOR THE DIFFERENCE BETWEEN INTERPLANETARY DUST AND METEORITIC MATERIAL

Option 1: Meteorites from asteroids; dust from comets

This requires that no cometary meteoroid has ever survived atmospheric entry to become a recognized meteorite. Asteroids at some time provided "warm" environments for the weathering processes that mobilized and redistributed Ca, S and Na in C1 meteorites. Fragile dust particles are not compacted in a regolith and contain ices in inter-grain spaces for most of the age of the solar system.

Option 2: Meteorites and dust from comets

The low-density dust is released during sublimation of ices. The observed ranges in dust porosity are caused by processes occurring near the comet surface. Meteorites are from compact residual cores of dead comets, crusts from aging comets or primordial compact masses within comets.

Option 3: Meteorites and dust from asteroids

Meteorites are from compacted asteroids and dust is from asteroidal materials too fragile to survive atmospheric entry and become meteorites.

INTERSTELLAR GRAINS AND COMETS (A CONJECTURE)

If interplanetary dust is presolar interstellar dust which accreted to form comets, then the results reported here may indicate that interstellar grains in the vicinity of the solar nebula were small, mundane particles not drastically different from particles which may have condensed from the solar nebula. The grains ranged in size from <100 Å to >50 μm with typical grains being on the order of 2500 Å. Some of the 1000 Å grains were aggregates of very much smaller grains. The diversity of grains indicates a variety of formational environments but the final product was a material fairly similar to carbonaceous chondrites with an average composition fairly close to solar. The grains ranged from high-temperature condensates like olivine and pyroxene to low-temperature materials like magnetite, iron sulfide, sometimes hydrated silicates and carbonaceous materials. A fraction of the material was amorphous, either the result of the mode of formation or subsequent radiation damage. In some cases whisker-like rods and euhedral crystals were formed but usually the grains were irregular. Some of the grains were highly rounded, possibly the result of sputtering. However the grains were formed, metallic iron was either rarely produced or it was later converted to sulfide or oxide. The major magnetic phases were Fe_3O_4 and $Fe_{1-x}S$. The accretional process that accumulated small grains must have been a very gentle process with very few high-velocity impacts.

DISCUSSION

The most detailed records of the formation of the solar system are recorded in asteroids, comets and possibly other yet undiscovered small solar system bodies. These records can only be adequately scrutinized in the laboratory where a remarkably versatile array of techniques can perform magic in extracting the many forms of information contained in solid primordial solar system matter. The samples that will be available in the foreseeable future are meteorites, interplanetary dust and samples from direct sample-return missions to a small number of asteroids and comets. The importance of interplanetary dust is that it provides samples of fragile materials which do not produce meteorites, and materials from a larger number of asteroids and comets than can be visited by spacecraft. It is clearly important to have samples of as many different types of primitive solar system bodies as possible. The mysterious existence and distribution of isotopic anomalies and short-lived radioactivities (Lee *et al.*, 1977; Clayton and Mayeda, 1977) within the region of the solar nebula which produced meteorite parent bodies, indicates that the processes which produced planetary bodies were highly complex.

Acknowledgments. E. Olszewski and D.A. Tomandl were instrumental in a variety of aspects of micrometeorite collection and analysis. The major part of this research was supported by the National Aeronautics and Space Administration.

REFERENCES

Anders, E. 1971. In *Physical Studies of Minor Planets* (T. Gehrels, ed.), p. 429. NASA SP-267. Washington, D.C.: U.S. Government Printing Office.
Bibring, J. P.; Brownlee, D.E.; and Maurette, M. 1978. Paper presented at 9th Lunar and Planetary Sci. Conf.
Brownlee, D.E. 1978. In *Cosmic Dust* (J.A.M. McDonnell, ed.); p. 295. London: Wiley.
Brownlee, D.E.; Ferry, G.V.; and Tomandl, D.A. 1976. Science 191: 1270.
Brownlee, D.E.; Tomandl, D.A.; and Olszewski, E. 1977. *Proc. Lunar Sci. Conf.* 8th. 149. New York: Pergamon.
Bunch, T.E., and Chang, S. 1978. Paper presented at the 9th Lunar and Planetary Sci. Conf.
Cameron, A.G.W. 1972. In *On the Origin of the Solar System* (H. Reeves, ed.), p. 56. Paris: C.N.R.S.
____. 1975. In *The Dusty Universe* (G.B. Field and A.G.W. Cameron, eds.), p. 1. New York: Neale-Watson.
Clayton, R.N., and Mayeda, T.K. 1977. *Geophys. Res. Lett.* 4:295.
Delsemme, A.H. 1977. In *Comets, Asteroids, Meteorites* (A.H. Delsemme, ed.), p. 3. Toledo, Ohio; University of Toledo.
Delsemme, A.H., and Rud, D. 1977. In *Comets, Asteroids, Meteorites* (A.H. Delsemme, ed.), p. 529. Toledo, Ohio: University of Toledo.
Dohanyi, J.S. 1976. In *Interplanetary Dust and Zodiacal Light* (H. Elsässer and H. Fechtig, eds.), p. 187. New York: Springer-Verlag.
Flynn, G.; Fraundorf, P.; Shirck, J.; and Walker, R.M. 1978. Paper presented at 9th Lunar and Planetary Sci. Conf.
Fuchs, L.H.; Olsen, E.; and Jensen, K.J. 1973. *Smithsonian Contrib. Earth Sci.* 10, Washington, D.C.: Smithsonian Institution Press.
Gaffey, M.J., and McCord, T.B. 1977. In *Comets, Asteroids, Meteorites* (A.H. Delsemme, ed.), p. 199. Toledo, Ohio: University of Toledo.
Goswami, J.N.; Hutcheon, I.D.; and Macdougall, J.D. 1976. In *Proc. Lunar Sci. Conf.* 7th. 543. New York: Pergamon.
Harwit, M. 1963. *J. Geophys. Res.* 68: 2171.
Hemenway, C.L., and Soberman, R.K. 1962. *Astron. J.* 67:256.
Hemenway, C.L.; Hallgren, D.S.; and Schmalberger, D.C. 1972. *Nature* 238:256.
Holweger, H. 1977. In *Comets, Asteroids, Meteorites* (A.H. Delsemme, ed.), p. 385. Toledo, Ohio: University of Toledo.
Horz, F.; Brownlee, D.E.; Fechtig, H.; Hartung, J.B.; Morrison, D.A.; Neukum, G.; Schneider, E.; Vedder, J.F.; and Gault, D.E. 1975. *Planet, Space Sci.* 23:151.
Jambor, B.J. 1976. *The Study of Comets* (B. Donn, M. Mumma, W.M. Jackson, M. Ahearn and R. Harrington, eds.), p. 943. NASA SP-393. Washington, D.C.: U.S. Government Printing Office.
Kresák, L. 1976. *Bull. Astron. Inst. Czechoslovakia* 27:35.
Lee, T.; Papanastassiou, D.A.; and Wasserburg, G.J. 1977. *Astrophys. J.* 211:L107.
Mason, B. 1971. *Handbook of Elemental Abundances in Meteorites.* New York: Gordon and Breach.
McSween, H.Y., and Richardson, S.M. 1977. *Geochim. Cosmochim. Acta.* 41:1145.
Millman, D.M. 1976. In *Interplanetary Dust and Zodiacal Light* (H. Elsässer and H. Fechtig, eds.), p. 359. New York: Springer-Verlag.

Newman, M.J., and Talbot, R.J. 1976. *Nature* 262:559.

Paddack, J.J., and Rhee, J.W. 1976. In *Interplanetary Dust and Zodiacal Light* (H. Elsässer and H. Fechtig, eds.) p. 453. New York: Springer-Verlag.

Parkin, D.W., and Tilles, D. 1968. *Science* 159:936.

Rajan, R.S.; Brownlee, D.E.; Tomandl, D.A.; Hodge, P.W.; Farrar, H.; and Britten, R.A. 1977. *Nature* 5607:133.

Rosen, J.M. 1969. *Space Sci. Rev. 9:58.*

Wasson, J.T. 1974. *Meteorites.* New York: Springer-Verlag.

Whipple, F.L. 1951. *Proc. Nat. Acad. Sci.* 36:687.

____. 1976. In *Interplanetary Dust and Zodiacal Light* (H. Elsässer and H. Fechtig, eds.), p. 403. New York: Springer-Verlag.

Zook, H.A., and Berg, O.E. 1975 *Planet. Space Sci.* 23:183.

PART III

Clouds and Fragmentation

STAR FORMATION IN MOLECULAR CLOUDS

N. J. EVANS II
The University of Texas at Austin

There is now considerable supporting evidence for the contention that stars and planets form in molecular clouds. These clouds are cold ($T_K \sim$ *10-80 K*), *neutral* [$n(e)/n < 10^{-5}$ *and probably* $< 10^{-8}$], *and gravitationally bound. We have divided clouds into two groups based on the maximum temperature in the cloud. The clouds with maximum temperatures above 20 K tend to be associated with infrared sources and compact H II regions, as well as decreased abundances of complex molecules. The Jeans mass in the dense cores of both groups is* $\sim$*2–4* $M_\odot$, *indicating that low-mass stars can form in both types of clouds, and that the cloud mass is much greater than the Jeans mass. A scenario for star formation in these clouds is presented and compared to observations. The possibility of direct detection of low-mass protostars in the infrared is considered and found feasible out to* $\sim$*1 kpc. Objects that may in fact be low-mass protostars have been found in clusters of infrared sources alongside much more luminous and hence more massive protostars. Surprisingly, no strong candidates for low-mass protostars have been found outside these clusters which exist in dense, hot cores of molecular clouds.*

The objective of this chapter is to present a general overview of the findings of molecular line and infrared astronomy regarding star formation. The general picture and some of the detailed conditions should be borne in mind by those working on the origin of planetary systems. Given the objective, I have refrained from detailed expositions of techniques and analysis, for which references are given, and have instead presented a simplified picture. Particular attention has been drawn to conclusions which should bear on the central question, "Which clouds and stars form planets?"

If a slight rearrangement of this question is permitted so that we ask, "Which clouds form stars and planets?," then the answer can be given with some certainty, namely that stars and planets form from *molecular* clouds. To the extent that planet formation is a natural adjunct of star formation, the question of where planets form reduces to the question of where stars form. An abundance of evidence now indicates that stars form from clouds which are primarily molecular. First, there is the confirmation by *Copernicus* observations (Spitzer *et al.*, 1973) of the hypothesis that clouds will change from atomic H to molecular H_2 once the column density is sufficient for self-shielding (Hollenbach, Werner and Salpeter, 1971). Thus, it is expected that all clouds will become primarily molecular before star formation occurs. Second, many molecular clouds have actually been observed (c.f., Zuckerman and Palmer, 1974). These clouds cover a wide range in mass – from about 20 $M_\odot$ up to at least 10^5 $M_\odot$. Third, the close association of compact H II regions and infrared sources with the dense portions of molecular clouds indicates that star formation *is* presently occurring in molecular clouds (c.f., Wynn-Williams and Becklin, 1974; Beckwith *et al.,* 1976).

My answer to the rearranged question has some important consequences. The most immediate of these is that the conditions in molecular clouds are input parameters for theoretical studies of star formation. For example, the ionization as measured by the electron abundance is quite low; $n(e)/n$ is certainly $<10^{-5}$, and probably $<10^{-8}$ (Guelin *et al.,* 1977). This fact may have serious implications for the role of magnetic fields.

Another consequence is that a nominal star-formation rate of 1-10 $M_\odot$ yr^{-1} is easily accounted for. Essentially all molecular clouds have more gravitational potential energy than thermal energy; and, in the absence of other support forces, they should be collapsing. We *observe* clouds which *should* be forming stars. In fact, we have an embarrassment of riches; if the observed clouds were all in free-fall collapse, the rate of star formation would be too high (Zuckerman and Palmer, 1974). Further discussion of this problem may be found in Section VII.

To make a direct attack on the original question of which clouds form planets is beyond our present capabilities. If any star can form a planet, then any molecular cloud can result in stars with planets. If, for the sake of argument, we take the conventional view that only low-mass stars form planets, then the question becomes "Which clouds form low-mass stars?" To examine even this much simpler question, we must first have a way of classifying clouds.

I. CLASSIFICATION OF MOLECULAR CLOUDS

Traditionally, molecular clouds have been divided into two general categories. One is called dark dust clouds or dark clouds based on their optical appearance; Bok globules belong in this category (see Bok, 1978). Dark clouds are generally nearby and are thought to have modest size, mass and density. The other category is less well defined; they are sometimes called massive (or giant) molecular clouds – reflecting the fact that they were found by molecular spectroscopy rather than by optical means. These clouds are thus more distant, and generally considered more massive. They contain regions of higher temperature and density, which we will call dense cores. The rest of the cloud is sometimes called a CO envelope (c.f., Zuckerman and Palmer, 1974).

For a number of reasons this categorization is not entirely satisfactory. Kutner *et al.* (1977) present a CO map of the archetypical massive molecular cloud, the one stretching across the constellation of Orion. They also show a schematic picture of the optical dust clouds in this area. The overall agreement is striking; this cloud could be put in both categories. Second, a study of molecular clouds associated with Sharpless H II regions (Sharpless, 1959) indicates that many have masses between those of "dark clouds" and "massive molecular clouds" (Blair, 1976). The mass distribution, though poorly known, is probably continuous, making the term "massive" a poor qualitative distinction. Finally, regions of high temperature and density are sometimes found in classical "dark clouds" – for example, the ρ Oph cloud (Encrenaz, Falgarone and Lucas, 1975). To avoid such confusion, we shall use a single parameter to classify clouds. Neither size nor distance is fundamental. Density is hard to determine and is not constant through the cloud. Mass is probably the most interesting parameter, but it is very hard to determine. The kinetic temperature, in contrast, is easily determined from CO. We shall divide clouds into Group A, those with temperatures less than 20 K throughout, and Group B, those where the temperature rises above 20 K somewhere in the cloud. In this scheme the ρ Oph cloud moves to Group B, leaving its fellow "dark clouds" behind in Group A. Although this categorization is *ad hoc*, we shall see that it has some useful features.

We now ask whether the division applies to any other properties. It would be particularly interesting to see if Group B clouds are *always* more massive, but this cannot yet be stated with assurance. The only direct measure of the density comes from studies of molecular excitation, notably that of H_2CO. These studies are restricted to the denser regions of both groups. So far, the highest densities found in Group B clouds exceed the highest densities found in Group A clouds (Evans *et al.*, 1975*a*; Evans *et al.*, 1975*b*; Evans and Kutner, 1976). The infrared sources and compact H II regions, indicative of massive stars forming, are generally restricted to Group B clouds. More studies are needed to see how general these trends are. A very striking difference between Group A and Group B clouds has been recently uncovered. The dense portions of Group A clouds have a higher relative abundance of complex molecules (notably HCO^+, ^{13}CO, H_2CO, HCN, and HNC) than do the dense parts of Group B clouds (Wootten *et al.*, 1977*a*). As shown by Wootten *et al.* (1977*b*), the abundance varies by up to three orders of magnitude, while the uncertainties are no more than about one and a half orders of magnitude. Thus there seems to be a chemical difference between Group A and Group B. These abundances are only determined in the dense cores; it would be very interesting to find the abundances in the less dense envelopes of the clouds. The ρ Oph cloud has low abundances; on this basis too, it is distinct from other dark clouds.

II. PHYSICAL PROPERTIES IN MOLECULAR CLOUDS

In Table I, the physical properties of the two groups are summarized, dividing each into an envelope and a core. As can readily be seen, our knowledge of densities in the envelopes is poor, and our knowledge of masses is also poor, especially in Group A. The Jeans mass (M_J) is the same (2-4 $M_\odot$) in the cores of both Group A and Group B clouds. At least in Group B cores, M_J is also much less than the total core mass. The presence or absence of strong infrared sources is subject to selection effects. Some of these effects have recently been removed by surveys of the Group B envelopes and of Group A clouds, which have not detected strong infrared sources, but more work is needed on this subject. As an example of the abundance variations, the average HCO^+ abundance is given for three Group A cores and for ten Group B cores; as noted above, no evidence on the envelopes is available.

TABLE I

Properties of Molecular Clouds[a]

	Group A		Group B	
	Envelope	Core	Envelope	Core
T_K(K)	10	10	10-20	20-70
n(cm^{-3})	10^2-10^3	10^4	10^3 ?	10^5 -10^6
$M(M_\odot)$	20-?	?	10^3-10^5	10^2-10^3
$M_J(M_\odot)$	8-24	2	?	2-4
Strong IR	no	no	no	yes (always?)
$\frac{n(HCO^+)}{n}$	?	2×10^{-9}	?	2×10^{-11}

[a]In all cases where a range of entries is given, the range represents values from sample clouds, not uncertainties in determinations in individual clouds. The typical uncertainties vary from ±10% (T_K), to factors of ~3 in n and M_J, up to factors of 5-10 in M and $n(HCO^+)/n$.

III. SCENARIO FOR STAR FORMATION IN MOLECULAR CLOUDS

To proceed further, we need at least a provisional scenario to show how star formation proceeds. Based on an amalgam of theory and observation, the following picture emerges.

For a variety of reasons, it is unlikely that massive clouds are collapsing uniformly at free fall (Zuckerman and Palmer, 1974; Zuckerman and Evans, 1974). However, we can identify one or more regions of higher density and/or temperature in, for example, the Orion cloud (Kutner *et al.*, 1977; Kutner,

Evans and Tucker, 1976). A simpler region is found associated with the S 140 Hα emission region (Blair *et al.*, 1978). Here the strong temperature peak is also a density peak and contains a strong infrared source. As this is a common pattern, it is natural to identify these cores as the sites of present star formation. Therefore, we begin with the first statement that massive clouds form stars first in one part, then perhaps later in another part (Zuckerman and Evans, 1974). For a smaller cloud, like a Bok globule, the collapse may be more regular and apply to the whole cloud. Unfortunately the small size of most globules compared to the resolution of current millimeter-wave telescopes makes it difficult to verify this hypothesis. The next step is to identify a number of stages in the evolution of the core of a massive cloud or perhaps the whole of small clouds.

1. First Stage (cool and dense). The cloud begins to contract; in the case of a massive cloud, a core becomes distinct from the rest of the cloud. The density rises, but the kinetic temperature remains low because of the effective cooling by molecular lines, first by CO and later perhaps by a combination of molecules (Goldsmith and Langer, 1977). As the density rises, collisions with dust become rapid and the kinetic temperature becomes coupled to the dust temperature (Goldreich and Kwan, 1974). The sub-millimeter or very-far infrared radiation by the dust could keep the gas cool even at very high densities (c.f., Larson, 1973). Further fragmentation presumably occurs, until fragments with the masses of individual stars become distinct.

2. Second Stage (hot and dense). If a massive protostar forms, its radiation will heat the remaining dust in the core. Since the density is still high enough to couple the dust and gas temperatures, the kinetic temperature will also rise, producing a CO hot spot which we have called a Group B core. It is sometimes suggested that this rise in temperature due to massive star formation inhibits the formation of low-mass stars by raising the Jeans mass. As we have seen, the Jeans mass in these regions is $2\,M_\odot$, about the same as in low-temperature clouds. This issue cannot presently be resolved without understanding the role of further fragmentation.

3. Third Stage (hot and rare; perhaps ionized). In this stage, which may only occur if a very massive star forms, a region of low density, but of high temperature is produced. For example, a substantial H II region may form and expand, a supernova may disrupt the core, or the new stars may simply accrete most of the matter in the core while keeping it hot.

Some examples may assist in clarifying these stages. The Orion nebula would represent a late Stage 3 object, while the Orion Molecular Cloud (OMC1) dense core behind it would be a late Stage 2.

IV. OBSERVATIONAL TESTS

What tests can be made regarding the above scenario? By examining temperature peaks in Group B clouds, we have found that many of these are dense and contain infrared sources. Thus, objects like OMC 1, OMC 2 (Gatley *et al.*, 1974), Mon R2 (Beckwith *et al.*, 1976), S 140 (Blair *et al.*, 1978), S 255 (Evans, Blair and Beckwith, 1977), etc. can be identified as cores in the second stage. A detailed study of two of these supports the hypothesis that the dust heats the gas; the requirement that the dust is at least as hot as the gas led to predictions for the far-infrared radiation, which were subsequently confirmed (Evans *et al.*, 1977; Blair *et al.*, 1978; Harvey *et al.*, 1978). Thus the fact that Group B cores contain strong infrared sources is now understood; the infrared sources cause the high T_K.

It is harder to locate cores in the first stage, dense and cold; without a CO peak to attract attention, they are unlikely to be observed in the H_2CO lines which are harder to detect. The Group A cores do have densities which are enhanced over those of the envelope, but the highest density measured in these regions is about 10^4 (Evans and Kutner, 1976). It would be very interesting to locate regions which are cold but have densities comparable to those in the Group B cores.

With the exception of obvious H II regions, cores in the third stage are also difficult to identify, because they may still be surrounded by dense regions. However, there are several Group B cores that do not have 2-mm H_2CO emission, suggesting that they have relatively low density.

Work by Sargent (1977) on the molecular cloud associated with the Cepheus OB3 association provides a nice example of the three stages. Within the extended cloud, she finds three regions of particular interest. One region is cold and dense, one is hot and dense, and one is hot and less dense since 2-mm H_2CO is not detected. There is some evidence that an early B star has formed in the hot and less dense region. Thus, cores in all three stages appear to be present in this cloud, in addition to the OB subgroups themselves.

V. FORMATION OF LOW- MASS PROTOSTARS

Let us return now to the question: Where do low-mass stars form? We begin by asking other questions. Is the formation of low-mass protostars inhibited by the formation of massive stars? If so, what determines whether a massive star forms? For example, do low-mass clouds produce only low-mass stars?

Based on a comparison of Jeans masses, we have *no* basis for arguing that low-mass stars *cannot* form in the same regions in which massive stars have just formed. Indeed, Larson (1977) found in a theoretical simulation that the number of subcondensations formed in a collapsing cloud is roughly equal to the total mass divided by the Jeans mass. Thus, in the group B dense cores we would expect 50-500 stars to form. Larson also found that the smallest subcondensation was 1/5 the initial Jeans mass, or about 0.5 - 1 $M_{\odot}$ for typical cores. If a massive star forms in a substantial H II region, it might rarefy most of the cloud, but fragments that are already dense are probably preserved, perhaps even compressed further. On the other hand, Sargent (1977) remarks that the masses of the dense cores in the Cepheus OB3 cloud are sufficient only to form the high-mass stars of an OB subgroup and she suggests that low-mass stars are formed elsewhere. More studies of this problem are needed.

VI. INFRARED DETECTION OF LOW-MASS PROTOSTARS

A more direct attack on this problem could be made by detecting the low-mass protostars in the infrared. Theoretical studies (Larson, 1972; Appenzeller and Tscharnuter, 1975) indicate that a 1 $M_{\odot}$ protostar produces 10-30 solar luminosities for a period of $\sim 10^5$ years. While the detailed shape of the emitted energy distribution depends on time and on the surrounding density and temperature structure, calculations (Larson, 1972; Bertout, 1976) tend to support the common sense view that low-mass protostars would look essentially like high-mass ones, only with lower luminosity. By analogy then, we would look for sources peaking strongly, longward of 10 μm with a luminosity of 10-30 $L_{\odot}$. At one kiloparsec, a 30 $L_{\odot}$ protostar would be magnitude +1.5 or about 7 Jy at 10 μm, well within detection limits.

In Larson's (1972) model, the luminosity drops later in the evolution; as the peak of emission moves to 2 μm, the luminosity is 3 $L_{\odot}$. Searches at 2 μm have discovered many objects which peak at 2 μm (c.f., Grasdalen *et al.*, 1973; Vrba *et al.*, 1975; Strom *et al.*, 1976*a*; Strom *et al.*, 1976*b*); unfortunately it is very hard to prove that any of these are protostars instead of reddened main-sequence stars, late-type giants, etc. Objects peaking strongly longward of 10 μm are more clearly associated with dust emission. While there is still controversy here, a source peaking strongly, longward of 10 μm and residing in a molecular cloud is a strong candidate for a protostar.

Have any such objects been found? The cores of Group B clouds often contain luminous sources of this type, such as the one in S 140 (Blair *et al.*, 1978). While these luminous sources usually receive most of the attention, there is often a cluster of infrared objects, some of which have much lower luminosity. For example, IRS4 and IRS5 in Mon R2 have luminosities less than 100 and 200 $L_{\odot}$ (Beckwith *et al.*, 1976). IRS3 in OMC2 produces $\sim$20 Jy at 10 μm (Gatley *et al.*, 1974). At the 500 pc distance of the Orion cloud, a 30 $L_{\odot}$ protostar would produce $\sim$30 Jy. Likewise IRS3 in OMC1 (Wynn-Williams and Becklin, 1974) has about the right flux. While converting these luminosities to masses is not straightforward, we seem to have evidence here that relatively low-mass stars form in the same regions where massive stars are forming.

Do they form anywhere else? Two places where we would like to look are Group A clouds (especially nearby dark clouds) and the extended envelopes of Group B clouds, both requiring that large areas be searched. This makes 10-μm observations extremely time consuming, as background noise restricts one to small beams. The only such 10 μm survey that I know of was made for other reasons and concentrated on more distant clouds (Nadeau, personal communication, 1977). The survey of the Air Force Geophysical Laboratory is not sensitive enough. However, the survey resulting from the Infrared Astronomy Satellite, together with much detailed follow-up work, may provide a breakthrough in this area.

What about searching at 2 μm where sensitive systems with large beams are available for sources which peak at 10 μm? Multicolor photometry would then be needed to select the very red sources. At 1 kpc the 30 $L_{\odot}$ protostar produces only ~0.003 Jy at 2 μm if it radiates like a blackbody. Even if it has an energy distribution like the observed more luminous sources, the flux would be only ~0.02-0.07 Jy. Many sources exist at this level and sifting through them would be very tedious; restrictions to more nearby clouds would probably be necessary. 2-μm surveys of large areas in nearby dark clouds have been made by Strom and co-workers at Kitt Peak (c.f., Grasdalen *et al.*, 1973; Vrba *et al.*, 1975; Strom *et al.*, 1975; Strom *et al.*, 1976*a*; Strom *et al.*, 1976*b*), while our group has surveyed large areas of extended envelopes of Group B clouds. So far, all the strong protostar candidates appear to exist in Group B cores.

VII. OLD CONTROVERSIES AND NEW DEVELOPMENTS

My objective in this chapter has been to present a general overview of the results of molecular line and infrared astronomy which bear directly on star formation. To complete this review, several controversies should be mentioned, as well as some detailed scenarios for star formation. The purpose of this section then is to discuss these areas more fully.

The first Pandora's box to be opened is the question of cloud collapse. Arguments for or against collapse in individual clouds have been presented elsewhere (Zuckerman and Evans, 1974; Snell and Loren, 1977; Leung and Brown, 1977) and need not be reviewed. The more general question of cloud collapse and star-formation rate is worth more discussion. Zuckerman and Palmer (1974) pointed out that if all molecular clouds were collapsing at free fall, the star-formation rate would be 30-100 times higher than accepted values. This problem has been exacerbated by the results of CO surveys of the galactic plane which indicate that the total mass in molecular clouds, interior to the galactic orbit of the sun, is $\sim 4 \times 10^9 M_{\odot}$ (Solomon, Sanders and Scoville, 1977). Even if the average density in these clouds is extremely low (~20 cm^{-3}) so that the free-fall time $\tau_{ff} \sim 10^7$ yr, the star-formation rate would be 400 $M_{\odot}$

yr^{-1}. The following set of suggestions are intended to cover all possible explanations of this situation:

1. The masses of molecular clouds have been grossly overestimated.
2. The clouds are supported against collapse so that the time scale for star formation is $\gtrsim 100\ \tau_{ff}$.
3. Star formation is cut off after only $\lesssim$ % of the cloud's mass has formed into stars.
4. This is an unusual time in galactic history; we are experiencing a burst of star formation much greater than the average rate.

To this list, some may immediately want to add the "explanation" that star formation is inefficient; on closer examination, this statement explains nothing and may be reduced to numbers 2 and 3 above. Explanation number 4 is clearly anti-Copernican, but cannot otherwise be ruled out. My own suspicion is that numbers 1-3 will all play roles in the eventual resolution of this dilemma (see also Field's chapter). The fact that associations and clusters have $\lesssim 10^3\ M_\odot$ in stars indicates that number 3 must play some role in the case of very massive ($\geqslant 10^5\ M_\odot$) clouds.

Another question which deserves more discussion is the role of external triggers in causing star formation. The fact that great difficulty is encountered in trying to provide adequate support forces to prevent clouds from collapsing at free fall makes it seem somewhat counterproductive to add external triggers to the already potent gravitational forces. On the other hand, if one bravely postulates that *some* support force must be preventing cloud collapse, then external triggers may provide a means of initiating star formation in a fraction of clouds. Furthermore, observational evidence exists which indicates that such triggers as cloud collisions (Loren, 1976), shock fronts from OB stars (Elmegreen and Lada, 1977; see Lada *et al.* chapter), and supernova remnants (Herbst and Assousa, 1977, see also their chapter; Wootten, 1977, 1978) do play a role in star formation by compressing molecular clouds. In a somewhat different sense, shocks from galactic density waves may initiate star formation by compressing diffuse clouds into the gravitationally bound molecular clouds that we observe (c.f., Woodward, 1976; Bash and Peters, 1976), although coalescence through cloud collisions (c.f., Penston *et al.*, 1969; Oort, 1954) or collection in magnetic field valleys (Mouschovias, 1974, see also his chapter) may also be responsible for the formation of molecular clouds.

The suggestion by Elmegreen and Lada (1977) that shocks from one group of OB stars initiate subsequent star formation is particularly appealing because it offers an explanation of age sequences of OB subgroups, provides a means of partial cloud disruption, and allows for differing initial mass functions. The shock both compresses and heats a portion of the molecular cloud adjacent to the H II region and thereby causes rapid star formation. The OB stars formed in the compressed layer ionize it and send shocks into the next layer of the cloud; this explains the age sequence of OB subgroups and the piecewise dissipation of the cloud. The elevated temperature, and perhaps turbulence, in the shocked layer raises the limiting mass which is unstable to further fragmentation (Silk, 1977), resulting in the formation of relatively high-mass stars.

The sequential formation mechanism of Elmegreen and Lada (1977) finds observational support in the examples of M 17 and W 3, but Sargent's (1977) study of the Cepheus OB3 association reveals some discrepancies with the detailed predictions of the model. Some of these discrepancies could be explained by geometry or irregularity in the cloud, but they may argue as well that gravitational fragmentation is also playing a role. It may be very difficult to distinguish these two mechanisms. Kutner *et al.* (1976) have argued that the dense condensation connecting Orion A and NGC 1977 may represent a pattern of fragmentation with the two most evolved fragments (Orion A and NGC 1977) at either end. Both these H II regions appear to be adjacent to a heated region in the dense condensation, suggesting that shocks from H II regions indeed affect the surrounding cloud, regardless of whether the detailed model of Elmegreen and Lada is correct. One prediction of their model which differs from the scenario presented in Sec. III is that the cloud may be heated by the shock before any massive protostars form. A systematic examination of temperature peaks in molecular clouds in both the near- and far-infrared should be able to test this prediction. Caution must be exercised in ruling out the presence of protostars on the basis of surveys even at 10 μm, however. A new class of very cool infrared sources was reported by Beichman *et al.* (1977). While still compact ($\lesssim 5''$), they have color temperatures ~100 K, and are not readily detected by surveys even at 10 μm.

The role of supernova remnants as triggers for star formation is of particular interest because the isotopic anomalies found in meteorites have been interpreted in terms of a supernova trigger for the collapse of our protosolar nebula. Lada's argument (see his chapter in this book) that only high-mass stars require a trigger would suggest that most low-mass stars would not form in this way. In addition, the picture of a supernova shock hitting an isolated protosolar nebula seems rather naive; a more relevant calculation would consider a supernova shock hitting a 10^3-10^5 $M_\odot$ molecular cloud. Under these conditions one wonders if the grains that carry the isotopic anomalies can get very far into the cloud.

VIII. SUMMARY

Beginning with the statement that stars, and therefore planets, form from molecular clouds, we have argued that the growing body of data on conditions in molecular clouds should serve as input parameters for theoretical calculations of star and planet formation. The fact that the gravitational potential energy of molecular clouds far exceeds their thermal energy implies that the required rate of star formation is easily explained. Indeed the most important problem in this field is that the rate of star formation predicted from molecular cloud studies is too high. Until this problem is understood, additional modes of inducing star formation, while they may be occurring, only add to our difficulties.

To approach the problem of which clouds form planets we have made the conventional assumption that only low-mass stars form planets. If this assumption is wrong, then essentially any molecular cloud can form planets. The traditional division of molecular clouds into "dark clouds" and "massive (or giant) molecular clouds" has been criticized; a new classification based only on the maximum kinetic temperature was proposed. Other properties of the cloud, like molecular abundances and the presence of strong infrared sources, seem to support the division of clouds into two groups, as shown in Table I. Indeed, the strong infrared sources are often the cause of the higher kinetic temperatures.

A scenario for star formation was presented in which three stages in the evolution of a dense core were identified. Observational tests of this scenario were also discussed. While the scenario has much observational support, there are various aspects which require further work.

Returning to the question of where low-mass stars form, we conclude that the physical conditions in molecular clouds provide *no* basis for arguing that low-mass stars *cannot* form in the same regions where high-mass stars are forming. The limited number of actual observations of low-luminosity protostars leads one to the same conclusion.

Acknowledgment. This work was supported by the National Science Foundation, National Aeronautics and Space Administration, and the Research Corporation.

REFERENCES

Appenzeller, I., and Tscharnuter, W. 1975. *Astron. Astrophys.* 40: 397.
Bash, F. N., and Peters, W. L. 1976. *Astrophys. J.* 205: 786.
Beckwith, S.; Evans, N. J., II; Becklin, E. E.; and Neugebauer, G. 1976. *Astrophys. J.* 208: 390.
Beichman, C.A.; Becklin, E. E.; Capps, R. W.; and Dyck, H. M. 1977. *Bull. Amer. Astron. Soc.* 9: 606.
Bertout, C. 1976. *Astron. Astrophys.* 51: 101.
Blair, G. N. 1976. Ph.D. Dissertation, The University of Texas at Austin.
Blair, G. N.; Evans, N. J., II; Vanden Bout, P.A.; and Peters, W. L. 1978. *Astrophys. J.* 219: 896.
Bok, B. J. 1978. *Moon and Planets* (special Protostars and Planets issue). In press.
Elmegreen, B. G., and Lada, C. J. 1977. *Astrophys. J.* 214: 725.
Encrenaz, P. J.; Falgarone, E.; and Lucas, R. 1975. *Astron. Astrophys.* 44: 73.
Evans, N. J., II, Blair, G. N.; and Beckwith, S. 1977. *Astrophys. J.* 217: 448.
Evans, N. J., II. and Kutner, M. L. 1976. *Astrophys. J.* 204: L131.

Evans, N. J., II; Zuckerman, B.; Morris, G.; and Sato, T. 1975*a*. *Astrophys. J.* 196: 433.
Evans, N. J., II; Zuckerman, B., Sato, T.; and Morris, G. 1975*b*. *Astrophys J.* 199: 383.
Gatley, I.; Becklin, E. E.; Matthews, K.; Neugebauer, G.; Penston, M. V.; and Scoville, N. 1974. *Astrophys. J.* 191: L121.
Goldreich, P., and Kwan, J. 1974. *Astrophys. J.* 189: 441.
Goldsmith, P., and Langer, W. 1977. Preprint.
Grasdalen, G. L.; Strom, K. M.; and Strom, S. E. 1973. *Astrophys. J.* 184: L53.
Guelin, M.; Langer, W. D.; Snell, R. L.; and Wootten, H. A. 1977. *Astrophys. J.* 217: L165.
Harvey, P. M.; Campbell, M. F.; and Hoffman, W. F. 1978. *Astrophys. J.* 219: 891.
Herbst, W., and Assousa, G. E. 1977. *Astrophys. J.* 217: 473.
Hollenbach, D.; Werner, M. W.; and Salpeter, E. E. 1971. *Astrophys. J.* 163: 165.
Kutner, M. L.; Evans, N. J., II; and Tucker, K. D. 1976. *Astrophys. J.* 209: 452.
Kutner, M. L.; Tucker, K. D.; Chin, G.; and Thaddeus, P. 1977. *Astrophys. J.* 215: 521.
Larson, R. B. 1972. *Mon. Not. Roy. Astron. Soc.* 157: 121.
______. 1973. *Ann. Rev. Astron. Astrophys.* 11: 219.
______. 1977. Preprint. Yale University Observatory.
Leung, C. M., and Brown, R. L. 1977. *Astrophys. J.* 214: L73.
Loren, R. B. 1976. *Astrophys. J.* 209: 466.
Mouschovias, T. C. 1974. *Astrophys. J.* 192: 37.
Oort, J. H. 1954. *Bull. Astr. Inst. Neth.* 12: 177.
Penston, M. V.; Munday, A.; Stickland, D. J.; and Penston, M. J. 1969. *Mon. Not. Roy. Astron. Soc.* 142: 355.
Sargent, A. I. 1977. Ph.D. Dissertation (unpublished), California Institute of Technology, Pasadena, Calif.
Sharpless, S. 1959. *Astrophys. J. Suppl.* 4: 257.
Silk, J. 1977. *Astrophys J.* 214: 718.
Snell, R. L., and Loren, R. B. 1977. *Astrophys. J.* 211: 122.
Solomon, P. M.; Sanders, D. B.; and Scoville, N. A. 1977. *Bull. Amer. Astron. Soc.* 9: 554.
Spitzer, L.; Drake, J. F.; Jenkins, E. B.; Morton, D. C.; Rogerson, J. B.; and York, D. G. 1973. *Astrophys. J.* 181: L116.
Strom, S. E.; Strom, K. M.; and Grasdalen, G. L. 1975. *Ann. Rev. Astron. Astrophys.* 13: 187.
Strom, K. M.; Strom, S. E.; and Vrba, F. J. 1976*a*. *Astron. J.* 81: 308.
______. 1976*b*. *Astron. J.* 81: 320.
Vrba, F. J.; Strom, K. M.; Strom, S. E.; and Grasdalen, G. L. 1975. *Astrophys. J.* 197: 77.
Woodward, P. R. 1976. *Astrophys. J.* 207: 484.
Wootten, H. A. 1977. *Astrophys. J.* 216: 440.
______. 1978. *Moon and Planets* (special Protostars and Planets issue). In press.
Wootten, H. A.; Evans, N. J., II; Snell, R. L.; and Vanden Bout, P. A. 1977*a*. *Bull. Amer. Astron. Soc.* 9:514.
Wootten, H. A.; Snell, R. L.; Blair, G. N.; Evans, N. J., II; and Vanden Bout, P. A. 1977*b*. Proc. 21st Liège Symposium – Small Molecules in Astrophysics. In press.
Wynn-Williams, C. G., and Becklin, E. E. 1974. *Publ. Astron. Soc. Pacific.* 86: 5.
Zuckerman, B., and Evans, N. J., II. 1974. *Astrophys. J.* 192: L149.
Zuckerman, B., and Palmer. P. 1974. *Ann. Rev. Astron. Astrophys.* 12: 279.

THERMAL EFFECTS IN NEARLY SELF-GRAVITATING DIFFUSE CLOUDS

M. JURA
University of California at Los Angeles

A necessary though not sufficient condition for interstellar clouds to collapse is for gravity to overpower thermal pressure. To determine the thermal pressure it is necessary to calculate the cloud temperature. It is likely but not certain that diffuse clouds are mainly heated by the photo-effect off grains and cooled by collisional excitation of the C^+ fine structure. In this picture, clouds become self-gravitating when they are sufficiently large that dust attentuation reduces the heating rate.

Stars differ presumably in part because of variations in the initial conditions in the clouds from which they form. It is therefore important to identify the types of clouds that may collapse into stars and to attempt to infer the physical conditions in these clouds (c.f., Elmegreen, 1978). The important question of the origin of clouds is not considered here.

In the simplest picture of star formation (although perhaps a misleading picture; Alfvén, personal communication, 1978), clouds collapse when gravity overpowers thermal pressure. It is therefore essential to calculate cloud temperatures as a function of different physical conditions. Interstellar clouds may be categorized into diffuse clouds and dark clouds. If $N_H = N(H) + 2\,N(H_2)$ denotes the total column density of hydrogen nuclei through a cloud, then diffuse clouds have $N_H \leqslant 2 \times 10^{21}$ cm^{-2} while dark clouds have $N_H \geqslant 2 \times 10^{21}$ cm^{-2} Although somewhat arbitrary, these definitions are useful in the sense that diffuse clouds are warm ($T \approx 80$ K; Spitzer and Jenkins, 1975), mostly atomic except for appreciable amounts of H_2 and contained by the pressure of an intercloud medium. In contrast, dark clouds are cold ($T \approx 10$ K; Dickman, 1975), mostly molecular and probably contained by their own self-gravity.

In this chapter I discuss the heating and cooling in diffuse clouds with particular regard to the issue of determining when clouds are unstable to collapse. Reviews of both diffuse clouds (Spitzer, 1978; Spitzer and Jenkins, 1975) and dark clouds (Heiles, 1971; Zuckerman and Palmer, 1974; Bok 1978; Evans, chapter in this book) exist while heating and cooling in diffuse clouds have been previously reviewed by Dalgarno and McCray (1972) and Silk (1973).

PHYSICAL CONDITIONS IN DIFFUSE CLOUDS

Measurements of the brightness temperature of atomic hydrogen at 21 cm and of the populations of the lower rotational levels of H_2 indicate an average temperature for diffuse clouds of about 80 K (Spitzer and Jenkins, 1975). There may be clouds at temperatures a factor of two below this average (Savage *et al.*, 1977), but there is as yet no clear correlation of the temperature within a diffuse cloud with any other parameter.

At the moment, the best estimates of cloud densities have come from analysis of the populations of the rotational levels of H_2 (e.g., Black and Dalgarno, 1973, 1977; Jura, 1975*a, b*). With *Copernicus* it will be possible to measure column densities of C I in the ground state and excited fine structure levels to determine densities from these ratios. However, at present, cloud densities are not determined to a factor of better than two. Much of the uncertainty is caused by the fact that for many clouds the density estimate depends inversely upon R, the rate for H_2 formation on grains. There is at least a factor of two uncertainty in the average value of R, and it is quite possible that R varies among clouds. Cloud densities may be as low as $n = 15$ cm^{-3}, where $n = n(\text{H}) + 2n(\text{H}_2)$, but most diffuse clouds have densities greater or equal to 20 or 30 cm^{-3}. Densities as high as 100 cm^{-3} are probably relatively common, and diffuse clouds with densities even as high as 1000 to 3000 cm^{-3} are also known as in the well-studied cloud toward ζ Oph (Black and Dalgarno, 1977). However, it appears that most clouds with densities greater than 100 cm^{-3} are near early-type stars (Jura, 1975*a, b*; Spitzer and Morton, 1976) including the cloud toward ζ Oph (Black and Dalgarno, 1973, 1977; Crutcher 1977; Jura, 1975*b*; Wright and Morton, 1978).

The electron density in diffuse clouds depends upon the hydrogen ionization rate and the abundance of minor constituents, such as carbon, that are photo-ionized by starlight. Although the hydrogen ionization rate may vary, it seems that it is usually less than 10^{-16} sec^{-1} (Spitzer and Jenkins, 1975). Consequently, in general $n_e/n \lesssim 10^{-3}$ and there may be clouds where it is 10^{-4} cm^{-3}. That is, the contribution to n_e from H^+ and C^+ are comparable. One consequence of this relatively low fractional ionization is that collisional excitation by electrons is less important in cooling than collisional excitation by hydrogen atoms.

The solar carbon abundance relative to hydrogen is uncertain by about a factor of two (Mount and Linsky, 1975; Lambert, 1978). In addition, it is not well known how much carbon is contained within interstellar grains. Although carbon resonance lines can be observed with *Copernicus* (e.g., Morton, 1975), they are so saturated that it is very difficult to infer carbon column densities. Some *Copernicus* observations suggest that $n(\text{C}^+)/n$ is as low as 4×10^{-5} (e.g., Morton, 1975; Snow, 1976), but other *Copernicus* observations of the carbon lines are consistent with values of $n(\text{C}^+)/n$ as high as 4×10^{-4} (York, 1975).

COOLING

In previous reviews, the cooling rate in diffuse clouds has been estimated to be on the order of $10^{-26}\, n$ erg sec^{-1} (Dalgarno and McCray, 1972; Silk, 1973;

Salpeter, 1976). Because of the data acquired with *Copernicus,* I suggest that a higher cooling rate, perhaps as high as 10^{-25} n erg sec^{-1} may be required to account for cloud temperatures.

Dalgarno and McCray (1972) have extensively reviewed the cooling processes in diffuse clouds. In the very hottest clouds (those with temperatures greater than 100 K), collisional excitation of the H_2 rotational levels begins to be important (Dalgarno and Wright, 1972); but in most clouds the main cooling process is collisional excitation of the fine structure level of C^+ by atomic or molecular hydrogen. Bazet *et al.* (1975) calculated the rate for collisional excitation of C^+ by atomic hydrogen to be a factor of two slower than the rate listed by Dalgarno and McCray (1972). Launay and Roueff (1977) have again computed this rate and found essential agreement with the results listed by Dalgarno and McCray. While there is still an uncertainty of about a factor of two, at 80 K the cooling rate in diffuse clouds is probably about 6 x 10^{-24} $n(H)$ $n(C^+)$ erg cm^3 sec^{-1}.

Rates for collisional excitation of C^+ by H_2 have been computed by Chu and Dalgarno (1975). Toshima (1975) and Flower and Launay (1977). These calculations do not agree very well with each other, but it appears that C^+ excitation by H_2 is somewhere between a factor of 2 and 10 less rapid than C^+ excitation by H.

In diffuse clouds with $n(H) = 50$ cm^{-3} and $n(C^+)/n(H) = 10^{-4}$, the cooling rate from C^+ excitation by H is about 3 x 10^{-26} $n(H)$ erg sec^{-1}. It seems very unlikely that the typical cooling rate in diffuse clouds is less than 10^{-26} $n(H)$ erg sec^{-1}. However, it is quite possible that it could be as high as 10^{-25} $n(H)$ erg sec^{-1}.

An argument for the higher cooling rate is given by Jura (1976*a*). In some clouds, if the hydrogen density is low, the sodium is only slightly depleted. If the hydrogen density is high, there may be some sodium depletion. Although few results are available, it seems that if the hydrogen density is low, there also cannot be much carbon depletion. By this argument, it appears that cooling rates as high as 10^{-25} $n(H)$ erg sec^{-1} may be appropriate for the typical diffuse cloud.

An upper limit to the cooling rate within diffuse clouds can be derived from the result that there are essentially no diffuse clouds with temperatures much in excess of 100 K (Savage *et al.*, 1977). Since the cooling rate increases with temperature, for temperatures above 100 K, cooling by collisional excitation of the H_2 rotational levels begins to become important. If the calculations of Dalgarno and Wright (1972) are modified to agree with the observed ratio of ortho H_2 to para H_2 it appears that cooling rates greater than about 10^{-25} n erg sec^{-1} do not occur within diffuse clouds.

HEATING

Historically, one of the outstanding problems of diffuse clouds has been to explain why clouds have temperatures as high as 80 K. Because the cooling time of clouds is short, it seems that some sort of external heating source is required. One of the most promising efforts to explain the observations was by Field *et al.* (1969) who postulated a very high rate of cosmic ray ionization with a correspondingly large amount of heating. However, observationally it now appears that the cosmic ray ionization is no greater than 10^{-16} sec^{-1} (Spitzer and Jenkins, 1975) and in some clouds it is probably as low as 10^{-17} sec^{-1} (Black and Dalgarno, 1977; Hartquist, Black and Dalgarno, 1978). With a heating input of about 4 eV for every ionization (Dalgarno and McCray, 1972), this cosmic ray ionization rate produces heating of less than 10^{-27} n(H) erg sec^{-1} which is at least a factor of 10 too small to account for the observed temperature in clouds.

There is a large number of possible heating sources, but most of them seem to be insufficiently powerful (Silk, 1973). The possibility of heating the gas by starlight is attractive because enough energy is observed to be absorbed within interstellar clouds even if only a small fraction of this absorbed energy is converted into heating the gas. Also, it appears that the heating rate in dark clouds (c.f., Evans *et al.*, 1977; Scalo, 1977) is probably lower than in diffuse clouds. That is, if the heating rate in dark clouds were 10^{-25} n erg sec^{-1}, equilibrium temperatures as high as 80 K might result, given cooling calculations for CO excitation (Goldreich and Kwan, 1974; Scoville and Solomon, 1974; de Jong *et al.*, 1975). (Cooling by grains is only dominant for densities greater than about 10^4 cm^{-3}; e.g., Scalo, 1977.) Even if the heating rate in dark clouds is 10^{-26} n erg sec^{-1}, temperatures >20 K would be common as may be extrapolated from the work of Leung and Liszt (1976) or Elmegreen *et al.* (1978). Since dark cloud temperatures are usually close to 10 K (Dickman, 1975), the heating rate in dark clouds is probably considerably less than in diffuse clouds. This contrast between dark clouds and diffuse clouds can be taken as indirect evidence that starlight heating of diffuse clouds, for example by the photoeffect off grains, may be important since starlight does not penetrate into dark clouds. It should be recognized that cosmic rays penetrate further into clouds than ultraviolet photons so that dark clouds may be mainly heated by cosmic rays.

Interstellar clouds absorb both in lines and in the continuum. Absorption in atomic lines is in fact only scattering and seems unable to inject any significant amount of kinetic energy into the gas. Absorption by H_2 can lead to dissociation of this molecule and appreciable heating of the gas may result either when the molecule is destroyed or when it reforms. Milgrom, Panagia and Salpeter (1973) have noted that photo-dissociation of H_2 through absorptions in the Lyman lines will produce non-thermal H atoms with an average energy deposition into the gas of about 0.25 eV (Stephens and Dalgarno, 1973). Also, a significant fraction of the hydrogen molecules in the cloud may be destroyed by chemical reactions (Dalgarno and Oppenheimer, 1974), and there may be as much as 4 eV of heating deposited within the gas for every molecule so destroyed. Finally, there is some possibility that a newly formed hydrogen molecule leaves a grain with an appreciable amount of kinetic energy (Spitzer and Cochran, 1973), perhaps as much as 2 eV (Glassgold and Langer, 1974; Barlow and Silk, 1976). If R is the H_2 formation rate defined above, the total heating from formation and destruction of H_2 may be as much as $10^{-11} R\, n(\mathrm{H})\, n$ erg cm^3 sec^{-1}. With $R = 3 \times 10^{-17}$ cm^{-3} sec^{-1} (Jura, 1975*a*) this amounts to $3 \times 10^{-28}\, n(\mathrm{H})\, n$ erg cm^3 sec^{-1}. If carbon is depleted by a factor of 10, the cooling rate at 80 K would be $3 \times 10^{-28}\, n(\mathrm{H})\, n$ erg cm^3 sec^{-1} and cloud temperatures could be explained. However, as noted above, if R is large, it appears likely that carbon is undepleted (Jura, 1976*a*). While further data are required to demonstrate this argument conclusively, at the moment it appears that heating from the formation and destruction of H_2 can produce at most only about one tenth of the heating required to explain cloud temperatures.

Essentially all of the continuous absorption observed in the interstellar medium is produced by grains, and most of the absorbed energy is reradiated by the grains in the infrared. However, kinetic energy may be injected into the gas by the photo-effect, and as first emphasized by Watson (1972), this process may very well cause the major heating rate within interstellar clouds. The composition and the photo-yield of interstellar grains are not well known. Consequently, the magnitude of photo-heating is uncertain. In any case, as shown by Watson (1972), the photo-yield of small grains is probably larger than the photo-yield of bulk materials since it is easier for an excited electron to reach a surface of a small grain. Therefore, the exact amount of photo-heating is quite uncertain.

Watson (1972) proposed a photo-heating rate of about $10^{-26}\, n(\mathrm{H})$ erg sec^{-1}. De Jong (1977) and Draine (1978) also derive similar heating rates. As discussed above, it is not clear that such a heating rate is sufficient to explain cloud temperatures. Nor is it obvious that a simple adjustment of the parameters, assumed to calculate the photo-heating, may produce a much higher heating rate. For example, an increase in the photo-yield leads only to a higher positive charge on the grain and not to more heating. One way around this difficulty has been proposed by Jura (1976*a*). In this model, grains are sufficiently small that photo-excited electrons are usually not scattered before they exit a grain. Consequently, the average energy of a photo-excited electron from a small grain may be considerably higher than from bulk materials. In this model, heating rates of about $10^{-25}\, n$ erg sec^{-1} can be explained. In any case, although photo-heating is very promising, it should be emphasized that we do not know that the photo-effect operates with sufficient power to explain diffuse cloud temperatures.

DISCUSSION

If the photo-effect is the main source of heating within diffuse clouds, several consequences follow. For example, as discussed above, it is possible to understand the observed difference in the temperatures between diffuse clouds and dark clouds.

More generally, if the photo-effect off grains is the dominant source of cloud heating, it is possible to estimate a minimum mass when clouds become self-gravitating. For example, in the calculations of Jura (1976*b*) for spherical clouds, when p/k has its typical interstellar value between 10^3 and 10^4 cm^{-3} K (where k is the Boltzmann constant and p is the intercloud pressure, inferred from observations; Jura, 1975*b*), it seems that the critical mass for collapse is between 10^4 and 10^2 $M_\odot$. The clouds that may collapse have column densities on the order of 10^{21} cm^{-2}, and it therefore seems important to study the heating and cooling rates in diffuse clouds to understand the conditions for the initial stages of cloud collapse. However, since the magnetic field (c.f., chapter by Mouschovias in this book) and turbulence have been omitted, this discussion yields only a minimum mass.

Pressures as high as $p/k = 10^5$ cm^{-3} K are observed to occur within the interstellar medium as would be expected near supernova explosions or stellar winds from early-type stars (Castor *et al.*, 1975; Weaver *et al.*, 1977). At such high pressures, clouds could be cold. Therefore, star formation near supernova remnants, for example, is a possibility (e.g., chapter by Herbst and Assousa in this book), and this could possibly be relevant to the formation of the solar system (c.f., chapters by Cameron, Reeves and Schramm; Clayton, 1978; Wassenburg, personal communication, 1978).

It is also possible to imagine physical conditions very different from those prevailing in the solar neighborhood. For example, in elliptical galaxies, the ultraviolet radiation field is probably low because there are few early-type stars. Consequently, clouds may always be cold so that low-mass clouds would collapse to form stars and only low-mass stars might form. Therefore, one can imagine a self-consistent scenario where no massive stars form because no massive stars are present (Jura, 1977).

Of course much of the above discussion is oversimplified and speculative. However, it seems at least possible that the basic processes that control diffuse cloud temperatures are understood, and this is an important step toward understanding the initial stages of star formation.

Acknowledgments. I thank B. Zuckerman for the very helpful discussions on interstellar matter. This work has been partly supported by the National Science Foundation and by a Research Fellowship of the Alfred P. Sloan Foundation.

REFERENCES

Barlow, M. J., and Silk, J. 1976. *Astrophys. J.* 207: 131.
Bazet, J. F.; Harel, C.; McCarroll, R.; and Riera, A. 1975. *Astron. Astrophys.* 43: 229.
Black, J. H., and Dalgarno, A. 1973. *Astrophys. J.* 184: L101.
______. 1977. *Astrophys. J. Suppl.* 34: 405.
Bok, B. J. 1978. *Moon and Planets* (special Protostars and Planets issue). In press.

Castor, J.; McCray, R. M.; and Weaver, R., 1975. *Astrophys. J.* 200: L107.
Chu, S. I., and Dalgarno, A. 1975. *J. Chem. Phys.* 62: 4009.
Clayton, D. D. 1978. *Moon and Planets* (special Protostars and Planets issue). In press.
Crutcher, R. M. 1977. *Astrophys. J.* 217: L109.
Dalgarno, A., and McCray, R. M. 1972. *Ann. Rev. Astron. Astrophys.* 10: 375.
Dalgarno, A., and Oppenheimer, M. 1974. *Astrophys. J.* 192: 597.
Dalgarno, A., and Wright, E. L. 1972. *Astrophys. J.* 174: L49.
de Jong, T. 1977. *Astron. Astrophys.* 55: 137.
de Jong, T.; Chu, S. I.; and Dalgarno, A. 1975. *Astrophys. J.* 199: 69.
Dickman, R. L. 1975. *Astrophys. J.* 202: 50.
Draine, B. T. 1978. *Astrophys. J. Suppl.* 36: 595.
Elmegreen, B. G. 1978. *Moon and Planets* (special Protostar and Planets issue). In press.
Elmegreen, B. G.; Dickinson, D. G.; and Lada, C. J. 1978. *Astrophys. J.* 220: 853.
Evans, N. J.; Blair, G. N.; and Beckwith, S. 1977. *Astrophys. J.* 217: 448.
Field, G. B.; Goldsmith, D. W.; and Habing, H. J. 1969. *Astrophys. J.* 155: L149.
Flower, D. R., and Launay, J. M. 1977. *J. Phys. B.* 10: 3673.
Glassgold, A. E., and Langer, W. D. 1974. *Astrophys J.* 193: 73.
Goldreich, P., and Kwan, T. 1974. *Astrophys. J.* 189: 441.
Hartquist, T. W.; Black, J. H.; and Dalgarno, A. 1978. *Mon. Not. Roy. Astr. Soc.* In press.
Heiles, C. 1971. *Ann. Rev. Astron. Astrophys.* 9: 293.
Jura, M. 1975*a*. *Astrophys. J.* 197: 575.
______. 1975*b*. *Astrophys J.* 197: 581.
______. 1975*c*. *Astrophys. J.* 200: 415.
______. 1976*a*. *Astrophys. J.* 204: 12.
______. 1976*b*. *Astron. J.* 81: 178.
______. 1977. *Astrophys. J.* 212: 634.
Lambert, D. L. 1978. *Mon. Not. Roy. Astr. Soc.* 182: 249.
Launay, J. M., and Roueff, E. 1977. *J. Phys. B.* 10: 879.
Leung, C. M., and Liszt, H. S. 1976. *Astrophys. J.* 208: 732.
Milgrom, N., Panagia, N.; and Salpeter, E. E. 1973. *Astrophys. Lett.* 14: 73.
Morton, D. C. 1975. *Astrophys. J.* 197: 85.
Mount, G. H., and Linsky, J. L. 1975. *Astrophys. J.* 202: L51.
Salpeter, E. E. 1976. *Astrophys. J.* 206: 673.
Savage, B. D.; Bohlin, R. C.; Drake, J. F.; and Budich, W. 1977. *Astrophys. J.* 216: 291.
Scalo, J. M. 1977. *Astrophys J.* 213: 705.
Scoville, N. Z., and Solomon, P. M. 1974. *Astrophys. J.* 187: L67.
Silk, J. 1973. *Pub. Astron. Soc. Pacific.* 85: 704.
Snow, T. P. 1976. *Astrophys. J.* 204: 759.
Spitzer, L. 1978. *Physical Processes in the Interstellar Medium* New York: John Wiley.
Spitzer, L., and Cochran, W. D. 1973. *Astrophys. J.* 186: L23.
Spitzer, L., and Jenkins, E. B. 1975. *Ann. Review Astron. Astrophys.* 13: 133.
Spitzer, L., and Morton, W. A. 1976. *Astrophys. J.* 204: 731.
Stephens, T. L., and Dalgarno, A. 1973. *Astrophys J.* 186: 165.
Toshima, N. 1975. *J. Phys. Soc. Japan.* 38: 1464.
Watson, W. D. 1972. *Astrophys. J.* 176; 103.
Weaver, R.; McCray, R.; Castor, J.; Shapiro, P.; and Moore, R. 1977. *Astrophys. J.* 218: 377.
Wright, E. L., and Morton, D. C. 1978. Preprint.
York, D. G. 1975. *Astrophys. J.* 196: L103.
Zuckerman, B., and Palmer, P. 1974. *Ann. Review Astron. Astrophys.* 12: 279.

FRAGMENTATION OF MOLECULAR CLOUDS

JOSEPH SILK
University of California at Berkeley

Fragmentation and complementary physical processes are studied, with the aim of elucidating how stars of solar mass can form, and more generally, how the initial stellar mass function originates. The evidence for fragmentation of molecular clouds is briefly discussed. The role of opacity in determining the minimum fragment size is reviewed, and the more realistic case of magnetic flux-limited fragmentation is described. The possibility of fragmentation initiated by molecule formation is also considered. The non-linear stages of fragmentation are examined in detail by considering various possible modes of fragment interactions. These include accretion, fragment coalescence, heat input by newly formed protostars, and the formation of binary systems by capture. Arguments are given that can account for the apparent exclusion of massive stars from dark cloud interiors. Probabilistic theories of fragmentation are also discussed. A final section describes observational clues to the nature of the star formation process which may help to discriminate between the various theoretical models.

Fragmentation is generally considered to be the initial process that a molecular cloud must undergo before stars can form. Yet its role in determining the final mass spectrum remains obscure. A thorough review of theoretical processes in star formation has been written by Mestel (1977), and the present discussion is restricted to the issue of fragmentation.

Numerical studies of the gravitational collapse of an interstellar cloud have in the past avoided the issue of fragmentation largely because a three-dimensional hydrodynamical code is required. Discussions of fragmentation are generally of a more or less qualitative nature, in order to cope with the complexity of a realistic collapse situation. This has not prevented rather diverse conclusions from being drawn about the role of fragmentation in a diffuse interstellar cloud which is triggered to collapse by a variety of possible mechanisms, ranging from passage of a spiral density wave to the formation of a nearby massive star or interaction with a supernova remnant. Those viewpoints can be summarized as follows:

1. Collapse rapidly becomes non-homologous, and no fragmentation occurs.
2. Magnetic stresses can strongly inhibit fragmentation.
3. Fragment interactions dissipate kinetic energy of random motions, preventing fragments from surviving as individual objects.

On the other hand, we are left with the firm impression that stars (and possibly planets) conspire to form during this collapse. Fragmentation must evidently have occurred, since the critical mass for gravitational collapse of a diffuse interstellar cloud exceeds $10^4\ M_\odot$. One can attempt to reconcile these possibilities as follows:

(A) The initial conditions of the collapse are crucial to the determination of the mass of the resulting protostars. An opaque core eventually forms, whose mass is relatively insensitive to initial conditions, but which grows by accretion. The amount of accreted material is determined by the Jeans mass M_J when the core initially develops, and by competition for this material with neighboring condensations. Fragmentation is important in understanding the range of possible initial values for M_J. Similar remarks apply to ring formation, as found in two-dimensional collapse studies, the ring masses being limited by accretion.

(B) The role of angular momentum is closely allied to that of the magnetic field. When they are approximately perpendicular to each other, Mestel (1977) argues that the tendency to fragmentation enhances the magnetic and centrifugal stresses, and that these forces will determine the ultimate fragment mass that can form. However, when angular momentum and magnetic field are approximately parallel, collapse can proceed along the field lines, and fragmentation can occur. The collapse is anisotropic, but the fragment size should ultimately be limited by thermal pressure which can only develop when the opacity of an individual fragment is appreciable. The discussion of (A) above suggests that these opaque cores continue to grow by accretion. One might expect that the magnetic stresses will play a crucial role in limiting the resulting protostellar masses.

(C) Head-on fragment collisions will probably result in coalescence, while glancing collisions will lead to the acquisition of non-radial motions. The former process may be self-limiting, leading to the development of a mass spectrum (Nakano, 1966; Arny and Weissman, 1973), while the randomization of the overall collapse by the latter effect is crucial to the formation of star clusters.

It is apparent from these remarks that gravitational fragmentation, considered as a unique process, is unsatisfactory. The classical fragmentation scenario developed by Hoyle (1953) and Hunter (1962) is almost certainly not valid (Layzer, 1963). Other important processes are occurrring simultaneously, and may even conspire to mask the role of fragmentation. This chapter will therefore discuss both fragmentation and complementary physical processes. One of the principal aims will be to indicate how stars of solar mass (and more generally, how the initial mass spectrum of stars) can form.

The principal topics to be covered in this review include a brief discussion of the evidence for fragmentation (Sec. I) and the role of opacity (Sec. II). The more realistic case of magnetic flux-limited fragmentation is treated in Sec. III, and effects of H_2 molecule formation are in Sec. IV. Various types of interactions are described in subsequent sections, including protostellar heat input (Sec. V), fragment coalescence (Sec. VI) and gas accretion (Sec. VII). The formation of binary systems by capture is discussed in Sec. VIII. Other plausible theories of the initial stellar mass spectrum are described in Sec. IX and Sec. X, and a final section attempts to make comparison with observable phenomena. Throughout this review emphasis will be placed on the origin of the initial stellar mass function, the explanation of which must be regarded as the ultimate goal of a satisfactory theory of star formation. It should be noted that according to some observers (e.g., Freeman, 1977) there appears to be no unique initial mass function: significant variations seem to be found when different clusters are compared. Thus flexibility must be a feature of a successful theory, a challenge (perhaps the only one) which most theories manage to meet.

I. EVIDENCE FOR FRAGMENTATION

Several arguments suggest that fragmentation may be occurring in molecular clouds at densities comparable to those observed. One of the major obstacles to this has been due to the influence of the magnetic fields; however, recent studies of dark clouds suggest that the fractional ionization may be much lower ($x < 10^{-8}$) than previously believed (Guelin *et al.*, 1977; Snyder *et al.*, 1977). In such regions, the molecular hydrogen density exceeds 10^4 cm^{-3}. For example, the time scale for the neutral particles to diffuse with respect to the magnetic field and the ionized component in a magnetically supported fragment (with magnetic flux near the virial theorem limit) is (Mestel, 1977)

$$t_d \approx \frac{5\, x \langle \sigma v \rangle}{4\pi\, G\, m_h\, (1{+}4y)} \tag{1}$$

where $y = n_{He}/n_{H} = 0.1$, G is Newton's gravitational constant, m_h is the mass of a hydrogen atom, and $\langle \sigma v \rangle$ is the ion – H atom collision rate. From Spitzer (1968), one finds that $\langle \sigma v \rangle = 2.2 \times 10^{-9}\,\mathrm{cm^3\,sec^{-1}}$, whence $t_d = 1.3 \times 10^{14} \times$ yr $< 10^6$ yr. This time scale is sufficiently rapid that one may expect flux loss to be an important process. Consequently, the dynamical time scale for cloud contraction would not be the free-fall time scale ($\sim 10^7 n^{-\frac{1}{2}}$yr), but the slower diffusion time scale.

Confirmation that fragmentation must be occurring at moderate densities comes from comparison with the average densities of typical bound star clusters. The gravitational energy released in cloud collapse must be preserved as the kinetic energy of random motions if star clusters are to form. As emphasized by Hoyle, fragmentation provides the most plausible way of avoiding excessive energy dissipation. It can be argued (Sec. VIII) that the high incidence of binaries suggests that fragmentation or fission continues until a relatively high average density is attained. Thus, a wide range of physical conditions must be spanned in a theory of fragmentation. The smallest fragment masses to form may be related to the masses of planets (or even to planetesimals).

Additional evidence for the widespread occurrence of fragmentation in molecular clouds comes from interpretations of CO line widths (Leung, 1978) and the excitation of NH_3 (Matsakis *et al.*, 1977). While not conclusive, and admittedly indirect, these results lead one to assert that molecular clouds may contain dense fragments moving at supersonic velocities relative to one another in a more diffuse medium.

In what follows it will therefore be assumed that collapsing molecular clouds are undergoing fragmentation. The hierarchical nature of fragmentation will now be explored.

II. OPACITY-LIMITED FRAGMENTATION

The initial stages of collapse are approximately isothermal, as the cloud can radiate freely via optically thin molecular transitions and thermal grain emission. Stability analyses (Hunter, 1962) have indicated that, at least for a homologous spherical collapse, scales containing more than a Jeans mass

$$M_J = \rho \left(\frac{\pi k T}{\mu G \rho} \right)^{\frac{3}{2}} \underset{\sim}{\propto} \rho^{-\frac{1}{2}} \tag{2}$$

will fragment. Consequently, smaller and smaller mass scales can separate out as the collapse proceeds to higher densities. The fragmentation terminates when individual fragments become opaque, for the optical depth across a Jeans length is proportional to $\rho^{\frac{1}{2}}$ (at constant temperature), and must eventually become appreciable. Once this occurs, the fragments heat up and pressure gradients inhibit any further fragmentation.

Consider the idealized spherical collapse of a molecular cloud (the role of magnetic fields will be considered subsequently). At appreciable densities, the cooling function is dominated by grain cooling, for which

$$\Lambda = A(T)\,\rho\,T^4 \text{ erg cm}^{-3}\text{sec}^{-1} \tag{3}$$

where ρ is the gas density, T is the grain temperature (here assumed equal to the gas temperature) and

$$A(T) = 3\,\sigma q\,Z\rho_g^{-1}\,T^{\delta} \equiv \alpha\,T^{\delta}. \tag{4}$$

The function $A(T)$ parametrizes the likely deviations from a Planckian cooling function for realistic grain models. The parameters q and δ are tabulated by Silk (1977*a*) for different grain models, σ is the Stefan-Boltzmann constant, Z is the grain mass fraction, and ρ_g is the grain material density. One can argue that fragmentation will proceed during cloud collapse provided that the cloud can radiate the compressional energy acquired during collapse, $\frac{\rho kT}{\mu}\frac{dln\rho}{dt}$, where k is Boltzmann's constant and μ is the mean molecular weight per gram. Adopting the rate appropriate to a spherically symmetric isothermal collapse, $dln\rho/dt = (16\,\pi G\rho)^{\frac{1}{2}}$, the condition that fragmentation occurs can be expressed in terms of the minimum fragment mass, equal to the Jeans mass, as

$$M_J = \frac{4\,\pi^2}{\alpha G^{\frac{3}{2}}}\left(\frac{k}{\mu}\right)^{\frac{5}{2}} T^{-\frac{3}{2}-\delta} \equiv \beta\,T^{-\frac{3}{2}-\delta}. \tag{5}$$

The optical depth, τ of a fragment can be expressed in terms of M_J as

$$M_J = \frac{\alpha}{8\sigma\tau}\left(\frac{\pi k}{\mu G}\right)^2 T^{\delta+2} \approx \tau^{-1}\,T^6{}_{100}\,M_\odot \tag{6}$$

for silicate grains, where $T_{100} = T/100$ K. This leads to a minimum value of M_J at optical depth unity of about 0.01 $M_\odot$. The insensitivity of M_J to the assumptions about opacity can best be seen by combining Eqs. (5) and (6) to write it in the form $M_J \propto T^{\frac{1}{4}}$, with no explicit dependence on grain parameters. The proportionality constant can be expressed as $\sim 10\,(kT/\mu c^2)^{\frac{1}{4}}\,M_{ch}$, where M_{ch} is the Chandrasekhar mass (Rees, 1976; Silk, 1977*a*).

What this argument purports to show is that the *mininum* masses of opaque self-gravitating fragments are on the order of 0.01 $M_\odot$. This argument is sensitive to the assumption of spherical collapse, and a more realistic collapse model can be formulated, as is shown in the following section.

III. MAGNETIC FLUX-LIMITED FRAGMENTATION

Consider now a model of anisotropic collapse in which gas can flow along the magnetic field lines, but collapse is inhibited across the field. The scenario is that outlined by Mestel (1965) in which, once collapse is initiated along the field lines, a flattened disk of material will form. This will in turn be unstable to further fragmentation parallel to the field. The reason for this is that the condition necessary for the initial contraction from a radius R is that the column density of material exceeds a critical value that depends only on the magnetic field strength. Now the magnetic field B remains unchanged, and in the course of the flattening the new fragment forms with radius roughly given by the disk thickness. Consequently, mass conservation yields ρR = constant, and the contraction criterion, if satisfied initially, is always satisfied subsequently. Fragmentation will cease only when the fragments become pressure supported, which again requires that the opacity becomes significant (assuming that the magnetic field effectively remains frozen in).

A characteristic mass scale also results from this repeated sequence of anisotropic collapse and fragmentation. To demonstrate this, note that the collapse rate of an oblate spheroid is increased over that of a sphere by a factor $\sim (\rho/\rho_0)^{1/2}$, where ρ_0 is the initial density and the spheroid is initially assumed to be moderately flattened (eccentricity $\sim$ 0.5). Comparison of cooling and collapse time scales for the collapse into a disk of an oblate spheroid of mass, M_0 and temperature, T_0 leads to the formation of fragments of mass M_1 and temperature T_1 where

$$M_1^2 = \beta M_0 \, T_0^{-\frac{3}{2}} \, T_1^{-\delta} . \tag{7}$$

The fragments can in turn collapse along the field lines; again one may assume that they initially are moderately flattened. This process can be extended to a fragmentation hierarchy, with the mass of a fragment at the nth stage satisfying

$$M_n = \frac{\beta^{\frac{1}{2}+\frac{1}{4}+ \cdots + 2^{-n-1}}}{T_0^{\frac{3}{2}\left(\frac{1}{2}+\frac{1}{4}+ \cdots + 2^{-n-1}\right)} \, T_1^{\delta/2} \cdots T_n^{\delta 2^{-n}}} > \beta \, T_0^{-\frac{3}{2}} \, T_n^{-\delta} \tag{8}$$

(where we have assumed that $T_n > T_{n-1}$). If the final fragment stage is denoted by $n = f$, one infers that M_f is a factor $(T_f/T_0)^{\frac{3}{2}}$ larger than the minimum fragment size obtained for spherical collapse (Eq. 5). Typical values are $T_0 \approx 5$ K for a dark cloud and $T_f \approx 20$ K; consequently these results suggest that the minimum fragment size will be increased in magnetic flux-limited fragmentation by an order of magnitude, to $\sim 0.1\ M_{\odot}$. One might expect a similar argument to be applicable to a collapsing, rotating cloud, which undergoes successive stages of ring formation and fragmentation, until opaque fragments are ultimately formed (c.f., chapter by Bodenheimer and Black in this book).

IV. FRAGMENTATION INDUCED BY MOLECULE FORMATION AND EXCITATION

A number of attempts have been made to demonstrate that molecule formation and excitation may result in thermo-chemical instability and subsequent fragmentation of molecular clouds. In the case of gas phase formation of such molecules as CO, one can show that the high optical depth in the cooling transitions of this molecule tends to quench any thermal instability that might otherwise occur.

The situation with regard to H_2 formation is not subject to this difficulty. Other problems arise, however. H_2 is formed on grains when the grain temperature drops below a critical value T_{cr}, where absorbed atoms stay long enough on the grain surface to encounter other H atoms. However, a considerable range in values of T_{cr} is expected, due to variations in grain composition and surface defects. Thus the H_2 formation rate is likely to be insensitive to variations in the grain optical depth as a consequence of the wide dispersion in T.

This conclusion already runs contrary to one of the principal claims of Reddish (1975) in his study of fragmentation by H_2 formation. Perhaps a fatal criticism of this aspect of the theory is that an incorrect form for the H_2 formation rate was assumed, which yielded on H_2 formation rate that greatly exceeded the supply of atoms to the grains. (The cited experiments that measure an increased rate of transfer of recombination energy to the grain surface as a function of molecule coverage can be explained by an increased efficiency of absorption of H_2 recombination energy due to the presence of adsorbed H_2 near newly formed molecules.) Correction for this drastically reduces the derived mass range of the fragments.

Nevertheless, there are physical processes, hitherto not adequately explored, that may lead to fragmentation of molecular clouds. For example, in the presence of an embedded infrared source, the grain temperature can exceed the gas temperature, and collisional coupling between gas and grain provides a substantial heat input. Such a situation is likely to be unstable to the formation of dense and relatively opaque clumps of cold matter, the characteristic clump size being largely determined by the grain opacity. A detailed calculation including radiative transfer of both continuum and line radiation for a mixture of gas and grains is evidently required.

V. PROTOSTELLAR HEAT INPUT

The extreme sensitivity of the mass of a self-gravitating fragment of unit optical depth to temperature (Eq. 6) suggests that, whatever the dynamical time scale associated with the fragmentation (which could be dependent on such effects as flux diffusion and dissipation or magnetic braking rather than free-fall), heat produced by the first protostellar fragments to form may affect the masses of future generations of fragments. The non-homologous nature of the collapse indicates that runaway cores form while much of the matter is diffuse, and magnetic and centrifugal forces will further inhibit the amount of material that can easily condense. If the cloud is opaque to the energy input from these first protostars, the energy feedback to the diffuse matter could have a significant effect on its ability to condense. Clearly, one would have to form a large number of, say, 0.1 $M_\odot$ protostars before the temperature of newly condensed material is raised to the point at which more massive protostars will form.

It has been proposed that this type of negative feedback could yield a mass function dN/dm (Silk, 1977*b*). The argument is simple; the radiation field, $u_{\rm rad} \sim \int \frac{dN}{dm} L(m)\, dm$, must maintain the gas temperature according to $u_{\rm rad} \sim \sigma T^{4+\delta}$. However, Eq. (6) relates the masses of newly formed fragments to their temperature; consequently, if the protostellar luminosity $L \propto m^3$, one must have that

$$\frac{dN}{dm} \propto T^{4+\delta} (mL)^{-1} \propto m^{-2\, -\, \delta/(2+\delta)} \,. \tag{9}$$

While this argument is suggestive in providing a negative feedback mechanism that yields a mass spectrum of negative slope, it clearly omits many physical effects. Such dynamical processes as coalescence of fragments and accretion of gas may be dominant. Nevertheless, it is useful to remark that at least one level of dynamical complexity, namely the effects of protostellar-driven mass motions and turbulence, can be incorporated into the foregoing scheme. This type of kinetic energy input will tend to drive up the gas temperature. Similar feedback arguments apply, except that now the coupling of gas and grain temperatures must be examined. The coupling begins to break down at $T \gtrsim 100$ K, when the critical mass of newly formed fragments exceeds 1 $M_\odot$. The grains are heated by collisions with gas molecules and they cool by radiation. This results in a considerable increase in sensitivity of M_J to grain temperature, yielding a somewhat steeper functional dependence for dN/dm at larger masses.

VI. FRAGMENT COALESCENCE

The ability of newly formed fragments to survive has been questioned on the grounds that they will retain an appreciable cross-section for collision with other fragments (Layzer, 1963). Such collisions are likely to be highly inelastic, resulting in coalescence. This need not be fatal to star formation, since one can readily imagine that the collisions will be self-limiting, the collision rate decreasing as coalescence continues. Indeed, numerical studies of fragment coalescence (Nakano, 1966; Arny and Weisman, 1973) have found that a unique spectrum of masses may be produced.

The process of fragment coalescence can be represented by the coagulation equation, for which exact solutions exist for simple choices of the collision rate. To demonstrate that collisions are indeed frequent, note that the probability of a newly formed fragment of radius r_f undergoing a collision within an initial free-fall time (which must be similar to the cluster crossing time) can be expressed as $\sim N\,(r_f/r_{cl})^2$ where N is the number of cluster stars and r_{cl} is the cluster radius. We may take r_f to be the initial Jeans length, or $10^{17}\,T_{10}^{-1}$ for a $1M_\odot$ fragment at temperature $T_{10} \equiv T/10$ K; hence collisions will be frequent provided $N \gtrsim 10^3$ and $r_{cl} \sim 1$ pc. Once fragments become appreciably condensed, their effective collision cross-section is greatly decreased. However, this takes roughly an initial free-fall time ($\sim 10^5$ yr), and somewhat longer if the fragments have retained (or acquired) significant angular momentum.

To illustrate the resulting form of the mass spectrum, consider now the coagulation equation:

$$\frac{\partial N}{\partial t} = \tfrac{1}{2}\int_0^m N(m-m')\,N(m')\,\alpha(m-m',m')\mathrm{d}m' - N(m,t)\int_0^\infty N(m',t)\,\alpha(m,m')\mathrm{d}m' \quad (10)$$

where $\alpha(m,\ m')$ denotes the velocity-averaged collision rate between fragments of mass m and m'. The asymptotic solutions to this equation are (Trubnikov, 1971), with $n(t) = \int_0^\infty N(m,t)\mathrm{d}m$

(a) if $\alpha(m,m') = \alpha_0$

$$N\,(m,t) \sim \frac{n^2\,(t)}{\rho}\,\exp\,[\frac{-m\,n(t)}{\rho}]$$

$$n(t) = n_0\,(1+\tfrac{1}{2}\,\alpha_0 n_0 t)^{-1} \quad (11)$$

(b) if $\alpha(m,m') = \alpha_0\,(m + m')$

$$N\,(m,t) \sim \frac{n(t)}{\langle m^2 \rangle_0}\left(\frac{m_0}{m}\right)^{\frac{3}{2}} \exp\left[-\frac{m}{2}\,\frac{m_0}{\langle m^2 \rangle_0}\,\frac{n^2(t)}{{n_0}^2}\right]$$

$$n\,(t) = n_0\,\exp\,(-\,\alpha_0\,\rho\,t)$$

where

$$m_0 = \rho / n_0$$

$$\langle m^2 \rangle_0 = \frac{1}{n_0} \int_0^\infty m^2 \, N(m,0) \, \mathrm{d}m$$

$$n_0 \equiv n(0). \qquad (12)$$

A realistic expression for α may be expected to have a mass-dependence somewhere between $m^{\frac{2}{3}}$ (proportional to the geometrical cross-section) and $m^{\frac{4}{3}}$ (due to gravitational focusing of the orbits of colliding particles). The case $\alpha \propto m^{\frac{2}{3}}$ has an asymptotic mass-dependence roughly as m^{-1} exp $(-\beta m)$ (Nakano, 1966). Evidently, a mass spectrum $\sim m^{-1}$ exp $(-\beta m)$ or $m^{-\frac{3}{2}}$ exp $(-\beta m)$ with an exponential decline above some $m_{\mathrm{max}} \equiv \beta^{-1}$ is predicted if coalescence is important during the fragmentation process. One can show that, when the non-equilibrium nature of the collapse is taken into account, m_{max} is roughly equal to the average number of collisions per fragment multiplied by the mean initial fragment mass m_0. The process must terminate after a finite time, determined by other considerations; for example, the fragment cross-section will be drastically reduced as it contracts to form a protostar. Hence, one might expect that if fragmentation first occurs at relatively low density, m_{max} would be less than if fragmentation occurs at high density. Such differences could plausibly arise, since the ability of a cloud to fragment depends on the efficiency with which it can diffuse through, or contract along, the magnetic field. A sudden compression, induced for example by a nearby supernova explosion, might result in fragmentation at high density and a large value for m_{max}, while fragmentation in a quiescent region could proceed at lower density and yield a lower value for m_{max}.

VII. ACCRETION

The importance of continuing accretion in enabling fragments to grow has been demonstrated in numerical studies of cloud collapse and star formation. The result seems to be that whatever can accrete will accrete; all the material within an initial Jeans length of a newly formed core tends to be accreted (c.f., Larson, 1978).

One might expect accretion to modify the effects of coalescence. One can attempt to examine this question by including an accretion term $\frac{\partial}{\partial m} [N(m,t) \frac{\mathrm{d}m}{\mathrm{d}t}]$, where $\mathrm{d}m/\mathrm{d}t$ is the rate at which a fragment of mass m accretes ambient gas, onto the left-hand side of the coagulation equation (Eq. 10). A plausible form for the accretion rate of a fragment moving subsonically at velocity v would be

$$\frac{\mathrm{d}m}{\mathrm{d}t} \approx \pi \left(\frac{GM}{c_s^2} \right)^2 \rho \, v \, . \tag{13}$$

One immediate result is that the total number density of fragments, $n(t) = \int N\,(m,t)\,\mathrm{d}m$, evolves in a way that is independent of the addition of the accretion term. To demonstrate this, one can simply integrate the coagulation-accretion equation with respect to mass, to yield an equation for $n(t)$. While exact solutions for $N(m,t)$ are not readily obtainable without numerical integration, a general property seems to be that accretion results in a general shift to larger masses. A characteristic mass spectrum still emerges at a finite time, determined now by exhaustion of the supply of accreting material.

VIII. BINARY FORMATION

A recent observational study (Abt and Levy, 1976; see Abt's chapter) has indicated that the difficulties implicit in star formation, if angular momentum is conserved during collapse from interstellar densities (Mestel, 1965), can be overcome as a consequence of binary formation. Provided that magnetic braking occurs during the early collapse, it appears possible to account for the observed range of binary periods below ~ 100 yr (Mouschovias, 1977). Abt and Levy argue that the flat frequency distribution of secondary masses in this range is consistent with the fragmentation hypothesis.

Longer period binaries ($\tau_b \gtrsim 100$ yr) possess a van Rhijn distribution of masses (independently for both primary and secondary masses), suggesting that a different binary formation process may be operative. Fragment interactions provide a clue to the nature of this process, namely binary formation by capture. The arguments of Sec. VII that assume coalescence are valid for collisions while the protostars are relatively diffuse and extended; one might expect indirect collisions of more centrally condensed protostars to result in binary formation. A necessary condition for this to occur is that there be a dissipation mechanism, and the presence of extended accretion envelopes around the protostars provides a medium in which viscous drag will be important. Thus, as long as such envelopes are present, expected to be for a period roughly equal to the initial free-fall time, binaries can, at least in principle, form by capture.

The plausibility of capture leading to a stable binary (rather than to merger and coalescence) can be examined by estimating the time t_s to substantially decelerate a newly captured fragment (mass m_s, velocity v, core radius r_s) from radius r ($> r_s$), namely

$$t_s = m_s \, [\pi \, r_s^2 \, v \, \rho(r)]^{-1} \tag{14}$$

where $\rho\,(r)$ is the local density. This expression for t_s can be written

$$t_s = \frac{Gm_s}{\pi^2 v c_i} \left(\frac{r}{r_s}\right)^2 \approx 500 \, (r/r_s)^2 \text{ yr} \tag{15}$$

where c_i is the sound velocity at the onset of fragmentation and the numerical value assumes $c_i = 0.3$ km sec^{-1}, and $m_s = 1 M_\odot$. Since the time scale for infall $\sim r_i/c_i \sim 10^5$ yr, this result suggests that fragment capture to form binaries can occur to initial radii $r \gtrsim 10\ r_s$. Because of orbital decay, one would expect the resulting distribution of binary periods to flatten at much smaller separations. Moreover, the distribution of binary member masses should reflect the initial fragment distribution, as seems to be the case for binaries with periods $\tau \gtrsim 100$ yr (Abt and Levy, 1966).

The range of binary periods resulting from fragment captures can be estimated as follows. The characteristic size of an opaque core resulting from a spherical collapse is $r_0 \sim 2 \times 10^{14}$ cm (Silk, 1977*a*). While the eventual size of these cores may be greatly augmented by accretion, this should provide a measure of the effective collision cross-section of a fragment. Collisions with a distance of closest approach of less than $2\ r_0$ will result in coalescence; more distant encounters may lead to the formation of binary systems if sufficient dissipation to ensure trapping of the colliding fragments occurs. For solar mass fragments, a separation of $2\ r_0$ reproduces a binary system of period ~100 yr, and provides an estimate of the minimum periods binaries one might expect to form initially as a result of capture. Of course, orbital decay may act to further reduce the period.

The longest period binaries produced by capture could have a separation as large as the initial Jeans length r_f appropriate to an individual fragment (this length scale giving a measure of the dimension of the effective density enhancement). Taking $r_f = 10^{17}\ (T/10\ \text{K})^{-1}$ cm for $1\ M_\odot$ yields $\tau_b \sim 10^6$ yr.

One difficulty with binary formation during the early stages of star formation may occur with tidal disruption (Kumar, 1972). For a binary pair of solar mass protostars, the Roche limit yields a minimum density of $\sim 10^{-10}\ (100\text{yr}/\tau_b)^2$ g cm^{-3}, implying that opaque fragments of the minimum critical mass, with density $\sim 10^{-12}$ g cm^{-3} (Silk, 1977*a*), can only form binaries with $\tau_b \gtrsim 10^3$ yr. Capture theory with some subsequent orbital decay should suffice to account for the long period binaries whose distribution of secondary masses satisfies the van Rhijn function.

IX. WHY ARE MASSIVE STARS EXCLUDED FROM THE INTERIORS OF DARK CLOUDS?

Infrared and radio maps of dark clouds have revealed that while T Tauri and low-luminosity stars are found throughout the cloud volume, massive stars and associated H II regions are generally confined to the cloud surface (Elmegreen, 1978; Lada, 1978). Evidently the star formation process is selective in choice of sites. It is of interest to consider how fragmentation theories can account for this phenomenon. In particular, two possible explanations appear tenable.

Fragmentation subject to an external compression

The maximum mass of stars forming as a result of coalescent encounters depends on the mean number of collisions suffered by a fragment. If the density of fragments were to be markedly increased, as might happen in a region where fragmentation is instigated by the passage of a strong shock incident on the surface of a dense cloud, the resulting mass function could extend to a considerably larger mass than in a more quiescent region.

One might suppose, for example, that in the cloud interior, pre-existing fragments in pressure equilibrium with a more diffuse and warmer ambient medium would occasionally coalesce, become Jeans unstable, and form a star. Incidence of a shock, by locally increasing the density of fragments, would accelerate the coalescence process and result in the formation of more massive stars. The shock could be associated with the H II region of a nearby massive star or be due to the interaction with a supernova remnant.

Fragmentation in a non-uniform cloud

Consider a cloud at the onset of fragmentation, that possesses a density gradient, as would be expected to develop if the cloud is collapsing. If the cloud is isothermal, the Jeans mass increases towards the exterior. In the fragmentation model of Reddish (1975) the fragment mass is determined by fixing the optical depth due to grain opacity, varies inversely as the square of the fragment gas density and so yields a strong increase towards the cloud exterior. Reddish used this result to infer the resulting mass function. One can readily adapt this approach to infer the initial mass function if either the Jeans criterion or the critical mass inferred in the presence of magnetic fields determines the local fragment size. In general, a fragmentation condition will yield a density dependence $M \propto \rho^{-\alpha}$. Take the cloud density to be $\rho \propto r^{-\beta}$. One now finds that the initial mass function

$$N(M) = \frac{4\pi \rho r^3}{M^2} \left(\frac{\mathrm{d}\, ln\, r}{\mathrm{d}\, ln\, \rho}\right)_M \left(\frac{\mathrm{d}\, ln\, \rho}{\mathrm{d}\, ln\, M}\right)_r \propto M^{\frac{3-\beta}{\alpha\beta} - 2} \tag{16}$$

Massive stars preferentially form in the outer regions if $\alpha > 0$ and $\beta > 0$, as expected in most physical models. Adopting $\beta = 2$, as inferred from isothermal collapse calculations, one obtains $N(M) \propto M^{-1}$ if $\alpha = ½$, appropriate to Jeans fragmentation or to magnetic fragmentation if $\beta \propto \rho^{\frac{1}{2}}$ (c.f., Mouschovias, 1976). This mass spectrum appears to be too flat compared to the initial stellar mass function inferred in the solar neighborhood, and would give an excess of massive stars. However, other physical effects may intervene to steepen the mass function, most notably due to the energy input by more massive stars.

X. PROBABILISTIC THEORIES

Various attempts have been made to discuss fragmentation as a series of random events in space and time (Auluck and Kothari, 1954, 1965; Kruszewski, 1961; Larson, 1973). While none of these theories are acceptable because of the lack of any real physical basis, it nevertheless may be premature to dismiss this approach to fragmentation.

To illustrate the possible advantages that may derive from treating fragmentation as a random process, I have constructed a simple example of how a theory of spontaneous fragmentation can be constructed. Let $C(m')$ $f(m)$ be the probability per unit time that a fragment of mass m' will subfragment into a piece of mass m. To conserve mass, write

$$pm' = \int_{m_0}^{m'} C(m')\, f(m)\, m \, \mathrm{d}m \tag{17}$$

where m_0 is a lower limit on the fragment distribution and p is a constant which may be interpreted as the rate at which the mass m' is fragmenting (in principle, p could be a function of mass m'). The mass spectrum $N(m,t)$ will satisfy an equation of the form

$$\frac{\partial N}{\partial t} = -pN + \int_{m}^{\infty} C(m')\, f(m)\, N(m', t)\, \mathrm{d}m'. \tag{18}$$

This equation has the steady state solution $N(m,t) = N(m)$, where

$$N(m) = \frac{Ap}{m^2} \quad \frac{\mathrm{d}\, ln\, [m/C(m)]}{\mathrm{d}\, ln\, m}. \tag{19}$$

For example, if $f(m) = m^{-n}$, one obtains $N \propto m^{-2}$; if $f(m) = \exp(-\beta m^2)$, one has $N \propto e^{-\beta m}(1-e^{-\beta m^2})^{-1} \underset{\sim}{\propto} m^{-2} \exp(-\beta m^2)$. This solution is valid over a finite mass range; mass is assumed to be fed in from some larger mass-scale and the resulting fragmentation spectrum maintains the specified form. For a simple choice of $f(m)$ (such as a power law), an exact time-dependent solution can be obtained, which has the property of relaxing to the steady state solution with an exponential decay rate $\sim \exp(-pt)$.

This approach seems promising insofar as functions suggestive of the initial stellar mass function emerge on the basis of rather general considerations. Further effort is needed, however, to incorporate realistic physical processes and parameters, perhaps derived from computer simulations of fragmentation, into his framework.

XI. CONCLUSIONS

There seems little doubt that fragmentation is occurring in molecular clouds. Radio and infrared observations with high spatial resolution are required to map out clumpiness on scales corresponding to aggregations of stellar mass. The original hierarchical fragmentation process outlined by Hoyle (1953) now seems very much an oversimplification. Some of the complexities attainable in a realistic situation include magnetic inhibition of collapse and flow along field lines, gas accretion, protostellar heat input, and fragment interactions leading to coalescence or capture. It seems encouraging that an initial mass function, resembling the observed distribution, nevertheless emerges (see Scalo's chapter). The details of the theories outlined above are not to be taken to apply to an actual cloud, but it seems entirely plausible that similar physical processes, notably of fragmentation, accretion, and coalescence, should dominate the collapse of an actual cloud.

There are at least three observational clues to the nature of the star formation process. The implications of one of these, the tendency for massive stars not to form in the interiors of dark clouds, have already been noted. A second clue is that studies of young galactic clusters indicate that the rotation axes of stars appear to be randomly oriented. If magnetic or centrifugal stresses were to dominate the fragmentation processes, one would expect a general alignment. Computations hitherto performed indicate that collapse of a non-magnetic rotating cloud generally results in ring formation. The rings are unstable to fragmentation, although relatively large perturbations must be applied to the initial configuration in order to instigate eventual fragmentation of the rings (Tohline, personal communication, 1978). Typically, he finds that the rings form two or three fragments, which retain about 10% of the original specific angular momentum (see chapter by Bodenheimer and Black). It is conjectured that these in turn will collapse and form rings that fragment, the net result being a stellar cluster containing close and distant stellar pairs whose rotation axes are parallel to the initial angular momentum vector. It seems more likely that the actual collapse and fragmentation will be a far more irregular process, with interactions between neighboring fragments playing a prominent role. While the extreme case of colescence described above is unlikely to apply in detail, similar processes are nevertheless likely to occur and result in a random distribution of rotation axes.

A third constraint on the nature of the star formation process that deserves attention is the fact that star formation evidently lasts for a considerable time. The presence of both massive young stars and low-mass stars already on the main sequence in specific associations requires a duration for continuing star formation that exceeds 10^7 yr. One can speculate that massive stars have formed only recently, and that the preceding phase of such associations was dominated by low-mass star formation just as in the interiors of dark clouds. Since the free-fall time of such a cloud is relatively short ($\sim 10^5$ yr), it seems likely that magnetic stresses must be playing a major role in inhibiting fragmentation and accretion processes that would otherwise result in the formation of massive stars in these regions.

Acknowledgment. This research has been supported in part by the National Aeronautics and Space Administration.

COMMENTS

C. Lada: Could you be more specific about your suggestion that observers search for fragmentation in molecular clouds? There is abundant evidence for fragmentation over a large range in scale in molecular clouds. For example, we have many observations of groups of embedded stars in high-density regions of molecular clouds; there is also evidence for fragmentation in OH maser sources on scales on the order of 20 – 1000 AU. In addition many large molecular clouds ($\sim$ 50 pc) with average densities around 10^3 cm^{-3} contain 1 – 3 pc fragments of densities $\sim 10^4$ cm^{-3}. Finally some clouds such as M 17 exhibit fragmentation on scales of the order 10 – 20 pc.

J.I. Silk: Fragmentation theory suggests that molecular clouds should exhibit small-scale structure on scales containing 1 $M_\odot$ or less. The principle spatial probe hitherto available on such a fine scale has been due to OH and H_2O masers and these necessarily are excited in unusually hot and dense regions. Interferometric studies of molecular lines, using lines of varying degrees of excitation, should be capable in the near future of searching for structures on scales of $\sim$ 0.1 pc in large molecular clouds. Detection of clumpiness on this scale at densities $\sim 10^4$ cm^{-3} could provide a vital link between molecular clouds and the star-formation process.

L. Mestel: Spitzer and I, and later Gaustad, estimated the limiting mass by an argument somewhat different from Hoyle's. We supposed that the opaque fragment is in hydrostatic support with an internal temperature gradient and demanded that the contraction rate due to heat leakage by Eddington's theory be just equal to the free-fall rate. This gives a mass larger (by a factor ten) than that from Hoyle's procedure. I was surprised by the large discrepancies. Have you any comments on this?

J.I. Silk: It cannot be overemphasized that the Jeans mass arguments of Hoyle *et al.* provide only a *lower limit* on protostellar masses. Many subsequent numerical studies have found that accretion plays a major role in increasing the mass of the initial opaque protostellar core. Coalescence may also be significant, and core growth by one or even two orders of magnitude in mass seems by no means improbable. Requiring that the Kelvin-Helmholtz contraction time scale be equal to the free-fall time scale indeed determines a limiting mass at the point of grain evaporation, although perhaps in the opposite sense to Hoyle's criterion. This is because more massive fragments are capable, in principle, of continued subfragmentation. However, because of the many complexities of a realistic fragmentation model, these limiting masses should be regarded as providing us with no more than a crude guide as to what is likely to emerge.

REFERENCES

Abt, H.A., and Levy, S.G. 1976. *Astrophys. J. Suppl.* 30: 273.
Arny, T., and Weissman, B. 1973. *Astron. J.* 78: 309.
Auluck, F.C., and Kothari, D.S. 1954. *Nature* 174: 565.
____. 1965. *Zs f. Ap.* 63: 15.
Elmegreen, B.G. 1978. *Moon and Planets* (special Protostars and Planets issue). In press.
Freeman, K.C. 1977. In *The Evolution of Galaxies and Stellar Populations* (B.M. Tinsley and R.B. Larson, eds.), p. 133. New Haven, Conn: Yale University Press.
Guelin, M.; Langer, W.D.; Snell, R.L.; and Wootton, H.A. 1977. *Astrophys. J.* 217: L165.
Hoyle, F. 1953. *Astrophys. J.* 118: 513.
Hunter, C. 1962. *Astrophys. J.* 135: 594.
Kruszewski, A. 1961. *Acta Astron.* 11: 199.
Kumar, S. 1972. *Astrophys. Space Sci.* 17: 453.
Lada, C.J. 1978. *Moon and Planets* (special Protostars and Planets issue). In press.
Larson, R.B. 1973. *Mon. Not. Roy. Astr. Soc.* 161: 133.
____. 1978. Preprint.
Layzer, D. 1963. *Astrophys. J.* 137: 351.
Leung, C.M. 1978. *Bull Amer. Astron. Soc.* 9: 590.
Matsakis, D.N.; Brandshaft, D.; Chui, M.F.; Cheung, A.C.; Yngvesson, K.S.; Cardiasmenos, A.G.; Shanley, J.F.; and Ho, P. 1977. *Astrophys. J.* 214: L67.
Mestel, L. 1965. *Quarterly J. Roy. Astr. Soc.* 6: 265.
____. 1977. In *Star Formation,* IAU Symp. No. 75, (T. de Jong and A. Maeder, eds.), p. 213. Dordrecht: Reidel.
Mouschovias, T. 1976. *Astrophys. J.* 207: 141.
____. 1977. *Astrophys. J.* 211: 147.
Nakano, T. 1966. *Prog. Theor. Phys.* 36: 515.
Reddish, V. 1975. *Mon. Not. Roy. Astr. Soc.* 170: 261.
Rees, M.J. 1976. *Mon. Not. Roy. Astr. Soc.* 176: 483.
Silk, J. 1977*a*. *Astrophys. J.* 214: 152.
____. 1977*b*. *Astrophys. J.* 214: 718.
Snyder, L.E.; Hollis, J.M.; Buhl, D.; and Watson, W.D. 1977. *Astrophys. J.* 218: L61.
Spitzer, L. 1968. *Diffuse Matter in Space.* New York: Interscience.
Trubnikov, B.A. 1971. *Soviet Phys. – Doklady* 16: 124.

MAGNETIC FIELD STRUCTURES AND STRENGTHS IN DARK CLOUDS

E. J. CHAISSON
Harvard-Smithsonian Center for Astrophysics

and

F. J. VRBA
U. S. Naval Observatory, Flagstaff Station

We review recent observational results concerning magnetic field structures and strengths in regions where star formation is currently believed to be taking place (dense molecular-dust clouds and very dense regions associated with H II regions). New optical and infrared linear polarization measurements of sources in and behind massive dust clouds continue to indicate that field strengths of several tenths of a milligauss may be necessary for grain alignment by the Davis-Greenstein mechanism. However, observations of OH Zeeman splitting at radio wavelengths on the interior regions of these clouds generally do not detect fields of this strength. The very dense molecular clouds associated with H II regions appear to be able to support much stronger fields. In particular, the protostar sites within the Orion molecular cloud infrared nebula almost certainly have a magnetic field of several milligauss strength, while a number of other protostellar galactic regions possibly have field strengths of this magnitude. Furthermore, polarization measurements of compact sources associated with the H II regions suggest that milligauss field strengths may permeate much of the parent molecular cloud in which the protostars are embedded. Recent observations of the compact infrared sources show that their polarization is a function of their age/size.

Numerous reviews have been written about indirect and direct studies of galactic magnetism. The most complete of these are by Aannestad and Purcell (1973) and Heiles (1976). We limit our discussion here to the detection of magnetic fields in the most dense portions of the interstellar medium, the regions currently thought conducive to star formation.

MAGNETIC FIELD DETECTION

Radio Observations

Unlike other astronomies, through which magnetism can only be inferred, radio astronomy provides a technique capable of directly measuring magnetic field strength. Since a magnetic field can alter the energy-level structure of paramagnetic atoms and molecules, the observational characteristics (namely splitting and polarization) of radio frequency spectral-line features can, in principle, be used as probes of the extent of alteration and hence of magnetism. This phenomenon, known as the Zeeman Effect after the Dutch physicist who first studied it during the 1890's displays appreciable energy-level splittings capable of astronomical observation only in atoms and molecules having reasonably large Landé *g*-factors.

Unfortunately, among the 40 or so interstellar chemicals now known to exist in galactic clouds (see Table I in E. Herbst's chapter), only two have appropriately large Landé *g*-factors capable of producing measurable Zeeman splitting in typical galactic regions. The substances are the hydrogen (H) atom and the hydroxyl (OH) molecule, both of which radiate in the microwave part of the spectrum. Other molecules observed largely in the millimeter-wave region require unrealistic fields of a few tens of gauss to produce splittings comparable to, or larger than, the observed line widths.

The energy-level splitting of the ground hyperfine $^2S_{\frac{1}{2}}$ state of hydrogen produces a pair of transitions at the approximate frequency of 1420 MHz. The extent of the splitting obviously depends on the magnetic field strength and can be calculated from the relation

$$\delta E = \mu_0 \, B \, g \, m_F \quad . \tag{1}$$

Here, B is the magnetic field, μ_0 (= 9.27×10^{-21} erg G^{-1}) the Bohr magneton, g the Landé factor, and m_F the magnetic quantum number. The extent of Zeeman splitting then amounts to 0.59 km sec^{-1} mG^{-1}.

The Zeeman Effect of the OH molecule is considerably more complex. Figure 1 shows the energy-level splitting of the ground rotational $^2\pi_{\frac{3}{2}}$, $J = 3/2$ state of OH, derived from the general description given by Townes and Schawlow (1955). The currently accepted frequencies of each of the four

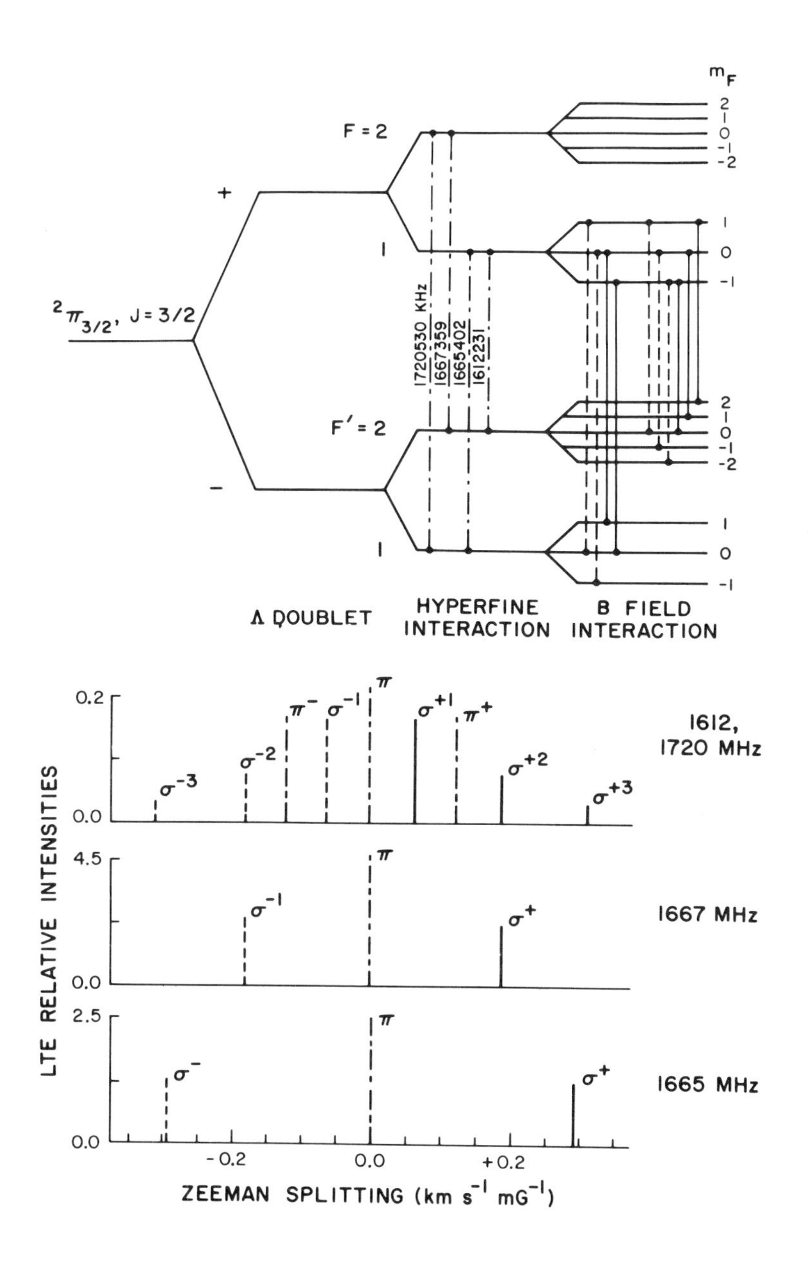

Fig. 1. Energy-level diagram for ground-state OH and its expected longitudinal Zeeman pattern of right (dashed) and left (solid) circularly polarized radiation. Linearly polarized radiation is shown by the dash-dot line, though, for clarity, only the circularly polarized transitions are shown at the top of the figure for the F = 1→1 ($\sigma^{\pm}$; 1665 MHz) and the F = 1→2 ($\sigma^{\pm1,2,3}$; 1612 MHz) radiation.

18-cm transitions are labeled vertically (ter Meulen and Dymanus, 1972). At the far right, all degeneracies are shown removed via a magnetic-field interaction. The magnitudes of the energy-level splittings (not drawn to scale) and neglecting the interaction between the magnetic field and the nuclear magnetic moment, are given by

$$\delta E = \frac{\mu_0 g_J m_F \, B(\mathbf{J}^2 + \mathbf{F}^2 - \mathbf{I}^2)}{2\mathbf{F}^2} \quad . \tag{2}$$

Here, **I**, **J**, and **F** are the usual angular momentum vectors, and g_J (=0.935, Radford, 1961) the Landé factor. Evaluation of this expression for the vector model and the weak-field approximation yields the distribution of circularly and linearly polarized features shown in the lower half of Fig. 1. The dashed bars of Fig. 1 refer to the σ^{-1}, σ^{-2}, and σ^{-3} transitions (Δm_F=+1) that produce right circularly polarized radiation about the longitudinal component of a magnetic field directed toward Earth; the solid bars similarly represent left circularly polarized radiation. Linearly polarized features (π transitions, Δm_F=0) are denoted by dash-dot bars. The relative intensities of the individual components of each Zeeman pattern are shown under the assumption of local thermodynamic equilibrium (LTE). Specifically, the sums of the intensities of all the components in each Zeeman pattern at 1612, 1665, 1667, and 1720 MHz vary according to 1, 5, 9, 1, respectively. Anticipating quantitative use of specific splittings later in this paper, we note that the separations between opposite circularly polarized components of the 1665- and 1667-MHz doublets are 0.59 and 0.35 km sec^{-1} mG^{-1}, respectively; similarly, the separations between corresponding pairs of opposite circularly polarized 1612-MHz components are 0.12, 0.37 and 0.61 km sec^{-1} mG^{-1}.

Optical and Infrared Observations

Polarization of the starlight of normal type stars (first observed by Hall [1949] and Hiltner [1949]) is known to be due to aligned, non-spherical, foreground interstellar grains. For an overview of optical and infrared polarization techniques and results, see Gehrels (1974). While a number of exotic mechanisms have been devised to explain this alignment, only the basic idea of a magnetic torque acting upon grains with paramagnetic impurities, first proposed by Davis and Greenstein (1951), now appears plausible. This torque will act to damp out rotation along axes perpendicular to the magnetic field direction and will act on a time scale

$$\tau_B = \frac{\gamma I}{V B^2 K} \tag{3}$$

where γI is the moment of inertia of the grain about the axis transverse to the symmetry axis, V is the grain volume, B is the magnetic field and K is the imaginary part of the grain's magnetic susceptibility divided by the rotation frequency. This alignment effect tends to be destroyed by randomizing collisions with surrounding gas atoms. This randomization occurs on a time scale

$$\tau_{\text{coll}} = \frac{C \gamma I}{\langle v \rangle \rho_{\text{gas}} V^{\frac{4}{3}}} \tag{4}$$

where $\langle v \rangle$ is the mean gas velocity, ρ_{gas} is the gas mass density, and C is a shape dependent parameter. Thus, an orientation parameter may be defined as

$$\delta = \frac{\tau_{\text{coll}}}{\tau_B} = \frac{CK\, V^{-\frac{1}{3}} B^2}{\langle v \rangle \rho_{\text{gas}}} . \tag{5}$$

Jones and Spitzer (1967) showed the effects of grain and gas temperatures (T_{gr} is the grain internal temperature and T is the gas kinetic temperature) on magnetic alignment noting that $T_{\text{gr}} \neq T$ is a necessary condition for net alignment. If $T_{\text{gr}} < T$, the normal Davis-Greenstein alignment occurs (spin axis perpendicular to the long axis of the grain producing linear polarization pseudo-vectors in the direction of the transverse magnetic field component) while if $T_{\text{gr}} > T$, the opposite alignment occurs. For the conditions found in interstellar space, diffuse clouds, and the peripheries of dense clouds, $T_{\text{gr}} < T$. However, for the conditions thought to be appropriate for the innermost regions of dense clouds, the value of T_{gr} will be near that of T if not higher (c.f., Goldreich and Kwan, 1974).

Due largely to the work of Purcell and Spitzer (1971) and Cugnon (1971), Eq. (5) may be related to an observable quantity such as $\Delta m_P / \Delta m$ (the ratio of linear polarization to absorption at a given wavelength) so that B may be solved for, provided that the following can be estimated: surrounding gas temperature and density, grain temperature, size, shape, and composition. While many of these quantities may be estimated with reasonable certainty, their combined uncertainties make magnetic grain alignment theories useful only as an order of magnitude estimate of magnetic field strength.

Linear polarization, however, is a very useful tool in producing maps of the structure of the transverse B field component by observations of many intrinsically unpolarized stars that are within, or in the background of, the region of interest. This technique has been used to map the local galactic magnetic field structure (c.f., Axon and Ellis, 1976; Fig. 12 of Field's Chapter) and field structures within individual star clusters. Very little optical work of this kind has been done in the dense cloud regions of interest here, principally because apparent magnitudes of background stars are very faint due to heavy extinctions. For central regions of these clouds, absorptions are so large ($10 < A_v < 40$ mag) that polarization work has had to await the development of sensitive near-infrared detectors in order to observe the embedded stars there. Fortunately, this review comes at an opportune time since the first extensive polarization work in dense clouds has recently become available.

DENSE CLOUD REGIONS

Should the magnetic field be coupled ("frozen-in") to the gas, we might expect that clouds of total densities on the order of 10^4 cm^{-3} would be permeated by milligauss magnetic fields. This suggestion is based on the conservation of magnetic flux per unit mass which, from simple geometric considerations, gives $B \propto n^{\kappa=\frac{2}{3}}$. However, during cloud collapse, matter is able to escape along the field lines and the value of κ becomes much lower. Recent theoretical work by Mouschovias (1976; see also his chapter in this book) suggests that $1/3 \leqslant \kappa \leqslant 1/2$, indicating that milligauss fields would only be expected for densities of $10^6 - 10^9$ cm^{-3}.

Milligauss magnetic fields should indeed be detectable via the Zeeman Effect provided that H and OH exist in those portions of the cloud permeated by the field. Verschuur (1970) attempted to detect Zeeman splitting of a deep self-reversed H feature toward the Taurus dark cloud, a region thought to be characterized by a total density of several times 10^3 cm^{-3}. His failure to detect it implied an upper limit of 10 μG, although he and others (c.f., Verschuur, 1974) had earlier succeeded in obtaining a positive measurement of hydrogen Zeeman Effect in a few other interstellar regions. These regions are characterized by narrow H absorption features arising in rather tenuous (10-100 cm^{-3}) H I clouds along the line of sight to the radio emitting supernovae Taurus A and Cassiopeia A, as well as the H II regions, Orion A and M 17. The measured values of 3-70 μG demonstrate directly the existence of galactic magnetism and are not inconsistent with the expected strengths given the gas density and the notion of magnetic freezing. However, the rather tenuous nature of such H I clouds and their association with H II regions imply that these clouds are probably not in the act of gravitational collapse; thus, further discussion of them in this, a review of protostars, will not be undertaken.

The implied upper limit of 10 μG in the relatively denser Taurus cloud probably does not reflect the physical conditions inside the dust cloud but only at the periphery where atomic hydrogen is prevalent (Heiles, 1971). Indeed, modern 21-cm studies have shown a definite anticorrelation between the dust and neutral hydrogen gas. Consequently, the fact that Zeeman Effect of the H atom has not yet been observationally demonstrated in dense

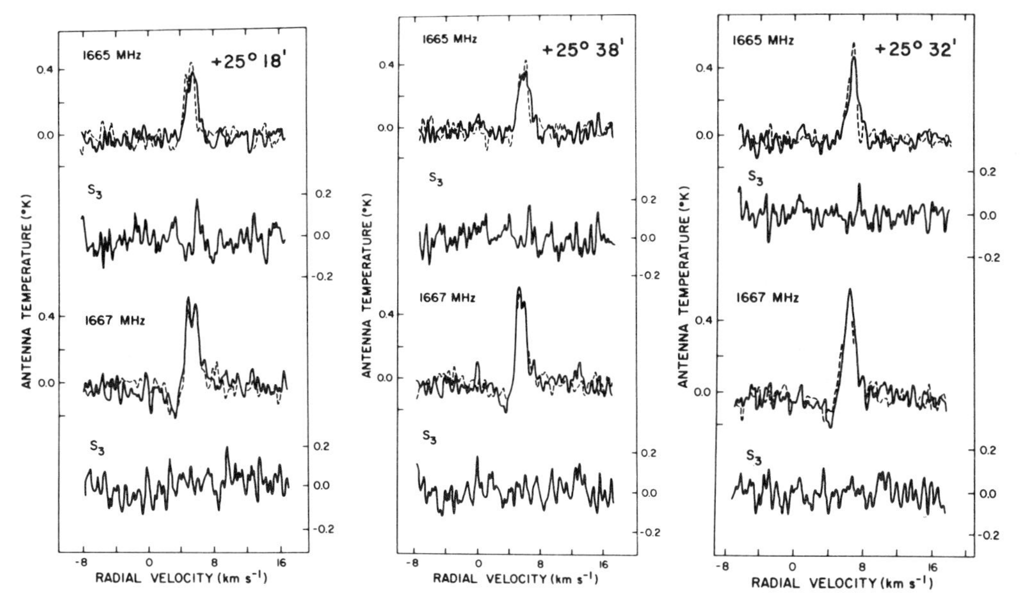

Fig. 2. OH radiation toward the Taurus dark cloud with 1.0 KHz resolution. The first and third spectra from the top of each frame represent superpositions of radiation observed with the right circular feed (solid line) and that observed with the left circular feed (dashed line) at each of the two main lines of the ground state. The second and fourth spectra illustrate the difference between radiation observed with feeds sensitive to right and left circular signals (the S_3 Stokes parameter). The radial velocity scale is referenced to the local standard of rest; the upper right-hand corner gives the declination of each observation and may be compared with those listed in Table I.

galactic clouds is not regarded as a problem, but only as an indication that atomic hydrogen is not coextensive with the molecules in the bulk of the clouds. Most of the hydrogen there is undoubtedly in molecular form, H_2.

The OH molecule, however, is regarded as a better probe of dense cloud interiors. Indeed, observation of OH constituted the first direct probe of the physical conditions in such clouds (c.f., Heiles, 1971). A few attempts have been made to measure the Zeeman splitting of OH in dark clouds, especially the main-line 1665 and 1667 MHz transitions which would show the largest effect (Turner and Verschuur, 1970; Crutcher *et al.*, 1975; Chaisson, 1975).

Figure 2 shows representative spectra observed toward a few positions in the Taurus cloud. At the top of each frame the left circularly polarized radiation (dashed curve) is superposed on the right circularly polarized radiation (solid curve) at 1665 MHz. The third spectrum from the top of each frame shows similar observations of the 1667 MHz circularly polarized radiation. The size, shape, and centroid of the OH signals can be examined as accurately as possible by forming the S_0 Stokes parameter (the sum of right and left circularly polarized antenna temperature), which contains 40% greater signal-to-noise ratio than illustrated for the individual right and left circularly polarized radiation. The S_3 Stokes parameter (the difference of right and left circularly polarized antenna temperature), which would be

TABLE I

Summary of OH Polarization Measurements in Dark Clouds

Cloud	α(1950)	δ(1950)	B(mG)[a]	Reference[b]
ρ Oph	$16^h23^m25^s$	$-24°16'$	0.5	1
ρ Oph	16 24 08	-24 28	0.05	2
ρ Oph	16 25 29	-24 31	0.4	3
ρ Oph	16 22 30	-24 22	0.4	1
Taurus	04 38 30	+25 18	0.4	1
Taurus	04 38 30	+25 38	0.4	1
Taurus	04 36 30	+25 30	0.3	1
Taurus	04 30 30	+26 12	0.05	2
NGC 1333	03 25 56	+31 10	0.6	1

[a]Upper limits only.
[b]References: (1) Chaisson and Beichman (unpublished); (2) Crutcher *et al.* (1975); (3) Chaisson (1975).

expected to show a horizontal S-shaped curve if the left and right circularly polarized spectra were appreciably Zeeman split, is shown as the second and fourth spectrum of each frame.

The observations shown in Fig. 2 are taken from an unpublished experiment by Chaisson and Beichman, but they are similar to other attempts (referenced above) to measure circularly polarized OH radiation from dense, dark clouds. In each of these experiments, the right and left circularly polarized signals of main-line OH were measured simultaneously by two spectrometers that operated in parallel, but which were slaved to the same synthesizer. This operational procedure, as well as one which carefully measures any imbalance in the gain of the antenna feeds, then assures minimal error in the alignment of the spectrum of the S_3 Stokes parameter.

No circular polarization of OH has yet been measured in dark clouds devoid of gaseous emission nebulae. That is, none of the S_3 spectra observed to date show any evidence for a signal, S-shaped or otherwise. Table I summarizes the negative results. For each cloud, the columns list Right Ascension and Declination for epoch 1950.0, the upper limit on the inferred magnetism, and the reference.

The first three sampled positions in the Taurus cloud are coincident with those of the strongest OH line emission. The $16^h\ 23^m\ 25^s$ position in the Rho Ophiuchi cloud is near the peak of millimeter-wave line and infrared continuum emission, while the $16^h\ 25^m\ 29^s$ position is at the cloud center (as determined from the distribution of grain density [Bok, 1956]), where Carrasco *et al.* (1973) found a high degree of optical polarization. The NGC 1333 region is a rich complex of young stellar objects including Herbig-Haro objects (see chapter by Böhm in this book), infrared and T Tauri stars (see chapter by Rydgren in this book; also Strom *et al.*, 1974); the position

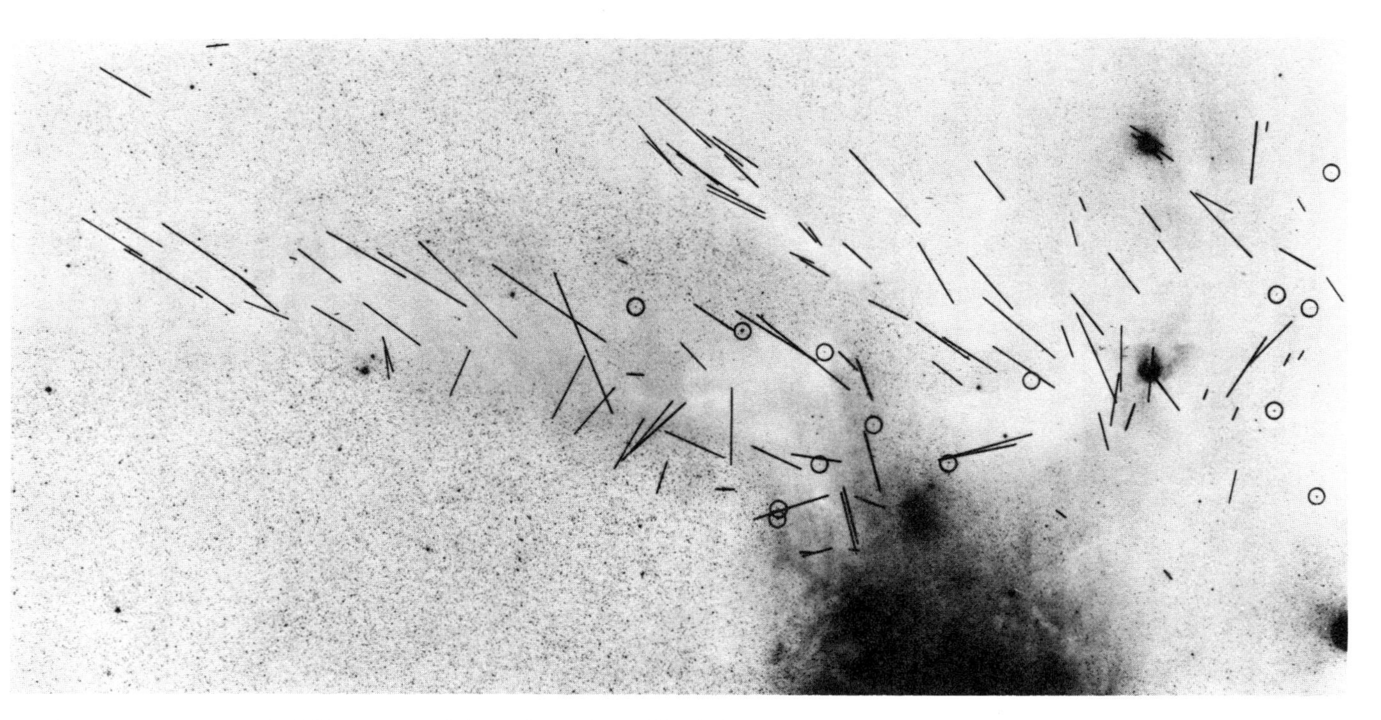

Fig. 3. The overall magnetic field structure of the dark cloud near Rho Ophiuchus. Lines represent polarizations measured for stars behind the cloud, with the line lengths proportional to the polarization amount and plotted in the direction of polarization position angle (presumably the same direction as that of the magnetic field). (From Vrba *et al.*, 1976.)

reported here is in the direction of the infrared star closest to H-H 12 in a region of 10^5 cm^{-3} mean total density. In summary, repeated failure to detect circular polarization in the main-OH spectral lines strongly suggests that uniform milligauss fields do not permeate typical dark clouds.

Optical and infrared results give a somewhat different picture of magnetic fields in such clouds. Vrba *et al.* (1976) carried out the first extensive mapping of the magnetic field structure in several massive dark clouds (Rho Ophiuchi, R Coronae Austrina, Lynds 1630, Lynds 1450 [NGC 1333], and Lynds 1551) by measurement of the optical polarization of stars which are background to the peripheries of the clouds. In Fig. 3 we show the magnetic field structure found for the Rho Ophiuchi cloud. The remarkable degree to which the magnetic field aligns with the eastern streamers implies that magnetic fields have played a major role in the evolution of this cloud. Close connection between the field and cloud structures was also found for the R Coronae Austrina cloud.

In order to deduce magnetic field strengths, Vrba, Coyne, and Tapia (1978) have made observations of polarization-wavelength dependence and polarization-to-absorption-ratio for many of the stars that had been observed by Vrba *et al.* (1976) as background to the Rho Ophiuchi cloud. Thus far, the only region of the cloud to have been adequately surveyed is the main eastern streamer, where $B \simeq 70$ μG is implied. With a total gas density in this region of 2×10^3 cm^{-3} (Vrba, 1977) this implies $\kappa \simeq 0.4$, which is consistent with the results of Mouschovias (see his chapter). These observations, however, do not serve as a confirmation of low magnetic field strengths throughout the cloud as they were obtained far from the dense star-formation region where the above OH surveys were carried out.

However, Vrba *et al.* (1976) made a 2.2 μm polarimetry survey of infrared sources (Vrba *et al.*, 1975) in the central region of the Rho Ophiuchi cloud (the region coincident with the radio OH data). They found the polarization position angles of these sources to lie predominantly in two directions, implying a two-component model for this region. More recently, Wilking *et al.* (personal communication, 1978) have confirmed this model by 2.2 μm polarimetry of additional infrared sources in Ophiuchus. In addition, Wilking *et al.* have found similar high degrees of 2.2 μm polarization ($1 < P < 4\%$) for sources embedded in Lynds 1630 and NGC 1333. Such polarizations certainly require field strengths of tenths of a milligauss, while, if $T_{gr} \gtrsim T$ as is theoretically expected for such dense regions, milligauss fields are required if our understanding of magnetic grain alignment is correct.

The discrepencies between the more stringent radio observations and the optical and infrared polarization observations imply one or more of the following:

1. Our understanding of magnetic grain alignment due to paramagnetic relaxation is incomplete. If ferromagnetic or "super-paramagnetic" effects are present in grains, field strengths several orders of magnitude

less would be required to produce the observed alignment.

2. Although the dark clouds have total densities generally between 10^3 and 10^4 cm^{-3}, embedded within them are small pockets of higher density. The OH observations use a relatively large antenna beamwidth (generally 18 arcmin) which may integrate over a tangled field within the beam, while the polarization observations essentially have an infinitely small beam.
3. Clouds which show a high degree of optical or infrared polarization must have a large transverse magnetic field component with, perhaps, only a small longitudinal component. (This is statistically an unlikely possibility.)

We stress that the negative results of the OH observations do not necessarily rule out milligauss field strengths in dark clouds as inferred from the degree of grain alignment. Physically, the observations imply that in at least some regions of dark clouds, either the magnetic field is not frozen in the gas (possibly because the fractional ionization is too small) or the magnetic field, if frozen in the gas, increases with cloud collapse at a rate $\kappa < 2/3$.

One other recent result of magnetic field structure studies should be mentioned at this point. Appenzeller (1971) noted that the Alpha Persei (Per OB3) star cluster was situated at the bottom of a magnetic well as inferred by polarimetric observations of surrounding stars. This result has recently been confirmed by Markkanen (1977), who further showed that the well only exists at the distance of the Alpha Persei star cluster. Vrba *et al.* (1975) found a similar structure associated with the R Coronae Austrina dark cloud and, more recently, Marraco and Forte (1978) have seen such a structure surrounding the cometary globule associated with the reflection nebulosity NGC 5367. If many more such structures exist (and they may, since little effort has thus far gone into determining polarization structure due to the large amounts of telescope time needed), this would indicate that magnetic fields commonly play a major role in the formation of dark clouds and associated star formation.

VERY DENSE REGIONS

The OH molecule also emits strong microwave radiation from small dense pockets of what seems to be protostellar-like gas. The compactness (diameter $\lesssim 10^{16}$ cm) and strength (brightness temperatures in the range $10^9 - 10^{12}$ K) imply that the source is a non-LTE maser of high density ($10^6 - 10^{10}$ cm^{-3}). This type of emission is almost invariably associated with H II regions or infrared emission sources, and contrasts sharply with the near-LTE OH emission in dark clouds.

Several attempts to measure Zeeman splitting in such sources have been made by several groups. Usually, strong, circularly polarized, main-line OH

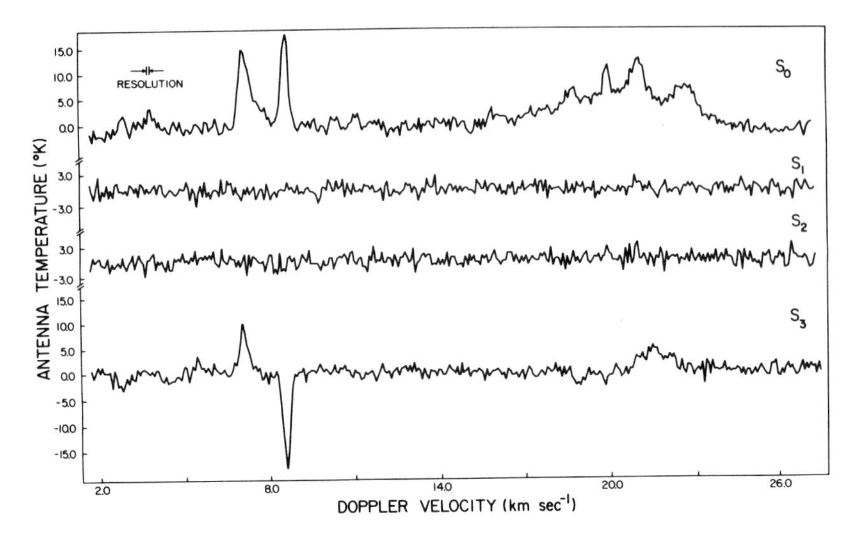

Fig. 4. Illustration of all four Stokes parameters for the 1665 MHz OH radiation in the direction of the Orion infrared/molecular cloud. Spectral resolution is 500 Hz; radial velocity is referenced to the local standard of rest.

features are indeed observed, although it is in all cases difficult, if not impossible, to pair them properly into a classical Zeeman pattern. Tentative interpretations of milligauss magnetic fields have, nevertheless, been made on the basis of ground-state and excited-state OH observations toward many galactic OH regions (c.f., e.g., Davies *et al.*, 1966; Rogers and Barrett, 1966; Palmer and Zuckerman, 1967; Zuckerman *et al.*, 1972; Davies, 1974; Beichman and Chaisson, 1974; Lo *et al.*, 1975). Controversy exists, however, in view of the fact that very long baseline interferometry (VLBI) sometimes shows that suspected Zeeman pairs arise in regions having large spatial separations (Moran *et al.*, 1968).

One region, however, has attracted considerable attention, and is almost certainly the site of a milligauss magnetic field. This is the infrared nebula/molecular cloud to the rear of the Orion A H II region. Figure 4 shows the entire run of 1665-MHz Stokes parameters observed with 500-Hz resolution in May 1974 with the 43-meter antenna of the National Radio Astronomy Observatory. This is an unpublished spectrum made by Chaisson and Beichman using the radio-astronomical system described by Chaisson and Beichman (1975). Several groups (Palmer and Zuckerman, 1967; Weaver, Dieter and Williams, 1968; Manchester, Robinson and Goss, 1970) have previously observed this OH spectrum toward Orion, but with inferior spectral resolution. The OH signals near the local-standard-of-rest (LSR) velocity of 21 km sec^{-1} seem hopelessly confusing. But the features near 8 km sec^{-1} are of greater interest not only because they are clearly separated but also because this is the approximate velocity at which virtually all other

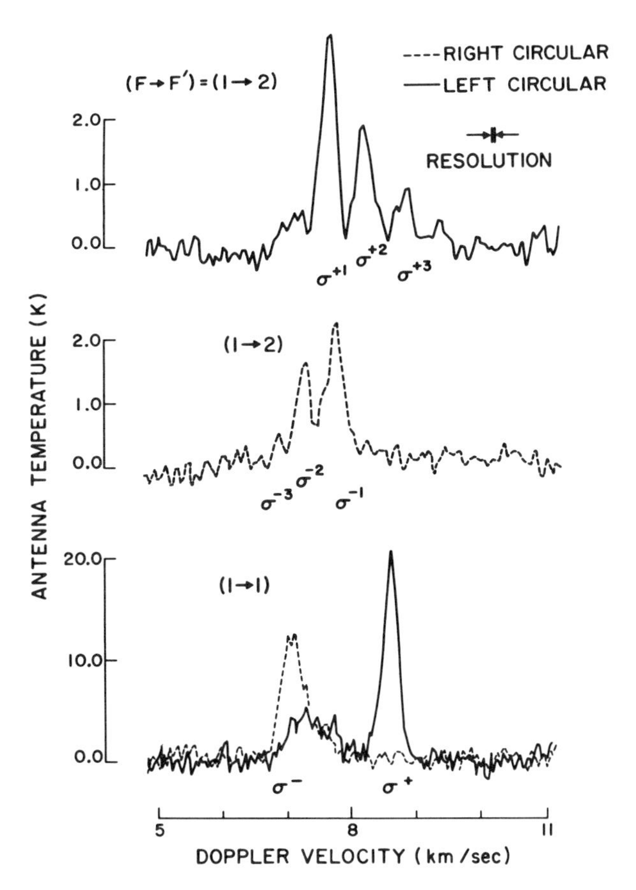

Fig. 5. Observations of circularly polarized OH signals at 1612 and 1665 MHz toward Orion. Spectral resolution is 250 Hz.

molecules emit in the Orion region.

The bottom of Fig. 5 shows observations of the 1665-MHz OH features near 8 km sec^{-1} with even better spectral resolution and sensitivity. The top of the figure shows observations by longer integration of the 1612-MHz satellite transitions of OH. All the observations in Fig. 5 were made with the highest spectral resolution available (250 Hz); they have been previously reported in Chaisson and Beichman (1975). As suggested in that communication, all these right and left circularly polarized σ features at 1665 and 1612 MHz are very close to, although not precisely, that expected in Fig. 1. The two triplet patterns at 1612 MHz slightly overlap, and the linearly polarized π components are missing. The lack of π components can be explained on the basis of arguments given by Goldreich, Keeley, and Kwan (1973) who have noted that, even for unsaturated amplification, the effects of Faraday depolarization and of angular-dependent masing can completely suppress linearly polarized features. (The near-LTE conditions implied from detailed analysis of the relative line intensities, despite masing, can be explained if the **F** = 1 level is at least partially saturated [Litvak, 1970].)

The unexpected overlap of the 1612 MHz triplets, however, is probably the result of two maser sources operating in magnetic fields having slightly different longitudinal components – either two separate condensations or two parts of a single region with a small velocity gradient – each producing Zeeman split radiation in only one sense of polarization. In other words, all the observations shown in Fig. 5 (1665 and 1612 MHz) are consistent with emission of all the right circularly polarized features from a region having an LSR velocity of 7.4 km sec^{-1} and a magnetic field of 3.5 mG, and all of the left circularly polarized features from a region with an LSR velocity of 8.0 km sec^{-1} and a magnetic field of 4.5 mG. The preferential emission of only one sense of circularly polarized radiation per region can be explained by the presence of a radial velocity gradient and a magnetic field gradient such that only one σ component has a long enough pathlength for coherent amplification; the π component can also be suppressed in such schemes (Cook, 1970; Litvak *et al.*, 1966; Shklovskii, 1969).

Indeed, VLBI observations of the 1665 MHz OH features near 8 km sec^{-1} support this Zeeman interpretation. Hansen et al. (1977) have demonstrated that there is a slight separation, no larger than 0.06 arcsec (20 AU), between the approximately 0.05 arcsec diameter sites of left and right circularly polarized emission. However, since the apparent maser sizes are thought to be considerably smaller than the true size of their parent condensation, both 1665-MHz features can now be regarded as part of the same compact region, very near the IRC 4 source and approximately 12 arcsec south of the infrared object of Becklin-Neugebauer (Becklin and Neugebauer, 1967; Wynn-Williams and Becklin, 1974). Because of a probable velocity offset ($\simeq$0.5 km sec^{-1}) of each region of maser emission, the inferred magnetic field should not really be derived from the 1665-MHz doublet; the more reliable magnetic field estimate is probably obtained from the 1612 MHz pattern as discussed above.

Hence, a genuine Zeeman interpretation of a milligauss magnetic field in a dense, protostellar-like cloud appears to require a special set of circumstances including relative suppression of the π components and partial saturation of the OH maser. The only alternative model would require six 1612-MHz OH-emitting clouds arranged so as to give the observed pattern of near classic relative intensities, equal widths, consistent velocity separation, and 100% circular polarization of the σ components of each triplet.

The angular coincidence of pairs of right and left circularly polarized ground-state OH features has also been observed toward several other sources, specifically the H II regions NGC 6334(N), W3(OH), W51, and NGC 7538(N) (Lo *et al.,* 1975). In addition, Moran and Reid (personal communication, 1978) have found several pairs of left and right circularly polarized features in the excited-OH spectrum of W3(OH) to be (virtually) spatially coincident. All of these OH splittings typically imply magnetic fields on the order of 10^{-3} G. It will be important to continue such VLBI observations, especially on the weaker, isolated OH spectral features.

Further, single-dish, high-sensitivity, high-resolution observations of satellite OH features toward additional H II regions would also be of considerable interest, although Chaisson and Beichman (unpublished) have not found as clear a pattern as that toward Orion in any of the following sources; IRC-20197, W28(A1), W28 (A2), OH1649-41, W43, IRC+10011, 42 Ori, OH0739-14, WX Ser, S269, S Cr Bor, OH1735-32, W3A, W3(OH), W51, NGC 7538, ON-3, Sgr B2, W44, and IRC 50137. Some of these sources, nonetheless, contain suggestions of triplet patterns at either the 1612 MHz or 1720 MHz satellite frequencies.

A further aspect of the suggestive Orion magnetic field results comes from studies of infrared polarization. A clear correlation between the extent of linear polarization and the depth of the 9.7 μm silicate absorption feature toward the Orion infrared nebula/molecular cloud (Loer, Allen and Dyck, 1973; Dyck *et al.*, 1973) implies that the grains responsible for the absorption of radiation are also responsible for the polarization of infrared radiation. Dyck and Beichman (1974) and Beichman and Chaisson (1974) used analytic and Monte Carlo treatments of the paramagnetic grain alignment mechanism (referenced in the dark cloud section) to suggest an order-of-magnitude transverse field of about 10 mG. Should the grain temperature be relatively close to the gas temperature, as suggested observationally by many researchers and theoretically by Goldreich and Kwan (1974), the field strength necessary to explain the infrared polarization could be larger. With the temperatures of the grain ($\simeq$70 K) and the gas ($\simeq$20 K) within a factor of 2 or 3 of each other, which is a representative case for a $10^3 - 10^6$ cm^{-3} region, a field no larger than several tens of milligauss would be required, as noted by Chaisson and Beichman (1975). Naturally, these field estimates suffer from the same uncertainties in grain parameters as noted for grain alignment in the above section on dark clouds; the derived estimate for magnetic fields is quite uncertain.

The infrared observations of polarization do seem to imply strongly, nonetheless, that magnetism is present throughout the Orion infrared nebula/molecular cloud. No other mechanism capable of aligning grains over such a large interval of space seems tenable at this time. Nevertheless, the relationship, if any, of this uncertain transverse milligauss field throughout the $10^3 - 10^5$ cm^{-3} infrared nebula and of the more definite longitudinal milligauss field at the OH site of approximately 10^8 cm^{-3} remains uncertain.

The field value of probably 3-5 mG for the Orion OH site is just about that expected if the field is frozen into the gas during collapse of regions that have total densities of about 10^8 cm^{-3} (c.f., Mouschovias, 1976).

Two other experiments have recently addressed the question of magnetism in the Orion region. Claiming that the SO profile has an anomalously large width because of Zeeman Effect, Clark and Johnson (1974) have argued that much of the Orion molecular cloud is embedded in a field of approximately 10 G. However, there are several other molecules, such as HCN (Snyder and Buhl, 1973), SiO (Dickinson, 1972), and H_2S (Thaddeus *et al.*, 1972), that have Landé *g*-factors nearly a thousand times smaller than that for SO and therefore should not be "magnetically broadened"; yet these emit lines having widths similar to SO. Other molecules, such as CN (Penzias *et al.*, 1974) and C_2H (Tucker, Kutner and Thaddeus, 1973), are observed to have relatively narrow widths despite the fact that they ought to be broadened similarly to SO if Gauss-strength fields really exist there. Millimeter-wave data are clearly suggestive of molecular emission from two regions toward Orion – a cool and/or perhaps less turbulent region that emits relatively narrow ($\simeq$4 km sec^{-1}) millimeter-wave lines and a warmer and/or perhaps more turbulent region emitting broader features ($\simeq$30 km sec^{-1}). Some aspects of this criticism are discussed by Zuckerman and Palmer (1975).

More fundamentally, the energy density of a Gauss-strength magnetic field, $B^2/8\pi \simeq 1$ erg cm^{-3}, exceeds the kinetic energy density, $\rho v^2 \simeq 10^{-9}$ erg cm^{-3}, by some nine orders of magnitude. On the other hand, equipartition of energy is almost achieved for the previously discussed milligauss-strength field, especially in the dense, protostar regions. We strongly believe that interstellar magnetic fields anywhere near Gauss strength present formidable difficulties as to their origin, their apparent incompatibility with other energy densities, and their physical maintenance throughout typical interstellar clouds.

Secondly, Troland and Heiles (1977) have recently reported upper limits to the magnetic field between 0.2 and 2 milligauss for several H II regions. These limits, the lowest of which applies to Orion A, are based on their failure to observe a Zeeman splitting of radio recombination lines. Their suggestion – that these limits, like those of the un-ionized dark clouds, are considerably less than those expected if the nebulae were formed via isotropic contraction of interstellar gas with a frozen-in magnetic field – is based upon the assumption that most of the recombination-line emission arises in dense ($\simeq 10^{4.5}$ cm^{-3}) ionized knots. However, this view, discussed by Lockman and Brown (1976), has been challenged by Chaisson and Dopita (1977), who claim that the bulk of the nebular radio recombination-line emission arises in regions of considerably lower ($\simeq 10^{3.3}$ cm^{-3}) density. Magnetic field reversals in the radio beam or magnetic field amplification at a rate slower than $\kappa = 2/3$, as predicted by theory, may also serve to diminish the expected fields in ionized regions, as was suggested earlier for dark clouds. At any rate, it presently remains uncertain whether or not the magnetic field limit for the Orion H II region has any bearing upon what magnetic field may permeate the Orion infrared nebula/molecular cloud. An upper limit of 0.2 mG in the

H II region certainly does not preclude the suggested milligauss fields of the molecular cloud.

Troland and Heiles (1978) have reported positive evidence of the Zeeman Effect in the H 90α recombination line from the Orion ionized nebula. Their derived value, 0.4 ± 0.1 mG, is larger than their previously reported upper limit, probably because of the superior angular resolution of the more recent experiment. The 70 μG field of the foreground H I cloud near Orion (Verschuur, 1970), the 0.4 mG field of the ionized H II region (Troland and Heiles, 1978), and the approximately milligauss field of the nearby molecular cloud (Chaisson and Beichman, 1975) suggest that the entire Orion environment may be magnetically enhanced.

Appenzeller (1974*a*) has noted, from the general interstellar magnetic field structure as determined from polarization observations of surrounding intrinsically unpolarized stars, that the magnetic fields in the vicinity of large H II regions are frequently distorted. Normally, this can be ascribed to the expansion of the H II region due to internal gas pressure and radiation pressure from the ionizing stars and their consequent interaction with an initially uniform magnetic field. However, in the case of the Orion nebula/molecular cloud complex, Appenzeller (1974*b*) shows that the observed surrounding interstellar polarization geometry cannot be fit by this simple model. Instead, it appears that the interstellar gas in Orion is suspended in an interstellar magnetic pocket in much the same manner as found for dark clouds in the previous section. This observation lends further weight to the idea that such magnetic wells are a common structure in our galaxy.

Dyck and Capps (1978) have reported near-infrared polarimetric measurements for 15 compact infrared sources associated with H II regions. As shown in Fig. 6, they find a strong inverse correlation between the linear diameter of the infrared sources and their 2.2 μm polarizations. Additionally, the sources which show the highest polarizations are also those that have absorption bands at the 9.7 μm silicate feature, show little radio continuum emission, and are associated with molecular clouds. These properties are thought to be those of the youngest protostar objects (Werner, Becklin and Neugebauer, 1977). Conversely, those objects with the least amount of polarization are more extended and have H II regions associated with them and are thus thought to be more evolved objects. From these observations it cannot be ascertained whether the polarizations are due to foreground grains aligned by a magnetic field or to polarization mechanisms intrinsic to the sources. Whatever the source of the polarization, it is safe to say that the conditions needed to produce the polarization in such objects are destroyed as the objects evolve.

Acknowledgments. The work of E.J.C. is supported in part by a Research Fellowship of the Alfred P. Sloan Foundation.

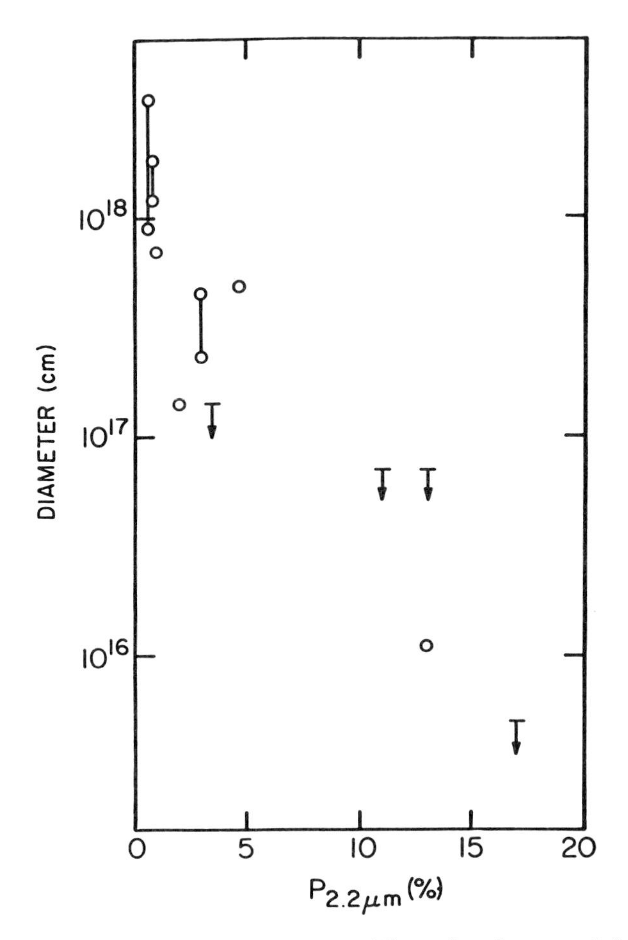

Fig. 6. 2.2 μm polarization versus estimated linear size of compact infrared sources that are associated with H II regions. (From Dyck and Capps, 1978.)

COMMENTS

K. R. Lang: Have you used polarization measurements to check ordering of magnetic fields on large angular scales?

F. J. Vrba: There is no way to get a large angular size to the polarimetric beam because it is possible only to sample the material in the line of sight between the observer and the star observed, no matter how large the aperture used. It might be possible to measure the polarizations of a number of stars closely grouped together. However, suitable clustering is rarely, if ever, seen due to the large foreground absorptions. The amount of time with large telescopes necessary to carry out such mapping would be prohibitive.

M. Werner: With regard to the high polarization seen at 2 μm in the ρ Oph cloud, Elias at the California Institute of Technology has found that some of the infrared sources in this cloud are embedded in infrared reflection nebulosity. Scattering by the nebulosity may be responsible for some of the observed polarization and thereby for the high inferred value of *B.* The possibility of such reflection effects is another indication that optical and infrared polarization measurements of stars in dust clouds must be interpreted carefully.

F. J. Vrba: The reflection nature of these nebulosities has not yet been demonstrated since there is no color information available. Furthermore, even if they are reflection nebulosities their ability to significantly influence the observed large 2.2 μm polarization is unclear due to their relative faintness and uncertain scattering properties at 2.2 μm.

T. Ch. Mouschovias: I would like to make an appeal to observers. Measuring the magnetic field in very dense clouds is indeed useful but not as important, from a theoretician's point of view, as it is to obtain contours of equal magnetic field strength for diffuse clouds. The distribution of magnetic flux during the formation of a cloud determines, to a large extent, the subsequent evolution of the cloud and also what kind of a stellar system could ultimately form.

REFERENCES

Aannestad, P.A., and Purcell, E.M. 1973. *Ann Rev. Astron. Astrophys.* 11: 309.
Appenzeller, I. 1971. *Astron. Astrophys.* 12: 313.
____. 1974*a*. *Estrato del Mem. della Soc. Astronomica Italiana* 45: 61.
____. 1974 *b*. *Astron. Astrophys.* 36: 99.
Axon, D. J., and Ellis, R. S. 1976. *Mon. Not. Roy. Astron. Soc.* 177: 499.
Becklin, E. E., and Neugebauer, G. 1967. *Astrophys. J.* 147: 799.
Beichman, C. A., and Chaisson, E. J. 1974. *Astrophys. J.* 190: L21.
Bok, B. J. 1956. *Astron. J.* 61: 309.
Carrasco, L.; Strom, S.E.; and Strom, K.M. 1973. *Astrophys. J.* 182: 95.
Chaisson, E.J. 1975. *Astrophys. J.* 197: L65.
Chaisson, E.J., and Beichman, C.A. 1975. *Astrophys. J.* 199: L39.
Chaisson, E.J., and Dopita, M.A. 1977. *Astron. Astrophys.* 56: 385.
Clark, F.O.; and Johnson, D.R. 1974. *Astrophys. J.* 191: L87.
Cook, A.H. 1970. *Nature* 211: 503.
Crutcher, R.M.; Evans II, N.J.; Troland, T.; and Heiles, C. 1975. *Astrophys. J.* 198: 91.
Cugnon, P. 1971. *Astron. Astrophys.* 12: 398.
Davies, R. D. 1974. *I.A.U. Symp.* No. 60, p. 275.
Davies, R. D.; de Jager, G.; and Verschuur, G. L. 1966. *Nature* 209: 974.
Davis, L., Jr., and Greenstein, J.L. 1951. *Astrophys. J.* 114: 206.
Dickinson, D.F. 1972. *Astrophys. J.* 175: 143.
Dyck, H.M., and Beichman, C.A. 1974. *Astrophys. J.* 194: 57.
Dyck, H.M., and Capps, R.W. 1978. *Astrophys. J.* 202. In press.
Dyck, H.M.; Capps, R.W.; Forrest, W.J.; and Gillett, F.C. 1973. *Astrophys. J.* 183: L99.
Gehrels, T., ed. 1974. *Planets, Stars and Nebulae Studied with Photopolarimetry.* Tucson, Arizona: The University of Arizona Press.
Goldreich, P.; Keeley, D. A.; and Kwan, J. Y. 1973. *Astrophys. J.* 179: 111.
Goldreich, P., and Kwan, J. 1974. *Astrophys. J.* 189: 441.
Hall, J. S. 1949. *Science* 109: 166.
Hansen, S. S.; Moran, J. M.; Reid, M. J.; Johnston, K. J.; Spencer, J. H.; and Walker, R. C. 1977. *Astrophys. J.* 218: L65.
Heiles, C. 1971. *Ann. Rev. Astron. Astrophys.* 9: 293.

———. 1976. *Ann. Rev. Astron. Astrophys.* 14: 1.
Hiltner, W. A. 1949. *Science* 109: 165.
Jones, R.V., and Spitzer, L., Jr. 1967. *Astrophys. J.* 147: 943.
Litvak, M.M. 1970. *Phys. Rev. A* 2: 2107.
Litvak, M. M.; McWhorter, A. L.; Meeks, M. L.; and Zieger, H. J. 1966. *Phys. Rev. Lett.* 17: 821.
Lo, K. Y.; Walker, R. C.; Burke, B. F.; Moran, J. M.; Johnston, K. J.; and Ewing, M. S. 1975. *Astrophys. J.* 202: 650.
Lockman, F. J., and Brown, R. L. 1976. *Astrophys. J.* 207: 436.
Loer, S. J.; Allen, D. A.; and Dyck, H. M. 1973. *Astrophys. J.* 183: L97.
Manchester, R. M.; Robinson, B. J.; and Goss, W. M. 1970. *Aus. J. Physics* 23: 751.
Markkanen, T. 1977. *Univ. Helsinki Report* 1/1977.
Marraco, H. G., and Forte, J. C. 1978. In preparation.
Moran, J. M.; Burke, B. F.; Barrett, A. H.; Rogers, A. E. E.; Ball, J. A.; Carter, J. C.; and Cudaback, D. D. 1968. *Astrophys. J.* 152: L97.
Mouschovias, T. Ch. 1976. *Astrophys. J.* 207: 141.
Palmer, P., and Zuckerman, B. 1967. *Astrophys. J.* 148: 727.
Penzias, A. A.; Wilson, R. W.; and Jefferts, K. B. 1974. *Phys. Rev. Letters* 32: 701.
Purcell, E. M., and Spitzer, L., Jr. 1971. *Astrophys. J.* 167: 31.
Radford, H. E. 1961. *Physical Rev.* 122: 114.
Rogers, A. E. E., and Barrett, A. H. 1966. *Nature* 210: 188.
Shklovskii, I. S. 1969. *Soviet Astron. – AJ* 13: 1.
Snyder, L. E., and Buhl, D. 1973. *Astrophys. J.* 185: L79.
Strom, S. E.; Grasdalen, G. L.; and Strom, K. M. 1974. *Astrophys. J.* 191: 111.
ter Meulen, J. J., and Dymanus, A. 1972. *Astrophys. J.* 172: L21.
Thaddeus, P.; Kutner, M. L.; Penzias, A. A.; Wilson, R. W.; and Jefferts, K. B. 1972. *Astrophys. J.* 176: L73.
Townes, C. H., and Schawlow, A. L. 1955. *Microwave Spectroscopy.* New York: McGraw-Hill.
Troland, T. H., and Heiles, C. 1977. *Astrophys. J.* 214: 703.
———. 1978. *Bull. Am. Astron. Soc.* 9: 611.
Tucker, K. D.; Kutner, M. L.; and Thaddeus, P. 1973. *Astrophys. J.* 186: L13.
Turner, B. E., and Verschuur, G. L. 1970. *Astrophys. J.* 162: 341.
Verschuur, G. L. 1970. *Astrophys. J.* 161: 867.
———, 1974. *Galactic and Extragalactic Radio Astronomy* (G. L. Verschuur and K. I. Kellermann, eds.) p. 179. New York: Springer Verlag.
Vrba, F. J. 1977. *Astron. J.* 82: 198.
Vrba, F. J.; Coyne, G. V.; Tapia, S. 1978. In preparation.
Vrba, F.J.; Strom, K.M.; Strom, S.E.; and Grasdalen, G.L. 1975. *Astrophys. J.* 197: 77. 77.
Vrba, F. J.; Strom, S. E.; and Strom, K. M. 1976. *Astron. J.* 81: 958.
Weaver, H.; Dieter, N.H.; and Williams, D.R.W. 1968. *Astrophys. J. Suppl.* 16: 219.
Werner, M.W.; Becklin, E.E.; and Neugebauer, G. 1977. *Science* 197: 723.
Wynn-Williams, C.G. and Becklin, E.E. 1974. *Publ. Astron. Soc. Pacific* 86: 5.
Zuckerman, B., and Palmer, P. 1975. *Astrophys. J.* 199: L35.
Zuckerman, B.; Yen, J. L.; Gottlieb, C. A.; and Palmer, P. 1972. *Astrophys. J.* 177: 59.

FORMATION OF STARS AND PLANETARY SYSTEMS IN MAGNETIC INTERSTELLAR CLOUDS

TELEMACHOS CH. MOUSCHOVIAS
University of Illinois at Urbana-Champaign

We view a planetary system as a special case of a binary star system. General dynamical principles can be used to understand the formation of binary stars through collapse and fragmentation of magnetic interstellar clouds. The early, more diffuse stages of interstellar clouds may hold the key to the resolution of fundamental problems of star formation, such as the angular momentum problem and that of providing support against gravity for some of the dense clouds. Cloud formation therefore is an integral part of a theory of star formation.

The role of magnetic fields in the formation, equilibrium and collapse of interstellar clouds is reviewed with emphasis on unsolved problems. The magnetic field is important in resolving the angular momentum problem by the time a cloud's density reaches $10^3 - 10^6$ cm^{-3}*. New results for the time scale of magnetic braking of a cloud's rotation are presented. The generally accepted notion that flattening along field lines reduces* κ *(the exponent in the relation between the field and the gas density* $B \propto \rho^{\kappa}$*) below its isotropic contraction value of 2/3 is shown to be false. The contraction of a self-gravitating magnetic cloud is highly nonhomologous and nonisotropic, and causes* κ *to vary from its value of 1/3 – 1/2 at the core to about 2 – 3 toward the poles;* κ *vanishes toward the equator. The implications of these new results (on support against gravity, rotation of dense clouds, fragmentation, and heating of clouds, for example) are discussed.*

A scenario for star formation in magnetic clouds is presented. A single mechanism can account for the formation of all binary stars, single stars, the sun-Jupiter and similar planetary systems.

From a dynamical point of view, we regard a planetary system as a special case of a binary star system. Both are presumed to form through the collapse of an interstellar cloud because observations show a strong spatial correlation between young stars and dense concentrations of interstellar matter. The per-unit-mass distributions of angular momentum, magnetic flux, thermal and gravitational energies and their mutual nonlinear interactions determine the dynamical evolution of a cloud and thereby the nature of the objects that may form during collapse. Since at least the initial distribution of these quantities in a cloud is determined by the mechanism responsible for cloud formation, it follows that, in order to formulate a complete theory of star formation, cloud formation must first be understood. Whether a cloud will give birth to multiple or binary star systems, to single stars or planetary systems may very well be decided by the residual angular momentum and magnetic flux at the later stages of collapse and by the value of the gas density at which angular momentum and magnetic flux reach their residual values.

Thermal and gravitational effects and, to a lesser degree, rotation have received much attention in calculations aiming at understanding the formation of clouds and the physical processes involved in cloud collapse. It is also generally recognized that the interstellar magnetic field, the energy density of which is comparable with any other interstellar energy density, must play a role in the dynamics of cloud formation, equilibrium and collapse. Yet, mathematical difficulties have prevented the formulation and solution of self-consistent problems relating to star formation that include the effects of magnetic fields in any but the linear regime.

Any rigorous calculation including the magnetic field requires as input the manner in which mass is distributed in the flux tubes in interstellar space or, at least, in clouds. This function largely determines the evolution of a cloud and whether or not it can collapse. It could be obtainable in principle if the mechanism responsible for the generation of the interstellar field were known. That not being the case, we must look to observations for at least a hint; we find none. What is required is a map of a cloud which, in addition to the usual isodensity contours, also exhibits isopedion contours (these are contours of equal magnetic field strength). Until this information becomes available, an indefensible arbitrariness will exist in any theoretical calculation. It is imperative that interferometric 21-cm Zeeman observations be undertaken for relatively massive H I clouds, such as the one in Cassiopeia A. It may be more attractive to search for Zeeman splitting in OH masers, but masers are generally associated with H II regions and current observational evidence therefore suggests that masers follow, rather than precede, star formation. Measuring the magnetic field in such objects can shed light on its role in the dynamics of the interaction of young stars with the surrounding gas, but, unless masers can be shown to be protostars (e.g., that they have appropriate masses, and line profiles characteristic of a collapse velocity field), such measurements can imply little about the process of star formation itself. It is much more useful for understanding the dynamical effects of magnetic fields on clouds and star formation to obtain isopedion contours for at least some relatively diffuse clouds and, if possible, for molecular clouds that have not yet given birth to stars.

In Sec. I we summarize the evidence, observational and theoretical, that points to the importance of magnetic fields in star formation and, in fact, in all large-scale interstellar gas dynamics. This is not intended to be a complete review or to contain the most recent observations, which are reviewed in the chapter of Chaisson and Vrba. Older work, recent advances and the problems remaining unsolved are summarized in Sec. II. In Sec. III new effects of the magnetic field (relating, for example, to fragmentation and the spatial separation of low- and high-mass stars) are described and effects previously proposed are further quantified. We point out where rigorous calculations are still lacking. Finally, Sec. IV summarizes some of our conclusions in the form of a scenario for star formation in magnetic clouds.

There are several excellent reviews of past work on the role of magnetic fields in interstellar gas dynamics and star formation (e.g., Mestel, 1965; Spitzer, 1978, Ch. 6; Parker, 1969; Field 1970; Mestel, 1977). In this chapter we concentrate on more recent work and bring controversial issues to the fore.

I. MAGNETIC FIELDS IN INTERSTELLAR GAS DYNAMICS AND STAR FORMATION

A. Observational Evidence

A magnetic field permeates the interstellar medium. *Synchrotron radiation* accounts for a major fraction of the background radio continuum emission in our galaxy (e.g., see Spitzer, 1968 and references therein; compilation of observations from 10 MHz to 400 MHz by Daniel and Stephens, 1970). The *polarization of starlight* from distant stars (Hall, 1949; Hiltner, 1949) and its correlation with interstellar reddening led to the generally accepted hypothesis that it is produced by elongated dust grains aligned dynamically by a magnetic field (Davis and Greenstein, 1951; Davis, 1958; Miller, 1962). The topology of the field deduced from polarization measurements exhibits an orderly large-scale behavior with field lines forming "arches" over distances of a few hundred parsecs (Mathewson and Ford, 1970; Davis and Berge, 1968; for reviews of interstellar polarization see Gehrels, 1974). Observations at 21 cm (Heiles and Jenkins, 1976) show an intimate association between the interstellar gas and the magnetic field. Large-scale condensations lie in valleys of the field lines, and where the gas rises high above the galactic plane in the form of filamentary arches, the filaments are aligned with the field lines.

Reliable measurements of *Faraday rotation* for hundreds of extragalactic radio sources (Wright, 1973) and tens of pulsars (Manchester, 1974) have yielded a mean strength (weighted by the electron density along the line of sight) of 1 – 4 microgauss for the magnetic field over distances of about 100 pc to a few kiloparsecs. Thus, the energy density in the magnetic field is at least comparable with that in thermal, turbulent and ordered motions, radiation, cosmic rays, and gravitational fields. The length scales over which this comparison is valid are large enough to be relevant to such processes as the formation of interstellar clouds, cloud-cloud collisions, motion of matter through spiral density waves, all of which are thought to be related to star formation. One would, nevertheless, like to establish observationally the relevance of magnetic fields to processes occurring on a much smaller scale (for example, that of an individual interstellar cloud) since it is over such a relatively small scale that the process of star formation may reveal itself most directly.

Zeeman observations on the 21-cm line implied magnetic fields with strengths ranging from a few to about 50 μG in different normal H I clouds (Verschuur, 1971 and references therein). The enhancement of the field over its background (intercloud) value of 1 – 4 μG is, presumably, due to the fact that the field is frozen in the matter (e.g., see Mestel, 1965) because of the very high electrical conductivity of the interstellar gas. It was expected that, if a cloud of density n_c formed by spherical isotropic contraction of intercloud material of density n_{ic}, the enhancement of the magnetic field would be $(B_c/B_{ic}) = (n_c/n_{ic})^{\frac{2}{3}}$ because of conservation of both mass and magnetic flux. This implied that, since cloud densities in the range 30 – 1000

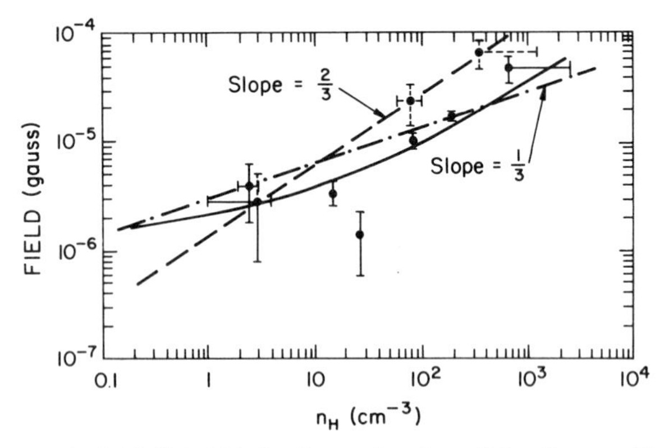

Fig. 1. Magnetic field, B, in H I clouds as a function of density, $n_{\rm H}$. The data are taken from Verschuur (1970). The broken error bars indicate uncertain values. Most densities are only estimates (Clark, 1965). The dashed curve (slope = 2/3) was drawn in by Verschuur. The dash-dot curve has a slope of 1/3 and fits the data at least as well; moreover, it yields a value of $B \approx 2$ microgauss at $n_{\rm H} \approx 0.2$ cm^{-3}, the presumed density of the intercloud medium. The solid curve with increasing slope, as suggested by our calculations, fits the data even better.

cm^{-3} are typical and $n_{ic} \lesssim 0.2$ cm^{-3}, the range 85 – 300 μG should also be typical. As Fig. 1 shows, much weaker fields were actually revealed by Zeeman observations. Was the assumption of flux-freezing incorrect? Mestel (1965) had earlier suggested that the exponent κ in the relation $(B_c/B_{ic}) = (n_c/n_{ic})^\kappa$ may be less than 2/3 due to flattening parallel to the field lines; such flattening increases the gas density without a corresponding increase in the field strength. Unfortunately, no quantitative estimate for the departure of the actual value of κ from its isotropic-contraction value of 2/3 had been obtained for comparison with observations.

The discrepancy between observations and quantitative theoretical predictions was even worse in the case of Zeeman splitting of OH lines in dense clouds. For example, Lo *et al.* (1975) used interferometric observations to measure a splitting of the 1720 MHz line that implied a field of a few milligauss in a maser region of linear extent less than 10^{16} cm. The authors argued that, since $B \propto n^{\frac{2}{3}}$, the density in the observed region must be 10^4 – 10^5 cm^{-3}, which is certainly too small for a maser. Earlier observations (e.g., Beichman and Chaisson, 1974) led to a similar paradox.

B. Theoretical Evidence

The observed parameters of the interstellar medium indicate a high electrical conductivity (e.g., see Spitzer, 1962, 1968) even in dense molecular clouds, down to a degree of ionization $\sim 10^{-9}$. The energy density in the magnetic field is at least comparable with the other energy densities in the interstellar gas and magnetic field, B, in the vertical galactic gravitational field, dynamics of the interstellar medium, from scales of kiloparsecs to protostellar sizes.

The possible role of the magnetic field in the formation of clouds was explored by Parker (1966). He showed through a linear stability analysis of the magnetohydrodynamic equations that the interstellar gas, which is partially supported by the magnetic field and cosmic-ray pressure gradients against the vertical galactic gravitational field, could be unstable with respect to deformations of the field lines. The gas tends to accumulate in "valleys" of the field lines under the action of the galactic gravitational field. Strittmatter (1966) used the virial theorem and Mestel (1966) a spherical isotropic model to obtain critical masses for the gravitational collapse perpendicular to the field lines of cold, uniform magnetic clouds. Field (1965) studied the effect of the magnetic field on thermal instability. These studies established that the magnetic field certainly introduces quantitative changes to the conclusions of previous calculations that ignored it. For example, the field increases the cloud masses required for gravitational collapse, and suppresses the thermal instability in directions perpendicular to field lines. Qualitative new effects, often unpredictable, are introduced when one considers the magnetic field.

An otherwise stable interstellar gas, for example, resting in the vertical galactic gravitational field, can be rendered Rayleigh-Taylor unstable by a frozen-in magnetic field.

II. RECENT THEORETICAL ADVANCES

A. Formation and Equilibria of Diffuse Interstellar Clouds

Parker's (1966) suggestion regarding the formation mechanism of interstellar clouds led us to calculate final equilibrium states for the interstellar gas and magnetic field, **B**, in the vertical galactic gravitational field, **g** (Mouschovias, 1974). These states differ from previously calculated ones (e.g., Lerche, 1967; Parker, 1968*a*) both quantitatively and qualitatively. First, they can be reached from Parker's initial state through continuous deformations of the field lines. Second, they include the effect of thermal pressure. The final equilibrium states represent large-scale ($\gtrsim$100 pc for the observed parameters of the interstellar medium), diffuse condensations of interstellar gas in valleys of the field lines – not "standard" interstellar clouds. The predicted variation of column density of hydrogen in these condensations (see Mouschovias, 1974, Fig. 3) was confirmed by 21 cm observations performed along spiral arms of the galaxy M 81 (Rots, 1974). Unexplained older observations that showed spiral arms of our own galaxy broken up along their length at regular intervals (Westerhout, 1963; Kerr, 1963) found a natural explanation. The magnetic Rayleigh-Taylor instability, which has a wavelength along the field lines of about $2\pi H \approx 1$ kpc corresponding to maximum growth rate (Parker, 1966) could be triggered by a spiral density wave, thus leading to the observed structures. The quantity H is the vertical scale height of the gas, ~ 160 pc. We suggested that this phenomenon is also responsible for the observed alignment of cloud complexes, giant H II regions, and OB stellar associations along spiral arms like "beads on a string" separated by regular intervals of ~ 1 kpc (Mouschovias, Shu and Woodward, 1974).

The development of the magnetic Rayleigh-Taylor instability behind a galactic shock also led to a natural qualitative explanation of the lack of strong enhancement of synchrotron radiation in spiral arms of external galaxies as compared with the radio emission in interarm regions. (As Piddington [1973] and Mathewson *et al.* [1972] pointed out, one-dimensional compression behind a galactic shock predicts too high an enhancement of radio emission in spiral arms.) The "buckling" of field lines in the vertical direction that characterizes the instability increases the volume of each flux tube after the instability enters the nonlinear regime (Mouschovias, 1974). Consequently, the cosmic-ray electrons, which tend to equalize their pressure along field lines rather quickly, "squirt out" of the region of compression (valleys of field lines) into the region of inflation, where the field is weak. That is, the relativistic electrons tend to avoid the regions of strong field, thereby reducing the radio emission below that predicted by one-dimensional compression, without vertical buckling. An estimate of the height above the galactic plane at which the tension of a typical inflating field line stops its expansion led to the suggestion that a fat, quasi-static radio disk of half thickness $\sim$ 1 kpc forms in this manner (Mouschovias, 1975*a*). The concept of a radio halo (e.g., see Shklovsky, 1953*a,b*; Pikelner, 1953; Baldwin, 1954; Biermann and Davis, 1958, 1960; Parker, 1968*b*) or, alternatively, of a radio disk (e.g., see Mathewson *et al.*, 1972) is a very old one. The question, however, of what the precise distribution of synchrotron radiation in and above spiral arms is, has not been answered theoretically yet.

The magnetic Rayleigh-Taylor instability leads only to the formation of the largest and most massive clouds (if the parameters of the interstellar medium are as we know them today). The thermal instability is capable only of forming clouds with masses less than a few tenths of a solar mass and sizes $\lesssim$ 0.1 pc (see Appendix A to this chapter; or Mouschovias, 1975*b*, pp. 5-12). The manner in which interstellar clouds form therefore remains an outstanding problem in theoretical astrophysics. This problem will become even more difficult if it turns out that the "third phase" of the interstellar medium (e.g., see McKee and Ostriker, 1977) occupies a large fraction of interstellar space. Yet, unless cloud formation is understood theoretically, the initial and boundary conditions needed in cloud collapse calculations and in formulating a theory of star formation will be arbitrary indeed.

Suggestions that expanding H II regions or supernovae are responsible for the formation of clouds miss the essential point, namely, that cloud formation must have preceded such events through a different mechanism because the existence of clouds is a prerequisite for star formation and, therefore, for H II regions and supernovae.

Our calculations of final states for the Parker instability left two questions unanswered. *We noted that, if the appropriate finite-amplitude perturbations are available in nature, the individual condensations in a "final" equilibrium state will tend to coalesce along field lines; they thus form another equilibrium state with a horizontal wavelength (i.e., spacing between valleys of the field lines) which is twice its original value.* The argument was based on the fact that we found the state with the longer wavelength to be lower in total energy (Sec. VIc of Mouschovias, 1974).

The first remaining question therefore is the following: Is there a general class of perturbations which would make such a coalescence inevitable? Whatever the nature of these disturbances, coalescence between neighboring large-scale condensations may not proceed to completion because of the very long times required ($\sim 10^8$ yr). The shearing effects of galactic rotation, not to mention supernova explosions, may also limit this process (Shu, 1974; Mouschovias, Shu and Woodward, 1974).

The second outstanding problem concerns the configuration of the large-scale condensations in three dimensions. Parker (1967) showed that the assumed initial stratified state is also unstable with respect to perturbations in the "third direction", i.e., the direction defined by $\mathbf{g} \times \mathbf{B}$. In this direction, wavelengths in the range H/25 – 20 H can grow with virtually identical growth rates. Parker chose to emphasize the shorter of these wavelengths and he suggested that the instability in the third direction produces small-scale hydromagnetic turbulence. We suggested that the extent of a condensation in the third direction is determined naturally by an external agent, the spiral density wave, and it is about 10^2 pc. Moreover, the maximum gas density and maximum field strength in the three-dimensional equilibrium may not be very different from the corresponding values of the two-dimensional states. The reason is that motions in the third direction, which are the only new ones that could enhance the maximum density and field strength of the 2-D equilibrium, tend to be inhibited by the inevitable compression of the magnetic field. These statements, however, can be further quantified only by exact three-dimensional calculations.

B. Nonhomologous Contraction and Equilibria of Self-Gravitating Clouds

The new formulation and method of solution of the problem of finding equilibrium states of the interstellar gas and magnetic field in the galactic gravitational field opened the way to the study of another problem more directly related to star formation. We followed the nonhomologous contraction (through a series of equilibrium configurations) of finite-temperature, self-gravitating, magnetic clouds embedded in a hot and tenuous intercloud medium (Mouschovias, 1976*a,b*). In general, the clouds are highly centrally condensed and flattened along the magnetic field. In most cases, the inner isodensity contours are flatter than the outer ones, which often obtain nearly spherical or cylindrical shapes. A parameter study of the solutions showed that, as the total mass-to-flux ratio, M/ϕ_B, of a cloud increases, the more compressed and the flatter a cloud becomes; beyond some critical value, $(M/\phi_B)_{crit}$, no equilibrium states are possible and collapse ensues. Similarly, a cloud of fixed mass-to-flux ratio gets more compressed and flatter as the intercloud thermal pressure increases relative to the magnetic pressure; for a fixed M/ϕ_B, a critical intercloud pressure exists beyond which no equilibrium is possible. Critical masses for gravitational collapse were thus obtained for magnetic clouds based on self-consistent models for the first time. *Because of flattening and the development of a central condensation, critical masses can be as low as 15% of those predicted by the virial theorem under identical conditions* (Mouschovias and Spitzer, 1976). A typical cloud on the verge of

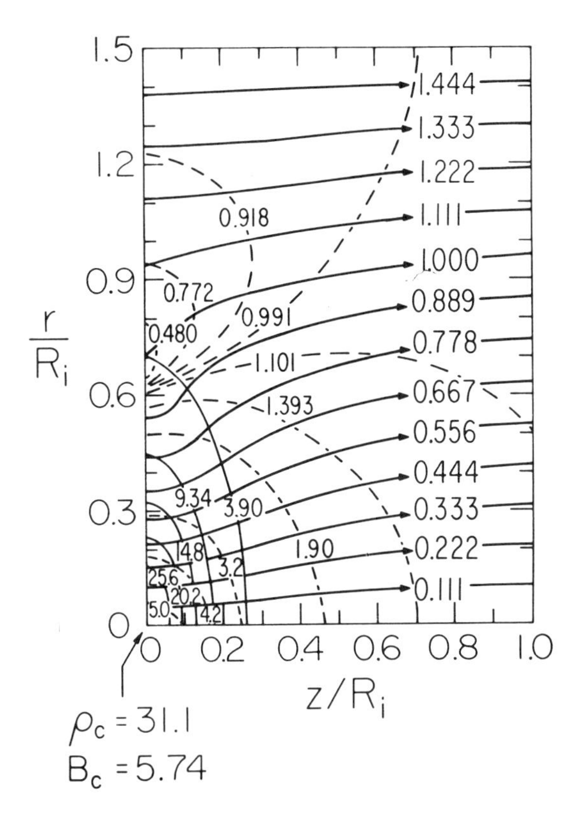

Fig. 2. An equilibrium state of a cloud on the verge of gravitational collapse. There is axial symmetry about the z-axis and reflection symmetry about the equatorial plane, $z = 0$. Both axes are labeled in units of the radius of the cloud, R_i, in a uniform, spherical initial state. The field lines are denoted by solid lines bearing arrows, and each is labeled by its r-coordinate in the initial state. The isodensity contours are the solid, oblate curves; each is labeled in units of the total uniform density of the initial state, n_i. The dashed curves are isopedion (equal magnetic field strength) contours and are labeled in units of the initial, uniform magnetic field strength, B_i.

collapse is shown in Fig. 2. The cloud contracted from an initially spherical, uniform configuration in which the ratio α_i of the magnetic field and gas pressures was taken to be 0.5 in the cloud and only 0.13 in the intercloud medium just to illustrate the point that even a weak field can cause considerable flattening. Initially the ratio of gravitational and magnetic energies was 2.30. For the canonical values $T = 50$ K and $P_{ext} = 1800\ k$ erg cm^{-3} (where k is the Boltzmann constant), the initial state was characterized by $n_i = 9.2$ cm^{-3}, $R_i = 10.9$ pc, $M = 1554\ M_\odot$, and $B_i = 0.9\ \mu$G. Thus, for the final state of Fig. 2 we have that $n_{center} = 285$ cm^{-3}; $n_{surface} = 36$ cm^{-3}; $R_{eq} = 7.7$ pc; $R_p = 2.8$ pc; and $B_{center} = 5.2\ \mu$G. Densities scale as $P_{1800\,k}/T_{50}$; lengths as $T_{50}/\sqrt{P}_{1800k}$; the magnetic field as $\sqrt{P}_{1800k}$; hence $M \propto T_{50}{}^2/\sqrt{P}_{1800\,k}$.

For typical interstellar conditions, critical masses for H I clouds range from 1200 to 2500 $M_\odot$, while for dark clouds they are in the range 15-100$M_\odot$. (The wider interval in the case of dark clouds is the result of observational uncertainties in the input parameters describing the environment of dark clouds and the particular way in which field lines are loaded with mass.) These results imply that, if dark clouds form from normal H I clouds through contraction during which conversion of atomic to molecular hydrogen takes place, magnetic fields are unlikely to prevent the collapse of the most massive dark clouds ($\gtrsim 10^2 M_\odot$); see below, however, for a discussion of the ability of magnetic fields to stop the collapse of the outer layers of a cloud. The possibility that the more massive dark clouds are stabilized by rotation is discussed by Field in this book.

It is emphasized that the above critical masses were calculated on the assumption that the distribution of mass in the flux tubes of a cloud is appropriate to that of an initially uniform, spherical cloud. This need not be so in reality. The assumption was adopted for two reasons: to enable us to compare results quantitatively with previous (not self-consistent) calculations, and to reduce a free function (i.e., the manner in which field lines are loaded with mass) to a free parameter (i.e., the ratio of total mass and total flux). A distribution of flux strongly concentrated in a small cylindrical region threading the cloud core can support significantly larger cloud masses than those given above. Conversely, a magnetic flux mainly threading a cloud's outer layers can allow a relatively low-mass core to collapse as long as it satisfies the Bonner-Ebert condition, $M_{BE} \gtrsim 1.2\, C^4/(G^3 P_0)^{\frac{1}{2}}$; where C is the isothermal speed of sound in the gas and P_0 the surrounding pressure. Clearly, when observations determine the distribution of mass in the flux tubes of clouds, the calculation of critical masses must be repeated in the light of the new observational evidence.

In addition to setting theoretical upper limits on masses of clouds which can exist in stable equilibrium, the states on the verge of collapse can be used as initial conditions for collapse calculations.

Our calculations quantified the relation between the magnetic field and the gas density, and predicted the exponent κ in the relation $B \propto n^\kappa$. We found *at the center of a cloud* that

$$\frac{1}{3} \lesssim \kappa \lesssim \frac{1}{2} \,. \tag{1}$$

Moreover, *the value of* κ *depends on the stage of contraction of the cloud as well as on the relative strength of magnetic and gas pressures in the initial state.* At low densities, $n_{\rm H}$ increases without a corresponding increase in B because self-gravity is not important and gas motions occur preferentially along field lines (see Fig. 1). The scatter of points about the solid curve of Fig. 1 could be caused either by different orientations of the clouds' magnetic fields with respect to the line of sight or by a different relative strength between magnetic and pressure forces in different clouds. The lower limit on κ is set by observations; no theoretical lower limit can exist without a detailed theory of how the interstellar flux is generated. With the above values of κ, the discrepancy between theory and the Zeeman observations described in Sec. I.A was removed both for H I and molecular clouds (Mouschovias, 1977*b*). If one insists on using a constant value of κ despite our results, the relation

$$\frac{B}{B_0} = \left(\frac{n}{n_0}\right)^{\kappa} \tag{2}$$

is a reasonable approximation, with κ given by Eq. (1) in the cloud core, B_0 = 3 μG, and n_0 (the mean density of gas in the interstellar medium) equal to 1 cm^{-3}. Equation (2) can also be applied to a self-gravitating fragment within a cloud, but different values of B_0 and n_0 should then be employed. It follows from Eq. (2) that *milligauss fields can only be found in regions of density* $10^6 - 10^9$ cm^{-3}, *while strict flux-freezing holds.* We discuss the exponent κ at points other than the cloud center in Sec. III.A.2.

The extended envelopes left behind around dense, oblate cores, in conjunction with the lack of equatorial pinching forces, which were previously suggested by the spherical-contraction model of Mestel (1966), led us to propose a new sequence of events leading to star formation (Mouschovias, 1976*b*, 1977*a,b*). In this, the field of the cloud does not detach from that of the intercloud medium; it is unlike Mestel's (1965) scenario. *Analytical arguments based on a slow, magnetically diluted collapse show that an extended envelope will be supported by the tension of the field lines*[a]*; this is contrary to accepted notions that, if gravitational forces induce collapse by exceeding the magnetic forces at some time, the latter will never catch up.* (The calculations of Bodenheimer and Black [see their chapter] confirm our conclusion as well as that regarding the value of κ in the core of a contracting cloud.) A central core will keep contracting until ambipolar diffusion (Mestel and Spitzer, 1956) can separate the field and the ions from the neutral matter.

Magnetic forces act only on the ionized matter. The magnetic field is

[a]Our result refers to a *collapsing* cloud (i.e., one with $M > M_{crit}$) while the work to which Mestel (1977, p. 218) refers concerns only clouds with $M < M_{crit}$. The older work is simply equivalent to the statement that the field will not allow a uniform cloud with $M < M_{crit}$ to collapse; the cloud will only contract to some equilibrium state. Thus the two results differ both qualitatively and quantitatively.

coupled to, or "frozen in," the neutral matter only through frequent collisions between the ionized and neutral components. When the degree of ionization becomes very low, the efficiency of collisional coupling between the two species is reduced and, it is commonly believed, magnetic forces can drive the field lines and the ionized matter through the neutrals, thus decoupling the field from the neutrals. This diffusion is misnamed "ambipolar." Another interpretation of this process is discussed in Sec. III.C.

The observed inefficiency of the star formation process (there are clouds with masses up to 10^6 $M_\odot$, but open stellar clusters seldom exceed 1000 $M_\odot$) *may thus be understood as a purely magnetic phenomenon.* Besides, we showed that neither an O5 star nor a supernova explosion, which have been suggested as responsible for this inefficiency, have enough *momentum* to disperse a dense, massive cloud. Since the magnetic field can be responsible for the inefficiency of star formation through the support of *relatively low-density* envelopes, it also makes it possible later on for a supernova or a group of O5 stars to disperse the weakly bound envelopes and expose the newly formed stars.

A crucial question, however, has not been addressed by these calculations: *Just how slowly does the boundary of a cloud contract during collapse in the presence of a magnetic field?* If its speed exceeds the local Alfvén speed, v_A, the ability of the magnetic field to prevent the envelope from collapsing and to transport angular momentum away from the cloud (see next paragraph) will be impaired. It is important therefore that a dynamical calculation resolve this question. Appendix B to this chapter has a semiquantitative kinematical argument which suggests that the contraction velocity of the boundary will be much less than v_A.

C. Magnetic Braking

There exist three theoretical possibilities (and several variations of them) for resolving the angular momentum problem without invoking magnetic fields.

1. The original angular momentum of the cloud is put in the orbital motions of stars (rather than in their individual spins) about the original axis of rotation of the cloud (e.g., see Bodenheimer, 1978). This, however, has the implication that all stellar clusters must rotate and therefore flatten, which is contrary to observations.

2. Somehow, fragments within a relatively diffuse (say $n_{\rm H} \sim 10$ cm^{-3}) cloud begin to collapse independently, but the cloud as a whole does not contract. This process maintains a small angular velocity of the resulting stellar cluster and therefore there is no significant flattening. A serious difficulty, however, lies in the fact that at a density of 10 cm^{-3} a blob of 1 $M_\odot$ is not self-gravitating. Moreover, the envisioned formation of a star of 1 $M_\odot$ would deplete of matter a spherical volume around it of diameter 1.8 pc. This

would have to be the distance between neighboring stars, and a series of dubious arguments would necessarily follow in order to explain binary and multiple star systems through captures. Yet another difficulty lies in the fact that, if stars formed independently at such low densities, the resulting mean density of young stellar clusters would be several orders of magnitude smaller than that observed.

3. The density of an initially uniform, spherical cloud can increase indefinitely, while angular momentum is conserved, if the cloud attains a nonaxisymmetric configuration (Weber, 1976). If such a cloud of initial radius R_1 and density ρ_1 turns into a bar (of length l, density ρ_2, and circular cross section of radius r, $r \ll l$) without increasing its angular velocity, it follows that $l = (24/5)^{\frac{1}{2}} R_1$, and $\rho_2/\rho_1 = 0.127\, l^2/r^2$ (Mouschovias, 1977*a*). Clearly, the density enhancement can become arbitrarily large. Illstrative as this example may be, the process will take place only if it can satisfy the additional local constraints imposed by the force equation. Although a bar configuration would bypass the objections raised against possibility (2), the same observational objection faced by possibility (1) is still valid. A further difficulty is that the length of the bar remains almost exactly equal to the original diameter of the cloud. Consequently, to attain a moderate density enhancement of only 10^3, a cloud's cross sectional radius would have to be smaller than its length by a factor of 1.1×10^{-2}. Observations do not reveal any such clouds.

It was suggested by Mestel and Spitzer (1956) that a magnetic field permeating a cloud may transport angular momentum from the cloud to the surrounding medium in a characteristic time τ_J roughly equal to the travel time of an Alfvén wave across the cloud radius, R_{CL}. Gillis, Mestel and Paris (1974) studied magnetic braking for a spherical uniform cloud surrounded by a radial magnetic field, while Mestel and Paris (1978) pursued the time evolution of such a system using the tensor virial theorem. Ebert, Hoerner and Temesváry (1960) studied the loss of angular momentum by a spherical cloud of density ρ_{CL}, rotating uniformly about an axis parallel to a uniform magnetic field B. The external medium, of density ρ_{ext}, is initially at rest. We (Mouschovias, 1977*a*) expressed their result in the useful form

$$\tau_J = \frac{\rho_{\mathrm{CL}}}{15\,\rho_{\mathrm{ext}}} \frac{R_{\mathrm{CL}}}{v_{A,\mathrm{ext}}} \tag{3}$$

where $v_{A,\mathrm{ext}}$ is the Alfvén speed in the external medium near the cloud, given by $v_{A,\mathrm{ext}} = B/(4\pi\rho_{\mathrm{ext}})^{\frac{1}{2}}$. [The factor 8/15 in Eq. (3) above was inadvertently written as 8/5 in Eq. 1 of Mouschovias (1977*a*). The conclusion of that paper, that magnetic braking is very efficient, is further strengthened by this correction.] For typical parameters of H I clouds and of the intercloud medium (i.e., $n_{CL} = 20\ cm^{-3}$, $n_{ext} = 0.2\ cm^{-3}$, $R_{CL} = 5$ pc, $B = 3$ μG), we find that $\tau_J \approx 2 \times 10^7$ yr. This being smaller than the period of galactic rotation at the sun's galactocentric distance ($\tau_G \approx 2 \times 10^8$yr), we suggested that *H I clouds are in synchronous orbits about the galactic center.* This conclusion is valid even if the lifetime of an H I cloud is less than τ_G. The reason is that the magnetic energy of a cloud is typically two orders of magnitude larger than its rotational kinetic energy; hence, magnetic braking is very effective. In fact, we may estimate the number of rotations, of a cloud about its own axis, within which a cloud will lose its angular momentum to the surrounding medium. This is τ_J/τ_{rot}, where τ_{rot} is the period of rotation.

For the geometry under consideration, we may write Eq. (3) in the form

$$\tau_J = \frac{8}{15}\left(\frac{\rho_\phi}{\rho_{ext}}\right)^{\frac{1}{2}} \frac{R_{CL}}{v_{A,CL}} \tag{4}$$

where $v_{A,CL}$ is the Alfvén speed *in the cloud.* The rotational energy density is $u_{rot} = (4\pi^2/5)\rho_{CL}R_{CL}{}^2\ \tau_{rot}{}^{-2}$. With the magnetic energy density being $u_m = B^2/8\pi$, Eq. (4) becomes

$$\frac{\tau_J}{\tau_{rot}} = \frac{2}{3\pi}\left(\frac{2}{5}\right)^{\frac{1}{2}}\left(\frac{\rho_{CL}}{\rho_{ext}}\ \frac{u_{rot}}{u_m}\right)^{\frac{1}{2}}. \tag{5}$$

For the standard parameters adopted following Eq. (3), the result $\tau_J/\tau_{rot} \approx 0.1$ is recovered for a cloud in a synchronous galactocentric orbit; hence the claimed efficiency of magnetic braking for H I clouds.

The core, or a fragment, of a molecular cloud of density $n_c = 2 \times 10^4$ cm^{-3} and radius $R_c = 0.5$ pc embedded in a CO envelope of density $n_{env} = 2 \times 10^3\ cm^{-3}$ and radius $R_{env} = 5$ pc (see Zuckerman and Palmer, 1974) will lose its angular momentum to the envelope, in accordance with Eq. (3) and $B \propto \rho^{\frac{1}{2}}$, in only a time $\tau_J \approx 6.7 \times 10^4$ yr. If the core can be better approximated by an oblate spheroid of major and minor radii R_{eq} and R_p, respectively, Eq. (6) is then applicable instead of Eq. (3), and τ_J is further reduced by the factor R_p/R_{eq}. Since the free-fall time, which is meaningful only in the absence of pressure and magnetic fields, is $\tau_{ff} = (3\pi/32G\rho)^{\frac{1}{2}} = 3.1 \times 10^7\ n_{H_2}{}^{-\frac{1}{2}}$ yr (for $n_{He}/2n_{H_2} = 0.1$), it follows that $\tau_{ff} = 2.2 \times 10^5$ yr for a molecular-hydrogen density $n_{H_2} = 2 \times 10^4\ cm^{-3}$. Thus, magnetic braking has plenty of time to act if only the field remains frozen in the matter. We shall see below that the ambipolar diffusion time is also larger than τ_J, so that the magnetic field can remain frozen in the matter long enough to resolve the angular momentum problem for such cores. The remaining question concerns the efficiency with which the extended envelopes themselves and the relatively compact dark clouds can transfer their angular momenta to the external medium.

To account for flattening, which has been demonstrated by detailed calculations of self-gravitating magnetic clouds (Mouschovias, 1976*a,b*), we modify Eq. (3) to apply to a uniform oblate spheroid of equatorial and polar radii R_{eq} and R_p, respectively, and with its minor axis along the field and angular momentum vectors. One can show (Mouschovias, 1978) that

$$\tau_J = \frac{8}{15} \frac{\rho_{CL}}{\rho_{ext}} \frac{R_p}{v_{A,ext}} \tag{6a}$$

$$= \frac{8}{15} \left(\frac{\rho_{CL}}{\rho_{ext}}\right)^{\frac{1}{2}} \frac{R_p}{v_{A,CL}} . \tag{6b}$$

The analogue of Eq. (5) is

$$\frac{\tau_J}{\tau_{rot}} = \frac{4\sqrt{2}}{15\pi} \frac{R_p}{R_{eq}} \left(\frac{\rho_{CL}}{\rho_{ext}} \frac{u_{rot}}{u_m}\right)^{\frac{1}{2}} . \tag{7}$$

Equations (6*a*), (6*b*) and (7) apply equally well to dark clouds and to the cores and envelopes of molecular clouds. Although a uniform density was assumed, these equations are useful because they allow us to compare τ_J with the ambipolar diffusion time τ_B, which is obtained (e.g., see Spitzer, 1968, p. 239) under the same assumption. In any case, one should bear in mind the approximate nature of both τ_J and τ_B.

Equation (6*b*) is best suited for our purposes. We scale the field with the gas density as in Eq. (2) with $\kappa = 1/2$. Conservation of both mass and flux imply a unique relation between the equatorial and polar radii of the uniform spheroid; namely, $R_p/R_{eq}{}^2$ = constant. This is consistent with the rapid establishment and preservation of near hydrostatic equilibrium along the field lines as the cloud contracts in the lateral direction. In fact it is known (Mestel, 1965) that a uniform oblate spheroid, which has achieved hydrostatic equilibrium along its axis of symmetry, contracts with an ever increasing oblateness. We showed that Eq. (6*b*) obtains a remarkably simple form under these conditions (Mouschovias, 1978), namely

$$\tau_J = \frac{8}{15} \left(\frac{2}{G}\right)^{\frac{1}{2}} \frac{C}{B_i} \tag{8a}$$

$$= 8.24 \times 10^5 \frac{(T/20°\mathrm{K})^{\frac{1}{2}}}{B_i/3 \text{ microgauss}} \text{ yr} \qquad (8b)$$

The quantity $C \equiv (kT/\mu m_H)^{\frac{1}{2}}$ is the isothermal speed of sound in the cloud; T is the cloud temperature in degrees K; m_H is the mass of a hydrogen atom; μ is the mean mass per particle in units of m_H ($\mu = 2.33$ for molecular clouds); and k is the Boltzmann constant. The cloud was assumed to have formed out of the external medium of density ρ_{ext} at an initial magnetic field strength B_i.

Since Eq. (8) is valid only when self-gravity is important, it should not be applied to "standard" H I clouds, in which thermal energy exceeds the gravitational potential energy by about two orders of magnitude. The second validity condition is that the magnetic field be frozen in the matter.

It follows from Eq. (8*b*) that a dark cloud ($T \approx 20$ K) will lose its angular momentum in a characteristic time 8.2×10^5 yr, while a CO molecular envelope ($T \approx 50$ K) will need 1.3×10^6 yr.[a] Although in molecular envelopes (typical density $\approx 2 \times 10^3$ cm^{-3}) ambipolar diffusion cannot decouple the field from the matter in such a short time, in the densest of dark clouds (whose densities may reach 10^5 cm^{-3}) decoupling is likely so that, for these objects, Eq. (8*b*) gives only a lower limit for τ_J. Angular velocities exceeding that of local galactic rotation are expected for the densest clouds. The critical density above which magnetic braking becomes inefficient is determined by the requirement that τ_J be equal to the ambipolar-diffusion time, and is roughly given by

$$n_{crit} \approx 2.4 \times 10^3 \text{ cm}^{-3} \qquad (9)$$

for dark clouds and cores of molecular clouds; for molecular envelopes n_{crit} is considerably larger (see Sec.II.D below).

For the oblate cloud under consideration we find at all stages of contraction

$$\frac{R_p}{R_{eq}} = 0.1 \left(\frac{R_{eq}}{1 \text{ pc}}\right)\left(\frac{T}{20°\mathrm{K}}\right)\left(\frac{10^2 M_\odot}{M}\right). \qquad (10)$$

---

[a] In a strict sense, in some parts of a molecular envelope $\kappa > 1/2$ while in others $\kappa < 1/2$ (see Sec. III.A.2 below). We take $\kappa = 1/2$ as a representative "mean" value for the present discussion.

This ratio increases by a factor of about 3 when the condition of uniformity is relaxed and the density is calculated in a self-consistent manner from exact hydrostatic balance along field lines (Mouschovias, 1978). Thus dark clouds are expected to be characterized by a ratio $R_p/R_{eq} \approx 0.3$ due to the presence of magnetic fields alone. Observations indicate that this is typical for dark clouds (see chapter by Field in this book). Molecular envelopes are expected to have a ratio R_p/R_{eq} closer to 0.1 if thermal pressure provides the only force against gravity along field lines.

If indeed the magnetic field is *crucial* in resolving the angular momentum problem at an early stage during the contraction of a cloud, the conclusion would follow that *no stars can form in the absence of magnetic fields.* Our present state of knowledge does not allow us to draw such a bold conclusion. Given, however, that magnetic fields are observed virtually everywhere in the interstellar medium, the seemingly crucial role of magnetic fields in star formation may eventually become an established fact. One still wonders, nevertheless, whether magnetic fields were also important when globular clusters were formed.

Recent observational papers have reported the detection of systematic velocity gradients (the signature of rotation) in some dense clouds (see Field's chapter). What is notable is not that these clouds rotate, but that they rotate so slowly compared to expectations based on conservation of angular momentum for clouds that formed out of a medium of mean density ~ 1 cm^{-3}. In other words, by the time densities representative of molecular clouds have been reached during the contraction process, the angular momentum problem is almost completely resolved. Thus, observations reinforce the need, emphasized earlier on theoretical grounds, to understand, within the context of a theory of star formation, the cloud formation mechanism and the process of cloud contraction during the early, diffuse stages.

In case a cloud is seen to rotate in a fashion consistent with conservation of angular momentum, it would mean that either it formed in a region of unusually low field or that the field decoupled from the matter at a relatively low density, by means yet unknown (Mouschovias, 1977*a*).

D. Loss of Magnetic Flux: A Single Mechanism for the Formation of All Binary Stars

If the magnetic field remained frozen in the matter throughout collapse, a typical star would have a field on the order of 10^8 G, contrary to observations. Mestel and Spitzer (1956) suggested that ambipolar diffusion decouples the field from the neutral matter at some stage during collapse. We have shown that the envelope of a cloud, which has exceeded the critical mass-to-flux ratio for gravitational collapse, will be prevented from collapsing by the tension of the field lines; ambipolar diffusion will set in only at a cloud's dense core (Mouschovias, 1976*b*). From the moment at which decoupling of the field from the matter has taken place, angular momentum is nearly conserved. (We say "nearly conserved" because the field cannot decrease below its background value in the cloud's envelope, and so magnetic braking can still go on but at a reduced efficiency.) *There is, therefore, a one-to-one correspondence between the density at which the field decouples from the matter and the angular momentum left in a collapsing blob.* This residual angular momentum is available to a binary star system that may form. The observed range of periods (10 hr – 100 yr), and therefore angular momenta, of binary stars can be accounted for if the field decouples from the matter at densities in the range $7.5 \times 10^3 - 2.2 \times 10^6$ cm^{-3} (Mouschovias, 1977*a*). Infrared sources embedded in CO envelopes are indeed characterized by such densities (e.g., see Zuckerman and Palmer, 1974). Ambipolar diffusion can decouple the field from the matter in a time

$$\tau_B = 2.0 \times 10^6 \frac{n_i/10^{-2}\ \mathrm{cm}^{-3}}{(n_c/2 \times 10^4\ \mathrm{cm}^{-3})^{\frac{2}{3}}} \left(\frac{M_c}{10^2 M_\odot}\right)^{\frac{2}{3}} \mathrm{yr}, \tag{11}$$

where the quantities n_c and M_c are the neutral density and mass of the core, respectively, and n_i is the ion density in the core. Equation (11) can also apply to dark clouds. Once again we took $\kappa = ½$ in Eq. (2). For the above range of densities, the field can decouple from the neutral matter in at most $9 \times 10^4 - 4 \times 10^6$ yr – a significantly short time. In fact, this time may be an overestimate by as much as one order of magnitude because Oppenheimer and Dalgarno (1974) give $n_i \simeq 10^{-3}$ cm^{-3} as a more realistic, but still uncertain value for the above conditions. On the other hand, if a dense massive cloud fragments before the field decouples from the matter, ionizing high-energy cosmic rays may penetrate the individual fragments and the core more easily. This effect would maintain a relatively high ion density, and the time scale for ambipolar diffusion could become proportionally longer.

Whatever the case, the above estimate of the characteristic time for ambipolar diffusion, as in the derivation of Spitzer (1968), neglects the internal structure of the cloud. It assumes that the field strength varies over a characteristic length equal to the core radius. It also considers the core of a cloud as characterized by a single density. It is clear that the detailed structure of the cloud can significantly affect the characteristic time for ambipolar diffusion. *Given that ambipolar diffusion is, at least so far, the most likely phenomenon responsible for the reduction of magnetic flux in a cloud, the detailed manner in which it takes place must be understood within the context of a theory of star formation.*

Since magnetic braking becomes inefficient when $\tau_J \approx \tau_B$, we can use Eqs. (8*b*) and (11) to obtain the "critical" density n_{crit} at which this condition is satisfied and the cloud's rotation begins to speed up. We find that

$$n_{\text{crit}} = 2.4 \times 10^3 \left(\frac{M_c}{10^2\, M_\odot}\right)\left[\frac{n_i/10^{-3}\ \text{cm}^{-3}}{(T/20\ ^\circ\text{K})^{\frac{1}{2}}}\ (B_i/3\mu\text{G})\right]^{\frac{3}{2}} \text{cm}^{-3} \qquad (12)$$

and for a typical dark cloud $n_{\text{crit}} = 2.4 \times 10^3\ \text{cm}^{-3}$. The same value of n_{crit} is obtained for a typical core of a molecular cloud ($T = 50$ K, $M = 200\,M_\odot$). For a molecular envelope of mass $2 \times 10^4\ M_\odot$, $n_{\text{crit}} = 2.4 \times 10^5\ \text{cm}^{-3}$. This larger value corresponds to the fact that ambipolar diffusion needs a longer time to operate over the larger length scales involved and, to reduce τ_B to a value equal to τ_J (which does not depend on length), the density has to be considerably larger. The critical density calculated above is *not* the density at which the field will necessarily decouple from the matter within a relatively short time; it is only the density at which the time scales τ_J and τ_B are equal *irrespective of their numerical values.*

The new scenario for star formation also predicted a single maximum in the period distribution of binary (and multiple) star systems (Mouschovias, 1977*a*). Older observations, nevertheless, showed two maxima in the number of binaries corresponding to short-period spectroscopic ones and wide visual pairs. It turned out that the "double hump" was the result of observational selection, and the recent observations by Abt and Levy (1976; see Abt's chapter in this book) on the multiplicity among solar-type stars showed a single maximum in the period distribution for the 88 available systems.

III. NEW RESULTS ON THE MAGNETIC FIELD, AND THEIR RELEVANCE TO STAR FORMATION

Numerical calculations of cloud collapse in the absence of magnetic fields have been made by, among others, Larson (1969*a,b*, 1972*a,b*), Ferraioli and Virgopia (1975), Kondo (1975), Appenzeller and Tscharnuter (1974, 1975), Tscharnuter (1975), Fricke *et al.* (1976), Black and Bodenheimer (1976), Nakazawa *et al.* (1976), Westbrook and Tarter (1975), Yorke and Krügel (1977), Black and Wilson (personal communications 1976), Shu (1977); also see the chapter of Bodenheimer and Black in this book. These calculations include spherical and axisymmetric collapse, with or without rotation, and even preliminary results from three-dimensional models. Since they invariably begin with a blob of high enough density and, when rotation is included, of low enough angular velocity, they do not address the problem of angular momentum (and magnetic flux) as described in the previous section. They have nevertheless shed light on some aspects of the dynamics as well as the thermodynamics of collapse. Important questions remain unanswered within the framework of non-magnetic calculations (e.g., just how much energy is radiated away during protostar formation).

Some issues on which non-magnetic calculations disagree among themselves may become moot in the presence of magnetic fields. For example, our discussion in Sec. II indicates that the importance of rotation and the consequent formation of rings in a collapsing blob might have been overemphasized by non-magnetic calculations. In addition, ring formation would tend to be inhibited by magnetic forces since it would involve compressing the field. Whether local enhancements in density would be large enough for gravitational forces to overwhelm magnetic forces is not clear *a priori.* This question is tied to the important question of whether flattening can produce fragmentation, as originally suggested by Mestel (1965). It is also possible that the field, which is tied to the external medium, may slow down the contraction of the envelope sufficiently to prevent the formation of an accretion shock.

Since the relation between the field strength and the gas density during the contraction of a cloud has important bearing on fragmentation, we begin with a closer examination of its meaning and origin.

A. The Slope of log B versus log ρ

1. General Considerations. In presenting values for the quantity κ ($\equiv d \log B / d \log \rho$) in Sec. II.B the discussion was restricted to the center of a cloud. (The values of B and ρ entering the definition of κ must refer to one and the same fluid element during its motion as the cloud contracts.) We now consider a conducting fluid element of density $\delta\rho$, of cross-sectional area δA normal to the local magnetic field, **B**, and of length δs along the field, at any arbitrary position in the cloud. Since the mass and flux of the fluid element are independently conserved during its motion, it follows that $\mathrm{d}(\rho\, \delta s / B)/\mathrm{d}t = 0$, so that

$$\frac{B}{B_0} = \frac{\rho}{\rho_0} \frac{\delta s}{\delta s_0} . \tag{13}$$

The subscripted quantities refer to some time t_0 in the past history of the same fluid element. *In general, therefore, the relation between B and ρ depends not only on the particular fluid element under consideration – the mass-to-flux ratio can be very different for different fluid elements – but also on the amount of stretching or compression suffered by a fluid element along field lines.* One can show that, since for uniform contraction normal to the field and for spherical isotropic contraction δs varies as ρ^0 and $\rho^{-\frac{1}{3}}$, respectively, Eq. (13) reduces to familiar expressions.

A general implication regarding the value of κ follows from Eq. (13). When gravitational forces dominate, the variation of δs is determined by the local interaction of gravity with magnetic and pressure forces. At the diffuse stages, however, in which gravity is not important and contraction is initiated by other means (see Sec. IV.A below), the variation of δs will be largely determined by the *velocity field* imposed externally on the cloud. We see that, until memory of the initiation of the collapse process is lost through takeover by gravity, the exponent κ will be determined by causes external to the cloud and may be very different for identical clouds placed in different environments.

2. *The Exponent κ as a Function of Position within a Cloud.* So far we have been concerned with the exponent κ as a function of time for one and the same fluid element as the contraction evolves. Nevertheless, there is an additional question concerning κ which bears directly on star formation. It concerns the variation of κ with position within the cloud. It is conceivable that $\kappa \ll 2/3$ in some region but $\kappa \gg 2/3$ in another so that fragmentation into spherical blobs will be facilitated in the former region but expressly forbidden in the latter. We studied the dependence of κ on position for sequences of equilibrium configurations. The detailed results depend on the physical parameters of a cloud – there are three free parameters (see Mouschovias 1976*a*, p. 763 for an intuitive explanation). Yet, some general statements can be made.

We recall first that at the center of a cloud κ is, in most cases, in the approximate range 1/3 – 1/2 for realistic sets of physical parameters. It becomes equal to 2/3 if the magnetic pressure is initially small compared to the thermal pressure (a detailed discussion is given by Mouschovias, 1976*b*). In general, as we move from the center to the pole of a cloud along the axis of symmetry, κ *increases.* In most cases, it reaches values between 2/3 and 2 at a distance $z = 0.8\ R_p$, where R_p is the polar radius of the cloud. The opposite behavior is observed as we move away from the center toward the equator in the equatorial plane: κ *decreases* slowly with distance from the axis of symmetry. It vanishes at, very roughly, $r \approx 0.8\ R_{\text{eq}}$, where R_{eq} is the equatorial radius.

The results just described may seem to violate one's intuition, which dictates that, along the axis of symmetry of the cloud, κ should be smaller than 2/3 because of flattening along field lines. Such flattening, it is argued, will decrease κ because it increases the density without a corresponding increase in the field strength. This is a popular misconception based on the implicit assumption that the contraction is uniform, i.e., that the density is constant throughout. In fact, *the contraction of a self-gravitating, magnetic cloud is highly nonhomologous and nonisotropic* (Mouschovias 1976*a,b*)*; a central concentration develops with a density often between one and two orders of magnitude larger than at the cloud surface even while the cloud is still in equilibrium.* Nonhomologous contraction along field lines decreases the density, say at $z = 0.8\ R_p$, well below the value it would have during uniform contraction. (Because of the general contraction of the cloud, however, the density may still be larger than the value it had in the uniform initial state.) The magnetic field at the same point, on the other hand, increases sometimes more rapidly than the density because it is tied to the dense central core and to the external medium. In other words, a fluid element on the axis of symmetry, say at $z = 0.8\ R_p$, is stretched along the field lines because of the non-homologous contraction; at the same time, however, the gravity of the dense core induces a compression of this fluid element normal to the field lines. The net result is that the density, which varies as $\delta A^{-1}\ \delta s^{-1}$, increases slightly, if at all, while the magnetic field, which varies as δA^{-1}, increases more rapidly. We see once again that arguments based on mean values of physical quantities or on the virial theorem and which ignore the internal gradients of the cloud can be very misleading. *Flattening along field lines does not by itself imply a reduction in the value of κ below its isotropic-contraction value of 2/3.*

The decrease of κ in the equatorial plane with increasing distance from the axis of symmetry is more easily understood intuitively. The gas simply moves into the "valleys" of the field lines under the action of the cloud's gravitational field (see Fig. 2).

3. Implications for Fragmentation. The implication of these results regarding fragmentation is evident. If conditions in a self-gravitating, magnetic cloud permit *fragmentation* at all, it *should proceed most easily in the equatorial plane,* where κ has its smallest values. The observable consequence is that local enhancements of molecular emission lines should be seen in the equatorial planes of dense clouds in which fragmentation has taken place. Observations by Morris *et al.* (1974) and by Clark *et al.* (1977) show such evidence. Dense clouds that have not yet given birth to OB stars would give the most unambiguous evidence for the relative location of fragments; if OB stars are turned on, their H II regions may alter the geometrical appearance of a cloud significantly to mask the location of the equatorial plane.

The extreme nonhomology introduced by the magnetic field also implies that low-mass stars should preferentially form first, and perhaps only, in the core of a dense cloud. This is so for the following reason. *Provided that the surrounding pressure is large enough,* a spherical fragment can separate out and collapse on its own if its mass exceeds

$$M_{\mathrm{crit}} = 5.04 \times 10^5\ \frac{(B/3\ \mu\mathrm{G})^3}{(n_0/1\ \mathrm{cm}^{-3})^2}\ M_\odot. \tag{14}$$

The quantity n_0 is the density of *protons* in the cloud ($n_0 = 2n_{H_2}$). The numerical constant in Eq. (14) is taken from exact calculations of critical masses (Mouschovias and Spitzer, 1976). Using Eq. (2) we write Eq. (14) in terms of the density or the field alone as

$$M_{crit} = 5.04 \times 10^5 \, n_0^{\,-(2-3\kappa)} \, M_\odot \tag{15a}$$

$$= 5.04 \times 10^5 \left(\frac{B}{3\,\mu G}\right)^{(3\kappa-2)/\kappa} M_\odot . \tag{15b}$$

Thus a blob of 1 $M_\odot$ can separate out at molecular cloud densities in the range $2.5 \times 10^5 - 1.3 \times 10^{11}$ cm^{-3} for κ in the respective range $1/3 - 1/2$. The smaller the value of κ the smaller the density at which fragmentation into blobs of solar mass can be achieved. (The Jeans mass for a non-magnetic cloud varies only as $\rho^{-\frac{1}{2}}$.) Although the lower values of κ found in the equatorial plane away from the cloud center would imply that low-mass stars can form there even more easily, another effect comes into play that may invalidate such a conclusion. The density in the envelope is lower than that in the core because of support of the envelope by the tension of the field lines. Additional self-consistent calculations including the magnetic field are clearly needed in order to resolve questions concerning fragmentation. Nevertheless the point can be made here that *solar-mass blobs can separate out at densities comparable with those in molecular clouds.* This is in agreement with observations, which show that the mean density of open clusters is on the order of that of molecular clouds. Also, a spatial separation between low- and high-mass stars seems to have been observed (W. Herbst, 1977).

Incidentally, Eq. (14) shows that the field required to stabilize a cloud of $M = 10^5\ M_\odot$ and $n_0 = 10^4$ cm^{-3} is equal to 0.81 milligauss.

B. Breakdown of the Force-Free Approximation for the Intercloud Medium

Although our formulation of the problem of the equilibrium of a self-gravitating, magnetic cloud was very general (Mouschovias, 1976*a*), we produced solutions only in the approximation that the intercloud medium is force-free. For the clouds considered, the scale height of the intercloud medium ($T \sim 10^4$ K) in the gravitational field of the cloud is about 10^2 times larger than the size of a typical cloud ($\simeq$ 5 pc). The approximation, therefore, that the intercloud pressure (P_{ext}) is constant along field lines, is excellent. This, coupled with uniformity at "infinity" (i.e., at a large distance from the cloud), would imply that P_{ext} = constant everywhere.

Since pressure is continuous across a cloud surface, the density just inside the cloud surface will remain constant (and equal to P_{ext}/C^2, where C is the isothermal speed of sound in the cloud) as long as the cloud temperature remains constant. The magnetic field strength, however, is not constrained to any particular value at the cloud surface. Even a small change in B will result in κ becoming $+\infty$ or $-\infty$ depending on whether B has increased or decreased, respectively. Our calculations showed that this actually happens at the cloud surface, but there is no cause for alarm because κ is not a physical quantity.

In the case of massive ($\sim 10^5\ M_\odot$), dense ($\sim 10^5\ \mathrm{cm}^{-3}$) clouds, the force-free approximation for the intercloud medium breaks down. The scale height of the intercloud medium ($T \sim 10^4$ K) in the gravitational field of the cloud is somewhat smaller than 1 pc, which is less than the size of the cloud. The external pressure at the cloud surface will consequently increase upon contraction, thus causing an increase in the density just inside the cloud surface which will eliminate the singularity in the values of κ.

An important consequence of the relaxation of the force-free approximation for the intercloud medium is that the magnetic field at the equator of the cloud may be more easily compressed, in which case its ability to prevent the collapse of the envelope may be impaired. This possibility should be further investigated.

C. Conversion of Gravitational Energy Into Heat Through Magnetic Fields: A Conceptual Re-examination of Ambipolar Diffusion

As a cloud contracts it compresses its frozen-in magnetic field, thus increasing the energy stored in the field. For a slow contraction, so that kinetic energy in ordered motions can be ignored, one can estimate from the virial theorem that the energy per gram stored in the magnetic field is on the order of the gravitational potential (with G as the gravitational constant), i.e.,

$$v_A^2 \sim \frac{GM}{R} \tag{16}$$

where v_A is the Alfvén velocity, M is the mass of the cloud, and R its radius. For a cloud of $M \sim 10^4\ M_\odot$ and a density $n \sim 10^4\ \mathrm{cm}^{-3}$, i.e., $R \approx 1.6$ pc, Eq. (16) gives an energy density $\sim 5 \times 10^{-9}$ erg cm^{-3}, which is two orders of magnitude larger than the thermal energy density at $T \approx 50$ K. If the gravitational energy which has been stored in the compressed magnetic field could be released somehow, it could represent an enormous source of heat for at least the dense clouds.

Fowler and Hoyle (1963) suggested that ambipolar diffusion serves to convert magnetic into thermal energy at high densities, when the ambipolar diffusion time becomes relatively short. Scalo (1977) used Spitzer's (1968) estimates for the ambipolar diffusion time to put an upper limit on the field strength in dense clouds by requiring that the heat released does not exceed that required to balance the heat lost from the cloud at the observed temperature. His reasoning is based on the assumption that ambipolar diffusion determines the exponent κ. We have seen, however, that long before the density becomes high enough for the process to set in, the dynamics of contraction (and in particular the inherent anisotropy of the magnetic force that vanishes in a direction parallel to the field) causes flattening in the cloud and a reduction in κ at least in the cloud core.

It is meaningless to write down the relation $B \propto \rho^{\kappa}$ if the magnetic field is not frozen in the matter, for Eq. (13) breaks down. To claim, as Scalo does, that once ambipolar diffusion sets in (i.e., once flux-freezing breaks down) it determines the exponent κ, is therefore incorrect. In fact, as a principally neutral fluid element diffuses toward the cloud center because of gravity, it finds itself in a region of stronger field, and it is stretched by differential gravitational forces. Altogether, for one and the same fluid element, κ can be larger than unity even though ambipolar diffusion is in progress and the local field has little effect on the dynamical evolution of the fluid element. As we have said above, κ is meaningful only while flux-freezing holds.

Scalo also assumed that the entire cloud ($M \sim 10^4\ M_\odot$) is characterized by a single value of the density and magnetic field as well as that κ is independent of position in the cloud, in contrast to our discussion in Sec. III.A above. It is a misconception to think *a priori* that magnetic forces drive the ions through the neutrals and that the field is expelled from a cloud, thereby releasing magnetic energy, during ambipolar diffusion. It is gravitational forces that drive the neutrals through the ions and the field. The structure and strength of the magnetic field may remain unaffected; thus the magnetic energy content of a cloud may suffer little change, if any, due to ambipolar diffusion. It is the gravitational energy released by contraction which gets converted into heat through neutral-ion collisions while the ions are "held in place" by the magnetic field, at least in the quasi-steady process envisioned by Spitzer (1968, pp. 239 – 240) in which magnetic forces nearly balance gravitational forces. In a cloud's envelope the degree of ionization remains high enough for ambipolar diffusion to be insignificant. In the core, the drift of the neutrals toward the cloud center has the effect of increasing the mass-to-flux ratio in the flux tubes threading the core, so that collapse becomes easier. In other words, *ambipolar diffusion achieves a redistribution of mass in the central flux tubes of a cloud, but it does not necessarily alter the total magnetic-energy content of the cloud.* The distinction, therefore, between a drift of the ions through the neutrals on the one hand and the neutrals through the ions on the other, is not merely a semantic one. It has additional implications for star formation.

If, driven by magnetic forces, the ions drift through the neutrals, thus reducing the field strength to its background value in a region about to give birth to stars, one should expect only very weak fields in H II regions, in which the background field gets refrozen in the matter. On the contrary, *if the neutrals diffuse through the ions, H II regions should be characterized by field strengths corresponding to the densities at which ambipolar diffusion became effective.* These fields are in the range of 1/4 – 4 mG, depending on the locally available ionizing radiation and, therefore, on the location of a collapsing fragment within a cloud. Observations support the latter interpretation of ambipolar diffusion. A field of 0.4 ± 0.1 mG has been detected very recently in the Orion A H II region (Heiles, personal communication, 1978).

We now estimate the heating rate of the cores of molecular clouds by ambipolar diffusion. Gravity acting on the neutrals provides the driving force (per unit volume) and causes a relative drift velocity between neutrals and ions, v_D. In the quasi-steady state suggested by Spitzer (1968), this force is transferred to the ions, of density n_i, which are held in place by the field, through collisions with the neutrals. The neutrals are assumed to be H_2 molecules of density n_{H_2}. If we balance the gravitational force per unit volume on the neutral particles with the opposing drag (per unit volume), exerted by the ionized particles on the neutrals, we find

$$2\pi G(1.4\, n_{H_2}\, m_{H_2})^2 r = n_{H_2}\, n_i \langle \sigma u_{H_2} \rangle m_{H_2}\, v_D. \tag{17}$$

In Eq. (17) we adopted Spitzer's "infinite, uniform cylinder approximation" and we have assumed that the ions are much heavier than the neutrals. The quantities σ and u_{H_2} are the collision cross section between H_2 molecules and ions, and the random speed of H_2 molecules relative to ions, respectively; the distance from the symmetry axis is denoted by r. A ten percent He abundance has been accounted for. The angular bracket denotes an average over all speeds. Equation (17) is valid as long as $v_D \ll u_{H_2}$, a condition that is usually fulfilled.

The neutrals "see" the ions drifting through them with a velocity $-v_D$. The rate of energy transfer to the neutrals is simply the product of the rate of collisions ($n_{H_2} \langle \sigma u_{H_2} \rangle$) and the energy transfer per collision (approximately $\frac{1}{2} m_{H_2} v_D^2$). Then the heating rate per unit volume is given by

$$\mathcal{H} = \frac{7.68\, \pi^2\, G^2\, m_{H_2}{}^3}{\langle \sigma u_{H_2} \rangle} \; \frac{(n_{H_2}\, r)^2}{x} \;\; \mathrm{erg\ cm^{-3}\ sec^{-1}} \tag{18a}$$

$$= 4.84 \times 10^{-21} \left(\frac{n_{H_2}}{10^5\ \mathrm{cm^{-3}}}\right)^2 \left(\frac{r}{0.2\ \mathrm{pc}}\right)^2 \left(\frac{10^{-8}}{x}\right) \mathrm{erg\ cm^{-3}\ sec^{-1}} \tag{18b}$$

where v_D has been eliminated through Eq. (17) and $x \equiv n_i/n_{H_2}$. We have also used $\langle \sigma u_{H_2} \rangle \approx 1 \times 10^{-9}\ \mathrm{cm^3\ s^{-1}}$ (Herbst and Klemperer, 1973). Clearly, the heating rate increases with the column density ($n_{H_2}\, r$) of the core. This is so because a larger column density implies stronger gravitational forces and, therefore, a larger v_D (and more energy transfer per collision) at a given x. The dependence $\mathcal{H} \propto x^{-1}$ at a given column density reflects the fact that the fewer the ions the more easily neutrals diffuse through and the higher the drift velocity is. Equation (18) is not valid at arbitrarily small x because, at some point v_D becomes larger than u_{H_2} and one of our assumptions is violated. One can show that for $x < 10^{-8}$ the heating rate actually decreases for scales larger than about 0.1 pc.

When x becomes even smaller and v_D much exceeds u_{H_2}, the predominantly neutral core contracts virtually unhindered and can in fact collapse if its mass exceeds the Bonner-Ebert value. A free-fall velocity field would tend to be set up, but because of the anisotropic nature of the collapse the gas pressure becomes increasingly more important and the possibility of shock formation arises; this would represent another source of heat for the small and dense cloud core.

Since at densities characteristic of dense cores gas-grain collisions dominate the cooling processes, one may equate the above heating rate with the cooling rate due to gas-grain collisions (Spitzer, 1949; Scalo, 1977),

$$\mathcal{L}_{\text{g-gr}} \approx 2 \times 10^{-33} \, n_{H_2}{}^2 \, T^{\frac{1}{2}} (T - T_{gr}) \text{ ergs cm}^{-3} \text{ sec}^{-1} \tag{19}$$

to determine the gas temperature. For $T_{gr} \approx 10$ K and the typical conditions used above, we find that the gas temperature will be maintained at $T \approx 46$ K. Even though this is in agreement with observations, the conclusion should be regarded only as tentative because of the approximate nature of the calculation and of the uncertainties in the grain parameters, which enter the derivation of Eq. (19). Detailed dynamical calculations are needed to understand the precise manner in which the magnetic field decouples from the matter and the consequent implications on the dynamic and thermodynamic future of a cloud in which such a process has taken place.

It remains now to piece together the discussion in this and the previous sections into a plausible scenario for star formation in magnetic clouds.

IV. A SCENARIO FOR STAR FORMATION IN MAGNETIC CLOUDS

A. Initiation of Cloud Collapse

Observations show that normal H I clouds have a magnetic energy comparable with thermal energy but two orders of magnitude larger than either gravitational or rotational energy. The most likely candidate for pushing such a cloud over the brink of gravitational instability is an increase in the intercloud pressure. Such an increase can be effected, for example, by passage through a spiral density wave, by a nearby supernova explosion, or heating due to nearby OB stars. Other chapters in this book discuss the possibility that any one of the above mechanisms gives rise to a shock that propagates through the cloud and forms gravitationally bound objects, as originally suggested by Dibai (1958). Shock calculations that include the magnetic field have not been performed yet. Here we simply note that shocks, although helpful, are not necessary for initiating cloud collapse.[a]

Cloud-cloud collisions and coalescence could also, in principle, lead to gravitational collapse. It is not clear at present, however, whether coalescence or dispersion of the clouds will be the result of such a collision. The effect of magnetic field have not been performed yet. Here we simply note that shocks, points to observational evidence that cloud velocities are highly organized with respect to the magnetic field, and that one gains the impression that the gas is moving along the magnetic field. If such is the case, cloud-cloud collisions may be much less frequent than commonly assumed.

B. Loss of Angular Momentum: Contraction at Nearly Constant Angular Velocity

Because of efficient magnetic braking, during the initial stages of compression a cloud maintains a nearly constant angular velocity, comparable to that of galactic rotation. Thus, the kinetic energy of rotation decreases as R^2. Flattening along field lines takes place. Once gravitational forces come into play, a dense central core will develop and further contraction will be highly nonhomologous. Molecules form over time scales relatively short compared to the dynamical time scale once the density becomes large enough. Since magnetic braking is less efficient at cloud densities larger than $\sim 10^4$ cm^{-3}, a dark cloud (and a molecular cloud with a dense envelope) may speed up noticeably above the rate of rotation that it would have if it remained in a synchronous orbit about the galactic center at all stages of contraction.

If the mass-to-flux ratio of the cloud exceeds some critical value, the cloud will collapse; otherwise, it will be set into oscillations, with nonradial

---

[a]It is often the case that mechanisms are proposed, or simply revived, for initiating gravitational collapse in *molecular* clouds. The problem with molecular clouds is not *what causes them to collapse,* but *why they do not all collapse.* (The free-fall time at $n_{H_2} = 3 \times 10^4$ cm^{-3} is only 2.2×10^5 yr.) Although magnetic fields could, in principle, stabilize a molecular cloud against gravitational collapse (see discussion following Eq. [14]), conclusions based on such an assumption should be avoided until the required fields are observed over extended regions in a molecular cloud – not merely in localized ($< 10^{16}$ cm) regions of OH masers.

It is sometimes argued that it is justifiable to begin one's formulation of a theory of star formation from molecular clouds (and thus avoid facing the angular momentum problem) because, after all, they are observed to exist. One can only point out that stars are also observed to exist.

velocities, about stable equilibrium configurations, such as the ones we have calculated. These oscillations are damped only over the time scale of ambipolar diffusion and could possibly explain the large linewidths of molecular clouds (Mouschovias, 1975*b*).

C. Formation of Extended Envelopes and Fragmentation

Even for a collapsing cloud, the magnetic field, which is tied to the external medium, can halt the collapse of the outer layers, thus forming an extended envelope. This only boosts flattening and enhances the nonhomologous nature of the contraction. The core of the cloud (where $1/3 \lesssim \kappa \lesssim 1/2$; $B \propto \rho^{\kappa}$) continues to contract, however, and blobs of solar mass can separate out even at as low a density as 2.5×10^5 cm^{-3} (for $\kappa = 1/3$) or about 10^{11} cm^{-3} (for $\kappa = 1/2$). Fragmentation is favored in the equatorial plane, where κ is small. Low-mass stars can form first (and perhaps only) in the core.

D. Loss of Magnetic Flux in the Core: Formation of Binary Stars

At even moderate densities, roughly $1 \times 10^4 - 2 \times 10^6$ cm^{-3}, ambipolar diffusion can significantly decrease the magnetic flux threading the core, or a fragment, of the cloud within a relatively short time ($10^5 - 4 \times 10^6$ yr). From that point onward, angular momentum is essentially conserved. For the above range of densities, the range of residual angular momenta in any blob that is destined to form a binary star system is exactly what is required to explain the entire range of periods of binary stars, from 10 hours to 100 years. A single mechanism, that is, can be responsible for the formation of all binary stars, in agreement with observations by Abt and Levy (1976). Single stars can form if decoupling of the field from the matter occurs at somewhat higher densities.

E. The Sun-Jupiter "Binary"

If a planetary system, such as the sun-Jupiter pair, is formed through the above scenario, Jupiter's angular momentum, which is essentially that of the entire solar system, is accounted for if the magnetic field decoupled from the matter at a density of about 10^9 cm^{-3}. This value of the density is not unreasonably high for decoupling if the parent solar fragment was located in the envelope of a dense cloud. Ionizing high-energy cosmic rays could then maintain the degree of ionization high enough for ambipolar diffusion to become effective only at this higher density.

Acknowledgments. I have benefited from numerous conversations on problems relating to star formation with a number of scientists. These include L. Spitzer, Jr., E. B. Jenkins, R. M. Kulsrud, J. P. Ostriker, G. B. Field, F. H.

Shu, W. D. Watson, and B. J. Bok. This chapter might not have begun without the hospitality of the Aspen Center for Physics in June 1977. Part of this chapter was completed while I was visiting the High Altitude Observatory in December 1977 and January 1978. This work has received partial support from a grant by the National Science Foundation.

APPENDIX A.

The Inability of Thermal Instability to Form Interstellar Clouds

It is often stated that a thermal instability is responsible for the formation of interstellar clouds. First, Field (1965) showed that a magnetic field as weak as 1 μG suppresses the instability in all directions except parallel to the field lines. Second, although perturbations with a broad range of wavelengths may grow at almost the maximum growth rate, these wavelengths have an upper bound determined by the fact (consistent with the force equation) that the "condensation mode" (Field, 1965) evolves almost isobarically. This means that the upper bound on the fastest growing wavelengths of a perturbation is approximately that distance within which a sound wave can establish pressure equilibrium in a time not exceeding the cooling time of the medium. Typical cooling times for an H I intercloud medium are less than 10^6 yr and become shorter as the gas density increases (Spitzer, 1968; Jura and Dalgarno, 1972). Therefore, with a sound speed smaller than 10 km sec^{-1}, the wavelengths that can grow at a rate near maximum will be less than 10 pc. Since the density of the intercloud gas is $\lesssim$ 0.2 cm^{-3}, the resulting condensation must have a size of about 0.01 pc if it is to reproduce the observed cloud densities ($\sim$ 30 cm^{-3}).

Aiming at obtaining larger condensations, Goldsmith (1970) considered the growth of perturbations with wavelengths considerably larger than the fastest growing ones. He chose $\lambda = 300$ pc, corresponding to an *e*-folding time of about 10^7 years; but, even so, the final condensation had an extent of only 0.13 pc, at $T \approx 20$ K – still a dwarf cloud. In any case, the thermal instability for a perturbation with a large enough wavelength to involve a sufficiently large mass will evolve more slowly than the magnetic Rayleigh-Taylor instability, in which the magnetic field is instrumental, rather than a detriment, in the formation of large-scale condensations.

Schwartz *et al.* (1972) proposed that inertial effects will maintain the flow set up by a thermal instability and therefore a condensation will continue to grow for a long time after the instability shuts off. Nevertheless, the final size of a condensation does not usually exceed 1/2 of the wavelength of the perturbation that initiated the instability. Since the observed dimensions of clouds (e.g., see Heiles, 1974; Heiles and Jenkins, 1976) are often larger than the wavelengths which can grow with an *e*-folding time less than about 10^7 yr, and since $n_{\rm cloud} \sim 30$ cm^{-3}, the thermal instability will not account for the formation of these condensations even if inertial effects are included.

Similar difficulties exist with a statistical model (Oort, 1954; Field and Saslaw, 1965), in which more massive clouds presumably form through collision and agglomeration of less massive ones (detailed discussion is given by Mouschovias, 1975*b*).

APPENDIX B

The Contraction Velocity of the Cloud Boundary Compared to the Local Alfvén Velocity

Consider a cloud of fixed mass M and radius R_{init} in which gravitational forces have just exceeded magnetic forces and collapse began. Initially the Alfvén velocity v_A is given by

$$v_A^2 \approx \frac{GM}{R_{\text{init}}} . \tag{B1}$$

The contraction velocity of the cloud's boundary is expected to be largest at the magnetic poles, where motions are along field lines and consequently are opposed only by thermal pressure forces. Let this velocity be denoted by v_p. A strict upper limit on v_p can be obtained if thermal pressure is ignored; this is just the free-fall velocity, v_{ff}, so that *at all stages* of the contraction we must have

$$v_p^2 \leqslant v_{\text{ff}}{}^2 = \eta \frac{GM}{R_{\text{eq}}} < \frac{GM}{R_p} , \tag{B2}$$

where η is a factor of order unity that depends on the degree of flattening, and R_{eq} and R_p are the equatorial and polar radii, respectively. One expects that near-hydrostatic-equilibrium will be established fairly rapidly along field lines and that contraction will thereafter proceed only as rapidly as gravitational forces can cause contraction *perpendicular* to the field lines. Since Eqs. (B1) and (B2) show that $v_p < v_A$ (and possibly $v_p \ll v_A$) initially, it remains to show that v_A^2 at the poles increases at least as rapidly as $R_p{}^{-1}$ during contraction.

If the external thermal pressure is fixed, continuity of pressure across the cloud boundary and isothermality, which is an excellent approximation for the relatively low densities at this stage of contraction, imply that the gas density at the cloud surface remains constant. It follows then that $v_A^2 \propto B_S^2$, where B_S is the field at the surface. Our calculations showed that the field at the poles, B_p, increases more rapidly than $R_{\text{eq}}{}^{-2}$, which is the expected dependence from conservation of total flux; let the actual dependence be

$R_{eq}^{-(2+a)}$, where $a > 0$. This is so because the tension of the field lines prevents R_{eq} from decreasing very much, while field lines passing near the poles are tied to the dense, nonhomologously contracting core (Sec. III.A.2). On the other hand, it is known (Mestel, 1965) that contraction of the kind we are considering proceeds in such a manner that the ratio R_p/R_{eq}^2 remains fixed. It follows that $v_A^2 \propto B_p^2 \propto R_{eq}^{-(4+2a)}$. Altogether then we have that

$$\left(\frac{v_A}{v_p}\right)^2 \propto R_p^{-(1+a)}, \quad a > 0. \tag{B3}$$

Since we have shown that, barring the compression of a cloud by a hydromagnetic shock, $v_p < v_A$ initially, the contraction velocity of the cloud boundary is unlikely to catch up with the local Alfvén speed.

REFERENCES

Abt, H.A., and Levy, S.G. 1976. *Astrophys. J. Suppl* 30: 423.
Appenzeller, I., and Tscharnuter, W. 1974. *Astron. Astrophys.* 30: 423.
_____. 1975. *Astron. Astrophys.* 40: 397.
Baldwin, J. 1954. *Nature* 174: 320.
Beichman, C.A., and Chaisson, E.J. 1974. *Astrophys. J.* 190: L21.
Biermann, L., and Davis, L., Jr. 1958. *Zs. f. Naturforsch.* 13a: 909.
_____. 1960. *Zs. f. Astrophys.* 51: 19.
Black, D.C., and Bodenheimer, P. 1976. *Astrophys. J.* 206: 138.
Bodenheimer, P. 1978. Preprint, *Lick Obs. Bull.* No. 804.
Clark, B.G. 1965. *Astrophys. J.* 142: 1398.
Clark, F.O.; Giguere, P.T.; and Crutcher, R.M. 1977. *Astrophys. J.* 215: 511.
Daniel, R.R., and Stephens, S.A. 1970. *Space Sci. Rev.* 10: 599.
Davis, L., Jr. 1958. *Astrophys. J.* 128: 508.
Davis, L., Jr., and Berge, G.L. 1968. In *Stars and Stellar Systems,* Vol. 7, *Nebulae and Interstellar Matter* (B. Middlehurst, and L.H. Aller, eds.) p. 755. Chicago: Univ. of Chicago Press.
Davis, L., Jr., and Greenstein, J.L. 1951. *Astrophys. J.* 114: 206.
Dibai, E.A. 1958. *Soviet Astron. – AJ* 35: 469.
Ebert, R.; Hoerner, S. von; and Temesváry, S. 1960. *Die Entstehung von Sternen durch Kondensation diffuser Materie.* Berlin: Springer-Verlag.
Ferraioli, F., and Virgopia, N. 1975. *Mem. Soc. Astr. Italiana* 46: 313.
Field, G.B. 1965. *Astrophys. J.* 142: 531.
_____. 1970. In *Proc. 16th Liège Astrophys. Symp.*, p. 29. Liège: Institut d' Astrophysique.
Field, G.B., and Saslaw, W.C. 1965. *Roy Obs. Bull.* No. 67.
Fowler, W.A., and Hoyle, F. 1963. *Roy Obs. Bull.* No. 67.
Fricke, K.J.; Möllenkoff, C.; and Tscharnuter, W. 1976. *Astron. Astrophys.* 47: 407.
Gehrels, T., ed. 1974. *Planets, Stars and Nebulae, Studied with Photopolarimetry.* Tucson: Univ. of Ariz. Press.
Gillis, J.; Mestel, L.; and Paris, R.B. 1974. *Astrophys. Space Sci.* 27: 167.
Goldsmith, D.W. 1970. *Astrophys. J.* 161: 41.
Hall, J.S. 1949. *Science* 109: 166.
Heiles, C., 1974. In *Galactic Radio Astronomy* (F.J. Kerr and S.C. Simonson, III, eds.), p. 13. Boston: Reidel.
Heiles, C., and Jenkins, E.B. 1976. *Astron. Astrophys.* 46: 333.
Herbst, E., and Klemperer, W. 1973. *Astrophys. J.* 185: 505.

Herbst, W. 1977. *Astron. J.* 82: 902.
Hiltner, W.A. 1949. *Science* 109: 165.
Jura, M., and Dalgarno, A. 1972. *Astrophys. J.* 174: 365.
Kerr, F.J., ed. 1963. *The Galaxy and the Magellanic Clouds.* p. 81. Dordrecht: Reidel.
Kondo, M. 1975. *Publ. Astron. Soc. Japan* 27: 215.
Larson, R.B. 1969*a*. *Mon. Not. Roy. Astr. Soc.* 145: 271.
____. 1969*b*. *Mon. Not. Roy. Astr. Soc.* 145: 297.
____. 1972*a*. *Mon. Not. Roy. Astr. Soc.* 156: 437.
____. 1972*b*. *Mon. Not. Roy. Astr. Soc.* 157: 121.
Lerche, I. 1967. *Astrophys. J.* 149: 395.
Lo, K.Y.; Walker, R.C.; Burke, B.F.; Moran, J.M.; Johnston, K.J.; and Ewing, M.S. 1975. *Astrophys. J.* 202: 650.
Manchester, R.N. 1974. *Astrophys. J.* 188: 637.
Mathewson, D.S., and Ford, V.L. 1970. *Mem. Roy. Astr. Soc.* 74: 143.
Mathewson, D.S.; van der Kruit, P.C.; and Brouw, W.N. 1972. *Astron. Astrophys.* 17: 468.
McKee, C.F., and Ostriker, J.P. 1977. *Astrophys. J.* 218: 148.
Mestel, L. 1965. *Quart. J. Roy. Astron. Soc.* 6: 161, 265.
____. 1966. *Mon. Not. Roy. Astr. Soc.* 133: 265.
____. 1977. In *Star Formation* (T. de Jong and A. Maeder, eds.) p. 213. Boston: Reidel.
Mestel, L., and Paris, R.B. 1978. *Mon. Not. Roy. Astr. Soc.* In preparation.
Mestel, L., and Spitzer, L., Jr. 1956. *Mon. Not. Roy. Astr. Soc.* 116: 503.
Miller, C.R. 1962. Ph.D. Dissertation, California Institute of Technology.
Morris, M.; Palmer, P.; Turner, B.E.; and Zuckerman, B. 1974. *Astrophys. J.* 191: 349.
Mouschovias, T. Ch. 1974. *Astrophys. J.* 192: 37.
____. 1975*a*. *Astron. Astrophys.* 40: 191.
____. 1975*b*. Ph.D. Dissertation, University of California, Berkeley.
____. 1976*a*. *Astrophys. J.* 206: 753.
____. 1976*b*. *Astrophys. J.* 207: 141.
____. 1977*a*. *Astrophys. J.* 211: 147.
____. 1977*b*. In *Star Formation* (T de Jong and A. Maeder, eds.), p. 235. Boston: Reidel.
____. 1978. In preparation.
Mouschovias, T. Ch.; Shu, F.H.; and Woodward, P.R. 1974. *Astron. Astrophys.* 33: 73.
Mouschovias, T. Ch., and Spitzer, L., Jr. 1976. *Astrophys. J.* 210: 326.
Nakazawa, K.; Hayashi, C.; and Takahara, M. 1976. Preprint.
Oort, J.H. 1954. *B.A.N.* 12: 177. No. 455.
Oppenheimer, M., and Dalgarno, A. 1974. *Astrophys. J.* 192: 29.
Parker, E.N. 1966. *Astrophys. J.* 145: 811.
____. 1967. *Astrophys. J.* 149: 535.
____. 1968*a*. *Astrophys. J.* 154: 57.
____. 1968*b*. In *Stars and Stellar Systems, Vol. 7, Nebulae and Interstellar Matter* (B. Middlehurst and L.H. Aller, eds.) p. 707. Chicago: Univ. of Chicago Press.
____. 1969. *Space Sci. Rev.* 9: 651.
Piddington, J.H. 1973. *Mon. Not. Roy. Astr. Soc.* 162: 73.
Pikelner, S.B. 1953. *Doklady Akad. Nauk.* (U.S.S.R.) 88: 229.
Rots, A.H. 1974. Doctoral Dissertation, Groningen University.
Scalo, J.M. 1977. *Astrophys. J.* 213: 705.
Schwartz, J.; McCray, R.; and Stein, R.F. 1972. *Astrophys. J.* 175: 673.
Shklovsky, I.S. 1953*a*. *Doklady Akad. Nauk.* (U.S.S.R.) 90: 983; 91: 475.
____. 1953*b*. *Soviet Astron. – AJ* 30:15.
Shu, F.H. 1974. *Astr. Astrophys.* 33: 55.
____. 1977. *Astrophys. J.* 214: 488.
Spitzer, L., Jr. 1949. *Astrophys. J.* 109: 337.
____. 1962. *Physics of Fully Ionized Gases,* 2nd ed. New York: Interscience.
____. 1968. *Diffuse Matter in Space* New York: Interscience.
Strittmatter, P.A. 1966. *Mon. Not. Roy. Astr. Soc.* 132: 359.
Tscharnuter, W. 1975. *Astron. Astrophys.* 39: 207.

Verschuur, G.L. 1970. In *Interstellar Gas Dynamics* (H.J. Habing, ed.), p. 150. Dordrecht: Reidel.
____. 1971. *Astrophys. J.* 165: 651.
Weber, S. 1976. *Astrophys. J.* 208: 113.
Westbrook, C.K., and Tarter, C.B. 1975. *Astrophys. J.* 200: 48.
Westerhout, G. 1963. *The Galaxy and the Magellanic Clouds* (F.J. Kerr, ed.), p. 78. Dordrecht: Reidel.
Wright, W.E. 1973. Ph.D. Dissertation, California Institute of Technology.
Yorke, H.W., and Krügel, E. 1977. *Astron. Astrophys.* 54: 183.
Zuckerman, B., and Palmer, P. 1974. *Ann. Rev. Astr. Astrophys.* 12: 279.

CONDITIONS IN COLLAPSING CLOUDS

GEORGE B. FIELD
Harvard-Smithsonian Center for Astrophysics

Because the masses of many dark interstellar clouds greatly exceed the critical Jeans mass for gravitational instability, it is often assumed that they are collapsing. This conclusion is suspect, both observationally and theoretically. While some clouds seem to show evidence for collapse in their molecular spectrum, the interpretation of the data is not unambiguous. Further, if all dark clouds were collapsing, there would be a discrepancy of a factor of 100 with the observed rate of star formation.

An alternative view is that dark clouds are supported against gravity by turbulence, rotation, or magnetic fields. Turbulence cannot provide hydrostatic support on the basis of a general theoretical argument. Magnetic fields are considered by Mouschovias in his chapter.

It seems possible that many dark clouds are in fact supported by rotation. The Hopper-Disney effect (alignment of elongated dark clouds with the galactic plane) can be interpreted in this way. Spectroscopic observations reveal that many dark clouds are rotating with angular velocities of order 1 km sec^{-1} pc^{-1}, about what is required for support against gravity, although only upper limits have been obtained in certain cases. Some insight into the data is provided by the theory of magnetic braking. The angular momenta observed are of interest for the theory of planetary system formation.

In this chapter we discuss dark interstellar clouds, and in particular the question whether or not most of such clouds are collapsing.

Observers often compare the masses of dark clouds with the Jeans mass M_J calculated for the observed temperature T and gas density ρ:

$$M_J = \frac{\pi^{\frac{5}{2}}}{6}\left(\frac{RT}{\mu G}\right)^{\frac{3}{2}} \rho^{-\frac{1}{2}}$$
$$= 3\left(\frac{T}{20\text{ K}}\right)^{\frac{3}{2}}\left[\frac{2n(H_2)}{10^5\text{ cm}^{-3}}\right]^{-\frac{1}{2}} M_\odot \qquad (1)$$

(Here we express the density in terms of the density of H_2 molecules, as most dark clouds are believed to be composed of molecular, not atomic hydrogen.) In the absence of significant turbulent stresses, magnetic pressure, or centrifugal force associated with rotation, a cloud with $M > M_J$ must gravitationally collapse, its internal gas pressure being inadequate for hydrostatic support. If one ignores turbulence, magnetic clouds, and rotation, one would expect most dark clouds to be collapsing, since their densities normally exceed $2n(H_2) = 10^3\text{ cm}^{-3}$, and their masses normally exceed the corresponding Jeans mass, $M_J = 30\ M_\odot$. As we shall see, it is by no means clear from the observations that most dark clouds *are* in fact collapsing. Since at least some are rotating so fast that centrifugal force is important in supporting them, it is possible that the majority of dark clouds are supported by rotation and are not collapsing. The amount of angular momentum involved has implications for the formation of stars and planetary systems.

OBSERVATIONS OF DARK CLOUDS

If clouds are collapsing, one can search for the predicted Doppler shifts of the molecular lines which they emit. The free-fall velocity acquired in collapsing from infinity to the observed radius R and density $n(H_2)$ is

$$v_{ff} = \left(\frac{2GM}{R}\right)^{\frac{1}{2}} = 10\left[\frac{2n(H_2)}{10^5\text{ cm}^{-3}}\right]^{\frac{1}{2}}\left(\frac{R}{1\text{ pc}}\right)\text{km sec}^{-1} \quad . \qquad (2)$$

Almost always the calculated free-fall velocity exceeds 1 km sec^{-1}. To observe it, one must be able to distinguish it from other Doppler velocities. A minimum value for the latter is the thermal velocity of the molecule in question (usually CO). The full-width at half maximum thermal velocity for CO at temperature T is

$$\Delta v_{th} = 2.4\left(\frac{kT}{m}\right)^{\frac{1}{2}} = 0.19\left(\frac{T}{20\text{ K}}\right)^{\frac{1}{2}}\text{ km sec}^{-1} \qquad (3)$$

a value which is so small that it does not interfere with the observation of collapse or other systematic velocities.

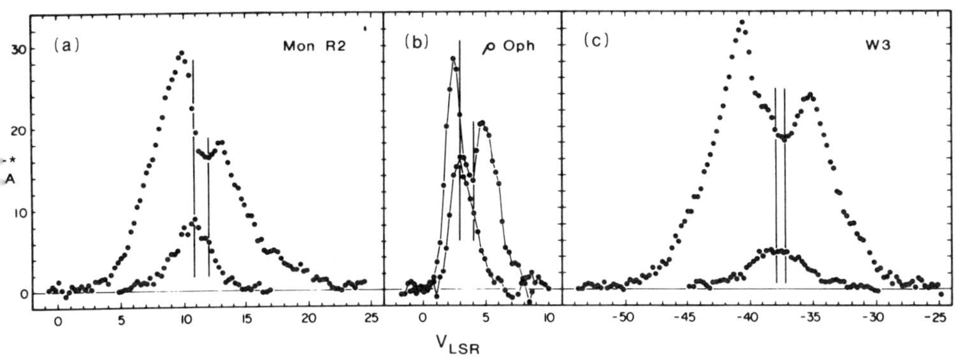

Fig. 1. Profiles (temperature vs velocity in km sec^{-1}, with respect to the Local Standard of Rest) of ^{12}CO emission (upper curve) and ^{13}CO emission (lower curve) toward the dark clouds Mon R2, ρ Oph, and W3. The optically-thin ^{13}CO profile has a single peak, marked by the left-hand vertical line. The optically-thick ^{12}CO profile shows a self reversal at a velocity about 1 km sec^{-1} greater than the ^{13}CO peak. The authors interpret this effect as absorption by infalling gas on the near side of the cloud (Snell and Loren, 1977).

In fact, almost all CO observations show velocity widths of several km sec^{-1} so we are sure that *some* kind of nonthermal velocities are present in most clouds. According to Goldreich and Kwan (1974), free-fall collapse is the most probable explanation.

A number of observers have attempted to test the collapse hypothesis. For example, Snell and Loren (1977) studied the velocity profiles of the Mon R2, W3, and ρ Oph dark clouds (Fig. 1). They find that the profile of ^{13}CO, which because of the low abundance of ^{13}C is believed to be optically thin, is symmetric, with a single peak at a definite velocity we denote by ν_{13}. On the other hand, ^{12}CO, which is believed to be optically thick because of its greater abundance, shows a double-peaked profile, with a dip at velocity ν_{12} which in each case exceeds ν_{13} by a small amount (~ 1 km sec^{-1}). Snell and Loren interpret these observations as follows. Since ^{13}CO is optically thin, one is observing all the gas present, at both the front and back of the cloud. Whether the cloud is collapsing or not, ν_{13} should therefore correspond to the center-of-mass velocity of the cloud as a whole. On the other hand, because ^{12}CO is optically thick, the ^{12}CO line is emitted by gas primarily on the front side. If the outer part of the cloud is somewhat cooler than the gas in the center of the cloud, one can understand the double-peaked profile as due to an absorption line formed in the cooler foreground gas. The fact that $\nu_{12} > \nu_{13}$ therefore means that the foreground gas is moving toward the center of mass of the cloud, or, in other words, that the cloud is collapsing (at ~ 1 km sec^{-1}).

Snell and Loren model a typical such cloud in detail, assuming a number of different variations of density, temperature, and collapse velocity with radius. Their best fits are with ν proportional to $R^{-\frac{1}{2}}$ (consistent with free-fall velocity if the mass is concentrated at the center). The resulting model

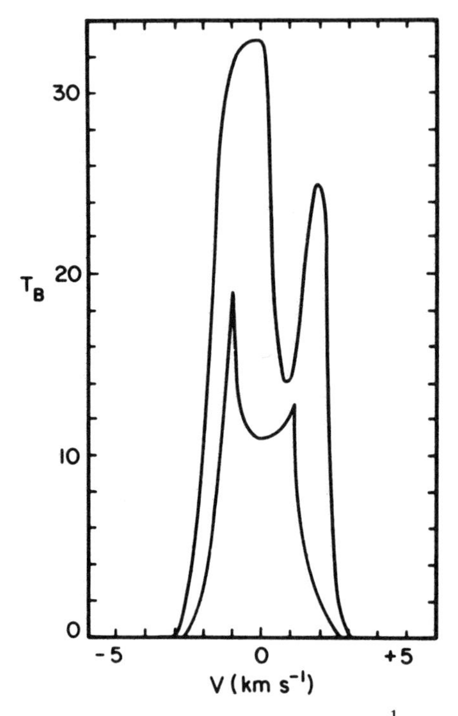

Fig. 2. Calculated profiles for a collapsing cloud ($v \propto r^{-\frac{1}{2}}$) with no turbulence; ^{12}CO is the upper curve, and ^{13}CO is the lower. Compare with Fig. 1 (Snell and Loren, 1977).

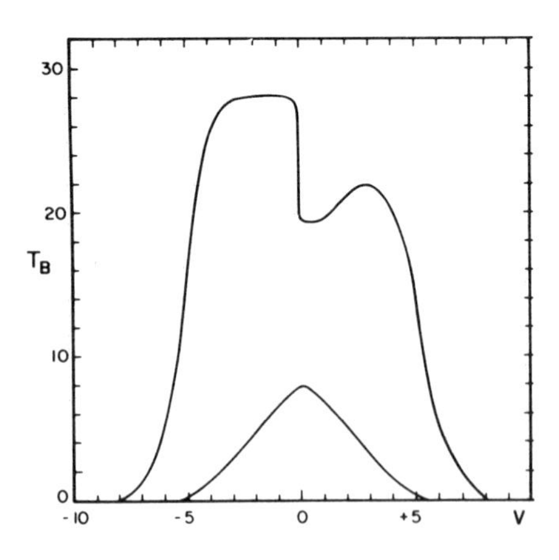

Fig. 3. Calculated profiles for a collapsing cloud ($v \propto r^{-\frac{1}{2}}$) with a turbulent velocity Δv_t = 3 km sec^{-1}. The cusps in Fig. 2 are smoothed out, but the displaced self reversal remains. Compare with Fig. 1 (Snell and Loren, 1977).

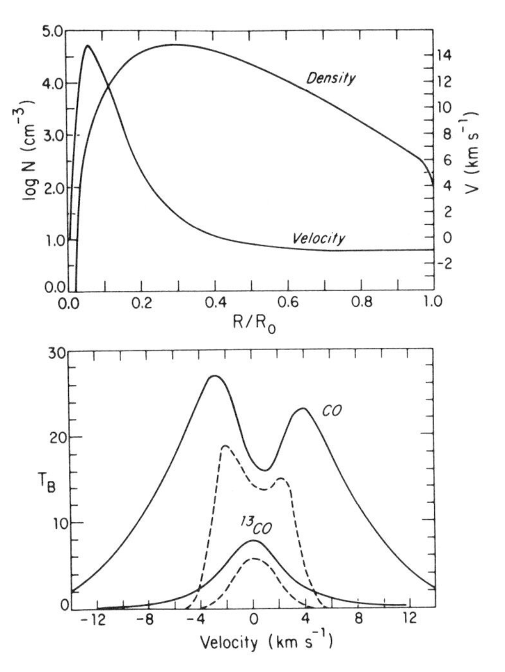

Fig. 4. The assumed density and velocity distribution in a dark cloud (upper panel) and resulting ^{12}CO and ^{13}CO profiles (lower panel). The solid lines refer to a line of sight through the cloud center (as does Fig. 1), and the dashed lines refer to a line of sight 0.75 pc from the center. Note that the outer part of the cloud is assumed to be collapsing at ~ 1 km sec^{-1}. Compare with Fig. 1. (From Leung and Brown, 1977.)

profiles show cusps which are not observed (Fig. 2). Snell and Loren find that they must introduce turbulence with a velocity spread $\Delta v_t = 3$ km sec^{-1} to get acceptable agreement with the data (Fig. 3).

Leung and Brown (1977) have criticized Snell and Loren's conclusion that the observed clouds as a whole are collapsing, pointing out that the assumed turbulent velocity in fact dominates the assumed systematic collapse velocity over almost the whole cloud in their model. Not only does this cast doubt upon the assumption of collapse (because large turbulent velocities would tend to support the cloud), but in the view of Leung and Brown, it vitiates a basic assumption made in constructing the model. Snell and Loren calculated the emergent radiation field using the Sobolev approximation, which requires that turbulent velocities v_t be small compared to systematic collapse velocities.

Leung and Brown calculated models without the Sobolev approximation, so their results are not subject to this criticism. Their best-fit velocity model requires that the exterior of the cloud is collapsing, but that the interior is actually expanding. Their fits to the observed profiles are good (Fig. 4).

From this discussion I conclude that the interpretation of CO profiles is controversial at this time. It is therefore not certain that the clouds observed by Snell, Loren and others are in fact collapsing throughout the bulk of their mass.

From a different point of view, the hypothesis that most dark clouds are in fact collapsing leads to difficulties in the theory of star formation (Zuckerman and Palmer, 1974). The free-fall time for a cloud of density ρ is

$$t_{ff} = \left(\frac{3\pi}{32G\rho}\right)^{\frac{1}{2}} = 1.4 \times 10^5 \left[\frac{2n(H_2)}{10^5\,cm^{-3}}\right]^{-\frac{1}{2}} yr. \quad (4)$$

Almost all dark clouds have $2n(H_2) > 10^3$ cm^{-3}, so $t_{ff} < 1.4 \times 10^6$ yr. According to Scoville and Solomon (1975) there are roughly $2.5 \times 10^9\ M_\odot$ of gas within interstellar CO clouds in the galaxy, which are believed to be dark clouds of the type under discussion. If all such clouds were collapsing, very dense interstellar matter would be generated at a rate exceeding $10^3\ M_\odot\ yr^{-1}$. As it is believed that only $\sim 1\ M_\odot\ yr^{-1}$ goes into forming new stars, one must ask what is the fate of the great bulk of the collapsing gas, which does not form stars. According to Vrba (1977) observations indicate that the mass of stars produced is typically 10% of that of the parent dark cloud. This would leave a large factor (~ 100) still to be accounted for. One possible explanation is that most dark clouds are not in fact collapsing, but are prevented from doing so by turbulence, magnetic fields, or rotation. Whether this is possible is the subject of the next section.

SUPPORT OF DARK CLOUDS

From the virial theorem we know that there are three ways to support a cloud against gravitational collapse: thermal motions, nonthermal motions (including turbulence, rotation, and various types of waves), and magnetic fields. Thermal motions are unimportant if the mass exceeds the Jeans mass, as is true for most dark clouds. According to Goldreich and Kwan (1974), turbulence cannot support clouds for the following reason. In order to be effective, the turbulent stress $\rho(\Delta v_t)^2$ must greatly exceed the thermal pressure $\rho RT/\mu$ (because the latter provides inadequate support). Hence

$$\Delta v_t \gg \left(\frac{RT}{\mu}\right)^{\frac{1}{2}} = v_s \quad (5)$$

where v_s is the speed of sound. Hence turbulence to be helpful must be supersonic. Such turbulence will generate shock waves within a time interval

$$t_t = \frac{L}{\Delta v_t} < \frac{R}{\Delta v_t} \tag{6}$$

where L is the scale of the turbulence and R is the radius of the cloud. Since the gas is a good radiator, the shock energy will be radiated away immediately, and turbulent support can therefore be effective for at most a time equal to $R/\Delta v_t$. From the virial theorem

$$\Delta v_t \simeq \left(\frac{GM}{R}\right)^{\frac{1}{2}} \tag{7}$$

if the support is to be effective so

$$t_t < \frac{R}{\Delta v_t} = \left(\frac{R^3}{GM}\right)^{\frac{1}{2}} \simeq t_{\text{ff}} \tag{8}$$

from Eq. (4). Hence turbulence at most increases the lifetime of a cloud against gravitational collapse by a factor of order unity. A similar argument rules out acoustic waves, because they would have to be of large amplitude to be effective, and would therefore quickly form shock waves. That leaves rotation and magnetic fields (including the hydromagnetic waves such fields can support). Since the effects of magnetic fields will be discussed by Mouschovias in this chapter, I will concentrate on rotation in what follows.

ROTATION

If a cloud is rotating with its axis in any orientation other than along the line of sight, the observer will detect a corresponding systematic velocity shift of the spectral lines across the cloud. A number of such studies have been made, with the results summarized in Table I. The observed line-of-sight velocity gradient dv/dr is given in km sec^{-1} pc^{-1} = 3.2×10^{-14} radians sec^{-1} (henceforth called "units").

In addition to this information about specific clouds there is an interesting effect discovered by Hopper and Disney (1974), who examined over 200 compact dark clouds listed in Lynds' (1962) catalogue, many of which are probably globules. They selected clouds which are sufficiently elongated so that a position angle of the major axis could be measured, and found that the major axis tends to be parallel to the galactic plane (Figs. 10 and 11). On the other hand, when the direction of the major axis is compared with the direction of the interstellar magnetic field in the local vicinity estimated from the polarization vectors observed in nearby stars (Fig. 12), no

TABLE I

Observations of Dark-Cloud Rotation

Object	Type	dv/dr (km sec^{-1} pc^{-1})	Comments	Reference
B 361 (core)	Globule	2.9	Plane ⊥ to galactic plane; see Fig. 5	Milman (1977)
B 163	Globule	1.9	Plane ‖ to galactic plane; retrograde rotation; flattened 3:1	Martin and Barrett (1978)
B 163 SW	Globule	3.1	Plane ‖ to galactic plane; retrograde rotation; flattened 3:1	Martin and Barrett (1978)
Five objects	Globule	Undetectable		Martin and Barrett (1978)
Mon R2	Massive cloud	2.3	Plane ‖ to galactic plane	Kutner and Tucker (1975)
Mon R2	Massive cloud	0.4	Plane ‖ to galactic plane; collapsing	Loren (1977*a*)
NGC 7129	Massive cloud	0.4	Inclined 45° to galactic plane	Loren (1977*b*)
Taurus dark cloud	Massive cloud	2		Ho *et al.* (1977)
L 1641	Large cloud complex	0.14	10 × 60 pc; retrograde rotation; see Figs. 6 & 7	Kutner *et al.* (1977)
L 1641	Condensation in L 1641	1.2	1.3 pc size; see Fig. 8	Kutner *et al.* (1976)
NGC 2264	Massive cloud	Detectable	Flattened; elongated in north-south direction	Crutcher (1977)
NGC 2264	Cloud near 15 Mon	Detectable		Crutcher (1977)
M 17 cloud complex	Individual condensations in M 17 cloud complex	Undetectable; <0.05	See Fig. 9	Elmegreen *et al.* (1978)

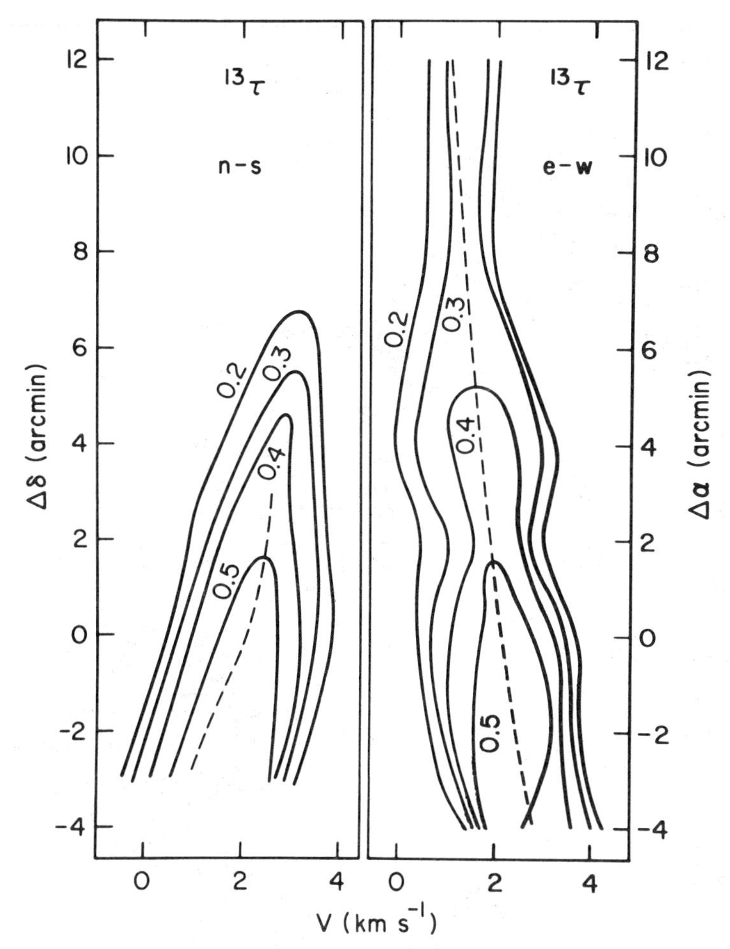

Fig. 5. Contours of ^{13}CO emission from the globule B 361, plotted against velocity and declination (left panel) and right ascension (right panel). The tilt of the contours indicates rotation at 2.9 km sec^{-1} pc^{-1} in a plane perpendicular to the galactic plane (Milman, 1977).

correlation was found. They examined various possible geometries (Fig. 13), and showed that the observations are best explained if the apparently elongated dark clouds are actually shaped like disks, with the disks tending to be parallel to the galactic plane. In a later paper (Disney and Hopper, 1975) they proposed that the clouds form disks as a result of the gradient in the gravitational force perpendicular to the galactic plane, which compresses the cloud in that direction. Heiles (1976) offers the alternative explanation that the Hopper-Disney clouds are rotating. In view of the direct observations quoted in Table I, this possibility seems worth pursuing.

To model the dynamics of such a situation we consider constant-density

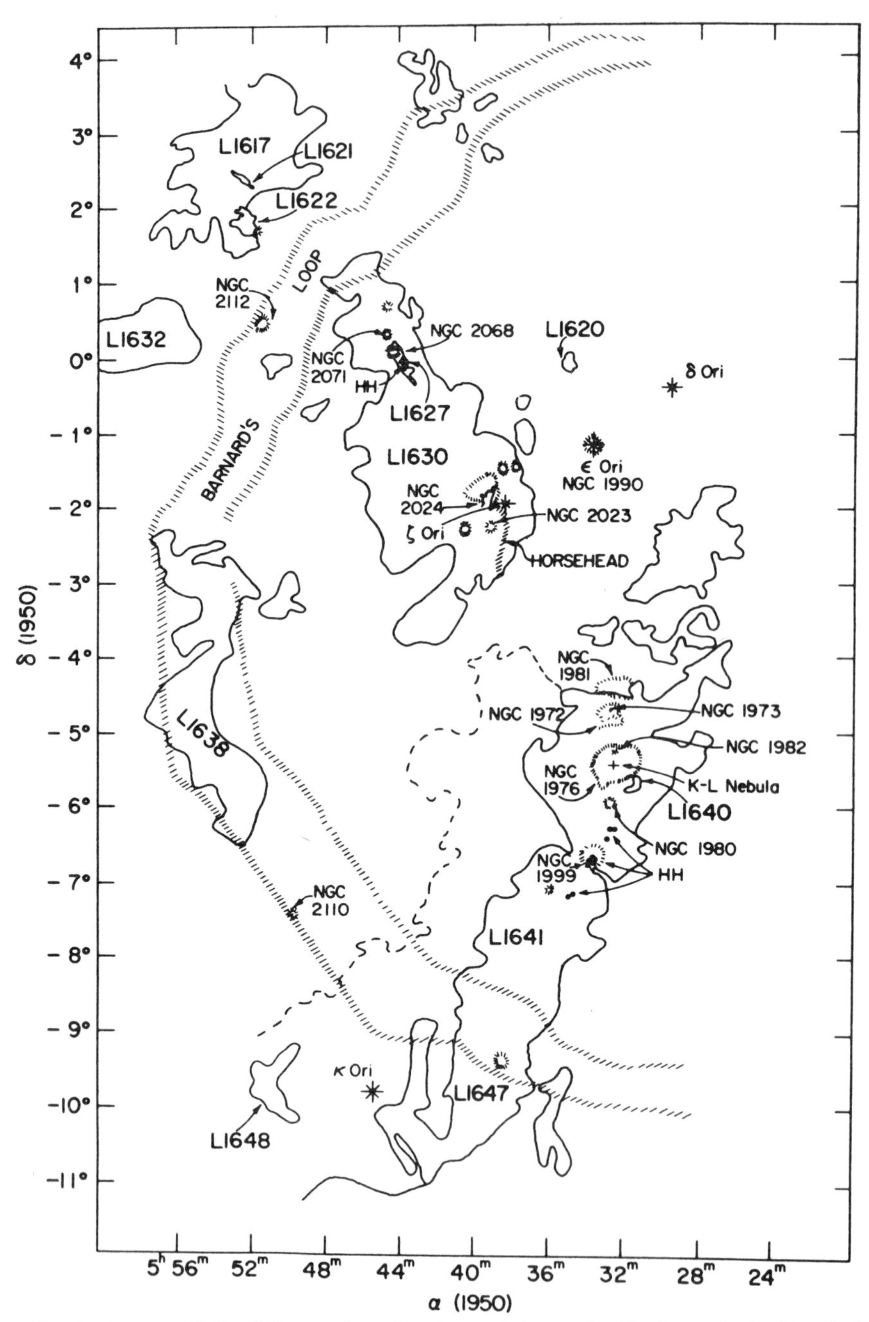

Fig. 6. A map of the Orion region, showing emission and reflection nebulae (hatched boundaries) and dark clouds (solid boundaries). The large cloud L 1641 is about 10 x 60 pc, parallel to the galactic plane. It is rotating as a whole in a direction opposed to that of galactic rotation (Kutner *et al.*, 1977).

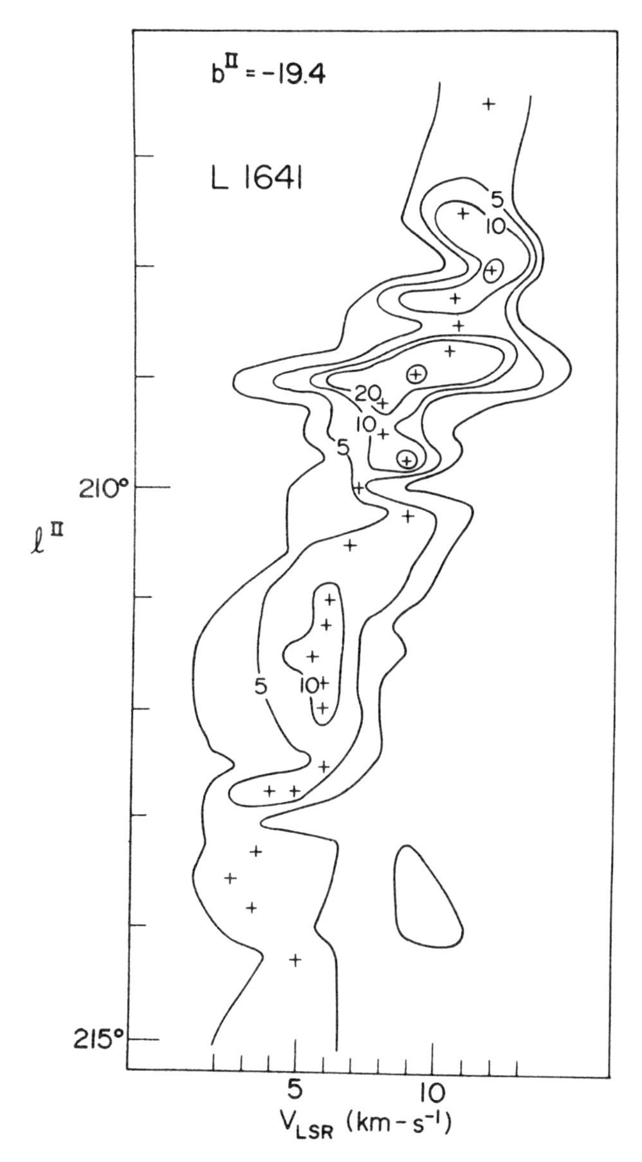

Fig. 7. The velocity of L 1641 plotted against galactic longitude. The inclination of this plot shows that L 1641 is rotating opposite to the galaxy at $\omega = 0.14$ km sec^{-1} pc^{-1} (Kutner *et al.*, 1977).

spheroids of radii a and b; to account for the observed flattening, we take $b = 1/3\ a$. Since authors often quote a single "radius" R, we take this to be the harmonic mean of a and b, so

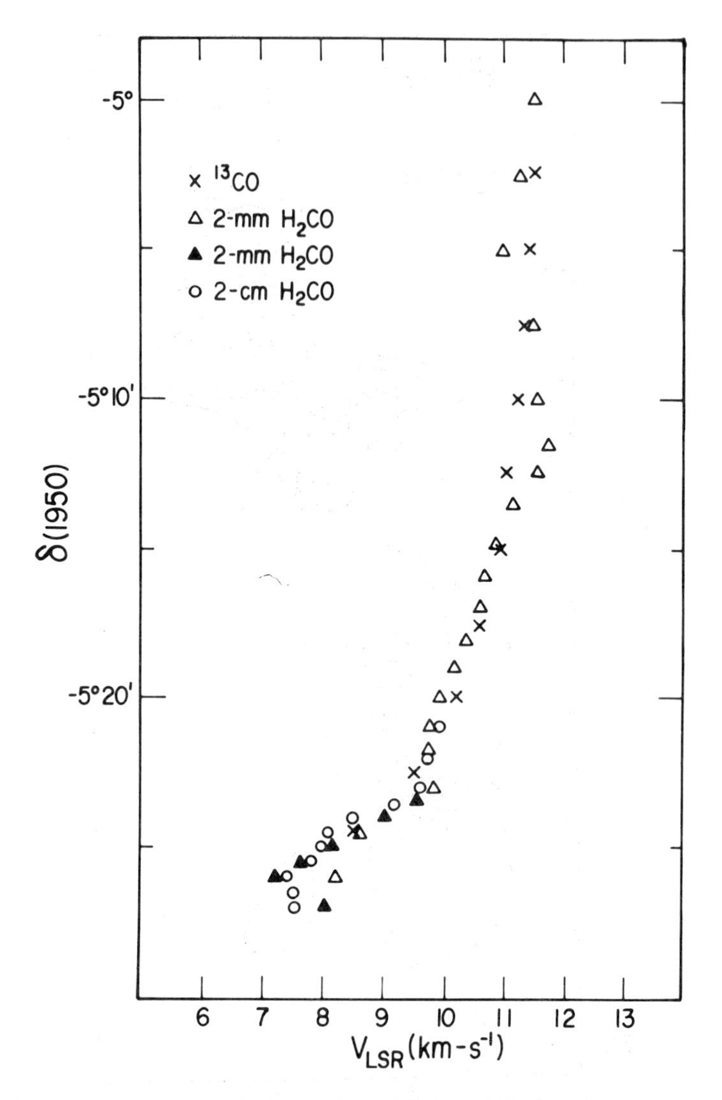

Fig. 8. The velocity pattern in the condensed region OMC-1/OMC-2 in Orion. The tilt of the central portion indicates rotation at about 12 km sec^{-1} pc^{-1} (Kutner *et al.*, 1976).

$$R = \frac{a}{\sqrt{3}} \quad . \tag{9}$$

In order to estimate the proton density $n = 2n(H_2)$ we use the relationship (Spitzer, 1978)

$$A_V = 3E(B{-}V) = 5.1 \times 10^{-22}\ N_H \equiv K N_H \quad . \tag{10}$$

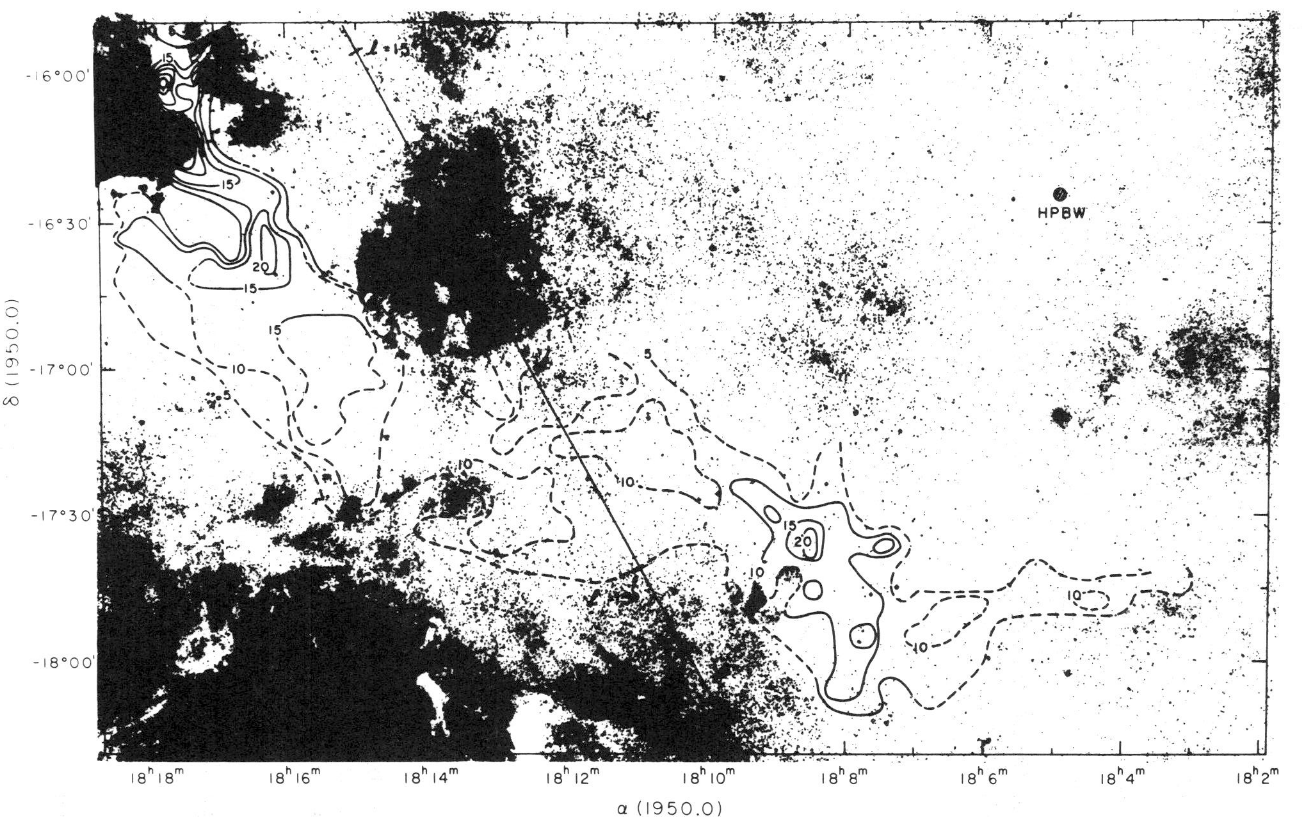

Fig. 9. The M 17 complex, showing CO contours superposed on a photograph of the region. Note the orientation of the galactic plane (straight line through the center of the figure). The complex is about 20 x 170 pc (Elmegreen *et al.*, 1978).

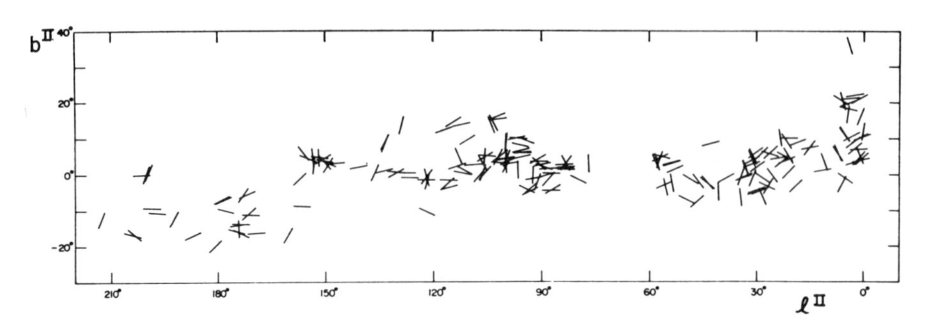

Fig. 10. The orientations of 213 dark clouds plotted in galactic coordinates (Hopper and Disney, 1974).

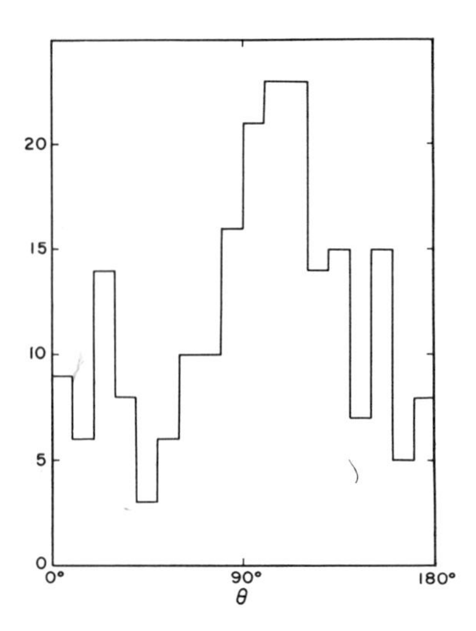

Fig. 11. Distribution of position angles θ of the dark clouds of Fig. 10. The angle θ is measured from the north galactic pole, so that 90° corresponds to alignment with the galactic plane (Hopper and Disney, 1974).

As we presumably see most of the disks edge on (to explain the Hopper-Disney effect), the mean extinction over the cloud is given by

$$A_V = KN_{\mathrm{H}} = \frac{KnV}{\pi ab} = Kn\,\frac{4\pi\sqrt{3}R^3}{3\pi R^2} = \frac{4}{3}\sqrt{3}\,KnR \tag{11}$$

so that

$$n = \frac{\sqrt{3}\,A_V}{4KR}\ . \tag{12}$$

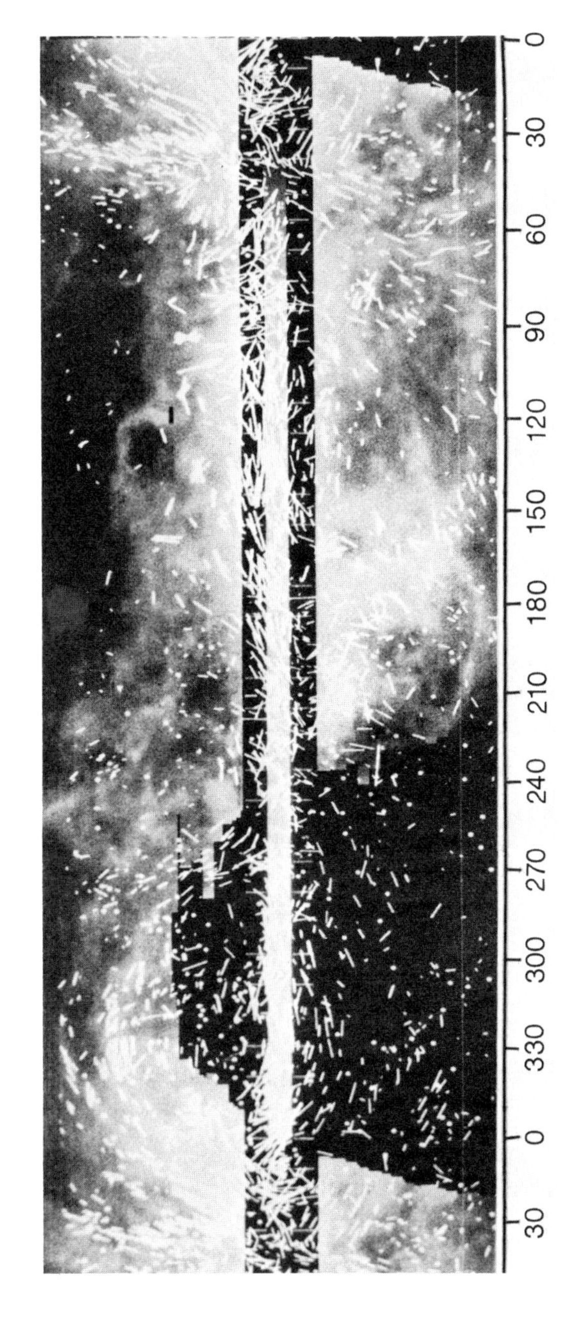

Fig. 12. A map of the polarization of starlight (Mathewson and Ford, 1970) superposed on a photographic representation of the column density of H I between −20 and +20 km sec^{-1} (Heiles and Jenkins, 1976). It is believed that the polarization vectors are parallel to the interstellar magnetic field.

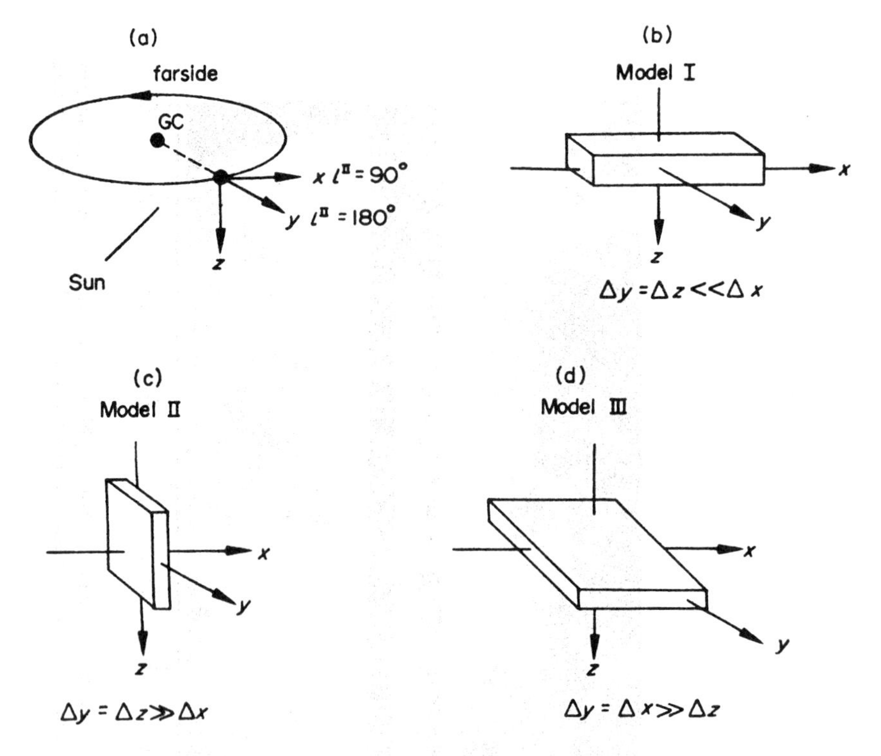

Fig. 13. Three models considered for explaining the data of Fig. 11. Model III, with disks parallel to the galactic plane, is the only one which fits the data (Hopper and Disney, 1974).

If $A_V = 10$ mag and $R = 0.5$ pc, we have

$$n = 5800 \text{ protons cm}^{-3} = 2900 \text{ H}_2 \text{ cm}^{-3} \tag{13}$$

and hence

$$\rho = 1.2 \times 10^{-20} \text{ g cm}^{-3} \tag{14}$$

for He/H = 0.1. Multiplying by the volume $4\pi\sqrt{3}\,R^3/3$, we get for the mass

$$M = \rho V = 3.2 \times 10^{35} \text{ g} = 160\, M_\odot \quad . \tag{15}$$

The values of R and M accord well with Heiles' (1976) estimated parameters for Hopper-Disney clouds, and also with Bok's (1977) parameters for large globules. In what follows we will assume that large globules and Hopper-Disney clouds are approximately the same thing.

The flattening may be due to rotation around the minor axis. As the spheroids are assumed to have constant density, the rotation would be solid-body rotation, with

$$\omega \simeq \left(\frac{3\pi GM}{4a^3}\right)^{\frac{1}{2}} = \left(\frac{\pi}{4\sqrt{3}}\frac{GM}{R^3}\right)^{\frac{1}{2}} = 1.6 \text{ km sec}^{-1} \text{ pc}^{-1} \ . \qquad (16)$$

Here we have used the formula for an infinitely thin spheroid, which is not far off for $b = a/3$. It is interesting that $\omega = 1.6$ units is of the same order of magnitude as the actual angular velocities observed for some of the globules in Table I; this is consistent with identifying globules with Hopper-Disney clouds. Not all the globules in Table I have detectable rotation. While in some cases this could be due to unfavorable projection effects, in other cases it may simply be that rotation of ~1 unit is present, but is undetectable with present techniques.

Heiles (1976) suggested that the rotation of the Hopper-Disney dark clouds can be understood if they formed via rapid compression of diffuse interstellar clouds, which are known from studies of the 21-cm line, optical absorption lines, and statistics of color excesses. Spitzer (1978) gives as typical parameters of such clouds $A_V = 0.2$ mag, $n = 20$ cm^{-3}, $R = 5$ pc and $M =$ 400 $M_\odot$. Since the estimated mass is comparable to what we have estimated for the Hopper-Disney clouds, it is instructive to ask what the parameters would be if the density were to increase a factor of 290, up to the value 5800 cm^{-3} observed for Hopper-Disney clouds. The mean radius would be $(290)^{-\frac{1}{3}} \times 5$ pc = 0.76 pc and the extinction would be 15 mag, not far from the observed values. If, as suggested by Heiles (1976) and Elmegreen (personal communication, 1977), angular momentum were conserved during the rapid compression, ω would increase by a factor of $R^{-2} = 44$.

What are reasonable values for the original angular velocity ω_0? As we shall see later, the large-scale galactic magnetic field should be effective in keeping diffuse clouds corotating with the galaxy, whence $\omega_0 = 8 \times 10^{-16}$ rad sec^{-1} = 0.025 units. An increase of 44 times would give $\omega = 1.1$ units, in rough agreement with what is calculated to be necessary to rotationally stabilize Hopper-Disney clouds. If the clouds are actually spun up in this way, it is plausible that their angular momentum vectors should be aligned with that of the galaxy, and that the clouds should therefore be aligned with the galactic plane.

What is not so readily understandable, however, is why in some cases globules appear to be aligned all right, but rotating in a retrograde manner. Here we must keep in mind Mestel's (1966) point that the direction of rotation (prograde or retrograde) of a cloud condensing from a differentially rotating disk depends upon its shape. In a disk like that of the galaxy, characterized by the differential rotation constants

$$\begin{aligned} A &= 0.015 \text{ km sec}^{-1} \text{ pc}^{-1} \\ B &= -0.010 \text{ km sec}^{-1} \text{ pc}^{-1} \end{aligned} \qquad (17)$$

a cloud which has an axial ratio (or radial extent ÷ azimuthal extent) greater than $[(A-B)/(A+B)]^{\frac{1}{2}} = 2.2$ would have retrograde rotation.

While the discussion up to now suggests that Hopper-Disney clouds may be stabilized by rotation, closer analysis makes this conclusion suspect. If we put aside turbulence as an important dynamical phenomenon for the reasons given earlier, and postpone discussion of the effects of magnetic fields, we are left with thermal pressure alone as a means of supporting the disks parallel to their axis of rotation. Leaving aside the question whether or not the temperature is adequate to do this (it appears not to be), we still must contend with the stability of the disk. It is known that disks (oblate spheroids) are unstable toward the formation of bars (prolate spheroids) if the kinetic energy of rotation exceeds 0.14 times the gravitational energy (Ostriker and Peebles, 1973). This can be translated into a requirement on the thickness of the disk by using the equations of Field (1975). A disk is unstable if $b/a < 0.52$. This criterion is violated by the globules B 163 and B 163 SW in Table I ($b \simeq 0.3a$) as well, presumably, as by many other elongated globules. Thus, it is conceivable that the observed clouds are more like bars than disks, and that in rotating end over end around their short axes they appear to the observer to be oriented parallel to the galactic plane. This would explain why some dark clouds appear to be spherical; we are looking along the major axis.

Even bar-like clouds may be gravitationally unstable on a smaller scale. To stabilize against such effects, it may be necessary to postulate that magnetic fields play a role; we defer such discussions to the next section. If for the moment we suppose that Hopper-Disney clouds are stabilized by rotation, we avoid a problem implicit in Bok's (1977) estimate that there are 25,000 large globules in the galaxy. If our estimate of 160 $M_\odot$ is correct, there are a total of 4×10^6 $M_\odot$ involved. The free-fall time according to Eq. (4) is 5.8×10^5 yr, so that if large globules were in fact collapsing in free fall, some 7 $M_\odot$ per year would be becoming superdense. Again, this rate is greater than the observed rate of star formation, although not by such a large factor as found earlier for dark clouds generally. On the other hand, globules, unlike dark clouds in general, show no signs of star formation, so we may be fairly sure that at most a tiny fraction of the total of 1 $M_\odot$ per year of interstellar gas going into stars does so in globules. Hence again there is a discrepancy of a factor of at least 10 with assuming that most globules are collapsing. While Bok (1977, 1978) believes they are, the Hopper-Disney result and the observation of rotation in several globules suggests that they are not.

MAGNETIC FIELDS

As Mouschovias' chapter discusses this topic in depth, I will confine my remarks to magnetic braking of cloud rotation. Fig. 12 demonstrates that magnetic fields influence the distribution of gas on the large scale. However, the same thing may not be true on the small scale of dark clouds. Two time scales are of interest in this connection. The first relates to the fact that the magnetic field is strongly coupled only to the charged-particle component of a cloud, affecting the motion of the dominant neutral component only through ion-neutral collisions. If the magnetic field inside a cloud is stronger than outside, magnetic stresses will force the ions to slowly diffuse through the neutrals, carrying the field with them and weakening it to the point that it has little dynamical effect. The diffusion time for a cloud of radius R containing an interior field B_i which is considerably larger than the external field B_e is (Spitzer, 1968)

$$t_D = \frac{8\pi R^2 n_i n_n \, m_{in} \, \langle \sigma v \rangle_{in}}{B_i^2} \quad . \tag{18}$$

Here n_i is the density of ions (assumed to be C^+), n_n is the density of neutrals (H I in diffuse clouds, H_2 in dark clouds), m_{in} is the reduced mass for the ion-neutral collision, and $\langle \sigma v \rangle_{in}$ is the corresponding collision rate.

Let us apply this to diffuse clouds. Although it is believed on the basis of interstellar polarization (Heiles, 1976) that such clouds have $B_i \simeq B_e = 3 \times 10^{-6}$ G, one can ask what would happen if B_i were slightly larger, say 5×10^{-6} G. If we take $n_n = 20$ cm^{-3} and $R = 5$ pc as typical (from Spitzer, 1978), $n_i = 6 \times 10^{-5}$ $n_n = 1.2 \times 10^{-3}$ cm^{-3} (corresponding to a five-fold depletion of carbon), and $\langle \sigma v \rangle_{in} = 2.2 \times 10^{-9}$ cm^3 sec^{-1} independent of temperature (Spitzer, 1978), we get

$$t_D \text{ (diffuse cloud)} = 6 \times 10^8 \text{ yr.} \tag{19}$$

If, on the other hand, we apply Eq. (18) to a globule, we have a larger neutral density, say $n_n = 2900$ cm^{-3}. Although the ion density is poorly known, an upper limit on $n_i/n_n = 10^{-8}$ has been obtained by Guélin *et al.* (1977) for five dark clouds. Thus in our case, $n_i < 2.9 \times 10^{-5}$ cm^{-3}. If we take $\langle \sigma v \rangle_{in} \simeq 1 \times 10^{-9}$ cm^3 sec^{-1}, appropriate for $C^+ - H_2$ collisions at 20 K (Herbst and Klemperer, 1973), and $R = 0.5$ pc, we get

$$t_D \text{ (dark cloud)} < 6 \times 10^7 \left(\frac{B_i}{3 \times 10^{-6} \text{ G}} \right)^{-2} \text{ yr.} \tag{20}$$

Although the field B_i is unknown, it is reasonable to suppose that it would exceed the intercloud value $B_e = 3 \times 10^{-6}$ G, because of compression of the field accompanying the formation of the cloud. Hence t_D (dark cloud) $< 6 \times 10^7$ yr.

If the field does not diffuse rapidly out of a cloud, radiation of Alfvén waves at speed v_A into the surrounding external medium (density ρ_e) will brake the rotation of the cloud in a time

$$t_M = \frac{1.6\rho_i R}{\rho_e v_A} = \frac{5.7\rho_i R}{B_e\sqrt{\rho_e}} \tag{21}$$

(Mouschovias, 1977), where ρ_i is the internal density. If we apply Eq. (21) to diffuse clouds ($\rho_i = 4 \times 10^{-23}$ g cm^{-3}, $R = 5$ pc, $B_e = 3 \times 10^{-6}$G, and $\rho_e = 2 \times 10^{-25}$ g cm^{-3}) we get

$$t_M \text{ (diffuse cloud)} = 8 \times 10^7 \text{ yr.} \tag{22}$$

Comparing Eqs. (22) and (21), we see that $t_M < t_D$ if the internal field exceeds 3×10^{-6}G somewhat. Even if $B_i = B_e = 3 \times 10^{-6}$G, t_M is smaller than the rotation period (2.5×10^8 yr) of the galaxy at 10 kpc distance from the galactic center. Hence our earlier assumption that diffuse clouds corotate with the galaxy seems reasonable.

Turning now to dark clouds, we may try to apply Eq. (21) to the Hopper-Disney clouds. Unfortunately, we do not know B_e or ρ_e for this case. If we take $B_e = 3 \times 10^{-6}$ G and $\rho_e = 2 \times 10^{-25}$ g cm^{-3} as for the diffuse clouds, $t_M = 2.6 \times 10^9$ yr, much larger than t_D. If this were the whole story, one would conclude that the majority of Hopper-Disney clouds probably lost their internal fields by diffusion before significant magnetic braking could occur; this would accord with the apparently high rates of rotation of such clouds. However, as Mouschovias (1977) has pointed out, B_e is probably much larger than 3×10^{-6} G as a result of the dragging-in of field lines demonstrated in his earlier work (Mouschovias, 1976). Some relevant limits on B_e have been obtained by Crutcher *et al.* (1975), who deduced $B < 5 \times 10^{-5}$ G in a dark cloud region from Zeeman splitting of OH lines.

In summary, then, magnetic braking should be effective for diffuse clouds, so they should corotate with the galaxy. In the case of Hopper-Disney clouds, however, it is not clear whether magnetic diffusion or braking is more rapid; the observed flattening, alignment with the galactic plane, and rotation of similar clouds suggests that magnetic braking is not very effective for many of these dark clouds, and that therefore they are stabilized by rotation.

None of this addresses the question of the condensations in giant molecular clouds where star formation is going on. The values of B_e and ρ_e are too uncertain to calculate t_M from Eq. (21). From Table I we see that the condensations in the M 17 complex are rotating with $\omega < 0.05$ units if at all, while in L 1641, a condensation is rotating at 1.2 units. Other massive clouds have intermediate values. It may turn out that magnetic braking is effective in some, but not all massive clouds.

Finally, we comment on the effect of Alfvén waves within dark clouds. Arons and Max (1975) suggested that the large observed widths of molecular lines may be due to the presence of such waves of large amplitude. If this were true, the associated wave motions would be a stabilizing term in the virial theorem, and would therefore be relevant to the question of whether such clouds are collapsing.

In their study of Alfvén waves in dark clouds, Arons and Max showed that waves of short wavelength are rapidly damped by ion-neutral friction; the minimum wavelength which can survive without excessive damping is proportional to n_n/n_i, a ratio they took to be 10^5. With the lower limit of 10^8 derived by Guélin *et al.* (1977), the minimum wavelength exceeds the radius of the clouds, so this idea must be discarded in its original form.

One might argue, however, that a cloud could undergo global oscillations supported by internal magnetic stresses. However, such oscillations would occur only if there is an equilibrium state to oscillate about, and this requires a magnetic energy comparable with gravitational energy. So this case reduces to that of a cloud stabilized by a static magnetic field (for a discussion of this problem see the chapter by Mouschovias in this book).

DISCUSSION

It has long been recognized that the evolution of the angular momentum and magnetic field of the solar nebula is crucial to understanding the planetary system. The data of Table I are suggestive in this connection. Consider a fragment of one solar mass within one of the condensations in the M 17 cloud complex. Since the mean density in such a condensation is estimated to be about 500 H_2 cm^{-3}, the radius r of such a fragment would be 0.2 pc at present. If it corotates with the condensation as a whole, its angular momentum per unit mass is

$$l < \frac{3}{5}\,\omega r^2 = 3.5 \times 10^{20}\ \mathrm{cm^2\ sec^{-1}} \qquad (23)$$

only a factor 3 greater than that of the solar system. Thus, as pointed out by Mouschovias (1977), magnetic braking early in the history of a condensation can result in angular momenta of the order observed in binary star systems and, as indicated here, in planetary systems. As more velocity maps of star-forming clouds become available, the question of angular momentum will become increasingly interesting.

Acknowledgments. Bruce Elmegreen and Telemachos Mouschovias were kind enough to provide helpful comments.

REFERENCES

Arons, J., and Max, C.E. 1975. *Astrophys J.* 196: L77.
Bok, B.J. 1977. *Publ. Astron. Soc. Pacific.* 89: 597.
____. 1978. *Moon and Planets* (special Protostars and Planets issue). In press.
Crutcher, R.M. 1977. *Bull. Am. Astron. Soc.* 9: 555.
Crutcher, R.M.; Evans, N.J. II; Troland, T.; and Heiles, C. 1975. *Astrophys. J.* 198: 91.
Disney, M.J., and Hopper, P. B. 1975. *Mon. Not. Roy. Astr. Soc.* 170: 177.
Elmegreen, B.G.; Lada, C.J.; and Dickinson, D.F. 1978. In preparation.
Field, G.B. 1975. In *Galaxies and the Universe* (A. Sandage, M. Sandage and J. Kristian, eds.), Ch. 10. Chicago: Univ. Chicago Press.
Goldreich, P., and Kwan, J. 1974. *Astrophys J.* 189: 441.
Guélin, M.; Langer, W.D.; Snell, R.L.; and Wootten, H.A. 1977. *Astrophys J.* 217: L165.
Heiles, C. 1976. *Ann. Rev. Astr. Astrophys.* 14: 1.
Heiles, C., and Jenkins, E.B. 1976. *Astron. Astrophys.* 46: 333.
Herbst, E., and Klemperer, W. 1973. *Astrophys J.* 185: 505.
Ho, P.T.P.; Martin, R.N.; Myers, P.C.; and Barrett, A.H. 1977. *Astrophys J.* 215: L29.
Hopper, P.B., and Disney, M.J. 1974. *Mon. Not. Roy. Astr. Soc.* 168: 639.
Kutner, M.L.; Evans, N.J. II; and Tucker, K.D. 1976. *Astrophys J.* 209: 452.
Kutner, M.L., and Tucker, K.D. 1975. *Astrophys J.* 199: 79.
Kutner, M.L.; Tucker, K.D.; Chin, G.; and Thaddeus, P. 1977. *Astrophys. J.* 215: 521.
Leung, C.M., and Brown, R.L. 1977. *Astrophys. J.* 214: L73.
Loren, R.B. 1977*a*. *Astrophys J.* 215: 129.
____. 1977*b*. *Astrophys. J.* 218: 716.
Lynds, B.T. 1962. *Astrophys. J. Suppl.* 7: 1.
Martin, R.N., and Barrett, A.H. 1978. *MIT Radio Astron. Contr. No. 3.* Submitted to *Astrophys. J.*
Mathewson, D.S., and Ford, V.L. 1970. *Mem. Roy. Astr. Soc.* 74: 139.
Mestel, L. 1966. *Mon. Not. Roy. Astr. Soc.* 131: 307.
Milman, A.S. 1977. *Astrophys J.* 211: 128.
Mouschovias, T.Ch. 1976. *Astrophys. J.* 206: 753; 207: 141.
____. 1977. *Astrophys J.* 211: 147.
Ostriker, J.P., and Peebles, P.J.E. 1973. *Astrophys. J.* 186: 467.
Scoville, N.Z., and Solomon, P.M. 1975. *Astrophys. J.* 199: L105.
Snell, R.L., and Loren, R.B. 1977. *Astrophys. J.* 211: 122.
Spitzer, L., Jr. 1968. *Diffuse Matter in Space.* New York: Interscience.
____. 1978. *Physical Processes in the Interstellar Medium.* New York: Wiley.
Vrba, F.J. 1977. In *Star Formation* (T. de Jong and A. Maeder, eds.), p. 243. Boston: Reidel.
Zuckerman, B., and Palmer, P. 1974. *Ann. Rev. Astr. Astrophys.* 12: 279.

THE STELLAR MASS SPECTRUM

JOHN M. SCALO
University of Texas at Austin

The most luminous stars in our galaxy and the Magellanic Clouds probably have masses in excess of 100 $M_\odot$. A number of processes have been proposed which may impose an upper limit to the masses of stars, but only one (radiation pressure on infalling material) has been treated in any detail, and so the dominant process remains to be identified. Opacity effects may impose a lower limit on stellar masses of greater than 0.005–0.01 $M_\odot$, but rotation and accretion of ambient gas should increase this limit. The average differential mass spectrum of open cluster stars has a mean slope of -2.5 ± 0.4 between 1 $M_\odot$ and 10 $M_\odot$. The mass spectra of associations cannot be determined. The initial mass spectrum of the average field star can be fit by a half-lognormal distribution between 0.1 $M_\odot$ and 50 $M_\odot$. The form of the stellar mass spectrum is unknown for $M \lesssim 0.1$ $M_\odot$, although it may be possible to infer the integrated number of low-mass stars. The evidence for spatial and temporal variations of the stellar mass spectrum is reviewed. The strongest evidence for an initial mass function that is non-universal is the deficiency of stars with $M \lesssim 1$ $M_\odot$ in some clusters, although this evidence is somewhat ambiguous. The mass spectra of field stars and open clusters do not differ significantly in the range 1-10 $M_\odot$. Comparisons of observed and theoretical photometric properties of globular clusters and galaxies have not given any firm evidence for mass spectrum variations. Interpretations in terms of fragmentation theories and fragment interaction models are briefly reviewed.

The form of the stellar mass spectrum is intrinsically interesting because it probably reflects the operation of particular processes of importance during star formation; it provides a basic constraint on theories for the formation and evolution of protostars. The stellar mass spectrum is also of prime importance in constructing models for the chemical evolution of the solar neighborhood and galaxies, and for calculating photometric properties of synthetic star clusters and galaxies. This chapter attempts to summarize our present knowledge concerning the observed stellar mass spectrum and, more briefly, how the observations relate to theoretical models.

The following definitions are used in this chapter. The *mass spectrum* $n(m)$ represents the number of stars per unit mass interval; $n(m)$ is the differential frequency distribution of masses. The *mass function* ξ $(\log m)$ is the number of stars per unit logarithmic mass interval, after Salpeter (1955). The mass spectrum and mass function are related by

$$\xi(\log m) = m\, n(m)/0.434. \tag{1}$$

We also define the *spectral index* γ as the logarithmic derivative of $n(m)$ with respect to m at a given value of m

$$\gamma \equiv \partial \log n(m)/\partial \log m \quad . \tag{2}$$

The quantity γ will be referred to as the index of the mass spectrum. Note that the logarithmic derivative of the mass function is $\gamma + 1$. It should also be noted that several papers exist which contain errors due to confusion between the differential mass spectrum and the mass function (e.g., Huang, 1961; Schlesinger, 1969; Butcher, 1976).

I. MASS LIMITS

A. Observational Evidence

The direct determination of the upper limit to stellar masses does not provide a feasible comparison with theory. The local luminosity function shows the rapidly decreasing probability of observing single stars of larger and larger masses, let alone very massive stars in binary systems with suitable mass ratios and orbital parameters. Because of these selection effects, the fact that the largest directly determined stellar mass is $M \sin^3 i \approx 37\ M_\odot$ for HD 228854 (Batten, 1973) should not be given much weight. The observed lower mass limit for luminous stars may be a more statistically meaningful quantity because of the large absolute space densities of low-mass stars, although their faintness severely limits the number which can be observed. Also, the accuracy of mass estimates for low-mass systems is unclear. (For an extreme example, compare the results for Luyten 726-8 A and B given by van de Kamp [1969] and redetermined by Luyten as given by van de Kamp [1975].) The resulting lower limit is about 0.06 $M_\odot$ for Ross 614 B (Lippincott and Hershey, 1972) and Wolf 424 A and B (Heintz, 1972).

The lower mass limit appropriate for comparison with gravitational fragmentation theories and galactic evolution studies may be much smaller, since there is strong observational evidence for "unseen companions" (see van de Kamp, 1975). The empirical lower limit to the masses of unseen companions is set by observational considerations. Besides, it is not known if these objects form in the same manner as stars, or by some accumulation process such as often invoked to explain our own planetary system.

For the high-mass limit an alternate approach is to search for very luminous stars and estimate their mass from a mass-luminosity relation or from evolutionary tracks. Stothers and Simon (1968) had estimated an upper limit of 60 $M_{\odot}$ using O stars and blue supergiants in our galaxy, but this limit now appears too small. Conti and Burnichon (1975) have compared the positions of O stars of known distances with evolutionary tracks in the Hertzsprung-Russell (HR) diagram and find strong evidence for stars more massive than 100 $M_{\odot}$. A comparison of these data with binaries of known mass and luminosity has been given by Hutchings (1976; see his Fig. 1). Humphreys (1977) finds even larger bolometric luminosities for the brightest O stars in our galaxy. Stothers and Simon (1968) have noted that there is a relatively sharp upper limit to bolometric luminosity in the Magellanic Clouds, indicating that the luminosity function and mass spectrum rapidly approach zero at this limit. Stothers and Simon estimated a limit of $M_{bol} \approx -9.5$, from which a mass of about 60 $M_{\odot}$ was deduced, but Osmer (1973) has re-examined the problem and finds a sharply defined upper limit to the luminosity $M_{bol} = -11 \pm 1$ in both the Large and Small Magellanic Clouds, which corresponds to a mass $\gtrsim 100\, M_{\odot}$ (see also Hutchings, 1976 and his Fig. 2). Humphreys (personal communication, 1977) finds $M_{bol} = -11.6$ for the brightest star in the Large Magellanic Cloud. Whether our galaxy also possesses a sharp cutoff is unknown. Sandage and Tammann (1974) demonstrated that the luminosity of the brightest star in a galaxy is correlated with the total luminosity of the galaxy, suggesting that the maximum luminosities observed in our part of the galaxy and in the Magellanic Clouds are only limited by the total number of stars. It may therefore be incorrect to assume that some physical process prevents stars more massive than, say, $100\, M_{\odot}$ from forming. All that can be said is that the most luminous stars in our galaxy and the Magellanic Clouds probably have masses in excess of $\sim 100\, M_{\odot}$.

B. Processes Controlling the Lower Stellar Mass Limit

It is generally believed that the physical mechanism which determines the lower mass limit of *luminous* stars is well understood. If the mass of a protostar is too low, the central regions become degenerate before the temperature becomes high enough for nuclear reactions to occur. As contraction proceeds, the gravitational energy is unable to supply both the radiated luminosity and the increasing kinetic energy of the degenerate electrons. The luminosity can then only be maintained at the expense of the ion thermal energy, so that the central temperature decreases and nuclear ignition is avoided. An interesting qualitative estimate of the minimum mass was made by Weisskopf (1975), and can also be derived by equating the de Broglie wavelength of an electron to the mean inter-particle spacing and assuming the star is in hydrostatic equilibrium at the hydrogen ignition temperature. The quantitative calculation is complicated by the equation of state, which must allow for effects like partial ionization in regions of partial degeneracy, Coulomb interactions, and pressure ionization. Grossman and Graboski (1971) have determined a lower mass limit of 0.08 $M_{\odot}$, in reasonable agreement with the mass of Ross 614B.

Of course none of the above discussion precludes the formation of low-mass objects which do not derive energy from nuclear fusion. An old idea (Hoyle, 1953; Fowler and Hoyle, 1963; Gaustad, 1963) which has received much recent attention (Lynden-Bell, 1973; Low and Lynden-Bell, 1976; Rees, 1976; Suchkov and Schevinov, 1976; Silk, 1977*b*) is that the minimum mass is the Jeans mass corresponding to an optical depth unity across a fragment, since opaque fragments contract adiabatically, increasing the Jeans mass and preventing any further fragmentation. These investigations indicate a minimum mass of around 0.005 – 0.01 $M_{\odot}$ at the present epoch (grain cooling dominates) for spherical fragments. Non-sphericity (Silk, 1977*b*) and accretion of ambient gas (Scalo and Pumphrey, 1978; chapter by Silk in this book) may increase this lower limit significantly.

Several other proposals for the process which sets the lower mass limit have appeared. For example, in the model of Nakano (1973), the minimum mass is set by the minimum wavelength relating to the instability of a uniformly rotating disk investigated by Goldreich and Lynden-Bell (1965). The result is about 0.04 $M_{\odot}$.

C. Processes Controlling the Upper Stellar Mass Limit

The condition of hydrostatic equilibrium can be used to show that the importance of radiation pressure must increase with stellar mass (Eddington's quartic equation). Furthermore the virial theorem, when generalized to include radiation pressure, shows that a star becomes only weakly bound in the limit of a radiation-dominated gas. The questions are: just how important must radiation pressure become in order to prevent the formation of a star, and what is the specific physical mechanism involved?

Ledoux (1941) showed by a linear stability analysis that stars more massive than 100 $M_{\odot}$ are pulsationally unstable to nuclear energy generation. Schwarzschild and Harm (1959) improved the input physics and revised this critical mass to 60 $M_{\odot}$, but further improvement by Ziebarth (1970) raised it to 90 $M_{\odot}$. The nonlinear problem is extremely difficult because the *e*-folding time scale of the instability is several orders of magnitude larger than the pulsation period. Various approaches to the problem have been given by

Appenzeller (1970*a*, *b*), Ziebarth (1970), Talbot (1971*a*, *b*) and Papaloizou (1973*a*, *b*). All these authors find that the amplitudes are damped, but there is still disagreement concerning the damping mechanism; for example, Appenzeller finds damping by mass ejection, while Papaloizou finds that energy is transferred from the fundamental mode to overtones. Nevertheless, it is generally agreed that vibrational stability does not set the upper limit to stellar masses.

Larson and Starrfield (1971) have discussed several mechanisms which might limit the masses of stars.

1. The time scale for accretion of the material remaining in a cloud may exceed the evolutionary lifetime of the stellar object already formed. In this connection it should be noted that a firm lower limit to stellar main-sequence lifetimes ($\sim 1 \times 10^6$ yr for normal hydrogen abundance) exists (see Stothers, 1972*b*).
2. The radiation from the protostar may heat the outer layers of the cloud to a point where gas pressure can overcome gravity. This effect is seen in the calculations of Appenzeller and Tscharnuter (1974), Westbrook and Tarter (1975).
3. Radiation pressure on dust grains, either due to stellar ultraviolet or reradiated infrared radiation, may reverse the collapse.
4. The formation of an H II region will ionize and heat the outer regions of the clouds, preventing further infall if the radius of the H II region is large enough to engulf a significant mass. The formation of an H II region in a protostellar cloud has also been investigated by Baglin *et al.* (1973) and Berruyer (1974). All these authors assumed, following Mathews (1969), that dust grains are destroyed by sputtering and do not impede the expansion of the H II region. However Barlow (1971) has pointed out that Mathews seriously overestimated the sputtering yield, so the rate of H II region growth will be reduced.

Unfortunately most of the mass limits estimated by Larson and Starrfield depend sensitively on the assumed initial conditions, and on the assumed density distribution, $\rho \propto r^{-\frac{3}{2}}$. Gerola and Glassgold (1978) have presented collapse calculations which include a careful treatment of chemical effects and heating and cooling mechanisms, and they find a much steeper density distribution. For example, such a steep density gradient would give very small equilibrium radii of H II regions, making Process 4 above much less effective.

It appears, then, that Larson and Starrfield (1971) have correctly identified the most important mechanisms for limiting stellar masses, but, except for one possible exception discussed below, reliable numerical estimates for the limiting mass have yet to be carried out.

Kahn (1974) treated the effect of radiation pressure on infalling material more quantitatively and showed that if the luminosity-to-mass ratio was too large, then accretion will be halted, either by ultraviolet radiation incident on the inner edge of the infalling dust for small infall rates, or by the reradiated infrared radiation trapped in the optically thick cocoon if the infall rate is large. Kahn estimated that stars more massive than about 40 $M_{\odot}$ cannot form, but his result is very sensitive to the unknown properties of interstellar dust, in particular to the assumption that most of the interstellar grains are graphite. Bedijn (1977) refined this calculation somewhat, using realistic graphite opacities, and found an upper mass limit of 60 $M_{\odot}$. It is not clear how an accurate treatment of the radiation field or uncertainties in the dust composition will affect this result. As discussed above, stars more massive than 100 $M_{\odot}$ probably exist.

Shields and Tinsley (1976) have pointed out that the process described above leads to an upper mass limit which increases with decreasing metal abundance Z, assuming the grain opacity is proportional to Z. In that case the upper mass limit in galaxies with Z-gradients should increase with distance from the nucleus, since Z is generally found to decrease with distance from the nucleus. The possibility of formation of very massive stars in the early history of our galaxy, suggested for other reasons by, e.g., Truran and Cameron (1971), is also implied, unless some other limiting mechanism comes into play.

For example, Elmegreen and Elmegreen (1978; see also Elmegreen and Lada, 1977; chapter by Lada *et al.* in this book) have summarized the observational evidence that star formation can be induced by shocks associated with spiral density waves, cloud collisions, H II regions, and supernova explosions and have derived the relation between growth rate and wavelength for a plane-parallel layer with external pressure. If the column density, σ of the layer is sufficiently large and/or the external pressure is sufficiently small, the fastest growing mode is unstable to gravitational collapse of a sphere, and stars may form. Elmegreen and Elmegreen make the plausible assumption that the maximum mass which can form is controlled by the condition that the growth time for this wavelength just be smaller than the duration of the shock, τ. The resulting maximum mass varies as $\tau^4 \sigma^3$ and therefore should vary as a function of time in a given region and also as a function of position in the galaxy, unless some other process (such as radiation pressure) sets a more stringent upper limit. The mass spectrum for this scenario has yet to be calculated.

II. OBSERVED MASS SPECTRUM

A. Open Clusters

Open clusters present a favorable case for determination of stellar mass spectra because all the members were formed at about the same time, so the mass spectrum below the main-sequence turnoff is the initial mass spectrum. This is in contrast to the situation encountered for field stars, where one must specify the history of the stellar birthrate function in order to account for stars which have died (see below). There are several problems which arise in determining the cluster mass spectrum: Membership is often difficult to establish, the number of stars which can be used in a given cluster is small, photometry is usually unavailable for the faintest stars, and one should in principle use only relatively young clusters, since the age distribution of clusters (Wielen, 1971) shows that cluster disruption processes, some of which may depend on stellar mass, are important for the older groups.

Early important work on the luminosity functions of open clusters was given by Sandage (1957), Jaschek and Jaschek (1957), van den Bergh and Sher (1960), van den Bergh (1957, 1961), and Walker (1957). Several of the luminosity functions of the first and third papers were converted to mass spectra by Nakano (1966). Some examples, using a more recent mass-luminosity relation, are shown in Fig. 1 and are discussed further below.

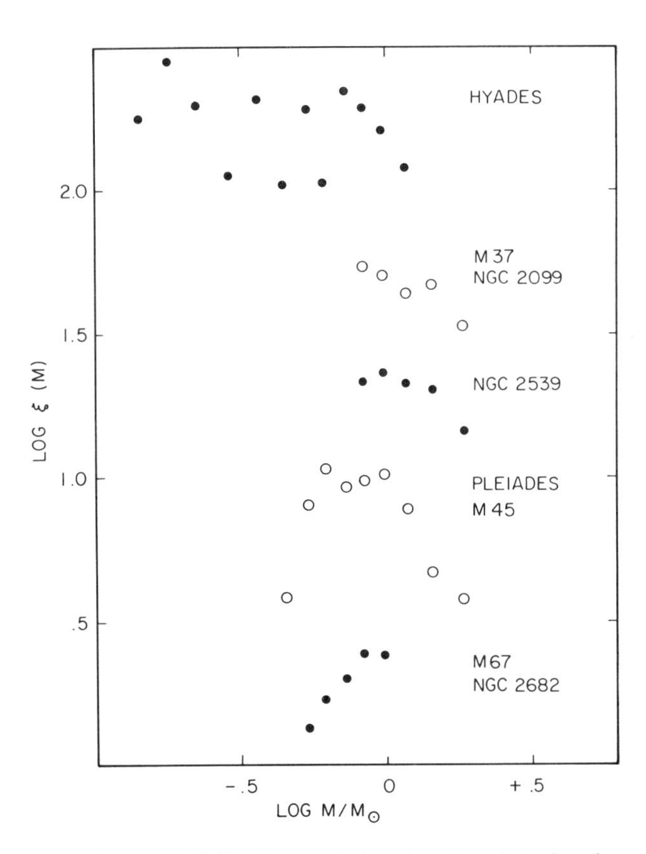

Fig. 1. Mass functions of individual open clusters, based on data given by van den Bergh (1961) except for the Hyades cluster, which is taken from van Altena's data as given by Pels *et al.* (1975).

Additional references can be found in Artyukhina (1972, 1973); Kholopov and Artyukhina (1972), Taff (1974), Frolov (1975), Archemashvili (1976), and Piskunov (1976).

Several papers have discussed the average mass spectrum of open clusters. Taff (1974) included 62 clusters after rejecting many more on the basis of number of members (if less than around 50), cluster age (to insure against cluster disruption effects), large differential reddening, and other factors. More than half of the 62 clusters contain over a hundred stars. A method for testing the hypothesis of universality of the luminosity functions of clusters is used which accounts for evolution at the bright end of the main sequence and incompleteness at the faint end. Taff concludes that the luminosity functions of clusters of different richness and concentration classes do not differ significantly for masses above about 1 $M_{\odot}$, except for 10 very rich clusters, and he attributes this difference to problems with the data. It should be emphasized that this statement only applies for $M \gtrsim 1\ M_{\odot}$. Taff's cluster mass function is shown in Fig. 2. The average mass spectrum between 1 and 10 $M_{\odot}$ is fit well by a power law with an index of about -2.7, although for such a small mass range other functions could be fit to the data. Uncertainties in the mass-luminosity relation have almost no effect on this result. Above 10 $M_{\odot}$ the spectrum flattens out ($\gamma \approx -1.8$), but the uncertainties are large because of the scarcity of high-mass stars. The spectrum apparently begins to flatten out below about 1 $M_{\odot}$, an effect also seen in the mass spectrum of field stars discussed below.

Burki (1977), apparently unaware of Taff's work, studied the bright end ($-6 < M_V < 1$) of the luminosity functions of 27 young ($\lesssim 15 \times 10^6$ yr) clusters. This luminosity range corresponds to a mass range of 2.5 $M_{\odot}$ to 60 $M_{\odot}$. Burki finds that the index of the mass spectrum is smaller for large clusters (diameter > 8 pc, $\gamma \approx -2.0$) than for small clusters (diameter < 4 pc, $\gamma \approx -2.7$) and notes that the theoretical mass spectrum of Larson (1973, see below) predicts the same sort of variation. Burki's result, if not affected by selection effects (e.g., corrections for field stars and/or differential reddening could depend on cluster size), would be an important constraint on theories of the stellar mass function. Taff (1974) found no clear systematic effects depending on richness or concentration class using more clusters and a larger average number of stars per cluster, but it is not clear whether cluster size is correlated with these parameters. The mean index for all the clusters was found to be about -2.2 by Burki, less steep than Taff's result. Piskunov (1976) finds a best fit to 61 clusters with a power law spectrum of index -2.3 over the range 1 $M_{\odot}$ to 25 $M_{\odot}$, with no evidence for large differences from cluster to cluster. We conclude that the average value of index γ of the mass spectrum in open clusters is -2.5 ± 0.4 between 1 $M_{\odot}$ and 10 $M_{\odot}$.

There is no assurance that the spectral index determined for open clusters applies to most stars in the galaxy or even to the solar neighborhood. First of all, most stars, if formed in groups, probably form as part of loose

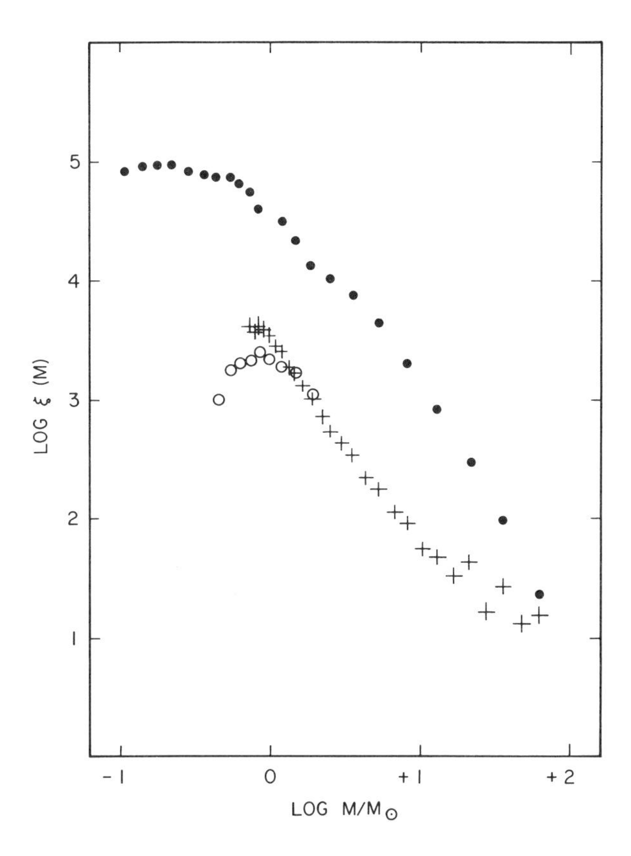

Fig. 2. Comparison of the initial mass function of field stars from Miller and Scalo (1978, filled circles) with the mean mass functions in open clusters of van den Bergh (1961, open circles) and Taff (1974, crosses). Normalization is arbitrary.

associations, not in open clusters (Ebert *et al.*, 1960; Wielen, 1971; Miller and Scalo, 1978). Secondly, the mechanism of star formation may be different in the two types of groups. It would therefore be useful to compare the mass spectra of associations, open clusters, and field stars. Unfortunately the luminosity function of associations is unknown except at the highest luminosities (Blaauw, 1964), because of the low star density, large angular size, and small proper motions, all of which makes membership studies difficult, and because often the lower luminosity stars are still approaching the main sequence. The comparison of the mass spectra of associations with other stars therefore appears out of reach at the present time.

The cluster mass spectrum below about 1 $M_\odot$ is more difficult to establish, and there is still considerable uncertainty in its form. The substantial evidence that large variations between individual cluster mass spectra exist is discussed in Sec. III below.

B. Field Stars

The second method for establishing the stellar mass spectrum uses the luminosity function of field stars, which is now fairly well established. A full discussion of methods for determining the luminosity function is given by McCuskey (1966), and more recent developments are summarized by Miller and Scalo (1978). The derivation of the present-day mass spectrum from the luminosity function, with an emphasis on the uncertainties, is discussed in detail by Miller and Scalo (1978). Luminosities are converted to masses using either an empirical spectral type-mass relation or a theoretical mass-luminosity relation. Space densities are converted to column densities in the galactic disk, which are more meaningful because the z-distribution (the distribution in the direction perpendicular to the galactic plane) of stars is a function of mass. The surprisingly large uncertainties in the z-density distributions of stars of various masses enters the determination of the mass spectrum at this point.

The present-day mass spectrum thus determined has been altered by the effects of possible changes in the birthrate of stars over the history of the galaxy, and because of the variation of stellar lifetime with mass. Hartmann (1970) has pointed out that the mass spectrum of field stars actually refers to discrete mass entities, whether or not found in multiple systems. The effect of unresolved multiple stars would be to effectively hide some low-mass stars, giving an apparent mass spectrum which is too flat at low masses (Whipple, 1977, personal communication). However Hartmann (1970) suggests that the effect is small by comparison of the luminosity function of nearby and distant stars.

For stars with main-sequence lifetimes $\tau_{ms}(M)$ greater than or equal to the age of the galactic disk T_0, the observed mass spectrum represents all such stars ever born, and therefore corresponds to the mass spectrum of a group of stars at birth. On the other hand, for stars with $\tau_{ms}(M) < T_0$ we only see those stars born later than time $T_0 - \tau_{ms}(M)$ after the beginning of star formation in the disk. The observed mass spectrum must then be increased to account for the stars which have died. This correction requires a knowledge of how the stellar birthrate $b(t)$ has varied with time in the solar neighborhood. As emphasized by Miller and Scalo (1978), the form of $b(t)$ and even the sign of db/dt (i.e., whether the birthrate increases or decreases with time) is essentially unknown, except that $b(t)$ cannot change too much over the age of the galaxy (c.f., Tinsley, 1977). In particular, there is essentially no basis for assuming that $b(t)$ is proportional to the first or second power of the gas density (c.f., Madore, 1977) an assumption which permeates the literature.

Given an assumed form for the birthrate, the observed mass spectrum can be corrected for evolution; the corrected spectrum, expressed per unit logarithmic mass interval, is referred to as the initial mass function (IMF).

Salpeter (1955) was the first to carry out this calculation, assuming a constant birthrate, and found that the IMF could be fit by a power law of slope −1.35 (i.e., index of the differential mass spectrum equal to −2.35). Although Salpeter's method for obtaining the IMF was correct in principle, revisions in main-sequence lifetimes, mass-luminosity relation, adopted age of the galaxy, and other input data substantially alter the result. This is important to note because of the widespread usage of Salpeter's result in the literature. Subsequent discussions were given by Schwarzschild (1958), Salpeter (1959), Schmidt (1959, 1963), Warner (1961*a*, *b*) and Limber (1960). A more recent discussion is given by Audouze and Tinsley (1976) and Tinsley (1977).

A redetermination of the field star mass spectrum including a discussion of the uncertainties in the input data, has been given by Miller and Scalo (1978). The resulting *initial mass function* for a constant birthrate is shown in Fig. 3 as the data points. The small bump at $\log M \approx -0.25$ and dip at $\log M \approx +0.3$ are not physical, but are due to uncertainties in the luminosity function and the fraction of stars which are main-sequence stars. Much of the uncertainty in the slope of the mass function at high masses has been eliminated by combining star counts according to luminosity with counts according to spectral type, and the use of both observed and theoretical mass-luminosity relations. On the other hand, it is possible that the number of massive stars has been underestimated if they remain shrouded in dust for a significant fraction of their lifetime. Also, the possible effect of mass loss from massive stars has been ignored. These points are briefly discussed by Wheeler, Miller, and Scalo (1978). At low masses the adopted luminosity function is a mean of those given by McCuskey (1966), Luyten (1968), and Wielen (1974), since there is no longer any evidence for an enhancement in the number of M dwarfs (see Faber *et al.*, 1976 and references therein). The spectrum shows no evidence for the kinks shown by Audouze and Tinsley (1976), which were mostly due to choice of luminosity functions and fitting noise, especially the use of the results of Ostriker *et al.* (1974) at high masses. The shaded region represents the maximum allowable range for the IMF, considering all uncertainties including the time-dependence of the birthrate.

The solid curve is a lognormal fit for a constant birthrate:

$$\xi = C_0 \exp\left[-C_1 (\log M - C_2)^2\right] \text{ pc}^{-2} (\log M)^{-1}, M > 0.1\, M_\odot \quad (3)$$

with $C_0 = 99.54$, $C_1 = 1.09$, $C_2 = -0.99$, and M in solar masses. Expressed as a power law, the mass spectrum $[n(M) = 0.434\xi/M]$ has a variable mass index

$$\gamma = -(1.94 + 0.94 \log M) \quad (4)$$

which varies from $\gamma = -1.0$ at $0.1\, M_\odot$, $\gamma = -1.9$ at $1\, M_\odot$, $\gamma = -2.9$ at $10\, M_\odot$, and $\gamma = -3.8$ at $100\, M_\odot$. Recall that the Salpeter value is -2.3, while $\gamma = -2.5$

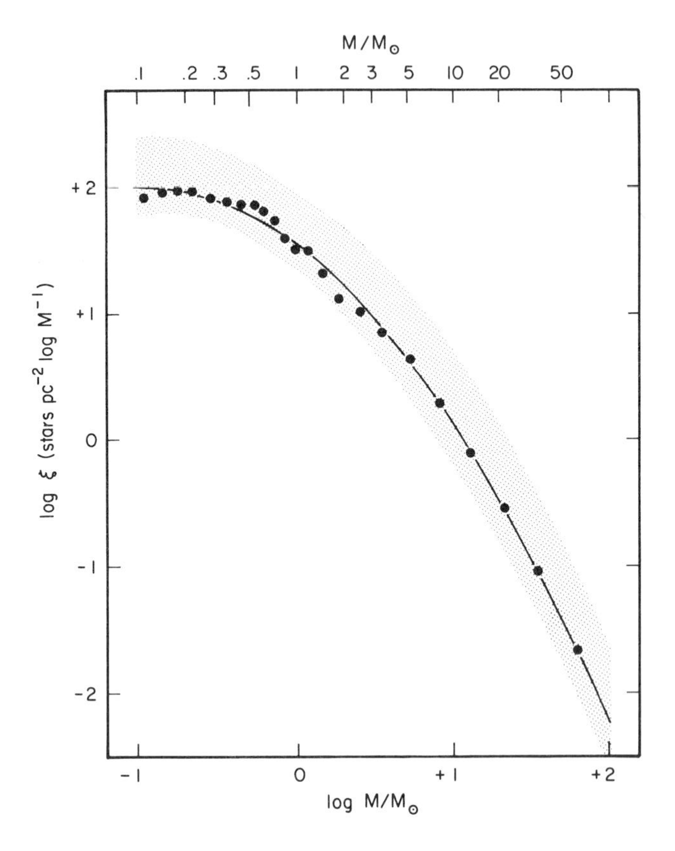

Fig. 3. Initial mass function ξ of field stars, given as the number of stars per square parsec per unit interval of log M. Data points were computed assuming a time-independent birthrate in the galactic disk. Solid line is a lognormal distribution. Shaded region shows the possible range for the initial mass function after estimating uncertainties in all input data. (From Miller and Scalo, 1978.)

$\pm$ 0.4 from 1 $M_\odot$ to 10 $M_\odot$ in open clusters.

An important point is that *the stellar mass spectrum is unknown for $M \lesssim$ 0.1 $M_\odot$*. Luyten (1968) has derived a luminosity function down to absolute photographic magnitude $\approx$ 24, but these data cannot be used to infer the mass spectrum for several reasons. First, the observational mass-luminosity relation is unknown for such faint stars. A theoretical mass-luminosity relation would be uncertain because of difficulties in model construction and large and uncertain bolometric corrections. Second, the scale height of faint dwarfs is unknown. Third, one does not know what fraction of these very faint stars are actually white (brown?) dwarfs rather than low-mass main-sequence stars. Luyten suggests the data are complete to $M_{pg} \approx 17$ but Hartmann (1970) has emphasized the uncertainty in this estimate. Also, since the stars were found in blink surveys, stars of low proper motion may have been missed. Therefore, although the mass spectrum at $M \approx 0.1 - 0.3\, M_{\odot}$ and Luyten's luminosity function suggest that the mass spectrum flattens or turns over for $M \lesssim 0.1\, M_{\odot}$, its quantitative form cannot be determined.

There are. however, a few indirect arguments related to the mass spectrum for $M < 0.1\, M_{\odot}$:

1. Assuming that binaries form by a capture mechanism, one can require that a random pairing of stars drawn from the mass spectrum match the observed distribution of mass ratios in binary systems. Warner (1961*b*) has used this method to infer a turnover in the IMF. Unfortunately the distribution of mass ratios is poorly known.

2. If we assume that the mass spectrum in open clusters is the same as that for the field, we can compare the integrated mass spectrum (extrapolated to low masses) down to, say, $0.1\, M_{\odot}$ with the observed dynamical masses for the few clusters with known virial theorem masses; any difference must be due to low-mass stars or remnants of high-mass objects. The results of Jones (1970) for the Pleiades and Praesepe clusters suggest that the fraction of mass in stars less massive than $\sim 0.1\, M_{\odot}$ is between 0 and 0.5. One of the main problems here is that the virial mass varies as the cube of the cluster distance. Also, this method only gives a lower limit to the mass fraction because of the evidence, summarized below, that open clusters are deficient in low-mass stars compared to the field.

3. We can add up the total amount of mass observed in the solar neighborhood, including main-sequence stars, white dwarfs, red giants, and gas, plus a correction for the expected mass of neutron stars and black holes, and compare with the "Oort limit" (the total mass required to account for the observed stellar motions). Any difference may be assigned to stars with $M < 0.1\, M_{\odot}$.

Previous discussions of this problem (e.g., Hartmann, 1970) have assumed that the mass function continues to increase with decreasing mass for very small masses, deriving a lower mass limit consistent with the missing mass. In contrast, the mass function of field stars shown in Fig. 3 is constant or may even be decreasing for $M \sim 0.1\, M_{\odot}$. Even if the mass function is constant for $M < 0.1\, M_{\odot}$ (i.e., $\gamma = -1$) down to $M = 0$, the total amount of mass contributed by "dark objects" would be far short of the usually quoted missing mass, although the total number of objects would about double for every decade in mass below $0.1\, M_{\odot}$. It is concluded that either (a) there is no missing mass problem (see Krisciunas, 1977 for discussion and references); (b) the slope of the present-day mass function is steeper than shown in Fig. 3 for $M \lesssim 0.1\, M_{\odot}$; or (c) the missing mass resides in some other form. Note that planetary systems associated with stars cannot be invoked for this purpose, since a very large mass of planets per star would be required. An additional means of possibly inferring variations in the relative amounts of mass locked up in stars with $M \lesssim 1\, M_{\odot}$ would be the detection of systematic variations in mass-to-light ratios in external galaxies (Audouze and Tinsley, 1976).

III. VARIATIONS IN THE STELLAR MASS SPECTRUM

Observational evidence for systematic variations in the form or limits of the stellar mass spectrum in space or in time could provide an important clue concerning the dominant processes in star formation. Evidence that the mass spectrum is "universal" in some sense would also constrain the theory. At present, much of the evidence regarding these points is ambiguous and/or very indirect. Theoretical models for the stellar mass spectrum are currently so rudimentary that no firm predictions of the dependence of the form of the mass spectrum on physical parameters such as metal abundance or turbulence can be given, although further investigations of the upper and lower mass limits along the lines of that given in Sec. I.C should be profitable. Recent theoretical discussions of star formation in metal-poor gas, along with further references, are given by Hutchins (1976) and Silk (1977*a*).

The most widely-discussed evidence for variations in the stellar mass spectrum concerns the lack of low-mass stars in open clusters. Van den Bergh and Sher (1960) studied the luminosity functions of 20 galactic clusters down to absolute photographic magnitudes varying between 5 and 11. Even after allowing for evolutionary effects, differences in the luminosity functions among clusters were apparent. The luminosity functions become either constant or decreasing for $M_{pg} \gtrsim 5$ in most of the clusters, unlike the "Salpeter (1955) function." The same effect has been noted by other authors (e.g., Roberts, 1958; Williams and Cremin, 1969) and can be seen in the more recent data on clusters M 11 and M 67 summarized by McNamara and Sanders (1977, 1978). Van den Bergh and Sher pointed out that this effect would not be due to evaporation of low-mass stars because the relaxation times are longer than the cluster ages, and they suggested that the mass spectrum of clusters is not the same as for field stars, implying that the physical conditions and/or processes involved in star formation are different for the two groups. This point was also emphasized by van den Bergh (1961) and others. However, it is clear that some process (e.g., tidal disruption or differential galactic rotation) disrupts most clusters in a fairly short time ($\sim 10^8$ yr), as can be seen from the cluster age distribution derived by Wielen (1971). In addition, radial mass segregation may occur and strong evidence for this effect in several clusters can be found in papers by Artyukhina (1972), Archemashvili (1976), and King and Tinsley (1976). Secondly, one must recall Hartmann's (1970) statement that "no reported turn-down in the mass function for low-mass stars can be naively assumed to be meaningful," because of problems with incompleteness. Nevertheless, a few clusters definitely show large deficiencies of low-mass ($\lesssim M_\odot$) stars which cannot be attributed to incompleteness (Walker, 1977, personal communication; Herbst,

1978) or to age (see Herbst, 1978 on NGC 6193).

A few of the better-determined mass functions of clusters are shown in Fig. 1, based on luminosity functions given by van den Bergh (1961), McNamara and Sanders (1977, 1978), and van Altena's (1969) data as given by Pels, Oort, and Pels-Kluyver (1975). It can be seen that even for these relatively nearby and/or rich clusters the data are fairly noisy. Some clusters have mass functions which become roughly flat for $M \lesssim 1\ M_{\odot}$. This behavior is entirely consistent with the field star IMF shown in Fig. 3. The difference between clusters and field arises only in cases like the Pleiades and M 67 which show definite turnovers in their mass functions below about 1 $M_{\odot}$. Herbst (1978) finds the same behavior for the young cluster NGC 6193.

Van den Bergh (1961) has given a mean luminosity function for $M \lesssim 2\ M_{\odot}$ in eight open clusters. These results have been converted to a mean cluster mass function, which is shown in Fig. 2 along with Taff's (1974) mass function for the higher mass stars and the field star mass function of Miller and Scalo (1978). The normalization is arbitrary. The important points to note are: 1. Within the uncertainties, Taff's (1974) average cluster mass function agrees with the field star IMF; 2. If Taff's and van den Bergh's (1961) average cluster mass spectra are forced to overlap at the high-mass end of van den Bergh's data, where incompleteness should not be a problem, then the two results are in disagreement for $M \lesssim 1-1.5\ M_{\odot}$, in the sense that van den Bergh's data show a deficiency of low-mass stars relative to Taff's result. Since Taff's mean includes many more clusters than does van den Bergh's, one must question whether the much-discussed turnover at low masses is an artifact resulting from a combination of incompleteness, difficulties in ascertaining cluster membership, segregation, and other factors. We tentatively conclude that differences exist between the mass spectra of field stars and some clusters for $M \lesssim M_{\odot}$, but a detailed re-examination of this problem is necessary before more quantitative statements can be made.

Concerning systematic differences among open clusters, no clear dependence on any cluster property has been found, except by Burki (1977), who finds that the slope of the upper mass spectrum is steeper for smaller clusters, as discussed earlier.

Evidence for or against a temporal variation of the mass spectrum of field stars in our galaxy is slim. Dixon (1970) divided a sample of late-type dwarfs into two age groups on the basis of their positions in the color-magnitude diagram and their space motions. A variation in the mass spectrum should manifest itself as a variation in the fraction of stars in each age group as a function of mass. No significant variation was found. The difficulty in accounting for the small number of metal-poor dwarfs in the solar neighborhood using simple chemical evolution models has led to several suggestions that the stellar mass spectrum was quite different early in the history of our galaxy. Schmidt (1959, 1963) and Truran and Cameron (1971) proposed an overproduction of high-mass stars, while Biermann and Tinsley

(1974) accomplish the same goal by not allowing low-mass ($\lesssim 1\ M_\odot$) stars to form until the metal abundance reached its present level. Enough alternate methods for explaining the number of metal-poor stars exist (e.g., infall, metal-enhanced star formation, comets, etc.; see Trimble [1975] and Audouze and Tinsley [1976] for reviews) so that the case for a variation in the mass spectrum cannot be considered very firm on the basis of this indirect argument.

There is somewhat ambiguous evidence for variations in the stellar mass spectrum in globular clusters and other galaxies. Limber (1960) has pointed out difficulties in accounting for the mass-to-light ratios of galaxies and globular clusters if a universal mass spectrum is assumed. Quirk and Tinsley (1973) invoked a variation of the lower mass limit in order to explain the mass-luminosity gradients in M 31 and M 33. Van den Bergh (1973) suggested that the different M/L ratios of dwarf ellipticals and giant ellipticals is evidence for a non-universal mass spectrum. However, more recent estimates of M/L ratios in ellipticals by Faber and Jackson (1976) and Sargent *et al.* (1977) and the possibility that the M/L ratios are affected by metallicity variations (Smith and Tinsley, 1976) make these conclusions doubtful. Schmidt (1975) has presented a luminosity function for halo dwarfs which may differ from that of disk stars, although Schmidt considers the result quite uncertain. Olson and Pena (1976) tried to account for the observed properties of the Magellanic Clouds with a closed one-zone galactic evolution model, and concluded that the mass spectrum must differ from that of our galaxy. Either the mass spectrum is steeper or the lower mass limit is much smaller in the Magellanic Clouds. However, these authors used a model with an age of 12×10^9 yr, while Butcher (1977) has shown that the difference between the observed luminosity functions of the Large Magellanic Cloud and our galaxy can be simply accounted for by the hypothesis that star formation has only occurred over the last $\sim 5 \times 10^9$ yr. Van den Bergh (1976*b*) notes that one could use a mass spectrum favoring high-mass stars toward the nuclear regions of galaxies in order to explain the observed metallicity gradients, although other explanations are possible.

The question of whether or not a mass spectrum enriched in low-mass stars is required to match the observed properties of elliptical galaxies has received much attention. Faber and Gallagher (1976) suggested that the lack of OB stars in elliptical galaxies is not due to a cessation of star formation, but to the fact that only low-mass stars form. Jura (1977) has pointed out that, because of the weak ultraviolet radiation field in ellipticals, gas clouds in these objects should be colder and contain more H_2 than clouds in, say, our own galaxy, and so the Jeans mass is very small ($\lesssim 10\ M_\odot$). Since these clouds may fragment during collapse, it is likely that only low-mass stars may form. However, Tinsley and Gunn (1976) concluded that a power law mass spectrum must have an index less steep than -2 in order to account for the strengths of the gravity-sensitive Wing-Ford band (FeH) and CO bands in elliptical galaxies, not inconsistent with the local field star mass spectrum. Whitford (1977) finds that the non-detection of the Wing-Ford band in giant ellipticals constrains power law mass spectra to have indices less steep than −3 at low masses. Since the mass spectrum of the field stars has an index of about −1 at the masses of interest, Whitford's result is not inconsistent with the proposed low-mass enriched spectrum, contrary to Whitford's own conclusion.

Spectacular evidence for a non-universal mass spectrum was provided by van den Bergh (1976*a*), who noted that the lack of H II regions in the spiral arms of M 104 implies that stars more massive than 10 $M_{\odot}$ are not currently being formed. However, recent work by Schweizer (1978) negates this conclusion. Sargent and Searle (1970) studied two metal-poor blue dwarf irregular galaxies and concluded that, if these galaxies are old and have had uniformly declining birthrates, their mass spectra must be enriched in massive stars. Later work by Searle and Sargent (1972) showed that a more consistent interpretation is that either the galaxies are young or that star formation has proceeded in bursts.

Searle (1973) pointed out that the fact that the average yield of metals in late-type spirals does not vary significantly implies that the mass spectrum cannot vary much. Sandage and Tammann (1974) find a good correlation of the luminosity of the brightest star in a galaxy with the total galactic luminosity, consistent with a universal upper mass spectrum. In a detailed comparison of synthetic stellar populations with colors of Sc and Irr I galaxies Searle, Sargent, and Bagnuolo (1973) found that a single power law for the mass spectrum led to acceptable agreement of models with observations if the time dependence of the birthrate was allowed to vary. Essentially the same conclusion had been reached earlier by Tinsley (1968).

In an extension of this type of work, Huchra (1977) has found that the colors and Hβ strengths of blue galaxies require a mass spectrum index less steep than −2.3 (Salpeter's IMF), assuming a power law IMF is applicable and that the upper mass limit is less than about 60 $M_{\odot}$. As was discussed earlier, it is likely that the upper mass limit exceeds 100 $M_{\odot}$, altering Huchra's conclusions based on Hβ strengths.

Da Costa (quoted by Freeman, 1977) has derived the mass spectra for three nearby globular clusters from $M_V \approx +5$ to $M_V \approx +9$ or +10, and finds a significant range in the spectral index, from ~ −2 to ~ −4. Unfortunately this result is strongly dependent on theoretical corrections for radial segregation according to mass. Since the observed luminosity functions only refer to the outer parts of the clusters, one expects an apparently steeper mass spectrum if segregation has occurred. The work of Artyukhina (1973) and Archemashivili (1976) indicates the severity of this effect for open clusters. The index of the mass spectrum for the local field stars is about −1.5 to −2 for the same M_V range as the globular clusters studied by Da Costa, so the range in γ derived by Da Costa indicates steeper mass spectra in some globular clusters. There is some suggestion of steepening with increasing metal abundance and/or increasing total mass. On the other hand, if the effects of radial segregation are more severe than the theoretical models indicate, the observations may be consistent with the field star IMF. The same remarks apply to inferences drawn from luminosity functions of other globular clusters.

Freeman (1977) has discussed preliminary results on the mass spectra of six young globular clusters in the Large Magellanic Cloud. Freeman finds that the cluster mass spectra are well-represented by a power law, and a range in γ from -1.2 to -3.5 among the clusters for $1\,M_\odot < M < 6\,M_\odot$. Three of the clusters have $\gamma \gtrsim -1.5$. Since the index of the local field star mass spectrum has an average value of about -2.2 in the same mass range, it appears that young globular clusters may possess steeper or flatter IMF's than the field stars in our galaxy. Uncertainties due to incompleteness at the faint end of the luminosity function or in ascertaining cluster membership were not discussed by Freeman, but assuming for now that these uncertainties are not large, Freeman's result is the strongest evidence to date that the shape of the IMF above $1\,M_\odot$ is not universal.

It is seen that much work remains to be done before anything can be said concerning systematic variations in the stellar mass spectrum. Although there is strong evidence that variations exist for $M \lesssim M_\odot$ in open clusters, there is no compelling evidence for variations above 1-2 $M_\odot$ or for variations in other galaxies, except perhaps for young globular clusters in the Large Magellanic Cloud (Freeman, 1977). This problem for external galaxies cannot be resolved until we have some means of disentangling the time dependence of the birthrate from the mass spectrum and its upper and lower limits. In addition, colors are simply not very sensitive to the shape of the mass spectrum.

IV. SUMMARY AND DISCUSSION

1. Observational evidence indicates that luminous stellar masses range from an upper limit above 100 $M_\odot$ down to below 0.1 $M_\odot$. The relative numbers of stars with $M \lesssim 0.1\,M_\odot$ and their mass spectrum are important unknowns.

2. A number of processes exist which would make it difficult for stars more massive than $\sim 10^2\,M_\odot$ to form. A quantitative study of several of these mechanisms is lacking; consequently the dominant process cannot yet be identified. The lower mass limit of fragments may initially be set at $10^{-2}\,M_\odot$ by opacity effects, but it is likely that interactions of fragments with each other and with ambient gas increase this limit somewhat.

3. The differences among the mass spectra of open clusters are small enough so that it is meaningful to define an average cluster mass spectrum in the range $1\,M_\odot < M < 10\,M_\odot$. The average spectral index is -2.5 ± 0.4 for this

mass range. The form of the mass spectrum of open clusters outside of these mass limits is still uncertain, but there are indications of variations among clusters at $M \lesssim 1\, M_{\odot}$.

4. The derivation of the initial mass spectrum of field stars is complicated by uncertainties in such quantities as the time dependence of the birthrate and stellar scale heights. The recent derivation by Miller and Scalo (1978) can be fit with a half-lognormal distribution from $0.1\, M_{\odot}$ to $\sim 50\, M_{\odot}$.

5. There is currently no unambiguous evidence for systematic variations of the stellar mass spectrum in space or time, except possibly for the dependence of IMF on cluster size found by Burki (1977). In particular, the mass spectrum of field stars (mostly *not* formed in open clusters) and open clusters do not differ significantly within the current observational uncertainties in the range 1-10 $M_{\odot}$. There is evidence for a deficiency of low-mass ($\lesssim 1\, M_{\odot}$) stars in some clusters, and this constitutes the strongest argument against a universal IMF. Comparisons of observed and theoretical photometric properties of globular clusters and external galaxies have not given any firm evidence for mass spectrum variations; photometric properties are not very sensitive to the mass spectrum and it is difficult to disentangle the history of the star formation rate from the mass spectrum. The only direct support for variations of the stellar mass spectrum in galaxies comes from Freeman's (1977) study of young globular clusters in the Large Magellanic Cloud.

Given our present knowledge of the stellar mass spectrum, can the theory of star formation offer any interpretation in terms of physical processes? Within the context of the theory that stars form by the fragmentation of interstellar clouds, there are essentially only two ways to interpret the observed stellar mass spectrum: either the mass spectrum reflects the manner in which mass is parceled out to fragments as they form (fragmentation models), or the spectrum reflects processes which add or subtract material from fragments after they have formed (fragment interaction models). These processes might include coalescence or disruption of fragments during encounters with each other and interactions with ambient gas through which fragments are moving, which can either be accreted or can strip material from the fragments. Fragmentation theories and fragment interactions are reviewed in Silk's chapter. The initial fragment mass spectra predicted by various fragmentation theories are compared graphically by Miller and Scalo (1978).

With a few exceptions the theories are ad hoc and highly simplistic, especially the geometrical and probabilistic theories. Nearly all the predicted mass spectra are power laws with indices between -2 and -3 and do not exhibit the flattening or turnover seen in the mass spectra of field stars and clusters. A notable exception is the probabilistic model of Larson (1973), which yields a lognormal mass spectrum. However, it is difficult for this theory to explain the degree of uniformity observed in cluster and field star mass spectra for $M \gtrsim M_{\odot}$ unless the fragmentation probability has almost

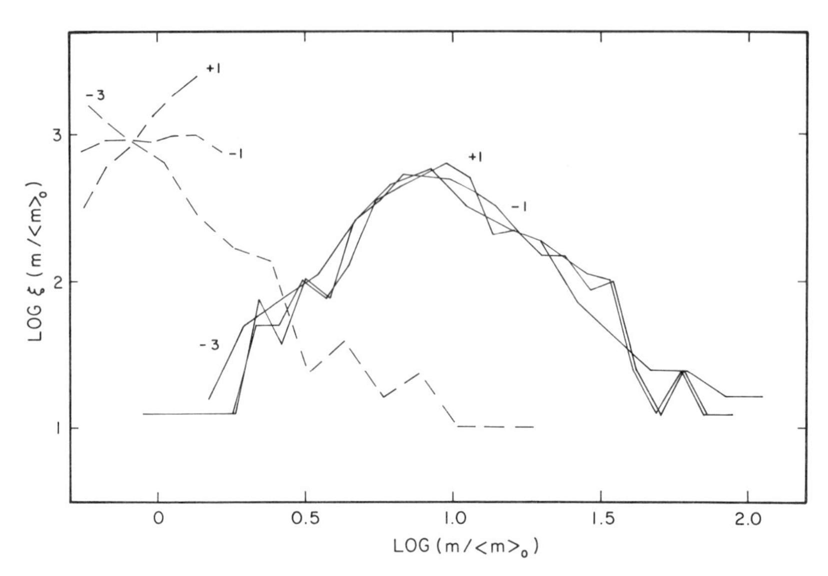

Fig. 4. Evolution of three different initial fragment mass functions due to fragment coalescence and accretion of ambient cloud gas, according to a 500-body simulation of Scalo and Pumphrey (1978). Masses are given in units of the mean initial fragment mass $\langle m \rangle_0$. Dashed and solid curves represent the initial and final mass functions, respectively. Each curve is labeled by the initial index γ of the mass spectrum. Coalescence dominates the spectra to the right of the peak while accretion dominates to the left.

precisely the same value independent of initial conditions.

Fragment interaction models have been discussed by Nakano (1966), Arny and Weissman (1973), Silk (this book), and Scalo and Pumphrey (1978). The latter paper shows by N-body simulations that when the fragment masses are changed by fragment-fragment encounters, the resulting mass spectrum is essentially independent of the initial fragment mass spectrum. At masses below some critical mass, accretion of ambient gas is more important than collisions, and the shape of the mass spectrum in this regime is dependent on the properties of the parent cloud, specifically the number of accretion time scales available. This model therefore can explain the uniformity of the observed mass spectrum above 1-2 $M_\odot$ and the variations observed in open clusters at lower masses. An example of the resulting mass spectra is given in Fig. 4 (Scalo and Pumphrey, 1978). This model is subject in principle to several observational tests involving correlations between features in cluster luminosity functions and cluster properties such as richness.

Acknowledgments. I am grateful to G. Miller and W. Pumphrey for numerous discussions, to B. Tinsley for her constructive suggestions, and to S. Murdock for her typing earlier and much longer drafts of this paper. The work of the author is supported in part by a research Fellowship of the Alfred P. Sloan Foundation.

REFERENCES

Appenzeller, I. 1970*a*. *Astron. Astrophys.* 5: 355.

_____. 1970*b*. *Astron. Astrophys.* 9: 216.

Appenzeller, I., and Tscharnuter, W. 1974. *Astron. Astrophys.* 30: 423.

Archemashvili, V. M. 1976. *Soviet Astron.–AJ.* 20: 116.

Arny, T., and Weissman, P. 1973. *Astron. J.* 78. 309.

Artyukhina, N. M. 1972, *Soviet Astron.–AJ* 16: 317.

Audouze, J., and Tinsley, B.M. 1976. *Ann. Rev. Astr. Astrophys.* 14:43.

Baglin, A.; Berruyer, N.; Morel, P. J.; and Pecker, J. C. 1973. *Astrophys. Lett.* 15: 9.

Barlow, M. J. 1971. *Nature Phys. Sci.* 232: 152.

Batten, A. H. 1973. *Binary and Multiple Star Systems.* New York: Pergamon.

Bedijn, P. J. 1977. Doctoral Dissertation, Sterrewacht te Leiden.

Berruyer, N. 1974. *Astron. Astrophys.* 30: 403.

Biermann, P., and Tinsley, B. M. 1974. *Astron. Astrophys.* 30: 1.

Blaauw, A. 1964. *Ann. Rev. Astron. Astrophys.* 2: 213.

Burki, G. 1977. *Astron. Astrophys.* 57: 135.

Butcher, H. R. 1976. *Astrophys. J.* 210: 489.

_____. 1977. *Astrophys. J.* 216: 372.

Conti, P. S., and Burnichon, M. L. 1975. *Astron. Astrophys.* 38: 467.

Dixon, M. E. 1970. *Mon. Not. Roy. Astr. Soc.* 150: 195.

Ebert, R.; von Hoerner, S.; and Temesvary, S. 1960. *Die Enstehung der Sternen durch Kondensation Diffuser Materie.* New York: Springer Verlag.

Elmegreen, B. G., and Elmegreen, D. M. 1978. *Astrophys. J.* In press.

Elmegreen, B. G., and Lada, C. J. 1977. *Astrophys. J.* 214: 725.

Faber, S. M.; Burstein, D.; Tinsley, B. M.; and King, I. R. 1976. *Astron. J.* 81: 45.

Faber, S. M., and Jackson, R. E. 1976. *Astrophys. J.* 204: 668.

Fowler, W. A., and Hoyle, F. 1963. *Roy. Obs. Bull.* 67: E299.

Freeman, K. C. 1977. In *The Evolution of Galaxies and Stellar Populations* (B. M. Tinsley and R. B. Larson eds.,) p. 133. New Haven: Yale Univ. Observatory.

Frolov, V. N. 1975. *Soviet Astron.–AJ Lett.* 1: 156.

Gaustad, J. E. 1963. *Astrophys. J.* 138: 1050.

Gerola, H., and Glassgold, A. E. 1978. *Astrophys J.* In press.

Goldreich, P., and Lynden-Bell, D. 1965. *Mon. Not. Roy. Astr. Soc.* 130: 97.

Grossman, A. S., and Graboske, H. C. 1971. *Astrophys. J.* 164: 475.

Hartmann, W. K. 1970. *Mém. Soc. Roy. Sci. Liège* 14: 49.

Heintz, W. D. 1972. *Astron. J.* 77: 160.

Herbst, W. 1978. Preprint.

Hoyle, F. 1953. *Astrophys. J.* 118: 512.

Huang, S. 1961. *Publ. Astron. Soc, Pacific.* 73: 30.

Huchra, J. P. 1977. *Astrophys. J.* 217: 928.

Hutchins, J. B. 1976. *Astrophys. J.* 205: 103.

_____. In *Structure and Evolution of Close Binary Systems,* (P. Eggleton, S. Mitton, and J. Whelan, eds.), p. 9. Dordrecht: Reidel.

Jaschek, C., and Jaschek, M. 1957. *Publ. Astron. Soc. Pacific* 69: 337.

Jones, B. F. 1970. *Astron. J.* 75: 563.

Jura, M. 1977. *Astrophys. J.* 212: 634.

Kahn, F. D. 1974. *Astron. Astrophys.* 37: 149.

Kholopov, P. N., and Artyukhina, N. M. 1972. *Soviet Astron.–AJ.* 15, 760.

King, I., and Tinsley, B. M. 1976. *Astron. J.* 81: 835.

Krisciunas, K. 1977. *Astron. J.* 82: 195.

Larson, R. B. 1973. *Mon. Not. Roy. Astr. Soc.* 161: 133.

Larson, R. B., and Starrfield, S. 1971. *Astron. Astrophys.* 13: 190.

Ledoux, P. 1941. *Astrophys. J.* 94: 537.
Limber, D. N. 1960. *Astrophys. J.* 131: 168.
Lippincott, S. L., and Hershey, J. L. 1972. *Astron. J.* 77: 679.
Low, C., and Lynden-Bell, D. 1976. *Mon. Not. Roy. Astr. Soc.* 176: 367.
Luyten, W. J. 1968. *Mon. Not. Roy. Astr. Soc.* 139: 221.
Lynden-Bell, D. 1973. *Dynamical Structure and Evolution of Stellar Systems* Saas Fee: Geneva Observatory.
Madore, B. F. 1977. *Mon. Not. Roy. Astr. Soc.* 178: 1.
Mathews, W. G. 1969. *Astrophys. J.* 157: 583.
McCuskey, S.W. 1966. *Vistas in Astronomy* 7: 141.
McNamara, B. J., and Sanders, W. L. 1977. *Astron. Astrophys.* 54: 56.
_____. 1978. *Astron. Astrophys.* 62: 259.
Miller, G., and Scalo, J. M. 1978. In preparation.
Nakano, T. 1966. *Prog. Theor. Phys.* 36: 515.
_____. 1973. *Publ. Astron. Soc. Japan* 25: 91.
Olson, G. L., and Pena, J. H. 1976. *Astrophys. J.* 202: 527.
Osmer, P. S. 1973. *Astrophys. J.* 186: 459.
Ostriker, J. P.; Richstone, D. O.; and Thuan, T. X. 1974. *Astrophys J.* 188: L87.
Papaloizou, J. C. B. 1973*a*. *Mon. Not. Roy. Astr. Soc.* 162: 143.
_____. 1973*b*. *Mon. Not. Roy. Astr. Soc.* 162: 169.
Pels, G., Oort, J. H., and Pels-Kluyver, H. A. 1975. *Astron. Astrophys.* 43: 423.
Piskunov, A. E. 1976. *Nauch, Informatsii* 22: 47.
Quirk, W. J., and Tinsley, B. M. 1973. *Astrophys J.* 179: 69.
Rees, M. J. 1976. *Mon. Not. Roy. Astr. Soc.* 176: 483.
Roberts, M.S. 1958. *Publ. Astron. Soc. Pacific.* 70: 462.
Salpeter, E. E. 1955. *Astrophys. J.* 121: 161.
_____. 1959. *Astrophys J.* 129: 608.
Sandage, A. 1957. *Astrophys. J.* 125: 522.
Sandage, A., and Tammann, G. 1974. *Astrophys. J.* 191: 603.
Sargent, W. L. W.; Schechter, P. L.; Boksenberg, A.; and Shortridge, K. 1977. *Astrophys. J.* 212: 326.
Sargent, W. L. W., and Searle, L. 1970. *Astrophys. J.* 162: L155.
Scalo, J. M., and Pumphrey, W. A. 1978. In preparation.
Schlesinger, B. M. 1969. *Astrophys. J.* 157: 553.
Schmidt, M. 1959. *Astrophys. J.* 129: 243.
_____. 1963. *Astrophys. J.* 137: 758.
_____. 1975. *Astrophys. J.* 202: 22.
Schwarzschild, M. 1958. *The Structure and Evolution of Stars.* New York: Dover.
Schwarzschild, M., and Harm, R. 1959. *Astrophys. J.* 129: 637.
Schweizer, F. 1978. *Astrophys. J.* In press.
Searle, L. 1973. In *Stellar Ages* (G. Cayrel de Strobel and A. M. Delplace, eds.), Obs. de Paris, Meudon.
Searle, L., and Sargent, W. L. W. 1972. *Astrophys. J.* 173: 25.
Searle, L.; Sargent, W. L. W.; and Bagnuolo, W. G. 1973. *Astrophys. J.* 179: 427.
Shields, G. A., and Tinsley, B. M. 1976. *Astrophys J.* 203: 66.
Silk, J. 1977*a*. *Astrophys. J.* 211: 638.
_____. 1977*b*. *Astrophys. J.* 214: 152.
Smith, H. A., and Tinsley, B. M. 1976. *Publ. Astron. Soc. Pacific* 88: 370.
Stothers, R. 1972. In *Stellar Evolution* (H.-Y Chiu and A. Muriel, eds.), p. 141. Cambridge, Mass: MIT Press.
Stothers, R., and Simon, N. R. 1968. *Astrophys. J.* 152: 233.
Suchkov, A. A., and Shchekinov, Y. A. 1976. *Soviet A. J.* 19: 403.
Taff, L. G. 1974. *Astron. J.* 79: 1280.
Talbot, R. J. 1971*a*. *Astrophys. J.* 163: 17.
_____. 1971*b*. *Astrophys. J.* 165: 121.
Tinsley, B. M. 1968. *Astrophys. J.* 151: 547.
_____. 1977. *Astrophys. J.* 216: 548.
Tinsley, B. M., and Gunn, J. E. 1976. *Astrophys. J.* 203: 52.

Trimble, V. 1975. *Rev. Mod. Phys.* 47: 877.
Truran, J. W., and Cameron, A. G. W. 1971. *Astrophys. Space Sci.* 14: 179.
van Altena, W. F. 1969. *Astron. J.* 74: 2.
van de Kamp, P. 1969. *Publ. Astron. Soc. Pacific* 81: 5.
____. 1975. *Ann. Rev. Astr. Astrophys.* 13: 295.
van den Bergh, S., and Sher, D. 1960. *Publ. David Dunlap Obs.* 2: 203.
____.1961. *Astrophys. J.* 134: 554.
____. 1973. In *Stellar Ages* (G. Cayrel de Strobel and A.M. Delplace, eds.). p. XL-1. Paris: Obs. de Paris-Meudon.
____. 1976*a*. *Astron. J.* 81: 797.
____. 1976*b*. *Ann Rev. Astr. Astrophys.* 13: 217.
van den Bergh, S., and Sher, D. 1960. *Publ. David Dunlap Obs.* 2: 203.
Walker, M. F. 1957. *Astrophys. J.* 125: 636.
Warner, B. 1961*a*. *Publ. Astron. Soc. Pacific* 73: 439.
____. 1961*b*. *Observatory* 81: 230.
Weisskopf, V. F. 1975. *Science* 187: 605.
Westbrook, C. K., and Tarter, C. B. 1975. *Astrophys. J.* 200: 48.
Wheeler, J. C.; Miller, G. E.; and Scalo, J. M. 1978. In preparation.
Whitford, A. E. 1977. *Astrophys. J.* 211: 527.
Wielen, R. 1971. *Astrophys. Space Sci.* 13: 300.
____. 1974. *Highlights of Astron.* 3: 395.
Williams, I. P., and Cremin, A. W. 1969. *Mon. Not. Roy. Astr. Soc.* 144: 359.
Ziebarth, K. 1970. *Astrophys J.* 162: 947.

NUMERICAL CALCULATIONS OF PROTOSTELLAR HYDRODYNAMIC COLLAPSE

PETER BODENHEIMER
Lick Observatory

AND

DAVID C. BLACK
NASA-Ames Research Center

Beginning in the 1960s significantly increased observational and theoretical understanding of the hydrodynamics of collapsing, self-gravitating objects have taken place. Advances on the theoretical aspects have come about primarily through numerical experimentation, and calculations of the spherically symmetric collapse of protostars have been carried out by a number of investigators with increasingly improved levels of physical approximation. The evolution of protostars of various masses has been followed, starting at the Jeans limit and ending when the entire object has gained hydrostatic equilibrium. Further development of hydrodynamic codes has made it feasible to undertake more complex, multi-dimensional numerical experiments, including the effects of rotation and magnetic fields. Axisymmetric experiments including rotation show that over a wide range of initial conditions a ring-like structure is formed rather than a central concentration of matter. Axisymmetric experiments including magnetic fields, but not rotation, show that the magnetic field increases in proportion to the square root of the density in the center of the collapsing cloud. The restriction to axial symmetry has been revised, and studies on the evolution of collapsing protostars and rings in three space dimensions have begun. Two- and three-dimensional calculations have general-

ly been limited to the earlier stages of protostellar collapse. Some of the classical problems of star formation, such as fragmentation, the angular momentum problem, and the relationship of star formation to the formation of planetary systems are discussed. Finally, the present results are summarized and future numerical experiments are suggested.

Hypotheses regarding the formation of planetary systems can be placed in one of two general categories, viz., those in which there is an intimate relation between the formation of a star and the formation of its planetary companions, and those in which the formation of a planetary system is for the most part independent of the formation of the central star. Currently adopted versions of the so-called nebular hypothesis (Cameron and Pollack, 1976; Cameron's chapter in this book; Safronov, 1972 and his chapter) fall into the former category, while the hypothesis discussed by Alfvén and Arrhenius (1976; see Alfvén's chapter) falls into the latter. If the nebular hypothesis is a correct description of the way planetary systems form, then the subject of the collapse of interstellar clouds is very relevant to obtaining an understanding of the processes involved. The collapse of the cloud establishes the form of the nebula and determines whether the physical conditions in it are suitable for the formation of planets.

Some of the principal questions which one hopes to address through numerical calculations on star formation are:

1. What are the dominant physical processes which control star formation?
2. What are the conditions in the circumstellar environment early in the history of a star?
3. How does a star evolve to the main sequence?
4. Is there any dependence of the answers to the first three questions on stellar type and if so what is that dependence?

The standard operational philosophy behind numerical studies which attempt to answer these questions is to isolate and study the effects of various physical phenomena (e.g., gas pressure, radiation, and rotation) on the dynamics of collapse. The approach is generally to study the collapse of a self-gravitating object in relative isolation. Most of the work which will be discussed in this review is concerned with objects of roughly one solar mass.

Protostars are generally thought to be objects with surface temperatures less than 3000 K which are evolving through the region of the Hertzsprung-Russell diagram that was found by Hayashi (1961) to be forbidden for stellar models in hydrostatic equilibrium. Two important mechanisms cause hydrodynamic instability for cool objects in this region. The first is the dissociation and ionization of hydrogen, and consequent reduction of the adiabatic exponent $\Gamma_1 \equiv [d(\log P)/d(\log \rho)]_s$ to values less than 4/3, in the temperature range 1800 K to about 3×10^4 K (Biermann and Cowling, 1939; Cameron, 1962). Collapse proceeds until enough gravitational energy has been released to dissociate and ionize the hydrogen and heat the object up to central temperatures on the order of 10^5 K, after which hydrostatic equilibrium is possible. The second mechanism occurs at an earlier stage when temperatures are about 10 K and densities are 10^{-13} g cm^{-3} or less. The protostar is transparent to its infrared radiation so that much of the gravitational energy released is lost through radiation and does not result in internal heating (Gaustad, 1963). Collapse occurs because the buildup of internal pressure is not as rapid as that of gravitational forces. One or both of these two basic physical mechanisms applies to all of the protostellar models considered in this review.

We structure the chapter by presenting studies in order of increasing complexity and/or decreasing restrictions on spatial symmetry. The status of one-dimensional (spherically symmetric) experiments is discussed in Sec. I. The status of two-dimensional (axially symmetric) experiments is reviewed in Sec. II, and the status of three-dimensional (symmetry about the equatorial plane) experiments is presented in Sec. III. We conclude in Sec. IV with a summary and with comments on the direction of future work.

It is important to recognize that numerical calculations are really numerical *experiments;* the computer code being analogous to apparatus, such as a mass spectrometer, used in laboratory experiments. As with laboratory apparatus, computer codes can be tested to see whether they obtain correct results on standard "samples." Numerical experiments also provide near-perfect aliquots for experiments by different investigators, a situation often not obtainable in laboratory experiments. Thus, when a given numerical code yields experimental results which are widely discrepant from results obtained with other, independent, numerical codes performing the same experiment, there is more justification in rejecting the discrepant result than there is in the analogous situation where a given laboratory experiment (on a meteorite say) yields results which vary from other similar experiments on the same meteorite. We point out this important aspect of numerical studies because of many questions and concerns on this point expressed by colleagues.

I. NUMERICAL EXPERIMENTS IN ONE SPACE DIMENSION

Restricting attention to the spherically symmetric case, in which self-gravity, internal pressure, and radiative transport of energy are the main physical effects considered, we generally start with a cloud of about 1 $M_\odot$ at the point of onset of collapse which occurs when the energy of self-gravitation becomes larger in absolute magnitude than the internal energy (usually referred to as the Jeans limit). If the temperature is 10 K, as it is suspected to be in dense interstellar clouds, the required minimum density is about 10^{-19} g cm^{-3}. This density is of course considerably higher than that of a typical interstellar cloud, so evolution up to this point would involve collapse of a more massive cloud, followed by fragmentation. Gerola and Glassgold (1978) have calculated the collapse of a cloud of 2.7×10^4 $M_\odot$, starting from a density of 3.4×10^{-22} g cm^{-3} and a temperature of 100 K, and including a detailed calculation of heating and cooling processes as well as of the chemical evolution of the more important molecules. Continuation of such studies with the inclusion of radiative transfer will provide important links between the theory and the observations of molecular clouds. It is not clear at what point fragmentation of such a cloud into isolated pieces of 1 $M_\odot$ occurs; thus the initial density of 10^{-19} g cm^{-3} that is generally chosen must be regarded as a parameter.

Starting from this point, evolution proceeds through the following main phases: (1) a transparent isothermal collapse during which cooling by infrared radiation from grains keeps the temperature near 10 K; (2) an adiabatic collapse, starting when the dust grain opacity results in an optical depth greater than 1 and leading, as a consequence of internal heating, to a temporary state of hydrostatic equilibrium in a small core of the protostar at temperatures of a few hundred K; (3) contraction of the core leading to further collapse induced by the dissociation of H_2 at the center, starting with a density $\rho = 10^{-7}$ g cm^{-3} and a temperature, $T = 2 \times 10^3$ K and ending with $\rho = 10^{-2}$ g cm^{-3} and $T = 3 \times 10^4$ K; (4) the formation of a second core in hydrostatic equilibrium containing only about 10^{-3} $M_\odot$; and (5) subsequent accretion of the remaining mass of the cloud onto the core through a standing shock at its surface. The first approximate calculation of the collapse of a protostar was made by Hayashi and Nakano (1965), starting in the opaque phase at $\rho = 10^{-13}$ g cm^{-3} and $T = 10$ K. They predicted the occurrence of a shock front and, later, a rapid flareup in luminosity resulting in a final equilibrium state for a 1 $M_\odot$ protostar which is fully convective with a radius of about 50 $R_\odot$ and a luminosity of 300 $L_\odot$.

These early estimates were followed by a number of numerical hydrodynamic calculations, the detailed physical approximations and results of which are discussed by Larson (1973*a,b*). Calculations of the isothermal transparent phase (Penston, 1966; Bodenheimer and Sweigart, 1968; Larson, 1969; and other references given by Larson, 1973*a,b*), starting near the Jeans limit for a 1 $M_\odot$ protostar, show the development of a highly centrally concentrated configuration, with collapse occurring on a very rapid time scale at the center as compared with that in the outer regions (time scales vary approximately as $\rho^{-\frac{1}{2}}$). Pressure effects retard the collapse to some extent, and as a consequence, the density distribution through much of the outer part of the cloud takes on the form $\rho \propto r^{-2}$, where r is the distance from the center (c.f., Shu, 1977). The time required to form the dense central condensation is about twice the free-fall time of the initial configuration. A relatively small mass fraction at the center (1%) enters the opaque (adiabatic) phase while most of the original material in the cloud is still transparent and at a density not much higher than the original value.

The first detailed calculations to be continued through the adiabatic and subsequent phases were those of Larson (1969). The short evolutionary time of the high-density "spike" at the center results in a rapid sequence of events; heating of the central region because of trapping of radiation, formation of a transcient hydrostatic core, contraction and further heating of the core, onset of collapse at the center caused by dissociation, and the formation of the small hydrostatic stellar core all occur before appreciable further evolution takes place in the outer transparent material. The final core remains in equilibrium during the process of the buildup of its mass by accretion. The accretion time is a few times the free-fall time of the bulk of the material external to the shock front, most of which is near its initial density of 10^{-19} g cm^{-3} at the time the core is formed. Once the accretion phase is well established, the density and velocity distributions in the inner part of the nearly free-falling envelope take on the form $\rho \propto r^{-\frac{3}{2}}$ and $v \propto r^{-\frac{1}{2}}$, respectively, while the outer region maintains $\rho \propto r^{-2}$. Shu (1977) has analyzed this form of the accretion flow by means of a similarity solution valid for isothermal collapse. Hunter (1977) describes additional similarity solutions and remarks that numerical calculations of collapse do not show a strong tendency toward agreement with such solutions.

During this relatively long period of accretion (about 10^6 yr for 1 $M_\odot$) the hydrostatic core is able to contract and release some gravitational energy. However, the bulk of the luminosity of the protostar is provided by dissipation of kinetic energy at the shock, which results in rapid heating behind the shock and subsequent escape of radiation through the shock. This radiation heats the pre-shock material to some extent, but most of it diffuses through the infalling envelope and is emitted in the infrared from the outer boundary of the protostar. Thus, the approximate protostellar luminosity during accretion is given by $L = G M \dot{M} / R$, where M and R are respectively the mass and radius of the core and $\dot{M}$ is the mass accretion rate. Up to the stage when about half of the mass of the envelope has been accreted, L increases with time due to the increase in M; beyond that stage L decreases because of the depletion of material in the envelope and the consequent reduction in $\dot{M}$. The point in the envelope where the optical depth is unity moves inward in radius and upward in temperature also as a consequence of the depletion of material. Calculation of the observable properties of the protostar indicates evolution through the forbidden region of the HR diagram, with the luminosity increasing to about 30 $L_\odot$, then decreasing again, and with the surface temperature gradually increasing from about 10 K to 3000 K (Fig. 1). By the time all the protostellar material has been accreted, the resulting star appears at the lower end of its Hayashi track with a radius of only twice its main-sequence value and a luminosity of only 1.3 $L_\odot$.

A number of other calculations have been made for the 1 $M_\odot$ case. The initial and final conditions pertaining to these calculations are summarized in

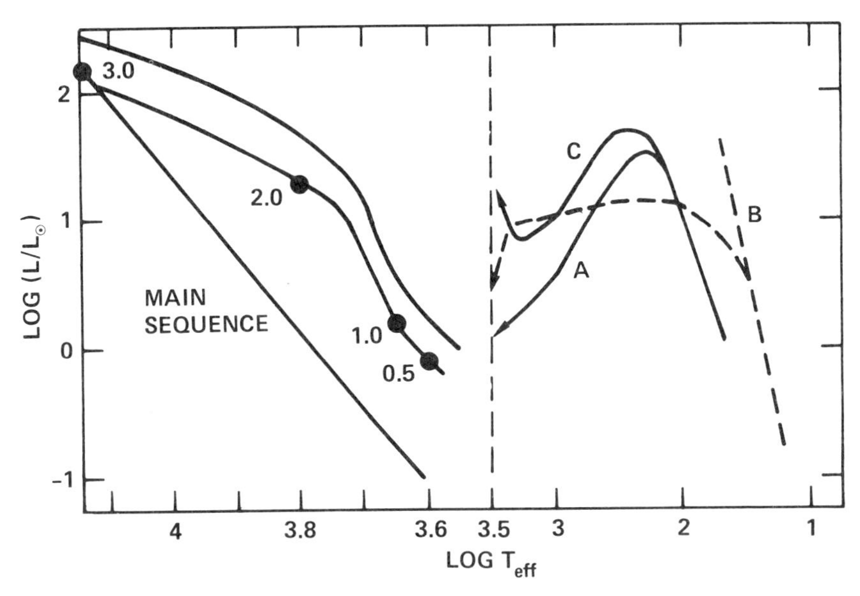

Fig. 1. Protostellar evolution in the spherical case. In the right-hand portion the evolutionary tracks are those calculated by (A) Larson (1969) for 1 $M_{\odot}$, (B) Appenzeller and Tscharnuter (1975) for 1 $M_{\odot}$, and (C) Larson (1969) for 2 $M_{\odot}$. The left-hand portion shows (solid line) the locus where protostars first appear in the Hertzsprung-Russell diagram as equilibrium objects (Larson, 1973*a*) assuming that collapse starts at the Jeans limit for each mass. Solid dots on this curve are labeled with masses in units of solar mass. The dashed curve in the right-hand portion gives the corresponding locus if the collapse is arbitrarily assumed to start at a radius of one-half the Jeans radius in each case.

Table I. Narita, Nakano, and Hayashi (1970) started from a polytropic (n = 4) density distribution in free-fall. Since the mean density ($\bar{\rho} = 10^{-11}$ g cm^{-3}) used by Narita *et al.* is well above that used by Larson, the mean free-fall time for their model is considerably shorter (by a factor of 10^4!) than the time scale for Larson's model. The hydrostatic core forms in the same manner as in Larson's calculation, but the subsequent accretion is extremely fast, resulting in a rapid flare-up of the star to the top of its Hayashi track (on a time scale of one day) when the shock reaches the optically thin outer layers. The qualitative difference in the two results is a consequence of the difference in the ratio t_D/t_F, where t_D is the radiative diffusion time from the shock front out to the boundary and t_F is the free-fall time of the envelope. In Larson's calculation the ratio is small so that the energy released at the shock has time to escape by diffusion to the surface. In the calculation by Narita *et al.* t_F is short and the ratio is large, so that the accretion is adiabatic and there is little energy loss until, of course, the layers ahead of the shock become so optically thin that photons from the shock can escape directly into space.

TABLE I

Comparison of Protostar One-Dimensional Calculations

Initial T (K)	Initial ρ ($g\ cm^{-3}$)	Final R ($R_\odot$)	Final L ($L_\odot$)	Evol. Time (yr)	Author
10	10^{-19}	2.0	1.5	1.2×10^6	Larson (1969)
10	10^{-16}	5.7	9.2	3×10^4	Larson (1969)
100[a]	7×10^{-9}[a]	118	10^3	10	Narita *et al.* (1970)
10	10^{-19}	4.5	2.5	4.3×10^5	Appenzeller & Tscharnuter (1974)
3	10^{-19}	90	165	2.7×10^5	Westbrook & Tarter (1975)
10	10^{-19}	3.0	20	3×10^5	Winkler (1976)

[a]at center

We conclude that the difference in results discussed above is due principally to the choice of initial density in the protostar. The reasons behind the large difference in radius and luminosity are explained in detail by Woodward (1978). Briefly, if the collapse is adiabatic and there is no appreciable energy loss by radiation, the binding energy of the final state is equal to the energy required to dissociate H_2 and ionize all the hydrogen and helium. The estimated final radius based on this energy argument is about 50 $R_\odot$ for 1 $M_\odot$, close to the value actually obtained by Narita *et al.* In the calculation by Larson the free-fall time is long so there is considerable energy loss by radiation, requiring more release of gravitational energy and therefore a smaller radius in the final state. Comparison of the first two lines in Table I confirms the conclusion that an increase in the initial density and a consequent reduction in the free-fall time results in a marked increase in the radius and luminosity of the final state.

More recent calculations have been started with initial conditions close to those used by Larson and have not always produced similar results. The principal differences between the calculations of Larson (1969), Appenzeller and Tscharnuter (1975), Westbrook and Tarter (1975), and Winkler (1978) are probably due to the treatment of the shock front and the radiative transfer in its vicinity. Larson used a shock-fitting technique and an approximate treatment of radiative transfer to join the solution in the hydrostatic core to that in the infalling envelope. Appenzeller and Tscharnuter used artificial viscosity to broaden the shock and used a diffusion approximation for radiative transfer. In spite of these differences, their final values of radius and luminosity differ by only about a factor of 2 from

Larson's and the evolution in the HR diagram is similar (see Fig. 1). Bertout (1976) has recalculated this evolution using an exact solution for the non-gray radiative transfer problem in the envelope, with little difference in the results compared with the more approximate Appenzeller-Tscharnuter calculation. Westbrook and Tarter obtain an evolution time not too different from that of Appenzeller and Tscharnuter, but their final results are closer to those of Narita *et al.* The cause of this discrepancy, as analyzed by Woodward (1978), appears to be related to the large artificial viscosity used by Westbrook and Tarter which resulted in the smoothing of the shock over a radius large compared to that of the core. The resulting re-adjustment of the density distribution increased the optical thickness of the layers in the vicinity of the shock and lengthened the photon diffusion time to the extent that relatively little energy could escape from the protostar during the collapse. The probable validity of the Larson and Appenzeller and Tscharnuter results has been recently confirmed by the work of Winkler (1978) who used an improved radiative transfer technique at the shock (c.f., Tscharnuter, 1977), and a movable numerical grid that gives very fine spatial resolution in the vicinity of the shock.

The evolution of protostars as a function of mass has been discussed by Larson (1972*b*) for the range $0.25 < M/M_{\odot} < 10$ and by Westbrook and Tarter (1975) for the range $0.1 < M/M_{\odot} < 50$. Appenzeller and Tscharnuter (1974) have calculated the collapse of a 60 $M_{\odot}$ protostar, and Yorke and Krügel (1977) have evolved massive protostellar clouds of 50 and 150 $M_{\odot}$. A number of the main points are discussed by Larson (1973*a,b*); here we emphasize more recent work.

As long as the collapse remains adiabatic, as it apparently does in the work of Westbrook and Tarter, the star arrives in hydrostatic equilibrium at the upper end of its Hayashi track, whatever its mass. However, if collapse starts at the Jeans limit for each mass and there is substantial energy loss during the accretion phase, as in Larson's work, the initial equilibrium state of the star depends strongly on mass. The main point is that the Kelvin contraction time for the stellar core decreases with increasing mass, as is well known from the theory of hydrostatic pre-main-sequence evolution. Furthermore, the density at the Jeans limit, at fixed temperature, goes as M^{-2}, so that collapse time scales ($\propto \rho^{-\frac{1}{2}} \propto M$) are longer the more massive the protostar is. The Kelvin time of the core and the free-fall time of the envelope are comparable for 1 $M_{\odot}$; however, as the mass increases the former shortens, the latter lengthens, the ratio t_D/t_F decreases, and the stage of evolution reached by the core becomes more and more advanced by the time infall of the envelope has been completed. For an object of 2 $M_{\odot}$ the evolution is well into the pre-main-sequence radiative contraction phase at the equilibrium point. The equilibrium core reaches the main sequence and starts nuclear burning while it still has an optically thick infalling envelope for masses $\gtrsim 5$ $M_{\odot}$. If the collapse starts at densities above the Jeans limit, the mass above which this effect occurs increases because of the shorter free-fall time.

At masses below 1 $M_{\odot}$ the free-fall time becomes relatively short so that a relatively small amount of energy is lost by radiative diffusion through the envelope during the accretion phase. Also, the Kelvin time in the core becomes long, so relatively little contraction takes place during accretion. As a result the low-mass stars come into overall hydrostatic equilibrium relatively high on their Hayashi tracks. Fig. 1 shows a line in the HR diagram which predicts, according to Larson's calculations, where a star should first appear as a "normal" hydrostatic object. Protostars with masses in the range $0.5 - 2 M_{\odot}$ become visible with stellar spectra when their radius is about 2 $R_{\odot}$, which is suspected to be the approximate radius of some of the T Tauri stars. Note that pre-main-sequence stars of masses greater than 3 $M_{\odot}$ would not appear in the HR diagram as equilibrium objects. Observations show, however, that young stellar objects do exist in this part of the diagram (Strom, 1977; Cohen and Kuhi, 1976; Cohen, 1978) from which we conclude that the density at the initiation of collapse may be higher than that given by the Jeans criterion at least in some cases (see also Sec. II).

An interesting development in the theory of massive protostars is the effect of gas pressure gradients (caused by radiative heating from the shock front) and of radiation pressure on dust grains in the infalling material. Westbrook and Tarter (1975) find that protostars of mass $\gtrsim 10 M_{\odot}$ eject a substantial fraction of their infalling envelopes during the accretion process. A protostellar cloud of 50 $M_{\odot}$ is estimated to lose about 65% of its mass by the time the remainder comes into equilibrium. In their calculation material is either ejected or accreted before the core reaches the main sequence. Appenzeller and Tscharnuter (1974) find that the core of a 60 $M_{\odot}$ protostar evolves to the main sequence; then rapid heating of the outer layers of the core results in the ejection of the envelope, leaving a final star of 18 $M_{\odot}$. A further effect is the ionizing radiation from hot, massive stellar cores which may result in the production of an H II region in the infalling cloud and subsequent halt and reversal of the collapse of the envelope (Larson and Starrfield, 1971); detailed hydrodynamic calculations have not been made.

Yorke and Krügel (1977) have calculated the collapse of massive protostars treating dust and gas as separate components coupled by friction. The dust is composed of pure graphite particles and of graphite cores with ice mantles. At temperatures above 150 K the ice mantles evaporate. Once a substantial (15 $M_{\odot}$) core develops and evolves to the main sequence, radiation pressure acts on the dust in the infalling material at distances from the center of $\sim 10^{14}$ cm, retards the collapse, and results in an optically thick layer at about 1000 K in which ultraviolet stellar radiation is converted to infrared. This inner cocoon (c.f., Kahn, 1974) is not ejected, but its radiation travels outward to the cool region where the ice-coated graphite grains exist. These grains at 10^{17} cm are not tightly bound gravitationally, and the infrared radiation pressure can result in their ejection; sufficient coupling exists to eject the gas as well. Original cloud masses of 50 and 150 $M_\odot$, starting at densities on the order of 10^{-19} g cm^{-3}, result in main-sequence stars of 16 and 36 $M_\odot$, respectively; these values are sensitive to assumed dust properties. The observable effects of radiative transfer in the envelope are discussed elsewhere in this book (see chapter by Bertout and Yorke). Continuation of calculations of this kind should establish a reliable upper limit to the mass of a spherically accreting object. Note however, that the results could be quite different if the density at the onset of collapse was considerably higher than 10^{-19} g cm^{-3}. In that case the core would not reach the main sequence before accretion was completed, the radiation field of the central core would be quite different, and the accretion and ejection process might be substantially altered.

II. NUMERICAL EXPERIMENTS IN TWO SPACE DIMENSIONS

Although numerical experiments in two space dimensions impose a degree of symmetry (typically axial symmetry) on the problem which is certain to be violated by real systems at some stage of their evolution, these experiments allow a detailed study of the important effects of rotation and magnetic fields on collapse behavior. Several individuals and groups have undertaken 2-D protostar collapse calculations, including effects of gas pressure, rotation and self-gravity (Larson, 1972*a;* Tscharnuter, 1975, 1978; Fricke, Möllenhof and Tscharnuter, 1976; Black and Bodenheimer, 1975, 1976; Deissler, 1976; Nakazawa *et al.*, 1976; Takahara *et al.,* 1977; Kamiya, 1977), and gas pressure, magnetic fields and self-gravity (Scott and Black, 1978). McNally and co-workers have carried out 2-D numerical experiments including rotation and self-gravity but not gas pressure (c.f., McNally, 1974). Here we limit our discussion to experiments which have included gas pressure. The numerical techniques employed by these investigators are many and varied, so that whatever agreement is found between experiments by different workers is likely to be due to the physics of the experiment rather than to an artifact of a particular numerical technique.

The experiments cited above employed differing boundary and initial conditions, the former usually being either constant volume or constant surface pressure. A rather wide range of initial conditions is represented, and there is little overlap in the extensive parameter space which specifies these experiments. This lack of overlap has served to obscure or make difficult meaningful intercomparison of results. There are, however, certain general results which have emerged from the experiments.

Experiments with Rotation, Gas Pressure and Self-Gravity:

Detailed studies (c.f., Larson, 1973*a*) of the thermal state of clouds in the density range $\sim 10^{-20} - 10^{-13}$ g cm^{-3} (the exact range depending on cloud mass and opacity) indicate that such clouds are, to a good approximation, isothermal at a temperature of ~ 10 K. The collapse studies referenced above all deal mainly with isothermal systems except that by Takahara *et al.* who study the adiabatic collapse that one would expect at densities higher than 10^{-13} g cm $^{-3}$.

Until recently, there has been controversy regarding the the collapse behavior of rapidly rotating, isothermal clouds, where "rapidly" means clouds characterized by an initial ratio β of rotational to gravitational energy $\gtrsim 10^{-3}$. The first numerical study of the collapse of a rotating isothermal cloud is that of Larson. He found that, as in the 1-D experiments, the collapse is non-homologous with a rapid increase in the ratio of central to mean density. After one free-fall time the calculation led to the formation of a ring-like structure in the interior of the cloud with the maximum density occurring in the ring rather than in the center of the cloud. The ring was massive enough so that the minimum in the gravitational potential was also located in the ring, causing material to move outward from the cloud center toward the ring. However, Larson's experiments were criticized because they were performed on a coarse (72 zones) numerical grid and they did not conserve angular momentum. The situation was clouded further when Tscharnuter (1975) published results from his experiments which were similar to Larson's in terms of initial conditions, but which did not give any evidence for the formation of rings. The first detailed experiments to show that rings form during the collapse of rotating isothermal clouds were those of Black and Bodenheimer (hereafter referred to as B^2). They used a finely zoned (1600 zones) moving grid, and angular momentum was exactly conserved during their experiments. Later studies by Nakazawa *et al.* confirmed the finding of B^2. Rings did not form during Deissler's experiments, but his calculations included the effects of turbulent viscosity and used an approximation for the gravitational potential that assumed a spheroidal mass distribution, so that his results are not directly comparable with those from the other experiments which explicitly calculate the gravitational field from the instantaneous mass distribution. Kamiya (1977) did not find rings. He used a Lagrangian grid for his calculations which should conserve angular momentum locally better than calculations which use Eulerian grids. However, many of Kamiya's initial models were not unstable to collapse and would thus not be expected to form rings. Further, his Lagrangian grid became highly distorted, raising doubts as to the accuracy of the finite difference scheme. More recently, Tscharnuter (1978) has done experiments and finds that rings form during the collapse of a cloud. However, the rings found by Tscharnuter form in a hydrostatic, opaque core and are thus not directly comparable to the rings found by other authors. In summary, it may be said with reasonable confidence that the collapse of a rapidly rotating, isothermal cloud leads to the formation of a ring in the interior of the cloud. The remainder of the discussion in this section will deal with a qualitative description of behavior found from 2-D experiments, with scaling laws for isothermal systems and with a comparison of the numerical experiments with observations of certain Bok globules (Bok, 1978).

The collapse of a rapidly rotating isothermal cloud has three identifiable phases, viz., (1) axisymmetric collapse leading to formation of a ring, (2) evolution of the ring to an equilibrium state, and (3) post-equilibrium evolution of the ring. The early phases of the collapse of a rapidly rotating cloud (i.e., during the first free-fall time of the initial cloud) proceed as one would expect *a priori.* The effect of rotation introduces an asymmetry into the dynamics such that the cloud becomes highly flattened in the direction parallel to the rotation axis. After approximately a free-fall time, a local maximum in the density begins to appear at a point in the equatorial plane away from the rotation axis. As the evolution proceeds, this protoring increases its mass and semi-major axis, d until it comes into mechanical equilibrium. That is, in the cylindrical coordinates R and Z the structure of the ring becomes static, with detailed force balance obtaining to a high degree throughout the ring structure. The integral properties and detailed structure of the equilibrium isothermal rings found by B^2 are in agreement with the analytic results for such rings obtained by Ostriker (1964). In particular, Ostriker demonstrated that equilibrium isothermal rings are characterized by a mass per unit length (i.e., $M_R/2\pi d$) which is uniquely determined by the temperature

$$\frac{M_R}{2\pi d} = \frac{2 R_g T}{\mu G} \tag{1}$$

where R_g is the gas constant, G is the gravitational constant, μ is the mean molecular weight and T and M_R are respectively the temperature and mass of the ring. In order to understand the post-equilibrium evolution of a ring, it must be recalled that the rings are not isolated configurations; they are embedded in the interior of a cloud. Material from the cloud is continuously being accreted by the ring at a rate $\dot{M}_R$ which is much greater than that which would allow Eq. (1) to be continuously satisfied. Thus the mass per unit length exceeds the equilibrium value and the ring collapses upon itself. None of the numerical experiments to date have followed this ring collapse to any extent because the rings very quickly become so small that they are contained entirely within a few zones of the computational grid, leading to a loss of numerical accuracy. It is reasonable to suppose that non-axisymmetric behavior will be important at this stage in the evolution of the cloud-ring complex, so that 2-D calculations are no longer valid. 3-D experiments are discussed in the following section, but it is worth noting here that the available evidence from 3-D numerical experiments indicate that even in the presence of non-axisymmetric initial conditions, a rapidly rotating isothermal cloud becomes axisymmetric to a very high degree, with ring formation proceeding essentially as indicated by the 2-D experiments.

The rings which form in most of the experiments are isothermal, having formed at low enough densities to be optically thin. There are, however, experiments in which the rings formed at sufficiently high densities for opacity effects to lead to heating of the ring (c.f., Cases 1A and 2B from B^2). For those situations where the cloud and ring are isothermal, one need not calculate models for all temperatures and masses. The initial models for those 2-D experiments can be characterized by two parameters, α and β (the parameter β was introduced above). The parameter α is defined as the ratio of the initial radius of the cloud to the Jeans radius corresponding to the temperature and mass of the cloud. (Note that $\alpha \sim$ thermal energy/grav. energy.) Given a set of α- and β-values, the results from one numerical experiment on a cloud of mass M and temperature T/μ may be used to obtain solutions for a test cloud of mass M' and temperature T'/μ' by means of the following scaling relations

$$
\begin{aligned}
L' &= L\left(\frac{M'}{M}\right)\left(\frac{\mu'}{T'}\right)\left(\frac{T}{\mu}\right) \\
\rho' &= \rho\left(\frac{M}{M'}\right)^{2}\left(\frac{T'}{\mu'}\right)^{3}\left(\frac{\mu}{T}\right)^{3} \\
\tau' &= \tau\left(\frac{M'}{M}\right)\left(\frac{\mu'}{T'}\right)^{\frac{3}{2}}\left(\frac{T}{\mu}\right)^{\frac{3}{2}}
\end{aligned}
\qquad (2)
$$

where L is any representative length scale in the reference cloud (e.g., its radius or the semi-major axis of a ring) and L' is the corresponding length scale in the test cloud of mass M', etc. The quantities τ and τ' are respectively time scales in the reference and test clouds. These scaling relations can be very useful in analyzing observed clouds in the context of numerical experiments.

Numerical experiments have not fully explored the (α, β) parameter space, but there are indications as to how ring properties vary as a function of α and β. The study of B^2 showed that as β is decreased (α fixed), the semi-major axis of the ring at equilibrium also decreases, the mass fraction in the ring decreases and its mean density increases. Qualitatively similar behavior has been found by other authors. There is not sufficient data at present to quantify a functional relationship, $d = d(\beta)$, but it must take a form such that in the limit $\beta \to 0$, $d \to 0$. Reducing α (β fixed) leads to a change in the detailed structure of a ring. In particular, rings formed from initial models with $\alpha = 0.05$ and $\beta = 0.32$ have been followed in their evolution to the point where they include most of the mass of the cloud. They tend to be highly flattened disk-like structures, while rings formed from initial models with $\alpha = 0.5$ and $\beta = 0.32$ have nearly circular cross-sections in a meridional plane. This strong dependence of ring structure on α may make it possible to observationally constrain the value of this important parameter, as we discuss below.

Numerical experiments to date have dealt with relatively simple physical systems. In particular, there have been no attempts to model anything as complicated as the Orion complex, that is, a cloud within which there are stars, ionization fronts, etc. There is a class of observable objects, namely Bok globules, (see Bok, 1978) which appear to be relatively simple systems. As such, one might hope to obtain meaningful comparisons between Bok globules and numerical experiments. The globule B 361 has been studied in some detail by Milman (1977) who reports data concerning the rotational velocity, collapse velocity, CO column density and radius of a dense inner core. In addition, there is a lower limit to the total gas density which is inferred from visual extinction measurements. The Bok globule B 163 (Martin and Barrett, 1977) gives evidence of rotation, orbital motion and possibly of fragmentation; it is further discussed in Sec. III.

Comparisons have been made between the observed properties of B 361 and the properties of 2-D models of collapsing, rotating clouds calculated with the B^2 numerical code. Numerical sequences were calculated as functions of total cloud mass M, and the ratios α and β. Details of those experiments as well as comparisons with other Bok globules will be presented elsewhere (Villere and Black, 1978), but we summarize here some of the salient results. There appears to be no match between the numerical experiments and Milman's data for *any* value of β or M if $\alpha = 0.5$ (i.e., for models which begin collapse near the Jeans limit). Fairly good matches to the observed properties are obtained for $\alpha = 0.05$, with a relatively narrow range in β $(0.02 \leqslant \beta \leqslant 0.08)$, and for *total* globule masses in the range $500 \lesssim M/M_\odot \lesssim 5000$. Note that a factor of 4 in β corresponds to a factor of 2 in the angular velocity Ω. The numerical experiments indicate that the mass in the dense core is $\sim 5\%$ of the total cloud mass. This comparison of non-spherical numerical experiments with observations apparently requires that α be small ($\lesssim 0.05$), indicating that clouds are compressed well beyond their Jeans radius before collapse ensues. If this result is found to hold in general, it will provide much needed insight into the initial conditions to be used for collapse calculations.

Experiments with Magnetic Fields, Gas Pressure and Self-Gravity

There have been many theoretical studies concerning the effects of magnetic fields on star formation (c.f., Mestel and Spitzer, 1956; Spitzer, 1968). Mouschovias (1976*a,b*) carried out the first self-consistent numerical studies on the structure of *equilibrium,* self-gravitating, magnetic clouds. He assumes that the clouds are isothermal, axisymmetric and have a "frozen-in" magnetic field. His findings generally support the earlier theoretical arguments which were based on the virial theorem. In particular, he finds that there exist "critical equilibrium states" such that any increase in either cloud mass or external surface pressure would cause a cloud to collapse. Mouschovias also finds that the strength of the frozen-in magnetic field, B_c at the center of the cloud varies with gas density, ρ_c at the cloud center according to a power law, viz., $B_c \propto \rho_c{}^n$, with $n \sim 1/3$ to $1/2$. This range of values for the exponent n is well below the value of 2/3 which is frequently (and erroneously) used. Although observational evidence on the strength of magnetic fields in dense clouds is equivocal on this point, there is evidence (Heiles, 1976) that the field strength in dark clouds is less than that which would be obtained from amplifying the background galactic magnetic field according to an $n = 2/3$ power law.

A recent numerical study by Scott and Black (1978) has examined the dynamics of *collapsing,* non-rotating, magnetic clouds. The numerical code used in that study is a modified version of the B^2 code, allowing for magnetic field effects in the equation of motion and for time dependence of the magnetic field. In order to test the modified numerical code and the results of Mouschovias' studies of equilibrium objects, a collapse experiment was carried out starting from a sphere of initially uniform density that was threaded by a uniform magnetic field. The sphere was immersed in a high-temperature, low-density medium which exerted a specified external pressure on the surface of the cloud throughout the experiment. The initial parameters of the experiment were consistent with the initial parameters of the model shown in Fig. 1*c* of Mouschovias (1976*b*). This initial uniform model was out of equilibrium and proceeded to undergo a dynamic relaxation toward an equilibrium state. The result of the experiment was that the cloud collapsed and subsequently "bounced" when gas and magnetic pressure in the interior of the cloud had increased sufficiently to overcome the cloud's self-gravity. The cloud settled into an equilibrium state with values of the central magnetic field strength and gas density which were within ten percent of the values for Mouschovias' corresponding equilibrium model. This rather harsh test of a hydrodynamic code corroborates the findings of Mouschovias on the structure of stable, equilibrium, magnetic clouds.

Of greater interest, however, is the collapse behavior of an unstable cloud. Again, the starting model used by Scott and Black was a uniform density sphere, but the initial values for this model were chosen to match one of Mouschovias' "critical models." As in the test case discussed above, the cloud underwent a dynamic relaxation toward the appropriate critical equilibrium model, but the ordered kinetic energy generated during the relaxation served to push the cloud into a parameter regime where there is no equilibrium for that particular cloud, and collapse due to gravitational instability occurred. As we are interested in the evolution of the cloud in the collapse phase of the experiment, we shall not describe the dynamic relaxation phase which led to the collapse.

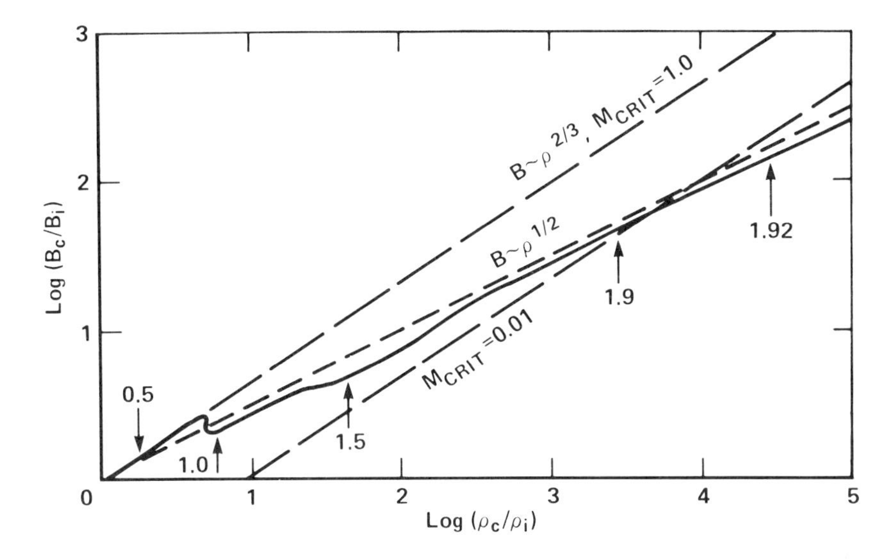

Fig. 2. Solid line shows central magnetic field versus central density, in units of the initial values of these quantities. Long-dashed lines correspond to $B \propto \rho^{\frac{2}{3}}$ and constant $M_{\rm crit}$ (see text). Short-dashed line corresponds to $B \propto \rho^{\frac{1}{2}}$. Arrows are labeled with time in units of the initial free-fall time. (From Scott and Black, 1978.)

There are two prominent features of the collapse phase which appear to be general characteristics of non-rotating, magnetic cloud collapse, viz.,

1. Steady core collapse with slowly flattening density contours.
2. Close approximation to a $B \propto \rho^{\frac{1}{2}}$ relation throughout much of the core of the cloud.

These characteristics are illustrated in Figs. 2 and 3. The flattening of a cloud is here defined as Z/R, where Z and R are respectively the intercepts of a given density contour on the rotation axis and on the equatorial plane. The initial magnetic field (as well as the background magnetic field) are parallel to the Z-axis. Characteristic (1) above implies that Z/R decreases as R decreases; the flattening is approximately proportional to R. Scott and Black argue that the near constancy of Z/R^2 during the collapse results from a natural feedback between gravitational and pressure forces in the Z-direction, and that the $B \propto \rho^{\frac{1}{2}}$ law follows from the same mechanism. We reproduce here the essential aspects of their argument as applied to an analogical model of a flattened, uniform density spheroid. Denoting the ratio of gravitational forces to pressure forces per unit volume in the Z-direction by Q, one has

$$Q = \frac{\text{Grav. Force/Vol.}}{\text{Pressure Force/Vol.}} \propto \frac{(ZR^4)^{-1}}{(Z^2R^2)^{-1}} \propto \frac{Z}{R^2} \tag{3}$$

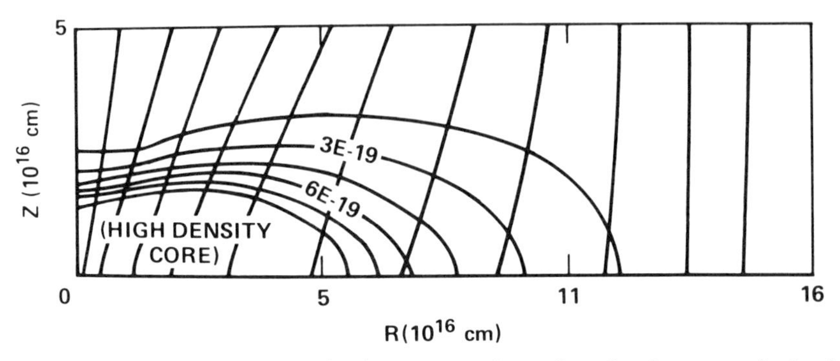

Fig. 3. Constant density contours for the outer envelope of a collapsing, magnetic cloud; density contours above 9×10^{-19} g cm^{-3} are not indicated. Also shown are typical magnetic field lines in the outer envelope of the cloud. (Data are from Scott and Black, 1978.)

where Z-gradients of a quantity are approximated by multiplying that quantity by Z^{-1}. If one examines the time dependence of Q, one has

$$\frac{d\,(\log Q)}{dt} = \left[\frac{d\,(\log Z)}{d\,(\log R)} - 2\right]\left[\frac{d\,(\log R)}{dt}\right]. \tag{4}$$

Taking the collapse along the equator as given (i.e., $d(\log R)/dt < 0$), one has

$$\frac{d\,(\log Q)}{dt} \propto (2 - k) \tag{5}$$

where $k \equiv d(\log Z)/d(\log R)$. When $k = 2$, Q (or Z/R^2) is constant. If $k < 2$, gravitational forces increase relative to pressure forces, causing k to increase until $k = 2$. If $k > 2$, pressure forces increase relative to gravitational forces, causing k to decrease until $k = 2$. Thus, there is a regulative mechanism operative to keep $k = 2$ and hence Z/R^2 constant. This argument is only valid if collapse in the R-direction is relatively insensitive to variation in k about $k = 2$, which Scott and Black have found to be approximately true. Physically this insensitivity derives from the fact that in the Z-direction pressure nearly balances gravity (hence the strong feedback), whereas pressure and magnetic fields balance only a small fraction of gravity in the R-direction and so variations in the ratio of gravity to pressure forces in the Z-direction induce little associated variation in $d(\log R)$. In the context of this analogical model, the condition of flux freezing ($B \propto R^{-2}$) and $\rho \propto (R^2 Z)^{-1}$ leads to

$$\frac{d\,(\log B)}{d\,(\log \rho)} = \left(1 + \frac{k}{2}\right)^{-1} \tag{6}$$

where k is defined above. For $k = 2$, one finds that $d(\log B)/d(\log \rho) = 1/2$, or $B \propto \rho^{\frac{1}{2}}$. Thus, it appears that the B-ρ relation found by Scott and Black in their numerical experiments is a consequence of the feedback mechanism which tends to keep $k = 2$.

The above results have implications regarding the question of fragmentation. The classic virial theorem arguments of Spitzer indicate that the critical mass for collapse in the presence of a magnetic field B and at a density ρ is given by

$$M_{\mathrm{crit}} = c_1 \frac{B^3}{G^{\frac{3}{2}} \rho^2} \tag{7}$$

where c_1 is a constant (Spitzer gives $c_1 = 0.0236$). If during the collapse $B \propto \rho^{\frac{2}{3}}$, as is frequently argued, then M_{crit} remains constant during evolution to higher densities. Put another way, as long as $B \propto \rho^{\frac{2}{3}}$ one would not expect a collapsing magnetic cloud to fragment into "blobs" of mass $M' < M_{\mathrm{crit}}$. However, if $B \propto \rho^{\frac{1}{2}}$, $M_{\mathrm{crit}} \propto \rho^{-\frac{1}{2}}$ so that M_{crit} decreases during collapse, it thereby raises the theoretical possibility of fragmentation during collapse. The experiments of Scott and Black showed no evidence for fragmentation even though the density increase was as high as 10^4 (M_{crit} was reduced by 10^{-2}). This lack of evidence for fragmentation is not conclusive; the numerical experiments may not have been carried far enough in their evolution for fragmentation to occur, or perhaps the time available is not sufficient for perturbations to develop even if the critical mass is exceeded. It should also be noted that the experiment carried out by Scott and Black was necessarily limited to axisymmetric fragmentation modes.

It was first suggested by Mouschovias, on the basis of his studies of equilibrium magnetic clouds, that a magnetic field would prevent much of the outer portion of a cloud from collapsing. This conjecture is confirmed by Scott and Black who find that magnetic forces stop the collapse of material near the magnetic equator. However, the fraction of the cloud mass which does not collapse is only ~10%, somewhat less than Mouschovias suggested. Some of the possible implications of this core-envelope "separation" for star formation are discussed in the chapter by Mouschovias in this book.

III. NUMERICAL EXPERIMENTS IN THREE SPACE DIMENSIONS

Because of the symmetry assumption that is made in the formulation of the 2-D problem, such calculations leave unanswered several critical questions. For a full understanding of fragmentation processes in interstellar clouds and of how binary, multiple, and planetary systems form we require calculations that take into account non-axisymmetric effects. Calculations of cloud collapse in three space dimensions are still rudimentary. Examination of the results of 2-D calculations suggests that initially the following specific points must be addressed:

1. Are axisymmetric rings unstable to fragmentation? If so, what is the dominant mode?
2. Do rings form at all in a collapsing cloud if the assumption of axial symmetry is relaxed?
3. What type of perturbation of an initial cloud is required so that fragmentation will eventually take place, either before or during a ring phase?

Calculations that are now underway include rotation, pressure, and self-gravity and are based on the following assumptions: a) symmetry about the equatorial plane; b) isothermal or adiabatic collapse; c) an ideal-gas equation of state; and d) conservation of angular momentum. As in the 2-D case, the initial conditions may be parameterized by α and β, as well as by the form of any initial perturbation. On the order of 10^4 grid cells are required for adequate numerical resolution; thus, even with the simplest assumed physics, considerable computing time and storage are required on the largest existing machines. Standard fluid-dynamic difference schemes as well as simulations employing finite particle techniques have been used to solve the equations of hydrodynamics.

Several authors (Miller, personal communication, 1977; Larson, 1977; Lucy, 1977; Gingold and Monaghan, 1977) have found that computing requirements can be lightened if the collapsing gas is represented by a set of individual "fluid elements," whose motions are followed by means of an N-body calculation. A grid system is not used; rather the Newtonian equations of motion for typically 100-150 such particles are solved (Miller uses 10^5 particles), taking into account mutual gravitational interactions. Pressure effects are simulated either through use of smoothing techniques or by introducing a repulsive force of the correct order of magnitude between a particle and its nearest neighbor(s). The work of Lucy and of Gingold and Monaghan has been applied primarily to the calculation of the fission of a rapidly rotating object in hydrostatic equilibrium and will not be further discussed here.

Larson has used his scheme to calculate the collapse of protostars, starting from initial conditions similar to those used in 2-D calculations and directed primarily toward an investigation of question (3) above. The particles are initially distributed randomly in a sphere and are given a uniform rotation but no random motions. A repulsive viscous force between particles is included to simulate dissipative effects, such as those resulting from shocks. In all cases a strong tendency toward fragmentation is observed. For the case $\alpha = 0.35$, $\beta = 0.19$, a dense concentration of matter collects in the center of the cloud, similar to that found in the non-rotating collapse calculations. No further fragmentation occurs and no ring is formed. The action of the built-in viscosity, which transports angular momentum outwards in the cloud, allows a significant amount of mass to accumulate at the center. With $\alpha = 0.25$ and $\beta = 0.30$, however, the system evolves to form two condensations of roughly equal mass in orbit about each other. When the thermal energy is still further reduced so that $\alpha = 0.075$, $\beta = 0.30$, a larger number (5 to 10) of particle clusters forms with masses ranging from 1 to 10% of the cloud mass. In one particular calculation a hierarchical triple system is formed, consisting of a close binary with a distant companion. The number of condensations formed in these various calculations is approximately equal, in each case, to the number of Jeans masses contained in the initial cloud; this number does not depend significantly on β. Once a condensation forms, it tends to accumulate more mass and it does not undergo further fragmentation. The mass of the largest condensation does depend strongly (inversely) on the assumed initial angular velocity. Larson suggests that a power law mass spectrum may occur as an end result of the fragmentation; however, the calculations are not sufficiently extensive to verify this conjecture.

Although these N-body experiments by Larson yield suggestive results, it should be stressed that they are *not* fluid dynamics experiments, but rather an attempt to simulate such experiments. The pronounced tendency for fragmentation, and coalescence of fragments of low α-values is a direct consequence of the strong (and *ad hoc*), viscous interactions between neighboring particles. This viscosity results in the outward transfer of angular momentum contained in the relative motion of neighboring particles, allowing them to coalesce. The absence of rings in Larson's experiments is also a direct consequence of these strong viscous effects. Unpublished controlled studies by one of us (DCB) have shown that 2-D numerical experiments with collapsing, rotating isothermal clouds do not yield rings if angular momentum is removed from the central regions of a cloud in the sense of violating local angular momentum conservation in these regions by $\gtrsim$ 10%. It is thus not surprising that Larson's experiments (particularly the one with $\alpha = 0.35$, $\beta = 0.19$) did not yield rings, since the strong viscous dissipation results in transfer of angular momentum away from the particle concentrations. We later discuss work by Tohline (1977) involving fluid dynamic experiments in three space dimensions in which angular momentum is conserved locally. These comments are not intended to be critical of Larson's work, but they should serve to point out that the experiments are approximate in nature and that an *ad hoc* viscosity plays an important role in the outcome.

The 3-D fluid dynamic studies, which generally use fluid-in-cell finite difference approximations to represent the equations of motion, energy, and continuity, have addressed the three questions posed above in a preliminary way. The first conclusion that has been reached is that isothermal ring structures of the type produced in the 2-D collapses are unstable to non-axisymmetric perturbations. Such instabilities had previously been suspected on the basis of a linear stability analysis of a constant density toroid (Wong, 1974). An extensive set of calculations has been carried out at Livermore by Norman and Wilson (1978) on a $40 \times 40 \times 26$ grid. The initial model, taken directly from Case 1D of B^2, was an isothermal ring, approximately in equilibrium, surrounded by the remainder of the collapsing cloud. The density in the entire system was perturbed, with an amplitude of roughly 10% and with an azimuthal dependence corresponding to a pure mode $m = 1,2,3,4$, or 6, or to a superposition of such modes.

In the pure-mode cases the ring breaks up into m blobs located at the density maxima of the initial perturbation. The growth rate is most rapid for $m = 2$ and in all cases fragmentation becomes well developed after half a rotation period of the ring. Initially the fragmentation time in the ring is somewhat longer than the time for accretion from the infalling cloud; thus the ring gains some mass before the increasing density shortens the fragmentation time and allows the perturbations to grow rapidly.

In the mixed-mode calculations the development of separate condensations again is rapid, with the lower-order modes ($m = 2$ and 3) dominating. Five calculations were performed, all starting from the same ring but with different phases and amplitudes for the various modes. In two cases, two objects formed in a binary system; in one case three objects formed, two of which seemed likely to coalesce; in one case four objects formed in pairs with an expected end result of coalescence into two final condensations in orbit; and in the final case three objects were formed nearly equally spaced around the original ring. The initial and final configurations in these five experiments are illustrated in Fig. 4. The masses of the condensations were roughly comparable and each equal to about 10% of the cloud mass. The spin angular momentum of the ring was largely converted into orbital motion; residual spin angular momenta of the blobs, while difficult to estimate because of the coarse grid in the azimuthal direction, were found to be 8 to 40% of their orbital angular momenta. In those cases where the end result was a binary system, the spin angular momentum of a condensation was roughly 20% of its orbital angular momentum.

The general conclusion that an isothermal ring is unstable has been confirmed in the pure-mode $m = 2$ case by Tohline (1977) using a $31 \times 31 \times 16$ grid and an initial condition comparable to that of Norman and Wilson (1978). However, he also found that other types of initial perturbations do not necessarily grow and result in fragmentation before the ring begins to collapse upon itself (see below). Additional experiments have been performed by Cook (1977) on a $10 \times 12 \times 5$ grid. He starts with equilibrium rings having a polytropic density distribution, and follows their evolution with an adiabatic rather than an isothermal pressure-density relation. Most of Cook's experiments start with pure-mode perturbations having $m = 2$ and a 1% amplitude. Toroids of various thicknesses all prove to be unstable to this

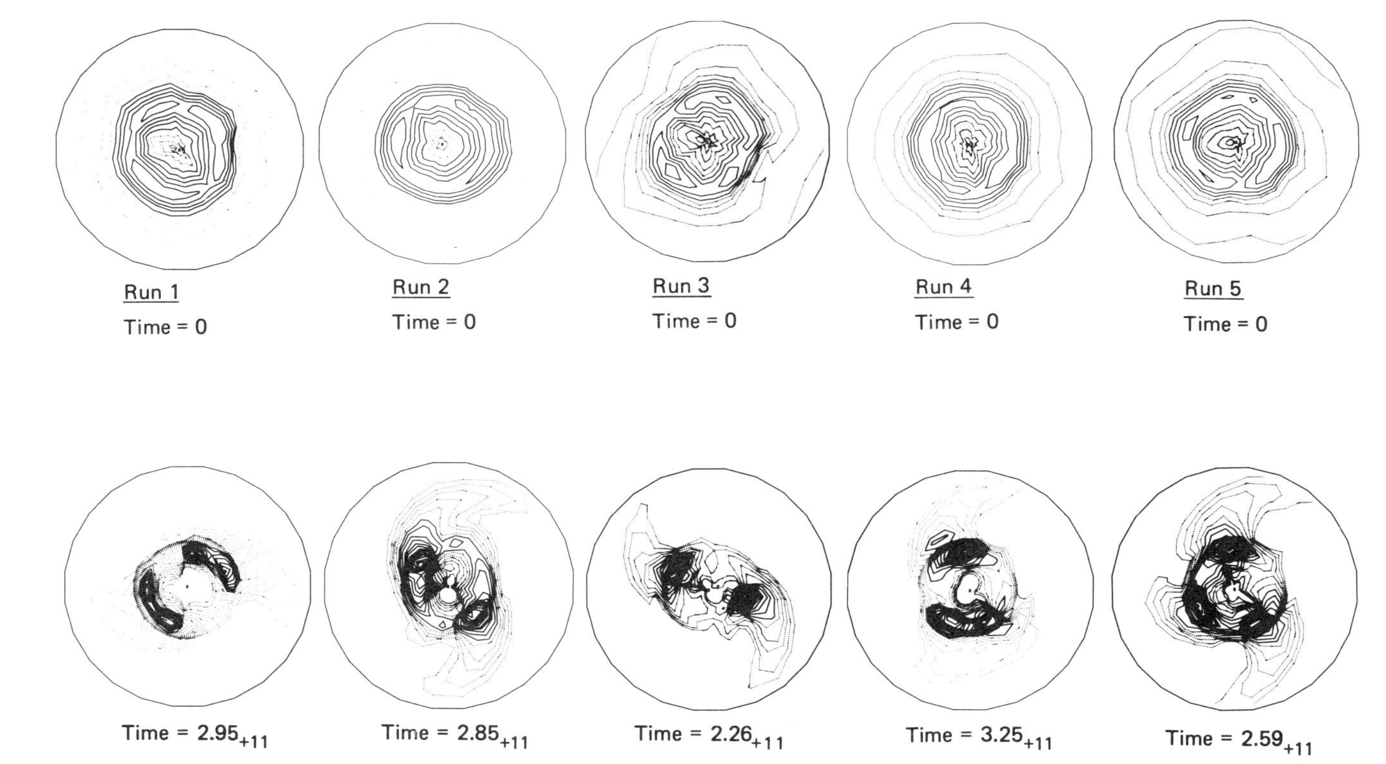

Fig. 4. Constant density plots of the initial and final mass distributions for the mixed mode runs by Norman and Wilson (1977). The plots of the initial distributions are a surface density projected on the equatorial plane. The times indicated in the figure correspond to evolutionary time in seconds for each of the runs (e.g., $2.95_{+11} = 2.95 \times 10^{11}$).

mode and they evolve to two well-defined condensations in orbit with roughly equal masses and with a small fraction of the ring's angular momentum remaining in spin.

The interesting ring results now lead us to the consideration of the second important question: do rings form at all in a non-axisymmetric collapsing cloud? The particle calculations discussed above suggest that ring formation is possible but that it may depend on initial conditions. The fluid-dynamics calculations to date show that ring formation does occur before there has been significant fragmentation over a wide range of initial conditions.

Calculations by Tohline, assuming isothermality and starting from a uniformly rotating, uniform density ($\rho = 1.4 \times 10^{-18}$ g cm^{-3}) sphere of 1 $M_\odot$ show that axisymmetric collapse and ring formation occur just as in the 2-D calculations of B^2. Three cases were run, with no initial non-axisymmetric perturbations, having respectively, $\alpha = 0.5$, $\beta = 0.02$, $J = 6 \times 10^{53}$ g cm^2 sec^{-1}; $\alpha = 0.5$, $\beta = 0.28$, $J = 2 \times 10^{54}$ g cm^2 sec^{-1}; $\alpha = 0.05$, $\beta = 0.28$. $J = 7.4 \times 10^{53}$ g cm^2 sec^{-1}, where J is angular momentum per unitmass. In the first case the central density increased by a factor of 10^5 during the experiment. Nevertheless, in the two or three free-fall times needed to form a ring, perturbations simply did not have a chance to develop, as one might expect as a consequence of the axisymmetric force field. A similar calculation was performed by Cook who assumed adiabatic collapse over a much smaller increase in density and showed that the cloud remained axisymmetric up to a point of incipient ring formation.

Even if significant non-axisymmetric perturbations are introduced into the initial cloud, ring formation still results in several cases that have been studied. Cook's adiabatic calculation with $\alpha = 0.2$, $\beta = 0.3$, and with an initial $m = 2$ perturbation of 1% amplitude in the azimuthal velocities, resulted in complete damping of the perturbation by the time the central density had increased by a factor of 4. Subsequent evolution was axisymmetric. When he reduced α to 0.1 (at constant β), however, the same initial perturbation grew and formed an incipient binary pair without the intermediate stage of ring formation. The components were of about equal mass and were spinning in synchronism with their orbital motion. The maximum density in the condensation was about 10 times the initial density in the cloud.

Initial perturbations introduced into isothermal collapses, however, do not produce analogous behavior. Three calculations by Tohline show the following:

(a) If $\alpha = 0.5$ and $\beta = 0.28$ initially, an $m = 2$ perturbation of the density with a maximum amplitude of 10%, localized in the region of the cloud where the ring is expected to form, damps out completely before ring formation occurs.

(b) If the pressure is reduced by taking $\alpha = 0.05$, $\beta = 0.28$, a similar perturbation damps initially, then begins to grow as the ring develops.

However, the ring reaches a stage when it collapses upon itself axisymmetrically on a free-fall time scale (the equilibrium mass given by Eq. (1) is exceeded). This collapse time becomes shorter than the fragmentation time and the non-axisymmetric effects do not develop further.

(c) If, for the case $\alpha = 0.5$, $\beta = 0.25$, a larger $m = 2$ perturbation is introduced that includes much of the mass of the cloud and has a maximum amplitude of 50%, the result is again initial damping, then growth of the perturbation, but no development of large non-axisymmetric effects before the onset of ring collapse.

The difference in ring fragmentation tendency between the experiments by Norman and Wilson (the results of which have been corroborated by Tohline), in which perturbations were applied only after an equilibrium ring had formed, and the experiments by Tohline, in which perturbations were introduced into the initial model, may be important in understanding the evolution of rings. The different results are *not* a consequence of perturbation amplitude. Norman and Wilson used 10% amplitude density perturbations in an equilibrium ring, whereas the amplitude of the $m = 2$ mode in Tohline's ring at the point where equilibrium was attained was as large as 20%. The difference in fragmentation tendency appears to result from phase effects in the perturbations. The experiments of Norman and Wilson involved perturbations which were intially in phase throughout the ring and surrounding material. As the experiment proceeds, differential rotation spreads the phase of the perturbation as a function of distance from the rotation axis, but the regions of space in which the amplitude of the perturbation is $\gtrsim$ 10% is concentrated in the ring and retains a strong collective behavior characteristic of the initial $m = 2$ mode (see Fig. 5*a*). As the experiment proceeds (Fig. 5*b*) more of the ring material shows large ($\gtrsim$ 10%) amplitude perturbation, but still retains the overall $m = 2$ character (180° symmetry). By contrast, perturbations in Tohline's experiment at stages in the ring evolution corresponding approximately with those shown in Fig. 5*a* and *b* are generally comparable in amplitude to those found by Norman and Wilson, but their relative phases are much different. In particular, there is no tendency toward a global $m = 2$ pattern for the perturbations. At a stage in Tohline's experiment (Fig. 5*c*) approximately corresponding to that shown in Fig. 5*b*, the larger amplitude perturbations in the ring are spread more uniformly throughout the ring than was the case in the Norman-Wilson experiment. Thus, it appears that the difference in ring fragmentation between the Norman-Wilson experiment and the Tohline experiment is due to the differences in relative phases of the perturbations at the equilibrium ring stage of evolution. Differential rotation arising during the phases of collapse leading up to ring formation is responsible for introducing the spread in phase.

We now consider the final question regarding what degree of asymmetry an initial cloud must have so that it will fragment. Larson's (1977) particle

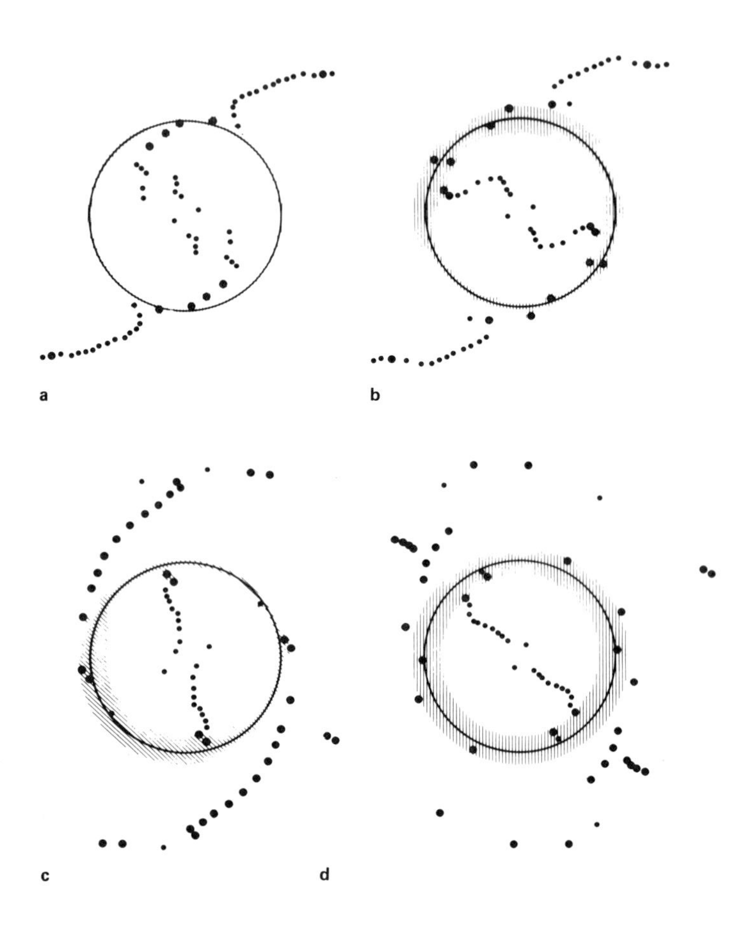

Fig. 5. Polar plot (in the equatorial plane) of the location of perturbations in the mass density distribution. The models shown have only a pure $m = 2$ perturbation (see text). Light dots indicate regions where the amplitude of the perturbation is significant, but small (between 1 – 10%); heavy dots indicate where the amplitude is large (greater than 10%). The ring is indicated by the stippled region and the center of the ring by the circle. Data in Figs. *a* and *b* are from an experiment in which a 10% amplitude perturbation was applied to the *equilibrium ring* in a phase coherent sense (see text), according to Norman and Wilson; Fig. *b* corresponds to a later stage in the evolution than does Fig. *a*. Data in Figs. *c* and *d* are from an experiment in which a 50% amplitude perturbation was applied to the *initial model;* the models are otherwise identical. Figs. *a* and *c* are at comparable stages of evolution, as are (at a later stage) Figs. *b* and *d*.

experiments indicate that fragmentation is inevitable and that it starts at a very early stage; however, the fluid-dynamic experiments support a different view. A wide variety of types of perturbed clouds remains to be considered, but it is becoming evident that (1) the initial perturbation must be strongly asymmetric, at least in the isothermal case, and (2) if pressure effects are reduced, the perturbation is less strongly damped. Point 2 has been demonstrated by the work of Cook and Tohline, and point 1 has been well illustrated in a calculation by Narita and Nakazawa (1977) who did a 3-D isothermal collapse with $\alpha = 0.3$, $\beta = 0.2$ and 0.05. A highly non-axisymmetric initial cloud was used by the latter authors that resembled an ellipsoid of non-uniform density with uniform angular velocity. Surprisingly, a ring does form that later breaks up into a binary system. The fragments, which include about half of the mass of the cloud, go into Keplerian orbit and do not subsequently accrete much of the remaining gas in the cloud.

Although detailed study of fragmentation is just beginning, it is already evident that a rotating interstellar cloud is likely to break up into two (or more) orbiting subcondensations by the time the density has increased by two to four orders of magnitude. The residual spin of a condensation will eventually halt its further collapse and result in a second stage of fragmentation. We now ask whether a process of hierarchical fragmentation on successively smaller and smaller scales can eventually result in the formation of stellar binary and multiple systems with observed properties. Estimates of fragmentation sequences, based on the general results of 2-D and 3-D numerical experiments, have been made by Bodenheimer (1978). It is assumed that fragmentation proceeds through a ring phase (although the argument works also without this assumption). At each stage the ring breaks up into two equal-mass fragments in Keplerian motion which conserve the total mass and angular momentum of the ring. The ratio of spin to orbital angular momentum in a fragment and the ratio of ring mass to total mass at the time fragmentation occurs are taken as parameters. A fragmentation sequence ends when the temperature (obtained from 2-D collapse experiments carries through the adiabatic phase) in the final fragments is high enough so that the objects can be in hydrostatic equilibrium, and when the angular momentum is low enough so that they can contract to the main sequence without undergoing further instability to break-up by fission.

The initial condition is taken to be a cloud of $10^4\ M_\odot$ at 75 K and a density of 10 particles/cm^3 rotating with an angular velocity of 10^{-15} rad sec^{-1} and having angular momentum per unit mass of 1.6×10^{24} cm^2 sec^{-1}. Based on a number of experiments starting with these initial conditions, we can conclude that reasonable values of the parameters lead to masses and angular momenta of the final fragments that lie in the observed range for main-sequence stars. For example, if fragmentation occurs when the ring mass is 1/4 of the total mass of the cloud (or cloud fragment) and if 10% of the ring's angular momentum remains in the form of spin, then the final fragment

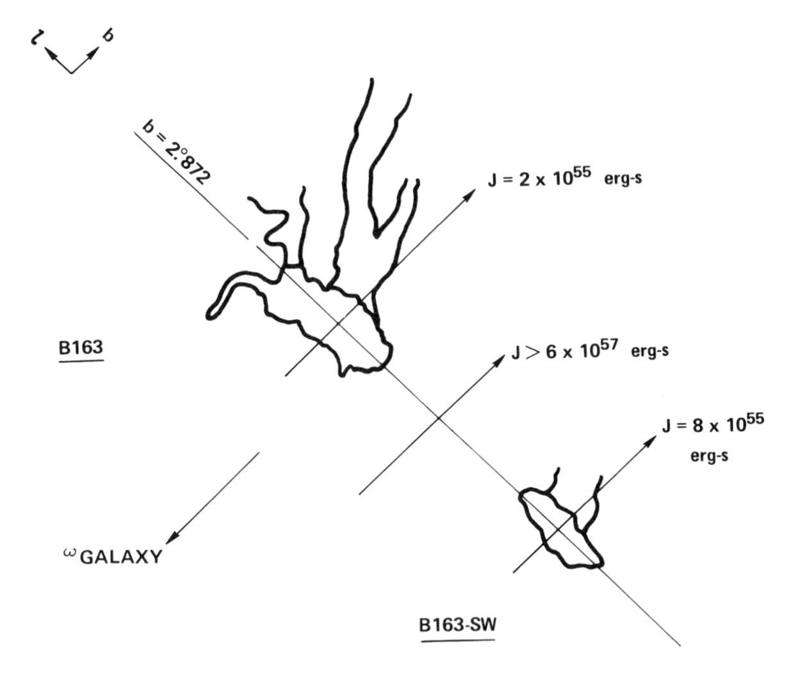

Fig. 6. Model for B 163 and B 163-SW. Minimum angular momentum associated with each cloud is given by *J*, and the directions of galactic coordinates are also shown. (From Martin and Barrett, 1977.)

has about 2 $M_\odot$ and is a member of a binary system with a period of 4 days. This binary system is one component of a quadruple system with a period of ~ 11 yr. These periods are characteristic of a number of observed stellar multiple systems, in particular, the well-observed χ Peg (Beardsley and King, 1976). The results are also consistent with observations reported by Abt and Levy (1976) which indicate that solar-type primary stars are likely to have both a close and a distant companion.

The rotational velocities of the sequence of fragments can be compared with observations of objects in two different phases of pre-main-sequence evolution. When a fragment first comes into hydrostatic equilibrium it has a mass, in several cases, of 0.5 to 2.0 $M_\odot$ and a radius of a few $R_\odot$. The rotational velocities at this point are calculated to be in the range 50-75 km sec^{-1}, in good agreement with those observed for T Tauri stars and other young contracting objects (Herbig, 1957; see chapter by Kuhi in this book). In the molecular cloud stage, where densities are the order of 10^4 particles/cm^3, the typical angular velocities (10^{-13} – 10^{-14} rad sec^{-1}) in a number of sequences are consistent with observations of rotation in, for example, B 361 (Milman, 1977) and B 163 (Martin and Barrett, 1977). The latter system provides further evidence for fragmentation at the molecular cloud phase, since it appears to consist of two rotating objects that are suspected to be in orbital motion with spin and orbital angular momentum vectors lined up in the same direction (see Fig. 6). It is also noteworthy that the predicted fragments at this stage have α and β of a few percent which are in the proper range that provides detailed agreement between B 361 and the 2-D experiments of Villere and Black (see Sec. II).

In conclusion, we see that 3-D hydrodynamical calculations indicate that a collapsing cloud will break up into two or more orbiting subcondensations with the possible subsequent development of a stellar multiple system. The formation of a "solar" nebula and an eventual planetary system is still a possibility; however, 3-D calculations to study such an event will be possible only with greatly improved spatial resolution over that possible at present.

IV. SUMMARY AND FUTURE WORK

The past decade has been one of many advances in the area of numerical experiments related to star formation and we have attempted to review results from these experiments. We summarize here the salient aspects of experiments in one, two and three space dimensions.

Experiments in One Space Dimension

Although 1-D (spherically symmetric) experiments of protostar collapse are highly idealized, they are the only ones which have been carried to a stage where a "stellar" object is formed. Experiments have shown that the parameters (e.g., radius and luminosity) of the visible stellar core are sensitive to the assumed initial conditions, particularly the initial density. If the initial density for 1 $M_\odot$ is such that the protostar is marginally Jeans unstable, the visible pre-main-sequence (PMS) stellar core is relatively small ($\sim 2\, R_\odot$) and is not very luminous ($\sim$ 10-20 $L_\odot$). However, if the initial density is significantly larger than that required to be Jeans unstable, the visible stellar core is much larger ($\sim 50\, R_\odot$) and more luminous ($\sim 10^3\, L_\odot$). An additional consequence of the assumed initial density is the dependence of the PMS evolution on mass. If the assumed initial density is near the Jeans density, the initial equilibrium state of the PMS object depends strongly on its mass, whereas if the assumed initial density is large enough to assure that collapse is adiabatic, the initial equilibrium state always lies at the top of the Hayashi track for the corresponding mass. A critical aspect of the evolution from protostar to a PMS object involves shocks formed by accretion of material onto an equilibrium stellar core. There have been significant advances, most notably by Winkler, in the modeling of such shocks and radiative transfer in the region of the shock, and it is now possible to predict with confidence the observable properties of PMS objects.

Presently, observational constraints on the properties of theoretical PMS objects are ambiguous. There do appear to be PMS stars occupying that region of the HR diagram where one expects to find stars $\sim 3\, M_\odot$, only if those stars began their collapse at densities well in excess of the Jeans density. However, it is premature to interpret the presence of these objects in terms of constraints on densities at the onset of collapse. If, as discussed by Cameron (see his chapter in this book) stars form as a consequence of dissipation in disks, the 1-D collapse experiments and the associated PMS evolutionary tracks may not be germane at all.

The principal focus of this book is on stars of roughly solar mass, but 1-D numerical experiments also provide information concerning the behavior of massive ($\gtrsim 10\ M_{\odot}$) stars. There is as yet no consensus regarding details of the PMS evolution of a massive star, but it is generally found that radiation and gas pressure around massive stellar cores limit and ultimately prevent accretion of circumstellar material onto such cores. This suggests that the upper limit to the mass of a star is determined by the properties of cores formed in the center of massive protostars. This subject is discussed in more detail in the chapter by Bertout and Yorke in this book.

Experiments in Two Space Dimensions

Two important physical parameters which could significantly affect star formation are rotation and magnetic fields. There have been several numerical experiments to determine the evolution of collapsing, rotating protostars. To date, none of these experiments has been followed to the stage where a "stellar" object forms. However, there is now a consensus that an important result of the collapse of a rotating protostar is the formation of a ring, and that the detailed structure of these rings depends on the dimensionless parameters α and β of the initial model for the collapse experiment ($\alpha \sim$ thermal energy/grav. energy; $\beta \sim$ rotational energy/grav. energy). In particular experiments with higher β-values (fixed α) yield larger more massive rings than do experiments with lower β-values; experiments with higher α-values (fixed β) yield rings with more nearly circular cross-section than do experiments with lower α-values, which yield more flattened rings. Unpublished studies by one of the authors (DCB) have shown that local conservation of angular momentum (to within 5-10%) in the central regions of a collapsing cloud appears to be necessary for ring formation. The axisymmetric nature of 2-D experiments limits the ability of these experiments to follow the evolution of rings, but it has been shown that in the absence of non-axisymmetric perturbations, rings formed from rapidly rotating ($\beta \gtrsim 10^{-3}$) protostars reach a short-lived equilibrium state (the properties of which agree with analytic results for such rings) and then because of mass accretion from their surroundings become unstable to collapse upon themselves. Non-axisymmetric effects are discussed below in connection with 3-D experiments.

Attempts to compare results from 2-D numerical experiments with observed parameters of Bok globules are just beginning. In particular, attempts to model the globule B 361 indicate that the initial model was one with a low α-value ($\alpha \sim 0.05$). Subject to the caveats that magnetic field effects are neglected in the numerical models and that detailed comparisons have only been made for one globule, this low α-value suggests that clouds evolve to a a density $\sim 10^3$ times the Jeans density before collapse ensues. Much more work needs to be done in this area.

The first self-consistent calculations on the structure of equilibrium self-gravitating magnetic clouds were made by Mouschovias. His results have been corroborated by Scott and Black, and the latter authors have done the first self-consistent experiments on the collapse of non-rotating, magnetized clouds. They find that the relationship between the magnetic field B and gas density ρ in the cloud core is a power law, viz. $B \propto \rho^{\frac{1}{2}}$. This law seems to hold for a variety of initial conditions. A potential consequence of this law is that the central regions of the cloud may be regions in which low-mass stars are more readily formed (see Mouschovias' chapter), because the critical mass for collapse ($M_{\rm CRIT} \propto B^3/\rho^{-2} \propto \rho^{-\frac{1}{2}}$) decreases in the cloud core.

Experiments in Three Space Dimensions

One of the major findings of 2-D numerical experiments is the formation of rings. A critical question concerns whether a 3-D experiment without the restriction of axial symmetry, would form rings. Studies by Tohline confirm that an initially axisymmetric cloud remains axisymmetric during collapse and rings form in qualitative and quantitative agreement with earlier 2-D experiments. Tohline has also shown that rings form even if there are substantial non-axisymmetric perturbations (50% amplitude $m = 2$) in the initial model. During the early stages of collapse, pressure effects tend to smooth out the initially non-axisymmetric perturbations. The question of ring evolution is still open. Experiments by Norman and Wilson show that phase coherent (see text) non-axisymmetric perturbations introduced into an equilibrium ring lead to rapid fragmentation of the ring, with a large fraction (60-90%) of the angular momentum of the fragments going into orbital angular momentum and the remaining 10-40% appearing as spin angular momentum of the individual fragments. However, experiments by Tohline show that fragmentation is inhibited if the perturbations are not coherent (see Fig. 5), and collapse of the ring upon itself occurs on a time scale comparable to or shorter than that of fragmentation. The role of pressure in smoothing out perturbations indicates that low-α initial models are more likely to retain non-axisymmetric effects than are high-α initial models. The possibility of fragmentation of rings into co-rotating blobs is suggestive as a mechanism for binary formation. Bodenheimer (1978) has shown that successive stages (~ 4) of ring formation and fragmentation, starting with a $10^4\ M_\odot$ cloud, could produce PMS stellar objects consistent with known properties of T Tauri stars. Studies of fragmentation, particularly fragment evolution, is difficult with finite difference numerical schemes. Larson has attempted to use a particle approach to model fluid behavior. While the technique has promise, limitations due to the number of particles (100) and to *ad hoc* representations of pressure and viscous effects leave doubt as to the validity of his results, particularly in the light of experiments by Tohline (using similar initial conditions and finite difference techniques) which did not yield even qualitatively similar results to Larson's particle experiments.

Future Work

Results from numerical experiments on star formation have produced exciting results, and raised many questions. Interdisciplinary interactions have also posed questions which can be answered in principle by numerical experiments. Some of the subjects for future study are discussed below.

It will be important to obtain model studies including more realistic opacities, both from grains and molecules and for continuum and line radiation. A preliminary step in this direction is discussed by Kondo (1978). A critical question concerns whether graphite grains exist in protostars. The highly refractory nature of such grains could have significant effects on radiative transfer in the region of shocks. It will also be important to understand and model the behavior of dust grains, particularily their growth and separation from the gas.

In addition to these specific effects, there is a need for a second generation of numerical experiments which will include improved and more complete physics and chemistry. Specific attention to the areas of radiative transfer, heating and cooling of clouds and cloud fragments and self-consistent models of cloud collapse including hydrodynamic, thermal and chemical effects will yield better numerical analogs for comparison with observations.

Magnetic fields and loss of angular momentum via magnetic fields may be important in the collapse of clouds and star formation. Numerical experiments, including both rotation and frozen-in magnetic fields will be carried out by the group at Ames Research Center. However, it will be very important to include effects due to ambipolar diffusion of magnetic fields from clouds as they collapse to higher density. In this connection, it will be very valuable to have improved observations relating to the strength of magnetic fields in clouds and cloud fragments of various densities.

Although we are far from understanding the often misused concept of fragmentation, several lines of future experimentation on this concept are clearly suggested. Much more work is needed on the fragmentation of rings, with emphasis on the distribution of angular momentum between spin and orbital motion. Further studies are also needed to determine whether fragmentation in large clouds, perhaps via some generalized Jeans fragmentation mechanism (see Silk's chapter), can occur in collapsing clouds. The availability of 3-D numerical codes offers the possibility of studying the dynamics of stellar fission – a mechanism often suggested as the means by which close binaries are formed (see Abt's chapter). Such studies could perhaps delineate the factors which control the mass and orbital period spectrum for close binaries.

It will also be valuable to obtain detailed evolutionary tracks for PMS objects, including more complete treatments of radiative transfer and accretion shocks. Preliminary efforts in this direction (Winkler, 1978) are underway. Improved observational constraints, discussed by Cohen (1978), will aid in defining the appropriate initial conditions for calculations concerning the final stages of star formation. In this context, it would be most valuable to obtain *observational* data concerning the masses of T Tauri stars. This would call for careful studies of T Tauri stars in binary systems where the mass of the other members of the binary is independently determinable. Finally, we urge observers to concentrate on cloud systems which are "simple." Complex structures, such as the Orion Nebula, are difficult to model numerically. The basic predictive aspects of numerical experiments on cloud collapse can best be tested against cloud structures where little or no star formation has yet occurred.

Acknowledgments. We thank M. Matthews and T. Gehrels for bringing people from diverse disciplines together to discuss the problems of planetary system formation. We acknowledge interchange of ideas with fellow numerical experimentalists who helped to clarify many detailed aspects of the work reviewed here. We thank E.H. Scott, M. Norman and J.E. Tohline for providing Figs. 3, 4, and 5, respectively. Partial support for one of us (PB) comes from an interchange agreement between NASA-Ames Research Center and San Francisco State University.

COMMENTS

L. Mestel: Even under the most favorable conditions, it is unlikely that braking by the galactic magnetic field will ever be so efficient that Type I protostars can reach main-sequence densities without centrifugal forces becoming important en route. Some further process of angular momentum loss must take place during the late phases of star formation. Conceivably a dynamo-built magnetic field may couple the star efficiently with a T Tauri phase stellar wind, or with surrounding protoplanetary material. Another possibility is that a rapidly rotating protostar flattens and divides into a binary system. (The argument that this is dynamically impossible seems misconceived; the process will be essentially dissipative, and appeals to the reversibility of the friction-free equations of motion therefore are beside the point.) If each component conserves its angular momentum of spin as it contracts, then the same centrifugal difficulties will quickly recur; we need an efficient process which converts spin angular momentum permanently into orbital.

Consider a binary system M_1, M_2, with centers of mass separated by a distance, d. The orbital angular velocity $\bar{\Omega}$ is given approximately by

$$\bar{\Omega}^2 d^3 = G(M_1 + M_2). \tag{1}$$

Let us suppose that some process enforces approximate synchronization between spin and orbital angular velocities until M_1 has contracted to a radius $\bar{R}_1$, and the subsequent contraction occurs with conservation of angular momentum of *spin.* Then at a subsequent radius R_1, the ratio of centrifugal force of spin to gravity is given by

$$\frac{\Omega_1^2 R_1^3}{GM_1} = \left(\frac{\Omega_1^2 R_1^4}{GM_1}\right)\frac{1}{R_1} = \left(\frac{\bar{\Omega}^2 \bar{R}_1^4}{GM_1}\right)\frac{1}{R_1} = \left(\frac{\bar{R}_1}{R_1}\right)\left(\frac{\bar{R}_1}{d}\right)^3 \left(1 + \frac{M_2}{M_1}\right) \tag{2}$$

and this does not become unity until

$$\frac{\delta(R_1)}{\delta(\bar{R}_1)} = \left(\frac{\bar{R}_1}{R_1}\right)^3 = \left(\frac{d}{\bar{R}_1}\right)^9 \frac{1}{(1 + M_2/M_1)^3}\,. \tag{3}$$

Thus, if $M_1 \simeq M_2$ and $(d/\bar{R}_1) \simeq 3$, $\delta(R_1) \simeq 2.5 \times 10^3\ \delta(\bar{R}_1)$; if $(d/\bar{R}_1) \sim 5$, $\delta(R_1) \simeq 2.5 \times 10^5\ \delta(\bar{R}_1)$. As in the discussion of magnetic braking, we again see how a synchronization process is equivalent to an enormously efficient transfer of angular momentum from spin to orbital motion.

If one or both of the stars in the binary system is largely convective (e.g., if it is in the Hayashi phase), then one can expect a large turbulent viscosity which will tend to destroy the tidal motions in a non-synchronized system. However, the rate of tidal friction $\propto (R_1/d)^6$ and so decreases rapidly as the stars contract. Magnetic coupling between the pair may stay efficient longer, but again as (R_1/d) decreases the number of field lines linking the two stars will decrease. A thorough re-appraisal of the synchronization problem seems overdue.

REFERENCES

Abt, H.A., and Levy, S.G. 1976. *Astrophys. J. Suppl.* 30: 273.
Alfvén, H., and Arrhenius, G. 1976. *Evolution of the Solar System* NASA SP-345 (National Aeronautics and Space Administration). Washington, D.C.: U.S. Government Printing Office.
Appenzeller, I., and Tscharnuter, W. 1974. *Astron. Astrophys.* 30: 423.
____. 1975. *Astron. Astrophys.* 40: 397.

Beardsley, W.R., and King, M.W. 1976. *Publ. Astron. Soc. Pacific* 88: 200.
Bertout, C. 1976. *Astron. Astrophys.* 51: 101.
Biermann, L., and Cowling, T.G. 1939. *Z. Astrophys.* 19: 1.
Black, D.C., and Bodenheimer, P. 1975. *Astrophys. J.* 199: 619.
____. 1976. *Astrophys. J.* 206: 138.
Bodenheimer, P. 1978. Preprint. Submitted to *Astrophys. J.*
Bodenheimer, P., and Sweigart, A. 1968. *Astrophys. J.* 152: 515.
Bok, B.J. 1978. *Moon and Planets* (special Protostars and Planets issue). In press.
Cameron, A.G.W. 1962. *Icarus* 1: 13.
Cameron, A.G.W., and Pollack, J. 1976. In *Jupiter* (T. Gehrels, ed.), p. 61. Tucson, Arizona: University of Arizona Press.
Cohen, M. 1978. *Moon and Planets* (special Protostars and Planets issue). In press.
Cohen, M., and Kuhi, L.V. 1976. *Astrophys. J.* 210: 365.
Cook, T.L. 1977. Los Alamos Publ. LA-6841-T: Ph.D. Dissertation, Rice University, Houston, Texas.
Deissler, R.G. 1976. *Astrophys. J.* 209: 190.
Fricke, K.; Möllenhoff, C.; and Tscharnuter, W. 1976. *Astron. Astrophys.* 47: 407.
Gaustad, J.E. 1963. *Astrophys. J.* 138: 1050.
Gerola, H., and Glassgold, A.E. 1978. *Astrophys. J. Suppl.* In press.
Gingold, R.A., and Monaghan, J.J. 1977. *Mon. Not. Roy. Astr. Soc.* 181: 375.
Hayashi, C. 1961. *Publ. Astr. Soc. Japan* 13: 450.
Hayashi, C., and Nakano, T. 1965. *Prog. Theor. Physics* 34: 754.
Heiles, C. 1976. *Ann. Rev. Astron. Astrophys.* 14: 1.
Herbig, G.H. 1957. *Astrophys. J.* 125: 612.
Hunter, C. 1977. *Astrophys. J.* 218: 834.
Kahn, F.D. 1974. *Astron. Astrophys.* 37: 149.
Kamiya, Y. 1977. Preprint. University of Tokyo.
Kondo, M. 1978. *Moon and Planets* (special Protostars and Planets issue). In press.
Larson, R.B. 1969. *Mon. Not. Roy. Astr. Soc.* 145. 271.
____. 1972*a*. *Mon. Not. Roy. Astr. Soc.* 156: 437.
____. 1972*b*. *Mon. Not. Roy. Astr. Soc.* 157: 121.
____. 1973*a*. *Fund. Cosmic Phys.* 1: 1.
____. 1973*b*. *Ann. Rev. Astron. Astrophys.* 11: 219.
____. 1977. In *Star Formation* (T. de Jong and A. Maeder, eds.), p. 249. Dordrecht: Reidel.
Larson, R.B., and Starrfield, S. 1971. *Astron. Astrophys.* 13: 190.
Lucy, L. 1977. *Astron. J.* In press.
Martin, R.N., and Barrett, A.H. 1977. *MIT Radio Astr. Contr.* No. 3.
McNally, D. 1974. *Irish Astr. J.* 11: 239.
Mestel, L., and Spitzer, L. 1956. *Mon. Not. Roy. Astr. Soc.* 116: 503.
Milman, A.S. 1977. *Astrophys. J.* 211: 128.
Mouschovias, T. Ch. 1976*a*. *Astrophys. J.* 206: 753.
____. 1976*b*. *Astrophys. J.* 207: 141.
Nakazawa, K.; Hayashi, C.; and Takahara, M. 1976. *Prog. Theor. Phys.* 56: 515.
Narita, S.; Nakano, T.; and Hayashi, C. 1970. *Prog. Theor. Phys.* 43: 942.
Narita, S., and Nakazawa, K. 1977. Preprint, KUNS 435. Kyoto University.
Norman, M.L., and Wilson, J.R. 1978. Preprint. Submitted to *Astrophys. J.*
Ostriker, J.P. 1964. *Astrophys. J.* 140: 1067.
Penston, M.V. 1966. R. *Obs. Bull.* No. 117.
Safronov, V.S. 1972. *Evolution of the Protoplanetary Cloud and Formation of the Earth and Planets,* translated from Russian (Israel Program for scientific translations).
Scott, E.H., and Black, D.C. 1978. In *Proceedings of 2nd Gregynog Workshop: Molecular Clouds.* In press.
Shu, F. 1977. *Astrophys. J.* 214: 488.
Spitzer, L. 1968. *Diffuse Matter in Space.* New York: Interscience.
Strom, S. 1977. In *Star Formation* (T. de Jong and A. Maeder, eds.), Dordrecht: Reidel.
Takahara, M.; Nakazawa, K.; Narita, S.; and Hayashi, C. 1977. *Prog. Theor. Phys.* 58: 536.

Tohline, J.E. 1977. *Bull Amer. Astr. Soc.* 9: 566.
Tscharnuter, W. 1975. *Astron. Astrophys.* 39: 207.
____. 1977. *Astron. Astrophys.* 57: 279.
____. 1978. *Moon and Planets* (special Protostars and Planets issue). In press.
Villere, K., and Black, D.C. 1978. In preparation.
Westbrook, C.K., and Tarter, C.B. 1975. *Astrophys. J.* 200: 48.
Winkler, K.-H. 1978. *Moon and Planets* (special Protostars and Planets issue). In press.
Wong, C. 1974. *Astrophys. J.* 190: 675.
Woodward, P.R. 1978. *Ann. Rev. Astron. Astrophys.* In press.
Yorke, H.W., and Krügel, E. 1977. *Astron. Astrophys.* 54: 183.

THE BINARY FREQUENCY ALONG THE MAIN SEQUENCE

HELMUT A. ABT
Kitt Peak National Observatory

Among the normal main-sequence B2-B5 and F3-G2 stars, there are two types of binaries in each group: (1) the binaries with periods less than 10 or 100 years have secondaries that decrease in frequency for smaller masses; while (2) among the long-period systems, both primaries and secondaries follow the van Rhijn distribution. We suspect that Group (1) is the result of bifurcation of rapidly-rotating protostars and Group (2) results from the capture of protostars. The frequencies and characteristics of binaries along the main sequence seem to be surprisingly similar. We observe half of the stars to be multiple systems. After correction for undetected companions, at least three fourths of the primaries have nearby (bifurcation) companions of some kind (secondary stars, black dwarfs, or planets). Whether or not all normal stars have nearby companions depends on a risky extrapolation of the frequencies of various secondaries.

Other kinds of stars have frequencies of short-period (less than a month or a year) binaries that vary from zero in some cases to nearly 100% in others. Perhaps all stars have the same frequency of long-period binaries. The Be stars have no short-period binaries. The magnetic Ap stars have a deficiency of short-period binaries but the non-magnetic Ap (Hg, Mn) stars have a normal frequency. Most or all Am stars are in short-period binaries but the normal A4-F1 stars of the same masses have no short-period binaries. High-velocity stars are deficient in short-period binaries. Some of these results are understandable in terms of the diffusion of elements in stars whose rotational velocities have been reduced by tidal interactions in short-period systems or by magnetic braking. On the other hand, emitting shells can occur around rapidly-rotating stars not found in short-period binaries.

This chapter is a summary of our current knowledge of the binary frequency and binary characteristics along the main sequence. This information is relevant to the occurrence of planetary systems in two ways:

(1) A planetary system can be considered to be a binary (or multiple) system in which the mass ratio is very large. By studying the frequencies of binary systems with different mass ratios, we can perhaps extrapolate as far as the mass ratios of star-planet systems.

(2) It may be that planets – or at least habitable planets – occur only around stars without stellar companions. Therefore searches for double and for single stars may have some bearing on the occurrence of (habitable) planets.

Studies of binaries among main-sequence stars quickly encounter an important effect whose relevance to planets is unknown. At many places along the main sequence there are several kinds of stars that are spectroscopically distinct and in several cases, at least, there is a close correlation between those characteristics and their differing binary frequencies. For example, in the region of the late A-type stars about 30% of the stars are "metallic-line" (Am) stars and the remainder are mostly normal. The Am stars have abnormal compositions, low rotational velocities, and are almost invariably in binaries; the remainder have normal compositions, high rotational velocities, and are never found in binaries with periods less than one year. Which kind of star could have planets?

Let us enumerate some of these cohabitors along the main sequence because the binary characteristics for some of these subgroups are known and will be described below. Among the O-type stars most have normal (solar) compositions but Walborn (1976) has specified cases that are rich in nitrogen (OBN) and others that are rich in carbon (OBC). Most, but not all, of his examples are for stars above the main sequence. Among B-type stars there are at least three distinct kinds: the normal B's, Be stars that have hydrogen emission lines, and various classes of peculiar stars (conventionally called Ap stars because they extend into the A-star region but most of the examples are B stars according to their temperatures). The Ap's have been reviewed by Preston (1974). Among the A-type stars there are the normal A's, an extension of the Ap stars, and the Am stars mentioned above. Among the F, G, and K stars most of the differences are minor: variations in the lithium abundance and in the strength of the Ca II K emission. However, there are high- and low-velocity stars that differ greatly in their composition. Finally, among the M's there are the normal dwarfs (dM) and those with hydrogen emission (dMe); the relative frequencies vary (Joy and Abt, 1974) from a few percent dMe's at M0 to 100% dMe's at M5.5. The characteristics of each of these subgroups will be discussed below.

In organizing this material one might think that the logical procedure is to start at one end of the main sequence and progress to the other end. Unfortunately our information about binaries among the subgroups is more complete for some than for others. Therefore we choose to discuss the better-studied subgroups first and then indicate whether the other subgroups show significant differences or not. Specifically, we shall start with the solar-type (F3-G2) stars first, then progress up the main sequence, and finally discuss the late-type stars briefly.

I. SOLAR-TYPE STARS

The most obvious difference among the intermediate-type stars is between the high-velocity stars (constituting roughly 10% of the nearby field stars) and the low-velocity ones. Of course there is a continuum of differences and it has been found (Strömgren, 1964) that the larger a star's space velocity relative to the local standard of rest, the lower its content of metals.

A. High-Velocity Stars

The frequency of binaries among high-velocity (Population II) field stars has been studied by Abt and Levy (1969) and Crampton and Hartwick (1972). They found that (1) the frequency of short-period ($P < 100^d$) binaries decreases drastically as one progresses to stars of higher space velocity, and (2) the high- and low-velocity stars do not seem to differ in the occurrence of long-period ($> 100^d$) systems.

If the deficiency of binaries was due to disruption in frequent near-collisions with disk-population stars, we would expect the long-period binaries to be lacking, rather than the short-period ones. Our only suggestion for the explanation is an initial low turbulent velocity in the nebular cloud, allowing immediate collapse, whereas the turbulent clouds resisted collapse until a later time.

These results may have a bearing on the question of the frequency of doubles in globular clusters. Eclipsing binaries are rare (Hogg, 1973) in globular clusters, but the stars surveyed are mostly giants; even among low-velocity giants, periods less than 100^d are rare (Scarfe, 1970) and therefore the probability of detecting eclipsing systems should be and is very small. But some stars (e.g., novae) that probably are invariably members of short-period binary systems occur in globular clusters and therefore, there must be some short-period binaries present, even if long-period binaries may have been disrupted in frequent stellar encounters.

B. Low-Velocity Stars

Earlier statistical estimates (Jaschek and Jaschek, 1957; Petrie, 1960; Jascheck and Gomez, 1970) indicated that about half of the low-velocity main-sequence solar-type stars have variable velocities attributable to duplicity. Trimble (1974) extended this study to show that the frequencies of various mass ratios (among binaries of all types) show two peaks at $m_2/m_1 = 1.0$ and 0.3, suggesting that there are two types of binaries present. But that study was based on binaries with published orbital elements and is subject to the strong selection effect that observers favor systems with double lines, large velocity ranges, and sharp lines.

An objective study of F3-G2 main-sequence stars was made by Abt and Levy (1976; see also Abt, 1977), who considered essentially all of those stars brighter than $V = 5.5$ mag and north of -20° declination; 123 low-velocity stars were included. The spectral range was selected to include sharp-lined stars later than the Am region and to include solar-type stars; there are very few stars later than G2 that are brighter than $V = 5.5$ mag. Many of those had been studied previously and all were well known to visual double-star observers. The basic results from this work yielded 88 periods due to spectroscopic binaries (*SB*) or published visual binaries (*VB*), and rough periods for common-proper-motion pairs (*CPM*). Branch (1976) pointed out that systems that were included in the $V = 5.5$ mag limit simply because they were double should be eliminated; this removed eight systems with 12 periods and reduced the totals to 36 *SB* (4 *SB*2, 32 *SB*1), 19 *VB*, and 21 *CPM*. The observed frequencies of singles:doubles:triples:quadruples is 45:46:8:1%

However, some binaries will not be detected because (1) the secondaries are of low mass, (2) the systems are so nearly pole-on that their radial velocities are not detectably variable, (3) the periods are long and the components are equal in brightness so that the double lines are not resolved, (4) the visual companions are faint and too close or faint and distant and lost among the background stars, etc. Attempts to allow for these selection effects were successful in most cases and led to the results shown in Fig. 1. The results shown differ slightly from the original in that the two corrections – the second was due to the fact that the first "bin" in secondary mass is narrower than the others – of Branch (1976) have been taken into consideration.

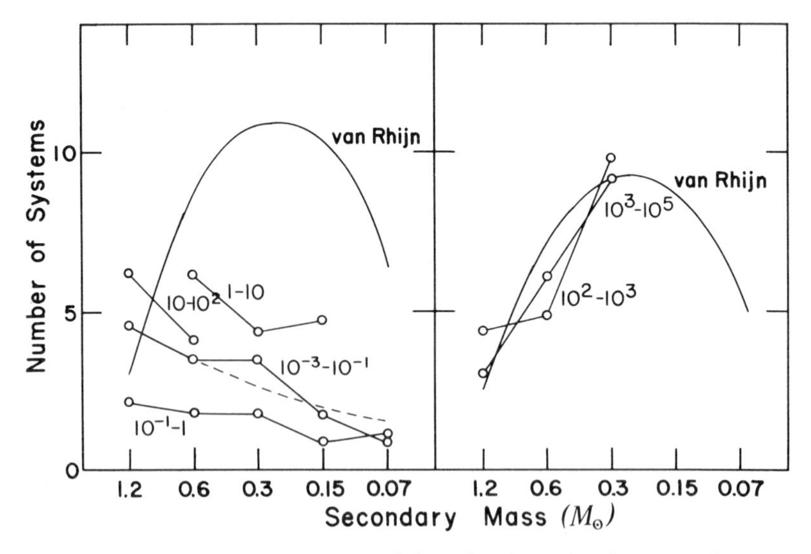

Fig. 1. The frequencies of systems with F3-G2 primaries and various secondary masses, grouped in bins differing by factors of 2. The segments are labeled with the orbital period ranges in years. The van Rhijn distributions are normalized to 3 secondaries of 1.2 solar masses in the left panel and 2.5 secondaries of $1.2 m_\odot$ in the right panel. The dashed line on the left shows the slope of a power law $N \propto m_2^{\,0.4}$.

Figure 1 shows the numbers of systems with various secondary masses, separated in bins differing by factors of 2.0. The period ranges in years are labeled on the line segments. The segments are, in some cases, incomplete because the incompleteness calculations were too uncertain. The systems with periods less than 100 years are shown on the left; those with periods greater than 100 years are on the right. The van Rhijn distributions are shown and are normalized to three stars in the first bin on the left and 2.5 stars in the first bin on the right. (The van Rhijn distribution is the relative frequencies per unit volume of space of primaries of various masses of luminosities.) The primaries would follow the van Rhijn distribution, although our sample extended over only a short range of types (F3-G2).

Notice that in the right panel of Fig. 1 the distant secondaries follow the van Rhijn function within the observational errors, which are at least the square roots of the numbers of systems. But in the left panel the close companions do not follow that distribution. Within the observational errors the curves in the left panel follow the slope of a power distribution $N \propto m_2^{0.4}$, which is shown as a dotted line.

We suspect that there are two types of binaries: the short-period systems are bifurcation or fission doubles in which a single protostar had too much angular momentum to form a single stable star, so it subdivided to form a close pair. When the angular momentum was divided between the members of the pair, the mass division favored equal pairs and smaller mass ratios (m_2/m_1) are increasingly improbable. On the other hand, the widely spaced pairs may be pairs of protostars that formed and collapsed separately and just happened to be gravitationally held together (as captures occurred). In those cases the mass distributions of both the primaries and the secondaries followed the van Rhijn distribution.

The dividing point between the two types at about 100 years probably tells us something about the dimensions (50 AU or 10^{10} km) of protostars at the time when the individual collapse starts.

It is of interest for the present discussion of protostars and planets to estimate the numbers of stellar and planetary companions to the solar-type primaries. Let us assume the 0.4 power law for the closely spaced systems, even though we have no assurance that this power is well determined or can be extrapolated well beyond the secondary mass of 0.07 $m_\odot$. Defining stars as objects with masses greater than 0.07 $m_\odot$, black dwarfs (Kumar, 1963) as objects with masses between 0.07 and 0.01 $m_\odot$, and planets as objects with masses less than 0.01 $m_\odot$, we find 54, 13, and 11%, respectively, as nearby companions. The sum is 78%, which is a departure from the 100% found in the original discussion (without the Branch corrections). This implies that perhaps not all primaries have nearby companions, even though it seemed originally that the universality of nearby companions would explain how the large amount of angular momentum in the initial clouds relative to that in stellar rotation could be absorbed. I suspect that the question of whether all or just most of the stars have nearby companions is still open. In addition, extrapolation along the van Rhijn curve shows that about 53% of the primaries have distant stellar companions; we know nothing about distant black dwarfs and planets because the van Rhijn distribution is not defined for masses less than 0.07 $m_{\odot}$.

This implication that roughly 11% of the solar-type stars have planetary secondaries is a first guess as to this frequency. Much better data should come from studies with more sensitive equipment, such as that of Serkowski's that can detect low-mass secondaries that perturb the primaries by amounts of the order of 10-100 m sec^{-1} (see Serkowski, 1977).

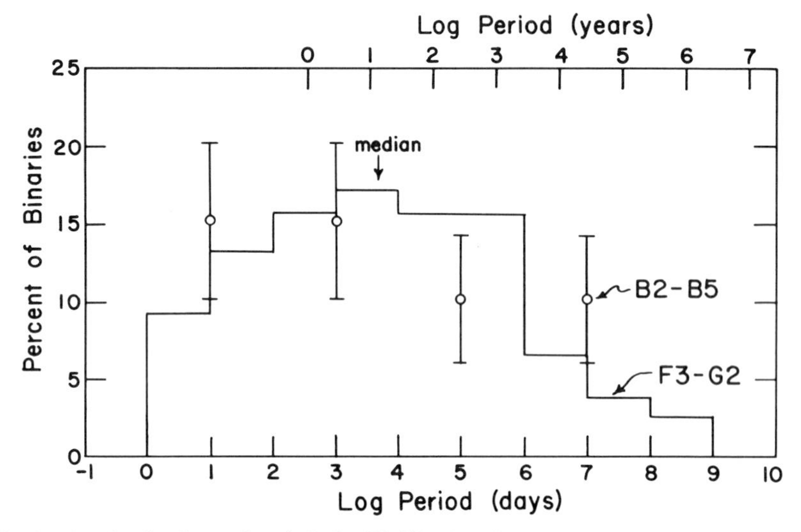

Fig. 2. The distributions of periods for F3-G2 primaries (solid line) and B2-B5 primaries (dots with error bars).

We have shown in Fig. 2 the distribution of periods (for both types of binaries). This distribution includes the observed spectroscopic, visual and *CPM* stars, but not the undetected pairs computed in the incompleteness calculations. It is perhaps strange that two mechanisms for the formation of binaries yield a single maximum, rather than a bimodal distribution. The median period is 14 years. The apparent termination of the curve at 1 day is not real; W Ursae Majoris binaries exist with periods as small as 0.25 days, but they are relatively rare (none occurred in this sample of 123 low-velocity stars), so they occur roughly at the 1% level. The other termination at about 10^6 years is an observational selection because no intensive study has been made for very distant *CPM* companions (>10 arc min or >10^4 AU. However, it is also likely that such widely spaced pairs would be disrupted by chance encounters and that the termination is real, at least after the 10^9 yr of existence of these binaries in the galactic plane.

C. Binaries and Calcium Emission

It seems likely (Young and Koniges, 1977) that stars with unusually strong Ca II H and K emission lines tend to be in closely spaced binaries of the λ And type (see Fig. 9 of Wilson and Bappu, 1957 for an illustration). Evidently tidal coupling causes enhanced chromospheric emission by increasing the density scale heights (Young and Koniges, 1977). This enhanced calcium emission should not be confused with the more frequent but less pronounced calcium emission associated with a small age (Wilson, 1963) among solar-type stars. Nevertheless, pronounced calcium emission seems to be an effective way to isolate short-period binaries.

II. A-TYPE STARS

The principal problem concerning binaries among the late A-type stars is the occurrence of two subgroups (mentioned earlier): the metallic-line (Am) stars and the normal A stars. The former have weak Ca II K and Sc II lines and strong lines due to iron-peak and heavier elements, all relative to the hydrogen line strength. On the basis of atmospheric temperatures or hydrogen strengths, the Am stars occur in the range A4-F1 IV,V (Abt and Bidelman, 1969) for the "classical" Am stars (those with at least a half spectral class difference between the spectral types from the K lines and from the metals), but for the "marginal" (Am:) stars with less than a half class range, the range can extend to A0 (like Sirius). Both kinds have about one-quarter of the rotational velocities of normal A stars (Abt and Moyd, 1973; Abt, 1975).

Abt (1961) found that essentially all (88%) Am stars are members of spectroscopic binaries, most (72%) of which have periods less than one year. On the other hand, the normal A stars in the same range of types have no known binaries with periods less than one year (Abt, 1965; Abt and Levy, 1974), although the two subgroups do not differ in the frequency of binaries of longer periods. The study of Am and Am: stars should be and is being repeated and extended by the author to answer many detailed questions (e.g., do Am and Am: stars differ in their binary frequencies; do all Am stars have short-periods after allowing for incompleteness?). But the contrast between the Am and normal A stars is striking.

The proposed explanation is that the short-period binaries have, by tidal interactions, the low rotational velocities that allow a diffusive separation of elements. The non-members of short-period binaries are rapid rotators that are well mixed and therefore have normal compositions.

Combining the Am's and A's together, the binary characteristics of the late A's do not seem to be different from those of the solar-type stars insofar as we know, but the measures are not as extensive and the analysis has not been carried as far as for the solar-type stars.

III. B-TYPE STARS

Among the B stars there are two types of studies that we wish to review: those of (1) normal B stars for a comparison with the solar-type stars and (2) the Be and Ap stars. These will be discussed in the next three subsections.

A. Normal B Stars

Abt and Levy (1978) made a study of 42 normal B2-B5 IV or V stars, searching for spectroscopic binaries and quoting data on visual binaries and *CPM* pairs. After eliminating three stars because their duplicity implied that the primaries were really outside the magnitude range ($B < 5.0$ mag), the statistics were limited to a sample of 39 stars. It was realized that this sample is too small to define the luminosity function of the secondaries, so a new study of 75 additional stars has been started.

The published study revealed a total of 15 spectroscopic binaries with orbital elements, one visual binary with known orbital elements, and seven *CPM* pairs. The observed ratios of single:double triple quadruple stars are 51:36:5:0%, which agree with the numbers (45:46:8:1%) for the solar-type stars within the statistical accuracy. After corrections, where they could be made, for incompleteness, the mass functions for the secondaries is shown in Fig. 3. Note that again the short-period systems have a mass function that decreases with decreasing secondary mass while the long-period systems follow the van Rhijn (or Salpeter, 1955) function. The breakpoint between the two types of binaries is given as 10 years, rather than the 100 years for the solar-type stars, but this is poorly determined because there was only one binary with a period between 10 and 100 years.

For the short-period (bifurcation) doubles, the mass function has a power-law slope of about one-quarter, but it could be the same exponent (0.4) found for the solar-type stars. After allowance for incompleteness has been made, 47% of the primaries have nearby secondaries with masses greater than 0.66 $m_{\odot}$. Extrapolation of the quarter-power relation suggests that 67% of the primaries have nearby stellar companions, 10% have black dwarf companions, and 16% have planetary companions, for a total of 94% having nearby companions of some sort. Again, it could be that all or most primaries have some companions, depending on the reliability of the extrapolation. At least 25% of the primaries have distant companions.

The distribution of periods for these B stars plus the Be stars (to be discussed below) is shown in Fig. 2 as dots superimposed on the distribution for the solar-type stars. Within the accuracy of the data, which is taken as the square root of the numbers involved, the distributions are the same.

Normally we would not expect any of these results for the B2-B5 primaries to be identical with those for the F3-G2 primaries because the primaries differ by a factor of 8 in mass, and this factor must have some effect on the distributions. But it is interesting to see that the general characteristics are the

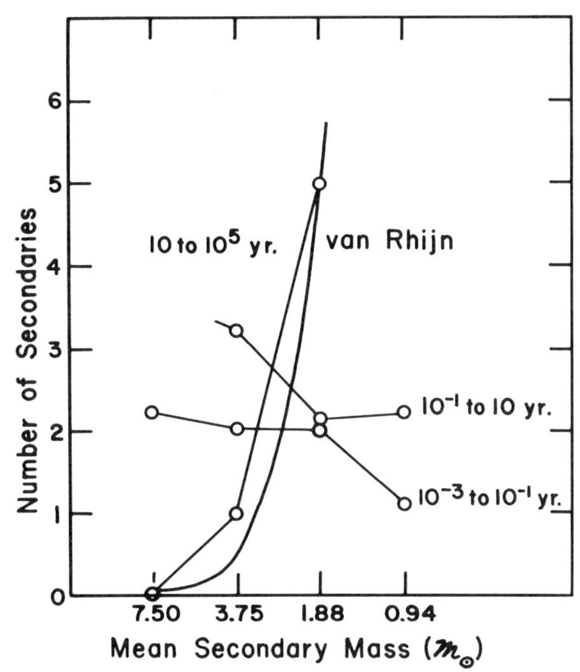

Fig. 3. The frequencies of systems with B2-B5 primaries and various secondary masses, grouped in bins differing by factors of 2. The orbital period ranges in years are labeled. The van Rhijn distribution, normalized to five stars in the third bin, is given for comparison.

same: (1) two types of binaries, (2) a period break between the two at 10-100 years, (3) the long-period secondaries obey the van Rhijn function, (4) the short-period binaries have a different relation with a fractional power law, (5) most of the primaries have some nearby companions and many have distant companions, and (6) the distribution of periods is similar for both sets of primaries and shows a single broad maximum, even though two types of binaries are represented.

A parallel study by Wolff (1978) concentrated on 83 normal late B-type main-sequence stars with sharp lines. The point of the study was partly to look for small velocity variations by virtue of their sharp lines. She found that for short periods, the frequency of various secondary masses decreases from a maximum at $m_2/m_1 = 1$ to 0.2. But surprisingly, the function shows a second maximum at $m_2/m_1 = 0.1$ or $m_2 = 0.25\ m_\odot$. That result is not based on derived periods but only on scatter in the radial velocities. Wolff argues that the scatter is not due to pulsation or other causes, but to perturbations by low-mass companions. The Abt-Levy data did not go to such low ($0.25\ m_\odot$) secondary masses because they were measuring mostly broad-lined stars. If the

Wolff interpretation is correct, it will not be valid to extrapolate the mass distributions below the regions from which they were derived. This would be unfortunate because it would mean that we could not predict the frequency of planets unless we actually measured their perturbations upon the primaries.

B. Be Stars

Among the early B stars (where the Ap or Bp stars are still infrequent), about one-quarter of the stars have hydrogen emission lines. Many Be stars have shells (Bidelman and Weitenbeck, 1976) and apparently all Be stars have rotational velocities near to the break-up speed (Slettebak, 1966). This probably implies that their rotational velocities have not been reduced by tidal interactions in closely spaced binaries. What are the binary characteristics of the Be stars?

Abt and Levy (1978) searched for binaries among 21 B2e-B5e main-sequence stars. They found no binaries with periods less than a month, although the frequency of binaries of longer periods did not differ from that for the normal B2-B5 stars. This substantiates the guess that these rapidly rotating stars do not occur in short-period binaries. There might be a way out of this conclusion if these stars are really "cocoon" systems – a close pair in a common opaque shell. That model may eventually be proven to be inappropriate. But then what is the difference between a normal B star in a long-period system and a Be star in a similar system? Is it just a matter of chance that the Be star ended with a rotational velocity near the break-up speed while the B star has a smaller rotational speed?

C. Peculiar A Stars

There are several families of Ap stars, although some of the differences are due to the surface temperature (or stellar mass) differences (e.g., Ap [Mn] stars are believed to be limited to the range 10,000-15,000 K in effective temperature [Preston, 1974], but at lower temperatures the manganese is neutral and the abnormality may not be recognized and other abnormalities would be obvious). Following Preston, let us discuss primarily the magnetic Ap stars (the Sr, Cr, Eu and the Si stars) and the non-magnetic ones (the Hg, Mn stars).

A study by Abt and Snowden (1973) of 62 Ap stars showed that the magnetic stars are deficient (but not totally absent) in binaries with periods less than about one year but normal with regard to long-period binaries. Evidently the low rotational velocities of the Ap stars are not due to tidal interactions in short-period systems, but rather to magnetic braking. There is synchronization between orbital and rotational motions for binaries with periods less than six days, but not for the longer-period binaries. Evidently magnetic Ap stars do not appear in short-period binaries very often, although some are known.

But Abt and Snowden (1973) and especially Aikman (1976) found that the binary frequency of the non-magnetic (Hg, Mn) Ap stars is the same as for normal B stars, both with regard to short and long periods. Ap (Hg, Mn) stars have rotational velocities that are much lower (Guthrie, 1965) than those of normal stars (i.e., $\langle V \sin i \rangle = 29$ km sec^{-1} compared with 178 km sec^{-1} [Abt, Chaffee and Suffolk, 1972]). How did the Ap (Hg, Mn) stars get such low rotational velocities without magnetic braking? Their very different rotational velocities are probably not attributable to their binary characteristics. This remains a problem unless it could be shown that the Ap (Hg, Mn) stars have strong internal magnetic fields despite having no external magnetic fields.

IV. O-TYPE STARS

The measurement of the radial velocities of O-type stars is complicated by (1) broad lines, (2) the effects of winds or mass loss in the photospheric regions (Conti, Leep and Lorre, 1977), causing different lines to have different velocities, (3) frequent effects of gas flows within binaries, and (4) much line blending due to the near coincidence of the He II and H I lines and to the fine structure in the He I lines. Therefore, although these stars tend to have massive secondaries that cause large velocity perturbations of the primaries, the measurement of true binary motion is difficult. Nevertheless, a systematic study has been started by Bohannan and Garmany, and a preliminary report (Bohannan and Garmany, 1978) indicates that some, but not many, binaries of all periods are being found. Conti, Leep, and Lorre (1977) found from a few spectra per star that about 58% of the O stars or 41% of the Of stars are probable binaries. We will presume that the binary frequency is similar to that of the B2-B5 stars until more complete results are available.

However, in addition to the normal O stars and those with some emission (Of), Walborn (1976) has pointed out some stars he calls OBN and OBC that are late O or early B stars – mostly supergiants – that have enhanced nitrogen or carbon lines. Among main-sequence stars these may constitute only a few percent of the total OB stars. Bolton and Rogers (1978) studied 11 OBN stars and 9 OBC stars for velocity variations. They found that between 50 and 100% of the OBN stars are members of short-period binaries but that none of the OBC stars are in short-period binaries. Again this seems to be a correlation between abundance peculiarities and binary characteristics. Bolton and Rogers favor, as an explanation for the correlation, the selective mass transfer of nitrogen in the (OBN) binaries and diffusion of core material in the evolved (OBC) single stars.

V. M DWARFS

The determination of the binary frequency of late-type dwarfs is hampered by instrumental problems. On the one hand, their masses are low and their orbital velocities are small (for a typical M2-M5 binary in a 10^d orbit, the primary velocity amplitude is only about 25 km sec^{-1}) and on the other hand, the stars are faint ($V \simeq 10$ mag for the brightest several dozen stars) so that high-dispersion spectra would require long exposure times. What is required is either a fast detector that will not degrade the usual coudé resolution or a special instrument that combines speed and spectral resolution. One can understand why preliminary results (three spectra per star) by Wilson (1967) on 300 dK and dM stars showed only 10% to be variable in velocity. In fact, Wilson felt that this (low) fraction of detectable binaries is roughly right if the binary frequency of dK and dM stars is normal.

Some M dwarf stars have hydrogen emission lines (dMe stars). The cause of the emission and the physical differences between dM and dMe stars is unknown. It has been thought that the dMe stars are younger, just like the solar-type stars with Ca II emission (Wilson, 1963) are younger than non-emission stars, but this is doubtful. The fraction of all M dwarfs that have emission spectra varies smoothly from about 3% at M0 to 100% at M5.5 (Joy and Abt, 1974).

There are preliminary statistics about the binary frequency among M dwarfs that are tantalyzing but inconclusive: all the (four) known M dwarfs that are double-lined have dMe spectra (Joy and Abt, 1974). They also have flare activity and are called BY Dra stars. Could it be that the emission phenomenon is associated with interacting binaries and the (absorption) dM stars are either single stars or widely spaced binaries? An objective study needs to be made.

VI. SUMMARY

We have looked at the fragmentary data on the frequency of binaries along the main sequence and described two types of results. First, we found that for most (but not all – e.g., the late A's) stars that are normal in composition and lack pronounced emission lines, the frequency of binaries is roughly the same and the binary characteristics do not seem to depend strongly on type or primary mass. The short-period binaries seem to be bifurcation doubles with secondary mass distributions that decrease in frequency with decreasing secondary mass. Extrapolations, if they are valid, to planetary masses suggest that roughly ten percent of the primaries have planetary secondaries but that most or all primaries have nearby companions of some sort (stars, black dwarfs, planets). The long-period binaries are probably gravitationally-held separate protostars in which each component follows the van Rhijn function.

Second, there is a variety of abnormal stars (and some normal ones) in which the binary frequency for closely spaced systems (periods less than one month or one year) is unusually high or low. A survey of these, plus the normal stars, is given in Table I. In that table the frequency of occurrence of each kind of star is given in percent. In some cases we think that we know why such correlations exist. We do not know whether such correlations have any bearing on the occurrence of planets, but they may be crucial in understanding the causes of abnormal spectra.

TABLE I

Summary of Binary Statistics

MAIN-SEQUENCE TYPE	FREQUENCY OF SHORT-PERIOD BINARIES LOW	NORMAL	HIGH
O	OBC? (few %)	Normal O (>90%)	OBN? (few %)
B0-A3	Be (~25%)	Normal B (~55%)	
	Ap (magnetic) (~10%)	Ap (Hg, Mn) (~10%)	
A4-F1	Normal A (70%)		Am (30%)
F3-G2	High-Velocity (10%)	Normal F,G (90%)	

COMMENTS

T. Ch. Mouschovias: The statement has often been made that binary stars are frequent, that more than 70% of all stars are binary systems. As previously emphasized (Mouschovias, 1977), this may be a simple consequence of the fact that it is easier to resolve the angular momentum problem by putting the angular momentum in the orbital motion than by forming very rapidly rotating stars. In this regard we should recall that most binary systems have an orbital angular momentum more than three orders of magnitude larger than that in the spins of the individual components of the binary system (Ambartsumian, 1956).

W. Kaula: This question relates to the problem of why relatively few T Tauri stars are members of binary pairs. Has anyone examined how close a dynamical system is constituted by the T Tauri stars and their associated clouds? Is it feasible to do this, or is the amount of mass in, or obscured by, the clouds too uncertain? Perhaps if the cloud were shrunk to a point mass, it would constitute an interactive dynamic system, such as is treated by Aarseth (1971) and others, which would evolve to several separate binaries and singles if the masses were comparable.

H. Abt: We know almost nothing about the binary frequency of T Tauri stars because these stars are faint ($V \geqslant 10$ mag) and their spectra are complex. But I would be surprised if the binary frequency were any different than for the main-sequence stars. Of course if you have a detailed dynamical model for a T Tauri binary system and its surrounding gas, you could study the orbital changes; but you need the model first.

K. R. Lang: There must be some selection effect and uncertainties in the observations of your second "long-period" class of binaries because only the obvious visual binaries could have been observed a hundred years ago, and it must at any rate be very difficult to measure periods of thousands of years.

H. Abt: Only visual binaries are used for one-hundred-year periods and the periods of longer period systems are very uncertain.

Z. Kopal: It should be kept in mind that by far the most numerous binaries in the sky are the short-period binaries of W UMa type – about 20 times more numerous than the rest.

H. Abt: The W UMa stars seem to be more numerous only because they are easy to discover. We know of many such systems of brightness V = 8-10 mag, but we know almost nothing about spectroscopic binaries in that magnitude range. In the complete sample of F3-G2 main-sequence stars brighter than V = 5.5 mag and north of -20° declination, there are no W UMa stars but there are dozens of spectroscopic binaries.

D. Dearborn: There should be large selection factors against detecting binaries with low-mass companions. This affects the power-law extrapolation you determined and therefore the number of black dwarfs and planets expected. What do you believe to be the magnitude of incompleteness?

H. Abt: The undetected systems with low secondary masses were, in some cases, recovered in the incompleteness calculations. But we included in Figs. 1 and 2 only those cases where the incompleteness did not exceed the numbers of measured systems. This generally took us down to $m_2 = 0.07\ m_\odot$ for the secondaries of solar-type stars, and it is all extrapolation after that.

L. Mestel: I believe that the Ap- and the Am-star phenomena can be qualitatively understood if one takes their star rotations as the essential pre-condition. Can you confirm that the close binaries have spins that are more or less synchronized with the orbital motions? If so, this could be evidence that e.g., tidal coupling is indeed significant.

H. Abt: For main-sequence binaries with periods less than six days, the synchronization of rotational with orbital motion is usually complete and for somewhat longer periods there is a tendency toward synchronization. There is a paper by Levato (1976) showing that older stars on the main sequence have gone further toward synchronization than younger ones.

B. G. Elmegreen: How do you know that the distant pairs are actually bound systems? Are their relative velocities less than the escape velocity?

H. Abt: For periods less than several thousand years there is a small relative motion observed during the past 100-200 years and preliminary orbital elements can be computed. For more widely spaced systems the relative motions are undetectable and we would see such pairs only as common-proper-motion pairs. We computed rough periods for these by assuming that their current separations are the sums of their semi-major axes.

J. I. Silk: Can you comment on whether OB associations are deficient in binaries?

H. Abt: There are many binaries known in OB associations and probably differences in the binary frequencies from one association to another. I would expect that the field stars, being escapees from associations, would be representative of the associations. But there might be selective effects in these escapes. The data to make such a comparison are partly available but that comparison has not yet been made.

REFERENCES

Aarseth, S. J. 1971. *Astrophys. Space Sci.* 14: 20.
Abt, H. A. 1961. *Astrophys. J. Suppl.* 6: 37.
______. 1965. *Astrophys. J. Suppl.* 11: 429.
______. 1975. *Astrophys. J.* 195: 405.
______. 1977. *Sci. Amer.* 236: 96.
Abt, H. A., and Bidelman, W. P. 1969. *Astrophys J.* 158: 1091.
Abt, H. A.; Chaffee, F. H.; and Suffolk, G. 1972. *Astrophys. J.* 175: 779.
Abt, H. A., and Levy, S. G. 1969. *Astron. J.* 74: 908.
______. 1974. *Astrophys. J.* 188: 291.
______. 1976. *Astrophys. J. Suppl.* 30: 273.
______. 1978. *Astrophys. J. Suppl.* In press.
Abt, H. A., and Moyd, K. I. 1973. *Astrophys. J.* 182: 809.
Abt, H A., and Snowden, M. S. 1973. *Astrophys. J. Suppl.* 25: 137.
Aikman, G. C. L. 1976. *Publ. Dom. Astrophys. Obs.* 14: 379.
Ambartsumiam, V. A. 1956. In *Vistas in Astronomy* (A. Beer, ed.) 2: 1708.
Bidelman, W. P., and Weitenbeck, A. J. 1976. In *Be and Shell Stars* (A. Slettebak, ed.), Dordrecht, Holland: D. Reidel Publ. Co., p. 29.
Bohannan, B., and Garmany, C. D. 1978. In preparation.
Bolton, C. T., and Rogers, G. L. 1978. *Astrophys. J.* In press.
Branch, D 1976. *Astrophys. J.* 210: 392.
Conti, P. S.; Leep, E. M.; and Lorre, J. J. 1977. *Astrophys. J.* 214: 759.
Crampton, D., and Hartwick, F. D. A. 1972. *Astron. J.* 77: 590.
Guthrie, B. N. G. 1965. *Publ. Roy. Obs. Edinburgh* 5: 1.
Hogg, H. S. 1973. *Publ. David Dunlap Obs.* 6: 1.
Jaschek, C., and Gomez, A. E. 1970. *Publ. Astron. Soc. Pacific* 82: 809.
Jaschek, C., and Jaschek, M. 1957. *Publ. Astron. Soc. Pacific* 69: 546.
Joy, A. H., and Abt, H. A. 1974. *Astrophys. J. Suppl.* 28: 1.
Kumar, S. 1963. *Astrophys. J.* 137: 1121, 1126.
Levato, H. 1976. *Astrophys. J.* 203: 680.
Mouschovias, T. Ch. 1977. *Astrophys. J.* 211: 147.
Petrie, R. M. 1960. *Ann. d'Astrophys.* 23: 744.
Preston, G. W. 1974. *Ann. Rev. Astron. Astrophys.* 12: 257.

Salpeter, E. E. 1955. *Astrophys. J.* 121: 161.
Scarfe, C. D. 1970. *Publ. Astron. Soc. Pacific.* 82: 1119.
Serkowski, K. 1977. *Astron. Quarterly* 1: 5.
Slettebak, A. 1966. *Astrophys. J.* 145: 121.
Strömgren, B. 1964. *Astrophysica Norwegica* 9: 333.
Trimble, V. 1974. *Astron. J.* 79: 967.
Walborn, N. R. 1976. *Astrophys. J.* 205: 419.
Wilson, O. C. 1963. *Astrophys. J.* 138: 832.
——— 1967. *Astron. J.* 72: 905.
Wilson, O. C., and Bappu, M. K. V. 1957. *Astrophys. J.* 125: 661.
Wolff, S. C. 1978. *Astrophys. J.* In press.
Young, A., and Koniges, A. 1977. *Astrophys J.* 211: 836.

PART IV

Associations and Isotopes

STAR FORMATION IN OB ASSOCIATIONS

CHARLES J. LADA
Harvard-Smithsonian Center for Astrophysics

LEO BLITZ
Columbia University

and

BRUCE G. ELMEGREEN
Harvard-Smithsonian Center for Astrophysics

We review the pertinent observations relating to the structure, content and star forming histories of OB associations. Studies of associations containing OB subgroups suggest that, once initiated, OB star formation may be sequential in nature proceeding through a cloud in a series of discrete bursts. Both the initial and subsequent bursts of star formation in associations could result from the application of external pressure to large molecular clouds. We discuss several mechanisms including ionization fronts, stellar winds, supernovae, spiral density waves and cloud-cloud collisions which have been proposed to supply such external pressure and thus to account for the formation of OB stars in associations. An observational test to evaluate the possibility that spiral density waves provide the initial trigger for the sequential formation process is proposed. In order to determine whether random supernovae in the galaxy may also initiate the first burst of star formation in OB associations, the maximum interaction distance between a supernova and a cloud surface at which a supernova must be able to trigger the formation of OB stars in the cloud is calculated. The results of a statistical argument show that if all supernovae within ~

10-30 pc of a cloud surface can trigger OB star formation in that cloud, then all OB associations in the galaxy could have been initiated by random (field) supernovae.

However, random galactic supernovae cannot account for the sequential star formation in associations. Once initiated, subsequent OB star formation is probably triggered by pressure forces intrinsic to the association (i.e., expanding H II regions, stellar winds and internal supernovae). We therefore study the dynamics and star-forming history of a molecular cloud complex adjacent to an OB subgroup in which one member becomes a supernova. Prior to the supernova blast, it is assumed that the OB subgroup generates an H II region which expands into the surrounding neutral material. We find that unless the supernova ignites at a very early epoch in the evolution of the OB subgroup (i.e., $< 10^6$ *yr), the explosion will have minimal effect on the overall morphology and star-forming history of the association. These results show quantitatively that ionization fronts dominate supernovae as the driving mechanism for the sequential formation process once it is initiated.*

The study of the birth of the solar system is ultimately tied to our understanding of star formation. In recent years an avalanche of radio, infrared and optical data has provided new insights into the formation process of massive O and B stars. Although, at present, it is unclear how these processes may relate to the formation of the solar system, what we have learned may represent a first step in this direction. In this chapter we review modern observational and theoretical work pertinent to the formation of OB stars.

Modern study of the formation of O and B stars probably began with the work of Ambartsumian (1947, 1955) who first recognized that associations – loose groupings of luminous blue stars – are very young compared to the lifetime of the galaxy. He based this conclusion on the fact that associations could not be gravitationally bound and he estimated expansion ages on the order of 10^7 yr for these stellar aggregates. The high luminosities and surface temperatures of these stars also implied relatively short lifetimes (a few million years) based on nuclear burning age estimates. Roberts (1957) examined the statistics of O and B stars in the galaxy and concluded that their number is consistent with the hypothesis that all of them were formed in associations. Understanding the birth of massive stars therefore requires an understanding of the star-forming processes in OB associations.

Numerous observations have greatly increased our knowledge of the structure and content of OB associations. In particular, new radio observations have for the first time enabled detailed examination of the gaseous component of these complexes. This is of great interest because, in general, most of the mass of a OB association-molecular cloud complex is in the gaseous or molecular component within which the newest stellar members of an association are being formed.

In Sec. I we discuss the observational picture of the structure and

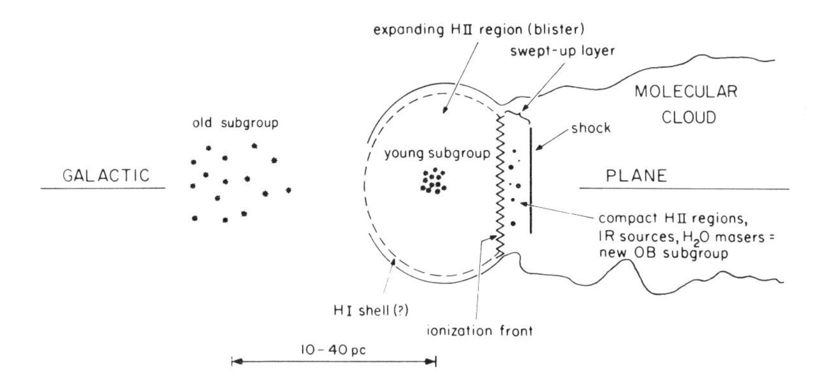

Fig. 1. Idealized representation of the structure of a typical OB association based on currently available radio and optical observations of nearby associations.

morphology of OB associations and their related gaseous components. These observations lead to the suggestion that the birth of massive stars is somehow induced by externally applied pressure acting on the protostellar gas. In Sec. II we summarize the mechanisms which have been proposed as triggers for OB star birth. In Sec. III we estimate the probability that random supernovae act as an initial trigger for OB star formation in all galactic associations. We assume that once started, subsequent star formation in an association is induced by intrinsic pressure sources (i.e., expanding H II regions, stellar winds, and supernovae igniting within the association). Using a simple model, we calculate the relative importance of expanding H II regions and "internal" supernovae to the dynamics and formation of new generations of stars in the gaseous components of the complexes.

I. THE STRUCTURE OF AN OB ASSOCIATION-MOLECULAR CLOUD COMPLEX

An OB association is a grouping of massive stars which contains (or once contained) at least one O-type star. R associations (loose clusters of reflection nebulae), containing B stars and T associations, containing T Tauri stars, are also found in some OB associations (Blaauw, 1964; Gratton, 1963; Racine, 1968; Herbst, 1975).

The structure of local ($\leqslant$ 1 kpc) OB associations, determined from optical observations of the most massive constituent stars (i.e., B2 and earlier) was summarized by Gratton (1963) and Blaauw (1964). Figure 1 shows an idealized representation of an evolved OB association. This diagram incorporates the essential features of the best-studied, nearby associations (e.g., Orion OB 1, Cep OB 3, M 17). The largest dimension of recognizable associations can reach sizes on the order of 50-100 pc and in one case (Sco-Cen) extends as far as 200 pc. Blaauw found that in nearby associations,

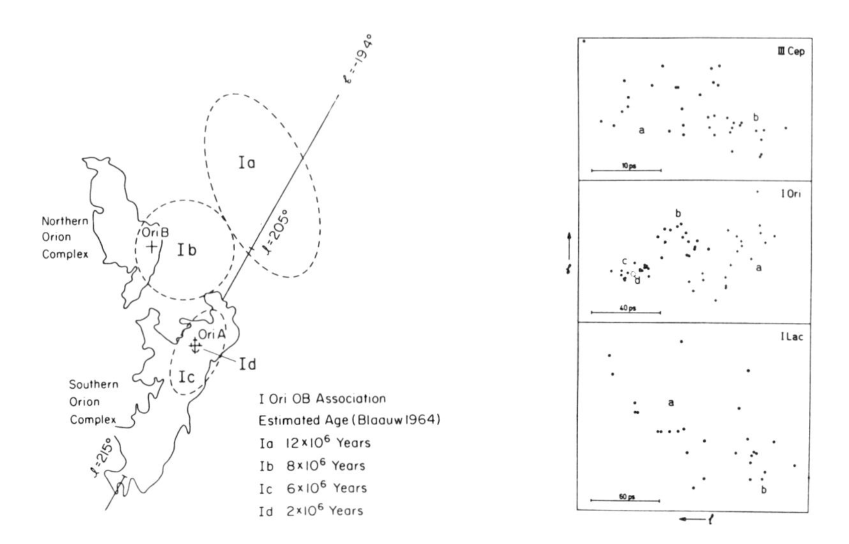

Fig. 2*a*. CO maps of the molecular clouds related to the Orion OB 1 association. Approximate extents of the subgroups identified by Blaauw (see Fig. 2*b*) are shown. (From Kutner *et al.*, 1977.)

Fig. 2*b*. Spatial distribution of stars in three nearby associations. Different symbols indicate the different subgroups *a,b,c* . . . The oldest subgroup in each association is designated by the letter *a*. (From Blaauw, 1964.)

the constituent stars were sometimes separated into subgroupings of discrete ages whose direction of separation was parallel to the galactic plane. Furthermore, Blaauw found that the spatial distribution of subgroups described a temporal sequence of ages starting with very old stars at one end of an association and ending with relatively young stars at the other. Usually the oldest subgroup has the largest angular size (i.e., is the most expanded) and is relatively free of interstellar material. On the other hand, the youngest subgroup has the smallest angular size and is usually associated with interstellar material of some form; Orion OB 1, shown in Fig. 2, is the prototype example of this phenomenon. It should be noted that whether an association contains one or more subgroups probably depends, among other things, on its age. Observationally, it is often difficult to distinguish subgroups even if they exist. Line-of-sight effects, cluster richness and distance all affect the determination of the number of subgroups in an association. It is not yet known what fraction of associations contains more than one subgroup, but Blaauw (1964) showed that more than one third of the local associations contain at least two subgroups.

Molecular-line radio observations of young OB subgroups in various associations have shown the young stars to be very near massive molecular clouds which themselves are often elongated parallel to the galactic plane (e.g., Lada *et al.*, 1976*b*; Lada, 1976; Elmegreen and Lada, 1976; Kutner *et* *al.*, 1977; Sargent, 1977; Cong, 1977; Blitz, 1978*b*; Baran, 1978). Such studies of the large-scale distribution of the gas in associations have indicated that in almost all cases the interstellar clouds adjacent to the associations have dimensions comparable to or greater than the associations themselves.

The close proximity of the youngest, least evolved, subgroups to the molecular clouds results in the creation of large, hot H II regions by the ionization of the molecular material from the Lyman continuum radiation of the O stars in the subgroup. These H II regions are ionization bounded by the molecular cloud in one direction and they freely expand into the relative vacuum of the association in the opposite direction. The H II region thus appears as a hot "firecracker" or "blister" at the edge of the large molecular cloud. Statistical studies of the comparative kinematics of H II regions and molecular clouds by Israel (1976) suggest that the "blister" model is generally applicable to the majority of galactic H II regions.

At the interface between the molecular cloud and H II region, ionization fronts propagate into the neutral gas, generating shock waves in the cloud (Kahn, 1954). In this way a swept up shell or layer of gas is driven into the molecular cloud. CO observations of IC 1805 (Lada *et al.*, 1978) suggest the presence of such a swept up layer in that source. Expanding shells have been observed toward many other sources (e.g., Brand and Zealy, 1975). Some of these shells could also be manifestations of the same phenomena, but not enough data are available to determine with any certainty the actual origin of these shells.

Detailed infrared and radio studies of the molecular clouds adjacent to the associations have provided much information concerning the most recent episodes of star formation in associations. Much of this information has been reviewed by Lada (1978). Studies of selected H II region-molecular cloud complexes have indicated that the sites of most recent star formation (as indicated by H_2O masers, compact continuum sources, CO bright spots, bright infrared sources, etc.) tend to occur in dense cloud fragments that are located near the surfaces of molecular clouds (i.e., within 1-4 pc of the surface) (Lada, 1978 and references therein). Furthermore, such signposts of recent OB star formation are found near the interface between the molecular cloud and the ionization fronts created by the subgroup adjacent to the cloud. Often these signposts are actually aligned along the ionization fronts (e.g., Sullivan and Downes, 1973; Gull and Martin, 1975; Lada *et al.*, 1976*a*; Elmegreen and Lada, 1977; Felli *et al.*, 1977; Lada *et al.*, 1978; Lada and Wooden, 1978; Elmegreen and Lada, 1978).

It is interesting to note that in the large molecular cloud complexes associated with Orion and IC 1805 (W4) there is no apparent evidence for O star formation beyond a few parsecs from the location of the presently visible H II regions. In Orion, O stars have been forming for the last 10^7 yr, but only in one end of the entire complex. During this period no O star formation is known to have occurred in the remaining 90 pc long, 10^5 $M_\odot$ clouds downstream of the H II regions Ori A and B. However, during this same period, stars of lower mass have apparently formed throughout the molecular gas, as the presence of T Tauri stars and Herbig-Haro objects testifies. Recently Myers (1977) surveyed 50 dark clouds for continuum emission from embedded sources. No such emission was found. Combining his observations with models predicting the fluxes and sizes of expanding, dusty H II regions, Myers concluded that the probability that an O star remains undetected is less than 2%.

The localization of newly formed O stars near the surfaces of molecular clouds adjacent to young OB subgroups suggests the validity of extrapolating the pattern of OB subgroup formation into the regions of most recent star formation within the molecular clouds. Apparently the formation of OB subgroups is an ordered, systematic process which starts at one end of a massive molecular cloud and then proceeds through the cloud in a series of consecutive bursts of star-forming activity. Blaauw found typical age differences between adjacent subgroups to be on the order of 2 to 4×10^5 yr, indicating the *delay* between successive star-forming events to be of the same order. The spatial separations of adjacent subgroups were found to be roughly 10 to 40 pc indicating that the OB star-forming process moves through a cloud at a velocity between 5 to 10 km sec^{-1}.

A dramatic observational example of the burst-like nature of OB star formation is illustrated in observations of the IC 1805 (W4) complex. Figure 3 shows a CO map of this region obtained by Lada *et al.* (1978). The O cluster IC 1805 (located in the center left portion of the field) contains nine O stars which ionize the large H II region, creating the extended ionization front which penetrates into the molecular cloud. The three bright CO peaks mark the locations of recent OB star formation in this region. All three strong peaks are found in a narrow but extended layer (6 pc $\times$ 36 pc) of high CO column density along the ionization front. This layer appears to be comprised of material swept up by the pressure forces in the H II region. The age of the exciting cluster is about 6×10^6 yr (Stothers, 1972), while the age of the most evolved neostellar objects in the swept-up molecular layer is only $\sim 10^5$ yr (Mezger and Smith, 1977). Star formation has occurred in two bursts, the most recent of which has happened nearly simultaneously in two widely separated locations in the high-density layer and within 10% of the dynamical history of the layer.

The relationship of the lower mass (i.e., spectral type later than B2) constituents of associations to this picture is unclear because of the relative difficulty involved in their observation. For a discussion of this point, we refer the reader to Elmegreen and Lada (1977).

II. TRIGGERING MECHANISMS

The observational picture discussed above is difficult to reconcile with

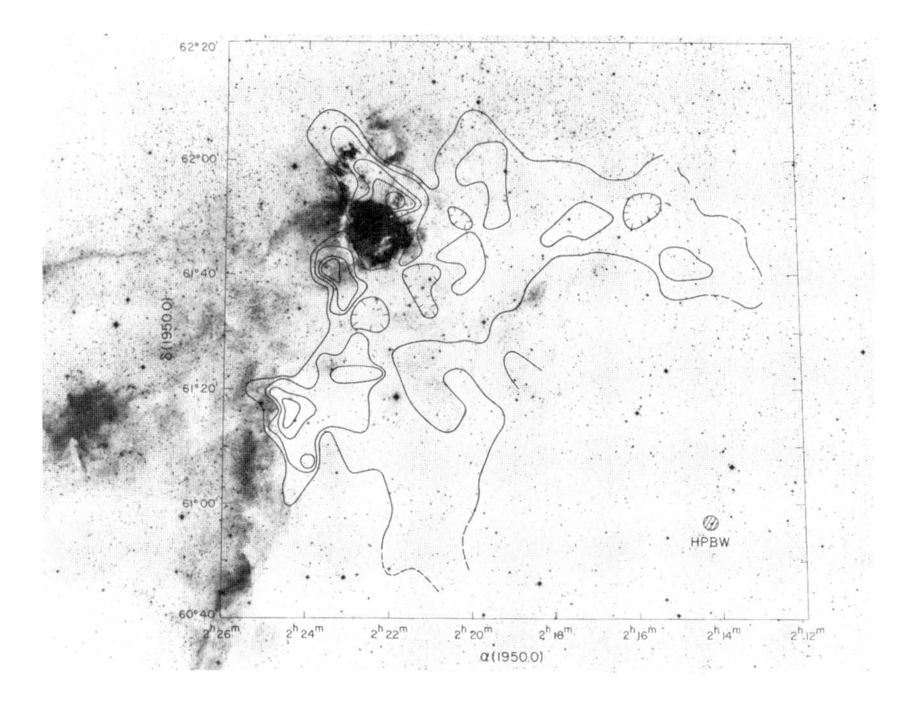

Fig. 3. CO contour map superposed on a red photograph of the IC 1805 (W4) region. The relatively evolved (6×10^6 yr old) O cluster IC 1805 is within the bright patch of nebulosity at the extreme left hand of the field. The brightest CO peak is associated with the radio source W3. All three bright CO peaks are found to be in a fragmented high-density layer approximately 36 pc in length and 6 pc in width (Lada *et al.*, 1978).

the idea that massive star formation is the result of a quiescent gravitational collapse occurring anywhere in a molecular cloud. One expects that massive stars would form preferentially in the deep interiors of the molecular complexes, and that the locations of the subassociations would be uncorrelated. Instead, it has been suggested by various authors that star formation has been induced by several triggering mechanisms which are reviewed below.

Ionization-Driven Shocks

Lada (1976) and Elmegreen and Lada (1977) suggested that the formation of OB subgroups could be triggered by pressure from ionization fronts due to O stars in a manner similar to that discussed by Oort (1954). The radiation from the O stars gives rise to an H II region which becomes ionization bounded in a direction towards a molecular complex. A shock front then separates from the ionization front creating a layer into which the ambient molecular material flows. The region between the fronts radiates efficiently producing a cool isothermal layer which is initially stable against perturbations until enough material accumulates to bring about the onset of a gravitational instability. The instabilities grow, ultimately forming another OB subgroup. The time scale for the collapse into stars was found to be typically 3×10^6 yr, a value concordant with the observed age differences between OB subgroups. Furthermore, the model predicts subgroup separations of 10-20 pc, a value which is also in agreement with the observations. This model also provides a mechanism which dissipates molecular clouds, aligns subgroups along the direction of the local magnetic field and may account for the observed expansion of OB associations.

Elmegreen and Elmegreen (1978) have investigated the growth of instabilities in an infinite plane-parallel pressure-bounded medium and have obtained expressions for the growth rate of instabilities as a function of the external pressure, sound speed, and initial central density. Their work implies that any shock which can accumulate a sufficiently large amount of gas may ultimately become unstable and form stars. None of the other mechanisms proposed as triggers for massive star formation have been analyzed in the detail that Elmegreen and Lada have analyzed their model. Nevertheless, any mechanism which can be treated by the analysis of Elmegreen and Elmegreen must be considered to be a plausible trigger. It is not to be supposed, however, that all such mechanisms are equivalent. Each must be analyzed in detail to see if it can produce stars, and if so, under what range of conditions.

Supernovae

Öpik (1953) suggested that the shell of material swept up by a supernova remnant could ultimately become unstable and form stars. This idea was attractive because it would explain the expansion of associations in a natural way. Another appealing aspect of the idea was that the death of one star in a supernova explosion would lead to the procreation of new stars. Öpik noted that this would result in a chain reaction of star formation, and the formation of each new subgroup would be delayed by a time equal to the nuclear burning time of the stellar progenitor. Although Blaauw's subsequent observations did confirm a burst-like nature in the formation of subassociations, the unidirectional temporal sequence of subgroups could not be explained in the model originally described by Öpik.

Ögelman and Maran (1976) revived Öpik's idea in a different context. Using subsequent theoretical and observational work, they constructed a model in which member stars of OB associations were formed by blast wave implosion of small independent interstellar clouds. They concluded that the "member stars (of an association) were not born collectively in a common interstellar cloud." However, they did not consider the observations of molecular clouds, and subsequently their hypothesis was not borne out by the detailed observations of Lada (1976), Sargent (1977), Cong (1977),

Kutner *et al.* (1977), Blitz (1978*b*) and Baran (1978).

Herbst and Assousa (1977) proposed that an old supernova remnant associated with CMa OB 1 and S-296 has induced the formation of the stars in CMa R 1. They argued that the 21-cm survey of Weaver and Williams (1973, 1974) shows a weak feature (at the 2σ level) at the position of the OB association which can be interpreted as the most negative velocity component of a supernova shell which is expanding with a velocity of 32 km sec^{-1}. Reynolds and Ogden (1978), however, have discovered clear evidence for the existence of an *ionized* shell of gas expanding at a velocity of 13 km sec^{-1} coincident with S-296. Reynolds and Ogden believe (see also Herbst and Assousa in this book) that the two observations could be consistent if the observed 21-cm emission is the blue edge of a turbulent H I shell and that most of the gas occurs at lower velocities. If that is the case, this gas should be observable above the background H I by the Weaver and Williams survey, but it was not found. If, on the other hand, the ionized gas results from particles streaming off the inside of an H I shell, it remains to be shown that the weak H I feature of Herbst and Assousa is consistent with the observations of Reynolds and Ogden. (Note that it was the H I shell properties which were used by Herbst and Assousa to derive the time since supernova outburst.)

Reynolds and Ogden (1978), and Blitz (1978*b*) considered the possibility that the *ionized* shell might be the result of either an expanding H II region, a supernova explosion, or a strong stellar wind. Considering all of the observations and analyses which have been done on this region to date, there appears to be no compelling reason for choosing one mechanism over the others. If the stellar wind theory of Weaver *et al.* (1977) is correct, the location of the stellar progenitor need not be correlated with any observed line splitting as implied by Herbst and Assousa (this book). If more than one star is responsible for the expansion of the H II region S-296 (two O stars in the association are projected on the shell), the observed morphology of the region can be consistent with an expansion velocity of 13 km sec^{-1}. The presence of the runaway O6.5 star does imply that a supernova is likely to have exploded in the region of CMa OB 1. If the ionized shell is a supernova remnant, its age is $\sim 10^6$ yr, but this value is very uncertain and depends on the outburst energy and the ambient interstellar density of the region.

The age of the OB association is given by Claria (1974) to be 3×10^6 yr, but this is probably an upper limit because there is no unambiguous evidence that stars have evolved off the main sequence. The age of the R association has been determined from the color-magnitude diagram by Herbst *et al.* (1978) to be 3×10^5 yr, but this is the age from the time a protostar has central temperatures of about 10^6 K (Iben, 1965), a state considerably evolved from the initial dark cloud material. The relevant age regarding the hypothesis of supernova-induced star formation is the time since the primitive molecular material was shocked by the remnant. For spherical collapse of a molecular cloud having uniform density, the free-fall time alone is 1.1×10^6 $n_3^{-\frac{1}{2}}$ yr where n_3 is the number density of the cloud in units of 10^3. Furthermore, Larson (1972) computed the evolution of protostars from dark cloud material using very simple assumptions, and concluded that "10^6 yr is probably the youngest cluster age or smallest age difference that can be meaningfully determined from cluster HR diagrams."

It may nevertheless be asked whether the stars in the R association are qualitatively younger than those in the OB association. Because the color-magnitude diagram of the OB association obtained by Claria (1974) did not take any special reddening corrections into account, it is directly comparable to the color-magnitude plot of the R association by Herbst *et al.* (1978) for a value of the total to selective extinction, R, of three. When such a comparison is made, the color-magnitude diagram of the OB association, which includes only six members of the R association, does not show significantly greater scatter than the one given for the R association. Furthermore, stars in the R association occur on the main sequence down to spectral type A0 (again, with R=3). This indicates that the OB association, the R association, and the shell (regardless of the mechanism which generated it) have ages of the same order of magnitude, with large uncertainties for each. Therefore, it does not seem warranted to argue that their ages provide evidence in favor of the hypothesis that a supernova triggered star formation in this region. As Herbst and Assousa point out in their chapter, the morphology of the region does suggest that there may be a causal relationship between the shell and the stars. However, as discussed above, there is no unique mechanism for the origin of the shell, and all of the proposed mechanisms are consistent with the observations.

Reeves (see his chapter) has speculated that 10-20 supernovae may occur during the 10^7 yr lifetime of an OB association. As we will show in Sec. III.A, however, many of the potential supernovae progenitors will probably migrate far away from the star-forming cloud before they explode. This may partially explain the lack of direct evidence for obvious supernovae remnants in OB associations. However, those supernovae explosions that occur close to a molecular cloud (perhaps 1 or 2 per subgroup) could possibly have an effect on star formation. In Sec. III.B we quantitatively consider this possibility.

The question of whether or not the sun could have formed by a supernova has recently been raised. Isotopic anomalies in meteorites have been interpreted to indicate the presence of a supernova in the immediate vicinity of the protosun. This might suggest that the formation of the sun was triggered by a supernova. However, such a conclusion must be regarded as very uncertain especially since Clayton (1978) has argued that the isotopic anomalies can be explained in other ways and has indicated problems concerning a supernova origin for the elemental enrichment of the meteorites.

Another observation concerning possible supernova-induced star formation was reported by Berkhuijsen (1974). She finds an apparent loop of radio continuum emission (the Origem Loop) whose properties are consistent with a supernova origin. The continuum emission in the loop, is dominated by a group of Sharpless (1959) H II regions which may be within the loop, suggesting that the expanding supernova perhaps triggered the formation of their exciting stars within the last 7×10^5 yr.

The uniqueness of the interpretation is, however, hard to determine with the presently available data. Studies of the ages of the exciting stars in the Sharpless H II regions and of the molecular clouds associated with these nebulae could possibly help in providing supporting evidence for the existence of supernova-induced star formation; as yet such information has not been available. Sancisi (1974) has presented evidence for the existence of expanding neutral shells of hydrogen gas toward Per OB2 and Sco OB2. These shells also appear to be related to recent OB star formation. The shells may have originated in a supernova explosion but as Sancisi points out, such an interpretation is not necessarily unique. A similar shell of possible supernova origin has also been found associated with Cep OB3 by Assousa *et al.* (1977).

Together these observations may be suggestive of an important role for the supernova as a trigger of OB star formation. In Sec. III we evaluate this possibility quantitatively.

Density Waves

It has long been noticed that in many external galaxies the OB associations delineate a regular large-scale pattern – usually the defining characteristic of a spiral galaxy. The regularity in the structure was difficult to understand because the galactic differential rotation should obliterate within a few rotation periods any pattern which is related to material spiral arms. To avoid this dilemma, Lin and Shu (1964, 1966) developed a theory in which a spiral wave pattern of enhanced density rotated around the center of the galaxy like a rigid body with a definite angular speed. The density wave could be characterized by three basic parameters: (1) the pattern rotation speed, Ω_p which determines (2) the pitch angle, i, relative to the circumferential direction, and (3) the amplitude of the spiral gravitational field relative to the smoothed axisymmetric field, F. From the work of Yuan (1969*a*,*b*) the best values of these quantities obtained from observations for the Milky Way are: Ω_p = 13.5 km sec^{-1} kpc^{-1}, $i = 6 - 7°$, and $F = 4 - 7\%$. At the position of the sun, the spiral pattern has a circular velocity of 135 km sec^{-1}, about half the 250 km sec^{-1} orbital velocity of the sun (see e.g., Burton, 1974). For a two-arm spiral, therefore, local star forming material passes through a spiral arm about once per solar rotation period.

Fujimoto (1966) has shown that the galactic spiral density wave will produce a large-scale shock in the interstellar medium. Roberts (1969) investigated the gas flow of a two-armed spiral gravitational field superimposed on the Schmidt (1957) model of the galaxy. He found that the galactic gas flows in closed, nearly concentric stream tubes passing through a two-armed spiral shock pattern. The typical gas compression along a stream tube is on the order of eight, and Roberts concluded that such a compression could cause the implosion and subsequent gravitational instability of some interstellar H I clouds. The triggering of the instability would typically require large masses ($\gtrsim 10^4\ M_\odot$), high densities ($\gtrsim$ 10-100 cm^{-3}), and low-velocity dispersions ($\lesssim 2$ km sec^{-1}). Woodward (1976) considered the implosion of an H I cloud with a radius of 15 pc, a density of 1.5 cm^{-3}, a temperature of 140 K, and a mass of 500 $M_\odot$ in a detailed numerical simulation. He found that such a cloud produced high density Rayleigh-Taylor tongues and a single dense clump near the center of the cloud which is nearly gravitationally bound. He then argued that scaling his results to larger initial masses would ensure that the central clump would be gravitationally bound and would form stars.

The density wave implosion model is attractive because it can explain the tendency of extreme population I objects to delineate spiral features in our own and in other galaxies. In addition, it could provide the initial trigger for the sequence of subgroups. The present theory does not, however, seem able to explain the existence of subassociations of different ages which are sometimes observed. An additional problem is that no calculation has yet been attempted of the effects of the passage of a galactic shock on a massive *molecular* cloud. Woodward's calculations may not scale for large clouds and his neglect of magnetic fields could significantly alter his final results. Therefore, although it is an attractive and plausible hypothesis, additional analysis must be performed to demonstrate that OB associations can indeed be formed as a result of density wave implosions.

Cloud Collisions

Collisions between interstellar clouds can form stars in two ways: (1) If two clouds with relative velocities significantly greater than their internal sound speeds collide, shocks can form which could induce the formation of stars. (2) Clouds which are initially gravitationally stable can coalesce by means of collisions until they have exceeded the Jeans mass. Kahn (1955) analyzed the collisions between clouds and concluded that typically, because the clouds are moving supersonically, collisions between them will be highly inelastic and could be a major source of heating for the clouds. Field and Saslaw (1965) investigated the collisions of interstellar clouds which coalesce to form larger masses and ultimately to form stars. While their results are in general agreement with some of the observations, their model does not explain either the formation of associations near molecular clouds nor Blaauw's results discussed above. Stone (1970*a,b*) calculated one- and two-dimensional models of colliding H I clouds with initial hydrogen densities of 10 cm^{-3} and 8 cm^{-3} and concluded that the growth time of the most unstable perturbation is greater than the duration of the compression, so that stars are not likely to form. His conclusions are likely to be different for molecular clouds where the initial densities are typically higher by two orders of magnitude than the ones he considered. Loren (1976) has suggested that the collisions between two molecular clouds induced the formation of massive stars related to NGC 1333. Blitz (1978*a*) has suggested that the giant OB molecular complexes are composed of smaller clouds. Collisions between them could form stars since the relative motions between them are generally supersonic, but it is difficult to see how they could be responsible for the sequential formation of associations.

Stellar Winds

Blitz (1978*b*) has suggested that because stellar winds from O and B stars (Morton, 1967; Lucy and Solomon, 1967, 1970) produce shells which have rapid cooling times (Castor *et al.*, 1975), they could sweep up enough material to be gravitationally unstable. Although the mass loss rates have been the subject of some controversy (see e.g., Lucy, 1975), the total wind energy emitted by an O star during its lifetime can be comparable to the energy expended by a Type II supernova or the expansion of an H II region (10^{50} – 10^{51} ergs). Because the effect of strong stellar winds is to increase significantly the radius of an H II region above its value in the absence of the wind (Weaver *et al.*, 1977), the winds can exert a singificant pressure on the ambient interstellar medium which is likely to affect the formation of stars.

In most cases, strong stellar winds are not expected to act in the absence of the radiation pressure which gives rise to ionization fronts. Because of the dependence of mass loss on luminosity class (see e.g., Hutchings, 1976), the details of the evolution of a wind and ionization-driven shock apropos the formation of stars is complex. Qualitatively, it is expected that the most important dynamical effect of the stellar wind is to increase the pressure on the shocked shell being swept up in a molecular cloud by an ionization-driven shock. This should be roughly equivalent to increasing the effective luminosity of the star giving rise to the H II region by an appropriate factor.

III. THE NATURE OF SUPERNOVA-INDUCED STAR FORMATION IN OB ASSOCIATIONS

Two problems are relevant to the formation of OB associations and of subgroups within the associations. We consider in Sec. III.A the possibility that a field supernova occurs at some distance from a dense cloud and the resulting shock travels through a low-density medium until it hits the cloud, where it eventually initiates star formation. The important quantity to determine is the maximum separation between the supernova explosion and the cloud that will allow the supernova shock to be of "sufficient" strength to induce star formation when it impinges on the cloud. This maximum distance will depend on the uncertain density of the intercloud medium, among other things. Although this maximum distance will be difficult to determine on theoretical grounds, we may estimate from statistical arguments what it would *have* to be in order that all OB associations in the galaxy are formed as a result of randomly occurring supernova shocks. These empirical estimates will provide a guideline for future theoretical work on one aspect of supernova-induced star formation.

A second case for considering the feasibility of supernova-induced star formation is in an existing OB association, where ionization pressure and/or pressure from stellar winds will have acted on an adjacent molecular cloud for some time before the supernova occurs. The supernova will then explode into an H II region, whose density can be estimated from Strömgren's formula (see Spitzer, 1968). It is important to determine in this second case the relative importance of supernovae and expanding H II regions on star-forming compression and large-scale mass motions in the OB association and molecular cloud complex. We use a simple model in Sec. III.B to illustrate the nature of star formation and cloud disruption in OB associations that experience a supernova explosion.

A. The Initiation of Star Formation in Giant Molecular Cloud Complexes by Random Occurrence of Nearby Supernovae

If supernova remnants from randomly occurring supernovae are responsible for the initiation of star formation in all OB associations, then the formation rate of OB associations is related to the supernova rate in the following manner:

$$\frac{R_{\mathrm{OB}}}{R_{\mathrm{SN}}} = n_c V_{\mathrm{int}}$$

where n_c is the number density of molecular cloud complexes within some appropriately specified galactic volume. V_{int} is the critical interaction volume defined as the volume which surrounds an individual molecular cloud complex and within which the detonation of a supernova can trigger the formation of OB stars in that cloud. To place the appropriate statistical constraints on the initiation of OB star formation in giant clouds we determine V_{int} which is related to D_{int}, the distance (from a cloud) within which any randomly occurring supernova will necessarily induce star formation.

Let us now consider the formation of OB associations and supernovae within 1 kpc of the sun. There are about 14 OB associations within 1 kpc of the sun (Alter *et al.*, 1968); the observable lifetime of an association is about 10^7 yr (Blaauw 1964; Gratton 1963) which results in a formation rate R_{OB} = 4.5×10^{-7} yr^{-1} kpc^{-2}. If the average supernova rate in the galaxy is (50

$\text{yr})^{-1}$ (e.g. Chevalier, 1977; see discussion below), then the local rate is probably half this value in order to account for the radial distribution of supernova remnants (Ilovaisky and Legueux, 1972*a*; Clark and Caswell, 1976). Therefore, $R_{SN} = 1.7 \times 10^{-5}\ \text{yr}^{-1}\ \text{kpc}^{-2}$. Nearly all OB associations which have been observed are associated with a giant molecular cloud (Blitz, 1978*a*); because the total number of such clouds in the galaxy (4000; estimated by Sanders and Solomon, 1978) is essentially equal to the total number of OB associations (4000; estimated by Blitz, 1978*c*), we assume the number of clouds locally is roughly equal to the number of associations or 14. The scale height (H_{SN}) of the supernovae locally is about 120 pc (Clark and Caswell, 1976), thus $n_c = 1.87 \times 10^{-8}\ \text{pc}^{-3}$ in the appropriate local galactic volume (i.e., $2\ H_{SN} \times 1\ \text{kpc}^2$) in which all the cloud-supernova interactions can occur. Therefore $V_{int} = 1.4 \times 10^6\ \text{pc}^3$ locally. To estimate D_{int} requires knowledge of the geometry of star-forming molecular cloud complexes. Observations of such complexes have been summarized by Blitz (1978*a*) who finds the average projected surface area A_{pj} of these complexes to be about $2 \times 10^3\ \text{pc}^2$, with little variation from cloud to cloud. To find D_{int} we set $V_{int} = (\sqrt{A_{pj}} + 2D_{int})^3$ with the result that $D_{int} \approx 34$ pc. It can be shown that this result is relatively insensitive to choice of geometry as long as such a choice is consistent with the observed projected surface areas. Random supernovae are capable of initiating star formation in all OB associations within 1 kpc of the sun only if every supernova within about 34 pc of a cloud can induce star formation in that cloud.

In a similar manner we can estimate D_{int} necessary to account for the initiation of star formation in all OB associations in the entire galaxy. Using 4000 as the value for the total number of clouds in the galaxy, 4000 as the total number of OB associations, $H_{SN} = 60$ pc (Ilovaisky and Lequeux, 1972*a*; Clark and Caswell, 1976) and a galactic supernova rate of $(50\ \text{yr})^{-1}$ (ibid.), we find $D_{int} \sim 12$ pc.

The most uncertain number in this calculation is the supernova rate. Combined evidence from observations of external galaxies and historical supernovae suggest a value of $(15^{+15}_{-5}\ \text{yr})^{-1}$ (Tammann, 1977) while the pulsar formation rate could be consistent with a value as high as $(6\ \text{yr})^{-1}$ (Taylor and Manchester, 1977). The rate of formation of supernova remnants, however, suggests a range from $(50)^{-1}\ \text{yr}^{-1}$ (Ilovaisky and Lequeux, 1972*b*) to $(150\ \text{yr})^{-1}$ (Clark and Caswell, 1976). This points to a serious discrepancy between the supernova rate and the rate of formation of supernova remnants. At least two possible explanations exist for this result. First, supernova remnants could be hidden if supernovae occurred preferentially in or near molecular clouds. To account for the observed difference requires that the progenitors of the supernova remnants be about 5 times more numerous within or near clouds than elsewhere in the galaxy; that is, they must be confined to 1/1000 the volume of the galactic disk! It is inconceivable that such a high density of O and B stars could be hidden within molecular clouds and have escaped detection in all the combined radio observations of such complexes (e.g., Myers, 1977). It also seems unlikely that such supernovae occurred within OB subgroups near molecular clouds. The death rate of all stars of spectral type A0 and earlier is 6.5×10^{-5} yr^{-1} kpc^{-2} (Ostriker *et al.*, 1974) which is a factor of four short of the upper limit to the pulsar formation rate. This means that the progenitors of pulsars must have spectral types of A0 or later. Consider that the lifetime of a B2 star is 2×10^7 yr. (Stothers, 1972). The expansion velocities of OB associations are believed to be ~ 5 km sec^{-1} (Blaauw, 1964), indicating that before a B2 star becomes a supernova it will be 100 pc from its protostellar cloud complex. A remnant from such a supernova should easily be detected. Obviously by the time A0 stars detonate, any trace of their relationship to an OB subgroup will have long disappeared. Apparently a supernova rate of (6 yr)$^{-1}$ cannot be explained by hiding their remnants within or near molecular cloud complexes.

Another explanation for the discrepancy is to have most supernovae occur far from the galactic plane. In this case the remnants would expand rapidly and might become unobservable on short time scales. However, since star-forming molecular complexes are closely confined to the plane ($H_{mc} \sim$ 50 pc; Cohen and Thaddeus, 1977), these supernovae would probably not interact effectively with clouds to form stars. Therefore, our use of a supernova rate of (50 yr)$^{-1}$ seems justified in the context of triggering OB star formation.

We conclude that randomly occurring supernovae can initiate star formation in all OB associations only if every supernova that detonates between $\sim 10 - 30$ pc from a cloud can induce star formation in that cloud. It remains to be shown either theoretically or observationally whether or not such supernova-cloud interactions can indeed produce stars.

The phenonemon of OB subgroups observed in the oldest nearby associations suggests that, once initiated, star formation may be a continual process within a triggered cloud. The fact that almost all associations contain O stars also supports this conclusion. The mean age of an O star is about 3×10^6 yr and the average difference in subgroup ages for the associations studied by Blaauw is also about 3×10^6 yr. There are at least 7 associations considered by Blaauw to have subgroups (in the solar neighborhood). With a subgroup formation rate of $(3 \times 10^6 \text{yr})^{-1}$ in each of these associations we estimate $D_{int} \sim 60$ pc. Thus the constraint on random supernovae triggering all subgroup formation is considerably more severe than it is for initiating the first epoch of star formation in each association. In any event, the observed unidirectional sequences of subgroup ages (see Fig. 1) strongly argues against random events of any kind as a trigger for sequential formation of OB stars. Supernovae can only play a role in sequential subgroup formation if their progenitors are themselves members of the association and are not random on galactic scales. In the following section we consider this in detail.

B. The Evolution of an OB Association in which a Supernova Explosion Occurs

We consider here the simple, two-step expansion of material surrounding a cluster of OB stars. Initially an H II region expands into a dense cloud, collecting neutral material into a dense shell (or partial shell) at the boundary between the ionized gas and the unperturbed cloud. We refer to this neutral shell as a circumionization shell, or CIS. After a time t_{SN}, a supernova explodes near the center of the exciting cluster and H II region. The supernova shock first propagates into the H II region. When it hits the inner boundary of the CIS, it will begin to sweep up the dense material that was previously shocked by the expanding H II region. If the supernova shell has enough momentum, it will continue to move through the CIS, picking up mass as it goes, until it sweeps up the entire CIS and emerges into the lower density, unshocked molecular cloud. Eventually the shock will either slow down to the preshock *rms* velocity, at which time the shell will begin to dissipate, or massive stars may form in the compressed matter and they will lead to its disruption.

To be specific, we consider a simple model where the Lyman continuum luminosity in the OB association is the constant value (typical of a single O5 star) $S = 5 \times 10^{49}$ photons $\sec^{-1}$ (even after the supernova occurs) and the unshocked cloud density is $n_c = 500\, H\ \mathrm{cm}^{-3}$.

Following Elmegreen and Lada (1977), we assume the H II region expands with a shock velocity of $v_{II} = (4P_{II}/3\rho_c)^{\frac{1}{2}}$ for $\rho_c = 1.4 m_H n_c$ and $P_{II} = 2.1\, n_{II}\, kT_{II}$ where

$$n_{II} = \left(\frac{3S}{4\pi R_{II}^3\, \alpha}\right)^{\frac{1}{2}} \tag{1}$$

for recombination coefficient $\alpha = 3.1 \times 10^{-13}$ $\mathrm{cm}^3\ \sec^{-1}$ at $T_{II} = 8000$ K. The density in the CIS is taken to be the pressure equilibrium value, $n_{CIS} = P_{II}/kT_{CIS}$ for $T_{CIS} = 100$ K.

For the expansion of a supernova into a uniform medium, we use the Sedov solution for an adiabatic blast wave (Woltjer, 1972) with a cooling function $\Lambda(T) = 6.2 \times 10^{-19}\ T^{-0.6}$ ergs $\mathrm{cm}^{-3}\ \sec^{-1}$ (Raymond *et al.*, 1976; c.f., McKee and Ostriker, 1977). In this way we determine that the supernova will become isothermal after it has expanded to a radius R_{rad}, in time t_{rad} to a velocity v_{rad} given by

$$\begin{aligned} R_{rad} &= 1.11 \times 10^{-5}\ \epsilon_0{}^{0.288}\ \rho_0{}^{-0.425} \\ t_{rad} &= t_{SN} + 2.78 \times 10^{-13}\ \epsilon_0{}^{0.219}\ \rho_0{}^{-0.562} \\ v_{rad} &= 0.4\, R_{rad}/(t_{rad} - t_{SN}). \end{aligned} \tag{2}$$

Here ϵ_0 is the supernova outburst energy and $\rho_0 = 1.4 m_H n_0$ is the preshock density. In all of the models presented here, the supernova became isothermal before it hit the CIS and $t_{\rm rad}$ was much less than $t_{\rm SN}$. This is why we can assume that the supernova explodes into a nearly uniform and constant density. Thus $n_0 = n_{\rm II}$ at the time of the supernova explosion $t_{\rm SN}$. Thereafter the remnant will expand isothermally, with a shock velocity v that satisfies three different equations at three epochs: In the short time (compared to $t_{\rm SN}$) before the supernova hits the CIS, momentum is conserved in a spherically divergent shell so

$$v(R) = v_{\rm rad}(R/R_{\rm rad})^{-3}. \tag{3}$$

For the short time (as calculated) during which the supernova shock is inside the CIS, momentum is conserved in a non-divergent shell, so

$$\frac{d\lambda}{dt} = \frac{v_{\rm hit} - v_{\rm CIS}}{1 + 3(n_{\rm CIS}/n_{\rm II})\lambda} \tag{4}$$

in the frame of reference moving at velocity $v_{\rm II} = v_{\rm CIS}$ with the CIS at time $t_{\rm SN}$. Here λ is the path length traveled by the shock inside the CIS ($\lambda \ll R$), and $v_{\rm hit}$ is the velocity at which the shock hits the CIS, $v_{\rm rad}\,(R_{\rm CIS}/R_{\rm rad})^{-3}$.

After emergence from the CIS, the supernova shell will carry its momentum further into the unshocked cloud. This phase may last for a relatively long time, $\gtrsim t_{\rm SN}$ so the additional pressure resulting from continued ionization at the inside of the shell and, possibly, from a hot low-density remnant must be considered. In this late phase, the shock velocity, v and its time derivative $\dot{v}$ satisfy the equation

$$\frac{1}{3}\rho_c R\dot{v} + \rho_c v^2 = P_{\rm II} + P_{\rm SN} \tag{5}$$

with an initial condition that v equals the velocity v_E at which the supernova shock emerged from the CIS:

$$v_E = v_{\rm II}(t = t_{\rm SN}) + (v_{\rm hit} - v_{\rm II})\left(1 + \frac{\rho_c}{\rho_{\rm II}}\right)^{-1}. \tag{6}$$

We assume that $P_{\rm II}$ continues to vary in proportion to $R^{-\frac{3}{2}}$ (since $n_{\rm II}$ at the ionization front is proportional to $R^{-\frac{3}{2}}$ in a spherically divergent shell) and that $P_{\rm II} = P_E$ and $R = R_{\rm CIS} = R_E$ at the time of emergence from the CIS. We also assume for some cases that the additional pressure from a hot remnant, P_{SN}, will be 2/3 times the energy density in the remnant at the cooling time, or $P_{rad} = (2/3) \times (\epsilon_o/2) \times (4\pi \, R_{rad}{}^3/3)^{-1}$, scaled down by adiabatic expansion as $\rho^{\frac{5}{3}} \propto (R/R_{rad})^{-5}$. Equation (6) then has the solution

$$v(R) = v_E \left(\frac{R}{R_E}\right)^{-3} \left\{ 1 + \frac{4P_E}{3\rho_c v_E{}^2} \left[\left(\frac{R}{R_E}\right)^{\frac{9}{2}} - 1 \right] + \frac{6P_{rad}}{\rho_c v_E{}^2} \left[\frac{R}{R_E} - 1 \right] \right\}^{\frac{1}{2}} . \tag{7}$$

For either the expansion of the CIS or the supernova shell we consider that stars have a potential for forming in the compressed (isothermal) shell if the mass column density in the shell, σ, and the pressure in the shell, P, (which equals $\rho_o v^2$ for preshock density ρ_o and shock velocity v) satisfy the relationship [referred to as "Condition (8)", below]

$$\frac{\sigma}{(P/\pi G)^{\frac{1}{2}}} \gtrsim 1 . \tag{8}$$

That such a relationship between σ and P at the onset of star formation should exist has been discussed by Elmegreen and Lada (1977) and Elmegreen (1978). The numerical constant on the right hand side should probably be determined empirically from observations of shock-induced star formation, but a value on the order of unity seems to be indicated both by the available observations (Elmegreen and Lada, 1977) and by a linearized stability analysis (Elmegreen and Elmegreen, 1978). In any case, this criterion will be useful here for a comparison of our results in different cases.

Here we discuss results for models with $\epsilon_o = 10^{51}$ ergs and with a variety of values of t_{SN}. We present the results in a manner which illustrates the differences between expanding shells around isolated H II regions, which have a radius R_{II} as a function of time t, and expanding shells around H II regions which contained a supernova explosion, for which the outer radius of the entire disturbance will be denoted by R_{SN}.

Figure 4 shows the fractional increase in the radius of the SN + H II region disturbance at time t over the disturbance produced by the H II region alone (i.e., if there is no supernova). This quantity is $[R_{SN}(t) - R_{II}(t)] / R_{II}(t)$. We plot it as a function of the time since the supernovae first emerged from the CIS, $t - t_E$, where t_E may be calculated from the equations of motion for the shells.

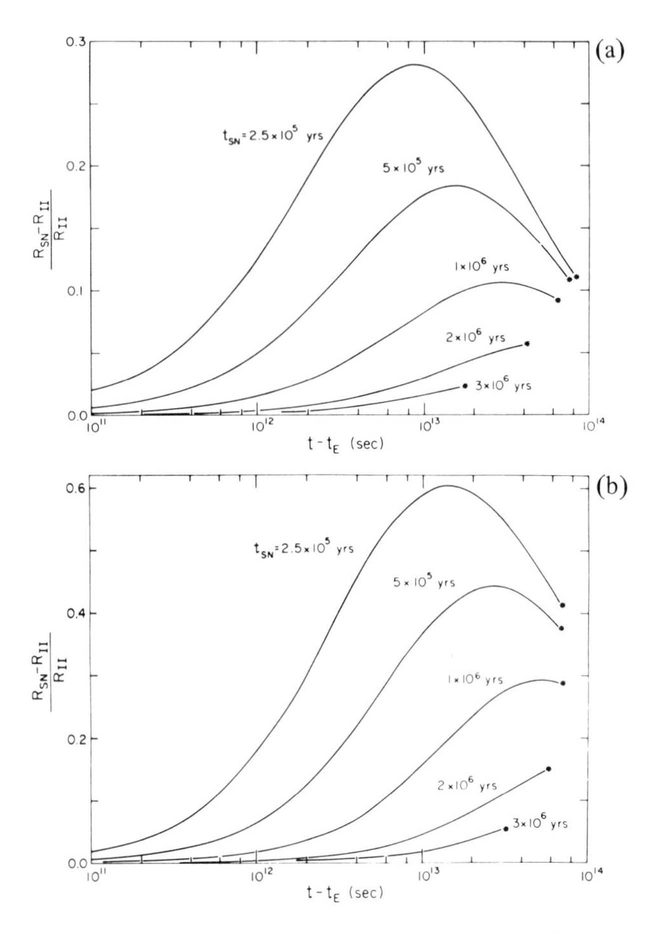

Fig. 4. The radius, R_{SN}, of the swept-up layer propagating into a dense cloud near an OB cluster in which a supernova ignited at time $t = t_{SN}$, is compared to the radius R_{II} of the layer that would have developed if no supernova occurred. Before the supernova occurs, ionization fronts will be the only source of pressure acting on the adjacent cloud, and a neutral shell (the CIS) will accumulate at the periphery of the expanding H II region. A short time after t_{SN}, the supernova blast wave will hit the CIS and the supernova shock will begin to compress and sweep up the neutral matter for a second time. At the time $t = t_E$ the supernova shock will have traveled all the way through the CIS (when $R_{SN} = R_{II}$) and will begin to move into the unshocked portions of the original cloud. Continued expansion will occur as the supernova shell plows into the remaining cloud and as pressures from the ionization fronts at the backside of this shell provide additional forces.

The effect of a supernova on the expansion of an H II region into a dense cloud is seen to be a relatively minor one, resulting in a small, temporary increase in the size of the cavity surrounding the OB cluster. The calculation terminated when Condition (8) was satisfied (see text) at which point star formation within the layer may become intense and disruptive. Figure 4*a* considers the limiting case where any pressure from the hot remnant interior is negligible, while Fig. 4*b* considers the opposite limit where the remnant interior is hot for all times after shell formation.

$$t_E = t_{SN} + (t_{rad} - t_{SN}) \frac{5}{8}\left[\frac{3}{5} + \left(\frac{R_{CIS}}{R_{rad}}\right)^4\right] + \frac{R_{CIS}}{3(v_{hit} - v_{CIS})} \frac{n_c}{n_{CIS}} \left[1 + \frac{n_c}{2n_{II}}\right]. \tag{9}$$

Here R_{CIS} and v_{CIS} are the radius and expansion velocity of the CIS at the time t_{SN} (obtained from the solution to the expanding H II region), and n_{II} is the proton density in the H II region at that time.

The additional pressure from the hot remnant, P_{SN}, will be important before the interior of the remnant cools. The time when this occurs cannot be obtained from existing hydrodynamic calculations for supernova remnants because in our case the remnant interior will be continuously filled in by the ionized gas that streams away from the inside of the shell. We might expect that after the crossing time for the ionized gas, R/c_{II}, the remnant will be filled in and cooled by the H II region; perhaps it will cool before this time. For the calculations shown here, R/c_{II} was less than the age of the supernova at the epoch of star formation for $t_{SN} > 2.5 \times 10^6$ yr; for smaller t_{SN}, the interior of the remnant may still be hot at the epoch of star formation.

Figure 4*a* shows our results for the limiting case when $P_{SN} = 0$ for all time; Fig. 4*b* shows the case when the interior of the remnant never cools, i.e., if P_{SN} is always included in Eq. (5). We might expect that for $t_{SN} \gtrsim 2.5 \times 10^6$ yr, Fig. 4*a* would be close to the true situation, while for $t_{SN} \lesssim 2.5 \times 10^6$ yr, Fig. 4*b* would apply. The difference between the two cases is approximately a factor of 2 in the quantity $(R_{SN} - R_{II})/R_{II}$ versus $t - t_E$.

The observed increase in $(R_{SN} - R_{II})/R_{II}$ at small $t - t_E$ is a result of the supernova blast depositing momentum into the CIS and causing it to move faster than it would have without the supernova. The decrease at larger $t - t_E$ occurs because R_{SN} (t) begins to move at the same (or smaller) velocity into the cloud as does R_{II} (t), since the ionization pressure in Eq. (5) becomes a strong driving force during the late phases in the expansion of the combined supernova – H II region disturbance. That is, the shell gains so much momentum as a result of this steady ionization-derived force that the momentum injected into the cloud by the supernova blast during its snowplow phase will become relatively insignificant. The calculation terminates when Conditions (8) is first satisfied, as indicated by dots in Fig. 4. In none of the cases shown was the shell velocity less than 1 km sec^{-1} at this point of termination; the final shell velocity ranged between 2 and 3 km sec^{-1} for t_{SN} between 0.25 and 3×10^6 yr.

If a supernova in an OB subgroup triggers the formation of the next subgroup, then t_{SN} will be the evolutionary or main sequence lifetime for the type of stars which become supernovae. This time is about a few million

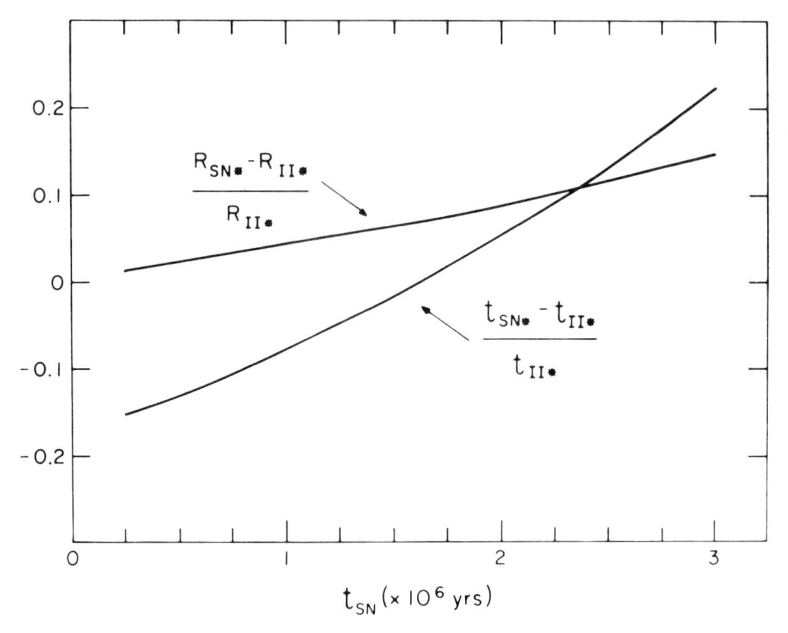

Fig. 5. The radii and ages of shocked layers are shown at the epoch when Condition (8) is satisfied (see text) and star formation within the layer may become intense. R_{SN*} is the star-forming radius of the cavity surrounding an OB cluster in which a supernova ignited at an earlier time t_{SN}, and R_{II*} is the star-forming radius the cavity would have had if no supernova occurred; t_{SN*} and t_{II*} are the corresponding times. The results show that a supernova occurring in a very young H II region may cause stars to form in the swept-up layer earlier than if no supernova occurred. Conversely a supernova may delay star formation if t_{SN} is large. The star-forming radius tends to increase when a supernova occurs because the additional supernova pressure on the swept-up shell inhibits its large-scale gravitational collapse, under the assumptions involved in formulating Condition (8). The same parameters were used as in Fig. 4*a*.

years. It may be seen from Fig. 4 that for such large values of t_{SN}, $(R_{SN} - R_{II})/R_{II}$ never exceeds 6% if $P_{SN} = 0$ (Fig. 4*a*) or 16% if P_{SN} is allowed to take the appropriate value for a hot remnant interior (Fig. 4*b*). Evidently the ionization front that pushed on the cloud during the 2×10^6 yr prior to the supernova outburst and during the snowplow phase is almost entirely responsible for the extent of cloud clearing. The difference a supernova makes in the extent of cloud disruption and compression will be very small. If t_{SN} happens to be small, e.g., 2.5×10^5 yr, the H II region will not have much time to expand before the supernova occurs and makes a 28 or 60% difference in the radius of the total cloud clearing as shown in Figs. 4*a* and 4*b*, respectively.

The influence a supernova might have on the place and time of formation of a second generation star cluster (i.e., the one that forms in the swept-up shell), is illustrated in Fig. 5 for the case when $P_{SN} = 0$. The radius and time at which the combined SN + H II region disturbance satisfies Condition (8) are denoted by R_{SN*} and t_{SN*}, respectively. Their values depend on t_{SN}. The radius and time at which the CIS satisfies Condition (8), which are calculated to be 14.5 pc and 3.4 × 10^6 yr for the present problem, are denoted by R_{II*} and t_{II*}, respectively. Figure 5 shows the fractional increase in radius $(R_{SN*} - R_{II*})/R_{II*}$ and time $(t_{SN*} - t_{II*})/t_{II*}$ at the epoch of star formation due to the occurrence of a supernova at time t_{SN}. The small fractional increase in radius is positive because the additional pressure from the supernova requires a correspondingly larger column density in the layer before Condition (8) is satisfied. The time it takes for the shell to satisfy this condition could be shorter if a supernova occurs at $t_{SN} \lesssim 1.4 \times 10^6$ yr than if no supernova occurs at all, since the critical column density would be swept up earlier for the higher average shell velocity in the case of a supernova. For larger t_{SN}, however, the supernova could delay the onset of star formation beyond the time when the H II region alone would cause stars to form, again because of the increased pressure in the shell. With the inclusion of P_{SN}, $(R_{SN*} - R_{II*})/R_{II*}$ ranges between 20 and 30% and $(t_{SN*} - t_{II*})/t_{II*}$ is between -34 and -40% for the same range in t_{SN}.

The reason that R_{SN*} or t_{SN*} can be larger than R_{II*} or t_{II*} is as follows. An increase in the pressure P of a layer results in a decrease in the thickness of the layer (as P^{-1}) and a corresponding decrease in the mass of the fragments that gravitationally condense out of the layer (roughly as P^{-2}). This decrease in fragment mass with increasing P is much more rapid than the decrease with P in the minimum unstable mass of a spherical condensate, which scales with $P^{-\frac{1}{2}}$. Thus, a large pressure causes the fragments that form out of a layer to be less massive than the minimum unstable mass if σ is small, and these fragments probably only form self-supporting spheres, which coalesce in time.

Several complications could change this picture of the epoch of star formation. If, for example, small globules form as the H II region expands (as is observed to be the case in many H II regions), then the supernova could compress these globules past the point of their stability and cause stars to form in them. A star formation mechanism similar to this has been proposed by Hjellming (1970) and others. However, in view of the numerous observations now available for large molecular fragments in regions where massive stars are forming (e.g., W3), it is most likely that this process will not be the primary means by which an entire OB subgroup is formed. The observations suggest quite a different picture (see Elmegreen, 1978*b*), namely that OB clusters form in 10^4 $M_\odot$ fragments which have been swept up as part of the large-scale expansion of neutral matter around pre-existing OB clusters, or perhaps around an explosion of a field supernova (see Sec. III.A). Although some star formation may occur as initial density perturbations are squeezed into collapse, the dominant process seems to be one where a layer forms first and then collapses under its own self gravity. In that case, a condition like that given by Eq. (8) is probably a reasonable indicator of when the next generation of large-scale star formation will occur in an expanding shell.

Finally we note that the results of our calculations presented here imply that the morphology and structure of the star-forming bursts in OB associations are relatively unaffected by the ignition of a supernova in the H II region. If the H II region were not present, then the supernova could have a significant impact on the cloud (as indicated by our low t_{SN} solutions), but the steady pressure from ionization fronts is likely to contribute more in the long run to cloud compression and possibly to star formation than will an occasional high-pressure burst from a supernova in the same region.

IV. CONCLUDING REMARKS

In this chapter we have attempted to synthesize a picture of the observed structure and content of OB association-molecular cloud complexes, based on currently available radio and optical observations. We have reviewed the evidence suggesting that OB star formation may be induced or stimulated by external pressure forces and we discussed the various mechanisms proposed to supply such external triggering.

It seems reasonable to assume that the sequential formation of OB subgroups is a phenomenon which arises from pressure forces intrinsic to the stellar component of the associations. At least three sources of such intrinsic pressure are available in an association. Expanding H II regions and stellar winds provide relatively *steady* sources of pressure, while supernovae, igniting within an OB subgroup, provide an *impulsive* source of pressure. The total energy expended by each of these pressure sources is roughly equal when integrated over the lifetime of the corresponding phenomenon. However, from a simple calculation, we have found that unless a supernova ignites very early during the evolution of the OB subgroup (i.e., $< 10^6$ yr) it will have a minimal effect on the dynamics of the incipient star-forming layer already expanding around the OB subgroup. Furthermore, a supernova may actually temporarily inhibit subsequent subgroup formation. Therefore, it would be very difficult to differentiate (from observations of star-forming layers alone) an association-cloud complex in which a supernova has exploded from one in which expanding H II regions and possibly stellar winds are the only dynamic forces. "Expanding" compressed layers of neutral gas, in which star formation may have recently occurred, will exist in both cases.

Expanding H II regions are observed in many associations and it is reasonable to conclude that the observed morphology and the sequential star-forming histories of these regions are essentially dominated by the dynamical influences of ionization fronts and perhaps stellar winds; supernovae may play a role, but only a supporting one. Our calculations do suggest that random field supernovae may act as an *initial* trigger for the OB star-formation sequence in massive clouds, as originally envisioned by Elmegreen and Lada (1977; see Sec. V), but as yet this is theoretically unconfirmed.

Another attractive initial triggering mechanism is that of spiral density wave implosion of the original clouds. An interesting observational test to evaluate the significance of such shocks on OB star formation is possible. If spiral density wave shocks trigger star formation, one would expect a correlation between the direction from which a spiral wave shock first hits a cloud and the location of the first epoch of star birth in the cloud. The existence of such a correlation would strongly implicate spiral density waves as the initial trigger of sequential OB star formation and in addition would provide a powerful tool for exploring the density wave structure in our galaxy. Cursory examination of the data of about a dozen associations is not inconsistent with the existence of such a correlation; however, not enough detailed data exist (especially in the optical domain) to permit a proper analysis of this test. Future observational attempts to search for such a correlation may be of great value to these problems.

Acknowledgment. We would like to thank L. Cowie, R. Chevalier, C. McKee and R. Stothers for interesting discussions. One of us (B.G.E.) is a recipient of a Junior Fellowship from the Harvard University Society of Fellows.

REFERENCES

Alter, G.; Ruprecht, J.; and Vanýsek, V. 1968. *Catalogue of Star Clusters and Associations.* Prague: Czechoslovak Academy of Sciences.
Ambartsumian, V.A. 1947. *Stellar Evolution and Astrophysics.* Armenian Academy of Sciences.
____. 1955. *Observatory* 75: 72.
Assousa, G.E.; Herbst, W.; and Turner, K.C. 1977. *Astrophys. J.* 218: L13.
Baran, G.P. 1978. Ph.D. Dissertation, Columbia University, New York. In preparation.
Berkhuijsen, E.M. 1974. *Astron. Astrophys.* 35: 429.
Blaauw, A. 1964. *Ann. Rev. Astr. Astrophys.* 2: 213.
Blitz, L. 1978*a*. In *Proceedings of the Gregynog Workshop on Giant Molecular Clouds.* In press.
____. 1978*b*. Ibid.
____. 1978*c*. Ph.D. Dissertation, Columbia University, New York.
Brand, P., and Zealy, W. 1975. *Astron. Astrophys.* 38: 363.
Burton, W.B. 1974. In *Galactic and Extragalactic Radio Astronomy* (G. Verschuur and K. Kellerman, eds.), p. 82. New York: Springer-Verlag.
Castor, J.; McCray, R.; and Weaver, R. 1975. *Astrophys. J.* 200: L107.
Chevalier, R.A. 1974. *Astrophys. J.* 188: 501.
____. 1977. *Ann. Rev. Astr. Astrophys.* 15: 175.
Claria, J.J. 1974. *Astron. Astrophys.* 37: 229.
Clark, D.H., and Caswell, J.L. 1976. *Mon. Not. Roy. Astr. Soc.* 174: 267.
Clayton, D.D. 1978. *Moon and Planets* (special Protostars and Planets issue). In press.
Cohen, R.S., and Thaddeus, P. 1977. *Astrophys. J.* 217: L155.
Cong, H.I. 1977. Ph.D. Dissertation, Columbia University, New York.
Elmegreen, B.G. 1978*a*. In *Proceedings of the Gregynog Workshop on Giant Molecular Clouds.* In press.

____. 1978*b*. *Moon and Planets* (special Protostars and Planets issue). In press.
Elmegreen, B.G., and Elmegreen, D.M. 1978. *Astrophys. J.* 220: 1051.
Elmegreen, B.G., and Lada, C.J. 1976. *Astron. J.* 81: 1089.
____. 1977. *Astrophys. J.* 214: 725.
____. 1978. *Astrophys. J.* 219: 467.
Felli, M.; Habing, H.J.; and Israel, F.P. 1977. *Astron. Astrophys.* 59: 43.
Field, G.B., and Saslaw, W.C. 1965. *Astrophys. J.* 142: 568.
Fujimoto, M. 1966. *Non-Stable Phenomena in Galaxies,* IAU Symp. No. 29, p. 453. Academy of Sciences of Armenian SSR. (In Russian.)
Gratton, L. 1963. *Proc. Int. Sch. Phys.* Course XXVIII, p. 243.
Gull, S.F., and Martin, A.H.M. 1975. In *H II Regions and Related Topics* (T.L. Wilson and D. Downes, eds.), p. 329. New York: Springer-Verlag.
Herbst, W. 1975. *Astron. J.* 80: 503.
Herbst, W., and Assousa, G.E. 1977. *Astrophys. J.* 217: 473.
Herbst, W.; Racine, R.; and Warner, J.W. 1978. Submitted to *Astrophys. J.*
Hjellming, R.M. 1970. *Evolution Stellaire,* 16 Liège Astrophys. Symp. (P. Ledoux, ed.), p. 105. Mém. Roy. Soc. Sci. Liège.
Hutchings, J.B. 1976. *Astrophys. J.* 203: 438.
Iben, I. 1965. *Astrophys. J.* 141: 93.
Ilovaisky, S.A., and Lequeux, J. 1972*a*. *Astron. Astrophys.* 18: 169.
____. 1972*b*. *Astron. Astrophys.* 20: 347.
Israel, F.P. 1976. Doctoral Dissertation, Leiden.
Kahn, F.D. 1954. *B.A.N.* 12: 187.
____. 1955. In *Gas Dynamics of Cosmic Clouds* (J.M. Burgers and H.C. van de Hulst, eds.), p. 60. Amsterdam: North Holland Publ. Co.
Kutner, M.L.; Tucker, K.D.; Chin, G.C.; and Thaddeus, P. 1977. *Astrophys. J.* 215: 521.
Lada, C.J. 1976. *Astrophys. J. Suppl.* 32: 603.
____. 1978. In *Proceedings of the Gregynog Conference on Giant Molecular Clouds.* In Press.
Lada, C.J.; Elmegreen, B.G.; Cong, H.I.; and Thaddeus, P. 1978. *Astrophys. J.* (Letters). In press.
Lada, C.J.; Dickinson, D.F.; Gottlieb, C.A.; and Wright, E.L. 1976*a*. *Astrophys. J.* 207: 113.
Lada, C.J.; Gull, T.R.; Gottlieb, C.A.; and Gottlieb, E.W. 1976*b*. *Astrophys. J.* 203: 159.
Lada, C.J., and Wooden, D. 1978. In preparation.
Larson, R.B. 1972. *Mon. Not. Roy. Astr. Soc.* 157: 121.
Lin, C.C., and Shu, F.H. 1964. *Astrophys. J.* 140: 646.
____. 1966. *Proc. Nat. Acad. Sci.* 55: 229.
Loren, R.B. 1976. *Astrophys. J.* 209: 456.
Lucy, L.B. 1975. *Mém. Roy. Soc. Sci. Liège, Ser. 6,* 8: 359.
Lucy L.B. and Solomon, P.M. 1967. *Astron. J.* 72: 310.
____. 1970. *Astrophys. J.* 159: 879.
McKee, C.F., and Ostriker, J.P. 1977. *Astrophys. J.* 218: 148.
Mezger, P.G., and Smith, L.F. 1977. In *Star Formation* (T. de Jong and A. Maeder, eds.), p. 133. Dordrecht: Reidel.
Morton, D.C. 1967. *Astrophys. J.* 147: 1017.
Myers, P.C. 1977. *Astrophys. J.* 211: 737.
Ögelman, H.B., and Maran, S.P. 1976. *Astrophys. J.* 209: 124.
Oort, J.H. 1954. *B.A.N.* 12: 177.
Öpik, E. 1953. *Irish Astron. J.* 2: 219.
Ostriker, J.P.; Richstone, D.O.; and Thuan, T.X. 1974. *Astrophys. J.* 188: L87.
Racine, R. 1968. *Astron. J.* 73: 233.
Raymond, J.C.; Cox, D.P.; and Smith, B.W. 1976. *Astrophys. J.* 204: 290.
Reynolds, R.J., and Ogden, P.M. 1978. Submitted to *Astrophys. J.*
Roberts, M.S. 1957. *Publ. Astron. Soc. Pacific* 69: 59.
Roberts, W.W. 1969. *Astrophys. J.* 158: 123.
Sancisi, R. 1974. In *Galactic Radio Astronomy* (F.J. Kerr and S.C. Simonson, eds.), p. 115. Dordrecht: Reidel.

Sanders, R.H., and Solomon, P.M. 1977. *Bull. Am. Astron. Soc.* 9: 554.
Sargent, A.I. 1977. *Astrophys. J.* 218: 736.
Schmidt, M. 1957. *B.A.N.* 13: 247.
Sharpless, S. 1959. *Astrophys. J. Suppl.* 4: 257.
Spitzer, L., Jr. 1968. *Diffuse Matter in Space.* New York: Wiley.
Stone, M.E. 1970*a*. *Astrophys. J.* 159: 277.
____. 1970*b*. *Astrophys. J.* 159: 293.
Stothers, R. 1972. *Astrophys. J.* 175: 431.
Sullivan, W.T., III, and Downes, D. 1973. *Astron. Astrophys.* 29: 369.
Tammann, G.A. 1977. *Supernovae* (D.N. Schramm, ed.), p. 95. Dordrecht: Reidel.
Taylor, J.H., and Manchester, A. 1977. *Astrophys. J.* 215: 885.
Weaver, H., and Williams, D.R.W. 1973. *Astron. Astrophys. Suppl.* 3: 1.
____. 1974. *Astron. Astrophys. Suppl.* 17: 1.
Weaver, R.; McCray, R.; Castor, J.; Shapiro, P.; and Moore, R. 1977. *Astrophys. J.* 218: 377.
Woltjer, L. 1972. *Ann. Rev. Astr. Astrophys.* 10: 129.
Woodward, P.R. 1976. *Astrophys. J.* 207: 466.
Yuan, C. 1969*a*. *Astrophys. J.* 158: 871.
____. 1969*b*. *Astrophys. J.* 158: 889.

THE ROLE OF SUPERNOVAE IN STAR FORMATION AND SPIRAL STRUCTURE

WILLIAM HERBST
AND
GEORGE E. ASSOUSA
Carnegie Institution of Washington

We discuss the astrophysical evidence that supernovae are involved in some star formation currently proceeding in the galaxy. In Canis Major R1 recent observations support the supernova-induced star formation scenario presented by Herbst and Assousa in 1977. It appears that density wave shocks cannot be responsible for any of the star formation occurring in the local spiral feature (this includes regions such as Cygnus, Cepheus, Orion, and Canis Major). The large-scale structure of this feature (and, presumably, similar features in other spiral galaxies) may be attributed to a supernova-cascade type process or to an underlying pattern in the gaseous component of the disk (e.g., an ultraharmonic resonance) which modulates the efficiency of localized triggering mechanisms such as supernovae. The meteoritic evidence suggesting that the solar system was formed as a result of a supernova event is, therefore, not in contradiction with what we know about star formation presently occurring in the vicinity of the sun.

The discovery of a ^{26}Mg anomaly in a portion of the Allende meteorite (Lee *et al.*, 1976) has shaken our conception of the early history of the solar system and led to the hypothesis that a supernova was directly involved in the sun's formation (see chapter by Schramm in this book). We review here the astrophysical evidence that present-day star formation in the galaxy is proceeding via this same mechanism. In particular we are concerned with the question of whether or not supernova-induced star formation may be a widespread phenomenon in the galaxy, and presumably in other spiral galaxies. The symbiotic nature of the relationship between this question and the meteoritic evidence concerning the formation of the solar system is apparent.

The idea that supernovae can cause star formation is not a new one. Öpik (1953) has written:

> "It would seem that the time has come to summarize observations referring to a number of remarkable phenomena: the expanding shells of supernovae . . . giant rings of hydrogen such as those in the constellation of Orion . . . loose stellar groups or associations . . . whose members are spread over too large a volume of space to be held together by gravitation . . . the expansion found for some of [these] indicating an origin from a common centre . . . . Theoretical difficulties in following up the phenomena are great, chiefly on account of the multitude of factors involved but the observational facts are so eloquent that a mistake in their general interpretation is hardly possible."

He went on to describe a picture of supernova-induced star formation very similar to the one discussed in this chapter.

The idea that supernovae are a trigger for star formation is not a radical one. Shock-induced star formation has been invoked by many authors in many situations. Examples are: spiral density waves (Roberts, 1969; Woodward, 1976), high velocity gas collisions in protogalaxies (Larson, 1976), and tidal interactions of galaxies (Larson and Tinsley, 1978). Supernovae are simply another mechanism for producing shocks in the interstellar medium. The source of the shocks, their strength, and their rate of production are reasonably well understood. There is no reason to believe that supernova shocks should be less efficient in inducing star formation than the other mechanisms mentioned. Woodward (1976), for example, has considered that his calculations of shock-driven implosions of clouds apply to supernova shocks as well as to density wave shocks.

A decade after Öpik's work, Bird (1964) further developed his idea, presenting a model of star formation in associations which depended on an explosion of the initial condensation. A somewhat different point of view, and perhaps the first suggestion that a supernova was involved in the formation of the solar system, was put forward by Brown (1971). He argued that instabilities in the expanding ejecta of a supernova would cause it to fragment into stellar-dimension condensations which would contract into solar systems. In fact, observations of the Cas A supernova remnant show that fragmentation has occurred; however, the physical conditions in the condensations (Chevalier and Kirshner, 1978) do not appear conducive to the formation of solar systems. In particular, the masses of the condensations are probably much less than 1 $M_\odot$.

Modern scenarios for supernova-induced star formation involve the interaction of the ejecta with the interstellar medium (aspects of this

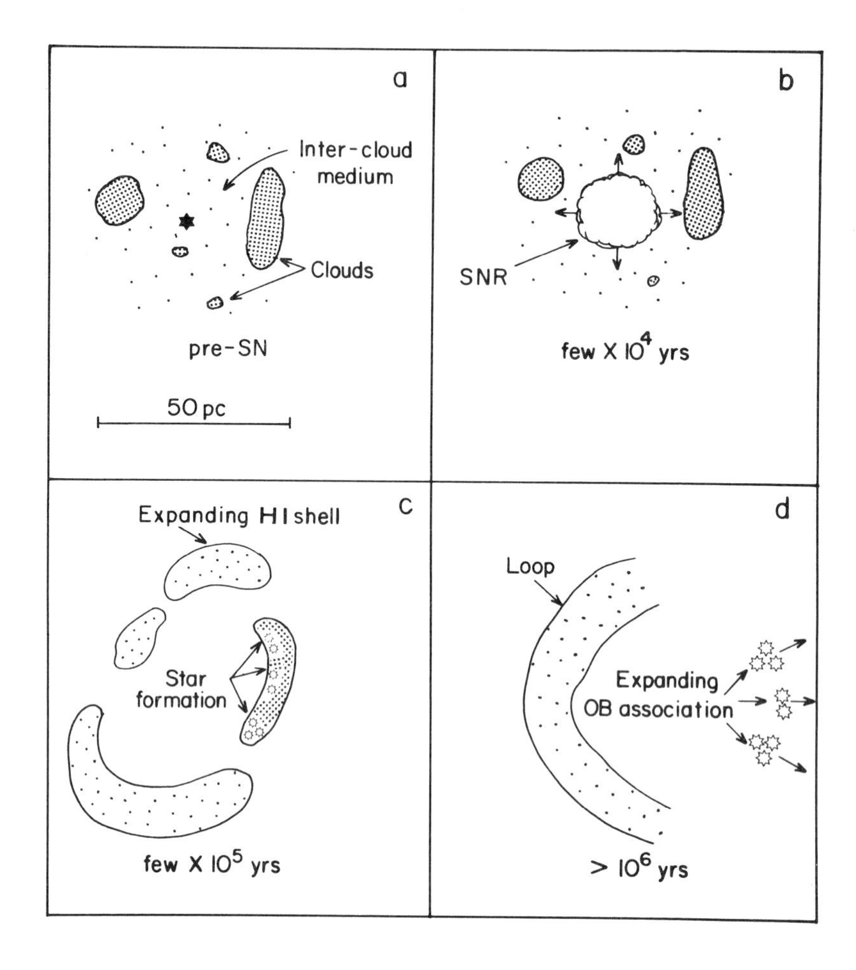

Fig. 1. A schematic representation of one possible means by which a supernova explosion might induce star formation. Fig. 1*c* depicts the present stage in CMa R1 and Fig. 1*d* the present stage in many OB associations (e.g., Cep OB 3) according to this scenario.

interaction have recently been reviewed by Chevalier, 1977). Two cases may be considered: the uniform medium and the cloudy medium. In the case of an initially uniform medium, the evolution of a supernova remnant is reasonably well understood (Woltjer, 1972; Chevalier, 1974). A characteristic of this evolution is the development of an expanding shell in which the density is several times the initial ambient density. Such shells have been observed around old supernova remnants (Assousa and Erkes, 1973; Assousa, Balick, and Erkes, 1974; Knapp and Kerr, 1974; Heiles, 1976; Giovanelli and Haynes, personal communication, 1978). Chevalier and Theys (1975) have speculated that star formation might occur within this shell. Another possibility is that the expanding supernova remnant encounters a cloud. In this case a shock front will develop and, as Woodward (1976) has calculated, cloud implosion and (very likely) star formation will occur. A schematic representation of this event is depicted in Fig. 1. A third possibility is that the radiation from a supernova remnant may affect the chemistry in nearby clouds so as to accelerate their collapse (Wootten, 1978). This effect might extend the influence of remnants beyond the volume directly affected by the shock.

Observational evidence that star formation occurs in expanding shells of neutral hydrogen was first presented by Sancisi (1974) and Sancisi *et al.* (1974). They detected shells in the Per OB2 and Sco OB2 associations and argued that some of the young stars were formed within these. They also showed that the shells could be interpreted as old supernova remnants. Berkhuijsen (1974) argued that the Origem Loop is an old supernova remnant and that several Sharpless H II regions were part of its shell. Williams *et al.* (1977) and Herbst and Havlen (1978) have discussed morphological evidence that star formation has occurred within large-scale shells.

Assousa, Herbst, and Turner (1977) re-analyzed the neutral hydrogen data in the direction of Cep OB3 and showed that a shell with a maximum expansion velocity of $\sim$35 km sec^{-1} exists around that association. They identify this as an old supernova remnant and show that the stellar ages and expansion motions are consistent with the hypothesis that at least the younger subgroup of Cep OB3 was formed as a result of this explosion. Perhaps the clearest example of star formation associated with an expanding shell, however, is the Canis Major R1 association (Herbst, 1976; Herbst and Assousa, 1977). Pertinent observations of this region are reviewed and updated in the next section.

I. CANIS MAJOR R1

The argument for supernova-induced star formation in CMa R1 presented by Herbst and Assousa (1977) has three main components. These are:

1. Star formation has recently taken place in CMa R1.
2. The emission ring adjacent to the association is a supernova remnant.
3. There is a causal relationship between the supernova event and the star formation.

Since their discussion a considerable amount of new data on this region has become available and in this section we review their implications.

Star Formation in Canis Major R1

The argument for recent star formation in CMa R1 has been strengthened by optical and infrared observations of its thirty members (Herbst, Racine and Warner, 1978). Many stars are found one to two magnitudes above the main sequence. The stars farthest from the main sequence generally have K-L excesses, indicative of circumstellar material. Later than spectral type B5, the stars are mostly well above the zero-age main sequence (ZAMS) and this is tentatively identified as the turn-off point of the association, suggesting an age of $\sim 3 \times 10^5$ yr. However, some of the later type stars are closer to the ZAMS then expected on the basis of a coeval formation hypothesis and Iben's (1965) pre-main-sequence models. More realistic models involving rotation and perhaps magnetic fields are necessary to estimate the age spread.

The Expanding Shell

The most exciting recent work on the CMa R1 ring is a Fabry-Perot spectrometer study by Reynolds and Ogden (1978). They have detected splitting of the [N II] line from the center of the ring, thereby confirming the existence of an expanding shell as inferred by Herbst and Assousa (1977) from the neutral hydrogen data. Both the front and back of the shell are detected allowing a good determination of the expansion velocity, 13 km sec^{-1}. This is considerably lower than that deduced from the neutral hydrogen data, 32 km sec^{-1}, probably for two reasons:

1. The H I value was based on the maximum velocity observed; the [N II] value was determined by fitting Gaussian curves to the two peaks and determining the separation of their maxima. It therefore allows for turbulence in the shell.
2. The ionized gas is presumably formed at the inner edge of the neutral shell and expands preferentially into the low-density, swept-out interior. The true expansion velocity of the shell therefore lies between the two values given above and is quite consistent with the scenario described by Herbst and Assousa (see Reynolds and Ogden, 1978). The major effect of the reduced expansion velocity on the picture of Herbst and Assousa is to increase the age of the remnant.

Another important result obtained by Reynolds and Ogden (1978) is that the [O III] line is enhanced in the vicinity of HD 54662 and at a velocity characteristic of the backside of the shell. This provides further confirmation of the star's association with the nebulosity and supports the suggestion by Herbst and Assousa that the star is a runaway from the supernovae event. (Since its radial velocity is +30 km sec^{-1} with respect to the association, we would expect it to be inside the back part of the shell at present.) Reynolds and Ogden also estimate the total emission measure from the CMa R1 ring and conclude that photo-ionization of the shell by the early-type stars can explain the observations.

Supernova Remnant or Stellar Bubble?

Perhaps the most controversial point in the interpretation of CMa R1 is whether the expanding shell of gas could have been produced without a supernova explosion. Blitz (1978) and Reynolds and Ogden (1978) have argued that one or more massive O type stars could have supplied the requisite energy. Since O type stars are commonly believed to end their lives as Type II supernovae (see glossary) (e.g., Tinsley, 1977), any hypothesized stellar sources for the shell should still be visible today. There are two sufficiently early-type stars in CMa OB1 that are of interest in this connection, HD 54662 (06.5 V) and HD 53975 (07.5 V). There are also a B0 giant (HD 53974; R association member No. 29) and a possibly luminous N type carbon star (W CMa; see Herbst *et al.,* 1977). Both of these are embedded in nebulosity, however, indicating that there is interstellar matter close to them. For this reason, it is difficult to see how either could have been involved in forming the expanding shell.

There are two possible mechanisms by which an early-type star might generate an expanding shell. First, by ionizing its surrounding interstellar medium, an O star creates a sphere of high-pressure gas which naturally expands into the neutral medium (see Mathews and O'Dell,1969). Lasker (1966, 1967) has numerically modeled the evolution of such an H II region in an initially uniform medium. He finds that the ionized sphere will expand, pushing ahead of itself a compressed neutral shell which may attain velocities comparable to the speed of sound in the ionized region (~10 km sec^{-1}). Observational confirmation of this model appears to come from Wade's (1958) observations of an H I shell expanding at 8 km sec^{-1} around the O star λ Ori. Riegel (1967), however, was unable to find further examples of this phenomenon in his 21-cm study of 27 galactic H II regions.

There is considerable difficulty in applying Lasker's models to the case of CMa OB1, as Blitz (1978) and Reynolds and Ogden (1978) have discussed. The observations show an expanding *shell* of ionized gas. According to Lasker's models only a slowly expanding (~4 km sec^{-1} maximum velocity) *sphere* of ionized gas exists, which would result in a flat-topped emission line rather than a double-peaked one. To account for the observations, one requires at least a two-star system, one to produce the shell and another to ionize it. What the evolution of such a system might be is unknown. A second objection has to do with the location of the O stars relative to the shell. It is important to realize that the ionized shell exhibits a high degree of symmetry (see the Fig. 2 of Herbst and Assousa, 1977). The coordinates of the center of the shell are rather well defined by the curvature of the emission ring. The O stars are projected on or very close to the edge of the shell, at least 30 pc from its center. The situation is, therefore, very unlike λ Ori, where the star is at the center of the shell, and unlike Lasker's models in which the star does not move relative to the gas. Finally, the earliest type star, HD 54662, is definitely moving supersonically with respect to the ionized gas in this picture. It has a radial velocity of +30 km sec^{-1} with respect to the mean association velocity and +24 km sec^{-1} with respect to the highest velocity ionized gas observed in the direction of the star by Reynolds and Ogden (1978). Lasker's models clearly do not apply in this case, and it is intuitively difficult to see how a well-formed expanding shell like the one in CMa R1 could result; we are not aware of any studies which address the question.

The second mechanism by which an O star might create an expanding shell is through the interaction of its stellar wind with the surrounding medium. This stellar "bubble" theory has been developed by Weaver *et al.* (1977) and is considered a possible mechanism operating in CMa R1 by Reynolds and Ogden (1978) and as the most likely source of the shell by Blitz (1978). It seems to us, however, that there are a number of considerations which make this hypothesis unattractive. First, we note that the stellar bubble theory has no unambiguous observational support. The O VI data and the Gum nebula observations which Weaver *et al.* (1977) claim support their theory, may also be explained as effects of supernovae (Reynolds, 1976*a,b;* McKee and Ostriker, 1977). The theoretical arguments for the formation of a bubble depend on a reasonably steady wind and a reasonably uniform ambient medium. It is well known that the interstellar medium is cloudy and real stellar winds are apparently not very steady. A secular variation occurs caused by evolution of the star off the main sequence. Luminosity Class V stars have little or no wind while Class I stars have strong winds; shorter term fluctuations may also occur as evidenced by the spread in wind parameters at a given spectral type (Hutchings, 1976). It must, therefore, be regarded as unproven that stellar bubbles even exist, let alone that they can explain the observations in CMa R1.

The star with the strongest stellar wind in CMa OB1 is presumably HD 54662. Its location within the shell at the present time does not necessarily argue against the stellar bubble mechanism. Weaver *et al.* (1977) claim that a bubble will form symmetrically around the location of the star when it "turns on" as long as it moves subsonically with respect to the hot, tenuous gas inside the shell having a sound speed of ~100 km sec^{-1}. What is unusual, however, is that Reynolds and Ogden (1978) do not find significant line splitting in the direction of HD 54662. (They find a dominant component of the [N II] line at +10 km sec^{-1} and an asymmetry of the line which indicates a weak component at +18 km sec^{-1}.) If it is the source of the 26 km sec^{-1} line splitting seen toward the center of the emission ring, why are the lines essentially single in its direction?

Finally, we point out that neither of the stellar source mechanisms can explain the presence of a runaway star (HD 54662) in CMa OB1. The star is moving at least 30 km sec^{-1} and probably over 40 km sec^{-1} with respect to the association (gas and other stars). Blaauw (1961) has shown that runaways can be formed from massive binaries in which one star becomes a supernova. This also explains why most if not all runaways (including HD 54662) are single stars. Herbst and Assousa (1977) have shown that the proper motion of the star is consistent with the hypothesis that it is a runaway from the same supernova event that produced the expanding shell and star formation in CMa R1. However, the proper motion data are not sufficiently accurate to prove that the star was once at the center of the emission ring. Without a supernova event it is difficult to account for this star's velocity. The "gravitational slingshot" mechanism in which three or more bodies interact so as to eject one, suffers from the lack of a suitable nearby binary or small cluster from which HD 54662 might plausibly have been ejected.

To summarize this discussion, the supernova scenario provides a simple, comprehensive explanation of the facts in CMa OB1/R1. The other scenarios discussed, while impossible to rule out entirely, require complicated histories or rely on observationally unconfirmed (stellar wind) theory, and leave the runaway star unexplained.

A Causal Relationship

Herbst and Assousa (1977) showed that the ages of the expanding shell and the CMa R1 members were consistent with the hypothesis of supernova-induced star formation hypothesis. The new data discussed here are in agreement with that conclusion. However, the strongest argument for a causal connection between the expanding shell and the star formation comes from the location of the newly-formed stars with respect to the shell. In Fig. 2 we show the locations of the CMa R1 members. Recent star formation has occurred over a projected linear extent of ~30 pc at or close to the ionization edge which forms the western boundary of the main emission ring. Note especially the two subgroups consisting of (16, 17, 18 and 19) and (9, 11, 12, 13 and 14). These lie almost right at the ionization boundary and each contains K-L excess stars which lie well above the zero-age main sequence (No. 9 = Z CMa, a Herbig emission star; see Herbig, 1960). As Woodward (1976) has shown, a shocked cloud is likely to form stars along the edge of the cloud facing the direction the shock came from. This is where most of the CMa R1 stars are found – along the concave edge of the dark cloud. We suggest that Fig. 1 is an appropriate schematic representation of the events occurring in CMa R1 with Fig. 1*c* representing the present configuration.

Earlier Stages of Supernova-Induced Star Formation

With currently available radio and infrared techniques it is possible to search for earlier stages of the star-formation process around old, but still clearly identifiable supernova remnants. It is important to realize that we should not expect to find such signs around every supernova remnant; many will be too young, and others may not be situated properly to interact with suitable clouds. Nonetheless, a survey by Wootten (1978) has produced positive results. His observation of a likely protostar in a cloud associated with the supernova remnant W 44 is particularly noteworthy. It may well be an example of an earlier stage of the same process which produced CMa R1. We feel that further molecular line mapping and infrared searches for

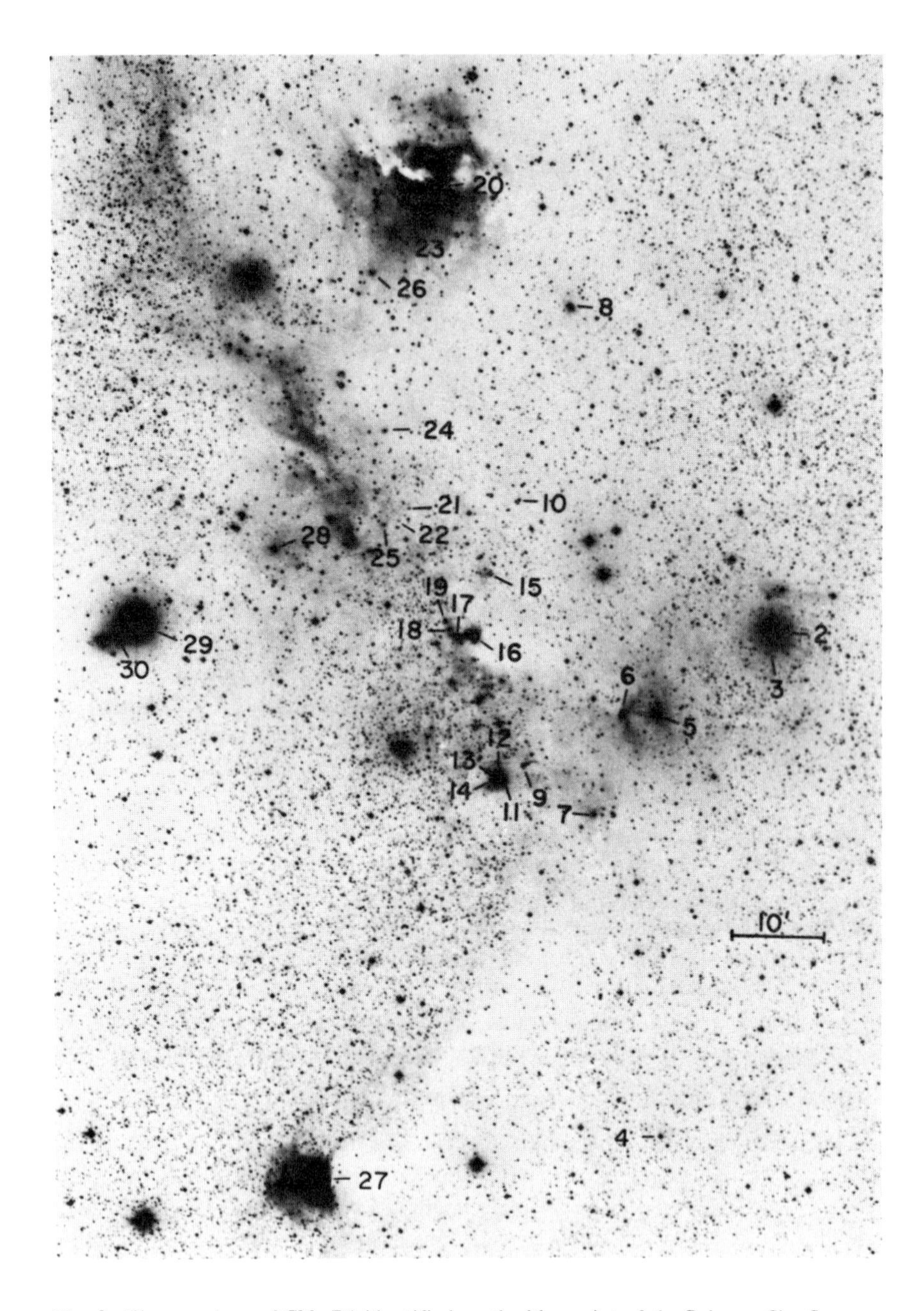

Fig. 2. The members of CMa R1 identified on the blue print of the Palomar Sky Survey. North is up; east is to the left. The angular scale is indicated and at the distance of the association 10′ corresponds to 3.3 pc. The Herbig emission stars are Nos. 9 (Z CMa) and 20 (HD 53367). (From Herbst, Racine and Warner, 1978.)

protostars in the vicinities of old supernova remnants will be a profitable direction for future research.

II. SPIRAL STRUCTURE

It is clear that recent star formation in external galaxies occurs mostly in a spiral pattern. There is every reason to believe that it does likewise in our galaxy, although the details of that pattern have not yet been established. The density wave theory as formulated by Lin and associates (e.g., Lin, 1967) has been most successful in accounting for this fact (however, problems remain; Toomre, 1977). The modification of shocks at the inner edge of the arms, introduced by Fujimoto (1968) and Roberts (1969), provides a triggering mechanism for star formation in the density wave theory. The major success of the shock theory is the explanation of narrow spiral structure, ordered on a galactic scale, such as that exhibited by many external galaxies. The other major successes of the density wave theory (which do not depend on the presence of shocks) are in explaining the concentration of H I to the spiral arms in M 81 (Rots and Shane, 1975) and the streaming motions detected across spiral arms in external galaxies (Rubin and Ford, personal communication, 1978).

Is there any room in this neat picture for widespread supernova-induced star formation? We believe so. In most spiral galaxies, some of the star formation occurs outside the main spiral arms in what have been called spurs or feathers. Some galaxies (NGC 2841 type; Sandage, 1961) seem to consist entirely of these short spiral pieces, with no global pattern. In some galaxies (e.g., Magellanic Clouds) the pattern is essentially chaotic with only a vague hint, if any, of spiral structure. It is difficult to see how the density wave theory with shocks can explain all of these features.

The Supernova Cascade Mechanism

One means of generating spiral-like patterns complete with spurs in differentially rotating galaxies is the supernova cascade mechanism originally proposed by Ögelman and Maran (1976). Their picture is that a supernova will result in the formation of stars with a range of masses at a distance of some 50 pc from the event. The most massive newly-formed star will soon become a supernova and cause another round of star formation, again displaced some 50 pc. In this way a chain of recent star-formation regions is created, lit up by the lower mass stars which evolve on a longer time scale. Mueller and Arnett (1976) studied numerical models of differentially rotating galaxies with star formation proceeding via this mechanism (or, alternatively, a stellar wind cascade mechanism) and found that, with an underlying spiral density enhancement they could produce a typical spiral galaxy, including spurs. Gerola and Seiden (1978) have improved on the Mueller and Arnett model by using a finer grid and adopting a probabilistic approach in which a

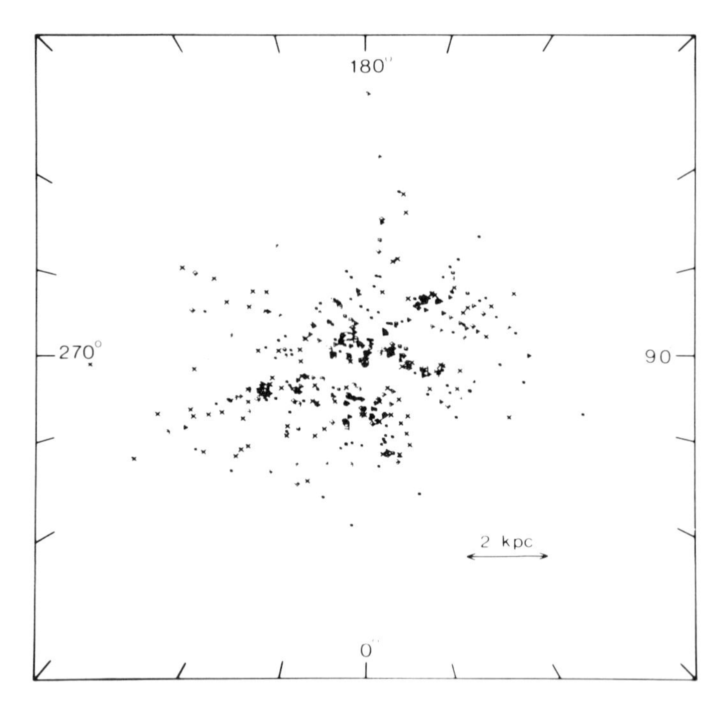

Fig. 3. The distribution of newly-formed stars in the vicinity of the sun. The sun is at the center of this diagram, along the inner edge of the local feature (also known as the Cygnus-Orion feature). The other well-known spiral features are the Perseus and Carina-Sagittarius arms. The gap in the distribution of young stars between the local and Carina-Sagittarius features is the best-established structure in this diagram. The galactic center is 8 to 10 kpc from the sun in the direction of 0° galactic longitude. (From Herbst, 1975.)

star-forming event does not always result in further star formation in an adjacent region. They are able to produce stable spiral structure without an underlying density enhancement and argue further that the type of resultant galaxy (Sb, Sc, etc.) depends only on the form of the rotation curve. While these results could be considered encouraging for the cascade mechanism, the simplistic nature of the models makes their application to real spiral galaxies doubtful.

The Local Spiral Feature

There is a considerable amount of star formation occurring within 1 kpc of the sun at the present time. In addition to the many well-known OB associations (e.g., Orion OB1, Scorpius OB2, Perseus OB2; Blaauw, 1964) there are T associations like the one in Taurus and R associations such as Mon R2. These regions contain stellar or protostellar objects with ages $\sim 10^6 - 10^7$ yr or less. The nearby associations are not randomly distributed over the galactic plane but show a large-scale order (see Fig. 3) which takes the form of a slightly arced chain of star forming regions about 7 kpc long and 0.5 kpc wide extending in the directions of galactic longitude $l \sim 70°$ (Cygnus) and $l \sim 270°$ (Vela), or possibly $l \sim 240°$. This is referred to as the local spiral feature. The sun is located near the inner edge (i.e., towards the galactic center, $l = 0°$). Note that the sun has made many circuits of the galaxy since its formation and was not born in the local feature. There are two other prominant "spiral" features in Fig. 3 – an outer one known as the Perseus Feature, seen most clearly around $l \sim 130°$ and an inner one, known as the Sagittarius-Carina Feature ($l \sim 280° - 25°$). This has been essentially the pattern seen in every optical map of star-forming regions since the work of Morgan *et al.* (1952, 1953).

Are these the density wave arms of our galaxy? Perhaps the Perseus and Sagittarius-Carina features are, but the local feature is not, according to Lin, Yuan, and Roberts (1978). They have analyzed stellar kinematical data and have shown that the disturbance in stellar orbits expected from the presence of a massive density wave arm is not seen in the local feature. There is some evidence that it is present in the Perseus and Sagittarius-Carina features. They conclude that the sun and all of the local star-forming regions are roughly midway between density wave arms. They consider the local Feature to be a spur. The presence of a "K-term" measuring expansion of the local set of stars at ~ 5 km sec^{-1} (e.g., see Frogel and Stothers, 1977) may be another indication that we are between density waves.

Shu *et al.* (1973) suggested an explanation for spurs in terms of ultraharmonic resonances which result in additional gas-density (but not stellar-density) enhancements outside of the density wave arms. (Ultra-harmonic resonances are secondary compressions of the gas [outside of spiral arms] which occur as a result of the resonant nature of non-linear gas flow in a disk galaxy within the spiral density wave [see Shu *et al.*, 1973].) This may well explain the large-scale order of the local feature, but it does not account for the "triggering" of star formation locally, since shocks cannot develop in ultraharmonic resonances (Woodward, 1975). Perhaps triggering is not necessary for star formation; it may proceed by simple gravitational instabilities (e.g., Larson, 1977). However, the observation that massive stars are invariably found at the edges of giant molecular clouds (Elmegreen and Lada, 1977) is an indication that a trigger mechanism is normally involved. Density wave shocks apparently cannot be the trigger for local star-formation regions. Ionization fronts or stellar winds also cannot provide the trigger for the initial round of star formation in a cloud since they require pre-existing massive stars. Supernovae, on the other hand, can provide this initial trigger, since they have a randomly distributed component, the Type I supernovae (Maza and van den Bergh, 1976). Type I supernovae have about the same outburst energies as the Type II (Chevalier, 1977) and occur with about the same frequency in spiral galaxies similar to our own (Maza and

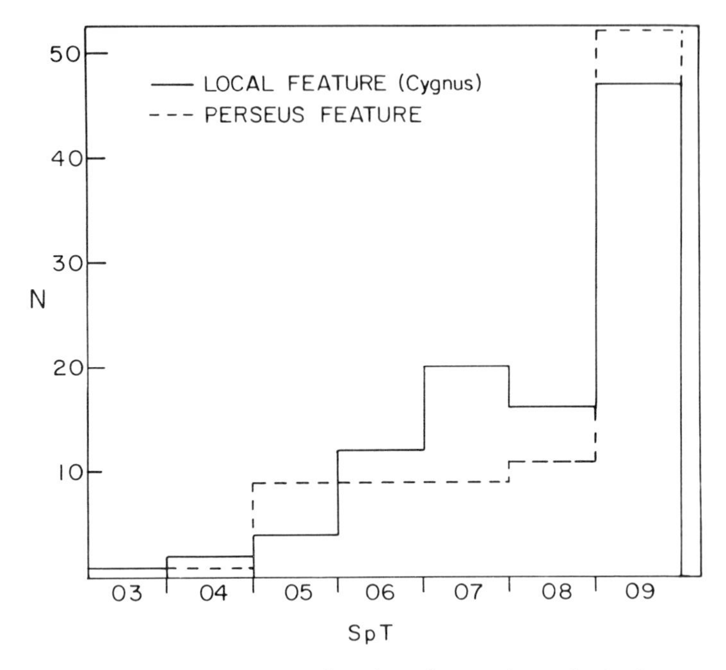

Fig. 4. The number of O stars as a function of spectral type in the Perseus and local (Cygnus portion only) features. No corrections have been applied for various selection effects which may influence this comparison; however, the preliminary result is that there is no gross difference in the rate at which young, massive stars are being formed in the two features. (From Herbst and Lockman, 1978).

van den Bergh, 1976). They do form remnants (Tinsley, 1977) and it is reasonable to suppose that they would be as effective at inducing star formation as the Type II. They seem to us to be the most likely source of the initial round of star formation in clouds within the local arm. The Type II supernovae, on the other hand, may be involved in the subgroup formation process in OB associations, in a role similar to that ascribed to ionization fronts by Elmegreen and Lada (1977). In Orion OB1, for example, Blaauw (1964) estimates that there have been five, or even more, Type II supernovae.

A question which naturally arises is: how representative is the local feature of star-forming regions in the galaxy. In particular, is the solar system likely to have formed in a similar region? No definitive answer can be given at the present time; however, one can get some information on this by comparing the star-formation rate in the local feature with those in the Perseus and Sagittarius-Carina arms. This work is in progress for the Perseus Feature (Herbst and Lockman, 1978) and a preliminary result is shown in Fig. 4. This histogram shows that the space density of O-type stars is not much different in the two features and that there are more stars of very early type in the local feature. The local feature may therefore be a spur but it is probably generating stars at least as rapidly as in what has been identified as a major density wave arm at about our distance from the galactic center.

III. SUMMARY AND CONCLUSION

The two major points addressed by this chapter are:

1. Supernovae are very likely responsible for at least some of the star formation presently occurring in the galaxy.
2. It is conceivable that they are the major "triggering mechanism" operating at the distance of the sun from the galactic center in our galaxy.

The evidence for point 1 was presented in our introduction and in Sec. I. It is further strengthened by Wootten's (1978) observation of interactions between molecular clouds and some old supernovae remnants such as the Monoceros Loop. These may be the sites of very early stages in the star-formation process and deserve a good deal of attention.

In Sec. II we pointed out that a density wave shock can probably not be invoked to explain star formation in the vicinity of the sun. Herbst and Assousa (1977) showed that the local supernova rate is sufficient to account for all the star formation in the local arm. However, other mechanisms, such as expanding H II regions (Elmegreen and Lada, 1977; also see their chapter), or simply large-scale gravitational instabilities (Larson, 1977) may also be important. An apparent advantage of the supernova mechanism is that expanding associations, which are the rule, may be expected quite naturally. More detailed studies of the interactions of supernova shocks and ambient clouds, and the evolution of OB associations are necessary to see if this apparent advantage is a decisive one. At the present time, the above points 1 and 2 are the strongest statements that can be made on behalf of this mechanism. From the point of view of the meteoritic evidence, we can phrase these conclusions by saying that it is no more *ad hoc* to suggest that the presolar cloud was brought to the point of collapse by a supernova shock than to suggest that a galactic density wave shock (or any other triggering mechanism) was responsible for this.

Acknowledgments. It is a pleasure to thank J. Lockman, R. Reynolds, and L. Blitz for stimulating discussions.

COMMENTS

Z. Kopal: Is there a chemical "signature" of the supernova events that occurred during the past one thousand years (Crab, 1572, 1604, Cassiopeia A) in the composition of lunar rocks?

J. R. Arnold: We do not see a direct signature, either chemical or isotopic, of the individual supernovae which you list. We should not expect to. These objects are far away and the cross-section of a lunar rock at that distance is exceedingly small.

The difficulty comes in understanding whether any signatures of supernovae of the last few million years might be recorded. Some of these must have been much closer than any of the recent ones. Modulation of the cosmic ray flux is to be expected – both increases and decreases seem possible.

REFERENCES

Assousa, G.E.; Balick, B.; and Erkes, J.W. 1974. *Carnegie Inst. Wash. Year Book* 73: 899.

Assousa, G.E., and Erkes, J.W. 1973. *Astron. J.* 78: 885.

Assousa, G.E.; Herbst, W.; and Turner, K.C. 1977. *Astrophys. J.* 218: L13.

Berkhuijsen, E.M. 1974. *Astron. Astrophys.* 35: 429.

Bird, J.F. 1964. *Rev. Mod. Phys.* 36: 717.

Blaauw, A. 1961. *B.A.N.* 15: 265.

____. 1964. *Ann. Rev. Astron. Astrophys.* 2: 213.

Blitz, L. 1978. In *Proceedings of the Gregynog Workshop: Giant Molecular Clouds.* In press.

Brown, W.K. 1971. *Icarus* 15: 120.

Chevalier, R.A. 1974. *Astrophys. J.* 188: 501.

____. 1977. *Ann. Rev. Astron. Astrophys.* 15: 175.

Chevalier, R.A., and Kirshner, R.P. 1978. *Astrophys. J.* 219: 931.

Chevalier, R.A., and Theys, J.C. 1975. *Astrophys. J.* 195: 53.

Elmegreen, B.G., and Lada, C.J. 1977. *Astrophys. J.* 214: 725.

Frogel, J.A., and Stothers, R. 1977. *Astron. J.* 82: 890.

Fujimoto, M. 1968. In *Non-stable Phenomon in Galaxies* (V.A. Ambartsumian, ed.), p. 453. Academy of Sciences, Armenian S.S.R. In Russian.

Gerola, H., and Seiden, P.E. 1978. Preprint. T.J. Watson Research Center of I.B.M.

Heiles, C. 1976. *Astrophys. J.* 208: L137.

Herbig, G.H. 1960. *Astrophys. J. Suppl.* 4: 337.

Herbst, W. 1975. *Astron. J.* 80: 503.

____. 1976. In *Supernovae* (D.N. Schramm, ed.), p. 143. Dordrecht: Reidel.

Herbst, W., and Assousa, G.R. 1977. *Astrophys. J.* 217: 473.

Herbst, W., and Havlen, R.J. 1978. *Astron. Astrophys. Suppl.* 30: 279.

Herbst, W., and Lockman, F.J. 1978. In preparation.

Herbst, W.; Racine, R.; and Richer, H.B. 1977. *Publ. Astron. Soc. Pacific* 89: 663.

Herbst, W.; Racine, R.; and Warner, J.W. 1978. *Astrophys. J.* In press.

Hutchings, J.B. 1976. *Astrophys. J.* 203: 438.

Iben, I. 1965. *Astrophys. J.* 141: 993.

Knapp, G.R., and Kerr, F.J. 1974. *Astron. Astrophys.* 33: 463.

Larson, R.B. 1976. In *Galaxies* (L. Martinet and M. Mayor, eds.), p. 67. Geneva: Geneva Observatory.

____. 1977. In *The Evolution of Galaxies and Stellar Populations* (B.M. Tinsely and R.B. Larson, eds.), p. 97. New Haven, Conn.: Yale University Printing Service.

Larson, R.B., and Tinsley, B.M. 1978. *Astrophys. J.* 219: 46.

Lasker, B.M. 1966. *Astrophys. J.* 143: 700.

____. 1967. *Astrophys. J.* 149: 23.

Lee, T.; Papanastassiou, D.A.; and Wasserburg, G.W. 1976. *Geophys. Res. Lett.* 3: 41.

Lin, C.C. 1967. *Ann. Rev. Astron. Astrophys.* 5: 453.

Lin, C.C.; Yuan, C.; and Roberts, W.W. 1978. Preprint. Massachusetts Institute of Technology.

Mathews, W.G.; and O'Dell, C.R. 1969. *Ann. Rev. Astron. Astrophys.* 7: 67.

Maza, J., and van den Bergh, S. 1976. *Astrophys. J.* 204: 519.

McKee, C.F., and Ostriker, J.P. 1977. *Astrophys. J.* 218: 148.

Morgan, W.W.; Sharpless, S.; and Osterbrock, D.E. 1952. *Astron. J.* 57: 3.

Morgan, W.W.; Whitford, A.E.; and Code, A.D. 1953. *Astrophys. J.* 118: 318.
Mueller, M.W., and Arnett, W.D. 1976. *Astrophys. J.* 210: 670.
Ögelman, H.B., and Maran, S.P. 1976. *Astrophys. J.* 209: 124.
Öpik, E.J. 1953. *Irish Astron. J.* 2: 219.
Reynolds, R.J. 1976*a*. *Astrophys. J.* 203: 159.
____. 1976*b*. *Astrophys. J.* 206: 679.
Reynolds, R.J., and Ogden, P.M. 1978. *Astrophys. J.* In press.
Riegel, K.W. 1967. *Astrophys. J.* 148: 87.
Roberts, W.W. 1969. *Astrophys. J.* 158: 123.
Rots, A.H., and Shane, W.W. 1975. *Astron. Astrophys.* 45: 25.
Sancisi, R. 1974. In *Galactic Radio Astronomy* (F.J. Kerr and S.C. Simonson, eds.), p. 115. Dordrecht: Reidel.
Sancisi, R.; Goss, W.M.; Anderson, C.; Johansson, L.E.B.; and Winnberg, A. 1974. *Astron. Astrophys.* 35: 445.
Sandage, A. 1961. *The Hubble Atlas of Galaxies*, p. 14. Washington, D.C.: Carnegie Inst. Washington.
Shu, F.H.; Milione, V.; and Roberts, W.W. 1973. *Astrophys. J.* 183: 819.
Tinsley, B.M. 1977. In *Supernovae* (D.N. Schramm, ed.), p. 117. Dordrecht: Reidel.
Toomre, A. 1977. *Ann. Rev. Astron. Astrophys.* 15: 437.
Wade, C. 1958. *Rev. Mod. Phys.* 30: 946.
Weaver, R.; Brand, P.W.J.L.; Longmore, A.J.; and Howarden, T.G. 1977. *Mon. Not. Roy. Astr. Soc.* 180: 709.
Weaver, R.; McCray, R.; Castor, J.; Shapiro, P.; and Moore, R. 1977. *Astrophys. J.* 218: 377.
Woltjer, L. 1972. *Ann Rev. Astr. Astrophys.* 10: 129.
Woodward, P.R. 1975. *Astrophys. J.* 195: 61.
____. 1976. *Astrophys. J.* 207: 484.
Wootten, A. 1978. *Moon and Planets* (special Protostars and Planets issue). In press.

SUPERNOVAE AND THE FORMATION OF THE SOLAR SYSTEM

DAVID N. SCHRAMM
The University of Chicago

A brief review is made of pieces of information which all seem to point toward the possible association of a supernova with the formation of the solar system. In particular, this information includes the discoveries over the past few years of isotopic anomalies in carbonaceous meteorites and the observation by W. Herbst and Assousa (see their chapter) of star formation associated with a supernova remnant near Canis Major R1. With regard to the isotopic anomalies, of utmost importance is ^{26}Al *(*$\tau_{1/2} = 7 \times 10^5$ *yr). The observations of Lee, Papanastassiou and Wasserburg show this nucleus to have been present when objects solidified and therefore* ^{26}Al *must have been synthesized within a few million years prior to such solidification. Arguments on energetics and the absence of other spallation-induced anomalies indicate that the* ^{26}Al *was probably produced by a supernova and not by spallation. Since* 10^6 *yr is less than the Kelvin-Helmholtz time of the sun, this argues that supernova shocks were of importance to the hydrodynamics at formation. Woodward and subsequently Margolis have shown that supernova shocks can cause compression of interstellar gas clouds. The question of injecting the anomalous material into the protosolar cloud can be answered with Rayleigh-Taylor instabilities at the boundary and/or the use of "shrapnel-like" grains made in the supernova. Lattimer, L. Grossman and Schramm have shown that the grain condensation sequences in supernova ejecta are very interesting with regard to this latter possibility. Margolis has investigated shock-cloud interfaces containing both gas and dust. Not only do grains have an easier time penetrating the protosolar cloud, but they also do* *not get their anomalies diluted away on the gas mixing time scale. (This point is particularly important to R. Clayton's ^{16}O anomaly.) However, vaporization of the injected grains during subsequent solidification seems necessary. A detailed review is made regarding the hydrodynamic calculations of the cloud-supernova interface. A consistent scenario which attempts to tie together the various pieces is presented.*

This chapter will review evidence that a supernova explosion may have triggered the formation of the solar system. This idea is not new (Hoyle and Öpik, for independent reasons, discussed such possibilities in the 1940's), but now strong observational and experimental evidence from several different areas seems to support this idea. This new "hard" evidence has stimulated many theoretical calculations, in attempts to develop logical, detailed scenarios.

To begin with we shall briefly review the observational and experimental evidence. It falls into two main classes:

1. Direct observation of supernova-stimulated star formation (Herbst and Assousa, 1977 and their chapter in this book) now going on in our galaxy: since this is being reviewed in detail in their own chapter, no further mention will be made. This, however, shows that the supernova trigger of the solar system was probably not a unique case.
2. Isotopic anomalies in meteorites imply that at least some material in the solar system had a different nucleosynthetic history than the bulk-average.

Although various aspects of these topics will also be reviewed elsewhere (see Clayton, 1978, and chapter by Reeves in this book), the fact that there are so many aspects with so many different implications, necessitates at least a brief review to focus on the relevant points. Of particular importance to the supernova-trigger hypothesis is the presence of ^{26}Al at the time of solidification of some objects in the solar system (Lee *et al.*, 1976, 1978). ^{26}Al has a half-life of 7×10^5 yr, and so a nucleosynthetic event (presumably a supernova) is needed within a few million years of solidification.

Once we have established that a supernova was necessary, we shall go into the consequences. In particular we shall review the hydrodynamic calculations (c.f., Woodward, 1976; Margolis, 1978), which show that a supernova shock can cause the compression of an interstellar cloud and thus induce star formation. We will not follow the scenario through to the formation of individual planets since that will be treated in other chapters in this book (see e.g., the chapter by Cameron). We will, however, discuss the important question of how material from this supernova might have been able to penetrate the presolar gas cloud. This problem is being treated in detail by Margolis.

Of particular importance for the penetration and for the retention of isotopic anomalies may be the formation of grains in the supernova ejecta. The details of grain nucleation and formation in supernova are presented elsewhere (Falk *et al.*, 1977). One important aspect of grain formation in the supernova ejecta is that the condensation sequence can be different from that in the isolated protosolar nebula. This point has been made by Lattimer *et al.* (1977, 1978). Such scenarios help in explaining the minerological locations of some of the anomalies. However, it should be remembered that even if grains carry anomalies into the protosolar nebula, those particular grains may no longer be present due to melting and recrystalization during the formation process of the solar system. It should be noted that the formation of grains in supernovae ejecta has important implications, not only for the supernova trigger, but also for the general problem of understanding the interstellar medium. This aspect of the problem is discussed by D. Clayton (1978) and in the chapter by Scalo among others. Some particular points to note are that the formation of grains in supernova ejecta can lead to a straightforward explanation for the depletion of heavy elements in interstellar gas since heavy elements are made in supernovae. In addition, it should be remembered that the average isotopic composition of the interstellar grains may be quite different from that of the gas. This point has been used by D. Clayton (1978) in an attempt to explain the anomalies in meteorites. However, even with such a possibility, one has a difficult time in understanding the ^{26}Al observations without a nearby supernova.

This chapter will end by presenting a scenario that attempts to tie the various pieces together. In particular, the scenario will relate the 10^8 yr xenon time scales with the galactic density wave, the 10^6 yr ^{26}Al time scale with the supernova trigger, and the O, Mg, Ca, Ba, Nd, etc., anomalies with the lack of complete mixing prior to the solidification of some samples in the early solar system.

ISOTOPIC ANOMALIES IN METEORITES

Isotopic abundance variations in meteorites due to the decay of radioactive species have been known for some time. The most interesting of the decade-old anomalies are the extinct activities ^{129}I and ^{244}Pu which decay to xenon (c.f., Hohenberg *et al.* 1967; Alexander *et al.* 1971). These indicate a time separation between the last *r*-process nucleosynthetic event and xenon retention in meteorites of $\sim 10^8$ yr (c.f., Hohenberg, 1969; Wasserburg *et al.*, 1969; the review by Schramm, 1974). In addition there are anomalies in the xenon gas composition in carbonaceous chondrites, known since the 1960's which have later been focused upon in detail by Lewis *et al.* (1975) but whose origins are still not clear.

The first anomaly for which the most reasonable explanation seemed to be primitive material of different isotopic composition than the solar system average was the Ne-E anomaly discovered by Black (1972). However, the study of rare gases is so complex, with neon having at least 5 different components contributing to the abundance of its 3 isotopes, that a definitive statement could not be made on the basis of the Ne-E anomaly alone.

The discovery of anomalies in oxygen by R. Clayton *et al.* (1973) initiated much of the present activity. Here was a major element showing isotopic variations in the primitive carbonaceous chondrites which could not be due to any simple mass fractionation and must have a nuclear origin different from that of the average solar system. The oxygen anomaly seemed to be best interpreted as a component of pure ^{16}O without ^{17}O and ^{18}O. This component represented up to ~ 5% of the oxygen in some mineral inclusions. The existence of the ^{16}O anomaly showed that when the carbonaceous chondrites solidified there were regions which did not have the same isotopic composition as others. This meant that the protosolar nebula was heterogeneous. Since gases would rapidly mix and dilute away any anomalies (unless solidification occurred on a gas mixing time), the simple explanation seemed to be that at least some primordial grains were present with different isotopic compositions from the bulk material of the nebula. The different isotopic composition in these grains could have been due to either of two reasons:

1. Interstellar grains in general having different isotopic composition from the gas.
2. Some grains coming from a nearby source and not yet completely mixed into the nebula. (This point could also work for gas from a nearby source but with more rapid mixing time scales.)

It is important to notice that the oxygen anomaly does not specify a time scale; it only tells us that the protosolar nebula was heterogeneous when the carbonaceous chondrites solidified.

The anomaly which does specify a time scale is $^{26}Mg - {}^{26}Al$. A Mg anomaly was discovered by Lee and Papanastassiou (1974) and Gray and Compston (1974) and was shown by Lee *et al.* (1976) to be due to decay of ^{26}Al. The key point is that for given rocks in carbonaceous chondrites, the ^{26}Mg abundance correlates with the Al/Mg ratio, showing that (1) it is due to the decay of ^{26}Al, and (2) that all of the minerals in the rock solidified with the same $^{26}Al/^{27}Al$ ratio ($\sim 5 \times 10^{-5}$).

The second point shows that ^{26}Al was still "live" when the rock solidified. Therefore, some ^{26}Al must have been synthesized within a few million years prior to the formation of the rock. Old primordial grains cannot explain the ^{26}Al.

We shall see later that it may be easier for grains to penetrate into the protosolar nebula. However, even if that were the case, the ^{26}Al isochron in the rocks studied implies that if grains were the carriers those grains were destroyed prior to the formation of the rocks under study.

It is interesting that Hutcheon *et al.* (1978) have found the same $^{26}Al/^{27}Al$ ratio as Lee *et al.* in other inclusions in the Allende meteorite. Although this anomaly was first studied in Allende, it has now been seen in other carbonaceous chondrites (Lorin *et al.*, 1978). It has not been seen in the less primitive ordinary chondrites and achondrites (Schramm *et al.*,

1970). Thus the ^{26}Al had decayed away by the time of their final solidification. One other point which Consolmagno and Jokipii (1978) emphasized is that the implied amount of ^{26}Al in carbonaceous chondrites causes significant ionization effects and a high magnetic Reynolds number.

Before going through the nucleosynthetic arguments which tend to point toward a supernova as the nearby source for ^{26}Al, it is important to note that a new set of anomalies has recently been found. To understand how these were found, let us remember that for the most part the ^{26}Al – ^{26}Mg anomaly and the ^{16}O anomaly do not correlate in the same minerals. This is probably due to the fact that the ^{16}O anomaly is only retained in very high temperature condensates whereas the ^{26}Al is present in rocks which were molten and most stable isotope anomalies would therefore have been diluted away. There do exist a few samples which simultaneously show peculiar O and Mg isotopic compositions (c.f., Lee, 1977). These samples are not merely a mixture of pure ^{16}O and normal oxygen but seem to show significant primordial fractionation as well, or else represent a mixture with another component. The magnesium composition associated with these samples show what at first appears like enormous fractionation effects, ~ 10 times that observed in other samples. However, on closer scrutiny these Mg effects cannot be fit with any standard fractionation formula, thus indicating that they may, in part, be nuclear in origin. Two inclusions were found with the above properties. The new result is that in addition to the above, one of these inclusions also showed anomalies in other elements, in particular, Ca, Ba, and Nd. In fact, in this one sample, *every element examined was found to be isotopically anomalous* to a few parts in a thousand (Lee *et al.,* 1978; McCulloch and Wasserburg, 1978).

NUCLEOSYNTHESIS OF ^{26}Al

As mentioned above, the anomaly which provides the most severe time constraint is ^{26}Mg – ^{26}Al. It tells us that only a few million years elapsed between the synthesis of ^{26}Al and the formation of rocks which are in carbonaceous chondrites. There are two ways ^{26}Al can be made (c.f., Schramm, 1971): by spallation of Si and other heavy elements, and by explosive and/or high-temperature hydrostatic carbon burning in (and just prior to) supernova explosions. In what follows we shall see that the spallation is probably not appropriate for the ^{26}Al seen in the carbonaceous chondrites and so one apparently is forced to the conclusion that a supernova occurred within a few million years of solidification.

The reasons for dismissing the spallation hypothesis are two-fold. One is the fact that a proton irradiation required to make the observed ^{26}Al/^{27}Al ratio runs into serious energetic problems. Ryter *et al.* (1971) have shown that any irradiation of the protosolar nebula prior to the formation of condensed objects and the loss of the gas is constrained by ionization loss in the hydrogen gas. If this constraint is applied to ^{26}Al, then it takes ~ 300 ergs for each ^{26}Al (Schramm, 1976). Since the maximum amount of energy available in the early solar system is the binding energy of the sun, the maximum ratio ^{26}Al/^{27}Al that could be produced throughout the solar nebula is $\sim 10^{-6}$ which is significantly less than the 5×10^{-5} observed.

The above constraint can be circumvented by only producing the ^{26}Al in a very small part of the nebula (c.f., Lee, 1977). However, such a local irradiation would produce other anomalies which as we shall see have not as yet been observed. Similarly if the irradiation took place after the hydrogen was dissipated, then again these other anomalies would be produced.

Any proton irradiation of the magnitude necessary to make the observed ^{26}Al/^{27}Al ratio must have $\gtrsim 10^{20}$ protons/cm^2 (Schramm, 1971). Such an irradiation will radically affect other species as well as ^{26}Al. In particular, it will produce the rare nuclei ^{40}K and ^{50}V. Anomalies have not been observed in these species (Burnett *et al.*, 1966; Balsiger, *et al.*, 1969). Similarly, for most cases, such a proton irradiation will also produce neutrons via (p, n) reactions. These neutrons will be thermalized and scour out certain isotopes of Gd which have high ($\sim 10^6$ barns) neutron capture cross sections. Analysis of Gd in various samples shows no such isotopic anomalies (Eugster *et al.*, 1970). Thus again *there seems to be no supporting evidence for any intense irradiation.* It has been shown by Lee (1977) that with enough free-mixing exposure parameters, a carefully selected spectrum and a sufficiently local irradiation, it may be possible to avoid conflict with the irradiation constraints. Such a solution seems to be sufficiently *ad hoc,* with more parameters than observations that one is tempted to state, if one must go to such extremes then, by Ockham's razor, this must not be the solution. Such a conclusion pulls us back firmly to a supernova as the source of the ^{26}Al.

It has been known for several years that ^{26}Al can be produced in explosive carbon burning which presumably takes place when the supernova shock traverses the carbon shell of the exploding star (see Fig. 1). In the calculations for explosive carbon burning of Arnett (1969) and Pardo *et al.* (1974) the ^{26}Al/^{27}Al ratio was $\sim 10^{-3}$ (see also Truran and Cameron, 1977). In calculations using revised reaction rates Woosley (1978, personal communication) has obtained ratios as high as 10^{-2}.

A new result is the realization that ^{26}Al can also be produced in the high-temperature hydrostatic shell burning phase (Wefel and Arnett, 1978), which immediately preceeds the passage of the shock. We must remember that nuclear reaction rates are sensitive functions of temperature. Thus the basic results of explosive carbon burning at $T \sim 2 \times 10^9$ K can be obtained at lower temperatures but with longer burning times. In particular, just prior to the explosion the carbon shell may reach temperatures of $\sim 1.2 \times 10^9$ K for a sufficiently long period that ^{26}Al/^{27}Al is comparable to that obtained in explosive burning. One further point of Wefel and Arnett is that in some cases the supernova shock may *not* heat up the material enough to explosively process it. Thus, in some cases, the only significant processing of the ejected

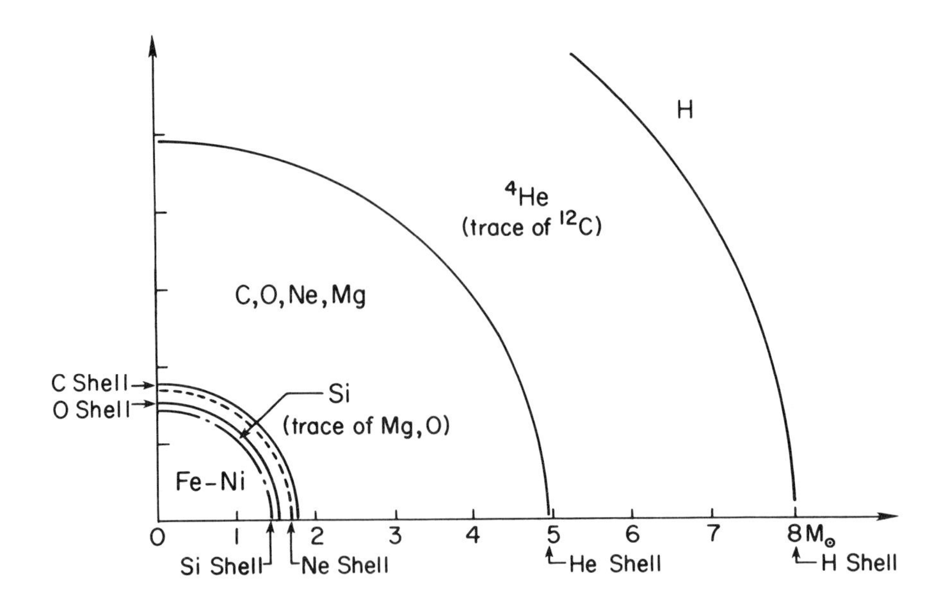

Fig. 1. Cross section of an evolved 22 $M_\odot$ star showing the 1.4 $M_\odot$ Fe-Ni core with the surrounding shells of Si, O, Ne, C, He and the H envelope.

material may be the hydrostatic burning at high temperature. The best models run to date seem to have only the inner part of the carbon shell explosively processed; however, these results are by no means conclusive.

In any case it is clear that ^{26}Al will be present in the ejecta in sufficient abundance to still be around in the observed amounts several million years after the explosion. If the ^{26}Al is from the carbon zone of a supernova, then we should also examine the composition of the other material in this zone. One important constituent is ^{16}O without any associated ^{17}O and ^{18}O. Thus this zone could also be the source of the anomalous oxygen. It should be remembered that one never sees 100% anomalous material, so some dilution with "normal" material must occur before solidification. Based on the maximum size of the ^{16}O anomaly (~ 5%), this means dilution of ~ 1:20, so the ^{26}Al/^{27}Al may also be diluted by this amount. Note also that one would not immediately expect to see a mineral with both ^{26}Al and ^{16}O anomalies since the ^{26}Al anomaly is only able to be seen in material with high Al/Mg whereas the oxygen anomaly, being due to stable isotopes whose effects can be diluted away, is largest in very refractory high-temperature condensates.

One important feature of explosive carbon burning is the production of large amounts of neutrons. These neutrons will alter the isotopic abundances of the heavy elements such as Ca, Ba, Nd, Sm, Gd, etc. This is particularly interesting with regard to the recently discovered anomalies of Lee *et al.* (1978) and McCulloch and Wasserburg (1978).

Detailed studies of these effects are still in progress (Blake, *et al.*, 1978); however, suggestive preliminary work is available (see also Howard *et al.*, 1971). The preliminary results of (Blake, *et al.*, show a depletion of the so called *s*-process isotopes and an enhancement of those previously attributed to the *r*-process. The carbon zone is thus a natural site for the *n*-process (Blake and Schramm, 1976*b*). We should remember that these effects would primarily involve stable isotopes; thus only early high-temperature condensates would retain anomalies which were not diluted away. It is also interesting to note that the calculations of Blake *et al.* do not significantly produce ^{129}I nor ^{244}Pu. Thus they do not effect these well established chronologies.

It should be noted that the carbon zone is toward the outer part of the core of a massive star with only He and H outside of it. It may be that only the material from these outer zones of the star was able to penetrate the protosolar gas cloud. We shall come back to this point later when we discuss the 10^8 yr xenon chronologies.

GRAIN FORMATION IN SUPERNOVA EJECTA

We have seen in the previous section that the interesting zones of a supernova have compositions which can be quite distinct from that of the average interstellar media or the bulk of the protosolar nebula. (Of course it may be that the nebula was not well mixed until the final condensation sequences.) Because of the different compositions as well as the different density-temperature evolution, the grains that form in the supernova ejecta will condense along a different sequence than the Grossman (1972) sequence normally attributed to the early solar system. These differences have been well studied by Lattimer, Schramm and Grossman (1977, 1978) and will not be examined in detail here. The prime differences between the solar system condensation sequence and that of the supernova ejecta are that in the interior zones of the supernova (1) there is no hydrogen, and (2) there can be in some stars C/O ratios which are $\gtrsim 1$ for some zones. The first difference means that for a given density and temperature the partial pressures of the heavy elements are significantly higher. The second effect means that for some zones all the O is locked up in CO and there is excess C as opposed to the normal solar system situation with excess oxygen.

An important result is that the carbon zone can have condensation sequences with spinel ($Mg Al_2 O_4$) coming out before melilite [$Ca_2 (Al_2, Mg Si) SiO_7$]. This is relevant because the solar sequence has melilite first and yet the ^{16}O anomaly is larger in spinel than in melilite. Such an explanation of supernova condensation for the ^{16}O does not provide the whole story because it is obvious that some melting, mixing and processing has taken place in forming the Allende inclusions, but the result is suggestive of a direction for scenarios.

The Lattimer *et al.* calculations assumes that grains are able to condense following an equilibrium distribution in unmixed shells from supernova ejecta. Whether or not grains are actually able to condense in supernovae is a complex question which has been addressed by Falk and Scalo (1975) and in more detail by Falk, Lattimer and Margolis (1977) who have shown that for many plausible supernova models grain nucleation and formation and condensation can indeed take place. Since this will be described in detail elsewhere (Falk, Lattimer and Margolis, 1977), the calculation will not be repeated here.

In addition to providing condensation possibilities, grains in supernova have the added advantage of providing a way of injecting the anomalies into the protosolar cloud (e.g., the "shrapnel theory"). This is true even if the supernova grains themselves are remelted prior to the final solidification of the meteoritic inclusions.

If a supernova did go off within a few million years of the condensation of objects in the solar system, then it seems reasonable to assume that the supernova shocks had significant hydrodynamic effects on the forming solar system. This is particularly obvious when it is remembered that the Kelvin-Helmholz time for the sun is $\sim 10^7$ yr. It also gains support from the observations of Herbst and Assousa (1977; see also their chapter).

It is known from calculations done at Livermore and elsewhere (c.f., Woodward, 1976) that for certain conditions a shock wave hitting an object such as an interstellar gas cloud can cause a compression of the cloud with the shock rapidly propagating around the cloud and compressing it because of the different velocities of sound internal and external to the cloud (see Fig. 2).

One question which arises is how far from the supernova must the cloud be in order for compression to occur. It is interesting to note that several different arguments give approximately the same answer. The arguments are as follows:

1. In order to cause compression, the supernova velocities must be much greater than the typical turbulent velocities in the interstellar medium.
2. If we want to make the 5% oxygen anomaly in this manner we must have the cloud close enough that dilution with the interstellar medium has not smeared out the anomaly to below the 5% level. If the supernova expanded uniformly and mixed continuously with the interstellar medium with densities corresponding to star-forming regions, then the supernova would have to be within ~ 0.1 pc. However, from looking at actual supernova remnants, we know that the expansion and mixing is not uniform but instead is filamentary and clumpy. One clump of supernova matter with 19 parts "normal" matter will still give a 5% anomaly.
3. Supernovae come from massive stars which evolve rapidly and therefore formed very recently. It is reasonable to assume that the protosolar cloud was not accidentally passing by, but was instead a cloud which was part of the star-forming region from which the supernova also descended.

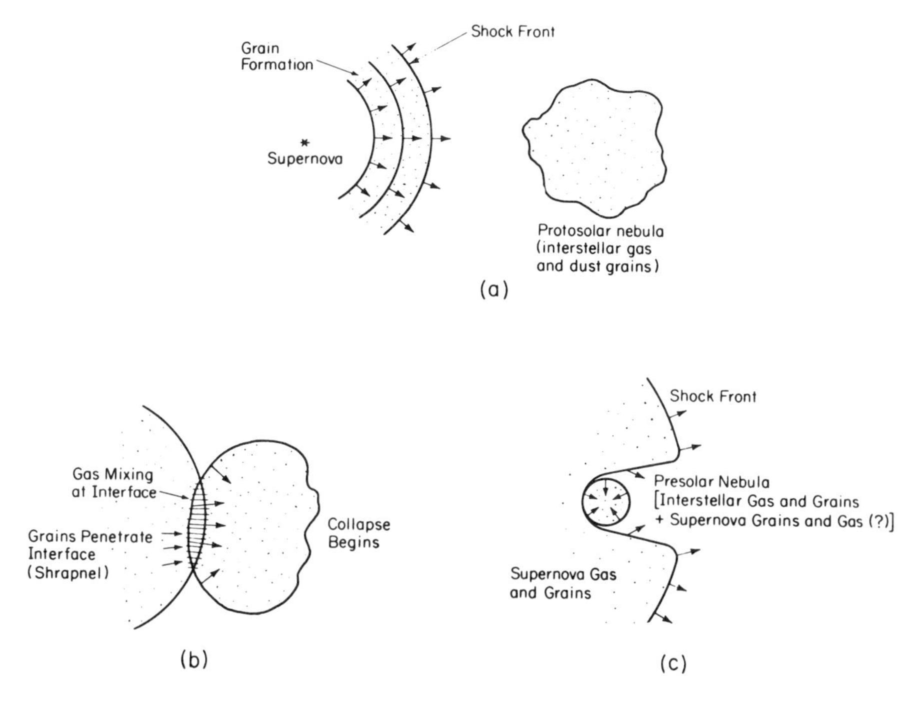

Fig. 2. Scenario for the formation of the solar system. Figure 2*a* shows the four components of presolar system material assumed in this paper: interstellar gas and dust in the protosolar nebula, and supernova gas and dust possibly formed in the expanding supernova ejecta. The ejecta encounter the protosolar nebula and trigger its collapse in Fig. 2*b*. Supernova gas mixes with nebular gas at the collision interface; supernova grains may penetrate into the protosolar cloud. The collapse of the protosolar nebula proceeds in Fig. 2*c*. The bulk of the supernova ejecta, continuing to expand, pass around the protosolar cloud.

Thus, the protosolar cloud may have been one which was slow to collapse, or was debris left over from the formation of other stars in the cluster. In any case, the separation should be not greater than the typical size of star-forming regions.

4. Herbst and Assousa (1977; see also their chapter) see star formation at the edge of the supernova remnant in Canis Major R1. The apparent distance from the location of star formation to where the supernova occurred agrees with the above arguments and is on the order of a few tens of parsecs.

Although none of the above arguments is rigorous, they all point to the same general conclusion.

An important problem in the interaction of the shock and the cloud is how to get the supernova material into the cloud. As mentioned before, grains seem to provide an important key. In the calculations of Woodward

(1976) which showed the shock triggering the collapse of the cloud, the matter behind the shock never penetrated the cloud but instead flowed around it. Margolis (1978) has re-examined this problem putting particular emphasis on the cloud shock interface and the boundary between cloud matter and supernova ejecta. In looking at the interface, Rayleigh-Taylor instabilities at the boundary were allowed. One particularly interesting aspect of Margolis' calculation is the use of a two-component fluid – gas and dust, with correspondingly different properties. Margolis ingeniously used a Lagrangian treatment for the gas and treated the dust in an Eulerian fashion with respect to the Lagrangian gas grid. As a result, he was able to see how the dust could move relative to the gas contact interface. Although the details of the calculation will be explained in more detail elsewhere (Margolis, 1978), the important points for this review are that he found that the dust was able to carry the anomalies and that the dust tended to congregate in the cooler outer regions of the collapsing protostar. This occurred because the shock transferred its momentum more efficiently to the gas of the protosolar gas cloud. The impacting dust embedded itself in the outer regions of the cloud and did not collapse into the central region. This implies that the anomalous inclusions condensed in this outer region. It also implies that the amount of anomalous material may have been small ($\ll 5\%$ of the protosolar cloud), but the effects were amplified because of the concentration in the outer region. In fact, Margolis' calculations show a definite gradient from normal well-mixed gas in the interior to highly anomalous dust at the outer parts of the cloud. (Such a picture makes a NASA mission to analyze Halley's Comet look very exciting!)

It should be noted that the condensation sequences in the high dust/gas mixture with anomalous material might be somewhat variable depending on the relative composition of C, O, etc. at each particular location (c.f., Lattimer *et al.*, 1978). These questions, as well as the question of grain growth, vaporization, etc. are being investigated in detail.

To summarize the hydrodynamic arguments, let us merely state that a supernova shock can cause compression of interstellar clouds out to a few tens of parsecs and that after the shock hits the cloud, fresh supernova material can become associated with the cloud through the existence of grains in the ejecta. This association seems to yield a gradient of composition from well-mixed in the center to very anomalous at the edges.

THE RESOLUTION OF THE 10^8 YR XENON AND 10^6 YR ^{26}Al TIME SCALES

Before going into a complete scenario it is necessary to address the meaning of the difference between the 10^8 and 10^6 yr time scales. Since this question is also addressed by Reeves elsewhere in this book and has been mentioned by Lattimer *et al.* (1977), these comments will be brief.

First let us remember that the xenon progenitors ^{129}I ($\tau_{\frac{1}{2}} \sim 17 \times 10^6$) and ^{244}Pu ($\tau_{\frac{1}{2}} \sim 82 \times 10^6$ yr) which give the 10^8 yr time scale are probably produced in the *r*-process. Let us also remember that the *r*-process makes $^{129}I/^{122}I \sim 1$ and $^{244}Pu/^{238}U \sim 1$, whereas it is observed that at the time of xenon retention $^{129}I/^{122}I \sim 10^{-4}$ and $^{244}Pu/^{238}U \sim 1/60$ (c.f., Hohenberg *et al.,* 1969; Podosek, 1970). Thus an event which makes both I and Pu needs to have $\sim 10^8$ yr before xenon retention. (In all of the following it should be remembered that there is no totally consistent astrophysical site for the *r*-process.

We have already seen that the explosive high-*T* hydrostatic burning event which made ^{26}Al must have taken place within a few million years before solidification. There are obviously several possible ways around the dilemma. They tend to dwell on either of the following cases:

1. The *r*-process production ratios in the last event may be non-standard and the contribution to the *r*-process for the ^{26}Al event may be very small ($\sim 10^{-4}$).
2. Xenon retention may be much later ($\sim 10^8$ yr) than Mg retention due to xenon being a rare gas and Mg being refractory.
3. The last *r*-process event contributing to the eventual solar system material may have taken place 10^8 yr before the last event which contributed ^{26}Al. This could have occurred because the ^{26}Al event did not have any *r*-process associated with it. (Perhaps *the* *r*-process only takes place in the jets resulting from collapsing, magnetic, stellar cores that are rapidly rotating according to LeBlanc and Wilson [1970]; or a neutron star mass-cut, where the *r*-process might take place, did not penetrate because only material from the outer carbon zone penetrated the protosolar cloud and the inner zones.)

The first of these possibilities has been utilized by Truran and Cameron (1977). However, there seem to be difficulties in getting any sort of neutron process to simultaneously produce the necessary abundance ratios (c.f., Blake and Schramm, 1976*a*). In addition, such a solution requires some extremely *ad hoc* selections for the mixing and dilution parameters in order to fit both the observations of I and Pu. (It is curious that the Ba anomaly [McCulloch and Wasserburg, 1978] is also $\sim 10^{-4}$ and so one might try to interpret it along with the I as an argument for Case 1. However, one should remember that $^{244}Pu/U \sim 1/60$, and that Ba chemistry, and therefore its dilution, is probably closer to that of Pu than I.) To most people such a selection is distasteful when it is remembered that $\sim 10^8$ yr may be the typical time associated between successive passages of spiral density waves in the galaxy. Since density waves are thought to stimulate star formation, it is natural to associate the *r*-process production with the passage of one wave, and the ^{26}Al supernova and subsequent solar system formation with the passage of the next. This would be the above Case 3. Case 2 also remains as a viable alternative. The resolution between Cases 2 and 3 can eventually be done with the ^{247}Cm/^{235}U chronology (Blake and Schramm, 1973); ^{247}Cm ($\tau_{\frac{1}{2}} \sim 15 \times 10^6$ yr) is also *r*-process produced and ^{235}U is refractory so it will be retained closer to the time of Mg retention rather than xenon retention. If Case 2 applies, then there should be extremely large ^{235}U anomalies whereas for Case 3 the ^{235}U anomaly would be $\sim 10^{-4}$ similar to the ^{129}I/^{127}I anomaly.

A SCENARIO

In reading the following scenario it should be remembered that all the caveats mentioned above apply and that where several possible options existed, somewhat arbitrary selections were made. Let us begin by noting that the galaxy had existed and been evolving for ~ 7 to 15×10^9 yr prior to the beginning of the scenario. Let us assume for simplicity that the interstellar medium consisted of well-mixed gas and dust with most of the heavy elements trapped in dust grains because they were ejected in that manner from supernova explosions. (It may be unnecessary to assume a well-mixed interstellar medium since mixing and vaporization in the early solar system may be sufficient to explain the bulk homogeneity.) We then have a galactic spiral wave go through a region of gas and dust about every 10^8 yr and stimulate some star formation. (It is intriguing to note that Gerola and Seiden [1978] were able to get spiral structure for the galaxy through supernova-induced star formation alone without need for a spiral density wave. However, for the present scenario we shall not go to such extremes.) When one of these waves passed our gas and dust $\sim 4.7 \times 10^9$ yr ago is produced one or more massive stars ($\gtrsim 8\ M_{\odot}$) which evolved rapidly ($\sim 10^7$ yr) and exploded, ejecting some *r*-process material, perhaps via a LeBlanc and Wilson (1970) jet (see also Meier *et al.,* 1976; Schramm and Barkat, 1972) from a rotating, magnetic collapse.

Thus, after this event the gas and dust which was eventually to become the solar system, had fresh *r*-process material mixed into it. When the next spiral wave hit this material 10^8 yr later ($\sim 4.6 \times 10^9$ yr ago), a new generation of stars formed. The massive ones evolved rapidly and became supernovae. One of the first to blow up would be the star which produced the observed ^{26}Al and stimulated the collapse of the solar system. It was presumably an early one because we will assume that no new *r*-process material was added to the protosolar gas and dust prior to its condensation. The triggering supernova added ^{26}Al along with ^{16}O and the other anomalous refractories. Its carbon zone had some "*r*-process-like" material (Blake *et al.,* 1978). Perhaps even some ^{22}Ne (Ne-E) was injected from the He burning zone. The anomalous material was concentrated in the grains at the outer part of the condensing protosun. Condensation, mixing, vaporization and recrystalization occurred. The ionization caused by the ^{26}Al would presumably effect the hydrodynamics of the mixing and accumulation. Most of the anomalies were mixed away. However, because of radioactive decay, anomalies such as ^{26}Al can still be retained following melting, and the stable ones that are trapped in high-temperature condensates (e.g., ^{16}O) may not be completely melted and mixed away. A few of the earliest condensing inclusions may still show some stable isotope anomalies. The assembly of large objects with ^{26}Al present will result in major melting and processing which will mix away any evidence for ^{26}Al in the later, more processed meteorite classes. Out of this mixture of small primitive bodies and large processed ones the solar system developed, presumably following some sort of disk scenario (c.f., chapter by Cameron).

SUMMARY

We are not able to say that the supernova trigger is absolutely necessary for planetary system formation; all we can say is that from isotopic evidence in meteorites it appears that a supernova did blow up within a few million years of the solidification of objects in the solar system. If such an event occurred, it is reasonable but not necessary to assume that the resultant supernova shock had a causal connection with the formation of the solar system and that a supernova may be one stimulus for formation of low-mass stars.

Acknowledgments. This work has profited by discussions and in some cases collaborations with the entire University of Chicago cosmochemistry and nuclear astrophysics groups, with particular thanks going to L. Grossman and J. Lattimer on the grain condensation, S. Falk on grain nucleation and related subjects, T. Lee on ^{26}Al and the "FUN" anomalies, H. Norgaard, J. Wefel and W. D. Arnett on carbon burning, R. Clayton on oxygen, E. Anders on xenon and S. Margolis on many hydrodynamic questions and on the subject of distances between the cloud and the supernova. In addition, information from former Chicagoans J. B. Blake and G. J. Wasserburg has been quite helpful.

This work was supported in part by NASA and NSF grants at the University of Chicago.

REFERENCES

Alexander, E. C.; Lewis, R. S.; Reynolds, J. H.; Michel, M. D. 1971. *Science* 172: 837.
Balsiger, H.; Geiss, J.; and Lipschutz, M., 1969, *Earth Planet Sci. Lett* 6: 117.
Black, D. C. 1972. *Geochim. Cosmochim. Acta* 37: 377.
Blake, J. B.; Lee, T.; Wefel, J.; and Schramm, D. N. 1978. In preparation.
Blake, J. B., and Schramm, D. N. 1973. *Nature* 243: 138.
____. 1976*a*. *Nature* 263: 707.
____. 1976*b*. *Astrophys. J.* 209: 846.
Burnett, D.; Lippolt, H. J.; and Wasserburg, G. J. 1966. *J. Geophys. Res.* 75: 2753.
Clayton, D.D., 1978. *Moon and Planets* (special Protostars and Planets issue). In press.

Clayton, R. N.; Grossman, L.; and Mayedo, T. K. 1973. *Science* 182: 485.
Consolmagno, G., and Jokipii, R. 1978. *Moon and Planets* (special Protostar and Planets issue). In press.
Eugster, O.; Tera, F.; Burnett, D. S.; and Wasserburg, G. J. 1970. *J. Geophys. Res.* 75: 2753.
Falk, S.; Lattimer, J.; and Margolis, S. 1977. *Nature* 270: 700.
Falk, S., and Scalo, J. 1975. *Astrophys. J.* 202: 690.
Gerola, H., and Seiden, N. 1978. IBM preprint.
Gray, G. M., and Compston, W. 1974. *Nature* 251: 495.
Grossman, L. 1972. *Geochim. Cosmochim. Acta* 36: 597.
Herbst, W., and Assousa, G. 1977. *Astrophys. J.* 217: 473.
Hohenberg, C. M. 1969. *Science* 166: 212.
Hohenberg, C.M.; Podosek, F.; and Reynolds, J. 1967. *Science* 156: 233.
Howard, M.; Arnett, W. D.; and Clayton, D. D. 1971. *Astrophys. J.* 165: 233.
Hutcheon, I.; Steel, I.; Solberg, T.; Clayton, R. N.; and Smith, J. V. 1978. *Meteoritics* 12: 3.
Lattimer, J.; Schramm, D. N.; and Grossman, L. 1977. *Nature* 269: 116.
____. 1978. *Astrophys. J.* In press.
LeBlanc, J., and Wilson, J. 1970. *Astrophys. J.* 161: 541.
Lee, T. 1977. Ph.D. dissertation, University of Texas.
Lee, T., and Papanastassiou, D. A. 1974. *Geophys. Res. Lett.* 1: 225.
Lee, T.; Papanastassiou, D.; and Wasserburg, G. 1976. *Astrophys. J.* 211: L107.
____. 1978. *Astrophys. J.* 220: L21.
Lewis, R.; Srinivason, B.; and Anders, E. 1975. *Science* 190: 1251.
Lorin, J.; Shimizu, N.; Michel-Levy, M. C.; and Allegre, C. J. 1978. *Meteoritics* 12: 299.
Margolis, S.H. 1978. *Moon and Planets* (special Protostars and Planets issue). In press.
McCulloch, M., and Wasserburg, G. J. 1978. *Astrophys. J.* In press.
Meier, D.; Epstein, R.; Arnett, W. D.; and Schramm, D. N. 1976. *Astrophys. J.* 204: 869.
Mueller, M. W., and Arnett, W. D. 1976. *Astrophys. J.* 210: 670.
Pardo, R.; Couch, R.; and Arnett, W. D. 1974. *Astrophys. J.* 191: 711.
Podosek, F. 1970. *Earth Planet. Sci. Lett.* 8: 183.
Ryter, D.; Reeves, H.; Gradsztajn, E.; and Audouze, J. 1970. *Astron. Astrophys.* 8: 389.
Scalo, J. 1978. In preparation.
Schramm, D. N. 1971. *Astrophys. Space Sci.* 13: 249.
____. 1974. *Ann. Rev. Astron. Astrophys.* 12: 383.
____. 1976. *Proc. Am. Geophys. U. Mtg.* Washington.
Schramm, D. N., and Barkat, Z. 1972. *Astrophys. J.* 173: 195.
Schramm, D. N.; Tera, F.; and Wasserburg, G. J. 1970. *Earth Planet. Sci. Lett.* 10: 44.
Truran, J., and Cameron, A. G. W. 1977. Center for Astrophysics preprint.
Wasserburg, G. J.; Schramm, D. N.; and Huneke, J. C. 1969. *Astrophys. J.* 157: L91.
Wefel, J., and Arnett, W. D. 1978. In preparation.
Woodward, P. 1976. *Astrophys. J.* 207: 484.

THE "BING BANG" THEORY OF THE ORIGIN OF THE SOLAR SYSTEM

HUBERT REEVES
Centre d'Etudes Nucléaires de Saclay

A typical protosolar cloud and the early solar system are affected by ten or more supernova remnants, of stars ranging from early O to B1 during the $\sim 10^7$*yr duration of an OB association. Coupled with the fact that massive supernovae yields are foreseen to be particularly rich in* ^{16}O *and Mg, contaminations by these supernovae remnants could be responsible for the oxygen and magnesium anomalies in various classes of meteorites.*

The high proximity of new born stars in the earliest subgroups of OB associations suggests that some exchange of matter between neighboring solar systems could have taken place. Combined with the first suggestion, this may account for some of the highly localized isotopic anomalies found in meteorites, particularly in the inclusions of Allende.

The prompt contamination of newly forming stars within an association might explain the paucity of low-metal stars. The various spurs and loops might well be the cumulative effects of the sequence of O star explosions in an association. Various observed phenomena in H II regions and nearby clouds seem to require the presence of fast moving matter which might find here a plausible origin.

In recent years, we have realized that stellar deaths and stellar births are intimately connected. The importance of this fact for our ideas on the formation of the solar system is only now emerging. The numerous theories of the origin of the solar system throughout the last hundred years or so have quite generally taken it for granted that the process of formation of the sun and the planets could be studied as a closed system having at best a weak electromagnetic link with the general galactic magnetic field. The so-called protosolar nebula was assumed to be composed of atoms formed long ago by multiple stellar deaths during the general nucleosynthetic activity of our galaxy. Well mixed by turbulent motions, the atoms of the nebula had later undergone a sequence of gravitational, thermodynamical and physico-chemical processes which had progressively transformed the nebula into the collection of planets, satellites, asteroids and comets that we observe today.

The discovery of important isotopic differences in the rare gases between various locations of the solar system came as the first hint that something could be wrong with the hypothesis of a well-stirred and well-mixed nebula. However, it could be argued that some isotopic separation could have been generated during the process of heavy chemical separation which these elements have undergone in stones, in atmospheres, etc. The isotopic anomalies were perhaps not reflecting genuine initial abundance differences.

The observations of isotopic differences in many other chemical elements, far less fractionated, such as oxygen, magnesium, calcium, etc., have made it clear that we are indeed dealing with real abundance effects. Next it was found that the magnesium isotopic anomaly had to be assigned to *in situ* decay of the isotope ^{26}Al, with a lifetime of 7.4×10^5yr; the implication is that at most a few million years had elapsed between the formation of the ^{26}Al (most probably in an evolved star; see appendix to this chapter for a detailed discussion) and its incorporation in the lattice crystal where we now find it. But, as Cameron and Truran (1977) formulated it, "Is it not highly unlikely that a supernova should have gone off close to the region of formation of the solar system within a few million years of the time the event occurred?" To erase the *ad hoc* character of the supernova hypothesis, these authors, following an idea of Öpik (1953), discuss the possibility that this supernova may be responsible for the trigger of the collapse of the protosolar nebula.

I will argue here that the supernova hypothesis is more than likely, and we do not need the trigger hypothesis to justify it; *the sun was most likely born amidst a fireworks of supernovae.* To defend this point I will first bring together the following propositions:

1. Star formation takes place in an extremely small fraction of the galactic volume.
2. The biggest stars have very short lives (a few million years) and explode before escaping their formation region.
3. Their supernova remnants expand far enough to contaminate the whole region of star formation.

The net result is that protosolar nebulae and early "solar" systems within these associations will be affected in many ways by these nearby violent events: thermodynamic perturbations and acquisition of newly made matter. Using recent calculations of the nucleosynthetic yields of O stars I shall discuss the various anomalies observed so far. At the end I will present a scenario describing how these anomalies could have come about.

STELLAR BIRTH OCCURS IN A VERY SMALL FRACTION OF THE GALACTIC VOLUME

A large fraction of stars with ages less than 10^7yr is found in OB associations. Blaauw (1964) has presented a detailed analysis of these associations in the solar neighborhood; the present chapter has been largely motivated by Blaauw's excellent review. He studied the ten or so associations which are within one kpc of the sun. Their dimensions vary from a few to a few hundred parsecs, their masses (in stars) from 300 to $10^4 M_{\odot}$; on various grounds, (kinetics or main-sequence departure) their lifetimes are estimated to be 10 to 15 $\times$ 10^6yr. The stellar mass of all the OB associations ($\sim$4 $\times$ $10^4 M_{\odot}$) within 1 kpc is about 10^{-3} of the total mass of gas in the same region (the total volume occupied is about 3 $\times$ 10^{-4} of the corresponding volume). With the lifetime quoted above, the lifetime for transformation of all the ages of OB associations into stars is 12 Myr/10^{-3} $\simeq$ 10^{10}yr (Blaauw, 1964), quite a reasonable number when compared with the age of our galaxy, taking into account the fact that the present fractional gas mass is about 0.1 in the solar neighborhood. This confirms the idea that OB associations are a major location of star formation. *One basic assumption of the present approach is that the sun was actually born in an OB association such as Orion.*

Studies of dense molecular clouds by far infrared and radio astronomy (Mezger and Smith, 1977; Elmegreen and Lada, 1977; Strom, Strom and Grasdalen, 1975; Zuckerman and Palmer, 1974; Zuckerman, 1975) show that within a given association stars with ages less than $\sim 10^6$yr are found near dense molecular clouds, H II regions, maser sources, etc., in a volume of a few pc^3, i.e., in a fraction of 3 $\times$ 10^{-4} of the whole association. Thus we have evidence that at any one time *a very small fraction* ($\simeq 10^{-7}$) of the galactic volume is producing *most* of the stars.

MORPHOLOGY OF A MEDIUM-SIZE ASSOCIATION (ORION)

I will describe here in some details the structure of the Orion association, a particularly well-studied example because of its proximity ($\sim$500 pc). However, the features reported here are not at all unique but are present in a number of other associations (Centaurus-Scorpius, W3, CMa R, Ophiuchus). The Orion association can be spatially subdivided into four subgroups of stars (Blaauw, 1964) which appear to have different ages (Table I and Fig. 1): Group *a*, the oldest group (Northwest) is also the largest. As we move from northwest to southeast, we find increasingly concentrated and younger subgroups, until we come to Group *d*, the Trapezium cluster, which we shall describe in more detail later. The general picture that emerges is that star formation behaves as an "infection" or a "forest fire". It starts in a given

TABLE I

Description of the Various Subgroups in the Orion Association[a]

Subgroup		Largest Projected Dimension (pc)	Ages (10^6 yr)
a	(Northwest)	50	12
b	(Belt)	25	8
c	(Outer Sword)	15	6
d	(Trapezium)	2	$\simeq 1$

[a]Table from Blaauw (1964).

region and moves progressively to adjacent areas (Elmegreen and Lada, 1977) which, in turn, are set afire.

In Orion, the wind blows from northwest to southeast. Abt and Levato (1977) and Levato and Abt (1976) have recently completed a new analysis of the four subgroups in Orion. In Table II, I have given the number of stars with spectral type earlier than a certain value, found in their survey. They also quote a recent redetermination of the ages and the four subgroups by Warren and Hesser (1978, personal communication), given in the last line in

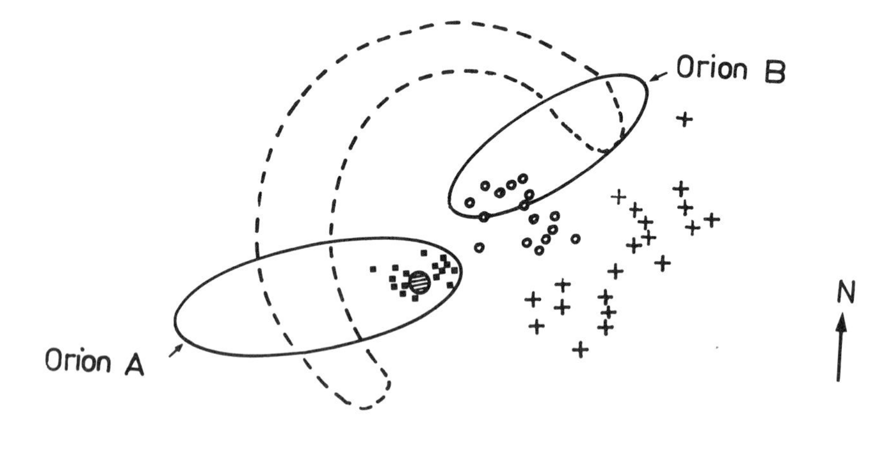

Fig. 1. The Orion OB association. Stars of Subgroup *a* (Northwest) are indicated by (+), Subgroup *b* (Belt) by (o) and Subgroup *c* (Outer Sword) by (■). The Orion Nebula, the formaldehyde cloud and the infrared cluster are all within the hatched circle, Subgroup *d*, enlarged in Fig. 2. Orion A and B are two large CO clouds. The dotted semi-ring is Barnard's Loop. The scale is shown at the bottom.

TABLE II

Stellar Statistics of the Orion Subgroups[a]

Subgroup	*a*	*b*	*c*	*d*
O6	0	0	0	1
O9	0	1	1	1
B0	0	4	2	2
B1	4	6	8	6
B2	13	12	14	8
Age (10^6 yr)	7.9	5.1	3.7	<0.5

[a]Data are from Levato and Abt (1976), Abt and Levato (1977) and Warren and Hesser (unpublished).

the table. Note how the subgroups are gradually depleted in early-type stars.

The motions of stars within the subgroups of various associations have been studied and appear to be coherent with a linear expansion from a point of convergence. The picture would then be that after birth in a very small area (less than a few pc^3), each subgroup would expand progressively with radial velocities of a few km sec^{-1}. Since this is more than the escape velocity of the group, the subgroup would, after $\sim 10^7$ yr, find itself dispersed by the general gravitational field of the galaxy.

The younger region, Trapezium, appears to be at the border of a large CO cloud of $\sim 10^5 M_\odot$ with dimensions of 30 pc which extends in the southeast of the stellar subgroups (Kutner *et al.*, 1977). This cloud contains two subunits of denser matter with $\sim 10^3 M_\odot$ which appear to represent two very early phases of star formation (Fig. 2). First, there is the optical H II region, the Orion nebula which is ionized by the brightest (O and B) members of the Trapezium Cluster. The Trapezium Cluster is a very young stellar group, less than 10^6 yr, containing several hundred stars, most of them confined within 10^5 AU (Blaauw, 1964). The mean distance between the stars is on the order of 10^4 AU. (For comparison, long period comets of our solar system have orbits extending to 5×10^4 AU.) Second, there is the dense molecular cloud, seen in formaldehyde, containing a number of infrared sources and masers of various types. According to Zuckerman and Palmer (1974) and many other authors, this cloud would represent an even earlier stage of stellar formation ($\leqslant 10^5$ yr), stars still deeply embedded in their "placental" cloud. The distance between the Trapezium and the formaldehyde cloud is estimated to be less than one parsec (Zuckerman and Palmer, 1974), and inside this cloud the distance between the various infrared sources and maser sources can be counted in thousands of AU. This may give us an idea of the degree of concentration of the youngest stellar objects that we know.

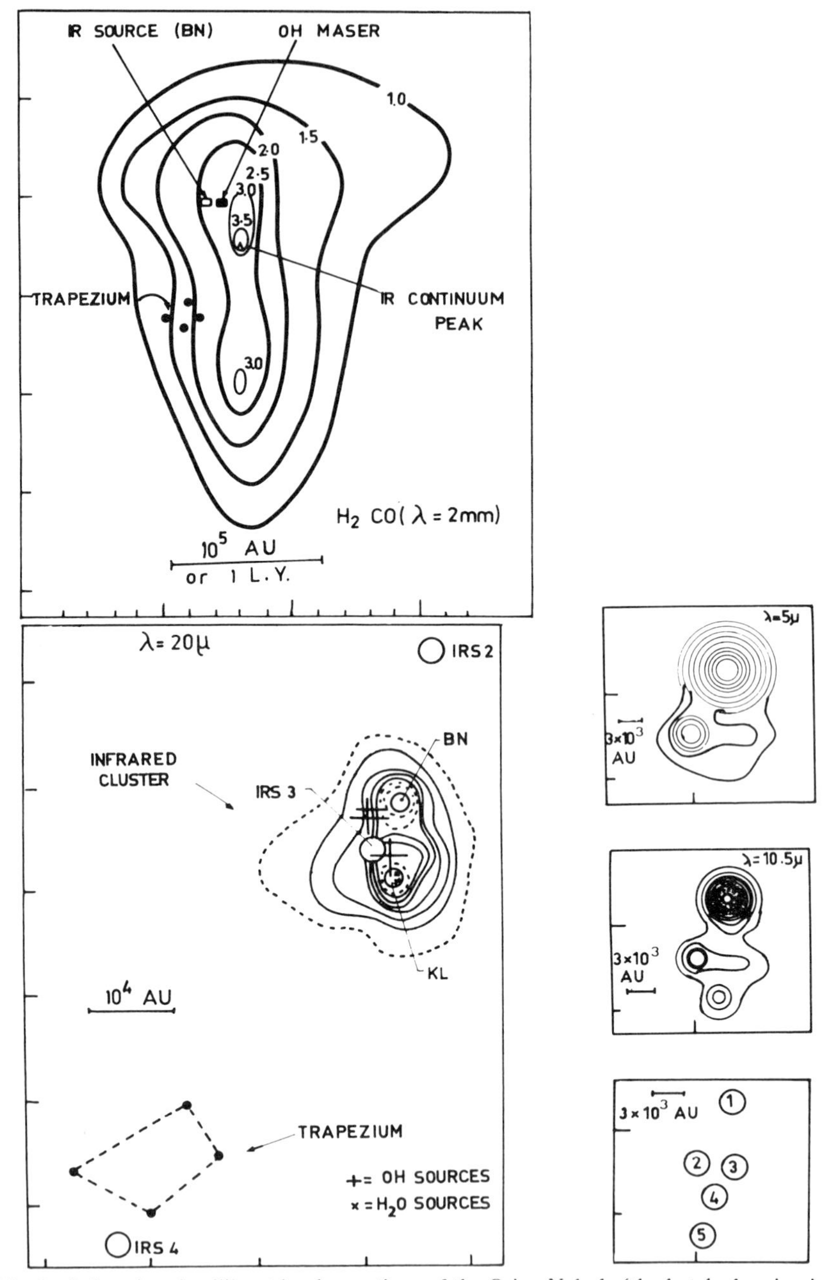

Fig. 2. Infrared and millimetric observations of the Orion Nebula (the hatched region in Fig. 1). The four brightest members of the Trapezium cluster, the stars responsible for the ionization of the nebula, are seen near a dense cloud, the formaldehyde cloud which contains young stellar objects. The upper diagram represents the profile of the formaldehyde cloud. Below, left is shown an enlarged view of the infrared cluster (BN) which is near the center of the formaldehyde cloud. Below, right, details can be seen of the infrared cluster; individual sources are resolved which are separated by a few thousand astronomical units.

HOW MANY STARS EXPLODE WITHIN AN OB ASSOCIATION?

Stellar evolution theory tells us that the $\sim 10^7$yr lifetime of an OB association is larger than the main-sequence lifetimes of stars earlier than B1 with $M = 16\ M_\odot$. (The pre-main-sequence and post-main-sequence lifetimes are expected to be even shorter.) A statistical survey showed that there are about 8 stars earlier than B1 ($M_V < -3$) per association (Blaauw, 1964). Furthermore, there are usually a few runaway stars, stars with high velocities, associated with each association. It is presently believed that runaway stars were former companions to larger stars which underwent supernovae explosions and thus freed themselves in a "sling-shot" fashion (Blaauw, 1964). Thus we would be witnessing the presence of already exploded O stars.

In this chapter I have used the very complete survey of O stars made by Cruz-Gonzales *et al.* (1974) to compute the number of large stars that should appear within a given association. I have used the estimates of main-sequence lifetimes quoted by Ostriker, Richstone and Thuan (1974; see Fig. 3) and the spectral type-mass relation given in their paper. (An important uncertainty is attached to the conversion of mass into main-sequence spectral type because of the unknown effects of mass loss.) Cruz-Gonzales *et al.* give the number of O stars per kpc^3 and also this number integrated over the galactic plane i.e., per kpc^2. In Table III I show the number of stars of mass larger than M_1 presently found per kpc^2 and the total number of these stars formed during 12 Myr i.e., the birth function multiplied by the lifetimes of an association. Up to 20 solar masses, the exponent of this initial mass function (IMF) is quite similar to Saltpeter's exponent (1.35). At higher masses the number of stars decreases much faster, although we cannot exclude the possibility that some of the largest stars never emerge from their opaque placental clouds; this would increase the O star contamination of associations. To obtain finally the number of stars within an association I adopt that there are about three associations per kpc^2. These numbers apply to ordinary H II regions. In giant H II regions they should be raised by an order of magnitude (Mezger and Smith, 1977).

One might claim that since about one half of the O stars are found outside of the associations, these numbers should be further divided by two. There are two objections:

1. The fraction of O stars found within an association varies with spectral class from a value close to one for the earliest type to less than one half at early B. This suggests that part of these later type stars simply survived the association in which they were born; this actually is one way by which the lifetimes of OB associations are obtained. (It is noted that Bash *et al.* (1977) find that CO emission is confined to star clusters with stars earlier than BO.)
2. A fair fraction of those stars which are outside associations is likely to

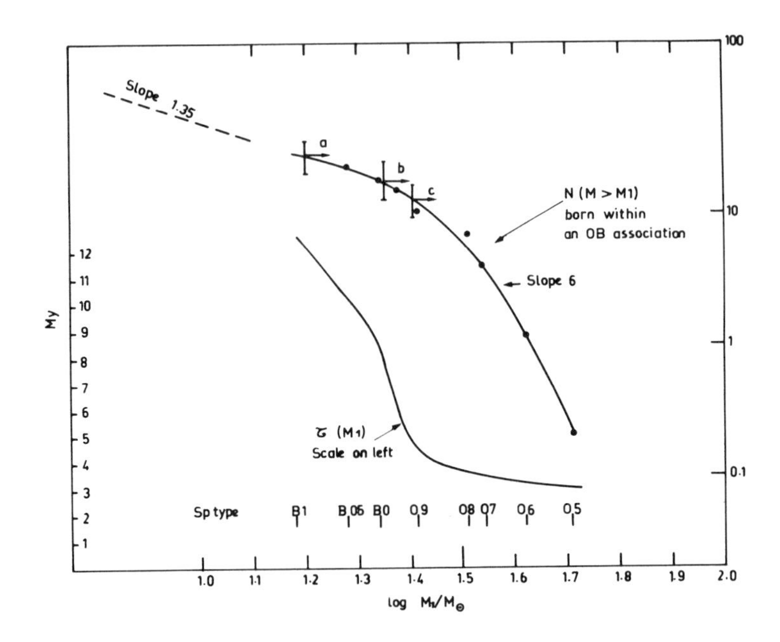

Fig. 3. Lifetimes and statistics of massive stars. The lifetimes, gathered by Ostriker *et al.* (1974), are plotted as a function of stellar mass. The horizontal lines with *a*, *b* and *c* describe the fact that in Orion, for instance, stars of the corresponding type are "ripe" to explode in subgroups, *a*, *b* and *c*.

have been ejected from inside. Cruz-Gonzales *et al.* (1974) find the fractional number of runaway stars to be as large as 30% of all O stars. Thus, the number of O stars born outside of associations appears to be very small.

According to stellar evolution theory, these massive stars are expected to end their lives in a catastrophic explosion associated with supernovae events. In this explosion, a large fraction of their masses is thrown away at velocities of 10^3 to 10^4 km sec^{-1}. These ejecta, which are expected to contain the nucleosynthetic products of the internal cooking of the star, are gradually decelerated by interaction with local interstellar matter to typical values of $\sim$10 km sec^{-1}. Calculations, based on Sedov's (1954) similarity methods, show that from 10^4 to 10^5 $M_\odot$ of interstellar matter is thus mixed with the ejecta.[a]

---

[a]Reeves (1972*b*) has discussed the existence of two time scales for mixing of the interstellar gas, one associated with the dilution of supernovae remnants ($\sim 10^6$yr) which could bring chemical inhomogeneities up to a few percent, and another associated with differential motions in the galaxy ($\sim 10^8$yr) which would wash out these inhomogeneities and justify the use of cosmochronologies.

The remnants are expected to extend several parsecs in ($n_H = 10^3$) or several tens of parsecs in the intercloud medium ($n_H = 1$) in 10^5 yr. Observed supernova remnants reach dimensions comparable to the whole OB association (Vela X has a diameter of 30 pc and the Cygnus Loop has 50 pc). For a detailed survey see the paper by Ilovaisky and Lequeux (1972).

In an OB association such as Orion we would expect the remnant to expand all the way to the dense molecular cloud and Trapezium region, where new-born stars are likely to be forming their solar systems from protosolar nebulae. We would expect this event to happen ten to twenty times during the life of the Orion association.

If so many supernovae have exploded around the stellar OB associations, why do we not see a number of supernova remnants, visible or radio, around these associations? According to the survey of Ilovaisky and Lequeux (1972), supernova remnants are not easily detected when they reach a diameter of 30 pc; the contrast with background vanishes as their surface brightness decreases. According to these authors, at least half of the remnants with more than 50 pc are missing. The time required to reach these dimensions varies with the energy input and the density of the medium but should be a few 10^5 yr. If an OB association records twenty supernovae in its 10^7 yr duration, the mean period between individual ones should be $\sim 5 \times 10^5$ yr. Thus we should not systematically expect to see one or more remnants around each OB association although some have already been seen.

Nevertheless, we might ask whether the cumulative effects of those numerous supernovae might not generate some observable features especially if we take into account the fact that we are dealing here with massive stars with presumably larger than average energy output (the nuclear energy content of exploding material in a 20 $M_\odot$ star is $\sim 10^{52}$ ergs). The whole Orion association is surrounded by a bright ring (Sivan, 1974) (Fig. 1), Barnard's Loop which several authors have considered as a possible supernova remnant (Herbst and Assousa, 1977; S. Peimbert, personal communication). A ring in neutral hydrogen gas (21-cm emission) around this association is reported and discussed in terms of supernova explosions by Heiles (1978). Its diameter is 100 pc and it has swept $\sim 7 \times 10^4$ $M_\odot$ of interstellar matter. In the same fashion the Scorpio Centaurus association is surrounded by a vast radio ring called Loop I. In general, there is quite a good correlation between the various radio spurs and loops and the location of the associations (Cassé and Reeves, 1978). Several authors have tried to explain the loops in terms of an individual supernova explosion. There is usually a problem of the lack of appropriate energy. This problem would, of course, be largely helped by appealing to several supernovae with larger than average energy content, exploding within the volume of an OB association. Ögelman and Maran (1976) present similar views.

These powerful supernovae may also be responsible for the observations of various high velocities and high-energy features around H II regions.

Heckman and Sullivan (1976) mention the possibility that the movements could be generated by supernovae (see also Baudry *et al.*, 1974). Traub *et al.* (1974) and Zuckerman (1975) mention the presence of a mass of $\sim 5\ M_{\odot}$ of fast moving gas (-60 to -100 km sec^{-1}) close to the Trapezium. Here there might be an explanation in terms of decelerated remnants.

DO SUPERNOVAE TRIGGER STAR FORMATION?

From Table I it appears that the time lag between the formation of the subgroups is approximately 3×10^6 yr, which is the lifetime of an O5 to O9 star. This may suggest that the "infection" of star formation is communicated by the explosion of these stars, an idea which has been rather popular recently (Cameron and Truran, 1977; Sgro, 1975; McKee and Cowie, 1975; Woodward, 1976; Margolis, personal communication, 1978; see Schramm's chapter). However, according to the analysis presented here the triggering process is not *required* by the observation of early stellar objects in supernovae remnants (see the chapter by Herbst and Assousa in this book) or by the detection of ^{26}Al in the early solar system (Lee *et al.*, 1976, 1978). Many supernovae remnants are expected to sweep across a typical association during the 10^7 yr of its life, even if none of them triggered anything.

ISOTOPIC ANOMALIES IN THE SOLAR SYSTEM

The various isotopic anomalies[a] that have so far been detected in our solar system can be classified according to their magnitude as follows.

Class A contains anomalies which are larger than one. We have neon – solar wind ^{20}Ne/^{22}Ne ~ 13.7 (Geiss, 1972), while in meteorites ratios of ~ 1.6 and 1.3 have been observed (Eberhardt, 1975; Niederer and Eberhardt, 1977); and helium – primordial ^{3}He/^{4}He varying from 4×10^{-4} to 2×10^{-4} in meteorites (Black, 1972; Frick *et al.*, 1978; Anders *et al.*, 1970).

Class B contains anomalies typically larger than 10% but usually smaller than one. We have here argon and xenon (Kuroda and Manuel, 1970; Frick *et al.*, 1978; Gros and Anders, 1977). Krypton has anomalies slightly less than ten percent.

Class C has anomalies typically larger than 1% but smaller than 10%. We have here the oxygen anomaly (R.N. Clayton, Onuma and Mayeda, 1976) which is best explained by assuming the existence of a pure ^{16}O component mixed with a "normal" oxygen component in proportions extending up to 5 or 6%. Then we have the anomalous ^{26}Mg (extending to 10%) which because of excellent correlation with Al can be associated with the decay of ^{26}Al (7.4×10^5 yr) (Lorin *et al.*, 1978; Gray and Compston, 1974; Lee *et al.*, 1976; Hutcheon *et al.*, 1977; Wasserburg *et al.*, 1977).

---

[a]Anomalies are defined in the following way: let R_n be the normal ratio to two nuclides and R_a the anomalous ratio, then $\delta = |R_n - R_a|/R_n$ is the magnitude of the anomaly.

Class D has anomalies between 10^{-3} and 10^{-2}. Here we have the calcium anomaly (Lee, Papanastassiou and Wasserburg, 1978) which apparently corresponds to variations of ^{42}Ca and ^{48}Ca.

Class E has anomalies between 10^{-4} to 10^{-3}. It contains barium and neodymium (McCulloch *et al.*, 1977), best ascribed to ^{135}Ba, ^{137}Ba and several Nd isotopes. To this class we are tempted to add strontium which shows variations up to 10^{-3} in the initial ^{87}Sr/^{86}Sr.

The first two classes contain *all* and *only* the rare gases, in order of increasing mass. In view of the fact that these elements are by far the most chemically fractionated in meteorites and that in various astrophysical locations they often show erratic behavior, which is not due to nucleosynthesis but to other non-nuclear effects (e.g., ^{3}He/^{4}He in the solar wind; ^{40}A/^{36}A at the surface of the moon), we are tempted to say that part of their anomalous behavior in meteorites may have no relation with initial solar system abundances. This view is also supported by other workers (Geiss, 1972; Wasserburg, personal communication, 1978). For the moment, we ignore classes *A* and *B* and begin our discussion with Class *C*.

WHAT COMES OUT OF A SUPERNOVA?

Arnett and Schramm (1973) and Arnett (1978) have tried to evaluate the mass yield of various elements from supernovae of various masses. Their results are given in Fig. 4 which shows the amounts of elements as a function of M_{core} (the mass of the stellar helium core). The top curve of the figure is the total mass of the star which they associate with the corresponding value of M_{core} of the appropriate spectral class. They point out that this last assignment is very uncertain, and because of the large uncertainties[a] in the nucleosynthesis calculations we should consider mostly the qualitative trend of the results.

We define g_i^M as the mass fractional amount of the nuclide, *i* in the ejecta from a star of mass *M* (Fig. 5), and $g_i^\odot$ is the same quantity in solar type matter. In Fig. 6 we plot $\epsilon_i^M = (Mg_i^M / M_{mix} g_i^\odot)$ the relative enrichment in the element, *i* brought about by the explosion of a star of mass *M* in a mass M_{mix} of 10^4 $M_\odot$ (ordinate on the left), or 10^5 $M_\odot$ (ordinate on the right) of mixed interstellar matter, neglecting here the mass of the stellar residue.

In the range B1 to O5 the strongest enrichment is in neon, followed by magnesium and by oxygen, in the form of ^{16}O. Carbon, silicon and iron are quite comparable, while helium is much less. O and Mg are the two elements

---

[a]According to stellar statistics, out of three stars two form a pair and one remains a single. Furthermore, for the case of O stars with high frequency of runaways we may often be dealing with massive close binaries. On the other hand, stellar evolution calculations have not yet reached the point of being able to compute the nucleosynthetic yields of binaries, which because of meridonal circulation may be quite different from the yields of isolated stars. I thank M. Arnold for this remark.

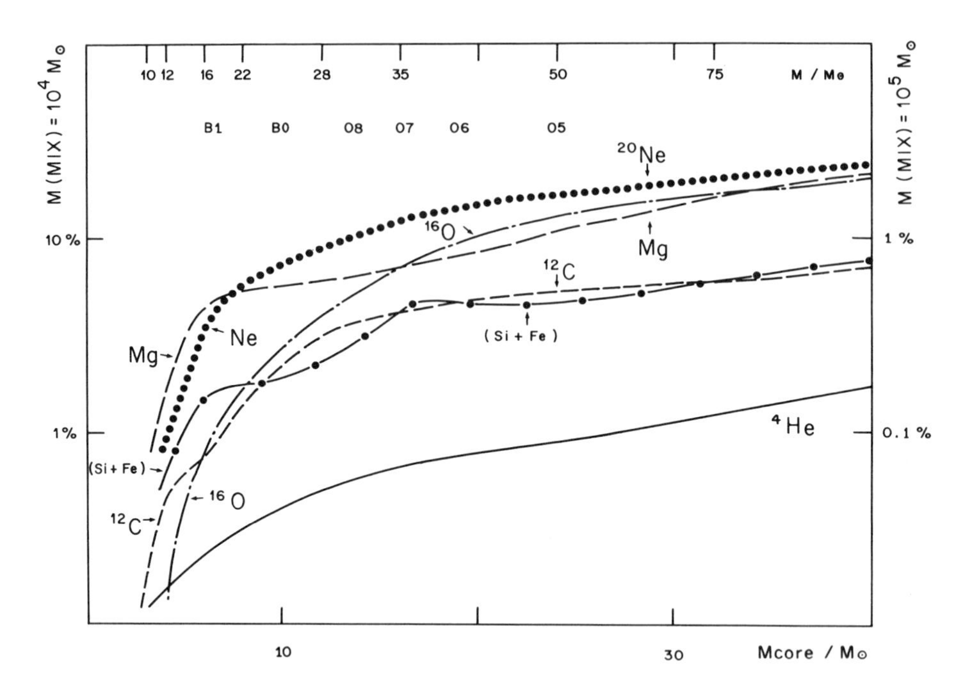

Fig. 6. The same as for Fig. 4 but showing the fractional mass contamination obtained by mixing the nuclear cooking of these stars with $10^4 M_\odot$ (ordinate on the left) or $10^5 M_\odot$ (ordinate on the right) of interstellar gas with solar composition. Note that, excepting rare gases, the largest contamination is in O, which appears here in the form of pure ^{16}O, and Mg, a fraction of what will appear in the form of $^{26}\text{Al} \rightarrow {}^{26}\text{Mg}$. (From Arnett and Schramm, 1973 and Arnett, 1978.)

appear at about one third of the O anomalies, i.e., around one percent. One candidate for observation is the $^{12}\text{C}/^{13}\text{C}$ ratio, since we do not believe ^{13}C to be generated in supernovae, but the fact that we have only two isotopes will make it difficult to eliminate mass fractionation effects. The Ca anomalies could result from the general process giving rise to Si + Fe as seen in Fig. 4, and is about the right order of magnitude. Another promising candidate would be ^{53}Mn ($\tau = 2 \times 10^6$ yr) which decays in ^{53}Cr. Thus, anomalies in the three Cr isotopes (^{52}Cr, ^{53}Cr, ^{54}Cr) should be expected at the ~1% level.

O STARS APPARENTLY GENERATE VERY LITTLE r-PROCESS ELEMENTS

Since very little is known about the astrophysical site for *r*-process formation we shall now take a different approach. We shall try to guess, from the anomalies themselves, what the B1 to O stars do generate, by assuming that the mixing arguments presented here provide a valid explanation for the observed isotopic anomalies.

Four points indicate that the "recent" additions of *r*-process elements in

TABLE III

Stellar Statistics of O-B1 Stars: Luminosity Function and Initial Mass Function in the Solar Neighborhood[a]

Spectral Range	Number of Stars (kpc^{-3})	Number of Stars Integrated over Galactic Plane (kpc^{-2})	Stellar Mass Range (M_2-M_1)	Stellar Lifetime for Lower Mass Range (M_1) τ_1 (10^6 yr)	Number of Stars with $>M_1$ (kpc^{-2})	Number of Stars with $>M_1$ per 12 Myr (kpc^{-2})	Number of Stars with $>M_1$ per Association
O3 -O4	1.3	0.2	∞-51	3.0	0.2	0.8	0.3
O5 -O5.5	4.3	0.68	51-42	3.1	0.9	3.4	1.2
O6 -O6.5	13	2.0	42-35	3.2	2.9	11	4
O7 -O7.5	20	3.0	35-32	3.6	5.9	21	7
O8 -O8.5	22	3.3	32-26	3.8	9.1	31	10
O9	31	4.7	26-24	4.2	13.9	45	15
O9.5-O9.7	36	5.5	24-22	6.4	19.4	55	18
B0		8.3[b]	22-19	8.8	28	66	22
B0.5		13.3[b]	19-16	10	41	82	28
B1		24[b]	16-12	12	65	106	36

[a]Stellar statistics from Cruz-Gonzales *et al.* (1974).
[b]The last three lines are estimates by combining Table 6 of Gonzales *et al.* (1974) with Table 6 of Torres Peimbert *et al.* (1974).

low iodine ratio observed. However, as pointed out by Clayton (1977) they would then run into serious difficulties with the $^{135}Cs - ^{135}Ba$ pair.

4. Some heavy element anomalies have been reported in ^{135}Ba and Nd at a level of 10^{-4} to 10^{-3} (McCulloch and Wasserburg, 1977). It is not yet clear which isotopes are responsible (*p, s* or *r* processes). We can nevertheless say that some recent addition of heavy isotopes has taken place with abundance ratios *different* from the solar ratios.

These anomalies, in the scenario presented here, offer new solutions to a number of old problems. The initial values of the ratio of $^{87}Sr/^{86}Sr$ obtained from internal isochrons show a spread of 10^{-3} (from 0.6985 to 0.6999) according to Gray, Papanastassiou and Wasserburg (1973); it has been usually interpreted as a time spread of $\sim 10^8$ yr between solidification of various minerals. The $^{129}I/^{127}I$ initial ratios show variations from 10^{-5} to 2×10^{-4} (Hohenberg *et al.*, 1967; Podosek and Lewis, 1972; Jordan *et al.*, 1977), implying a spread of 6×10^7 yr in their formation. It now appears that these variations are at the level of the heavy elements recent addition, as witnessed by Ba and Nd, and that their interpretation in terms of chronology alone may no longer be advisable. Clayton (1977) has reached the same conclusion from a model based on fractionation of dust from gas.

Variations at the same level should logically be expected in the $^{235}U/^{238}U$ ratio (see also Cameron, 1973). Chemical fractionation effects (Pellas and Storzer, 1975; Crozaz, 1974) would probably make it impossible to find such small variations associated with ^{244}Pu or ^{232}Th. However, the fossil products of all these radioactive isotopes including also ^{248}Cu may appear in the lead isotopic ratios. Difficulties in the search for the non-radiogenic isotopic ratio of lead (Tatsumoto, 1976) may find some explanation there.

One tentative conclusion of this study is that the O stars have a different nucleosynthesis yield than the mean of all processes responsible for the abundance of elements such as supernovae of all masses, novae, planetary nebulae, mass ejection of red giants, etc. They generate rather more of the so-called explosive carbon-burning elements and less of the *r*-process elements. (From nucleosynthetic calculations Hillebrand *et al.*, [1976] arrive at the conclusion that only one out of $\sim 10^4$ supernovae generates *r*-process elements.) Quantitatively we may say that the ratio of carbon-burning yield to *r*-process yield in early stars compared to the same ratio for the summation of all sources is $\geqslant 10^2$.

A corollary to this statement is that nature has been kind enough not to confuse the scheme of galactic cosmochronology (Fowler, 1972; Schramm, 1974; Reeves and Johns, 1976; Quirk and Tinsley, 1973; Talbot and Arnett, 1973; Truran and Cameron, 1971). In fact, one can write for the ratio of two elements a formula which takes into account both the galactic heritage (in a simple model of exponentially decreasing stellar activity) and a sum of recent additions

$$\frac{N_j}{N_i} = \frac{P_j \tau_j}{P_i \tau_i} \frac{(1 + \bar{\epsilon}_j T/\tau_j)}{(1 + \bar{\epsilon}_i T/\tau_i)} \qquad (2)$$

$$\bar{\epsilon}_i = \Sigma \frac{M g_i^M}{M_{\text{mix}}\, g_i^{\odot}}$$

where the sum is taken over all the stellar ejecta from stars of mass M with mass fraction g_i^M which have contaminated the protosolar nebula. M_{mix} is the mixing mass and T the period between the birth of the galaxy and the birth of the sun ($T \sim 10^{10}$ yr). From our previous discussion we expect $\bar{\epsilon}_i$ for r-process nuclides to be at most 10^{-3}. Thus, cosmochronoly based on ^{187}Re, ^{232}Th, ^{238}U, ^{235}U would still work perfectly well and even ^{244}Pu would be acceptable (Drozd *et al.*, 1977). At the same time this opens up the possibility of variations on the level of 10^{-3} for the uranium isotopic ratio.

ORBITS IN EARLY SOLAR SYSTEMS

The study of the early solar OB association may help to solve an acute problem of meteoritics. Wasserburg and his group have shown that anomalous isotopic effects of magnesium are uniform among distinct phases within all Allende inclusions, which however have distinct Ca and Ti abundances. This is one of the main difficulties with the supposition that anomalies have been carried into the solar system by interstellar grains, since one can hardly see how these grains could have received the complex chemical differentiation processes needed to explain their petrological aspect (Lewis *et al.*, 1975). Such differentiations are much better accounted for by processing in a parent body.

A first step toward a solution is to assume that all centimeter-size inclusions came into Allende as interstellar "marbles" [a] (Wasserburg, personal communication, 1978). We cannot rule out the possibility of a certain fraction of the interstellar dust consisting of "pebbles" or even "boulders" of all sizes. The travel time of these marbles should be obtained by comparing the irradiations of the whole meteorite with the irradiation ages of the inclusion. Drozd *et al.* (1977) find that inclusions have an age of 5(± 0.5) × 10^6 yr, while the age of the whole rock is 3.6(± 0.4) × 10^6 yr; they assign the difference to possible diffusive loss from the matrix. One could as well associate it with the life span of the inclusions after their births and before they were integrated in Allende (see the chapter in this book by Wasson for a

[a] As pointed out by P. Pellas, (personal communication) one still has to account for the oxygen isotopic differences.

of one lightyear in radius (the Trapezium Cluster for instance) has a total gravitational energy of 5×10^{47} ergs (~0.1 eV per nucleon). This is only 10^{-4} to 10^{-5} of the supernova energy yield and might suffice to initiate the expansion if an appropriate coupling mechanism between the gas and stellar motion is found, which is by no means evident. Perhaps one should consider a coupling of extended protostellar nebulae strongly threaded by magnetic fields; this would be an interesting problem in plasma astrophysics.

As the first subgroup expands, star formation is initiated a few parsecs away (in Orion we are now in Belt Region *b*). Whether this new phase is causally related to the explosion of the first stars in Region *a* we do not know, but the idea is attractive. Many authors (Wentzel, 1966; Chevalier, 1974; Sgro, 1975; McKee and Cowie, 1975; Mufson, 1975; Woodward, 1976; Chevalier and Theys, 1975; Cameron and Truran, 1977; Margolis, personal communication, 1978) have analyzed the conditions in which the shock wave induced by a stellar explosion could initiate nebular collapse in a neighboring area.

The processes described here, by which a newborn subgroup (here Subgroup *a*), first frees itself from its dark nebulosity by the effect of its H II regions and later initiates stellar birth (by its own supernova explosion?) in adjacent areas of the cloud, may yield an explanation to the puzzling fact that star formation seems to take place mostly at the *edges* and not uniformly *inside* the dark clouds. Star formation in this context grinds its way into the cloud and leaves behind itself a low-density interstellar medium (see Fig. 1).

In the newly "infected" belt area, Region *b*, the same sequence of events takes place as previously in Region *a*: infrared sources, development of H II regions, formation of protoplanetary nebulae and, later on, supernovae from the largest stars. Now emerges the simultaneous occurrence in the older Region *a* of explosions involving stars of somewhat later type (O7-O9) which have by now reached the end of their lives. Their specific ejecta will contaminate the newly formed stars and protostellar nebulae both in subgroups *a* and *b*. Contamination from these supernovae will tend to be smaller than before due to the increased distances that the remnants have to span, but the events will be more numerous due to the larger number of later type stars.

A few million years later, star formation has moved further along the cloud, initiating in turn the equivalent of subgroups *c* (Outer Sword) and *d* (Trapezium) in Orion. A typical protostellar nebulae in Subgroup *d*, such as the Kleinman-Low nebula will be made of gas consisting *primarily* of the original cloud matter with standard composition, but also of the recent additions of diluted supernova ejecta of type O to B1 received from the earlier subgroups during the last 10^{7} yr. Because they occupy different space locations and also because of their own physical structure (density, temperature, state of turbulence) the various stellar objects of a young subgroup will not be affected in the same manner by the incoming tidal waves of neighboring supernova remnants. Neighboring cloud subunits may receive somewhat different shares of newly made material resulting in anomalies differing by up to several percent.

A corollary is the possibility of observations of isotopic variations in the molecules responsible for radio emission (CO, H_2CO, HCN, etc.) inside a given molecular cloud (Herbig, personal communication). Dickman *et al.* (1977) have reported such variations which however need confirmation. In the same fashion this may give a partial explanation for the spread of about 50% in the abundance of heavy elements for stars of the same age (Henry *et al.*, 1977). Another long standing problem might perhaps also be solved, namely the paucity of low-metal stars. If the *in situ* supernovae of an OB association contaminate the whole cloud, and hence most of the stars born within this association, with up to several percent of heavy element abundance, we should expect very few stars, even first-generation stars, to have less than one percent metal abundance. Furthermore, stars with very small metal abundance should reflect the ejecta of the O stars responsible for their early contamination within the association. Spite and Spite (1977) and that stars with metal abundances below one percent of the solar ones have no Na or Al deficiency. This is to be expected if all these elements are produced by carbon or oxygen hydrostatic burning in massive stars. For the heavy elements, strong deficiency appears for Ba, mild deficiency for Y and apparently no deficiency for Eu. In the solar system Ba is believed to be 90% due to *s*-process and Y 80% *s*-process while Eu is largely *r*-process. According to the earlier finding that *r*-process elements are very poorly generated by O stars, we would have expected the opposite trend. One difficulty here is that we do not know the isotopic distribution in these old stars. Further observations such as those of Spite and Spite would be of great interest (see also Cohen, 1978).

What should be the life of a forming planetary system in the context of an OB association? The standard picture of a protosolar nebula as a closed system, evolving without gravitational or thermodynamical interaction with its surrounding, should be revised. Matter exchange between neighboring nebula could take place in a number of ways: gas motion induced by H II regions; gravitational tides caused by close encounters; supernovae explosions; and later on, collisions of solid bodies in a given system which would possibly eject whole pebbles to the nearby systems. As time goes on, the subgroups continue to expand away and reach, after 10^7yr, a situation where the general gravitational field of the galaxy becomes stronger than the field of the subgroup. Individual stars are freed from the cluster and move away in the galaxy where in their isolation they still remember, by their isotopic anomalies or fossil radioactivities, the turmoil of their youth.

Cosmologists have their "Big Bang" theory of the origin of the universe. In view of the supernovae fireworks accompanying the birth of stars, cosmogonists are perhaps entitled (just to add a little confusion to a subject

association. The ^{26}Al, whether present in our solar system or carried away from a nearby system, is created in this scheme by the supernovae exploding within the solar association and it was still alive since the whole association witnesses $\simeq$20 supernovae in 10^7yr or one in $\sim 5 \times 10^5$yr. After expansion of the remnant, these radioactive products are spread around and contaminate the whole association.

O. K. Manuel: I have three comments on your conclusion that anomalies in noble gases are not nucleogenetic. (1) There are excellent correlated variations between isotopic ratios of Ar, K and Xe and elemental abundance of He and Ne. If you are going to dismiss this as not being nucleogenetic, then you should give a clear definition of another mechanism which will explain these correlations. (2) You mentioned that Os in residues of Allende are isotopically normal as evidence that there are no nucleogenetic anomalies in those residues. But Anders and coworkers dismissed the Os in these residues as a fortuitous consequence of the insolubility of noble metals in their HC1-Hf acids. Thus, the Os is *not* indigeneous to the residues and the isotopic composition of this contamination is meaningless. (3) Arden reported in *Nature* (October 1977) that $^{235}U/^{238}U$ in the residues of Allende and a few other meteorites is enriched in ^{235}U by as much as 300%. Do you have an explanation for this?

H. Reeves: Regarding your first comment, my remarks were mostly to justify a strategy. I do not dismiss the rare gas anomalies as being at least partly nucleogenetic; I have reasons to believe that they are not entirely and perhaps not even mostly nucleogenetic. (The almost linear correlation between Xenon isotopes in average carbonaceous chondrites and in the atmosphere is another example.) Unfortunately, I have no non-nuclear mechanisms which I can defend properly, although like everybody I can think of many. Regarding your second point, I have no comment. In answer to your third comment, this result, if confirmed, is far too large to be accommodated in the scheme I have discussed. In fact it would be puzzling on many grounds.

L. Alaerts: Manuel's argument of the osmium isotopic normalcy in etched residues of Allende, being a fortuitous consequence of its insolubility in and HC1, is not valid. Realizing this problem Takahashi *et al.* etched the residue in a stepwise experiment. Part of the osmium dissolved readily by etching with cold HNO_3 together with mineral Q – the fortuitous Os. The rest of the Os would not be etched away by making conditions more severe. This indigenous Os in chromite is as normal as the fortuitous one. The still high Os content may be an indication that the chromite's parent material (Fe Ni Cr grains) condensed on a "Pt metals" early condensate.

REFERENCES

Abt, H.A., and Levato, H. 1977. *Publ. Astron. Soc. Pacific* 89: 797.
Anders, E.; Heymann, D.; and Mazor, E. 1970. *Geochim. Cosmochim. Acta* 34: 127.

Arnett, W.D. 1978. *Astrophys. J.* 219: 1008.
Arnett, W.D., and Schramm, D.N. 1973. *Astrophys. J.* 184: L47.
Bash, F.N.; Green, E.; and Peters, W.L. 1977. *Astrophys. J.* 217: 464.
Baudry, A.; Forster, J.R.; and Welch, W.J. 1974. *Astron. Astrophys.* 36: 217.
Blaauw, A. 1964. *Ann. Rev. Astron. Astrophys.* 2: 219.
Black, D.C. 1972. *Geochim. Cosmochim. Acta* 36: 347, 377.
Cameron, A.G.W. 1973. *Space Sci. Rev.* 15: 121.
Cameron, A.G.W., and Truran, J.W. 1977. *Icarus* 30: 447.
Cassé, M., and Reeves, H. 1978. In preparation.
Chevalier, R.A. 1974. *Astrophys. J.* 188: 501.
Chevalier, R.A., and Theys, J.C. 1975. *Astrophys. J.* 195: 53.
Clayton, D.D. 1977. *Icarus* 32: 255.
Clayton, D.D.; Dwek, E.; and Woosley, S.E. 1977. *Astrophys. J.* 217: 300.
Clayton, R.N.; Onuma, N.; and Mayeda, T.K. 1976. *Earth Planet. Sci. Lett.* 30: 10.
Clossa, B. 1978. Thesis, Université de Paris. In preparation.
Cohen, J.G. 1978. Preprint, Univ. California, Berkeley.
Crozaz, G. 1974. *Earth Planet. Sci. Lett.* 23: 164.
Cruz-Gonzalez, C.; Recillas-Cruz, E.; Costero, R.; Peimbert, M.; and Torres-Peimbert, S. 1974. *Rev. Mex. Astron. Astrofis.* 1: 211.
Dickman, R.L.; Langer, W.D.; McCutcheon, W.H.; and Shuter, W.L.H. 1977. In *CNO Isotopes in Astrophysics* (J. Audouze, ed.), p. 95. Dordrecht: Reidel.
Drozd, R.J.; Morgan, C.G.; Podosek, F.A.; Poupeau, G.; Shirck, J.R.; and Taylor, G.J. 1977. *Astrophys. J.* 212: 567.
Eberhardt, P. 1975. *Meteoritics* 10: 401.
Elmegreen, B.G., and Lada, C.J. 1977. *Astrophys. J.* 214: 725.
Epherre, M. 1972. Thesis, Université de Paris-Sud.
Fireman, E.L., and Goebel, R. 1970. *J. Geophys. Res.* 75: 2115.
Fontes, P.; Perron, C.; Lestringuez, J.; Yiou, F.; and Bernas, R. 1971. *Nucl. Physics* A165: 405.
Fowler, W.A. 1972. In *Cosmology, Fission and Other Matters* (F. Reines, ed.), p. 67. Boulder: Univ. Colorado Press.
Fowler, W.A.; Greenstein, J.L.; and Hoyle, F. 1962. *Geophys. J. Roy. Astron. Soc.* 6: 148.
Frick, U.; Moniot, R.K.; Neil, J.M.; Phinney, D.L.; and Reynolds, J.H. 1978. Preprint. Univ. California, Berkeley.
Geiss, J. 1972. In *L'origine du Système Solaire* (H. Reeves, ed.), p. 217. Paris: C.N.R.S.
Gray, C.M., and Compston, W. 1974. *Nature* 251: 495.
Gray, C.M.; Papanastassiou, D.A.; and Wasserburg, G. J. 1973. *Icarus* 20: 213.
Gros, J., and Anders, E. 1977. *Earth Planet. Sci. Lett.* 33: 401.
Hartmann, W.K. 1977. *Scientific American* 236: 84.
Heckman, T.M., and Sullivan, W.T. 1976. *Astrophys. Lett.* 17: 105.
Heiles, C. 1978. *Scientific American.* In press.
Henry, R.C.; Anderson, R.; and Hesser, J.E. 1977. *Astrophys. J.* 214: 742.
Herbig, G. 1976. In *Origin of the Solar System* (S.J. Dermott, ed.), New York: Wiley. In press.
Herbst, W., and Assousa, G.E. 1977. *Astrophys. J.* 217: 473.
Hohenberg, C.M.; Podosek, F.A.; and Reynolds, J.H. 1967. *Science* 156: 233.
Holweger, H. 1977. In *Comets, Asteroids and Meteorites* (A.H. Delsemme, ed.), p. 385. Toledo, Ohio: Univ. Toledo Press.
Hutcheon, I.D.; Steel, I.M.; Solberg, T.N.; Clayton, R.N.; and Smith, J.V. 1977. *Meteoritics* 12: 261.
Ilovaisky, S.A., and Lequeux, J. 1972. *Astron. Astrophys.* 18: 169.
Jordan, J.; Kirsten, T.; and Richter, H. 1977. *Meteoritics* 12: 269.
King, C.H.; Austin, S.M.; Rossner, H.H.; and Chien, W.S. 1977. Preprint, Michigan State Univ.
Kuroda, P.K., and Manuel, O.K. 1970. *Nature* 227: 1113.
Kutner, M.L.; Tucker, K.D.; Chin, G.; and Thaddeus, P. 1977. *Astrophys. J.* 215: 521.
Lee, T. 1978. Submitted to *Astrophys. J.*

Lee. T.; Papanastassiou, D.A.; and Wasserburg, G.J. 1976. *Geophys. Res. Lett.* 3: 109.
____. 1978. Preprint, California Inst. Technology.
Leich, D.A.; Niemeyer, S.; and Michel, M.C. 1977. *Earth Planet. Sci. Lett.* 34: 197.
Levato, H., and Abt, H.A. 1976. *Publ. Astron. Soc. Pacific* 88: 712.
Lewis, R.S.; Srinivasan, B.; and Anders, E. 1975. *Science* 190: 1251.
Lorin, J.C.; Shimizu, N.; Christophe-Michel Levy, M.; Allegre, C.J. 1977. *Meteoritics* 12: 299.
Mason, B., ed. 1971. *Handbook of Elemental Abundances in Meteorites.* New York: Gordon and Breach.
McCulloch, M.T., and Wasserburg, G.J. 1977. *Astrophys. Lett.* In Press.
McKee, C.F., and Cowie, L.L. 1975. *Astrophys. J.* 195: 715.
Meyer, J.P., and Reeves, H. 1977. *15th Intern. Cosmic Ray Conf.* 2: 137, Bulgarian Academy of Sciences.
Mezger, P.G., and Smith, L.F. 1977. In *Star Formation* (T. de Jong and A. Maeder, eds.), p. 133. Dordrecht: Reidel.
Mufson, S.L. 1975. *Astrophys. J.* 202: 372.
Niederer, S., and Eberhardt, P. 1977. *Meteoritics* 12: 327.
Ögelman, H.B., and Maran, S.P. 1976. *Astrophys. J.* 209: 124.
Öpik, E. 1953. *Irish Astr. J.* 2: 219.
Ostriker, J.P.; Richstone, D.O.; and Thuan, T.X. 1974. *Astrophys. J.* 188: L87.
Pellas, P., and Storzer, D. 1975. *Meteoritics* 10: 471.
Phinney, D. 1978. *Lunar Planetary Sci. Conf.* 9th. 2: 893.
Podosek, F.A., and Lewis, R.S. 1972. *Earth Planet. Sci. Lett.* 15: 101.
Quirk, W.J., and Tinsley, B.M. 1973. *Astrophys. J.* 179: 69.
Reeves, H. 1972*a*. In *L'origine du Système Solaire* (H. Reeves, ed.), p. 378. Paris: C.N.R.S.
____. 1972*b*. *Astron. Astrophys.* 19: 215.
Reeves, H., and Audouze, J. 1968. *Earth Planet. Sci. Lett.* 4: 135.
Reeves, H., and Johns, O. 1976. *Astrophys. J.* 206: 958.
Ryter, C.; Reeves, H.; Gradsztajn, E.; and Audouze, J. 1970. *Astron. Astrophys.* 8: 389.
Schramm, D.N. 1971. *Astrophys. Space Sci.* 13: 249.
____. 1974. *Ann. Rev. Astr. Astrophys.* 12: 383.
Sedov, A.I. 1954. *Methods of Similarity and Dimension in Mechanics.* Gostekhizdat.
Sgro, A.G. 1975. *Astrophys. J.* 197: 621.
Sivan, J.P. 1974. *Astron. Astrophys. Suppl.* 16: 163.
Spite, M., and Spite, F. 1978. Submitted to *Astron. Astrophys.*
Strom, S.E.; Strom, K.M.; and Grasdalen, G.L. 1975. *Ann. Rev. Astron. Astrophys.* 13: 187.
Takahashi, H.; Higuchi, H.; Gros, J.; Morgan, J.W.; and Anders, E. 1977. *Proc. Nat. Acad. Sci.* 73: 4353.
Talbot, R.J., and Arnett, W.D. 1973. *Astrophys. J.* 186: 69.
Tatsumoto, M.; Unruh, D.M.; and Desborough, G.A. 1976. *Geochim. Cosmochim. Acta* 40: 617.
Tobailem, J.; de Lassus St-Genies, Ch.H.; and Reeves, H. 1973. Rapport CEA-R-4441, CEN-Saclay, France.
Torres-Peimbert, S.; Lazcano-Araujo, A.; and Peimbert, M. 1974. *Astrophys. J.* 191: 401.
Traub, W.A.; Carleton, N.P.; and Hegyi, D.J. 1974. *Astrophys. J.* 190: L81.
Truran, J.W., and Cameron, A.G.W. 1971. *Astrophys. Space Sci.* 14: 179.
____. 1978. *Astrophys. J.* 219: 226.
Wasserburg, G.J.; Lee, T.; and Papanastassiou, D.A. 1977. *Geophys. Res. Lett.* 4: 299.
Weaver, T.A.; Zimmerman, B.; and Woosley, S.E. 1977. Preprint UCRL 80460, University of California.
Wentzel, D.G. 1966. *Astrophys. J.* 145: 595.
Woodward, P.R. 1976. *Astrophys. J.* 207: 484.
Zuckerman, B. 1975. In *Review of Orion A and Orion B in H II Regions and Related Topics* (T.L. Wilson and D. Downes, eds.), p. 360. Berlin: Springer Verlag.
Zuckerman, B., and Palmer, P. 1974. *Ann Rev. Astr. Astrophys.* 12: 279.

THE CHAOTIC SOLAR NEBULA: EVIDENCE FOR EPISODIC CONDENSATION IN SEVERAL DISTINCT ZONES

WILLIAM V. BOYNTON
University of Arizona

The fine-grained Ca-Al-rich aggregates in the Allende meteorite occur in two distinct groups. One group has rare earth element (REE ≡ lanthanides) abundances which are unfractionated relative to solar except for a depletion of the two most volatile REE, Eu and Yb. This REE pattern suggests isolation of the grains from the solar nebula before condensation of the REE was complete. The other group has a REE pattern which is very strongly fractionated by loss of the least volatile REE. Because the least volatile elements cannot be preferentially lost by partial vaporization, the grains in this group of aggregates cannot be unvaporized interstellar grains. It is more likely that they condensed from a gas which was previously fractionated by the condensation and removal of a more refractory phase. These two groups could not form in the same region of the nebula, but required different condensation zones, with episodes of condensation and isolation of grains from gas occurring in each zone. Although components of the meteorites show evidence for several condensation zones, the elemental abundances in the bulk meteorites are consistent with a simple continuous monotonic cooling sequence. This apparent contradiction suggests that the bulk meteorites may have acquired the average composition of a large region of the nebula by sampling many zones.

Most models for the condensation of the elements from the solar nebula assume that the nebula was homogeneous and cooled monotonically (e.g., Larimer, 1967; Lewis, 1972; Grossman and Larimer, 1974). The models assume that the condensates either maintained equilibrium with the gas until some low temperature or that they became isolated from the gas as soon as they formed. However, these assumptions only provide an idealized model of what may have occurred. Physical models of the formation of the solar nebula indicate that convection zones and turbulence may have been present (Cameron and Pine, 1973; Cameron, 1978, and see his chapter). If such zones did form on a large scale, they could have complicated the condensation sequence by moving material between regions of different pressure and/or temperature. Because we have no quantitative predictions of a reasonable scenario for solar system formation, we cannot form more elaborate condensation models without making them more *ad hoc.* In spite of these simplifications, the models have been successful in accounting for the elemental abundances in meteorites (Anders, 1968; Grossman and Larimer, 1974; Wai and Wasson, 1977; Anders, 1977). In fact, until recently, there was no compelling evidence to suggest that a relatively simple condensation sequence did not occur.

More recent meteorite data, however, suggest the presence of independent condensation zones in the nebula. In several zones condensation apparently started and stopped several times and solids were rather abruptly isolated from a gas of one composition and brought into contact with gas of another composition. Keeping in mind the wide scope of this emerging field of Protostars and Planets, I shall attempt to explain these data without using the jargon of the cosmochemist, and hope that modelists may incorporate the constraints of these data into more quantitative phsyical models in the future.

ABUNDANCES OF RARE EARTH ELEMENTS IN ALLENDE INCLUSIONS

The Allende meteorite, a large carbonaceous chondrite which fell in Mexico in 1969, contains a diverse mixture of chondrules, irregular aggregates and dark fragments of other rock types imbedded in a fine-grained matrix (Clarke *et al.,* 1970). Chondrules, after which chondrites are named, are spherical bodies generally composed of silicates, sulfides and metal, and are generally 1-2 mm in diameter. In the Allende meteorite, many of the chondrules are as large as 2 cm, and contain coarse-grained silicates and oxides of refractory elements Ca, Al and Ti. The aggregates are irregular in shape, approximately a centimeter in size, and are composed of many fine-grained bodies (Wark and Lovering, 1977). Many of the fine-grained aggregates are also composed of refractory element grains, but, in addition, contain other grains rich in volatile elements such as Na and Cl. The high abundances of refractory elements suggest that these Ca-Al-rich inclusions (CAI) may be condensates from the solar nebula (Marvin *et al.,* 1970; Grossman, 1973; Boynton, 1975) or residues of a partial vaporization process (Kurat, 1970; Chou *et al.,* 1976). Only the CAI will be the subject of this discussion.

Before one can consider the rare earth element (REE) abundances in the CAI, a brief review of the cosmochemistry of the REE is necessary. The REE, or properly, the lanthanides, are the group of elements in the periodic table ranging from lanthanum (La) to lutetium (Lu). They are characterized by having rather similar chemical properties. For most processes, the only property that distinguishes them is their size, which decreases smoothly with increasing atomic number. This property causes the REE to distribute themselves between different co-existing minerals in a way that is a smooth function of size and is characteristic of the particular minerals involved. The only exception is europium (Eu), which can exist in the divalent state, giving it different chemical properties from the other, trivalent REE.

For processes such as condensation or vaporization in thermodynamic equilibrium with the nebula gas, the abundances of the REE in the solid phase are controlled by the vapor pressures of the REE oxides. Vapor pressure is one property of the REE that varies dramatically between individual REE in a way that is *not* a smooth function of size. Figure 1 shows values of D, the solid/gas distribution coefficient, for the REE calculated from thermodynamic data (Boynton, 1975). The value of D is a measure of the relative volatility of the REE; the lower the value, the more volatile the element. The distribution coefficient is actually the ratio of the amount of REE in the solid to the amount in the gas. Because it is difficult to calculate the absolute value of this ratio, all values are taken relative to La. Note that the difference between individual REE are quite large and the total range spans a factor of 10^7.

The D values may change slightly, depending on the mineral in which the REE are condensing, because of the mineral's preference for REE on the basis of size. This mineral effect is expected to be small, perhaps a factor of 2 between adjacent REE and a factor of 10 to 100 maximum difference between the largest and smallest REE. Because the mineral effect is a smooth function of size (and atomic number), it will not significantly change the pattern of Fig. 1; it can only raise or lower one end relative to the other in a smooth and gradual way. It is clear, for example, that no mineral effect will eliminate the large volatility difference between ytterbium (Yb) and lutetium (Lu). The REE will probably substitute for the large Ca ion; thus, the smaller REE, to the right in Fig. 1, will be more strongly excluded from the mineral and will tend to remain in the gas phase. This effect will lower the D values of the smaller REE and make Yb more volatile than any other REE except for Eu. (See Boynton [1975, 1978] for a more rigorous discussion of the effect of mineral preference).

To understand the REE abundances in the Allende inclusions, one needs only a qualitative understanding of the REE volatility. The most volatile REE are Eu and Yb. These REE will be the last to condense into the solid and will be the most depleted relative to the other REE if the solid is isolated from

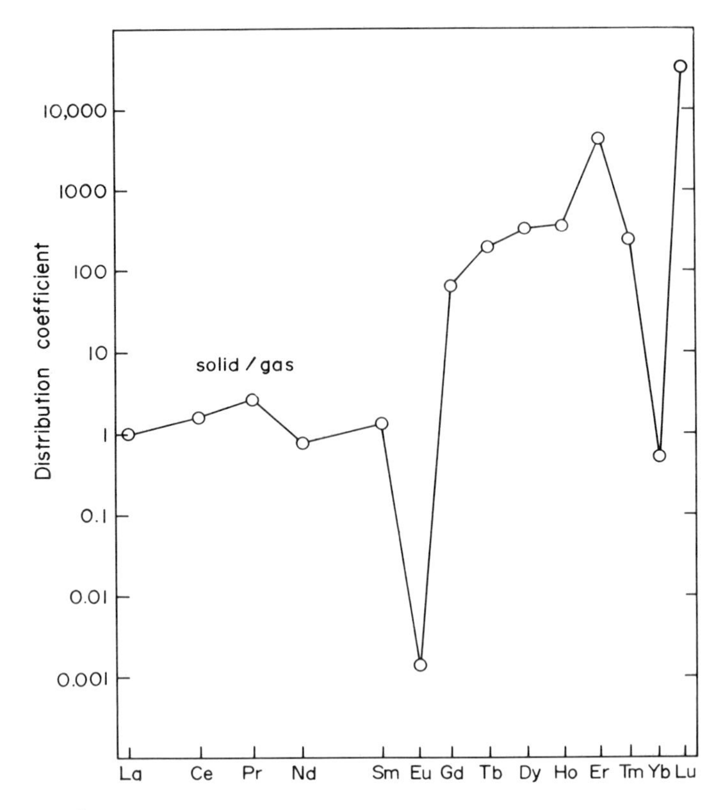

Fig. 1. Solid/gas distribution coefficients of the rare earth elements (REE) from Boynton (1975). The values are equal to the ratio of REE in the solid to REE in the gas relative to lanthanum. The volatility of the REE is inversely proportional to the distribution coefficients. The values of the distribution coefficients may change slightly in a smooth and gradual manner, depending on the mineral in which the REE are condensing.

the gas before condensation is complete. Lu and erbium (Er) are the least volatile, or most refractory, and they will be the first to condense and the most strongly enriched in the very first condensate.

The solar normalized abundances of the REE in the first condensate will have a pattern similar to that of Fig. 1. However, as condensation progresses with decreasing temperature, the pattern will change. This occurs because the most refractory elements soon become totally condensed and their abundances no longer increase, while the abundances of the more volatile elements are still increasing. As the more volatile elements continue to condense, the large fractionation between the different REE begins to diminish. Obviously, at the end of condensation the REE will have an unfractionated pattern. (See Fig. 2 of Boynton [1975] for abundance patterns at different stages of condensation.)

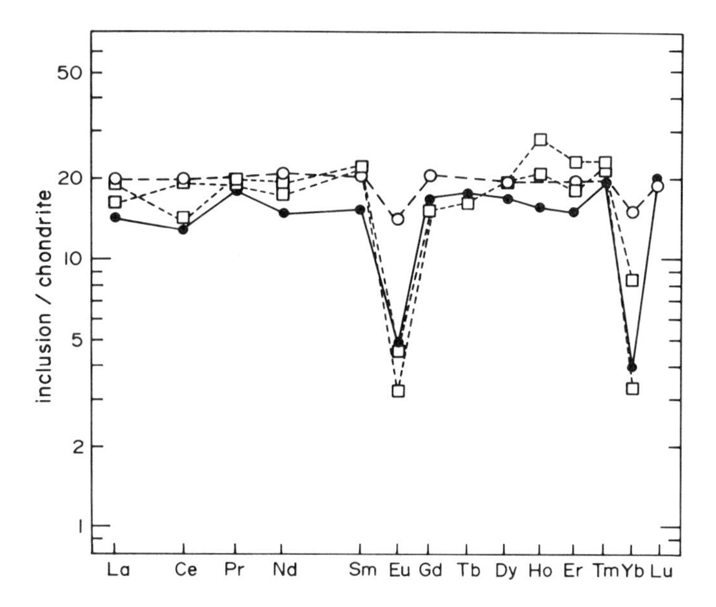

Fig. 2. Rare earth element (REE) patterns in Group III aggregates. The Group III pattern is unfractionated relative to chondritic (solar) abundances except for a depletion of the two most volatile REE, europium and ytterbium. This depletion suggests that the grains that make up the aggregates were isolated from the nebula gas before condensation of the REE was complete. Solid circles are data from Conard *et al.* (1975); large open circles from Tanaka and Masuda (1973); and open squares from Martin and Mason (1974).

If one is to gain any information from the REE about high-temperature condensation processes, it will be necessary to examine samples that preserve a fractionated REE pattern. Such material can only occur if a mechanism exists for isolating solids from the gas before condensation is complete. Some of the CAI found in Allende have, indeed, preserved evidence of gas/solid fractionation of REE.

The coarse-grained chondrules and fine-grained aggregates characterize the two major types of CAI (Gray *et al.*, 1973). The fine-grained aggregates are subdivided into two groups based on differences in their REE abundances (Martin and Mason, 1974). The REE patterns of the two groups, called Group II and Group III by Martin and Mason, are illustrated in Figs. 2 and 3.

The abundances are normalized to the abundances in chondrites, our best estimate of solar abundances. The Group III REE pattern shows a rather uniform enrichment at a factor of ~20 times chondrites except for a depletion in Eu and Yb. Since Eu and Yb are the two most volatile REE, this pattern suggests that the grains that make up these aggregates condensed from a gas of solar composition but were isolated from the gas before condensation of the REE was complete. Due to their greater volatility, significant amounts

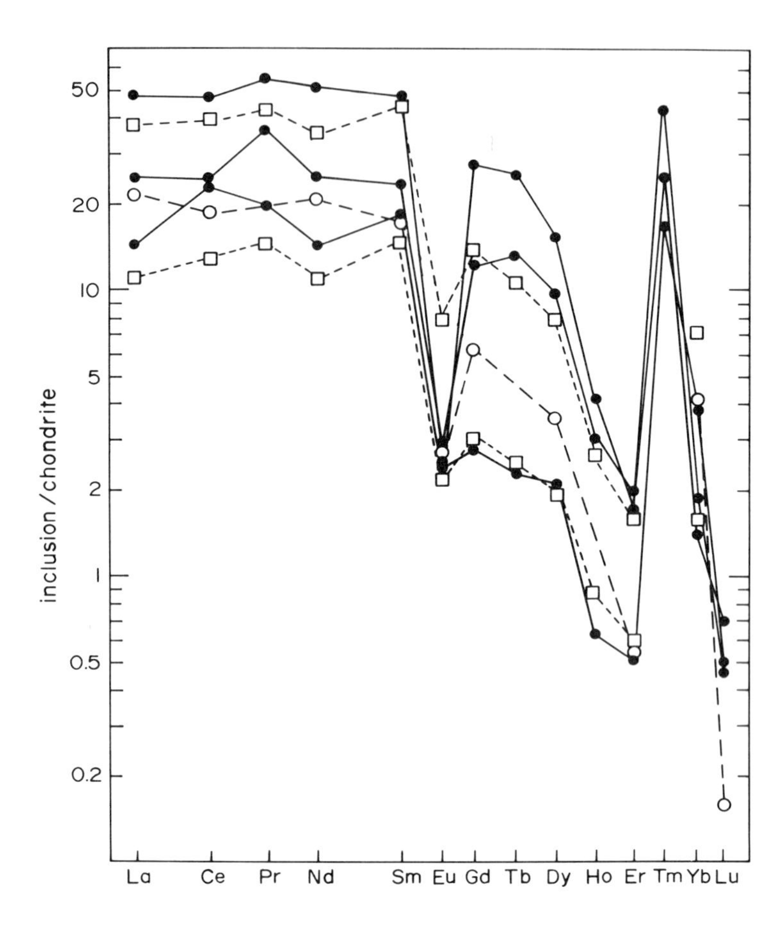

Fig. 3. Same as for Fig. 2 except showing rare earth element (REE) patterns in Group II aggregates. The Group II pattern is characterized by a large enrichment of light REE over heavy REE and contains abundance "anomalies" at europium, thulium and ytterbium where the pattern differs from a smooth curve. With the exception of europium, the REE can only be fractionated as a smooth function of size (and atomic number) during an igneous fractionation process. In particular, the enrichment of thulium over neighboring erbium and ytterbium cannot result from any igneous fractionation on a parent body, but can be accounted for by gas/solid fractionation in the nebula. The depletion of the least volatile REE, lutetium and erbium, suggests that the grains condensed from a gas that had previously been depleted in the missing REE by a more refractory condensate not sampled. Solid circles are data from Conard *et al.* (1975); large open circles from Tamaka and Masuda (1973); and open squares from Martin and Mason (1974). The data of Martin and Mason represent a range of five.

of Eu and Yb still remained in the gas phase. Europium and Yb, although the most volatile REE, are still much more refractory than volatile elements such as Na, K, Cl and zinc (Zn), which are also enriched in the fine-grained aggregates (Grossman and Ganapathy, 1976; Mason and Martin, 1977). The texture of the aggregates, "a loosely-packed, cavernous aggregate of crystals which project into the pore spaces" (Grossman *et al.,* 1975), rather strongly suggests that the Na- and Cl-rich minerals are also condensates. Thus, the Group III aggregates experienced at least two episodes of condensation: one in which the REE partially condensed and one in which the volatile elements condensed. The volatiles will be discussed in more detail in the next section.

The Group II REE patterns, Fig. 3, suggest a more complex history. The patterns are characterized by an enrichment of the light REE and a strong depletion in the heavier REE. In addition, there are "anomalies" at Eu, Tm and Yb. Such remarkable REE patterns could not be formed by any igneous process such as occurs on earth or on any meteorite body, because it is impossible for a mineral grown from a melt to preferentially incorporate thulium (Tm) over the adjacent Er and Yb. The REE patterns appear to require a gas/solid fractionation and, in fact, the entire REE pattern can be duplicated within experimental error of the REE solid/gas distribution coefficients (Boynton, 1978*a*).

Although these CAI have REE patterns which require a gas/solid fractionation, they *cannot* be the earliest condensate from a solar gas. The two most refractory REE, Er and Lu, are the most strongly *depleted* in these inclusions, but, as discussed earlier, they should be the most strongly *enriched* in the first condensate to form. In fact, except for Eu and Yb, the depletion is inversely related to the volatility, suggesting that the REE pattern is complementary to that expected for the earliest condensate (Boynton, 1975). Some earlier, more refractory condensate (or residue from an incomplete vaporization of pre-existing grains) must have formed, depleting the gas in the refractory REE. After this missing refractory material was isolated from the gas, condensation began again; the REE remaining in the gas condensed into the grains that made up the Group II fine-grained aggregates. The depletion of Eu and Yb, as in the case of the Group III aggregates discussed earlier, occurred presumably because the grains again became isolated before the condensation of the REE was complete.

Thus it appears that the REE depict a record of several distinct episodes of condensation and separation of gas from grains. The Group II and Group III aggregates have little mineralogic or textural difference (Martin and Mason, 1974; Grossman and Ganapathy, 1976), yet they must have formed under quite different circumstances. The Group III aggregates must have condensed from a gas having *solar* REE abundances and must have maintained contact with the gas until all the REE except Eu and Yb had fully condensed. The Group II aggregates, however, probably condensed from a gas having *fractionated* REE abundances. In the zones of the Group II aggregates, some very efficient means of removing grains that are refractory rich must have been present. Over 95% of the Lu and Er, and the grains containing them, not only were isolated from the gas but never were mixed back into the gas. Had they been mixed back before the micrometer-sized grains

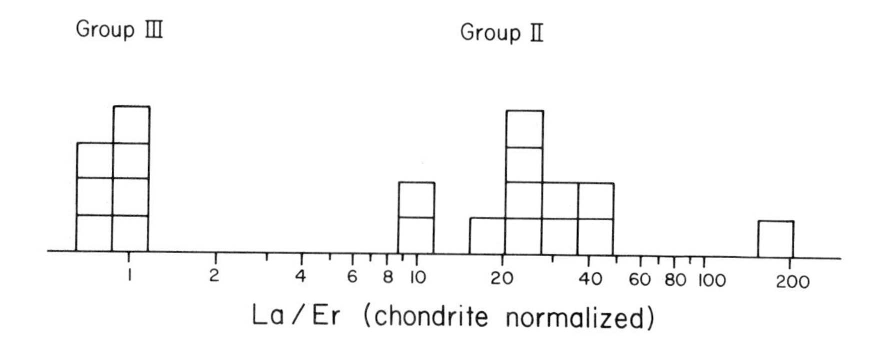

Fig. 4. Histogram of lanthanum/erbium ratios in Group II and Group III aggregates in the Allende meteorite. Data are from sources in Figs. 2 and 3 and Taylor and Mason (1978). The REE patterns clearly form two groups and do not represent a continuum.

agglomerated to form centimeter-sized aggregates, they surely would have been incorporated into the aggregates.

It is not likely that these two types of REE patterns represent a continuum. In the Group III patterns, the chondrite-normalized abundance of Lu and Er are the same as La, but in the Group II patterns they are always depleted by large factors. In Fig. 4 a histogram of La/Er ratios shows the clear distinction between the groups. The chondrite-normalized La/Er ratios cluster around 25 in Group II aggregates but are equal to unity in the Group III aggregates. There is a hiatus of a factor of ten between the two groups.

Although the aggregates do not represent a continuum of REE patterns, one must consider whether it is possible that they may represent two different samples of a continuum. A continuum might represent mixing between two (or more) different end members or a continuous change in some parameter such as temperature of isolation from the gas. For any type of continuous process, the Group III REE pattern is unlikely to represent an *intermediate* stage but is most likely an end member. If it were an intermediate stage product, highly fractionated material would have to be altered to yield the unfractionated pattern (except for Eu and Yb) of the Group III aggregates, and the process would have to be sampled at just that point where the product was unfractionated. This is a most unlikely occurrence.

The highly-fractionated Group II REE patterns may be a member of a continuum in which the amount of REE condensed into the missing highly refractory condensate varies in a continuous manner. However, there is no easy way to form the Group II patterns from material having a Group III REE pattern. As mentioned earlier, the Group II patterns cannot be formed by partially vaporizing material with unfractionated REE abundances. Because Lu and Er are depleted by over a factor of 20 relative to the light REE, it is not possible to generate Group II patterns by mixing any other material with Group III patterns unless the other material makes up at least 95% of the mixture and is itself nearly identical to the Group II material.

McSween (1977) in a study of the major element composition of CAI suggested that the inclusions followed a cooling sequence, with the coarse-grained chondrules condensing before the fine-grained aggregates. A similar suggestion was made by Grossman and Ganapathy (1976) based on trace elements. McSween, however, noted that although a monotonic cooling sequence under equilibrium conditions is adequate to explain the observations to a first approximation, strict adherence to equilibrium is not supported unless the inclusions are transported between sections of varying temperature, pressure or composition. The REE data in the fine-grained aggregates suggest that the aggregates could not have condensed in the same region as the coarse-grained chondrules. This is because the REE are expected to be strongly enriched in the earliest condensates. The coarse-grained chondrules, in fact, are significantly enriched in REE as expected, but if they formed first they should significantly deplete the gas in REE. It is unlikely that a sufficient amount of REE would remain in the gas to provide the enrichments of REE observed in the fine-grained aggregates. It is even less likely that the gas remaining would have an unfractionated pattern as observed in the Group III aggregates.

One is forced to conclude that the REE condensation process recorded in the Allende fine-grained aggregates requires two independent condensation zones in the solar nebula and that both zones require separation of gas from grains. Two episodes of separation occurred in the zone where Group II aggregates formed. The first episode, during which the highly refractory material was lost, must have been very efficient. The implications of this conclusion will be considered after the following discussion.

RIM SEQUENCE ON COARSE-GRAINED CHONDRULES

As mentioned earlier, the coarse-grained chondrules comprise the other major type of CAI. The chondrules, like the aggregates, are subdivided into two groups (Grossman, 1975), but this division is based on mineralogy, not on trace element content. The criteria for distinguishing the two groups, called Type A and Type B, is based on the mineralogy of the inclusions and is of no concern for this discussion.

Wark and Lovering (1977) found that these CAI are invariably coated by a rim containing layers of different minerals. Generally, the rims contain about six rather sharply defined layers. Wark and Lovering (1977) stated that the rim layers were almost certainly deposited while the CAI were still dispersed in the nebula, rather than after incorporation into the meteorite. One observation supporting this conclusion is that on broken CAI the rims occur only on the smooth rounded surfaces and not on the fractured surface, even though both surfaces are in contact with the same meteorite matrix material.

There are two remarkable features of these rims that are important to this discussion. The first is that the mineral sequence does not follow any simple condensation sequence. Layers of refractory grains such as Ti-rich silicates are found outside a layer of volatile-rich grains such as Na- and Cl-bearing silicates. It is not likely that any monotonic cooling sequence, even a rapid non-equilibrium sequence, could condense volatile elements followed by refractory elements. This sequence of layers suggests that the CAI, as complete rimmed bodies, were transported from a volatile-rich gas which was the source of the Na- and Cl-bearing silicates to a refractory-rich gas which deposited the Ti-rich silicates. The layers are sharp and well-defined, suggesting that this transportation was rather abrupt. The other noticeable feature is that all Type A CAI have one sequence of layers and all Type B CAI have a different sequence. The significance of these rim sequences will not be discussed in detail, since Wark and Lovering (1977) have provided such a discussion. The sequences do provide additional strong evidence for the presence of different rim sequences supports the conclusion based on REE data that episodic condensation occurred in several different regions of the nebula.

Wark and Lovering (1977) found both Type A and B sequences in the fine-grained aggregates. The aggregates themselves were not rimmed, but the bodies that composed the aggregate were made up of a similar sequence of minerals. In fact, Wark and Lovering suggested that the fine-grained aggregates are composed of pure rim material. Presumably, the volatiles in the fine-grained aggregates were added to the volatile-rich rim material, indicating that the episodes of condensation and isolation from the gas continued after the condensation of the REE.

IMPLICATIONS FOR MODELS OF SOLAR SYSTEM FORMATION

It should be clear from the preceding discussion that in at least several zones the nebula went through episodes of condensation and gas/grain separation. Four separate zones are required to make the two types of fine-grained aggregates and the two types of coarse-grained chondrules. Other types of inclusions in Allende, such as the olivine aggregates (Grossman and Steele, 1976), may well require additional condensation zones. All of these conclusions are based on data from one meteorite. Detailed examination of CAI in other meteorite types may yet reveal additional REE patterns that require zones distinct from those sampled by the Allende meteorite.

Although the details of the gas/grain separation mechanism in these zones is not well understood, it is clear that a single monotonic-cooling condensation sequence followed by isolation of grains from gas cannot account for these observations. This is, perhaps, surprising, since the equilibrium condensation model has had quite good success in describing meteorite properties (e.g. Anders, 1968; Grossman and Larimer, 1974; Wai and Wasson, 1977). One explanation for this success could be that most of the nebula did indeed obey the simple condensation sequence and that Allende, and presumably other meteorites of the same type, sampled a very peculiar region of the nebula. This is not likely, however, since the carbonaceous chondrites, a class to which Allende belongs, are probably the best examples of success of the equilibrium condensation model. Another explanation is that the nebula had many condensation zones, but the meteorites sampled a sufficiently large volume of material such that an average composition appropriate for a particular heliocentric distance was achieved. Thus, the simple condensation sequence would be able to describe the bulk properties of the meteorites if the condensation zones were small compared to the region that accreted to form the meteorite parent bodies.

The mechanism for establishing these zones is not understood, nor is it clear how grains can be isolated from the gas so efficiently. Any realistic explanation for these condensation zones must await more detailed physical models of processes in the solar nebula. It is hoped that this discussion of the meteorite data will encourage the development of such models.

Acknowledgments. The author is grateful to L. L. Wilkening for constructive comments and to S. R. Taylor and B. Mason for permission to cite their unpublished data. This work was supported in part by a grant from the National Aeronautics and Space Administration.

COMMENTS

D. Clayton: The literature shows that cosmochemists have regarded the early high-temperature gaseous nebula as a real event, not merely a convenient way of thinking. A major scientific objective should be that of ascertaining whether or not such an event did occur. In a model of accumulation processes within a cold molecular cloud, astonishing rare earth patterns are possible. The most refractory rare earths will be in the refractory interstellar grains, whereas the most volatile ones will be more in the gas phase. After accumulations with varying dust/gas ratios, closed-system chemical modification can result in rare earth patterns similar to those you describe.

W. Boynton: Although it may be possible for significant gas/solid fractionations to occur in condensation associated with environments other than the solar nebula, accumulation with variable dust/gas ratios will only give two component mixtures. It is not possible to generate the REE patterns observed in Allende CAI with just two components. However, the point you raise is interesting. If preserved interstellar grains can be found in meteorites, the REE may be quite useful in determining the environment in which they formed. The relative volatility of several of the REE is very sensitive to the oxidation conditions in the gas such that a fractionated REE pattern can distinguish environments such as an envelope of a carbon-rich star or a supernova from each other or from the solar nebula. This application has been discussed by Boynton (1978*b*).

REFERENCES

Anders, E. 1968. *Accounts Chem. Res.* 289.

____. 1977. *Earth Planet. Sci. Lett.* 36: 14.

Boynton, W.V. 1975. *Geochim. Cosmochim. Acta* 39: 569.

____. 1978*a*. *Earth Planet. Sci. Lett.* In press.

____. 1978*b*. *Lunar and Planetary Sci.* 9th. p. 120. Lunar Sci. Conf., Houston.

Cameron, A.G.W. 1978. *Moon and Planets.* In press.

Cameron, A.G.W., and Pine, M.R. 1973. *Icarus* 18: 377.

Chou, C.-L.; Baedecker, P.A.; and Wasson, J.T. 1976. *Geochim. Cosmochim. Acta* 40: 85.

Clarke, R.S., Jr.; Jarosewich, E.; Mason, B.; Nelen, J.; Gómez, M.; and Hyde, J.R. 1970. *Smithson. Contrib. Earth Sci.* 5.

Conard, R.L.; Schmitt, R.A.; and Boynton, W.V. 1975. *Meteoritics* 10: 384.

Gray, C.M.; Papanastassiou, D.A.; and Wasserburg, G.J. 1973. *Icarus* 20: 213.

Grossman, L. 1973. *Geochim. Cosmochim. Acta* 37: 1119.

____. 1975. *Geochim. Cosmochim. Acta* 39: 433.

Grossman, L.; Fruland, R.M.; and McCay, D.S. 1975. *Geophys. Res. Lett.* 2: 37.

Grossman, L., and Ganapathy, R. 1976. *Geochim. Cosmochim. Acta* 40: 967.

Grossman, L., and Larimer, J.W. 1974. *Rev. Geophys. Space Phys.* 12: 71.

Grossman, L., and Steele, I.M. 1976. *Geochim. Cosmochim. Acta* 40: 149.

Kurat, G. 1970. *Earth Planet. Sci. Lett.* 9: 225.

Larimer, J.W. 1967. *Geochim. Cosmochim. Acta* 31: 1215.

Lewis, J.S. 1972. *Icarus* 16: 241.

Martin, P.M., and Mason, B. 1974. *Nature* 249: 333.

Marvin, U.B.; Wood, J.A.; and Dickey, J.S., Jr. 1970. *Earth Planet. Sci. Lett.* 7: 346.

Mason, B., and Martin, P.M. 1977. *Smithson. Contrib. Earth Sci.* 19: 84.

McSween, H.Y., Jr. 1977. *Geochim. Cosmochim. Acta* 41: 1777.

Tanaka, T., and Masuda, A. 1973. *Icarus* 19: 523.

Taylor, S.R., and Mason, B. 1978. In preparation.

Wai, C.M., and Wasson, J.T. 1977. *Earth Planet. Sci. Lett.* 36: 1.

Wark, D.A., and Lovering, J.F. 1977. *Proc. Lunar Sci. Conf.* 8th. p. 95. New York: Pergamon.

NOBLE GASES IN METEORITIC GAS-RICH MINERALS: SOME IMPLICATIONS FOR THE FORMATION OF THE SOLAR SYSTEM

LEO ALAERTS
AND
ROY S. LEWIS
The University of Chicago

Noble gas abundance patterns in chondrites may provide direct clues to temperatures in the inner solar nebula. Due to large differences in heat of solution between heavy and light noble gases, their ratios are sensitive cosmothermometers which only need calibration by laboratory experiments with the appropriate gas-rich minerals.

Contrary to isotopic anomalies in O, Si, Ca, S, Ba and Nd, the carbonaceous chondrite fission-xenon anomalies, which occur only in secondary, low-temperature minerals of the most gas-rich meteorites from all classes, are best explained by local processes such as mass fractionation during trapping and fission of a volatile superheavy element.

A "planetary" noble gas component is present in all chondrites (Zähringer, 1962, 1968; Marti, 1967; Heymann and Mazor, 1967) as well as in the atmosphere of Earth (Signer and Suess, 1963) and Mars (Anders and Owen, 1977; Owen, 1978). He and Ne show very large but similar depletions relative to solar system abundances (Cameron, 1973), while the three heavier gases show progressively smaller depletions. The fact that such chemically different objects as carbonaceous and ordinary chondrites, and objects so different in size as meteorite parent bodies and the earth, contain noble gases in similar proportions suggests that this fractionated noble gas component was formed early in the history of the solar system. It seems likely that the fractionation occurred during trapping of the noble gases in dust grains growing from the solar nebula, before accretion of larger bodies. Since the trapping process probably depends on the pressure and the temperature, the gas abundance pattern is a potential clue to conditions in the solar nebula. One has to know, however, the actual host phases, and have the assurance that the pattern was not altered by later losses.

A significant advance in the study of the planetary noble gases in meteorites was achieved when Lewis *et al.* (1975) succeeded in chemically separating and partially identifying the major carrier phases in the Allende (C3V) carbonaceous chondrite. A small fraction (0.04% of the meteorite), insoluble in HF/HCl but soluble under oxidizing treatment, and probably an Fe, Cr sulfide (phase Q), contains 30-60% of this meteorite's primordial noble gases, of normal planetary isotopic composition. Smaller amounts of gas, of distinctive isotopic composition (carbonaceous chondrite fission [CCF] xenon [see below]), are contained in chromite and amorphous carbon (0.4% of the meteorite). Localization of the noble gases in single minerals appreciably clarified the noble gas record, compared to studies done by stepwise heating of bulk meteorites.

A study on LL chondrites of different petrologic type showed that these meteorites, too, contained the same principal noble gas carriers (Alaerts *et al.*, 1977). Furthermore, there seemed to be no evidence for metamorphic redistribution of noble gases in Q and chromite since Ne/Ar ratios decline with increasing Ar content, exactly the opposite of what one would expect for a diffusive loss process.

ELEMENTAL NOBLE GAS RATIOS AS COSMOTHERMOMETERS

Volatile trace metals such as Bi, Tl, In, and Pb have been used as condensation/accretion thermometers of the solar nebula (e.g., Keays *et al.*, 1970; and Laul *et al.*, 1973), with theoretical condensation curves (Larimer, 1967, 1973) as a standard. If we knew the condensation behavior of the planetary noble gases under the solar nebula conditions, then we might use them for the same purpose. Trapping in mineral grains growing from a plasma (Arrhenius and Alfvén, 1971), physical adsorption (Fanale and Cannon, 1972) and clathration in ices (Sill and Wilkening, 1978), all at extremely low temperatures (~100 K or lower), have been proposed as mechanisms to produce "planetary" noble gas patterns. A major difficulty with these models is then to explain the observed correlation of noble gases with volatile metals, which should be fully condensed by ~390 K. Larimer and Anders (1967) and Mazor *et al.* (1970), on the other hand, have suggested that the "planetary" noble gases could be trapped during the formation of secondary minerals at moderately low temperatures, by reaction of their parent materials with H_2S, H_2O and CO in the solar gas. So far, all of the major noble gas carriers seem to belong to this category of minerals (Anders *et al.*, 1975; Lewis *et al.*, 1977).

Although in theory a number of these reactions are accessible in the laboratory, such experiments could not be done until the relevant phases were known and even now slow rates at low temperatures are a major difficulty (Lancet and Anders, 1973; J. Yang, personal communication). Thus far, data have been published on only one reaction: trapping of noble gases in magnetite growing from μm-sized iron powder in an atmosphere of water and noble gases (Lancet and Anders, 1973). Two important findings came out of these experiments: the heats of solution for this type of trapping processes are negative, and they show a remarkable difference between the light and the heavy noble gases (He = -2.4, Ne = -2.2, Ar = -15.2, Kr = -13.0, Xe = -12.5 all in kcal/mole). Ratios such as $^{20}Ne/^{36}Ar$ in a single mineral therefore are sensitive functions of the formation temperature and can in principle be used as condensation/accretion thermometers. This noble gas ratio is a pressure independent cosmothermometer in the region where Henry's solubility law is valid (for magnetite this limit is $>$ 1000 times the noble gas partial pressures of the solar nebula [Lancet and Anders, 1973]).

Although magnetite is not a relevant mineral for ordinary chondrites, the basic difference in the trapping behavior of the light and heavy noble gases should depend mainly on the electronic properties of the atoms themselves and should therefore persist in minerals other than magnetite. A comparison between the Ne/Ar ratios in LL-chondrite residues and the data from the laboratory magnetite experiment is therefore useful. As seen in Fig. 1 (Alaerts *et al.,* 1977) the Ne/Ar ratio of the Krymka Q phase (most primitive) is approximately two orders of magnitude lower than in the St. Séverin Q phase (most metamorphosed). Based on the magnetite calibration this would imply that Krymka formed from material that equilibrated with the solar nebula at a temperature ~100°C lower than the material of St. Séverin.

The linear correlation of $^{20}Ne/^{36}Ar$ versus ^{36}Ar on a log-log plot, as seen in Fig. 1, is predicted by the solubility law of van't Hoff. Q necessarily has an operational definition as that portion of the residue soluble in HNO_3. If Q were a mixture of two phases, samples from different meteorites might well give a mixing line in Fig. 1. But such a mixing line, on this log-log plot, is concave and not straight as observed.

Laboratory trapping experiments with the relevant meteoritic minerals are needed to provide the absolute calibration of this noble gas condensation/accretion thermometer. However, limits can be set from the mineralogy of these meteorites. None of them contains magnetite which puts a lower limit at 400 K (Urey, 1952), while the constancy of the troilite content suggests almost complete condensation of S, putting an upper limit around 600 K (Larimer, 1967).

For the C1, where the evidence is most complete, Anders (1972) has noted that six other cosmothermometers (based on equilibrium and kinetic isotope effects, and on pressure-dependent and -independent chemical equilibria) all point approximately to the same temperature of 360 K. Assumptions are required for all these thermometers, but the variety of processes underlying them makes an accidental concordance improbable. It

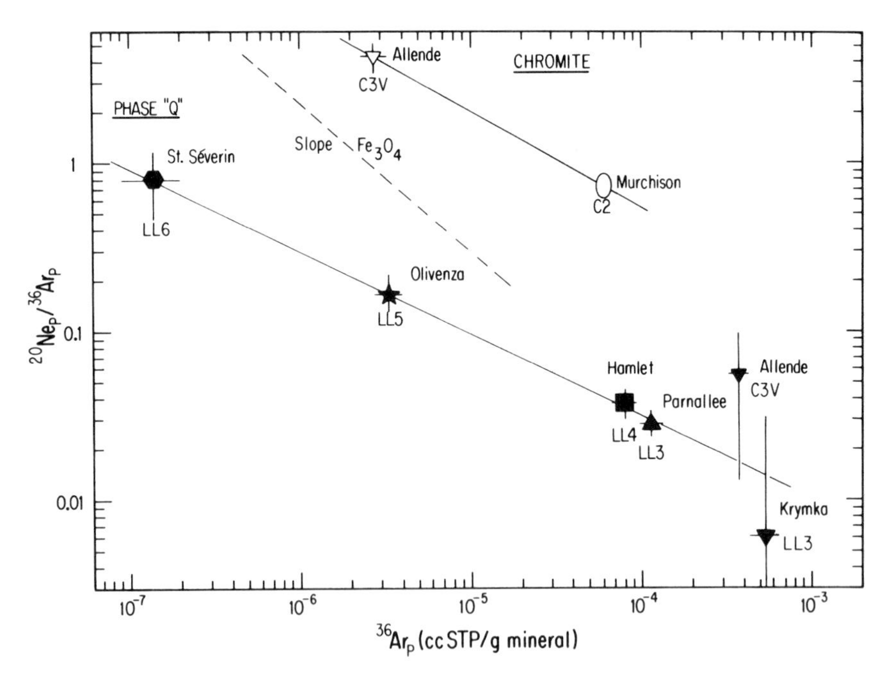

Fig. 1. The Ne/Ar ratio in meteoritic gas-rich minerals decreases with increasing Ar content, as expected for equilibrium trapping of noble gases in solids at decreasing temperatures. Compare trend of LL chondrite data with the dashed line for synthetic Fe_3O_4 (Lancet and Anders, 1973). Diffusion losses during metamorphism would have given the reverse trend and therefore seem to have been negligible.

seems that during accretion of the meteorite parent bodies, temperatures were considerably higher than the present 170 K in the asteroid belt.

Manuel and Sabu (1975, 1977) have noted correlations in our data between various elemental and isotopic ratios, and proposed that these derive from a variable sampling of material from a local nucleosynthetic event. If their hypothesis is true, then all of the foregoing (and also the following) discussion is faulty. We shall return briefly to this point at the end of this chapter, but for the present, we shall just agree with Podosek (1978) that such a hypothesis, which derives the bulk of the solar system from a single supernova remnant is a "rather extreme astrophysical scenario which is widely considered unacceptable."

NOBLE GAS ISOTOPIC ANOMALIES

There is general agreement that discoveries of isotopic anomalies in O (R. Clayton *et al.,* 1973), Si (R. Clayton *et al.*, 1978), Ca (Lee *et al.*, 1978), S (Rees and Thode, 1977) Ba and Nd (McCulloch and Wasserburg, 1978) in high-temperature inclusions of the meteorite Allende, all point to presolar material carrying the record of presolar nuclear processes. The temptation has therefore been great to attribute the CCF xenon component discovered by Reynolds and Turner (1964), and localized by the acid treatment technique (Lewis *et al.,* 1975), to such processes (Rowe, 1968; Manuel *et al.,* 1972; Black 1975; D. Clayton, 1975). Indeed NeE (Black and Pepin, 1969) as well as a newly discovered xenon component (Srinivasan and Anders, 1978) seem to be noble gas components with just such an origin. However, more difficulties than benefits accrue, in our opinion, to any similar attempt with the venerable CCF xenon.

CCF xenon is an isotopically anomalous component for which the enrichments of the heavy isotopes suggested a fissiogenic origin. As known fissioning elements were ruled out, three groups of authors (Srinivasan *et al.,* 1969; Anders and Heymann, 1969; Dakowski, 1969) simultaneously suggested that a superheavy element might have been the progenitor. Anders and Heymann (1969) noted that the correlation of CCF xenon with volatile trace elements pointed to the elements 112-119, the only transuranium elements expected to be volatile. Chromite, carbon and Q, an ill-determined sulfide, were found to be principal carriers of CCF xenon (Lewis *et al.,* 1965). This was done by first treating meteorite samples with HC1-HF, yielding a ~0.5% residue (chromite, carbon and Q) and then etching the residue in an oxidizing medium such as HNO_3. This last treatment removes less than 10% of the residue (Q), about one third of the residues' fission xenon but 95% of the trapped xenon. This leaves an etched residue (chromite and carbon) strongly enriched in CCF xenon.

It appears that a local origin for this CCF component, such as fission for the enrichment in the heavy isotopes and mass fractionation for the enrichment in the light ones, is the best hypothesis. However, we certainly do not claim that our favored model is without drawbacks. In particular, we would agree with Manuel *et al.* (1972) that the correlation of excess light isotopes, which are shielded from production by fission, with the excess heavy isotopes (putatively fissiogenic) presents a problem. First seen by Reynolds and Turner (1964) this correlation requires a coincidence in our model such that the strongest fractionation must occur in the same phase as will have, after our treatment, the greatest ratio of fissiogenic to trapped xenon. This has become a very entangled field and we refer the reader to the growing body of literature (e.g., Lewis *et al.,* 1977; Frick and Moniot, 1977; Manuel and Sabu, 1975; D. Clayton, 1975, and references cited therein).

Briefly the two main host phases of CCF Xe do not have any of the other isotopic anomalies expected for a presolar origin, nor a chemical composition consistent with it. Indeed, no anomalies were found in any element checked thus far, except for the noble gases. Q has normal, solar system Pb and a large complement of normal trapped noble gases. Chromite/carbon has essentially normal O, C, and Os. Especially the $^{12}C/^{13}C$ ratio of 90.4, that falls squarely within the terrestrial range, is strongly diagnostic of local material since such high ratios are very unusual in the galaxy (most interstellar clouds have ratios near 40 [Townes, 1977]). It may be easy to envision a presolar nucleosynthetic event which changed the $^{124}Xe/^{132}Xe$ ratio by at least 63%, but it is difficult to envision this anomalous xenon being packaged with such commonplace, O, C, and Os, having isotopic compositions within ±2% of solar. Of all the elements measured in these minerals, only the noble gases show large anomalies. Since noble gases differ from other elements only in their chemical, not their nuclear properties it seems likely that the origin of the anomalies must involve these chemical properties.

A most important observation in this context is that chromite and carbon contain noble gases of virtually identical isotopic ratios, elemental ratios and abundances. Lewis *et al.* (1977) proposed that the chromite and the carbon were formed simultaneously by oxidation of Fe, Cr grains with CO: $Fe + 2Cr + 4CO \rightarrow FeCr_2O_4 + 4C$. The noble gases, after adsorption on the Fe, Cr grains would then be trapped simultaneously in the two growing phases. This model readily explains the similarity of the noble gases in the very distinct phases, chromite and carbon, a fact difficult to account for in a supernova scenario.

With this model in mind, the monotonic trend of the isotopic composition of the noble gases in chromite/carbon relative to the solar ratios (δ%) ($^3He/^4He$ = −62%; $^{20}Ne/^{22}Ne$ = −36%; $^{36}Ar/^{38}Ar$ = −10%; $^{80}Kr/^{82}Kr$ = +2% and $^{124}Xe/^{132}Xe$ = +63%) suggests a chemical process, namely mass fractionation. The reversal of sign is no embarrassment, because a mechanism for such reversal has been in the literature for 17 yr. W.B. Lewis (1961) found both enrichments and depletions of −2 to +4% per mass unit in the trapping of Xe in UO_2 and proposed a two-step mechanism: adsorption, favoring the heavy isotope, and migration to a potential well, favoring the light one. If, for a particular solid, the first step is rate-determining for the lighter gases and the second step for the heavier gases, the reversal in sign would be accounted for.

All models for the origin of the CCF-xenon require two processes, one to enrich the light isotopes, another to enrich the heavy ones. A major advantage of a supernova origin for the CCF xenon is that the *p* and *r* processes are thought to occur there together. Hence the observed association of the light and heavy xenon enrichments would be quite natural. The proponents of a supernova (local or not) have as yet not explained how this highly anomalous xenon got trapped in isotopically normal chromite and carbon in such similar amounts. This is especially hard to account for since carbon, oxygen, Fe + Cr, the different noble gases and the nuclear processes which allegedly changed their isotopic composition, all occur in different zones of a layered supernova (Fig. 14 of Trimble, 1975).

It is instructive to contrast the CCF xenon with the anomalous ^{16}O rich component discovered by R. Clayton *et al.* (1973). Generally, large ^{16}O enrichments (up to 50‰) are found in high-temperature inclusions of some primitive meteorites, precisely the sort of refractory material most likely to have avoided complete mixing with the solar nebula. These are the same inclusions which sometimes have isotopic anomalies in the other elements cited in the first paragraph of this section. Smaller variations in the amount of the admixed ^{16}O component (a few $^{0}/_{00}$) separate the different classes of solar system material (R. Clayton *et al.*, 1976). The ^{16}O component, in this case, is well mixed with the rest of the oxygen. Separate carriers enriched in the ^{16}O component are not found. Rather, different phases show only fractionation trends on a $\delta^{17}O$ versus $\delta^{18}O$ plot. It is only the CCF xenon which is a large and unhomogenized anomalous component, carried in secondary low-temperature minerals, and never (as yet, at least) found to be associated with isotopic anomalies in elements other than the noble gases.

The previous ideas have been mainly based on work on Allende. Table I includes data from most of the meteorites studied by the acid dissolution technique at the University of Chicago thusfar. The meteorites are classified in order of increasing percent-mean-deviation (PMD) of the Fe content in the olivine, a measure of primitiveness. The last column demonstrates that only the petrologically more primitive meteorites show the 136 enrichment in the etched residues. High $^{136}Xe/^{132}Xe$ ratios also correlate with the content of volatile metals in the bulk meteorites and with the xenon content in the unetched residue. A low Cr/Fe ratio in the residues of the meteorites showing the ^{136}Xe enrichment, is indicative of Fe(III) substitution for Cr(III) in the chromite which again suggests a low formation temperature. The presence of phases other than chromite, mostly carbonaceous matter, is another characteristic of these gas-rich residues. All this evidence suggests that, if the heavy part of CCF xenon is made by fission, the progenitor has to be volatile. A more limited number of these correlations, some years ago, was the basis for the suggested volatile superheavy element origin of the heavy part of CCF xenon (Anders and Heymann, 1969). There are difficulties with the fission origin but weighing all the evidence, the other hypotheses have more problems (Anders, 1978).

As mentioned at the end of the previous section, Manuel and Sabu (1975, 1977) have laid great emphasis on their observation that the elemental ratios of the noble gases correlate with the xenon isotopic variations. They would argue that these correlations support a local supernova model in which the *r*- and *p*-process isotopes of xenon are made in regions of a supernova rich in the light elements. Various admixtures then of this component (Xenon-X) with a second noble gas component of normal xenon and no light elements (He and Ne burned up), would produce the observed correlations. Their hypothesis would clearly, if true, make all of the previous discussion in error. For our part, we see no need for such a global anomalous component. Our residues are binary mixtures of two noble carriers. Q (low $^{20}Ne/^{132}Xe$, low $^{136}Xe/^{132}Xe$) and chromite/carbon (high $^{20}Ne/^{132}Ne$, high $^{136}Xe/^{132}Xe$). Two processes, both favored by low temperatures, are responsible for the correlation. Low temperatures favor condensation of the volatile superheavy

TABLIE I

Correlation of Fission Xenon with Primitiveness of the Meteorites[a]

Class, Name	Oliv. PMD	Bulk parts per billion		Residue			Etched Residue 136/132
		Bi	Tl	132[b]	Cr/Fe	"C" %	
LL6 St. Séverin		–	–	2	1.7	2	0.315
LL6 Manbhoom		8.7	0.9	4	–	–	0.315
LL5 Olivenza		5.1	14.6	13	1.4[c]	10	*0.26*[c]
LL4 Hamlet	3.5	8.7	2.3	1110	1.8	24	0.329
LL3 Parnallee	19	36.4	1.5	1230	1.8	19	0.327
LL3 Chainpur	32	67	31	3200	–	–	0.359
LL3 Bishunpur	39	16.6	24	2850	–	–	0.352
LL3 Krymka	45	39.2	114	3190	1.3	53	0.421
C3O Lancé	63	49	84	3035	1.0	65	0.380
C3O Ornans	68	10.1	15.7	4624	1.0	50	0.356
C3O Kainsaz	70	42.8	41.0	3730	1.0	70	0.386
C3V Allende	86	44.5	59	2330	1.4	61	0.572
C3V Leoville	95	–	–	2320	1.2	84	0.381
C2 Murchison	100+	62	94	4010	1.7[c]	95	0.412

[a]Percent-mean-deviation (PMD) values from Dodd *et al.* (1967) and Van Schmus (1969). Bi and Tl values from Laul *et al.* (1973), Anders *et al.* (1976), Takahashi *et al.* (1978), Ikramuddin *et al.* (1977) Krähenbühl *et al.* (1973). Except for Allende (Lewis *et al.* 1975, 1977) and Murchison (Srinivasan *et al.* 1977), the other data are from two papers in preparation by Alaerts *et al.* (1978).

[b]10^{-10} ccSTP/g.

[c]Uncertain.

element and hence raise ^{136}Xe/^{132}Xe. They also change the composition of the chromite, which seems to favor trapping of He and Ne. Slightly different conditions for formation of the gas-rich minerals in different meteorites give variations by a factor of two in the slope of the mixing lines (Fig. 2).

Studying meteorites other than Allende has shown that the correlation between the ^{124}Xe and the ^{136}Xe enrichments is not as perfect as concluded from Allende data alone (Lewis *et al.*, 1978). Thus Manuel's Xe-X component (enriched in *p*- and *r*-process nuclides) is not as uniform as his supernova model seems to predict.

This area has become subject to strong feelings and divergent interpretations. What is obviously needed is more pertinent data. In particular, we see a need for a better characterization of the elusive Q phase, laboratory experiments on the trapping behavior of known noble gas carriers, a search for other host phases, an explanation for the fractionation trend

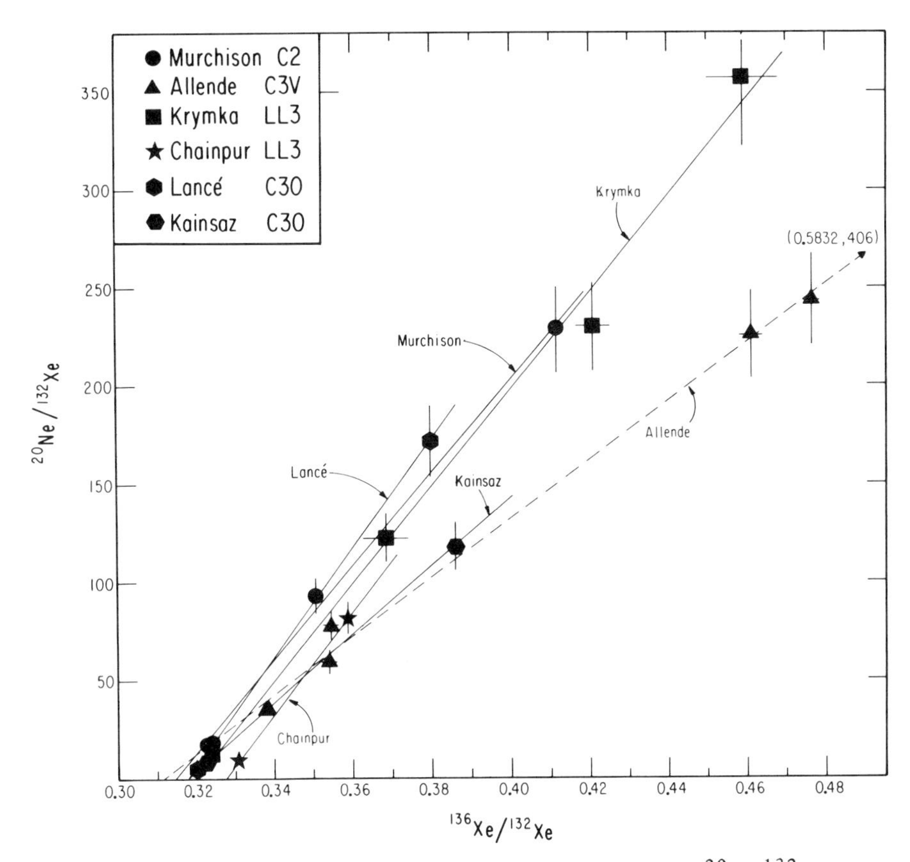

Fig. 2. Principal gas-bearing minerals in Allende are Q (low $^{20}Ne/^{132}Xe$, low $^{136}Xe/^{132}Xe$) and chromite/carbon (high $^{20}Ne/^{132}Xe$, high $^{132}Xe/^{132}Xe$). Allende samples are mixtures of these two phases and hence lie on a mixing line. Other meteorites have their own mixing lines, due to slight differences in condensation temperature (changing the elemental ratios) and/or to different amounts of fission xenon (changing the isotopic ratios).

from He to Xe and a testing of the predictions of the fission hypothesis through a search for anomalous fission barium in association with the anomalous xenon.

Acknowledgements. We thank E.J. Olsen of the Field Museum, Chicago, for meteorite samples and J. Gros and H. Takahashi for some of the chemical work. We express our gratitude for the guidance this work has received from E. Anders. This research has been supported in part by the National Aeronautics and Space Administration. One of us (L.A.) thanks the North Atlantic Treaty Organization for additional financial support in the form of a Research Fellowship.

REFERENCES

Alaerts, L; Lewis, R.S.; and Anders, E. 1977. *Science* 198: 927.
____. 1978. In preparation.
Anders, E. 1972. In *L'Origine du Système Solaire* (H. Reeves, ed.), p. 179. Paris: C.N.R.S.
____. 1978. *Robert A. Welch Foundation Conference, XXI. Cosmochemistry*. Houston, Texas: Welch Foundation. In press
Anders, E., and Heymann, D. 1969. *Science* 164: 821.
Anders, E.; Higuchi, H.; Ganapathy, R.; and Morgan, J.W. 1976. *Geochim. Cosmochim. Acta* 40: 1131.
Anders, E.; Higuchi, H.; Gros, J.; Takahashi, H.; and Morgan, J.W. 1975. *Science* 190: 1262.
Anders, E., and Owen, T. 1977. *Science* 198: 453.
Arrhenius, G., and Alfvén, H. 1971. *Earth Planet. Sci. Lett.* 10: 253.
Black, D.C. 1975. *Nature* 253: 417.
Black, D.C., and Pepin, R.O. 1969. *Earth Planet. Sci. Lett.* 6: 395.
Cameron, A.G.W. 1973. *Space Sci. Rev.* 15: 121.
Clayton, D.D. 1975. *Astrophys. J.* 199: 765.
Clayton, R.N.; Grossman, L.; and Mayeda, T.K. 1973. *Science* 182: 485.
Clayton, R.N.; Mayeda, T.K.; and Epstein, S. 1978. *Lunar and Planetary Sci. Conf. 9th.* (Abstract) 1: 186.
Clayton, R.N.; Onuma, N.; and Mayeda, T.K. 1976. *Earth Planet Sci. Lett.* 30: 10.
Dakowski, M. 1969. *Earth Planet. Sci. Lett.* 6: 152.
Dodd, R.T., Jr.; Van Schmus, W.R.; Koffman, D.M. 1967. *Geochim. Cosmochim. Acta* 31: 921.
Fanale, F.P., and Cannon, W.A. 1972. *Geochim. Cosmochim. Acta* 36: 319.
Frick, U., and Moniot, R.K. 1977. In *Proc. Lunar Sci. Conf. 8th.* p. 229. New York: Pergamon.
Heymann, D., and Mazor, E. 1967. *Science* 155: 701.
Ikramuddin, M.; Binz, C.M.; and Lipschutz, M.E. 1977. *Geochim. Cosmochim. Acta* 41: 393.
Keays, R.R.; Ganapathy, R.; and Anders, E. 1970. *Geochim. Cosmochim. Acta* 35: 337.
Krähenbühl, U.; Morgan, J.W.; Ganapathy, R.; and Anders, E. 1973. *Geochim. Cosmochim. Acta* 37: 1353.
Lancet, M.S., and Anders, E. 1973. *Geochim. Cosmochim. Acta* 37: 1371.
Larimer, J.W. 1967. *Geochim Cosmochim. Acta* 31: 1215.
____. 1973. *Geochim. Cosmochim. Acta* 37: 1603.
Larimer, J.W., and Anders, E. 1967. *Geochim Cosmochim. Acta* 31: 1239.
Laul, J.C.; Ganapathy, R.; Anders, E.; and Morgan, J.W. 1973. *Geochim. Cosmochim. Acta* 36: 329.
Lee, T.; Papanastassiou, D.A.; and Wasserburg, G.J. 1978. *Astrophys. J. (Letters)*. In press.
Lewis, R.S.; Alaerts, L.; and Anders, E. 1978. *Lunar and Planetary Sci. Conf. 9th.* (Abstract) 1: 645.
Lewis, R.S.; Gros, J.; and Anders, E. 1977. *J. Geophys. Res.* 82: 779.
Lewis, R.S.; Srinivasan, B.; and Anders, E. 1975. *Science* 190: 1251.
Lewis, W.B. 1961. *Atomic Energy Canada Ltd.*, Report AECL-1402.
Manuel, O.K.; Hennecke, E.W.; and Sabu, D.D. 1972. *Nature (London) Phys. Sci.* 240: 99.
Manuel, O.K., and Sabu, D.D. 1975. *Trans. Missouri Acad. Sci.* 9: 104.
____. 1977. *Science* 195: 208.
Marti, K. 1967. *Earth Planet. Sci. Lett.* 2: 193.
Mazor, E.; Heymann, D.; and Anders, E. 1970. *Geochim. Cosmochim Acta* 34: 781.
McCulloch, M.T., and Wasserburg, G.J. 1978. *Astrophys. J. (Letters)*. In press.
Owen, T. 1978. *Moon and Planets* (special Protostars and Planets issue). In press.
Podosek, F.A. 1978. *Ann. Rev. Astron. Astrophys.* In press.
Rees, C.E., and Thode, H.G. 1977. *Geochim. Cosmochim. Acta* 41: 1679.
Reynolds, J.H., and Turner, G. 1964. *J. Geophys. Res.* 69: 3263.

Signer, P., and Suess, H. 1963. In *Earth Science and Meteoritics* (J. Geiss and E.D. Goldberg, eds.), p. 241. Amsterdam: North-Holland.
Sill, G.T., and Wikening, L.L. 1978. *Icarus* 33: 13.
Srinivasan, B.; Alexander, E.C., Jr.; Manuel, O.K.; and Troutner, D.E. 1969. *Phys. Rev.* 179: 1166.
Srinivasan, B., and Anders, E. 1978. *Science.* In press.
Srinivasan, B.; Gross, J.; and Anders, E. 1977. *J. Geophys. Res.* 82: 762.
Takahashi, H.; Janssens, M.J.; Morgan, J.W.; and Anders, E. 1978. *Geochim. Cosmochim. Acta* 42: 97.
Townes, C.H. 1977. *Observatory* 97: 52.
Trimble, V. 1975. *Rev. Mod. Phys.* 47: 877.
Urey, H.C. 1952. *The Planets.* New Haven: Yale Univ. Press.
Van Schmus, W.R. 1969. In *Meteorite Research* (P.M. Millman, ed.), p. 480, Dordrecht: Reidel.
Zähringer, J. 1962. *Z. Naturforsch,* Teil A 17: 460.
____. 1968. *Geochim. Cosmochim. Acta* 32: 209.

PART V

Protoplanets and Planetesimals

PHYSICS OF THE PRIMITIVE SOLAR NEBULA AND OF GIANT GASEOUS PROTOPLANETS

A. G. W. CAMERON
Harvard-Smithsonian Center for Astrophysics

There is a growing list of elements in meteoritic inclusions which show isotopic anomalies, including evidence in magnesium for the presence of the extinct short-lived radioactivity ^{26}Al in the early solar system. Truran and the author have suggested that these anomalies represent an abnormally rich admixture of nuclides produced in a supernova which occurred nearby at the time of formation of the solar system. We further have suggested that the supernova triggered the collapse of an interstellar cloud, which fragmented into progressively less massive parts during its collapse. We have postulated that one of the ultimate fragments then formed the primitive solar accretion disk from which the solar system developed. The evolution of such a viscous accretion disk has been studied using a physical theory developed by Lynden-Bell and Pringle. There would be several vigorous stirring processes present in such a primitive solar nebula, and an optimistic view has been taken during the studies of the magnitude of the resulting turbulent velocities. Mechanical waves generated by the turbulence are expected to heat the upper layers of the primitive solar nebula, and, when there has been a sufficient diminution of the infall accretion pressure on these layers, a warm corona could form from which mass loss from the accretion disk will be quite rapid. Disk dissipation leads to an outward flow of angular momentum and energy, and to a mass flow which is inward near the center of the disk and outward at larger radii. The time behavior of characteristic quantities within the disk are presented for typical values of critical parameters. The results are found to be invalid in detail because of the neglect of global instabilities which will form rings in the primitive solar nebula. These rings would break up into giant gaseous protoplanets resulting in the formation of many such protoplanets; their number would be reduced by mutual collisions. The *hydrostatic evolution of giant gaseous protoplanets has been studied by DeCampli. Protoplanets with masses comparable to that of Jupiter or less are expected to have liquid droplets of rocky material in their interiors; the rainout of these rocky materials to form a core of the protoplanet would occur very rapidly, according to the work of Slattery. In the inner part of the solar system, protoplanets are expected to undergo tidal stripping which would leave only cores of condensed materials if such have formed. Giant gaseous protoplanets would then raise tides within the primitive solar nebula that would lead to rapid transport of angular momentum, an important process that must be included in the future studies of the evolution of the primitive solar nebula.*

THE SUPERNOVA TRIGGER

The discovery by Lee, Papanastassiou and Wasserburg (1976) that there were significant amounts of ^{26}Al present in the early solar system has had a profound effect upon thinking about the formation of the sun and its accompanying family of planets. This isotope has a half life of 0.72×10^6 yr. and the principal mechanism for its manufacture in nature is in the explosive carbon-burning process which occurs in supernovae. From this it follows that there was a supernova close to the solar system at the time of its formation. The probability that such a close proximity should be accidental is rather low, and Cameron and Truran (1977) therefore proposed that the supernova responsible for injecting ^{26}Al into the early solar system was in fact responsible for triggering the collapse of an interstellar cloud in order to produce a system of stars, one of which would be the solar system.

Truran and I have suggested that a variety of isotopic anomalies which had recently been discovered in meteorites might also have been produced by this triggering supernova. Elements in which such isotopic anomalies had been found include oxygen, neon, aluminum, iodine, krypton, xenon, lead, and plutonium. Some of the anomalies take the form of short-lived radioactivities, which would be present in the early system, but are extinct today. These anomalous isotopes might be manufactured in a variety of layers within the supernova when explosive nucleosynthesis occurred in such layers.

There has recently been an explosive increase in the rate of discoveries of isotopic anomalies, particularly associated with two inclusions in the Allende meteorite, EK1-4-1 and C1. Both inclusions have very large magnesium anomalies, but it has not yet been found whether ^{26}Al contributes to these. Other elements showing anomalies, particularly in EK1-4-1, are silicon (Clayton, Mayeda and Epstein, 1978; Yeh and Epstein, 1978), calcium (Lee, Papanastassiou and Wasserburg, 1978), strontium (Papanastassiou, *et al.*, 1978), barium and neodymium (McCulloch and Wasserburg, 1978), and samarium (Lugmair, Marti, and Scheinin, 1978).

A common feature of these anomalies is the change in the ratio of the abundances of the odd and even mass numbers within a multi-isotope heavy element. McCulloch and Wasserburg (1978) suggested that anomalies of this type might represent the addition or subtraction of *r*-process material relative to the solar system normal mixture of the products of nucleosynthesis. However, the ratio of odd to even mass number abundances can also change in the *s*-process if the temperature at which neutron capture occurs is changed. Therefore, it is premature to associate these anomalies with nucleosynthesis in any particular layer of a supernova, and the bearing of these anomalies on the supernova trigger hypothesis cannot yet be determined.

A supernova can be expected to act as a trigger in the following way: The explosion will occur when the presupernova star is at a random location in the interstellar medium (although very massive stars will not have moved very far from their places of formation, which are likely to be the rich cloud complexes in the spiral arms of the galaxy). Most supernovae will explode in the intercloud medium since clouds, although containing an appreciable fraction of the mass of the medium, occupy only a relatively small fraction of its volume. However, the distance to the nearest cloud is likely to be only a few parsecs. When the envelope is ejected from a star in a supernova explosion, the envelope gases are slowed by a shock which propagates through the intercloud medium. This converts the kinetic energy of the expanding gases and the swept-up interstellar medium to very high gas kinetic temperatures.

To a first approximation, the gas in the interstellar medium will tend to approach a common pressure throughout, being accelerated by pressure gradients wherever such pressure uniformity has not been achieved. The pressure in the material expanding away from a supernova explosion is considerably higher than the pressure of the medium into which the expansion occurs, at least until distances of a few tens of parsecs have been reached. When the expanding material encounters a cloud of much higher density, it finds this to be an obstacle around which it flows, exerting a higher pressure upon the boundaries of the cloud. This pressure will cause anisotropic compression of the cloud, possibly leading it to the threshold of gravitational collapse. Studies of the hydrodynamics of this process have been reported by McKee and Cowie (1975), Sgro (1975) and Woodward (1976). In addition, Gull (1973) has discussed the formation of Rayleigh-Taylor instabilities at the forward face of the interstellar cloud, leading to the penetration of tongues of the supernova ejecta which extend into the interior of the cloud, and may inject the products of the supernova nucleosynthesis into the interior of the cloud.

It is possible to offer some qualitative arguments suggesting that a cloud triggered into gravitational collapse in this way will undergo extensive fragmentation. The hydrodynamic studies cited above attempt to show that the collapse which occurs will be very anisotropic and the critical Jeans mass for the collapse is unlikely to be that of the cloud as a whole, but rather of a fairly small section of the cloud. Furthermore, the dynamic compression is likely to lead to the development of a vigorous turbulence within the collapsing gas. Such vigorous turbulence, when it involves velocities approaching sound speed, leads also to strong dissipation within the cloud with an accompanying set of density fluctuations. As the cloud collapses, the critical Jeans mass for growth of instabilities in a quiescent medium continually decreases, so that it seems likely that many of the density irregularities produced by dissipation of turbulence can become nuclei about which additional fragmentation will occur. After fragmentation has proceeded to the point where the ultimate fragments have masses comparable to that of the sun, it is expected that their further collapse will lead to the formation of stars. However, any fragment formed within a collapsing interstellar cloud will have internal velocities derived from the shear field of the turbulence, and hence the fragments will have a random angular momentum distribution derived from the turbulence. In most cases, the collapse of the fragment will then lead to the formation of a gaseous accretion disk which has dimensions large compared to those of a star in spherical hydrostatic equilibrium.

This is the rationale behind the assumption which I made everywhere here, namely that an appropriate initial condition from which to study the primitive solar nebula is that of a collapsing interstellar cloud fragment, which is highly turbulent and has been formed from a highly turbulent medium.

EVOLUTION OF THE PRIMITIVE SOLAR NEBULA

Interstellar cloud fragments of the type just described are likely to exhibit a spectrum of masses. In approaching the problem of the origin of the solar system, I have generally studied gas masses containing about two solar masses. The reason for this is that the process of forming stars could be a relatively inefficient one as far as the usage of the available mass is concerned. Thus, it appears that young stars may possibly lose some tens of percent of their initial mass in the course of their T Tauri mass loss. In addition, when the infalling gas forms a gaseous disk, not all of the mass in the disk can dissipate inwards into a star, since some must remain to absorb the angular momentum which is transported outwards. Thus I have taken two solar masses as the mass of the infalling material that was responsible for the formation of the solar system, but the amount could have been somewhat less or perhaps substantially more.

As mentioned above, the infalling gas is expected to be highly turbulent, and the two solar masses of gas can be expected to have a net angular momentum derived from a field of turbulence. I have crudely estimated that the angular momentum of this infalling matter should be about 8×10^{53} gm cm^2 sec^{-1}, and used this value throughout all the studies reported here (Cameron, 1973).

The first steps in studying the behavior of the primitive solar nebula formed by interstellar cloud collapse were to assume that the primitive solar nebula was completely formed by accretion, to examine the structure of the nebula, and to examine the rate of dissipation in the gas. This formed the work of Cameron and Pine (1973). The models computed by us proved to be rather fat disks close to the point of global instability according to a criterion depending on the ratio of thermal to rotational energy developed by Ostriker (1972). However, the most important results of the study by Cameron and Pine were the determination that a characteristic cooling time in the inner regions of the primitive solar nebula would be in the range $10^2 - 10^3$ yr, and the characteristic angular momentum transport time would be in the range $10^3 - 10^4$ yr. Angular momentum transport in these models of the primitive solar nebula was taken to depend upon meridional circulation currents only. Turbulence was not assumed to be present except where driven by thermal convection. These various dissipation times are short compared to the infall time of the gas in the last stage of collapse of an interstellar cloud fragment which probably lies in the range $10^4 - 10^5$ yr (Larson, 1969). From this study it became clear that a proper treatment of the evolution of the primitive solar nebula would have to include the effects of dissipation occurring within the solar nebula during the entire process of accretion from the interstellar cloud fragment. It also became clear that it was unrealistic to ignore the effects of turbulence that would be induced by the vigorous meridional circulation currents which were estimated to exist in the models constructed by Cameron and Pine.

The basic physical theory needed to treat the evolution of a viscous accretion disk was published by Lynden-Bell and Pringle (1974). I describe here the essential features of the calculation that was carried out for the formation of the sun from a viscous accretion disk by my application of the Lynden-Bell and Pringle theory (Cameron, 1978). It must be kept in mind that the details contained in these calculations are not in themselves important, for the results of these calculations have also revealed important deficiencies in the treatment which need to be corrected in subsequent studies.

When the study reported by Cameron (1978) began, it was clear that there would be two important driving forces which should induce vigorous turbulence within the primitive solar accretion disk. As indicated above, one of these is the meridional circulation currents which have been treated in some detail only for nearly spherical rotating stars (see Mestel, 1965 for a review). In a rotating star it is not possible to construct coincident surfaces on which thermodynamic variables and effective gravitational potentials are constant. Under these circumstances, the luminous flux escaping from the interior of a star exerts a small acceleration upon the matter, which sets up a circulation within the meridional planes of the star. As the ratio of the centrifugal to the gravitational acceleration increases, the velocity of the circulation current rapidly increases. Such effects have only been quantitatively studied for stars in which the centrifugal effects can be treated as perturbations on the stellar structure. It is clear that the velocities of the circulation currents must continue to increase as large rotational deformations are produced. On the other hand, in the other extreme where rotational flattening would produce a disk of infinitesimal thickness, the surfaces of constant thermodynamic variables and of effective gravitational potentials would once again coincide, and the circulation currents would vanish. Therefore, we have a qualitative expectation that the meridional circulation currents will have their greatest velocities for fat disks, in which the thickness is a substantial fraction of the radial distance. For a rough estimate of the circulation current velocity, one can use a simplification of an estimate given by Mestel (1965), in which I have set the centrifugal force equal to the gravitational force, which gives

$$v \sim L/Mg \tag{1}$$

where L and M are the luminosity and mass of a cm^2 column in the disk, and g is the local acceleration of gravity. A random sampling of disk conditions in the models produced by Cameron (1978) indicate that this value of the meridional circulation velocity is comparable to or somewhat greater than the local value of the sound speed in all of the sampled cases.

When flow velocities within a gas approach sound speed, very efficient dissipation takes place. This generally limits the value which can be obtained by the flow velocities to some number in the range of one-tenth to one-third of the sound speed, provided the driving force is very strong. The above estimates, using Eq. (1), indicate that the meridional circulation current probably does provide a strong driving force, and that the actual meridional circulation velocity may approach the speed of sound.

For the large dimensions which are characteristic of the primitive solar accretion disk, and the large velocities associated with the meridional circulation currents, the critical Reynolds number for the onset of turbulence is exceeded by many orders of magnitude in most parts of the disk. Only in the very outermost parts of the disk, where the gas density is so low that the disk becomes optically thin at the center, does the criterion for driving meridional circulation currents break down and invalidate these arguments. This criterion alone leads to the expectation that the primitive solar nebula should be nearly everywhere vigorously turbulent, with the velocities of the largest eddies approaching some substantial fraction of sound speed.

The second initial reason for expecting extensive turbulence in the primitive solar nebula is associated with the character of the infalling accreted gas. As discussed above, this gas is also likely to have a vigorous turbulence within it. This means that at any particular point on the accretion disk at which infall is occurring, the component parallel to the plane of the disk of the vector velocity of the infalling gas will undergo large temporal fluctuation; in other words, the angular momentum brought in by the infalling gas will undergo fluctuations in time which are substantial fractions of its mean value. Furthermore, the mean value of the infalling angular momentum per unit mass at any point on the disk will not be correlated with the angular momentum per unit mass within the disk itself, since substantial redistributions of angular momentum can occur within the disk on relatively short time scales. The upper layers of the atmosphere of the disk which absorb the infalling material will therefore be given substantial random velocities with respect to the mean motions. This should provide a vigorous stirring of the upper layers of the disk atmosphere, which might not be directly acted upon by the meridonal circulation currents.

During the course of the study carried out by Cameron (1978), it became evident that certain portions of the disk would have other driving forces which should induce turbulence. For portions of the disk with temperatures in excess of 2000 K it was found that there was a superadiabatic temperature gradient along the central plane of the disk in the radial direction. This superadiabatic gradient was very large, and would drive enormously strong thermal convection. In addition, as will be described below, it was found that the disk should become repeatedly unstable against gravitational instabilities which would form giant gaseous protoplanets. These massive condensations will stir the surrounding gas, and also they must be expected to raise tidal bulges within the surrounding gaseous disk, which can lead to a rapid transferal of angular momentum within the disk in a manner analogous to that proposed for binary systems by Papaloizou and Pringle (1977). The latter effect is an angular momentum transport process which is not included within the theory of Lynden-Bell and Pringle, and it should be explicitly treated in future studies of the evolution of the primitive solar nebula.

The calculations reported by Cameron (1978) have taken an optimistic view of the strength of the driving forces. I have assumed that the turbulent velocity in the largest eddies is one-third of the local sound speed. This means that the calculations present a limiting case in which the turbulent viscosity will be taken as large as it could possibly attain under any reasonable physical circumstances. Downward adjustment of the turbulent velocity in the largest eddies simply increases the strength of the conclusion reached below concerning gravitational instabilities in the primitive solar nebula.

The theory of Lynden-Bell and Pringle, which describes the dissipation of an accretion disk, does not make any assumptions about the character of the viscosity involved, but treats this quantity as a parameter to be supplied by the user. Consider two adjacent cylinders concentric about the spin axis within the primitive solar nebula, each of thickness one centimeter. Lynden-Bell and Pringle write an expression for the couple, g which is exerted at a radius, R of the inner material on the outer material in the form

$$g = 4\pi R^2 \nu\sigma A \tag{2}$$

where $A(R)$ is the local rate of shearing

$$A(R) = -\frac{1}{2}R\frac{\mathrm{d}\Omega}{\mathrm{d}R} \tag{3}$$

Ω is the angular velocity about the center, $\nu\,(R)$ is the viscosity and $\sigma\,(R)$ is the surface density. Utilizing the specific angular momentum h in the form

$$h = \Omega R^2 \tag{4}$$

Lynden-Bell and Pringle obtain a basic equation

$$F\frac{\mathrm{d}h}{\mathrm{d}R} = -\frac{\partial g}{\partial R} \quad . \tag{5}$$

In the interpretation of Eq. (5) it should first be noted that the specific angular momentum is a monotonically increasing function of the radial distance; therefore $\mathrm{d}h/\mathrm{d}R$ is always a positive quantity. The viscous couple g is small near the spin axis, increases steadily with increasing radial distance until it reaches a maximum, and then declines steadily until it vanishes at the outer edge of the disk. The details of this behavior depend upon the associated physics of the problem and the boundary conditions applied to the system in which the equation is being solved. However, it follows that the mass flux, F is inwards in the inner portion of the disk and outwards in the outer portion.

It is necessary to supplement Eq. (5) by an equation for the continuity of the flow. We may write this in the form

$$\frac{\partial\sigma}{\partial t} + \frac{1}{2\pi R}\frac{\partial F}{\partial R} - \frac{\phi\,(R)}{2\pi R} = 0 \tag{6}$$

where ϕ is the vertical flux accreted on a ring one centimeter wide at radial distance R due to the infall of material from the collapsing interstellar cloud fragment.

In order to apply the Lynden-Bell and Pringle theory to a disk structure, it is necessary to specify the vertical structure of the disk. Among other things, this will determine the area of the adjacent cylinders between which turbulent viscosity will provide friction. In the most general terms, one must expect that the pressure within the disk will decline between midplane and the high atmosphere. If one were to move a large fluid element from the region of the midplane to the high atmosphere, then to the degree that the fluid element expands and the expansion is adiabatic, one would expect it to cool. However, in a medium with such vigorous turbulence as is expected here, there is a very high efficiency for converting kinetic energy of turbulence into mechanical waves which will propagate in all directions through the disk. For the presently assumed conditions, such waves start with a large amplitude and they will steepen in amplitude to become shocks as they progress into a region of smaller density. Hence, the upper layers of the disk can be expected to be strongly heated by shock dissipation.

Clearly this is a very complicated situation, and I have made some simplifying assumptions with regard to it. During the time that the bulk of the infall accretion is occurring onto the disk from the collapsing interstellar cloud fragment, the pressure exerted on the upper atmosphere by the infalling material is quite substantial. I have assumed that the infalling material will form an accretion shock in the high atmosphere in which the potential energy released by the infall will be radiated away. I have taken the vertical structure of the disk to be isothermal and I have ignored the modification of the uppermost structure due to the infall accretion shock.

It may be expected that in the later stages of the evolution of the disk, infall accretion will have greatly diminished. Under those circumstances the shock heating of the higher layers of the disk atmosphere can be expected to make these layers much hotter than the midplane of the disk. At a sufficiently large distance from midplane, the density of the atmosphere will become sufficiently low so that the efficiency of radiative processes is very low. In the sun the deposition of energy in the upper atmosphere by mechanical waves produces a high-temperature plasma, in which the thermal conductivity is very high, so that most of the energy deposited in the chromosphere is conducted downwards to the base of the chromosphere where the energy is radiated into space. Therefore, only a relatively small fraction of the deposited energy is available for the expansion of the corona to form the solar wind. In contrast, the mechanical energy deposited in the uppermost layers of the disk atmosphere should go much more efficiently into the hydrodynamic expansion of those layers. The base of the solar corona must be raised to a temperature of at least 10^6 K in order to produce the solar wind flow, but in the primitive solar nebula the gravitational potential is generally much smaller and hence the temperature in the "disk corona" needed is typically only 10^3 K or less. Perhaps it would be better to describe such a disk corona as warm rather than hot. At these lower temperatures the thermal conductivity is very much less, and the scale height in the atmosphere is enormous compared to that in the solar chromosphere. Hence thermal conduction cannot be expected to play a significant role in the disk atmosphere.

I have made a very crude estimate that when mass loss is free to occur in the primitive solar nebula about ten percent of the energy produced in the disk, where the temperature at midplane is about 100 K, is used to energize the escape of mass from that point in the disk, and that the efficiency of use of the energy produced within the disk declines toward the inner region of the disk as the temperature at midplane of the disk increases. It should also be noted that the energy per gram required for mass loss increases as the inner portion of the disk is approached and the gravitational potential deepens. The actual empirical formula I used for estimating the mass loss under typical disk conditions was

$$\frac{\mathrm{d}M}{\mathrm{d}t} = - \frac{D}{R f_r T^{\frac{1}{2}}} \tag{7}$$

where D is the rate of energy production per cm^2 column in the disk, f_r is the radial force at midplane in the disk, and T is the temperature in midplane. I used a symbolic expression for the radial force within the disk because the distributed nature of the mass in the disk cannot in general be accurately represented by an analytical expression. I found that the general character of the evolution of a disk is not very sensitive to the precise form of this mass loss equation.

The energy generated at any point in the disk results from the viscous dissipation taking place within the disk. The expression given by Lynden-Bell and Pringle for the rate of dissipation per cm^2 column is

$$D = \frac{1}{2\pi R} g \left(- \frac{\partial \Omega}{\partial R} \right) \tag{8}$$

where Ω is the angular velocity within the disk. If we neglect the channeling of a small fraction of this dissipated energy into mass loss, and assume that the rate of radiation from the cm^2 column into space is equal to the rate at which energy is dissipated within the column, then the photospheric surface temperature is

$$T = \left(\frac{D}{2\sigma_0} \right)^{\frac{1}{4}} \tag{9}$$

where σ_0 is the radiation constant. The factor 2 which appears in this expression comes from the fact that any cm^2 column in the disk has 2 cm^2 of radiating area. With the assumption that the vertical structure in the disk is isothermal, it therefore follows that the temperature at midplane is also given by Eq. (9).

The equation of hydrostatic equilibrium in the vertical z direction is

$$\frac{dP}{dz} = -\frac{f_r z \rho}{R} \tag{10}$$

where P is the pressure and ρ is the density. These two quantities may be connected by an ideal gas equation of state

$$P = \frac{N_0 k T \rho}{\mu} \tag{11}$$

where N_0 is Avogardro's number, k is the Boltzmann constant, and μ is the mean molecular weight. Hence, Eq. (10) can be solved to give the vertical density structure

$$\rho = \rho_c \exp\left(-\frac{\mu}{2N_0 kT}\frac{f_r}{R}z^2\right) \tag{12}$$

where ρ_c is the density of midplane.

We define the principal scale height, H within the vertical structure to be

$$H = \left(\frac{2N_0 kT}{\mu}\frac{R}{f_r}\right)^{\frac{1}{2}} . \tag{13}$$

This allows the density variation to be expressed in the simple form

$$\rho = \rho_c \exp(-z^2/H^2) . \tag{14}$$

During the evolution of the primitive solar nebula the mass in the disk may be comparable to the mass in hydrostatic equilibrium on the spin axis at the center. As mentioned above, the radial force within the disk in general cannot be accurately expressed in analytical form, owing to the fact that the surface density, $\sigma(R)$ cannot be accurately expressed as an analytical function of R. However, in the course of the numerical work carried out in this study, I found that the surface density could very crudely be expressed in the form

$$\sigma = \frac{M_0}{2\pi R_0 R} \sin^{-1}\left(1 - \frac{R^2}{{R_0}^2}\right)^{\frac{1}{2}} \tag{15}$$

where M_0 is the mass of the disk and R_0 is the radius of the outer boundary of the disk. The properties of these disks have been extensively discussed by Hunter (1965). For such a disk accompanied by an additional mass, M at the center, the radial force is

$$f_r = \frac{GM}{R_2} + \frac{\pi}{2} \frac{GM_0}{R_0 R} \tag{16}$$

The specific angular momentum of the combination is

$$h = \left(GMR + \frac{\pi}{2} \frac{GM_0 R^2}{R_0}\right)^{\frac{1}{2}} \tag{17}$$

The remaining physical quantity that must be defined is the turbulent viscosity, $\nu(R)$. To the degree that moving blobs of turbulent gas can be thought of as equivalent to molecules in a gas, then the form of the turbulent viscosity can be written by analogy to that of the molecular viscosity. This means that one must take the turbulent viscosity to be one-third of the product of a characteristic velocity and a characteristic mixing length. The interaction between the largest eddies within the disk will therefore give the greatest contribution to the turbulent viscosity within the disk. I have written the turbulent viscosity in the following form:

$$\nu = \frac{2}{9} c_s H \tag{18}$$

where c_s is the sound speed given by

$$c_s = \left(\frac{\Gamma_1 N_0 k T}{\mu}\right)^{\frac{1}{2}} \tag{19}$$

where Γ_1 is the first adiabatic exponent of the gas.

These are the essential features which must be specified in addition to the basic equations of Lynden-Bell and Pringle in order to determine the evolution with time of the disk structure. It is also necessary to specify boundary conditions which the solutions must satisfy. One of these concerns the character of the infall accretion. I have made the crude simplifying assumption that the angular momentum distribution in the infalling material is given by assuming that this material is derived from the interior of a uniformly rotating cloud of uniform density having the assigned total angular momentum for two solar masses of material. I have made a further simplifying assumption that the infalling material lands on the disk at a point where its specific angular momentum matches that of the material in the disk. These assumptions avoided the necessity to treat the angular momentum redistribution which would actually occur due to the mismatch between the specific angular momenta of the fluids at the points of merger. I do not believe that they cause a gross distortion in the general character in the evolution of the disk.

Another boundary condition concerns the treatment of the outer radial edge of the disk. A straightforward utilization of the equations given above leads to the result that the temperature at midplane of the disk becomes extremely small at very large radial distances. This actually is a nonphysical situation, since a great deal of energy is released by viscous dissipation in the interior parts of the disk. This radiant energy slowly diffuses away from the primitive solar nebula through the infalling accreting material of the collapsing interstellar cloud fragment. This diffusion causes the outgoing radiation to heat the outer layers of the disk at quite large distances. A crude estimate of the magnitude of this effect caused me to prohibit the material in the outer portion in the disk from having a temperature below 20 K. This sets a lower limit on the sound speed in the outer portions of the disk. It was optimistically assumed that the outer edge of the disk could advance in the outward radial direction at one-third of the local sound speed during the time of accretion of material from the collapsing interstellar cloud fragment. During the subsequent mass loss phase, mass was also assumed to be lost through two vertical scale heights of the disk at the outer edge, and the outer edge was therefore assumed to retreat towards smaller radial distances at a rate which corresponded to this mass loss.

EVOLUTION OF THE PRIMITIVE SOLAR NEBULA

The paper of Cameron (1978) includes several evolutionary histories of possible primitive solar nebulae, in which various fundamental parameters have been varied. I show here the results for only one of the cases, utilizing values for the parameters which I thought to be most appropriate. These parameters include the expected angular momentum mentioned earlier for the central two solar masses of the infalling material, an infall rate of two solar masses in 3×10^4 yr, and a transition from accretion to mass loss when the temperature at the orbit of formation of the planet Mercury rises to 1500

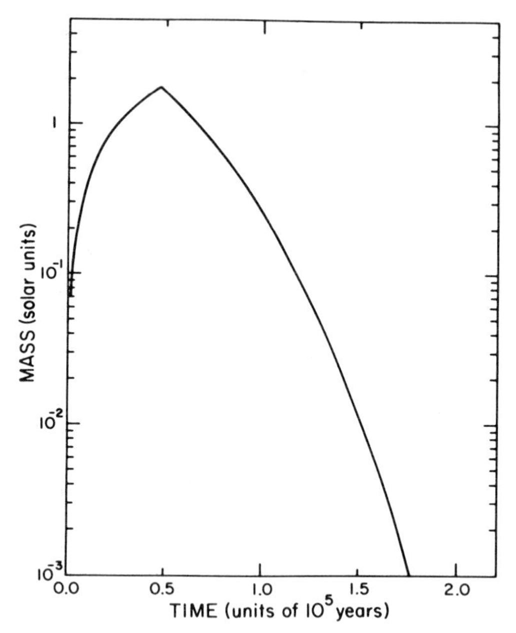

Fig. 1. Mass of the disk as it varies with time. Standard case.

K. Figure 1 shows the history of the mass in the disk. It may be seen that the disk mass becomes quite large, an indication that mass accumulates in the disk much faster than dissipation can move large portions of it into the center. When the disk mass approaches its maximum, only about 10% of the mass lies in the range of the present planets, from Mercury to Neptune, and the rest lies at much larger distances. The period of the accretion turns out to be slightly less than 5 x 10^4 yr, a relatively short period of time, but consistent with estimates of the dynamical collapse time of an interstellar cloud fragment in its late stages. Once mass loss begins from the disk, the mass in the primitive solar nebula becomes reduced to a relatively small remnant of the maximum amount on a similarly short time scale. In this particular example, of the maximum mass which exists in the disk at the time accretion ends, about 15% continues to flow into the sun and the rest is lost by hydrodynamic expansion from the warm corona.

Figure 2 shows the time history of the radius of the primitive solar nebula. It may be noted that the radius grows to a value of several hundred astronomical units during the mass accretion phase, after which it falls quite rapidly for a while during a period of rapid mass loss, and then it shrinks more slowly. The combination of a massive solar nebula at a large radial distance which disappears in a relatively short period of time provides the basis for a new theory of comet formation discussed by Cameron (1978).

It is evident that the accretion process leads to a primitive solar accretion disk which contains relatively little mass near the center in the early stages of

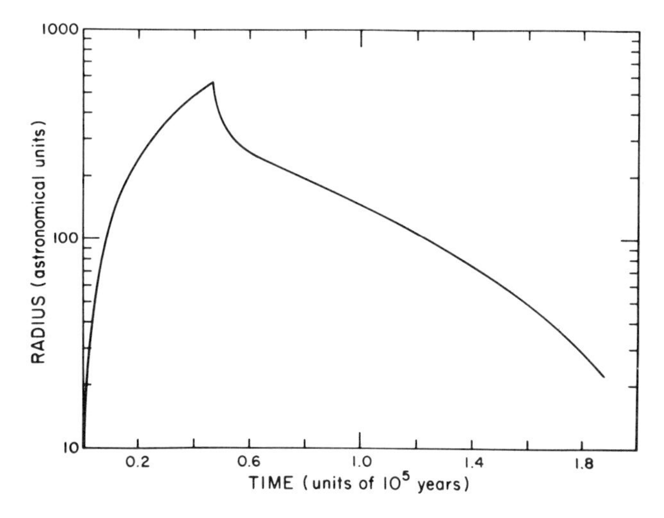

Fig. 2. Radius of the disk as it varies with time. Standard case.

the accretion. Let us define a region of planet formation to be a position in the disk having a given specific angular momentum equal to that possessed by the planet today. The radial distance of any large planetary body possessing this specific angular momentum would change as the body conserves its angular momentum while mass flows past it toward the center of the nebula during the course of the dissipation.

Figure 3 shows how the radial distances of the various regions of planet formation vary during the evolution of the solar nebula.

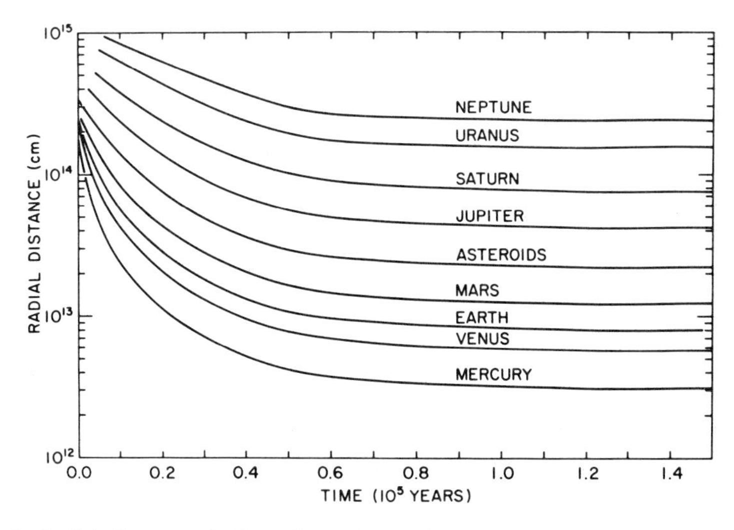

Fig. 3. Radial distances of planet formation regions as they vary with time, centered on specific angular momenta of present planets.

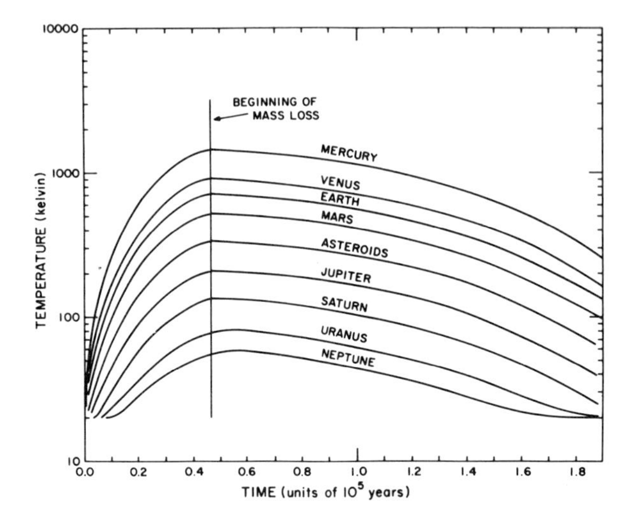

Fig. 4. Temperature in regions of planet formation as it varies with time.

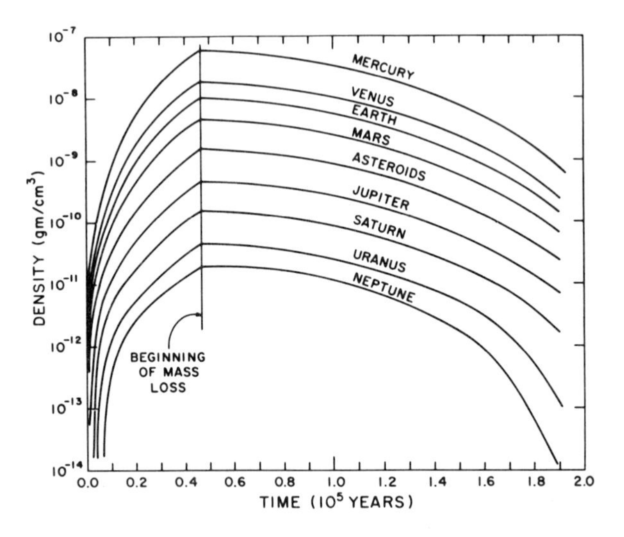

Fig. 5. Density at midplane in regions of planet formation as it varies with time.

Figures 4 and 5 show the variation with time of the temperature and density within the regions of planet formation. It may be noted that only near the orbit of Mercury would it appear that the temperature within the unperturbed primitive solar nebula may become high enough to vaporize completely the original interstellar grain material which would fall into the primitive solar nebula with the incoming accreting gas. Such interstellar grains would lose by evaporation their more volatile materials within the inner portion of the solar system, but a residual structure containing iron and silicates should remain to form the basis from which larger bodies can accumulate. These conclusions are not arbitrarily dependent upon the choice of the temperature at the region of formation of Mercury for the switch from the mass accretion to the mass loss. If mass accretion had been allowed to continue for a significant period of time beyond the chosen switching time, then the temperature in the regions of formation of the inner planets would indeed rise to the point where grains could completely evaporate, but then totally unrealistic quantities of mass would have been incorporated into the primitive solar nebula.

The reason for the low temperatures within the primitive solar nebula throughout its entire history is easy to understand. During the collapse of the interstellar cloud fragment, the energy released by the compression of the gas is readily radiated away, and most of the collapse of the gas cloud occurs with interior temperatures which are likely to be close to 10 K. When the material falls onto and merges with the primitive solar accretion disk, there is plenty of time for the infall energy of accretion to be radiated away into space; that energy is radiated from the accretion shock and has been neglected here. The energy radiated away in these calculations is the dissipation energy of the gas in the disk. The viscous friction within the disk not only transports outwards angular momentum but also the energy which is released by the friction and supplied from the gravitational potential well into which the material is moving. That means that the energy which is released at any point in the disk is for the most part generated at points much closer to the spin axis. The temperature in the disk can therefore be increased only if the disk contains much more mass or if the viscous dissipation per unit mass is increased. But I have already been as optimistic as is conceivably reasonable for the efficiency of the viscous dissipation. The mass is already as high as is reasonable. Since the viscous dissipation would probably be considerably less than shown here, it follows that the temperatures shown in Fig. 4 are probably overestimates.

INSTABILITIES IN THE PRIMITIVE SOLAR NEBULA

In the study by Cameron (1978) it was not possible to make an explicit test within each model to determine whether instabilities were present. Instead, two approximate criteria were calculated for each model, one being a rough test for a local patch instability, and the other being a test for a global instability of the axisymmetric type (ring formation).

The test for a local patch instability was based upon a criterion given by Goldreich and Ward (1973). Based upon a limiting wavelength for propagation of radial waves as determined from a dispersion relation, these authors define a critical surface density

$$\sigma^* = \frac{\kappa c_s}{\pi G} \tag{20}$$

where

$$\kappa^2 = 2\Omega \left(\Omega + \frac{d(R\Omega)}{dR} \right). \tag{21}$$

The derivation of this formula is correct only for an infinitessimally thin disk, for which a local patch instability will exist for surface densities larger than σ^*. The critical surface density for the instability will be increased for a more realistic disk of appreciable thickness. Goldreich and Ward estimate that a more appropriate criterion for the local patch instability would be that the surface density should exceed about three times that given by Eq. (20).

Figure 6 shows the local instability criterion for the accretionary phase of the primitive solar nebula model whose behavior has been discussed in the preceding section. It may be seen that the surface density, measured in the critical unit σ^*, rises rapidly in all of the zones of planet formation in the primitive solar nebula at an early time in the evolution, and then the surface density in these critical units starts to decrease at times comparable to or less than 2×10^4 yr, relatively early in the history of the accretion disk. The peak surface densities in the various regions of the disk lie between 1.1 and 2.6 critical units. Thus, it appears that the surface density in this model comes close to that required to induce local patch instabilities, but it is not evident that these actually occur anywhere. It is worth noting however that if I had used a smaller velocity for the turbulence in the largest eddies, then the disk would contain a larger surface density and it would be cooler; under such

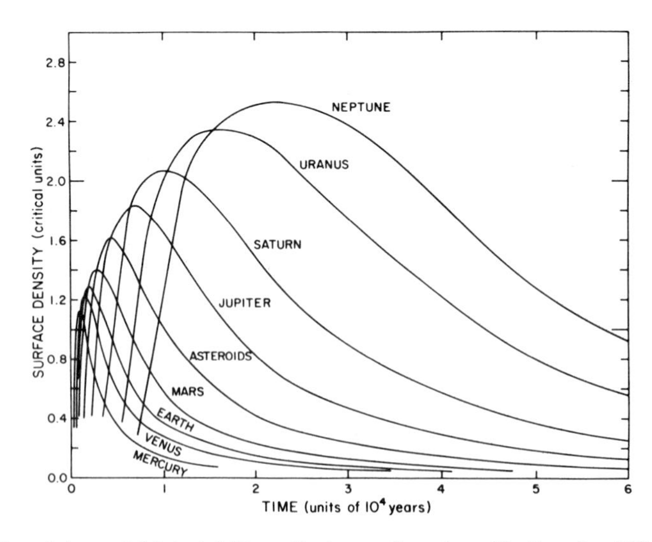

Fig. 6. Local (or patch) instability criterion as it varies with time for different planet formation regions. Standard case.

conditions it is probable that a calculation similar to this would indicate periods of local patch instability during the evolution. However, it is likely that the effects of the global instabilities, to be described next, will reduce the surface density in the disk below that shown in these calculations, thus rendering it unlikely that local patch instabilities will become superimposed upon global instabilities.

The instabilities of a global nature in the primitive solar nebula arise because of the inability of dissipation in the disk to transport mass quickly enough in toward the spin axis where it can accumulate onto a central object. It is therefore evident that the criterion for the onset of global instabilities of the ring formation type will depend in some way upon the ratio of the mass in the disk to that at the center. For a somewhat more quantitative estimate, I have depended upon an analysis by Yabushita (1966, 1969) who considered the stability of Saturn's rings against the formation of axisymmetric perturbations of the type considered here. For various values of the ratio of the outer to the inner radii of the ring-formation region, Yabushita determined critical values for the ring mass relative to the mass of Saturn at which instabilities would commence. Yabushita's analysis is also based upon the assumption that the ring-forming region of the disk is infinitesimally thick.

In order to apply Yabushita's criterion in an approximate way, it is clearly necessary to choose a ratio of ring radii large enough so that the difference in radial distances will be large compared to the thickness. The bulk of the mass comprising the surface density in the disk is contained within two scale heights as defined by Eq. (13), one on each side of the midplane. A typical value for these two scale heights in the models is about 10-15% of the radial distance. As a matter of convenience, I have therefore taken the ratio of the radial distances for the outer and inner edges of the rings to be equal to the ratio of radial distances between successive planets; this ratio averages about 1.7. I have also included with the central mass the mass in the disk out to the inner edge of the ring-forming region, for two reasons. The first reason is that according to an analysis by Mestel (1963), for a disk surface density variation which is approximately true in the models studied here, the angular motion of the disk material at any radial distance is approximately Keplerian in the force field produced when all the inner-lying matter in the disk is considered to be concentrated at the center. The second reason is that the inclusion of this portion of the disk mass with the central mass gives a more conservative criterion for the global instability to occur.

For the ratio of outer and inner radii of the ring-forming region set equal to 1.7, Yabushita's criterion indicates a critical mass ratio of ring material to central materials of $\sim 7 \times 10^{-3}$. If the amount of matter in the ring-forming region becomes larger than given by this criterion, then it is to be expected that the primitive solar nebula will become unstable against the formation of a ring in a place where this criterion is met. Figure 7 shows the mass ratios as

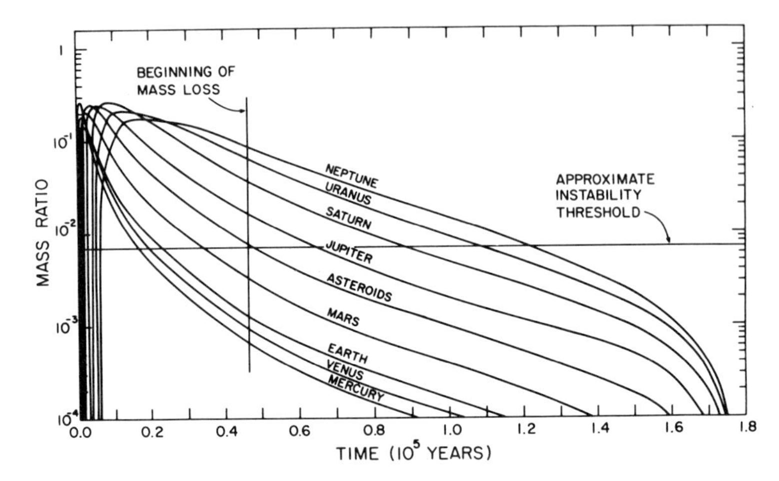

Fig. 7. Global (axisymmetric) instability criterion as it varies with time for different planet formation regions. Standard case.

a function of time in the various regions of planet formation for the models discussed in the preceding section. It may be seen that all of the regions of planet formation become unstable against ring formation very early in the evolution of the primitive solar nebula. The instability persists until late in the stage of dissipation of the disk, at least for the outer regions of planet formation.

I have looked for mechanisms which might be capable of suppressing this ring-formation instability, without success. It may be recalled that the turbulence within the disk was taken to be as vigorous as seemed to be reasonably possible. If the turbulence is assumed to be less vigorous within the primitive solar nebula, then the efficiency of the transport of energy, angular momentum, and mass within the nebula will be reduced, and the instability would be increased according to the criterion used here. Therefore, it does not appear to be possible to escape the conclusion that the primitive solar nebula became repeatedly unstable against the formation of rings early in its evolutionary history.

When a ring instability occurs, the initial behavior is a reduction in the cross-sectional area of the torus by contraction. When the torus becomes sufficiently thin, it must be expected that local instabilities along the line of the torus will cause a fragmentation into several pieces. Current 3-dimensional hydrodynamic calculations of ring instabilities suggest that two or three pieces are likely to be formed (J.E. Tohline, M.L. Norman and J.R. Wilson, personal communications, 1978; see the chapter of Bodenheimer and Black). Each of the pieces will contract upon itself to form a giant gaseous protoplanet.

These giant gaseous protoplanets are initially in an unstable orbital configuration. A certain amount of merging of protoplanetary bodies by collision is likely to take place. Other protoplanets may be perturbed into fairly elliptical orbits, which will later lead to collision with additional protoplanets, possibly those formed in the same ring, but also possibly those formed in another ring.

The thresholds for instability shown in Fig. 7 correspond initially to ring masses not very much greater than the mass of Jupiter. For this reason the most interesting range of protoplanetary masses to be investigated is that of a Jupiter mass, plus or minus a factor of a few, formed from the primitive solar nebula. Such giant gaseous protoplanets will cause very large perturbations in any realistic history of the primitive solar nebula which might replace that shown in the earlier figures of this paper. However, it is clear that a more realistic history cannot be constructed until a reasonable understanding of the evolutionary properties of giant gaseous protoplanets has been achieved.

Calculations by Larson (1972) and by Black and Bodenheimer (1976) indicate that the central region of a collapsing interstellar cloud fragment which forms the disk will become unstable against formation of a ring. Actually, it is more likely to become unstable against the formation of a bar. In either case, the net result is likely to be a pair of giant gaseous protoplanets. If the distance between these is too great, then presumably a binary system will form, and that is of no interest for the formation of the solar system. However, if the distance is not too great, then the work of Papaloizou and Pringle (1977), which discussed accretion disks in binary star systems, may be qualitatively extended to suggest that the rotation of the two protoplanets around their common center of gravity will raise substantial tides in the surrounding disk. This will lead to tidal dissipation in which the angular momentum associated with the mutual rotation of the protoplanets will be transported into the disk by the tidal couple, allowing the protoplanets to spiral together into a common object. D. Lin (personal communication) reports that he and Papaloizou are working on this problem, and have independently arrived at this same result.

There is an important extension of this to the formation of protoplanets by ring instabilities in the disk. Such protoplanets are likely to raise both inner and outer tidal disturbances within the disk. The inner tidal disturbance will result in the transport of angular momentum from the inner disk to the protoplanet, strictly analogous to the binary star models discussed by Papaloizou and Pringle. On the other hand, the interaction of the protoplanet with the outer tidal disturbance will transport angular momentum from the protoplanet into the outer disk. In this way, it is clear that the presence of giant gaseous protoplanets within the primitive solar nebula may be very effective in the transport of angular momentum within the disk. Such a mechanism is not included within the formulation of Lynden-Bell and Pringle, and the angular momentum transport, accompanied of course by mass and energy transports, are additional processes that must be taken into account in improved calculations of the evolution of the primitive solar nebula.

We shall see shortly that protoplanets are very large objects, initially having radii on the order of an astronomical unit. Such objects could not be formed in the inner solar system as it exists today, but it may be noted from Fig. 3 that at early stages in the formation of the disk the radii of planet-forming regions are very much larger than the present planetary radii. It may also be noted from Fig. 7 that the disk instabilities of the global type are expected to set in very early in the evolution of the primitive solar nebula. Therefore, at the time such instabilities may form, there is sufficient room for the objects to exist in the inner region of the disk. There then exists a race between the shrinkage of the protoplanet as a result of its evolutionary processes on the one hand, and the spiraling inward of the protoplanet as mass flows past it, to accrete at the center of the disk and strengthen the gravitational bonds between the forming sun and the giant gaseous protoplanet. If the shrinkage of the radius of the protoplanets loses the race, then the inner Lagrangian point between the giant gaseous protoplanet and the sun and solar nebula will pass inside the radius of the protoplanet, and the envelope of the protoplanet will be tidally stripped away. It appears that this happens in the inner regions of the solar nebula. On the other hand, there are very much larger distances available in the outer solar system, so that protoplanets formed there are in no danger of tidal stripping.

GIANT GASEOUS PROTOPLANETS

A first survey of the properties of giant gaseous protoplanets has been carried out by DeCampli (DeCampli and Cameron, 1978). The structure and evolution of a protoplanet is determined in precisely the same way as the structure and evolution of a star. We therefore start by writing down the familiar equations which govern stellar structure.

The equation of hydrostatic equilibrium for the interior of a protoplanet is

$$\frac{dP}{dr} = -\frac{GM(r)\rho}{r^2} \tag{22}$$

in which r is the radial distance in the interior of the structure, G is the universal gravitational constant, and $M(r)$ is the mass inside the radial distance r, given by

$$\frac{dM(r)}{dr} = 4\pi\rho r^2 \quad . \tag{23}$$

It may be noted that these equations assume spherical symmetry. It may also be noted that these equations for hydrostatic equilibrium require some additional conditions to establish a relation between the pressure, P and the density, ρ, i.e. an equation of state. This relationship will usually depend upon the local value of the temperature, and therefore it is not possible to solve the above two equations for the structure of a body without knowing the run of temperature through it. In order to obtain this, we write down an equation for the luminous flux in the interior, $L(r)$,

$$\frac{\mathrm{d}L(r)}{\mathrm{d}r} = 4\pi\rho r^2 \epsilon \tag{24}$$

in which ϵ is the rate of energy generation at a point in the interior. For the protoplanets under consideration here, ϵ will be determined by the local rate of energy release by gravitational contraction. With knowledge of the luminous flux, it is possible to calculate the temperature gradient required to transport this flux through the interior.

$$\frac{\mathrm{d}T}{\mathrm{d}r} = -\frac{3}{4ac}\frac{\kappa\rho}{T^3}\frac{L(r)}{4\pi r^2} \tag{25}$$

in which a is the radiation constant, c is the speed of light and κ is the opacity. In a simplified calculation one compares this temperature gradient with an adiabatic temperature gradient in the star (the temperature gradient which one would obtain in moving a very large blob of gas through the interior structure of the body quickly while adjusting the blob of gas everywhere to be in local pressure equilibrium with its surroundings). This adiabatic temperature gradient is

$$\frac{\mathrm{d}T}{\mathrm{d}r} = \left(1 - \frac{1}{\gamma}\right)\frac{T}{P}\frac{\mathrm{d}P}{\mathrm{d}r} \tag{26}$$

where γ is the ratio of specific heats.

For calculations deep in the interior of a star, it is customary to use the smaller of the two temperature gradients given by Eqs. (25) and (26). This recognizes that convection is an efficient way to transport energy, so that if the structure becomes unstable against convective transport, which occurs when the radiative gradient exceeds the adiabatic one, then it is expected that the actual temperature gradient will be very close to the adiabatic one. This assumption is also fairly good for the deep interiors of giant gaseous protoplanets. However, near the surface, the density becomes low and the efficiency of convective transport is significantly diminished. Under these circumstances it is necessary to use a more complicated formulation which will give a temperature gradient intermediate between those of Eqs. (25) and (26). This is usually done as some form of mixing length theory (Cox and Giuli, 1968). The actual version of the mixing length theory used in the calculations described here was that due to Henyey, Vardya and Bodenheimer (1965).

These structure equations must be solved subject to certain boundary conditions. The mass, M and the luminosity, L must vanish at the center. At the surface the total mass of the object must be obtained correctly and at a suitable optical depth (two-thirds in a gray-body approximation), the temperature of the structure must be precisely that required to radiate the luminosity into space. Compliance with these boundary conditions is usually achieved by dividing the structure of the body into concentric zones and iterating the conditions at all zone boundaries until the errors in matching the inner and outer boundary conditions have become suitably small.

The calculations on the structure of giant gaseous protoplanets which have so far been done are preliminary and can be used to discuss expected conditions in the primitive solar nebula only with the greatest caution. The reason for this is that the protoplanet models were considered to be the same kind of isolated bodies that stars are: luminous self-gravitating bodies surrounded by a vacuum. As will be seen, the resulting surface temperatures have proved to be only a few tens of degrees. However, the giant gaseous protoplanets were imbedded in the primitive solar nebula at a variety of locations, some of which have temperatures of hundreds of degrees. It is clear that a proper treatment of the structure and evolution of protoplanets under these conditions must make allowance for a continuous gaseous connection to the surrounding primitive solar nebula, and the temperature in the connecting gas must vary smoothly from the protoplanet to the surrounding nebula. This means that the mass associated with giant gaseous protoplanets is likely to change during the course of the evolution, due to gain or loss of mass from the surroundings, and the luminous flux coming from the interior of the protoplanet may undergo large changes, with consequent effects upon the lifetime of the evolution. Therefore, the results that are shown here should be considered to be a first survey of the situation to determine the character of the possibilities.

The most important and least certain of the physical quantities associated with the determination of the interior structure is the opacity, κ. At low temperatures the opacity depends almost entirely on the character of the small particles suspended in the gas. We have earlier seen that the temperature in the unperturbed solar nebula is not expected to rise high enough to evaporate completely the interstellar grains, except at very small distances from the spin axis. Therefore, the character of the particles suspended in the gas will depend upon the degree to which the interstellar grains have managed to clump together during the collapse of the interstellar cloud fragment. I have suggested that some degree of such clumping may be likely (Cameron, 1975). It seems reasonable for a first approach to assume that a substantial part of the interstellar grains have undergone little or no clumping. From the point of view of opacity calculations, the size distribution of the grains does not become important until the grains are large enough to exceed the wavelength of the light which is near the peak of the Planck spectrum.

Two sets of opacity relations have been used in these calculations. One of these sets, shown in Fig. 8, will be described as "low opacities". These opacities contain a peak at very low temperatures due to absorption by water ice. At temperatures of a few hundred degrees there is another peak due to iron and to some degree to rocky materials. For simplicity it was assumed that the opacity due to these condensed materials vanishes when a single evaporation curve, that due to iron, is exceeded by the temperature. The most important source of opacity remaining at that time is that due to water vapor. The water vapor opacity is more important at higher densities, because then the individual lines in the water vapor spectrum undergo greater pressure broadening. There is a further drop in opacity when water vapor becomes dissociated. At still higher temperatures the opacities once again begin to increase because the most important component of them is absorption of the negative hydrogen ion; the number of hydrogen ions increases as the temperature produces increased ionization of elements such as potassium, sodium, and aluminum. This accounts for the principal features shown in Fig. 8, in which the lines which are plotted are those for constant density.

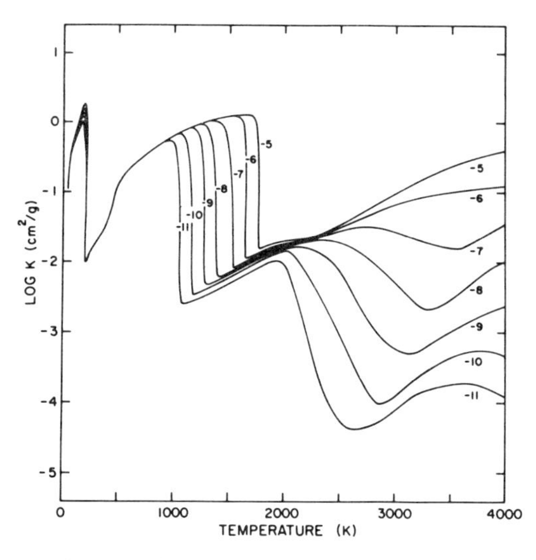

Fig. 8. "Low" opacities as a function of temperature for line of constant density (constant log ρ).

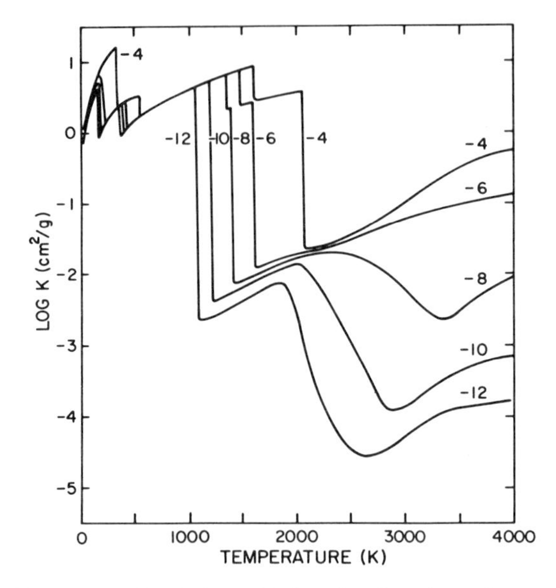

Fig. 9. "High" opacities of Pollack (preliminary) as functions of temperature for lines of constant density (constant log ρ).

Figure 9 shows what we describe as "high opacities". They include new and preliminary calculations by J. Pollack for the opacities due to all of the condensed substances. These opacities are substantially higher than those shown in Fig. 7 for temperatures $\lesssim$1000 K. Both sets of opacities have been used in computing the structure and evolution of protoplanets, owing to the great uncertainty in the opacities and in order to determine somewhat explicitly how sensitive various effects are to whether the opacity is chosen to be high or low.

From these preliminaries we turn now to the results which were obtained for the structure and evolution of giant gaseous protoplanets. Calculations have been carried out for protoplanetary masses ranging from 0.3 times the mass of Jupiter to nearly five times the mass of Jupiter. Such calculations typically started with small values for the central temperature of the model and continued until the temperature at the center had increased to significantly greater than 2000 K. At such temperatures molecular hydrogen starts to be dissociated into atomic hydrogen. This process absorbs much of the released gravitational potential energy, and leads to a dynamical instability in the model causing it to collapse until the hydrogen molecules have been dissociated and hydrogen and helium have been ionized throughout most of the model. Two of the protoplanetary models were followed through this hydrodynamic instability region using a hydrodynamic code.

It is usual to present the results of stellar evolution calculations in a Hertzsprung-Russell diagram, and I do so for protoplanets in Fig. 10. Both

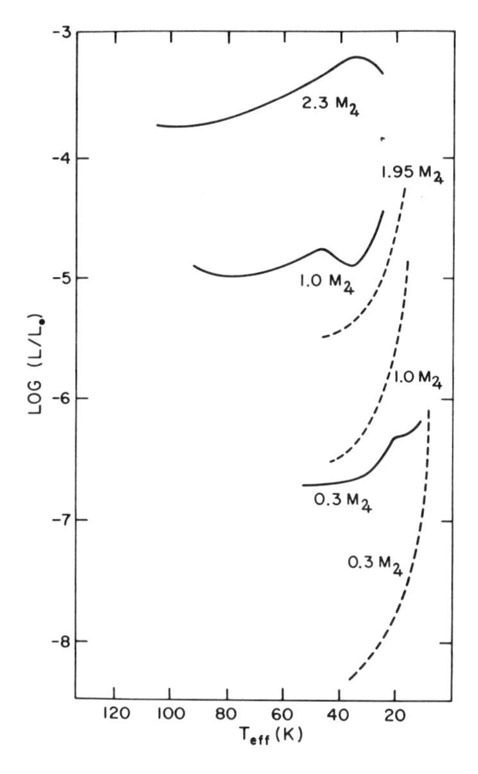

Fig. 10. Evolution of protoplanets showing the Hertzsprung-Russell diagram for giant gaseous protoplanets. Solid line represents the low-opacity case, dashed line the high-opacity case.

high- and low-opacity cases are shown in this figure. It may be noted that the essential difference between the evolution of models of about the same mass but differing opacities is that the luminosity for the low-opacity case is an order of magnitude greater than that of the high-opacity case. Therefore, it follows that the proper determination of the opacities to be used in the study of protoplanetary interiors is especially important. It may also be noted that the luminosities of the models are strong functions of their masses. This is also true for the evolution of stars. Consequently, the evolutionary times of the smaller masses are very much longer than those of the larger masses. These relative features associated with Fig. 10 are the more important ones; the absolute values of the surface quantities are expected to undergo large changes in more realistic models of embedded giant gaseous protoplanets.

Table I shows some characteristics of protoplanetary models with low opacities near the beginning of their evolution. Note the rather low values of the central temperatures and densities. The large radii of the protoplanetary models are particularly striking; the typical values of the radii on the order of an astronomical unit allow the existence of young protoplanets only very

TABLE I

Unevolved Protoplanets

M/M_J	**0.3**[a]	**1.0**	**1.95**
Age (yrs)	156	130	59
Radius (AU)	0.73	2.1	3.7
Central Density ($gm\ cm^{-3}$)	4.8×10^{-10}	6.8×10^{-11}	2.8×10^{-11}
Central Pressure (bars)	7.4×10^{-7}	1.2×10^{-7}	5.7×10^{-8}
T_{eff} (K)	13	17	18
% Convective	85	87	80
Surface Convection Speed/Sound Speed	0.04	0.07	0.096
Log $(L/L_\odot)$	-6.2	-4.8	-4.3

[a]We employed Bodenheimer's (1974) opacities in this model sequence.

early in the evolution of the primitive solar nebula, to judge from the then availability of large distances in the inner solar system as shown in Fig. 3.

Tables II and III show low- and high-opacity protoplanetary models for the most part near the end of their pre-collapse evolution. It may be seen that the central temperature has risen to the vicinity of or beyond 2000 K, and the radii have shrunk by about an order of magnitude. Consistent with the difference in luminosity, the lifetimes for protoplanetary models of about the same mass but different opacities differ by a factor of approximately 10. Again it must be cautioned that in more realistic calculations the masses of the protoplanets may change during the evolution, and the lifetimes will change in presently unknown ways.

Figures 11 and 12 show some gross structural features of the interiors of a Jupiter mass protoplanet in the low- and high-opacity cases, for a variety of models throughout the evolutionary sequence. The essential feature in these diagrams is the large extent of the interior which is convective during major parts of the evolution. This means that most of the mass of a protoplanet will be circulated through the region near the center in a time of a few years. Note also that near the surface the convective velocity becomes a significant fraction of local sound speed. This means that there should be a fairly high efficiency for generating mechanical waves near the surface, which can propagate into the upper atmosphere of the protoplanet, producing a warm corona. It is likely that there will be a substantial loss of mass from the protoplanets as the result of a protoplanetary wind generated by the hydrodynamic expansion of these warm coronas. However, no attempt has yet been made to estimate such potential mass loss.

TABLE II
Evolved Protoplanets
Low-Opacity Case: Models Near Hydrodynamic Collapse

M/M_J	**0.3**	**1.0**	2.3	**4.5**
Age (yrs)	880,000	72,380	8000	1025
Radius (AU)	0.031	0.06	0.163	0.304
Central Density ($gm\ cm^{-3}$)	1.2×10^{-3}	4×10^{-5}	2×10^{-6}	5×10^{-7}
Central Pressure (bars)	130.0	4.5	0.12	0.04
T_{eff} (K)	46	88	100	104
Log ($L/L_{\odot}$)	-6.7	-4.8	-3.7	-3.0

TABLE III
Evolved Protoplanets
High-Opacity Case: Models Near Hydrodynamic Collapse

M/M_J	**0.3**[a]	**1.0**	**1.95**
Age (yrs)	4,031,000	905,000	196,500
Radius (AU)	0.02	0.05	0.10
Central Density ($gm\ cm^{-3}$)	7.2×10^{-5}	4.1×10^{-5}	8×10^{-6}
Central Pressure (bars)	5.0	4.4	0.7
T_{eff}(K)	27	42	48
log ($L/L_{\odot}$)	-7.9	-6.5	-5.7
Mach number of surface convection	0.018	0.02	0.04

[a]Quantities for this model are evaluated about halfway (in time) to hydrodynamic collapse point.

Figure 13 shows the character of the evolution near the center of the protoplanet. Shown here on a plot of the temperature versus the logarithm of the density are several thermodynamic conditions. One of these is the line of demarcation marking the 10% dissociation of hydrogen molecules into atomic hydrogen. This is the approximate position at which the center of a protoplanet goes into hydrodynamic collapse. Most of the protoplanetary evolution has been calculated only to the vicinity of this line, but two of the protoplanetary models have been computed through the hydrodynamic collapse stage. Also in the diagram is the line showing the change from condensed iron to iron vapor. The region of condensed iron is further broken

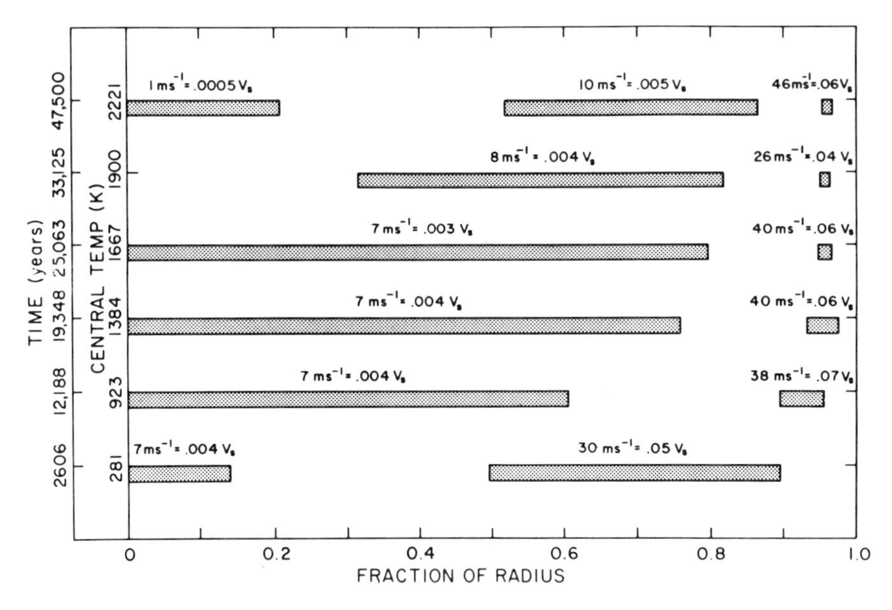

Fig. 11. Interior convective structure of a Jupiter mass protoplanet. The low-opacity case.

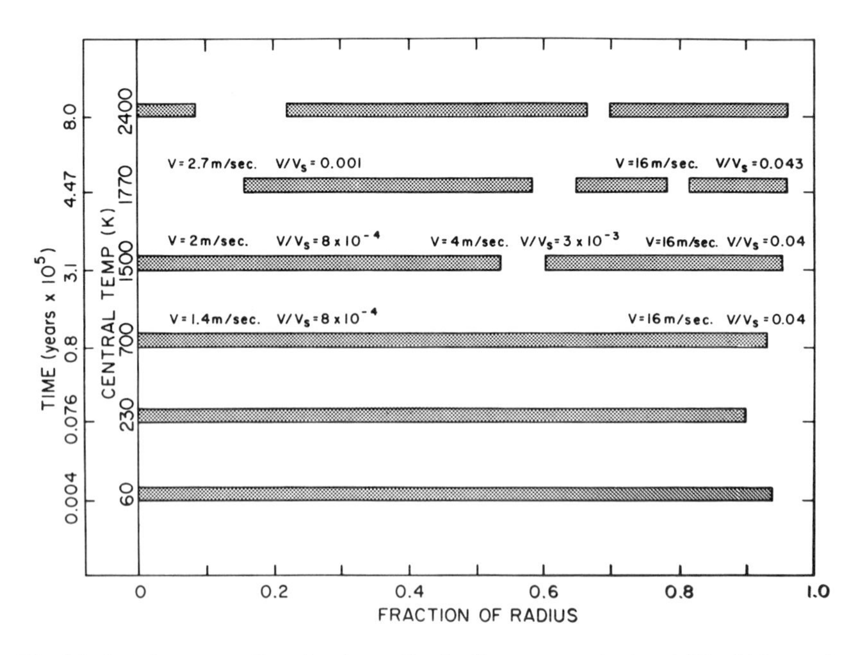

Fig. 12. Interior convective structure of a Jupiter mass protoplanet. The high-opacity case.

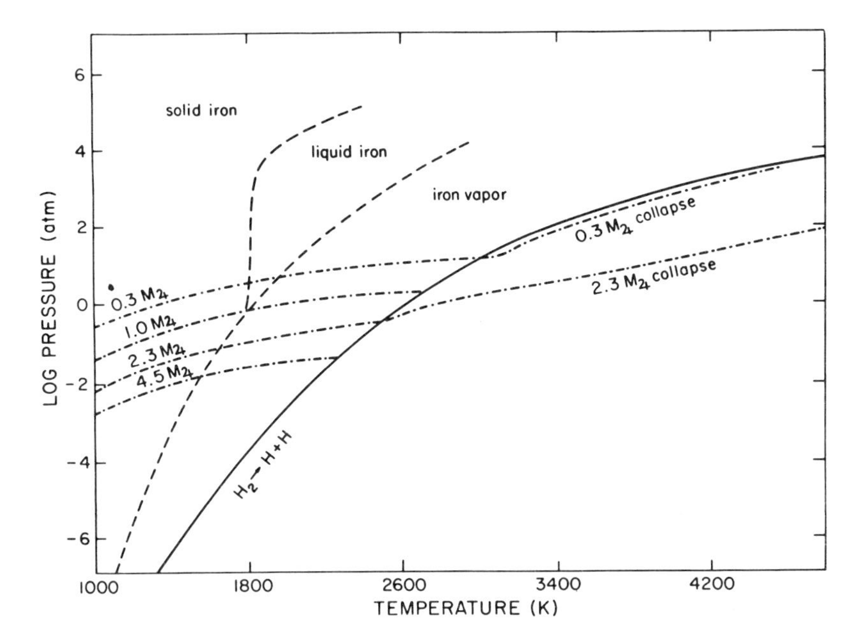

Fig. 13. Evolution of the centers of giant gaseous protoplanets showing the thermodynamic tracks for protoplanet centers in relation to iron and hydrogen phase transitions. Composition: X = 0.77; Y = 0.21; Z = 0.02; Fe = 1.13×10^{-3}.

down into a region of solid iron and one of liquid iron. It may be noted that the centers of protoplanets of about one Jupiter mass and less move through the region of liquid iron. Certain mineral phases present in meteorites will be liquid during a portion of the evolutionary tracks of protoplanets up to two Jupiter masses.

These regions in the interior of a protoplanet in which liquids will exist during a part of the evolutionary lifetime may be of great significance for the formation of planetary cores. There is no special reason to expect two small grains in the interior of the protoplanet to stick together when they collide, any more than two grains of sand would do. However, if the grains or some significant component of them become molten, then it is much more likely that they will stick together upon collision, just as do liquid rain droplets in the atmosphere.

The growth of liquid droplets by coalescence upon collision has been studied in a preliminary way by Slattery who adapted a theory originally devised for the terrestrial raindrop problem to the problem of condensed solids under protoplanetary conditions. He has found that a part of the mass in small particles develops into large particles quickly, in a period of just a few years, because the larger droplets fall more and more rapidly in the gravitational field of the protoplanet, sweeping up smaller droplets as they go. The growth ultimately tends towards saturation due to two effects.

Aerodynamic pressures cause a flattening of the droplets, so that the fall velocities tend toward an asymptotic limit. For these high velocities, very small droplets tend to be carried around the large droplet by the motion of the fluid in which they are suspended, so that in general they can be accumulated onto the large droplet only when the collision is a very central one which allows the small droplet to pass through the stagnation point of the flow about the large droplet. However, by the time these effects become important, the droplets have grown to a radius of many centimeters. Under such circumstances, it only takes them at most a few decades to fall to the center of the protoplanet.

Slattery has obtained his results for initial particle distributions extending to radii of about 10^{-3} cm. The growth times for particle distributions with a smaller size cutoff will clearly be lengthened, and those cases remain to be examined. Figure 13 shows a Jupiter mass protoplanet passing through the tip of the liquid stability region in the thermodynamic diagram. A small reduction in the mass is all that is required to make the track spend several thousand years or more in the liquid region. During this time the convection in the interior of the protoplanet will cause the bulk of the mass of the protoplanet to circulate through the central region, so that the entire mass of the protoplanet can contribute a significant fraction of its condensed solids for fallout to form a central core. A protoplanet with a Jupiter mass, initially formed with solar composition, contains about one Earth mass of condensed material of rocky composition within it. Therefore, it seems likely that a protoplanet close to a Jupiter mass will be able to rain out into the central core a substantial fraction of an Earth mass of refractory rocky materials. The core is quite stable against later re-evaporation into the main volume of the planet. This core formation process may have some relation to the formation of cores in the center of the giant planets, for which there is now some evidence. It may also have some relation to the formation of major portions of the terrestrial planets, if it can occur before the tidal stripping process which we consider next.

It has already been mentioned that there is an inner Lagrangian point between a protoplanet and the sun and solar nebula. After a protoplanet has formed, the dissipation processes within the primitive solar nebula will cause mass to flow past the protoplanet toward the center of the system where that mass will become a part of the sun. This strengthens the gravitational binding between the sun and the protoplanet, causing the protoplanet to spiral inwards. Figure 3 shows the rate at which this happens for the various regions of planet formation. The inner Lagrangian point therefore moves toward the center of the protoplanet. There is some interest to compare the motion of this point with the radius of the protoplanet as a function of time.

Figure 14 shows this comparison for the standard case of evolution within the primitive solar nebula as illustrated in the earlier figures of this chapter, and for a one Jupiter mass protoplanet in both the high- and

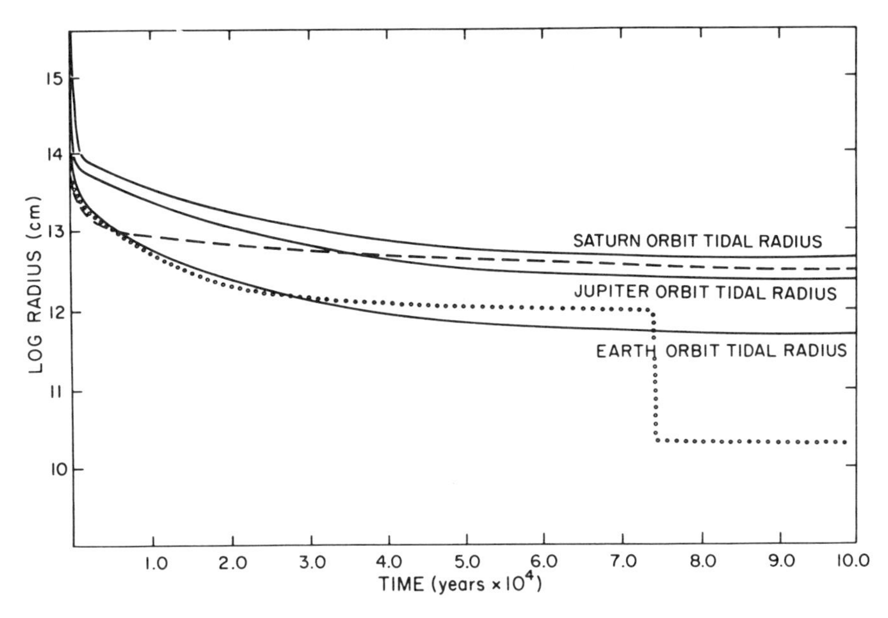

Fig. 14. Protoplanet radius showing the tidal stability of a Jupiter mass protoplanet. Dotted curve represents the low-opacity case, dashed curve the high-opacity case.

low-opacity cases. The tidal radius referred to in this figure is the distance from the center of the protoplanet to the inner Lagrangian point in the solar nebula model. The comparisons were made for the protoplanet placed at the orbits of Earth, Jupiter, and Saturn. It may be seen that the low-opacity model is stable at the position of the earth's orbit for nearly 3×10^4 yr, by which time a condensed core has formed at its center. After that the inner Lagrangian point moves inside the surface of the protoplanet, and the envelope will be stripped away, leaving behind a centrally condensed core if one has formed. This model will be stable at Jupiter's orbit long enough for the central hydrodynamic collapse to take place, after which it will be safely stable throughout the remainder of the evolution of the primitive solar nebula. On the other hand, the high-opacity protoplanet becomes stripped very quickly at Earth's orbit and before a central core has had a chance to form at Jupiter's orbit. All of the protoplanetary models are completely stable at Saturn's orbit.

Figure 14 should be considered principally as showing the main issues which must be resolved in the study of protoplanets and their relationships to the primitive solar nebula. The interesting time scales on which things happen are quite sensitive to the interior opacities. On the other hand, the time scales of the evolution are critically dependent upon the response of the protoplanet evolution to the proper surface boundary conditions, which have not yet been employed and which involve embedding the protoplanets within the primitive solar nebula. Until these issues are resolved by better calculations, it is best simply to note the possibility that tidally stripped cores of rocky materials may be formed in the inner solar system, and that giant gaseous protoplanets may evolve into the giant planets of the solar system with an early pre-collapse formation of rocky cores.

The principal conclusion to be drawn from all of the material presented in this chapter is that the primitive solar nebula was a rather chaotic place, highly turbulent, with the multiple formation of giant gaseous protoplanets. It seems likely that the evolution of the primitive solar accretion disk will depend critically upon processes connected with the presence of giant gaseous protoplanets. It also appears likely that the precise evolution of the giant gaseous protoplanets will depend on the thermodynamic condition in the primitive solar nebula to which their surface conditions must be matched. It is thus evident that a proper understanding of this early episode in the history of the solar system will require a coupled study of the evolution of the solar accretion disk interacting with giant gaseous protoplanets. Because of the stochastic nature of the formation of protoplanets and their subsequent orbital behavior, it would be unreasonable to expect that the present conditions of the planets in the solar system can be predicted entirely from first principles. However, it may become possible to determine well enough the range of possibilities so that examination of early solar system materials will suffice to select the actual situation from among these possibilities.

Acknowledgment. This work has benefited greatly from discussion with D. M. Hunten, W. DeCampli, W. Slattery, D. C. Black and S. J. Peale. It has been supported in part by a grant from the National Aeronautics and Space Administration.

REFERENCES

Black, D.C., and Bodenheimer, P. 1976. *Astrophys. J.* 206: 138.
Bodenheimer, P. 1974. *Icarus* 23: 319.
Cameron, A. G. W. 1973. *Space Sci. Rev.* 15: 121.
_____. 1975. *Icarus* 24: 128.
_____. 1978. *Moon and Planets.* In press.
Cameron, A. G. W., and Pine, M. R. 1973. *Icarus* 18: 377.
Cameron, A. G. W., and Truran, J. W. 1977. *Icarus* 30: 447.
Clayton, R.N.; Mayeda, T. K.; and Epstein, S. 1978. In *Lunar and Planetary Sci.* 9th p. 186. Lunar Planetary Institute, Houston.
Cox, J. P., and Giuli, R. T. 1968. *Principles of Stellar Structure.* Gordon and Breach: New York.
DeCampli, W. M., and Cameron, A. G. W. 1978. In preparation.
Goldreich, P., and Ward, W. R. 1973. *Astrophys. J.* 183: 1051.
Gull, S. F. 1973. *Mon. Not. Roy. Astr. Soc.* 161: 47.
Henyey, L. G.; Vardya, M. S.; and Bodenheimer, P. 1965. *Astrophys. J.* 142: 841.
Hunter, C. 1965. *Mon. Not. Roy. Astr. Soc.* 129: 321.
Larson, R. B. 1969. *Mon. Not. Roy. Astr. Soc.* 145: 271.
_____. 1972. *Mon. Not. Roy. Astr. Soc.* 156: 437.

Lee, T.; Papanastassiou, D. A.; and Wasserburg, G. J. 1976. *Geophys. Res. Lett.* 3: 109.
____ . 1978. *Astrophys. J.* 220: L21.
Lugmair, G. W.; Marti, K.; and Scheinin, N. B. 1978. In *Lunar and Planetary Sci.* 9th p. 672. Lunar Planetary Institute, Houston.
Lynden-Bell, D., and Pringle, J. E. 1974. *Mon. Not. Roy. Astr. Soc.* 168: 603.
McCulloch, M. T., and Wasserburg, G. J. 1978. *Astrophys. J.* 220: L15.
McKee, C. F., and Cowie, L. L. 1975. *Astrophys. J.* 195: 715.
Mestel, L. 1963. *Mon. Not. Roy. Astr. Soc.* 126: 553.
____. 1965. In *Stellar Structure* (L. H. Aller and D. B. McLaughlin, eds.), p. 297. Chicago, Ill.: University of Chicago Press.
Ostriker, J.P. 1972. In *L'origine du Systeme Solaire* (H. Reeves, ed.), p. 154. Paris: C.N.R.S.
Papaloizou, J., and Pringle, J. E. 1977. *Mon. Not. Roy. Astr. Soc.,* 181: 441.
Papanastassiou, D. A.; Huneke, J. C.; Esat, T. M.; and Wasserburg, G. J. 1978. In *Lunar and Planetary Sci.* 9th p. 859. Lunar Planetary Institute, Houston.
Sgro, A. G. 1975. *Astrophys. J.* 197: 621.
Woodward, P. R. 1976. *Astrophys. J.,* 207: 484.
Yabushita, S. 1966. *Mon. Not. Roy. Astr. Soc.* 133: 247.
____ . 1969. *Mon. Not. Roy. Astr. Soc.* 142: 201.
Yeh, H. W., and Epstein, S. 1978. In *Lunar and Planetary Sci.* 9th. P. 672. Lunar Planetary Institute, Houston.

MAXIMUM TEMPERATURES DURING THE FORMATION OF THE SOLAR NEBULA

JOHN T. WASSON
University of California at Los Angeles

With the possible exception of the CI and the fine-grained matrix of the CM chondrites, meteoritic evidence favors the conclusion that the chondritic meteorites formed in the solar nebula. The correlated fractionation of refractory elements and degree of oxidation among chondrite groups can be explained by plausible nebular fractionation processes. The alternative that these reflect pre-existing differences in the presolar matter is unlikely both because turbulent mixing would have tended to erase such differences, and because of the absence of isotopic effects that would have resulted from extended (~0.1 Gyr) irradiation of interstellar grains by cosmic rays.

The enstatite and ordinary chondrites have substantially lower refractory abundances than the carbonaceous chondrites; those in the latter appear to be representative of mean solar matter. Models to explain this fractionation require a refractory-rich component at the locations of enstatite and ordinary chondrite formation at a time when the major elements Mg, Si and Fe were in the gaseous phase. Similarly, the anomalously large size of the refractory inclusions in CV chondrites is best accounted for by incomplete volatilization of interstellar solids at temperatures above the condensation temperature of Mg. Boynton has shown that a peculiar rare-earth abundance pattern found in fine-grained refractory inclusions can only be explained by condensation from a gas phase fractionated by removal of the most refractory rare earths at nebular temperatures just below the condensation temperature of Ti (T=1400 K at $pH_2 = 10^{-6}$ atm). The absence of large refractory inclusions and of major O-isotope fractionations at the formation locations of the enstatite, ordinary and IAB chondrites, indicates that only very small amounts of presolar solids survived evaporation at these locations, and that maximum nebular temperatures were higher than the Ca, and probably higher than the Ti condensation temperature (~1550 K at $pH_2 = 10^{-4}$ atm).

At the present time most numerical models of cloud collapse yield the result that temperatures were never above about 1000 K $\geqslant$1 AU from the axis of the forming solar system. In contrast, most meteorite researchers hold that higher temperatures are necessary to account for a variety of elemental fractionations found between groups of meteorites, between members of a single group, and between components of a single meteorite. In this chapter I attempt to summarize the meteorite researchers' viewpoint.

There are two categories of meteorites: *chondrites* have solar abundances of non-volatile elements, and appear to have formed in the solar nebula, and *differentiated meteorites* which formed by melting and associated fractionation processes occurring in planet-like parent bodies. With regard to chondrites, although formation in a nebula is commonly invoked, more exotic models would lead to similar maximum temperatures.

Since it is likely that the chondritic meteorites were produced in the solar nebula, they can provide boundary conditions for models of solar system formation and evolution. In particular, numerous authors have demonstrated that chondrite compositions can be semiquantitatively explained by an approach to chemical equilibrium in a cooling gaseous nebula. Fractionations between classes can be plausibly explained by a combination of kinetic effects and mechanical separations (e.g., of magnetic particles from non-magnetic, or small particles from large).

It may seem tempting to ascribe some of the evidence for high temperatures to processes predating solar system formation, particularly now that there is isotopic evidence indicating that chondrites contain presolar materials that managed to avoid isotopic equilibration. For this reason, I first examine the evidence indicating that chondrites formed in the solar nebula.

MOST CHONDRITES ARE NOT AGGREGATES OF INTERSTELLAR GRAINS

Surprisingly, there seems to be no published discussion of the meteoritic evidence for or against this important conclusion. (Cameron [1973] argued that CI chondrites consist of collections of interstellar grains based on the fact that his collapse models failed to produce high nebular temperatures.) The following is an attempt at such a synthesis; suggested additions to the arguments would be appreciated.

As indicated in Table I there are 10 groups of chondrites. Some of these are closely interrelated, and the clusters of interrelated groups are designated "clans." Although essentially solar in composition, significant elemental fractionations between groups are observed. Particularly important are the fractionations in refractory element/Si ratios (Fig. 1), and in the oxidation state of Fe [measured by the FeO/(FeO+MgO) ratio]. These are summarized in Table I (Ca is an example of a refractory element). It is generally accepted that neither fractionation could have resulted from processes occurring on or

TABLE I

Properties of Chondrite Groups

Group[a]	Clan	Fe/Si (atom %)	Ca/Si (atom %)	FeO/(FeO+MgO) (mole %)	Density (g cm^{-3})	Fall freq. (%)
CV	Vigarano	73-79	9.1-9.9	~30	3.4	1.1
CI	Carbonaceous	86-93	6.7-7.4	~40	2.3	0.68
CM		80-85	6.4-7.0	~35	2.8	1.9
CO		81-85	6.5-7.1	~30	3.4	0.81
LL	Ordinary	50-56	4.5-5.0	27-32	3.5	7.2
L		55-62	4.7-5.1	21-25	3.6	39.0
H		75-83	4.8-5.2	16-19	3.7	32.1
IAB	IAB	50-80	3.9-5.0	4-7	3.8	1.0
EL	Enstatite	60-70	3.4-3.8	0.1-0.2	3.6	0.81
EH		90-100	3.3-3.7	0.5-1.0	3.8	0.68

[a]Group symbols are those proposed by Wasson (1974) except that the enstatite chondrites are now divided into high-Fe (EH) and low-Fe (EL) groups. Type-I or C1 carbonaceous chondrites are synonymous with group CI chondrites.

in a parent body, but rather that these predate the agglomeration of grains to form the chondritic rock (Wood, 1963; Anders, 1964; Dodd *et al.*, 1967; Larimer and Anders, 1970; Wasson, 1972). Support for this view is provided by the high abundance of chondrules (mm-sized spheroidal grains that appear to be frozen droplets of material having compositions similar to the bulk chondrite), the large (factor of 10) "disequilibrium" ranges in Fe/(Fe+Mg) ratios in the common silicate minerals of the most primitive chondrites, and the absence of petrographic evidence for planetary-type igneous processes.

If the observed fractionations did not occur in parent bodies, it seems likely that they are associated with formation at varying distances from the sun (Wasson, 1977). Members of each clan probably formed near one another, whereas large spatial separations are reflected by the distinct hiatus in certain properties between the different clans.

There appears to be no way to produce the observed spectrum of chondritic properties by the agglomeration of interstellar grains. It seems inconceivable that collapse of an interstellar cloud could lead to a 3-fold enrichment in refractory elements at the CV location relative to EH and EL. It also seems impossible that mechanical separation of two or more distinct populations of interstellar grains could result in EH and EL material being so reduced as to contain several percent metallic Si in the Fe-Ni metal, whereas

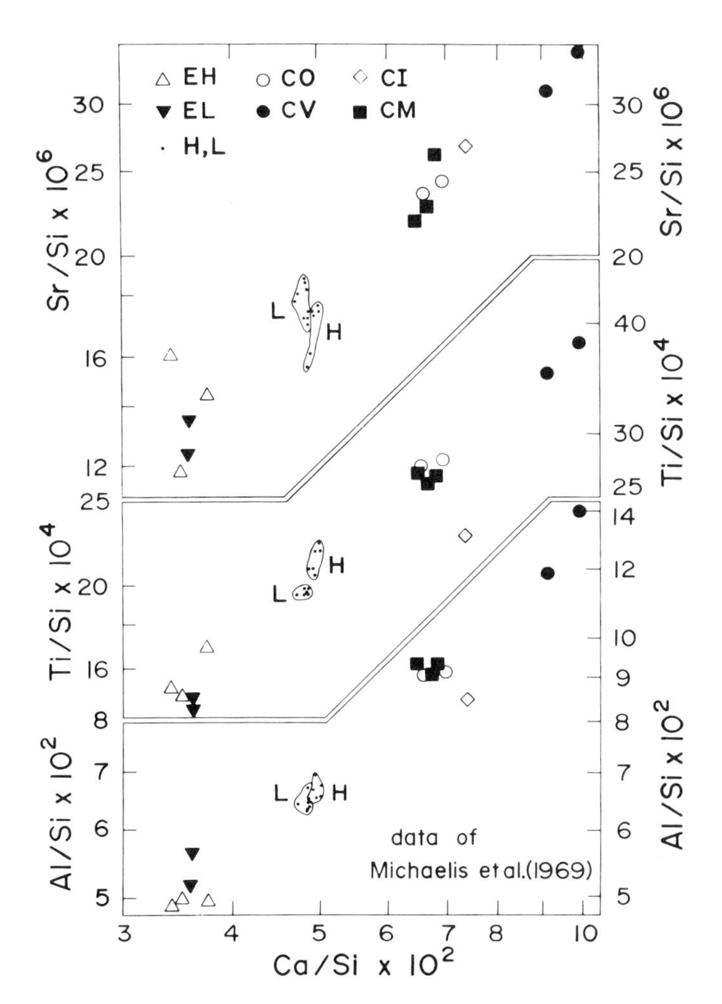

Fig. 1. Four elements (Sr, Ti, Al, Ca) that are refractory during nebular condensation show large fractionations among the chondrite groups, which occupy four distinct fields on this diagram. The four fields correspond to chondritic clans; IAB chondrites were not included in the study. Slopes are close to +1 on this log-log diagram, indicating that these refractory lithophiles are not fractionated from one another. (Data from Michaelis *et al.*, 1969.)

CI chondrites are so oxidized that no Fe-Ni metal exists, and roughly half the Fe is in the +3 (rather than the more common +2) oxidation state. Even if there were populations of interstellar dust grossly different in refractory element/Si ratios and in degree of oxidation, it seems probable that these would have been mixed by turbulence during the collapse phase, and that such differences were not preserved in the bulk matter at specific distances from the sun. Note that any such heterogeneity in the refractory contents of interstellar grains does not result from condensation in the various "onion shells" of a protosupernova (Clayton and Woosley, 1974; Lattimer *et al.*, 1978), since each such shell has enhanced abundances of only a few neighboring elements whereas the refractories range in atomic number from 13 to 92.

The above arguments and others of similar ilk indicate that accretion of interstellar dust to the solar nebula and agglomeration there cannot account for the large fractionations spanning the spectrum of chondrite groups. In contrast, reasonably plausible hypotheses based on the formation of chondrites at nebular locations that are widely separated and thus differing in pressure and degree of turbulence can account for the differences.

The question then arises whether it is still possible that one or a few groups could consist of agglomerated interstellar grains. If so, the most likely candidates are meteorites that appear to have formed farthest from the sun the CI chondrites, while the CM chondrites are also a possibility. In bulk CI and the "matrix" of CM the grain size is small (sub-μm) and the degree of disequilibrium between grains is higher than in other chondrite groups. The CI group also has the highest and the CM group the second highest concentration of volatile metals. As a result these are more nearly solar in composition than are other chondrites and are the only chondrite groups containing hydrated (layer-lattice) silicates. These features are commonly attributed to formation at large distances from the sun where nebular temperatures were lowest.

Could CI or CM meteorites consist largely of unvaporized interstellar grains? The most direct evidence comes from studies of cosmic-ray effects. No significant shielding of interstellar grains from cosmic rays is expected since the average grain spends 99% of its lifetime outside of dense clouds and in any case, Cesarsky and Völk (1977) have shown that the energetic cosmic rays responsible for most of the production of cosmogenic nuclides penetrate to the centers of dense clouds. In sub-μm grains the production rate of cosmogenic nuclides such as ^{3}He and ^{21}Ne would have been lower than that in a meter-sized meteorite because of (1) the greatly reduced number of secondary cosmic ray particles, and (2) the escape of a large fraction of cosmogenic products from the smallest interstellar grains as a result of nuclear recoil. The combination of these effects should not reduce the production rate by more than a factor of 100. Although the average recoil length resulting from interactions with GeV cosmic rays may be a few μm, a non-negligible fraction of these products will have much smaller values. Further, production by low-energy (MeV) particles and by cosmic ray gammas produces negligible recoil. Following recoil escape, some fraction of the condensable cosmogenic nuclides will re-adsorb on grains, but this will not occur for the light rare gases commonly studied.

The lowest cosmic ray ages of CI chondrites are about 3 Myr, but the cosmic ray exposure age of the Cold Bokkeveld CM chondrite is only 350 kyr; these "ages" are based on production rates calculated assuming meter-sized dimensions throughout the exposure period (Mazor *et al.*, 1970; Macdougall and Phinney, 1977). The activity of cosmogenic 740-kyr ^{26}Al in Cold Bokkeveld is that expected if the entire irradiation period immediately preceeded its fall (Heymann and Anders, 1967); thus, only a minor fraction ($\leqslant$10%) of the measured cosmogenic stable isotopes could have resulted from a presolar irradiation.

This conclusion has also been reached by Macdougall and Phinney (1977) on the basis of measurements of rare gases in olivine, a mineral which is relatively refractory, and in a fine-grained matrix that mainly consisted of hydrated layer-lattice silicates (phyllosilicates). The mean grain size of the olivine fraction was larger than that of the matrix. If these materials were all of interstellar origin, they should have initially contained relatively large amounts of ^{21}Ne produced by presolar cosmic ray interactions. The ^{21}Ne concentration in the olivine should have been higher because of a larger abundance of target elements and because a smaller fraction would have undergone recoil escape from the larger grains. During mild reheating in the solar system, diffusional escape of ^{21}Ne would have been more rapid from the matrix both because of the smaller size and the high diffusion coefficients for movement parallel to the lattice planes of the phyllosilicates. In fact, matrix and olivine in Cold Bokkeveld showed no resolvable difference in ^{21}Ne "age"; the observed differences in ^{21}Ne content were precisely those predicted on the basis of the differences of the composition-related production rate. The indicated conclusion is that essentially all the ^{21}Ne resulted from irradiation since the meteorite formed.

Is the absence of presolar cosmogenic rare gases in Cold Bokkeveld the result of heating to moderate temperatures during accretion to the solar nebula? According to Anders (1971) ~400 K is the maximum that would leave phyllosilicates unaltered. If we assume a mean olivine grain size of 1 μm and a ^{21}Ne diffusion coefficient of about 10^{-26} cm^2 sec at 400 K (Fechtig *et al.*, 1963), one arrives at a rough estimate of the duration of heating of 10^7 years in order to reduce the ^{21}Ne concentration by about 90%. It seems doubtful that low-temperature nebula models can account for such a long period of cooking at 400 K.

If we assume negligible loss during outgassing and a production rate that is 100 times lower in interstellar grains than in meteorites, the magnitude of an allowed presolar irradiation of CI chondrites might be as large as 10^8 yr. The probable value is much lower, since the dynamic processes that regulate the recent exposure of stony meteoroids lead to expected values in the 1-20 Myr range and could easily account for the entire measured amount of cosmogenic nuclides. The estimated limit on the presolar exposure of Cold Bokkeveld grains is a few Myr. Estimates of the mean lifetime of interstellar grains given by Salpeter (1977) have large uncertainties; it appears that the mean lifetime cannot be greater than 10^9 yr, and could be as low as 10^7 yr. The latter is based on poorly quantified processes occurring in molecular clouds and H II regions. If we accept the Cold Bokkeveld data as offering the most precise limit on presolar effects, the indicated conclusion is that most presolar grains were either vaporized in the solar system or were subjected to thermal effects severe enough to lead to near total outgassing of the most retentive site. It is to be hoped that future studies of cosmic ray produced nuclides or charged particle tracks will be focused on determining a more precise limit on the presolar irradiation of Cold Bokkeveld and other CI or CM chondrites having very low contents of cosmogenic nuclides. Until then, we must rely on plausibility arguments of the sort I have presented.

Note that the arguments in this section refer only to the bulk of the material in these chondrites classes. The isotopic data of R. Clayton, P. Eberhardt, G. Wasserburg and others make it clear that minor amounts of unvaporized interstellar solids are preserved in most classes. This may have consisted largely of materials accreted after nebular temperatures were significantly lower than those discussed in the following section.

MAXIMUM NEBULAR TEMPERATURES DURING THE FORMATION OF CHONDRITES

Table II lists equilibrium 50% condensation temperatures of Ti, Ca and Mg for nebular pressures ranging from 10^{-6} to 10^{-2} atm. These pressures bracket the range of reasonable nebular pressures for the solar system inside the orbit of Jupiter. A minimum mass (0.01 $M_{\odot}$) preplanetary nebula yields a pressure $\sim 2 \times 10^{-5}$ atm at 1 AU, $\sim 10^{-6}$ atm at 3 AU. More massive preplanetary nebulas give proportionately higher pressures.

Arrhenius and Alfvén (1971) have argued that the nebula was transparent to infrared radiation, and therefore that grains and gas were not in thermal equilibrium. More recently De (1977) has shown that these conditions would lead to (1) a condensation sequence essentially identical to those calculated for gas-grain equilibrium (Larimer, 1967; Grossman, 1972; Wai and Wasson,

TABLE II

50% Condensation Temperatures (K) in Equilibrium Formation

Elem.	Associated with formation of	Nebular pressures (atm) 10^{-6}	10^{-5}	10^{-4}	10^{-3}	10^{-2}
Mg	Forsteritic olivine Mg_2SiO_4	1203	1268	1340	1417	1494
Ca	Gehlenitic melilite $Ca_2Al_2SiO_7$	1382	1447	1518	1596	1683
Ti	Perovskite $CaTiO_3$	1420	1482	1549	1622	1703

1977), and (2) grain kinetic temperatures only about 100° lower than equilibrium values like those listed in Table II.

The large range of abundances of refractory elements among chondrite classes is demonstrated in Table I and Fig. 1. Larimer and Anders (1970) suggested that this resulted from differential settling of refractory-rich grains to the median plane of the nebula. Making the common (and plausible) assumption that the CI chondrites are representative of mean solar system matter, they concluded that the missing refractory material at the ordinary and enstatite chondrite formation locations settled to the median plane and agglomerated to larger bodies before the subsequent settling of the particles that agglomerated to form these classes of chondrites. Weighing against this model is the fact that their hypothetical early agglomerate has never fallen as a meteorite, despite the fact that, at the location of the enstatite chondrite formation there should be about the same amount of refractories as that present in the enstatite chondrites.

I consider an alternate process more proable, viz., that the missing refractories were mainly contained in very small condensate particles that were prevented by turbulence from settling to the median plane and were eventually removed from the solar system, perhaps by a T Tauri wind. Differential radial transport resulting from momentum exchange between grains and gas may also have been important (Weidenschilling, 1977). In any case, both models explaining the refractory fractionation demand the existence of solid refractories at a time when the bulk of the matter was in the gaseous state. Thus, the refractory fractionation among the different chondrite groups demands temperatures $\gtrsim$1200 K in the portion of the solar nebula inside 3 AU. Still higher temperatures are inferred below.

The highest refractory concentrations are observed in CV chondrites; as shown in Fig. 1, abundances are about 30% higher than solar abundances, assumed to be similar to those in CI or CM chondrites. This is the only group in which the refractories are present as large, 1-25 mm inclusions (Fig. 2). Refractory-rich clasts are also found in CO and CM chondrites, but the maximum size is about 1 mm. In the remaining meteorite classes refractory-rich inclusions are essentially nonexistent; a few small ($\leqslant$0.4 mm) materials rich in refractories also have high contents of the volatile element Na, indicating reaction with nebular gases at low temperatures ($\leqslant$1000 K; Noonan, 1975). Since the settling of large Ca-Al-rich clasts to the nebular median plane would have been more efficient than that for smaller particles (and especially those in the μm-sized matrix), it seems likely that this mechanism accounts for the fact that refractory abundances in CV chondrites are higher than those in mean solar system matter. As noted above, the inverse process probably is responsible for the low refractory abundance in E, IAB and ordinary chondrites.

Chou *et al.* (1976) pointed out that very special circumstances are needed if refractory minerals, which account for only about 4% of the mass

Fig. 2. Refractory inclusions in CV chondrites range in size from about 1 to 25 mm, the longest dimension of the large ovoid in this section of the Allende chondrite. Refractory abundances in CV chondrites are about 30% higher than those in the CI or CM chondrites, assumed representative of mean solar system matter, perhaps because of preferential settling of the large inclusions onto the nebular median plane. The large size of the refractory inclusions can be explained by partial vaporization of large presolar agglomerates at temperatures near the 50% condensation temperature of Ca (Table II). (Smithsonian Institution photograph.)

of condensible solar matter, are to form grain assemblages many times larger than those formed by the olivine, which accounts for about 80% of the mass. Condensation processes in a uniformly cooling nebula should produce the opposite phenomenon, namely, grains of olivine larger than those of refractory materials. They suggested that the observations were most simply explained by a model in which the large refractory inclusions formed by partial volatilization of pre-existing solids. These cm-sized precursors could have formed prior to collapse of the interstellar cloud (it appears that there is no optical evidence that would rule out the possibility that several tens of percent of the interstellar solids are so large), or as Cameron (1975) proposed, during the collapse of the cloud. If equilibrium prevails, the composition of a residue from partial volatilization is compositionally identical to that of a condensate formed at the same temperature. Although isotopic data indicate that equilibrium was merely approached, the low concentration of Mg in these Ca-rich inclusions in CV chondrites nonetheless provides firm evidence that temperatures were higher than the Mg but lower than the Ca condensation temperature (Table II).

More detailed and less model-dependent information regarding maximum temperatures at the location of CV formation is available. The 14 rare earths are a geochemically cohesive group of elements. With occasional exceptions involving Eu, igneous processes result in only modest fractionations that are correlated with variations in the ionic radii of the +3 oxidation-state which systematically decrease from La (~1.13 Å) to Lu (~0.94 Å). Tanaka and Masuda (1973) reported an erratically fractionated rare earth pattern in a fine grained refractory-rich inclusion from Allende, including a high abundance of $_{70}$Yb but order-of-magnitude lower abundances of $_{68}$Er and $_{71}$Lu. These inclusions are almost certainly of solar system origin, since they are highly fragile, and many show evidence of plastic flow during accretion. If they are presolar the whole rock is presolar, and if the rock is presolar, it should contain much larger amounts of cosmic ray produced nuclides.

Boynton (1975) showed that this distinctly non-igneous pattern could be explained by a nebular condensation model if a still more refractory rare earth bearing condensate had been removed previously from equilibrium with the gas. Fig. 3 compares a calculated pattern with one similar to that of Tanaka and Masuda (Boynton, 1978). Boynton inferred that the missing

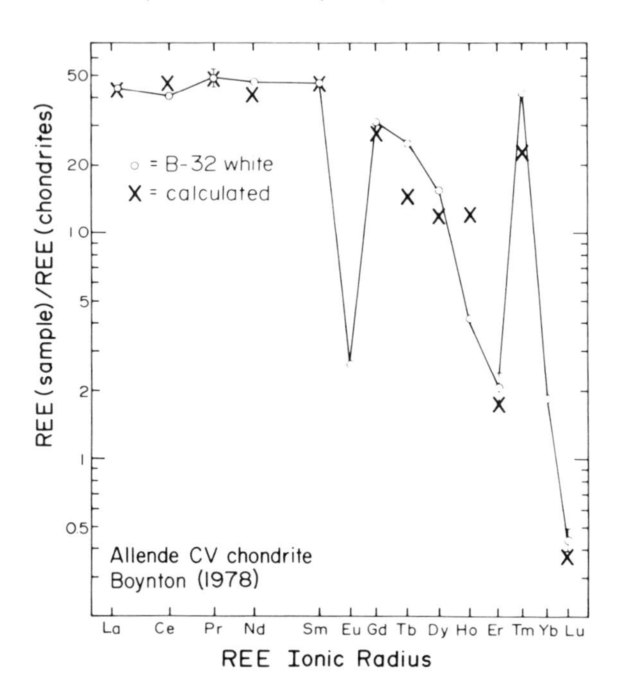

Fig. 3. Chondrite normalized rare earth abundances in a fine grained inclusion from the Allende CV chondrite. Boynton's fit is shown to the pattern generated by equilibrium condensation of rare earths from a nebular gas. This gas is partially depleted in the most refractory rare earths as a result of their removal together with Ti as perovskite $CaTiO_3$ at temperatures lower than the 50% condensation temperature of Ti but higher than that of Ca (Table II) REE ≡ rare earth elements.

refractory rare earths condensed into $CaTiO_3$. This implies nebular temperatures lower than the 50% condensation temperature of Ti and above the 50% condensation temperature of Ca (Table II; note that only 3.5% of the Ca can condense in $CaTiO_3$). If nebular pressures in this region were about 10^{-6} atm, a temperature of ~1440 K is inferred.

Blander and Fuchs (1975) assembled an impressive array of evidence indicating that many refractory-rich inclusions were molten at one time. Since these materials are not stable as liquids at nebular pressures, Chou *et al.* (1976) suggested that metastable liquids formed as a result of fluxing by volatiles during rapid heating of material accreting to the solar nebula. Whatever the mechanism, it is clear that these objects were solid before impacts with olivine and less refractory minerals could result in much "contamination".

Clayton *et al.* (1973, 1976) showed that the O-isotopic composition of each chondrite group is distinctive; Fig. 4 is a $\delta^{17}O$–$\delta^{18}O$ diagram summarizing the whole-rock data. Because whole-rock data are not available for CM, CO or CV chondrites, I have estimated the positions of the fields from data on separated fractions; the actual ranges are probably somewhat

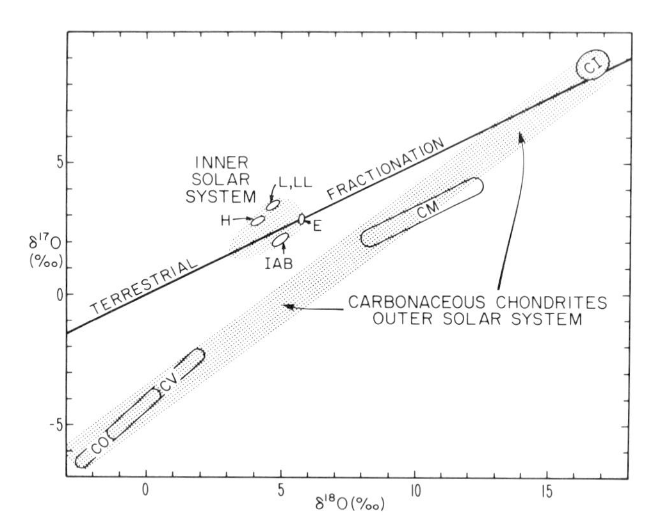

Fig. 4. It seems likely that the mean O-isotope composition of the solar system lies somewhere near the fractionation line linking terrestrial samples. Chondrite groups having significant fractions of unvaporized presolar solids (CM, CO and CV chondrites) have much larger contents of an anomalous ^{16}O-rich component; temperatures where these groups formed were high enough to vaporize Si, Mg and Fe but not refractories such as Ca and Al. At the inner solar system location where the ordinary, enstatite and IAB chondrites formed, vaporization of refractories was essentially complete, requiring temperatures comparable to the condensation temperatures of Ti (Table II). The delta notation shows normalized differences in isotope ratios between sample and standard, e.g.: $\delta^{17}O = [(^{17}O/^{16}O)_{sample} - (^{17}O/^{16}O)_{standard})] \, (^{17}O/^{16}O)_{standard}$.

smaller. Note that the CV and carbonaceous chondrite groups occupy fields lying along a diagonal band spanning the entire diagram, whereas the remaining chondrite groups occupy smaller fields near each other on either side of the slope 0.50 terrestrial fractionation line. (Modest amounts of fractionation produce twice as much change in the $^{18}O/^{16}O$ ratio as in the $^{17}O/^{16}O$ ratio, thus the slope of 0.50.) Upper mantle rocks offer the best estimate of the mean terrestrial O-isotope composition; although not shown, these occupy a small field just lower in ^{18}O than the E-chondrite field.

Although the detailed interpretation of the O-isotope data is far from clear, the following seems relatively certain:

1. The mean O-isotope composition of the preplanetary nebula occupied a position on or near (within about $\pm 1‰$ $\delta^{17}O$) the terrestrial fractionation line.
2. An anomalous, ^{16}O-rich component has been mixed with this normal material to produce the composition of the CM, CO and CV chondrites. This could have occurred in either of two ways: (a) at each location, pre-existing heterogeneities in the total nebular materials were partially preserved despite turbulent mixing during accretion of the solar nebula; or (b) the anomalous ^{16}O-rich material was preserved in incompletely volatilized interstellar grains that more efficiently settled to the nebula median plane, and thus were enhanced in abundance in chondritic solids relative to materials having normal O-isotope compositions.

Earlier in the chapter I outlined reasons for doubting that major compositional heterogeneities survived the anticipated turbulent mixing that occurred during collapse of the interstellar cloud to form the nebula. Above I noted the arguments indicating that significant portions of the larger refractory-rich inclusions in CV and the somewhat smaller inclusions in CM and CO chondrites consisted of unvaporized interstellar material. As a result, the second explanation involving unvaporized carriers of the high ^{16}O components of CM, CO and CV chondrites seems to fit much better with the facts. It then follows that the absence of large anomalies in the H, L, LL, EH, EL and IAB chondrites and in the earth indicate that the bulk of the ^{16}O-rich solids were vaporized at the solar system locations where the major components of these bodies formed, and that temperatures at these locations were higher than the 50% condensation temperatures of Ti (Table II). This conclusion is also supported by the Sr, Mg and PB isotopic evidence indicating that the CV chondrites formed (or cooled) earlier than the ordinary, IAB or enstatite chondrites (Wetherill *et al.*, 1973; Gray *et al.*, 1973; Chen and Tilton, 1976; Lee *et al.*, 1976; Tatsumoto *et al.*, 1976). If we assume a typical pressure of 10^{-4} atm where these latter chondrites formed, vaporization of $CaTiO_3$ requires maximum temperatures $\gtrsim 1550$ K. It is noted that this model implies that the ^{16}O-rich interstellar solids contributed only a minor fraction of the solar system O, but accounted for a substantial portion of the large ($\gtrsim 1$ mm) interstellar grains accreted to the solar nebula.

Alternative schemes can be devised to explain many of the facts cited in favor of high nebular temperatures. However, it appears difficult if not impossible to explain all these observations by a plausible model of nebula formation and early evolution which involves significantly lower nebular temperatures. Under the circumstances it would appear that satisfactory astrophysical models for the formation of the solar system must be able to generate high temperatures ($\gtrsim$1550 K) in the region 1-3 AU from the sun where enstatite, IAB and ordinary chondrites appear to have formed (Wasson, 1977), and temperatures near 1400 K in the regions (probably $>$3 AU) where the CV chondrites formed.

Acknowledgements. The chapter has benefited from discussions with numerous persons including J.R. Arnold, W.V. Boynton, D.D. Clayton, J.M. Greenberg, M. Jura, J. Kerridge, G. Morfill, H. Reeves and H. Völk. This research was mainly supported by the National Aeronautics and Space Administration.

COMMENTS

O.K. Manuel: You suggest that the large-grained, white refractory grains in Allende contain isotopic anomalies and might therefore be residues of incompletely vaporized presolar grains, but it should be noted that the isotopic anomalies seen in elements (Ba, Ca and Mg) of these inclusions are small (0.1-1%). Large isotopic anomalies, up to several hundred percent, are seen in extremely volatile elements of meteorites, such as Hg, Xe and Kr. The geochemical character of these elements is exactly opposite to that of the refractory elements which might survive in the partially vaporized presolar grains you mention.

J.T. Wasson: The Hg anomalies were not confirmed by Helden and Begemann (1976). The rare gas contents of these inclusions are very small, and may be present in carriers that account for only a tiny fraction of the mass of the analyzed material. One can think of various *ad hoc* explanations for the fact that this minor component failed to outgas completely.

REFERENCES

Anders, E. 1964. *Space Sci. Rev.* 3: 583.
____. 1971. *Ann. Rev. Astron. Astrophys.* 9: 1.
Arrhenius, G., and Alfvén, H. 1971. *Earth Planet. Sci. Lett.* 10: 253.
Blander, M., and Fuchs, L.H. 1975. *Geochim. Cosmochim. Acta* 39: 569.
Boynton, W.V. 1975. *Geochim. Cosmochim. Acta* 39: 1605.
____. 1978. Submitted to *Earth Planet. Sci. Lett.*
Cameron, A.G.W. 1973. In *Interstellar Dust and Related Topics* (J.M. Greenberg and H.C. van de Hulst, eds.), p. 545. Dordrecht: Reidel.

____. 1975. *Icarus* 24: 128.
Cesarsky, C.J., and Völk, H.J. 1977. *Int. Cosmic Ray Conf., Plovdiv* 1:61.
Chen, J.H., and Tilton, G.R. 1976. *Geochim. Cosmochim. Acta* 40: 635.
Chou, C.L.; Baedecker, P.A.; and Wasson, J.T. 1976. *Geochim. Cosmochim. Acta* 40: 85.
Clayton, D.D., and Woosley, S.E. 1974. *Rev. Mod. Phys.* 46: 755.
Clayton, R.N.; Grossman, L,; and Mayeda, T.K. 1973. *Science* 182: 485.
Clayton, R.N.; Onuma, N.; and Mayeda, T.K. 1976. *Earth Planet. Sci. Lett.* 30: 10.
De, B.R. 1977. *Proc. Lunar Sci. Conf.* 8th. p. 87. New York: Pergamon.
Dodd, R.T.; Van Schmus, W.R.; and Koffman, D.M. 1967. *Geochim. Cosmochim. Acta* 31: 921.
Fechtig, H.; Gentner, W.; and Lämmerzahl, P. 1963. *Geochim. Cosmochim. Acta* 27: 1149.
Gray, C.M.; Papanastassiou, D.A.; and Wasserburg, G.J. 1973. *Icarus* 20: 213.
Grossman, L. 1972. *Geochim. Cosmochim. Acta* 36: 597.
Helden, J. von, and Begemann, F. 1976. *Meteoritics* 11: 297.
Heymann, D., and Anders, E. 1967 *Geochim. Cosmochim. Acta* 31: 1793.
Larimer, J.W. 1967. *Geochim. Cosmochim. Acta* 31: 1215.
Larimer, J.W., and Anders, E. 1970. *Geochim. Cosmochim. Acta* 34: 367.
Lattimer, J.M.; Schramm, D.N.; and Grossman, L. 1978. *Astrophys. J.* 219: 230.
Lee, T.; Papanastassiou, D.A.; and Wasserburg, G.J. 1976. *Geophys. Res. Lett.* 3: 109.
Macdougall, J.D., and Phinney, D. 1977. *Proc. Lunar Sci. Conf.* 8th. p. 293. New York: Pergamon.
Mazor, E.; Heymann, D.; and Anders, E. 1970. *Geochim. Cosmochim. Acta* 34: 781.
Michaelis, H. von; Ahrens, L.H.; and Willis, J.P. 1969. *Earth Planet. Sci. Lett.* 5: 387.
Noonan, A.F. 1975. *Meteoritics* 10: 51.
Salpeter, E.E. 1977. *Ann. Rev. Astron. Astrophys.* 15: 267.
Tanaka, T., and Masuda, A. 1973. *Icarus* 19: 523.
Tatsumotio, M.; Unruh, D.M.; and Desborough, G.A. 1976. *Geochim. Cosmochim. Acta* 40: 617.
Wai, C.M., and Wasson, J.T. 1977. *Earth Planet. Sci. Lett.* 36: 1.
Wasson, J.T. 1972. *Rev. Geophys. Space Phys.* 10: 711.
____. 1974. *Meteorites-Classification and Properties.* New York: Springer.
____. 1977. In *Comets, Asteroids and Meteorites* (A.H. Delsemme, ed.), p. 551. Toledo, Ohio: University of Toledo Press.
Weidenschilling, S.J. 1977. In *Comets, Asteroids and Meteorites* (A.H. Delsemme, ed.), p. 541. Toledo Ohio: University of Toledo Press.
Wetherill, G.W.; Mark, R.; and Lee-Hu, C. 1973. *Science* 182: 281.
Wood, J.A. 1963. *Icarus* 2: 152.

CONCLUSIONS DERIVED FROM THE EVIDENCE ON ACCRETION IN METEORITES

J. M. HERNDON
University of California at San Diego

AND

L. L. WILKENING
University of Arizona

Stony meteorites contain both individual mineral grains and rock fragments. Some contain foreign fragments which differ from one another and the host meteorite in their isotopic, chemical, and mineral compositions. From investigations of such fragments it has been found that mixing and reprocessing of condensed matter took place during the early history of the solar system. Some mineral grains appear to have escaped extensive alteration and from detailed investigations of these grains conclusions can be drawn which bear on the nature of the physical and chemical environment prevailing at a time before accretion of the meteorites. We discuss the physical morphologies of the following three types of meteoritic grains, the evidence for their formation in space, and the conclusions which can be drawn from their existence: (1) Magnetite platelets; (2) Chondrules; and (3) Ca-Al-rich inclusions. The well-preserved, yet extremely fragile morphologies of these materials are indications of low relative velocities during accretion. From chemical considerations, reasons are given for the conclusion that at the time chondrules existed as molten droplets in space, a major portion of the gaseous components of solar matter had already been lost.

Studies of cosmological accretion, the process of growth or enlargment of aggregates of matter by the accumulation or adhesion of additional condensed matter, attempt to elucidate the physical conditions during the growth of planetesimals from individual grains to bodies hundreds of kilometers in diameter. The craters that exist on the surfaces of the terrestrial planets and their satellites are visible justification for theoretical studies of

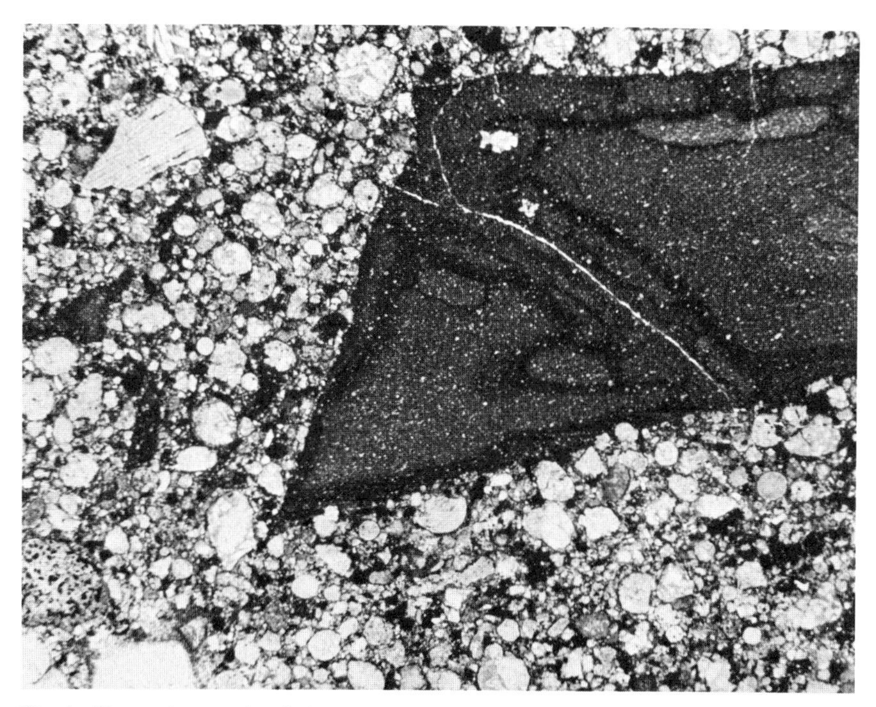

Fig. 1. Photomicrograph of the Sharps (H3) chondrite. The black angular fragment is about 0.5 cm wide. It is a carbonaceous chondrite (C2 or C3). The small fish-shaped inclusion in the corner of the carbonaceous fragment is a Ca-Al-rich inclusion. Note the chondrule-rich ground mass of Sharps.

the collisional accretion of kilometer-sized and larger bodies (e.g., Greenberg *et al.*, 1977, and their chapter in this book) and for a variety of investigations into aspects of planetesimal growth. Cratering studies have particular relevance for investigations of the final stages of accretion (e.g., Hartmann, 1966, 1975; Baldwin, 1974; Soderblom *et al.*, 1974). In this chapter we will be concerned with accretion of objects on a much smaller scale.

Meteorites are rocks of extraterrestrial origin which survived atmospheric entry and impact. Typically smaller than one meter in dimension, their arrivals were events in the accretion of the earth. Many meteorites are an agglomerate of individual grains and rock fragments and thus are themselves the result of accretion events. This is clearly the case for the example illustrated in Fig. 1. Rock fragments contained in meteorites are evidence of even earlier accretion, fragmentation, and mixing. It is well known from isotope studies that the mineral constituents of meteorites obtained their presently observed morphologies at an early time in the history of the solar system, $\sim 4.5 \times 10^9$ years ago. Prompt formation of individual mineral grains is evidenced by the presence of the decay products of extinct radioactivities (Reynolds, 1960; Rowe and Kuroda, 1965; Alexander *et al.*, 1971; Lee *et al.*,

1976; the chapters of Schramm, Reeves, and Alaerts and Lewis in this book). In the present chapter we point out materials which seem to have escaped extensive reprocessing after accretion and discuss what their existence implies about the conditions in the solar system during accretion of small objects.

MIXING AND REPROCESSING OF CONDENSED MATTER

Of the various types of meteorites listed in Table I, chondrites have elemental abundance ratios for certain condensable elements which are similar, although not identical, to corresponding ratios derived from the spectral analysis of sunlight; this is shown in Fig. 2. Groups of chondrites are chemically and isotopically distinct, as illustrated in Table I and Fig. 3 respectively. Urey and Craig (1953) first showed that the ordinary chondrites formed discrete compositional groups in terms of iron content and oxidation state. Although there is slight variation within each type, the classes H, L, LL, C, and E have unique chemical compositions and states of oxidation, Clayton *et al.* (1976) found that chondrites formed similar groups in terms of oxygen isotope compositions, except that L and LL groups are indistinguishable with present data. These observations indicate that the accretion sampled only restricted volumes of space which were characterized by different elemental and nuclear compositions.

In general it appears that mixing and reprocessing of condensed matter was extensive during meteorite formation. Certain types of meteorites, irons

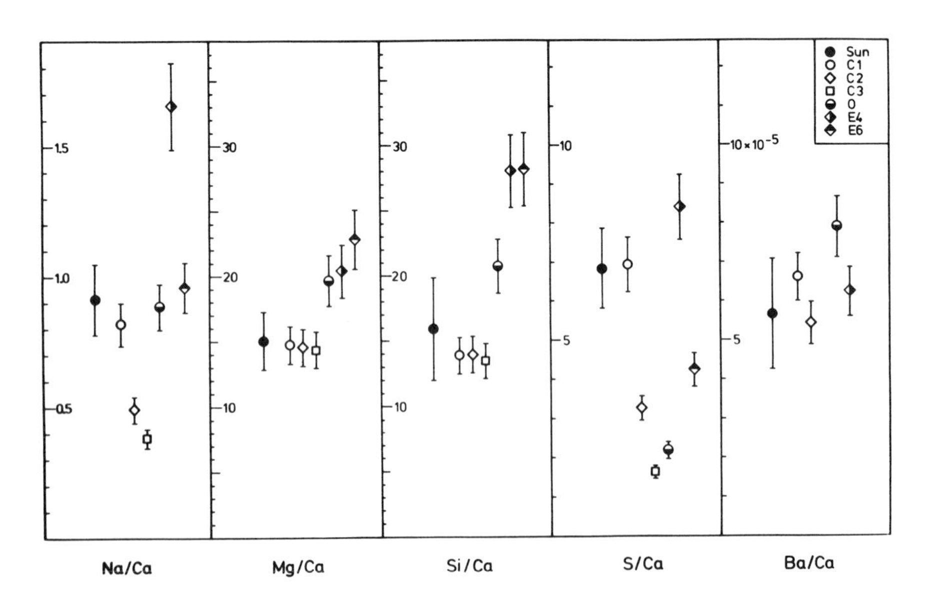

Fig. 2. Comparison of chondritic compositions with those of the sun (solid circles). For discussion of the technique see Holweger (1977).

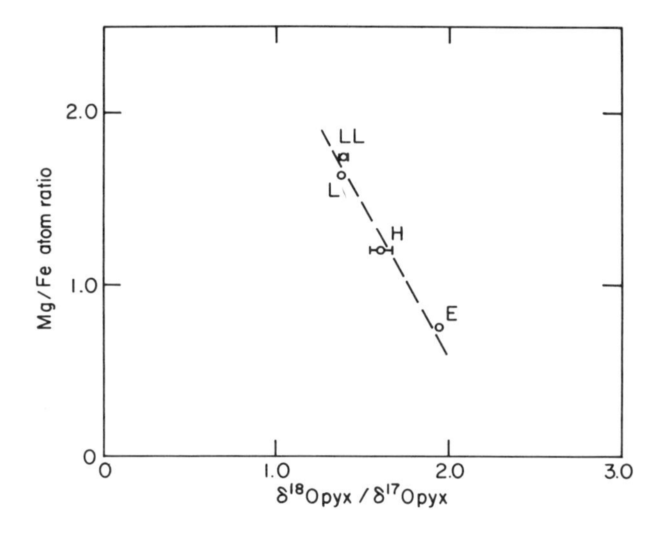

Fig. 3. The discrete chemical and isotopic compositions of the chondrites. The error bar on H shows the variation within H-group pyroxene oxygen isotope data. $\delta\,^{18}O_{pyx} \equiv \left[\frac{^{18}O_{pyx}/^{16}O_{pyx}}{^{18}O_{stan}/^{16}O_{stan}} - 1\right] \times 10^{-3}$. Variations in the Mg/Fe ratio are ten percent or less. Straight line relationship may be fortuitous.

and achondrites, have experienced melting and magmatic differentiation. The chondrites, which are considered the most primitive meteorites, are brecciated (broken, fragmented internal texture) rocks which display evidence of mixing, accretion, fragmentation and varying degrees of heating. The constituents of some brecciated meteorites have uniform chemical compositions. Other brecciated meteorites, such as the example shown in Fig. 1, contain assemblages of matter distinct and different in chemical and oxygen isotopic compositions. It has been established that mixing took place among different types of meteorites as well as within a given type (Wilkening, 1973, 1977; Wilkening and Clayton, 1974). These studies show that while initial accretion may have been largely restricted to pools of chemically and isotopically distinct matter, mixing must have taken place, at least at a later stage. Dynamical studies of circumstances under which such mixing could occur during planetary accretion have been made by Wetherill (1975; also see his chapter).

MATERIALS WHICH PARTICIPATED IN ACCRETION

Bearing in mind the lessons of the previous section about reprocessing of meteoritic materials, we consider a few of the components of meteorites which have been suggested primarily on the basis of morphology and

TABLE I

Meteorite Types and Fe/Fe+Mg Ratios[a]

		Classes	Fe/Fe+Mg
Primitive	Carbonaceous chondrites	C1 (CI)	–[b]
		C2 (CM)	–
		C3 (CV or CO)	6-23%
	Ordinary chondrites	H	16-19
		L	21-25
		LL	27-32
	Enstatite chondrites	E	0.04-1.4
Differentiated	Achondrites	Eucrites	50-67
		Howardites	25-40
		Diogenites	25-27
		Aubrites	<0.03
		Ureilites	10-25
	Stony-irons	Mesosiderites	23-27
		Pallasites	11-20
	Irons	Many types	–

[a]Mole % in silicates.

[b]Layer lattice silicates do not provide a useful comparison to other meteorite minerals. Fe is highly oxidized in these materials.

chemistry to have formed in space and persisted little-altered in at least some types of meteorites. The most useful evidence comes from those chondrites which consist of individual grains and rock fragments that were not strongly heated following their final accumulation.

Magnetite of C1 chondrites

The presence of magnetite in C1 chondrites had been known since the time of the original investigations of the Orgueil meteorite (Cloez, 1864). At low magnification the magnetite appears to be present mostly in the form of spherules. The work of Jedwab (1965, 1967, 1971) clearly showed that at higher magnifications the magnetite of C1 chondrites appears to exist in several distinct morphologies. In fact, the spherules observed at lower magnifications are found in many cases to be composites of even smaller spherules, crystals or stacks of platelets. Jedwab (1971) recognized the

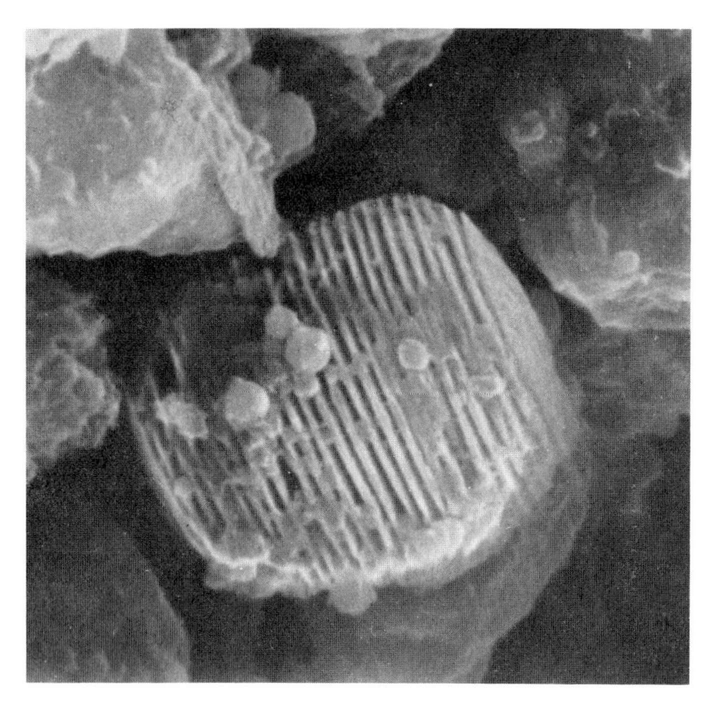

Fig. 4. Magnetite platelets in rounded stacks. The stack is ~6μm high. (Photograph by J. Jedwab.)

following morphological categories: nodules, which are the most abundant form; isometric crystals, which are scattered in the matrix and filling cavities; stacks of platelets (illustrated in Fig. 4); and framboidal aggregates of tiny crystals. A number of additional morphological forms including cubes and radially fibrous spheroids were also noted. The unusual magnetite morphologies, such as the stacks of platelets in Fig. 4, are unknown in terrestrial occurrences of that mineral.

The primitive nature of the carbonaceous chondrite magnetite has been the subject of debate. Its ancient age was demonstrated by I-Xe dating of magnetite separated from the Orgueil (C1) and Murchison (C2) meteorites (Herzog, *et al.,* 1973; Lewis and Anders, 1975). Herzog *et al.* found Orgueil to be ~2 Myr older than Karoonda (C4), the oldest meteorite that had at that time been determined by I-Xe dating techniques. However, Drozd and Podosek (1976) have determined that an ordinary chondrite, Arapahoe, appears to antedate Orgueil and Murchison magnetite so that the significance of the age measurement is clouded. Kerridge (1970) investigated the nickel content of the magnetite and found no evidence that this mineral formed directly from iron metal containing Ni. He concluded that at least two generations of magnetite are present in Orgueil. More recent analyses do not seem to support the latter conclusion, and he now thinks that all magnetite is a secondary alteration product formed in the meteorite parent body (Kerridge, personal communication). The morphological form characterized by the stacks of platelets shown in Fig. 4 is unique to C1 chondrites. It was, in fact, the observation of this particular morphology that led Jedwab (1965) to suggest that these were grown from a vapor phase in space and that their morphological forms resulted from crystal dislocations during growth from a vapor (Donn and Sears, 1963). The rather delicate nature of many of the morphological forms of magnetite observed in C1 chondrites speaks for extremely gentle physical circumstances from time of formation through final accretion. Relative velocities must have been quite low. In contrast, the sharp and angular shards, broken from olivine crystals, which are present in small amounts in the Orgueil C1 chondrite (Reid *et al.*, 1970), must have had their origins in different and more violent circumstances.

Chondrules

Chondrites, by definition, contain chondrules, nearly spherical objects approximately one mm in diameter. Chondrule sizes appear to have a narrow frequency distribution (Dodd, 1976; Martin and Mills, 1976). The chondrules are composed, in general, of the same minerals as the host meteorite. The chondrule content of chondritic meteorites varies. Not rarely, such as in the

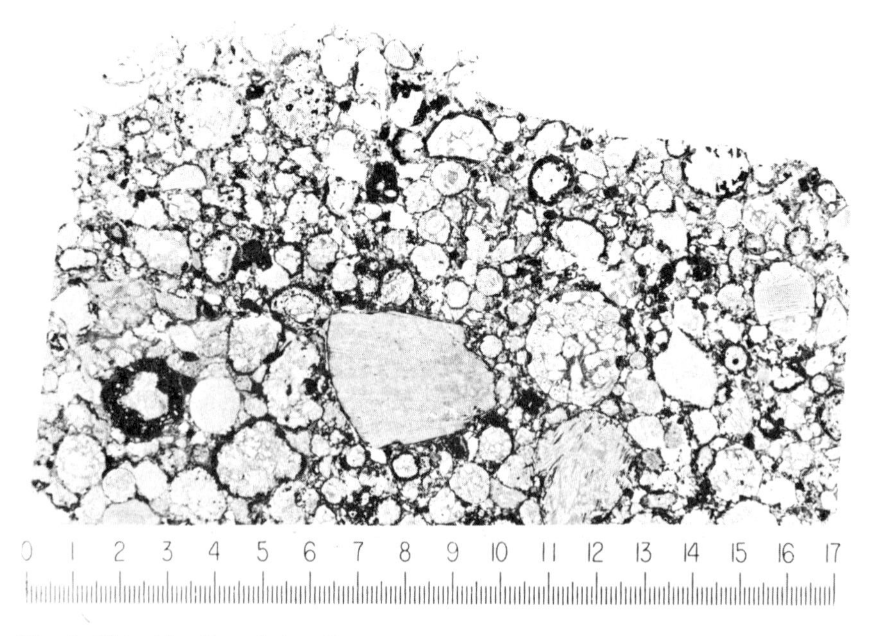

Fig. 5. This thin slice of the Chainpur meteorite shows a typical, highly chondritic structure of unaltered ordinary chondrites. One division is ~0.1 mm.

cases illustrated in Figs. 1 and 5, chondrules comprise approximately 70% of the meteorite.

Chondrules are a general morphological feature of all chondritic meteorites except C1 chondrites. For years scientists have intuitively considered chondrules to represent "frozen-droplets" that formed in a fiery rain by condensing out of a hot gas (Sorby, 1877). Many, although not all, chondrules bear the unmistakable signs of once having been molten droplets that obtained their nearly spherical shapes from the surface tension of the melt, which presumably solidified from a supercooled state. Suggestions have been made to the contrary, viz., that chondrules were produced *in situ* in meteorites, for example, by abrasion (Tschermak, 1875) or metamorphism (Mason, 1960). In experiments, however, Nelson *et al.* (1972) were able to obtain morphologies, remarkably similar to certain chondrules (Fig. 6), from the solidification of molten mineral droplets. Their experiments provide compelling evidence in support of Sorby's concept.

The frozen-droplet appearance of many chondrules has evoked various suggestions for their origins as molten objects (Urey and Craig, 1953; Ringwood, 1961; Wood, 1963; Reid and Fredriksson, 1967; Whipple, 1966; Wasson, 1972; Podolak and Cameron, 1974; Kieffer, 1975). The considerable diversity among chondrules with regard to mineralogy, composition, texture and shape, even within a single chondrite (Keil, 1962; Keil and Fredriksson, 1964) makes it difficult to discover what the exact origin of the chondrule is, if indeed, there was a single means of producing chondrules. However, the same frozen-droplet appearance makes it possible to derive from chemical considerations some conclusions regarding the chemical composition of the

Fig. 6. An artificial chondrule (b) Nelson *et al.*, 1972) is compared to a real chondrule from Khohar (a). (Photographs courtesy of K. Keil.)

ambient gas phase. This is because at the temperatures at which silicates are molten (~ 1500 K) chemical reactions between mm-sized droplets and the ambient vapor progress rapidly in the direction of equilibrium. An indication of the rapidity with which oxidation-reduction reactions of iron progress is evident from the experiments of Schmidt *et al.*, (1964). These authors impacted a steel ball into a steel target in air and found that the ejecta were spherules of magnetite (Fe_3O_4) indicating rapid reaction with the ambient atmosphere.

By calculating phase boundaries of the vapor-condensate using thermodynamic data, Wood (1963) was able to show that silicate liquid droplets could exist in equilibrium with vapor in a medium of solar composition at pressures $\gtrsim 10^3$ atmospheres. The composition of such a liquid (FeO-free magnesium silicates plus iron metal), however, is not that of the chondrules of ordinary chondrites which contain FeO-bearing silicates. Herndon and Suess (1977) calculated that an atmosphere depleted in hydrogen by a factor of $\sim 10^3$, but otherwise of solar composition, would permit the coexistence of this atmosphere with a liquid of the composition for ordinary chondrite chondrules, i.e., FeO-bearing magnesium silicates plus iron metal plus iron sulfide (Figs. 7 and 8). However, another condition is necessary; namely, that the ambient gas be deficient in oxygen by about half an order of magnitude relative to the solar abundance. Otherwise, iron oxide could form at the expense of both iron metal and iron sulfide (Herndon, 1977). Such an ambient vapor would be realized for chondrules formed from

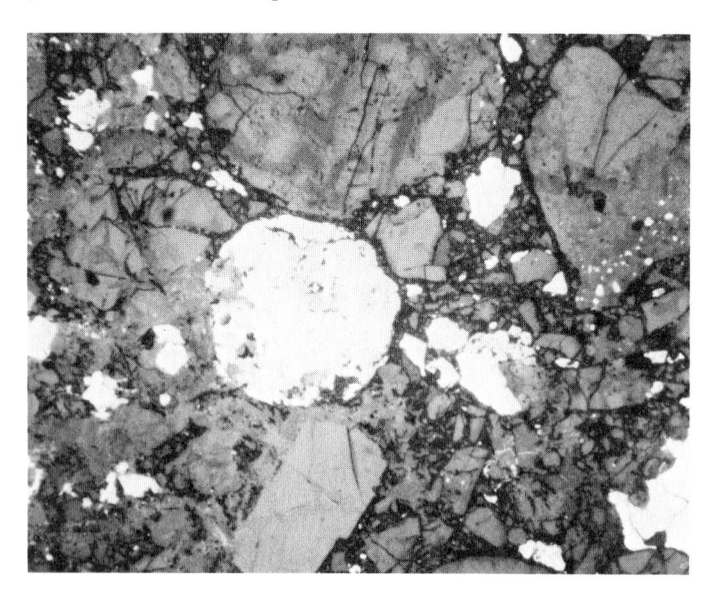

Fig. 7. A chondrule of metallic iron plus iron sulfide (0.5 mm diameter) co-exists with chondrules containing FeO-bearing silicates, metallic iron and iron sulfide in the H-group meteorite Weston.

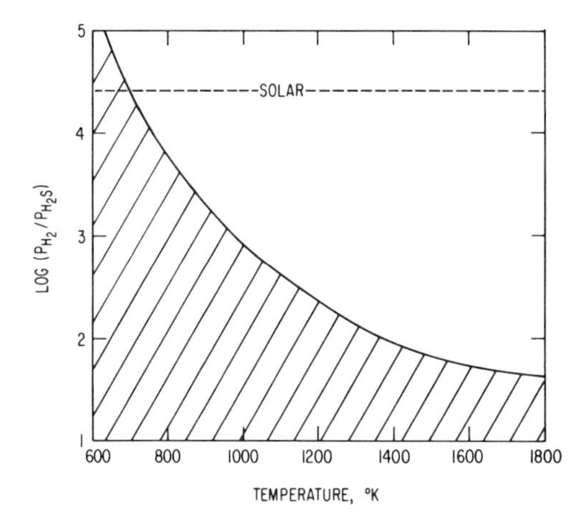

Fig. 8. Equilibrium P_H/P_{H_2S} ratio as a function of temperature at the Fe-FeS boundary. FeS is thermodynamically stable below the solid curve. The broken curve indicates the ratio for unfractionated solar matter.

matter, perhaps similar to that of the C1 chondrites, which was re-evaporated after first having been separated from a large fraction of the gaseous components of solar matter. It can be concluded that most, if not all, chondrule formation for the ordinary chondrites, and for C2 and C3 chondrites, occurred at a time after gas-condensate separation (Herndon, 1978). Consequently, it should be noted that in model calculations of meteorite accretion or of chondrule fractionation by aerodynamic sorting (Whipple, 1972; Dodd, 1976) the gas pressure should be reduced from the value for solar matter by several orders of magnitude.

Ca-Al-rich Inclusions

A great amount of scientific inquiry is directed toward the Ca-Al-rich inclusions in C2, C3 and in certain ordinary chondrites because of the discoveries of isotopically unusual elements in some samples from the Allende meteorite (Clayton *et al.*, 1973; Gray and Compston, 1974; Lee and Papanastassiou, 1974). Larimer and Anders (1970) and Marvin *et al.* (1970) first pointed out that the distinctive white or pink mm- to cm-sized inclusions in carbonaceous chondrites (C2 and C3) are made of the minerals predicted to be the first condensates from a gas of solar composition (Lord, 1965). However, Kurat (1970) has argued that they could be refractory residues rather than refractory condensates. Even though we cannot adequately review here the many attempts to resolve this and other questions about the Ca-Al-rich inclusions, a few conservative inferences can be drawn for our purposes. The distinctive isotopic and chemical attributes of the Ca-Al-rich inclusions preclude their formation in the meteorite matrix in which they are found.

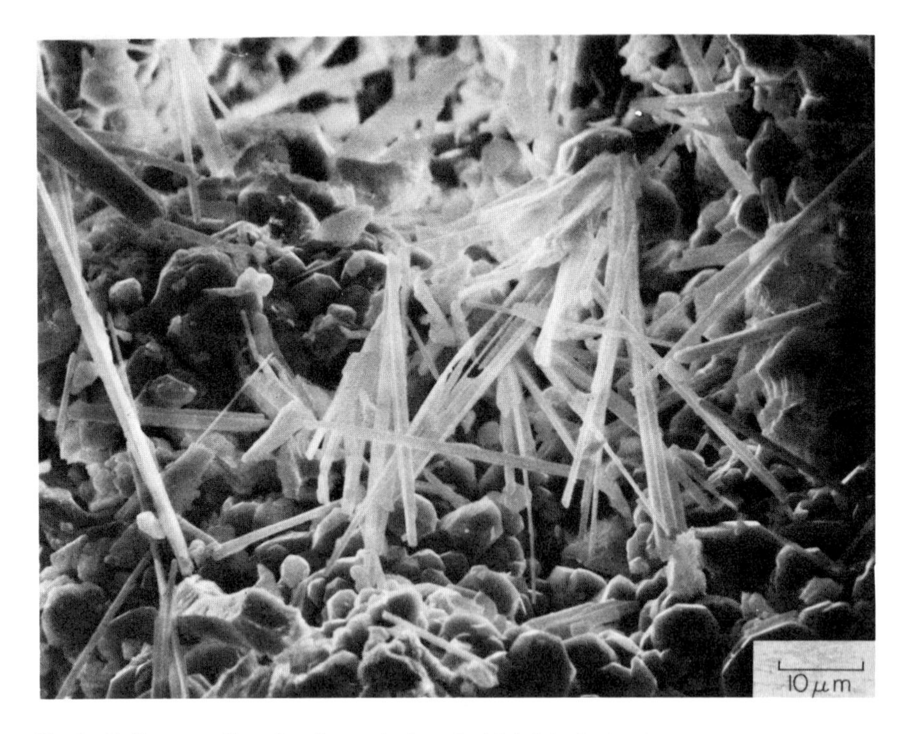

Fig. 9. Delicate needles of wollastonite in a Ca-Al-rich inclusion from Allende. (Scanning electron microscope photomicrograph by I. Hutcheon.)

This means they must have been mixed into the meteorites in an accretionary event. As was the case for magnetite, the morphology of some of the individual mineral grains comprising these inclusions testifies to low relative velocities during accretion. Fine-grained Ca-Al-rich inclusions yield examples such as that shown in Fig. 9 of delicate crystals projecting into cavities (Fuchs, 1971; Grossman *et al.*, 1975). Some layered rims surrounding coarse-grained inclusions have retained intact several layers of minerals of different compositions, but other Ca-Al-rich inclusion are totally lacking rims on angular, apparently broken surfaces (Wark and Lovering, 1977). Hence, accretion and, possibly, later processing were not uniformly gentle.

SUMMARY

The study of the evidence on accretion in meteorites shows that gross chemical features of chondritic meteorite types (C, H, L, LL, E) were established by accretion of materials with characteristic chemical and isotopic compositions. Some meteorites have been subjected to intensive mechanical and/or thermal processes. Mixing has occurred among meteorite types, presumably late in the accretion sequence. From the materials which appear to have escaped substantial modification the following conclusions can be drawn:

1. Relative velocities between accreting grains varied but in some cases were very low.
2. Accretion took place after most, if not all, of the gaseous components had been separated from the condensed portion of solar matter.

Acknowledgements. We thank H. Suess for his guidance and I. Hutcheon, J. Jedwab and K. Keil for providing unpublished photographs. This work was supported by the National Aeronautics and Space Administration.

COMMENTS

J. Wasson: Equilibrium calculations can indeed account for many of the compositional features and for some of the details (e.g., rare earth abundance patterns) of chondritic meteorites in terms of condensation in a cooling solar nebula. There is abundant evidence, however, that the chondrites that best preserve the conditions in the solar nebula are mixtures of grains that were never in equilibrium with each other. As you note, chondrules contain both high-temperature and low-temperature materials and there are major differences between one chondrule and the next. But it is precisely for this reason that most meteorite researchers now hold that the chondrules formed by flash heating of random mixtures of pre-existing grains followed by a very rapid cooling. I know of no evidence that would support your contention that chondrules were hot long enough to allow a significant amount of equilibration with an ambient gas if one were present.

J. M. Herndon: At elevated temperatures the chemical reactions between mm-sized droplets and ambient gas progress at extremely rapid rates. One piece of evidence is the steel ball impact experiment (Schmidt *et al.*, 1964) that we described and that indicates reaction with the ambient gas. Also, the experiments of Stephens and Kothari (1978) indicate extensive reaction between liquid drops and ambient gas on a time scale of milliseconds.

W. Boynton: The fine-grained aggregates all have fractionated rare earth patterns. The order in which the rare earths condense are a strong function of the H_2/O_2 ratio. The fractionated patterns are consistent with thermodynamic equilibrium with a solar H_2/O_2 ratio (Boynton, 1975) and are not possible in a strongly H_2 depleted region.

J. M. Herndon: The behavior of trace constituents is dominated by the behavior of the major phases in which they reside. It would be interesting if one could discern from trace constituents the behavior of the major phase host. Thermodynamic data is at the present time insufficiently precise to make this possible. It should, however, be possible to perform experiments to address the question.

J. F. Kerridge: There is no direct evidence that the magnetic platelets in C1 carbonaceous chondrites were formed in the nebula. The idea that these platelets represent condensation products arose because their morphology is unique among known magnetite occurrences and because platey morphologies have been predicted for vapor condensation phases. In reality, the chemical composition and petrographic associations of this magnetite (and most other phases in C1 meteorites) are consistent with an origin by secondary alteration on the meteorite parent body and may well require such an origin. The unusual morphology of the magnetite platelets was probably controlled by local growth conditions, specifically by intergrowth with another mineral species.

J. M. Herndon: Could one obtain such magnetite morphologies from laboratory experiments? Perhaps we might also learn something of the origin of this particular magnetite by looking at the isotopic compositions of Ni and Fe. It would be interesting to know whether any ^{60}Ni excess from the ^{60}Fe decay product exists in these.

H. E. Suess: There must have been more than one period of time when condensation, or rather interaction between condensates and a gas phase, occurred. An early first period established the elemental composition of the condensates, as investigated by the Chicago School (Anders, Grossman, etc.), and a second one established the degree of oxidation of the condensates. The first one determined the rare earth abundance distribution (see Boynton's chapter in this book), while the second is the one considered by Herndon and Wilkening. How these two stages of gas-condensate interaction might fit into a model for the formation of the solar system is another question that the astrophysicists have to resolve.

REFERENCES

Alexander, E.C., Jr.; Lewis, R.S.; Reynolds, J.H.; and Michel, M.C. 1971. *Science* 172: 837.
Baldwin, R.B. 1974. *Icarus* 23: 157.
Boynton, W.V. 1975. *Geochim. Cosmochim. Acta* 39: 569.
Clayton, R.N.; Grossman, L.; and Mayeda, T.K. 1973. *Science* 182: 485.
Clayton, R.N.; Onuma, N.; and Mayeda, T.K. 1976. *Earth Planet. Sci. Lett.* 30: 10.
Cloez, S. 1864. *Compt. Rend.* 59: 37.
Dodd, R.T. 1976. *Earth Planet. Sci. Lett.* 30: 281.
Donn, B., and Sears, G.W. 1963. *Science* 140: 1208.
Drozd, R.J., and Podosek, F.A. 1976. *Earth Planet. Sci. Lett.* 31: 15.
Fuchs, L.H. 1971. *Amer. Mineral.* 56: 2053.
Gray, C.M., and Compston, W. 1974. *Nature* 251: 495.
Greenberg, R.; Wacker, J.F.; Hartmann, W.K.; and Chapman, C.R. 1978. *Icarus,* in press.
Grossman, L.; Fruland, R.M.; and McKay, D.S. 1975. *Geophys. Res. Lett.* 2: 37.
Hartmann, W.K. 1966. Publ. No. 3 *Center for Meteorite Studies,* Tempe, Ariz: Ariz. State Univ.

____. 1975. *Icarus* 24: 181.
Herndon, J.M. 1977. *Comets, Asteroids, Meteorites.* (A.H. Delsemme, ed.), p. 537. Toledo, Ohio: Univ. Toledo.
____. 1978. Submitted to *Proc. Roy. Soc. London.*
Herndon, J.M., and Suess, H.E. 1977. *Geochim. Cosmochim. Acta* 41: 233.
Herzog, G.F.; Anders, E.; Alexander, E.C., Jr.; Davis, E.C.; Davis, P.K.; and Lewis, R.S. 1973. *Science* 180: 480.
Holweger, H. 1977. *Earth Planet. Sci. Lett.* 34: 152.
Jedwab, J. 1965. *C.R. Acad. Sci. Paris* 261: 2923.
____. 1967. *Earth Planet. Sci. Lett.* 2: 440.
____. 1971. *Icarus* 15: 319.
Keil, K. 1962. *J. Geophys. Res.* 67: 4055.
Keil, K., and Fredriksson, K. 1964. *J. Geophys. Res.* 69: 3487.
Kerridge, J.F. 1970. *Earth Planet. Sci. Lett.* 9: 299.
Kieffer, S.W. 1975. *Science* 189: 333.
Kurat, G. 1970. *Earth Planet. Sci. Lett.* 9: 225.
Larimer, J.W., and Anders, E. 1970. *Geochim. Cosmochim Acta* 34: 267.
Lee, T., and Papanastassiou, D.A. 1974. *Geophys. Res. Lett.* 1: 225.
Lee, T.; Papanastassiou, D.A.; and Wasserburg, G.J. 1976. *Geophys. Res. Lett.* 3: 109.
Lewis, R.S., and Anders, E. 1975. *Proc. Nat. Acad. Sci.* 72: 268.
Lord, J.C. 1965. *Icarus* 4: 279.
Martin, P.M., and Mills, A.A. 1976. *Earth Planet. Sci. Lett.* 33: 239.
Marvin, V.B.; Wood, J.A.; and Dickey, J.S., Jr. 1970. *Earth Planet. Sci. Lett.* 7: 346.
Mason, B. 1960. *Nature* 186: 230.
Nelson, L.S.; Blander, M.; Skaggs, S.R.; and Keil, K. 1972. *Earth Planet. Sci. Lett.* 14: 338.
Podolak, M., and Cameron, A.G.W. 1974. *Icarus* 23: 326.
Reid, M.; Bass, N.M.; Fujita, J.; Kerridge, J.F.; and Fredriksson, K. 1970. *Geochim. Cosmochim. Acta* 34: 12-53-1255.
Reid, A.M., and Fredriksson, K. 1967. *Researches in Geochemistry* (P.H. Abelson, ed.), vol. 2 p. 170. New York: Wiley.
Reynolds, J.H. 1960. *Phys. Rev. Lett.* 4: 8.
Ringwood, A.E. 1961. *Geochim. Cosmochim. Acta* 24: 159.
Rowe, M.W., and Kuroda, P.K. 1965. *J. Geophys. Res.* 70: 709.
Schmidt, R.A.; Keil, K.; and Gault, D.E. 1964. *Proc. 7th Hypervelocity Impact Symposium* 5: 1505. Tampa, Florida.
Soderblom, L.A.; Condit, C.D.; West, R.A.; Herman, B.M.; and Kreidler, T.J. 1974. *Icarus* 22: 239.
Sorby, H.C. 1877. *Nature* 15: 495.
Stephens, J.R., and Kothari, B.K. 1978. *Moon and Planets* (special Protostars and Planets issue). In press.
Tschermak, G. 1875. *Sitz. Ber. Math-Naturw. Kl., Akad, Wiss. Wein* Vol. 71: part 2, p. 661.
Urey, H.C., and Craig, H. 1953. *Geochim. Cosmochim. Acta* 4: 36.
Wark, D.A., and Lovering, J.F. 1977. *Proc. Lunar Sci. Conf. 8th,* p. 95. Pergamon.
Wasson, J.T. 1972. *Rev. Geophys. Space Phys.* 10: 711.
Wetherill, G.W. 1975. *Proc. Lunar Sci. Conf.* 6th, p. 1539. New York: Pergamon.
Whipple, F.L. 1966. *Science* 153: 54.
____. 1972. In *From Plasma to Planet* (A. Elvius, ed.), p. 211. New York: Wiley.
Wilkening, L.L. 1973. *Geochim. Cosmochim. Acta* 37: 1985.
____. 1977. In *Comets, Asteroids, Meteorites* (A.H. Delsemme, ed.), p. 389. Toledo, Ohio: University of Toledo.
Wilkening, L.L., and Clayton, R.N. 1974. *Geochim. Cosmochim. Acta* 38: 937.
Wood, J.A. 1963. *Icarus* 2: 152.

METEORITE MAGNETISM AND EARLY SOLAR SYSTEM MAGNETIC FIELDS

E. H. LEVY AND C. P. SONETT
University of Arizona

Remanent magnetization of carbonaceous chondrite meteorites apparently reflects the presence of a strong magnetic field during the formation of the solar system. This remanence implies that primitive carbonaceous chondrites have experienced a magnetic field having an intensity on the order of one gauss. The major possible models for the origin of this field are discussed. Magnetic fields of planetary-scale parent objects or the compressed interstellar field seem to offer the least likely explanations. An extended solar field and a nebular dynamo magnetic field offer possible explanations. These are discussed and the need for additional work is pointed out.

The principal aim of this chapter is to review the evidence for the presence of a substantial magnetic field during the formation of the solar system, some 4.6×10^9 yr in the past, and to review several possible models for the origin of the field. It is apparent that angular momentum transfer by magnetic fields plays a central role in the evolution of stellar spin rates (Struve, 1930; Kraft, 1972); the present slow rotation of the sun may be evidence for the influence of magnetic torques during the sun's formation and evolution (Mestel, 1968; Hoyle, 1960; Alfvén, 1954; Weber and Davis, 1967; Modisette, 1967). A strong early solar magnetic field is essential to the hypothesis that inductive heating by an early solar wind magnetic field was important to the evolution of solar system bodies. Among the objects which may have been heated in this way are asteroidal parent objects (Herbert and Sonett, 1978), possibly some of the planets such as Mercury (Sonett *et al.*, 1968), and the moon (Sonett *et al.*, 1975; Herbert *et al.*, 1977). Since a high solar spin rate is also required in order that heating by solar wind induction be efficient, the hypothesis implies that magnetic braking has diminished the sun's spin. Strong fields may also have had an important influence on the dynamics of a preplanetary nebula if the electrical conductivity was sufficiently high to couple the nebular gas to the field (Alfvén, 1954; Hoyle, 1960). In addition, it has been suggested that a magnetic field may have played a significant role in producing some separations of chemical species (Alfvén and Arrhenius, 1976; Jokipii, 1964).

Since the 1960s small-scale and regional fields have been discovered on the moon (e.g., Coleman and Russell, 1977, and Fuller, 1974 for reviews); Mercury (Ness *et al.*, 1974; Ness *et al.*, 1975) has been shown to have a global magnetic field; and the planets Mars (Dolginov, 1978*a,b*) and Venus (Russell, 1976) may have magnetic fields. In the case of the moon, magnetization of basalts implies the presence of a magnetizing field during the interval 3.2 – 3.9 $\times 10^9$ yr before the present. Although earlier highland magnetization may be tied to the primordial solar system, rather complex arguments are required to connect the basalt magnetization to events taking place 4.6 $\times 10^9$ yr ago. (Urey and Runcorn, 1973 give an example).

In the case of Mercury the contemporary field source is uncertain, but it is of sufficient intensity to create a weak magnetosphere about the planet. The most obvious model invokes a presently active hydromagnetic dynamo, similar to that which is thought to generate Earth's magnetic field. However, such a model depends strongly upon the thermal history of the planet and it is questionable whether a heat source is present sufficient to maintain the planet's core in a molten and convecting state (Cassen *et al.*, 1976). Stephenson (1976) has suggested that magnetization of the planet's crust could account for the magnetosphere, but such magnetization may have occurred relatively late in the planet's evolution.

The most direct evidence for the existence of strong magnetic fields during the early evolution of the solar system comes from the magnetic remanence of meteorites. Remanent magnetization is familiar through the everyday occurrence of permanent magnetism. As is well known from common experience, ferromagnetic substances may acquire or lose magnetization under a variety of circumstances. For example, ferromagnetic material which cools from a high to a lower temperature while held in a magnetic field is very efficiently magnetized. Magnetization acquired in this way is called *thermoremanent magnetization*. Similarly, some degree of magnetization may be imparted by immersion in a magnetic field at constant temperature. Such magnetization is referred to as *viscous remanence*, or *isothermal remanence*, depending on the detailed microscopic properties of the mineral grains. In this chapter we can only briefly define a few important magnetizing processes and refer the reader to the literature for more complete discussions (e.g., Stacey and Banerjee, 1974; Tarling, 1971). Two additional magnetizing processes, however, deserve mention here. The growth of mineral grains in the presence of a magnetic field produces *chemical remanent magnetization*; and the accretion or sedimentation of magnetic particles in the presence of a magnetic field produces *depositional remanence.*

The total magnetization of a natural object is in general a superposition of magnetization components acquired through several of the above processes, during various times in the object's history. Generally, the different components of magnetization have distinct properties because they are retained by mineral grains which differ in their compositions, structures, and dimensions. As a result, these different components can be disentangled, one from the other, and some details about the conditions under which each was acquired (blocked) can be identified, especially including the blocking temperature and the intensity of the magnetizing field. A number of laboratory techniques and procedures have been developed to accomplish this; descriptions can be found in the cited literature. In general the older and primary components of magnetization are hard, or stable, which is to say they are difficult to remove by demagnetization. More recently acquired, and extraneous components are generally soft, or easily removed by demagnetization.

Virtually all meteorites carry natural remanent magnetization. This is illustrated graphically in Fig. 1, which shows the natural magnetic moments of a large class of such objects. The most striking aspect of Fig. 1 is the strong linear relation between the specific magnetization and the specific magnetic susceptibility, viz the ferromagnetic carrier concentration, for a wide range of meteorites in several generic classes (Sonett, 1978). Fig. 1 also reflects the strong grouping within generic (and thus possibly genetic or evolutionary) types (Gus'kova and Pochtarev, 1959).

In the data shown in Fig. 1 no attempt has been made to assess the significance of the natural remanence or to separate the hard, and presumably primordial magnetization from the extraneous, soft components of presumably recent origin. This involves detailed analysis of individual objects which has been carried out only for a much smaller population of objects. Stacey and Lovering (1959) first demonstrated the presence of a hard component of magnetization in two ordinary chondrite meteorites and speculated that this might be the result of ancient magnetization.

The outer layers of meteorites are heated to the melting point upon entry into Earth's atmosphere. During the subsequent cooling in the latter stages of entry, thermoremanent magnetization is acquired in these layers. But Lovering *et al.* (1960) have calculated that only the outer 3 cm or so of an entering object is heated above 100°C, and as material is ablated the isotherm moves with the ablation to maintain this small thickness of heated region. Weaving (1962) has experimentally detected and accounted for "atmospheric entry" thermoremanent magnetization in the ordinary

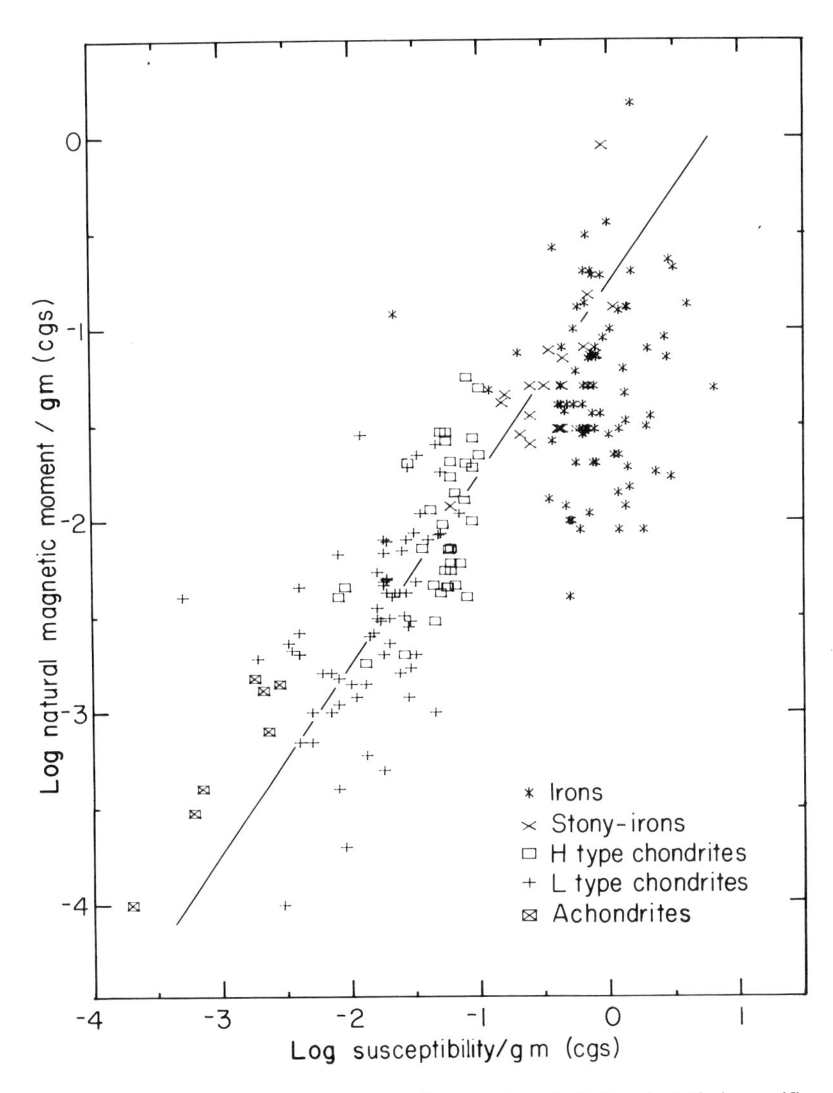

Fig. 1. Specific natural magnetic moments of meteorites plotted against their specific magnetic susceptibilities (Sonett [1978] from a compilation of measurements from Russian collections by Herndon *et al.* [1972]).

chondrite Brewster, confirming the calculation of Lovering *et al.* (1960). Generally, atmospheric entry affects a meteorite's magnetization only in the outer few centimeters and does not interfere with identification of primordial magnetization.

Under detailed investigation individual meteorites have usually been shown to possess a hard and stable residual magnetization (Stacey and Lovering, 1959; Stacey *et al.*, 1961; Weaving, 1962; Banerjee and Hargraves,

1971, 1972; Butler, 1972; Brecher, 1972; Brecher and Arrhenius, 1974; Brecher and Raganayaki, 1975), which appears to be of thermoremanent, chemical remanent, or mixed origin. The success with which these measurements can unambiguously be interpreted, as indicative of primordial magnetization, depends on the magnetic properties of the material and on its evolution subsequent to formation. The carbonaceous chondrites are especially suited to the present discussion as these meteorites appear to be among the least evolved and differentiated relics of the solar system's early years. Furthermore, their magnetic properties seem sufficiently well-behaved to have preserved a clear record of their early magnetic history. Consequently, in the remainder of this article we concentrate solely on the carbonaceous chondrites.

The next section briefly reviews the relevant properties of carbonaceous chondrites and summarizes the available information on their magnetic history. Following that, we discuss principal candidates for the origin of the magnetizing field. Our approach is restricted to the existing empirical basis of information; we also indicate where substantial work is necessary. The discussion of models relies as little as possible on *ad hoc* assumptions constructed solely to explain the magnetic measurements.

THE CARBONACEOUS CHONDRITES

The general physical appearance, the chemistry, and the mineralogy of carbonaceous chondrite meteorites suggest that these objects have remained largely unaltered since their formation in the primitive preplanetary nebula (Wasson, 1974). Information about meteorite ages is consistent with this idea; however, ambiguities still remain about the relative ages of the meteorite classes derived from isotope relative abundances (Reynolds, 1967). Wasson (1974) has classified thirty-five of the known meteorites as carbonaceous chondrites and of these he has isolated five as the apparently most primitive objects; these are called the CI 1 meteorites for cataloging purposes.

Careful measurements of the magnetic properties and the inferred primeval magnetizing fields have been carried out for six carbonaceous chondrites; these, together with their estimated magnetizing fields (based upon thermoremanent magnetization), are listed in Table I. Note that, of these six, two belong to the putative class of most primitive objects, the CI 1. (The most extensive magnetic measurements have been carried out on samples of the Allende meteorite, largely as a result of the great abundance of material available from this object.)

Estimates of the primordial magnetizing fields (viz., the fields responsible for the hard, stable component of remanence) have been obtained through the use of a variety of techniques. Some degree of ambiguity is introduced by the uncertainty that laboratory experiments always accurately simulate the actual magnetization processes. However, the striking agreement between

TABLE I

Estimates of Intensity of the Primordial Magnetic Field From Carbonaceous Chondrites

Object	Class	Magnetic Field (gauss)	References
Allende	CV3	1.1	Butler, 1972; Banerjee and Hargraves, 1972
		0.25 - 1.0	Brecher, 1972
Murchison	CM2	0.18	Banerjee and Hargraves, 1972
		0.5 - 3	Brecher, 1972
Orgueil	CI1	0.67	Banerjee and Hargraves, 1972
Renazzo	CV2	2	Brecher, 1972
Murray	CM2	0.7	Brecher, 1972
Ivuna	CI1	0.5	Brecher, 1972

measurements made on various objects and by various techniques suggests the qualitative validity of the inferred paleofield intensities. Indeed, the spread of inferred paleofields from one object to the next in Table I is comparable to what one might expect even if they were all magnetized by the same background magnetic field but on a variety of objects spinning with different orientations with respect to the field.

The inferred paleointensities are generally based on the assumption that the magnetization is thermoremanent. This is a generally conservative assumption since magnetization apparently is acquired more efficiently through a thermal process than via chemical changes, phase changes, or some combination of these and thermoremanence. Relaxing the assumption that the magnetization is of thermal origin would thus yield larger estimates of the primordial background field which is assumed to be responsible for the magnetization of meteorites. However, the precise nature of the magnetizing process – thermal, chemical, or mixed – should not significantly affect the subsequent discussion or the final conclusions.

The temperature history of the meteorites is of particular significance since it carries important information about conditions in the early nebula. The physical properties and chemistry of carbonaceous material (Wasson, 1974) and the character of the natural remanent magnetization (Banerjee and Hargraves, 1971, 1972; Butler, 1972; Brecher, 1972, 1977) suggest that carbonaceous chondrites have remained at relatively low temperatures ($<$ 200°C) since their formation. A spectrum of blocking temperatures is present in all objects, and a significant fraction of the thermoremanence seems to have been acquired at temperatures well below 150°C (Butler, 1972; Brecher,

1977). Further work (Lanoix *et al.*, 1977, 1978), however, indicates that individual chondrules from Allende may have been magnetized at a higher temperature (> 160°C) and in a more intense magnetic field (2 to 7 G) than the meteorite matrix.

At this time there are at least two additional pieces of information emerging which must have a substantial impact on our ideas about the early history of meteorites. These are the embedded noble gases, which seem to have impinged on meteoritic grains at energies on the order of 10^3 electron volts (comparable to the kinetic energies of solar wind particles); and the evidence of energetic particle fluxes, similar to those from solar flares, carried by the grains. Both indicate that meteoritic grains were exposed to solar emissions, unshielded by dense nebular gas, for at least a limited time before their final incorporation into compacted bodies.

INTERPRETATION OF THE MEASUREMENTS

We recall that the apparent primordial magnetization is homogeneous over macroscopic ("rock-sized") dimensions of meteoritic material. The most straightforward and conservative interpretation of these data implies that the remanent magnetization was imprinted after the accretion and compaction of the meteorite's parent bodies. Furthermore, the observed macroscopic-scale magnetization implies that meteorite parent bodies were immersed in a magnetic field having a component with non-zero mean in some direction viewed from a frame fixed in the object. This component must have had an intensity of the order of one gauss, and it must have persisted long enough for the remanence to be imprinted.

The time scale for acquisition of magnetic remanence is set by the time taken for parent objects to relax thermally through their blocking temperatures. Although blocking temperature spectra are available from laboratory studies, these provide no evidence of just how rapidly remanence was blocked into meteorites. The thermal and chemical evolution of small primitive bodies was most likely governed by the ambient conditions in the surrounding nebula. These conditions are unlikely to have changed substantially in only a few years. (There is of course the possibility that the remanence was imprinted by fast, or catastrophic, processes such as collisional shocks, in which case the time scale could be very short. This possibility deserves further scrutiny.)

ORIGIN OF THE MAGNETIZING FIELD

So far as we can see, there are four major candidates for the origin of the primordial magnetic field which produced the remanence in carbonaceous chondrites. They are:

1. Magnetic fields generated in very large meteorite parent bodies;
2. The interstellar magnetic field compressed to high intensity by the

inflowing gas;

3. A strong solar magnetic field permeating the early solar system;
4. A hydromagnetic dynamo field produced in the gaseous nebula itself.

We shall consider each of the these possibilities in turn.

1. Parent Body Magnetic Fields

A magnetizing field generated in a meteoritic parent object would require that carbonaceous chondrites originally resided on differentiated bodies large enough to have generated internal magnetic fields – certainly hundreds of kilometers in radius. However, the carbonaceous chondrites seem to be composed of essentially undifferentiated, primitive material and do not seem to show evidence of the violent processes which would be necessary to remove material from the surfaces of such large objects. Indeed various arguments have been advanced against a parent body origin for meteorite magnetization (Brecher [1973] gives references to arguments on both sides of this question). Thus, while it probably cannot be demonstrated conclusively at present, it is generally thought that the carbonaceous chondrites do not descend from the surfaces of very large objects. Altogether then we will not consider further the possibility of magnetization by a parent body field; but we caution the reader that additional work is needed to fully illuminate the difficulties and merits of this possibility.

2. Compressed Interstellar Field

Compression of the local interstellar magnetic field to an intensity of the order of one gauss in the preplanetary nebula requires that the electrical conductivity of the infalling interstellar gas remained high throughout the pre-nebular collapse phase. It is easy to see that if the conductivity did remain sufficiently high, then the field could in principle be compressed to the necessary intensity. For example, in isotropic collapse of a very highly conducting gas, preservation of the total embedded magnetic flux implies that $B \propto \rho^{\frac{2}{3}}$, where ρ is the gas density. Then compression of interstellar gas from an original density of one atom cm^{-3} to a final density of say 10^{16} atoms cm^{-3} would compress an initial 3×10^{-6} G magnetic field to 10^5 G. More complete calculations of the compression of a magnetic field by a collapsing gas cloud (e.g., Mouschovias, 1977) suggest that the field intensity more likely varies as $B \propto \rho^{\frac{1}{2}}$ which would give a compressed field intensity of several hundred gauss (see Mouschovias' chapter).

However, calculations of the ionization level in dense, interstellar, clouds (Nakano and Tademaru, 1972; Oppenheimer and Dalgarno, 1974) show that the electron and ion concentrations in these objects fall to very low values at relatively low gas densities. As a result the magnetic field and the neutral gas become uncoupled through the process of ambipolar diffusion whereby the field and the charged particles slip through the neutral gas. In consequence of this the interstellar field is thought to escape the collapsing pre-nebular gas (e.g., Mouschovias, 1977) without exceeding the milligauss range. This scenario seems to be consistent with the most intense, dark-cloud magnetic fields so far observed in interstellar space (see the chapter by Chaisson and Vrba in this book).

This model for the escape of the interstellar magnetic field from dense gas clouds is based on stationary hydromagnetic equilibrium calculations. It suggests that the interstellar field did not contribute substantially to the strong early magnetic fields inferred from meteorite remanence. However, this conclusion should be regarded as tentative, since the more complete dynamical calculations, necessary to firmly rule out the possibility of a more strongly compressed interstellar field, have not been carried out.

3. Solar Magnetic Field

We consider two possibilities for a solar origin of the magnetizing field. The first is that the field is centrally confined so that all of the field stress is taken up in the sun, with the exterior field in either a vacuum or force-free configuration. In this case the field strength must vary at least as rapidly as the inverse cube of the distance from the sun. Thus, a one gauss field at several hundred solar radii would imply an internal solar field in excess of 10^7 G. Several lines of argument suggest that such a large early solar field is unlikely. Observationally, no example is known of such an intense stellar field, except in association with collapsed, degenerate objects. The most intense stellar magnetic field known (HD 215441) is some three orders of magnitude smaller than this. In addition, since the resistive decay time of the dipolar mode in a sun-like object is on the order of 4×10^9 yr (Wrubel, 1952), such a strong early field should have persisted to the present day. The general poloidal field of the present sun is on the order of one gauss.

This simple model of the dissipation rate of an early strong solar field actually requires several modifications. A magnetic field of intensity 10^7 G will provide the dominant structural force in the outer layers of the sun. Under these conditions the field must take up a largely force-free configuration. One possibility is that in the outer layers the field adopts the geometry of a vacuum field so that all electrical currents as well as magnetic stresses are confined to the deep interior. If so, the initial dissipation time for the field may be appreciably shortened, since it is proportional to the square of the length scale characterizing the volume within which the electrical currents are confined. (See Mestel, 1967, for a more complete discussion of this point.) However, once the surface field intensity falls to several tens of thousands of gauss, the classical dissipation described above applies. Thus, we would expect to see a residual field of several tens of thousands of gauss left from a strongly magnetized early sun.

It is worth considering briefly the possibility that a very strong magnetic field has somehow become hidden beneath the sun's surface. While we see no reasonable scenario through which a strong field could have become buried within the sun, there is a fundamental reason why it would not remain buried. Regardless of the structure of a magnetic field, its net stress is always expansive. This is a straightforward result of the virial theorem. In consequence of this a magnetic field can be thought of, in crude terms, as a high-pressure, zero-density gas. Thus, a magnetic field embedded in a fluid is extremely buoyant. A field confined beneath the sun's surface, even if in equilibrium, would be extremely unstable through an analogue of the Rayleigh-Taylor instability. In a time short compared with the sun's age (Parker, 1974), large amounts of magnetic flux would poke through the surface and the field would make its presence known.

A major area of uncertainty in limiting the possibilities for a strong magnetic field in the early sun concerns the effect of vigorous convection during, say, a possible Hayashi phase (Hayashi and Nakano, 1965; Hayashi, 1966). Such convection may influence the field in two ways. The turbulence may dominate the transport of magnetic field lines and thus increase the field dissipation rate over that used in the previous discussion. In addition, the convection may produce a magnetic field through hydromagnetic dynamo action. To our knowledge, detailed consideration of these possibilities has not been carried out. However, a crude and tentative argument suggests that such effects do not have an important bearing on magnetization of the carbonaceous chondrites. In order for turbulent fluid motions to dominate the magnetic field transport and to tangle the field to smaller and smaller scales, thus speeding its dissipation, the kinetic energy density must be greater than the field energy density. Furthermore, in order for a hydromagnetic dynamo to be effective on the time scale characteristic of Hayashi phase evolution ($\sim 10^6$ yr), it is probably necessary that turbulence dominate the field's dissipation in order to provide sufficient simplification of the field's topology (Parker, 1971*b*; Rädler, 1968); in this situation the field energy density must again be less than the kinetic energy density of the fluid flow so that the field lines can be carried to small scales in a turbulent cascade. Thus, it is altogether plausible that a global magnetic field strongly influenced by a convective Hayashi phase did not have energy density in excess of the kinetic energy density of the convection.

A simple calculation suffices to show that if Hayashi's model is correct, any associated fields are probably too weak to account for the magnetization of parent objects. Recall that thermal properties of carbonaceous chondrite remanence suggest that the magnetization was acquired at relatively low temperatures, i.e., $\leqslant 150^\circ$C (Butler, 1972). If the meteorite parent objects had, as their main source of heat, direct insolation at a distance of 3 AU from the sun, then with an assumed albedo of the parent object albedo of about 10% at the time of magnetization, the solar luminosity must have been less than fifty times its present value. For a one solar mass object this implies a solar radius less than about fifteen times the present value (Hayashi, 1965). Referring to Hayashi's analysis of early stars, and using standard mixing length theory to obtain the convection velocity, we find that the magnetic field must be less than about 10^3 G if its energy density is to be less than the kinetic energy density of convection. If the field falls away as the inverse cube of distance from the sun, then at 3 AU the intensity will be on the order of 10^{-2} G, too small to account for meteorite remanence. This argument suggests that a possible Hayashi convective phase did not substantially affect the magnetic fields responsible for meteorite magnetization. At our present state of knowledge this discussion should probably be regarded as illustrative; it points up the need for a definitive understanding of meteorite thermal histories and the sun's early history.

Instead of the situation in which the nebular field stress is taken up wholly within the sun, the solar field may be extended throughout the nebula by a solar wind outflow. Sonett (1974) has previously suggested a similar field structure to produce an inductive heating episode in small solar system bodies. In this case, the field components in the nebula and near the solar equatorial plane are given by a well-known relation (Parker, 1958)

$$B_r = B_s R_s^2 r^{-2} \quad \text{and} \quad B_\phi = B_s \, \Omega \, R_s^2 \, (V_w r)^{-1} \tag{1}$$

where R_s is the sun's radius, B_s is its surface magnetic field strength, V_w the velocity of the solar wind, and Ω the sun's spin rate. Supposing, for the purpose of illustration, that the solar wind speed was 3×10^7 cm sec^{-1}, and that the solar spin rate was about a hundred times its present value (a reasonable supposition for the early sun, before it had been braked by a solar wind[a]), it is easily seen that a one gauss field at, say, 3 AU requires a solar surface field of about 10^3 G. We see no fundamental reason why such a field could not have been produced in the early sun. In fact, such a field could have been produced by a hydromagnetic dynamo in the solar convective zone. In this zone the hydromagnetic time scale is very short and we would see no residual today from such an early magnetic field.

In order that a magnetic field of one gauss should be carried to several astronomical units by the solar wind, it is necessary that the kinetic energy density of the flow exceed that of the field. With a solar wind velocity of about 300 km sec^{-1}, this requires that the mass density be about 9×10^{-17} gm cm^{-3}, which implies a mass-loss rate from the sun on the order of 10^{-6} $M_\odot$/yr. This is somewhat greater than the mass flows thought characteristic of vigorous T Tauri winds (Kuhi, 1964) and suggests that if such a strong flow did occur it must have lasted a relatively brief time.

---

[a]But which does not take into account a possible lower limit imposed by fission instabilities of the sun itself.

In addition to the potential difficulty with the plasma flow noted above (based upon interpretation of spectroscopic observation [Kuhi, 1964] and model calculations suggesting a T Tauri lifetime on order of 10^7 yr [Iben, 1965]), there are two additional difficulties with the extended "solar wind" geometry as applied to explain meteorite remanence. The extended field is configured in a three-dimensional archimedean spiral. A spinning and orbiting object will in general see no unidirectional component from such a field. Therefore, magnetization of meteorites by a solar wind magnetic field requires that the passage through magnetic blocking conditions be short in comparison with the orbit period of several years. As we mentioned earlier, this seems unlikely, unless the magnetization was imprinted as a result of fast processes, for example collisional heating.

The second difficulty is that repetitive changes in direction of the field are expected during one solar rotation, if the solar field is dipolar and drawn out by plasma flow. This supposition is based upon current models for the present epoch which show a sector structure thought to be due to the presence of a neutral sheet separating the northern and southern halves of the interplanetary field. However, as pointed out by Sonett *et al.* (1975) this objection can be resolved if the solar field also contains a quadrupolar component which will cause the net interplanetary magnetic field to have a dominant polarity at planetary distances.

4. Nebular Dynamo

Turbulent gas motions in a differentially rotating nebula (Cameron, 1978) can produce hydromagnetic dynamo action, and thereby generate a large-scale magnetic field, if the electrical conductivity, σ, of the gas is sufficiently high. Consolmagno and Jokipii (1978) have pointed out that the recent indications of an excess ^{26}Al abundance in the early solar system (Lee *et al.*, 1976) in the amounts that may have been present (Cameron and Truran, 1977) could produce a moderately high electrical conductivity by the emission of charged particles through radioactive decay. The precise value of the conductivity depends on several model-dependent variables which are discussed further by Consolmagno and Jokipii (1978).

Calculations of the character of dynamo magnetic field generation under the physical and geometrical conditions thought to be representative of a preplanetary nebula have not been carried out so far as we are aware. However, several illustrative remarks can be made on the basis of general considerations. Parker (1971*a*) computed some of the properties of dynamo magnetic field generation in an extended slab of gas, with specific reference to the galactic magnetic field.

The efficiency of magnetic field generation in a natural hydromagnetic dynamo depends on the rate of the fluid's nonuniform rotation, the intensity of its helical or cyclonic convection[a], and the electrical conductivity. The precise criterion for the growth of a magnetic field depends on the details of the fluid flow and on the geometry, but a general measure of efficacy for turbulent motions in a differentially rotating fluid is the dynamo number, which can be written (Parker, 1970; Levy, 1976)

$$N \equiv \frac{\gamma \Gamma \delta^2}{\eta^2} \tag{2}$$

where γ is a measure of the fluid's rate of global shear; Γ is the helical or cyclonic component of the turbulent convection; δ is scale of the magnetic field; and $\eta = c^2/4\pi\sigma$ is the magnetic diffusivity. Since the gas is nearly in Keplerian motion, we can write (Parker, 1971*a*)

$$\gamma \equiv \frac{\partial V_\phi}{\partial r} - \frac{V_\phi}{r} \simeq \frac{3}{2}\,\Omega \tag{3}$$

where r is the radial coordinate in a standard spherical coordinate system, V_ϕ is the azimuthal velocity, and $\Omega = (GM_\odot/r^3)^{\frac{1}{2}}$. Similarly for the cyclonic component of the turbulent motion, $\Gamma \simeq 0.25\, l\Omega$, where l is the scale of a large turbulent eddy (Parker, 1955, 1971*a*; Steenbeck *et al.*, 1966; Lerche, 1971; Moffatt, 1970; Levy, 1978). Numerically, in a nebula surrounding the sun $\gamma \sim 1.9 \times 10^{13} r^{-\frac{3}{2}}$ sec^{-1} and $\Gamma \sim 3.2 \times 10^{12}\, l r^{-\frac{3}{2}}$ cm sec^{-1} (Cameron, 1978). Then taking both the scale of the turbulent eddies, l, and the scale of the field, δ, equal to about half the disk thickness of $\sim 5 \times 10^{12}$ cm (Cameron, 1978), we find $N \sim 1.7 \times 10^{35}\, \eta^{-2}$ when $r \sim 4$ AU. Parker's (1971*a*) calculations for a roughly similar physical configuration found that a large-scale magnetic field is generated when $N \simeq 6$. This requires that $\sigma \geqslant 400$ sec^{-1}, a relatively low electrical conductivity. Consolmagno and Jokipii (1978) suggest that ionization produced by the decay of ^{26}Al can produce an electrical conductivity on the order of 10^3 sec^{-1} in a preplanetary nebula where the density is about 3×10^{-10} gm cm^{-3} and the temperature ~ 200 K. These conditions are comparable to those which may exist in such a nebula at several astronomical units from the center, roughly between the asteroid and the Jupiter regions (Cameron, 1978).

Recalling that meteorite remanence suggests the presence of a magnetic field of approximately one gauss in the early nebula, it is of interest to estimate the intensity of the field that might be expected. Since the Coriolis

---

[a]Efficient generation of magnetic fields generally requires fluid motion with helicity, i.e., $(\Delta \times V) \cdot V \neq 0$. (A review and references are in Levy, 1976.)

force is responsible for enforcing on a fluid motion the large-scale organized character that produces efficient magnetic field generation, the maximum characteristic intensity of such a field is given by a balance between the Lorentz and Coriolis forces. (Levy, 1976 has a more complete discussion of this question.) Then the global field satisfies $\rho V\Omega \sim \langle B_p B_\phi \rangle / 4\pi l$ where V is the typical convective velocity, ~ 0.4 km sec^{-1} and B_p is the poloidal magnetic field. Using the values quoted above we find that $\langle B_p B_\phi \rangle^{\frac{1}{2}} \sim 5 - 10$ G, so that characteristic maximum field strength to be expected is on the order of 5 – 10 G. The meteorite remanence implies that the primitive carbonaceous chondrites experienced a unidirectional field component on the order of one gauss; our crude analysis suggests that this is consistent with a possible nebular dynamo origin of the magnetizing field. These estimates are, of course, very rough. A more complete treatment will require taking thorough account of other dynamical effects such as ambipolar diffusion.

Among the most easily excited modes of a dynamo are the stationary, axisymmetric fields. Of these, the odd-symmetry, dipole-like fields should have a component which is normal to the nebular disk and which will therefore have a unidirectional projection as seen from a spinning and orbiting body. Thus, the nebular hydromagnetic dynamo can be expected to generate a global magnetic field with a morphology conducive to producing the observed meteoritic remanence. Active field generation need not take place at the location of the magnetized planetesimals. Dynamo generation at remote locations in the gaseous nebula may still produce a field in the vicinity of the meteorite parent bodies.

SUMMARY AND CONCLUSIONS

We have briefly surveyed the available information on the primordial magnetization of meteorites. The present discussion has concentrated on the carbonaceous chondrites because their magnetic properties seem well-behaved and because of remanence in these objects can be interpreted in the least ambiguous fashion. Furthermore, the carbonaceous chondrites are relatively primitive, according to the prevalent view, and their characteristics are thought to most directly reflect conditions in the preplanetary nebula. Measurements indicate that these meteorites were magnetized by a magnetic field of approximately one gauss.

Of the four possible origins for the magnetizing field which we considered, two are least likely: the internal magnetic field of a large parent body, and the compressed interstellar field. However, more complete dynamical calculations are needed to firmly rule out the latter. Magnetization by a solar magnetic field seems difficult. A field rooted only in the sun easily satisfies the geometrical requirements for meteorite magnetization but requires magnetic strengths which appear to be unrealistically large. On the other hand, a magnetic field extended through the solar system by a powerful solar wind more easily generates the required intensity, but there are difficulties with its geometry. Magnetization by such a solar-wind field probably would require a rapid traverse of the meteorites through their magnetic blocking conditions. Catastrophic processes, for example collisions, could satisfy this requirement; but the carbonaceous chondrites do not appear to be part of a population of heavily shocked material. Finally, field production by a hydromagnetic dynamo in the nebula gas offers an explanation but detailed calculations remain to be carried out in order to explore this possibility.

Of course our list of possible origins of the magnetizing field is not exhaustive. For example, one could imagine that the magnetization resulted from electrical discharges, or "lightning strikes", in the early nebula. Some intensely magnetized lodestones on Earth are thought to arise in this way. Such phenomena have occasionally been suggested to account for the melting of chondrules (Whipple, 1966). It has also been suggested that hypothetical, strong magnetic fields induced in cometary nuclei may have been responsible for imprinting meteorite remanence (Brecher, 1977). It seems unlikely to us, even if such strong fields are induced in comets, that their direction would remain sufficiently constant to account for the observed magnetization. We have confined our more detailed discussion to phenomena which are not *ad hoc* and follow in a natural way from the physical evolutionary states which the solar system may have traversed during its birth.

Magnetic fields of large-scale extent with an intensity of 1 – 10 G in the early solar nebula can have important dynamical consequences (Alfvén, 1954; Hoyle, 1960). As an illustration, consider the redistribution of angular momentum through magnetic torques in a gaseous nebula. Representing the nebula as a disk having a density of about 10^{-9} gm cm^{-3} and radial scale of three or four astronomical units (Cameron, 1978) orbiting the sun, it is straightforward to show (Levy and Rose, 1974 made an illustrative calculation) that the characteristic angular momentum redistribution time varies from 10^4 yr for a one gauss average field to 10^2 yr for a ten gauss average field. These are comparable to the evolutionary time scales for such a nebula.

REFERENCES

Alfvén, H. 1954. *On the Origin of the Solar System.* Oxford: Oxford Univ. Press.

Alfvén, H., and Arrhenius, G. 1976. *Evolution of the Solar System,* NASA SP-345. Washington, D.C.: U.S. Government Printing Office.

Banerjee, S.K., and Hargraves, R.B. 1971. *Earth Planet. Sci. Lett.* 10: 392.

_____. 1972. *Earth Planet. Sci. Lett.* 17: 110.

Brecher, A. 1972. In *Origin of the Solar System* (H. Reeves, ed.), p. 260. Paris: C.N.R.S.

_____. 1973. In *Evolutionary and Physical Properties of Meteoroids,* NASA SP-319. p. 311. Washington, D.C.: U.S. Government Printing Office.

_____. 1977. In *Asteroids, Comets and Meteorites* (A.H. Delsemme, ed.), p. 415. Toledo, Ohio: Univ. Toledo Press.

Brecher, A., and Arrhenius, G. 1974. *J. Geophys. Res.* 79: 2081.
Brecher, A., and Ranganayaki, R.P. 1975. *Earth Planet. Sci. Lett.* 25: 57.
Butler, R.F. 1972. *Earth Planet. Sci. Lett.* 17: 120.
____. 1978. *Moon and Planets.* In press.
Cameron, A.G.W., and Truran, J.W. 1977. *Icarus* 30: 447.
Cassen, P.; Young, R.E.; Schubert, G.; and Reynolds, R.T. 1976. *Icarus* 28: 501.
Coleman, P.J., Jr., and Russell, C.T. 1977. *Phil. Trans. Roy. Soc. London* 285A: 485.
Consolmagno, G., and Jokipii, J.R. 1978. *Moon and Planets* (special Protostars and Planets issue). In press.
Dolginov, Sh. Sh. 1978*a*. *Geophys. Res. Lett.* 5: 89.
____. 1978*b*. *Geophys. Res. Lett.* 5: 93.
Fuller, M. 1974. *Rev. Geophys. Space Phys.* 12: 1.
Gus'kova, E.G., and Pochtarev, V.I., 1959. *Geomag. Aeron.* 7: 245.
Hayashi, C. 1966. *Ann. Rev. Astron. Astrophys.* 4: 171.
Hayashi, C. and Nakano, T. 1965. *Prog. Theor. Phys.* 34: 754.
Hayashi, M. 1965. *Publ. Astron. Soc. Japan* 17: 177.
Herbert, F., and Sonett, C.P. 1978. *Astrophys. Space Sci.* (special volume commemorating the 70th birthday of Hannes Alfvén). In press.
Herbert, F., Sonett, C.P., and Wiskerchen, M.J. 1977. *J. Geophys. Res.* 82: 14.
Herndon, J.M.; Rowe, M.W.; Larson, E.E.; and Watson, D.E. 1972. *Meteoritics* 3: 263.
Hoyle, F. 1960. *Quart. J. Roy. Astr. Soc.* 1: 28.
Iben, I., Jr. 1965. *Astrophys. J.* 141: 993.
Jokipii, J.R. 1964. *Icarus* 3: 248.
Kraft, R. 1972. In *Solar Wind,* NASA SP-308. (C.P. Sonett, P.J. Coleman, Jr. and J.M. Wilcox, eds.), p. 276. Washington, D.C.: U.S. Government Printing Office.
Kuhi, L.V. 1964. *Astrophys. J.* 140: 1409.
Lanoix, M.; Strangway, D.W.; and Pearce, G.W. 1977. *Proc. Lunar Sci. Conf.* 8th. p. 689. New York: Pergamon.
____. 1978. *Lunar Planetary Sci. Conf.* 9th, (Abstract).
Lee, T.; Papanastassiou, D.A.; and Wasserburg, G.J. 1976. *Geophys. Res. Lett.* 3: 109.
Lerche, I. 1971. *Astrophys. J.* 166: 627.
Levy, E.H., 1976. *Ann. Rev. Earth Planet. Sci.* 4: 159.
____. 1978. *Astrophys. J.* 220: 325.
Levy, E.H., and Rose, W.K. 1974. *Astrophys. J.* 193, 419.
Lovering, J.F.; Parry, L.G.; and Jaeger, J.C. 1960. *Geochim. Cosmochim. Acta* 19: 156.
Mestel, L. 1967. In *Magnetism and the Cosmos* (W.R. Hindmarsh, F.J. Lowes, P.H. Roberts and S.K. Runcorn, eds.), p. 194. New York: Elsevier.
____. 1968. *Mon. Not. Roy. Astr. Soc.* 140: 177.
Modisette, J.L. 1967. *J. Geophys. Res.* 72: 1521.
Moffatt, H.K. 1970. *J. Fluid Mech.* 41: 435.
Mouschovias, T. Ch. 1977. *Astrophys. J.* 211, 147.
Nakano, T., and Tademaru, E. 1972. *Astrophys. J.* 173: 87.
Ness, N.F.; Behannon, K.W.; Lepping, R.P.; and Whang, Y.C. 1975. *J. Geophys. Res.* 80: 2708.
Ness, N.F.; Behannon, K.W.; Lepping, R.P.; Whang, Y.C.; and Schatten, K.H. 1974. *Science* 185: 151.
Oppenheimer, M. and Dalgarno, A. 1974. *Astrophys. J.* 192: 29.
Parker, E.N. 1955. *Astrophys. J.* 122: 293.
____. 1958. *Astrophys. J.* 128: 664.
____. 1970. *Astrophys. J.* 162: 665.
____. 1971*a*. *Astrophys. J.* 163: 255.
____. 1971*b*. *Astrophys. J.* 163: 279.
____. 1974. *Astrophys. Space Sci.* 31: 261.
Rädler, K.H. 1968. *Z. Naturforsch.* 23a: 1851.
Reynolds, J.H. 1967. *Ann. Rev. Nucl. Sci.* 17: 253.
Russell, C.T. 1976. *Geophys. Res. Lett.* 3: 589.
Sonett, C.P. 1974. In *Solar Wind Three* (C.T. Russell, ed.), Los Angeles: University of California, L.A.

____. 1978. *Geophys. Res. Lett.* In press.
Sonett, C.P.; Colburn, D.S.; and Schwartz, K. 1968. *Nature* 219: 924.
____. 1975. *Icarus* 24: 231.
Stacey, F.D., and Banerjee, S.K. 1974. *The Physical Principles of Rock Magnetism.* New York: Elsevier.
Stacey, F.D., and Lovering, J.F. 1959. *Nature* 183: 529.
Stacey, F.D.; Lovering, J.F.; and Parry, L.G. 1961. *J. Geophys. Res.* 66: 1523.
Steenbeck, M.; Krause, F.; and Rädler, K.H. 1966. *Z. Naturforsch.* 21a: 369.
Stephenson, A. 1976. *Earth Planet. Sci. Lett.* 28: 454.
Struve, O. 1930. *Astrophys. J.* 72: 1.
Tarling, D.H. 1971. *Principles and Applications of Paleomagnetism.* London: Chapman and Hall.
Urey, H., and Runcorn, S.K. 1973. *Science* 180: 636.
Wasson, J.T. 1974. *Meteorites.* Berlin: Springer-Verlag.
Weaving, B. 1962. *Geophys. J. Roy Astr. Soc.* 7: 203.
Weber, E.J., and Davis, L., Jr. 1967. *Astrophys. J.* 148: 217.
Whipple, F.L. 1966. *Science* 153: 54.
Wrubel, M.H. 1952. *Astrophys. J.* 116: 291.

FROM DARK INTERSTELLAR CLOUDS TO PLANETS AND SATELLITES

HANNES ALFVÉN
The Royal Institute of Technology, Stockholm

The common conclusion that magnetic fields necessarily counteract the collapse of an interstellar cloud is model dependent. In other and at least equally reasonable magnetic field models the magnetic field compresses the cloud. It is possible that dark clouds are formed and kept together by electromagnetic effects. Studies of the properties of dusty plasmas are essential for the understanding of the behavior of dark clouds. The motion of electrically charged small ($\lesssim 10^{-5} - 10^{-6}$ *cm) dust grains is controlled by electromagnetic effects; larger* ($\gg$*1 cm) solid particles move under the action of gravitation and inertial forces, whereas intermediate-size particles will settle under the action of gravitation and viscous forces. In a dusty cloud, gravitation collects the dust at the center of gravity of the cloud. A dust ball is formed which, when it has grown large enough, collects gas from its surroundings. This process may lead to the formation of a star. In a cloud with irregular structure a number of such dust balls may be formed which later join by a process which is similar to the "planetesimal" formation of planets and satellites around the sun. Such a "stellesimal" accretion may result in a body having the mass and angular momentum of the primeval sun. The process of star formation by dust accretion would be effective even for a cloud as small as one solar mass.*

Specific conclusions about the physical and chemical state in the region surrounding a forming star emerge from the model of star formation discussed here. They can also be deduced by extrapolation back in time from the present state of the planet and satellite systems. Both approaches lead to a consistent picture of the structure of the early solar system.

The period during which matter became emplaced in the region of space which developed into our solar system can be characterized as an intermediate stage in an evolutionary chain, beginning with the formation and evolution of an interstellar source cloud, and ending with the planet and satellite systems observed today. Two independent methods are used to reconstruct this intermediate stage, one which makes use of data on the solar system and the other which uses data on the interstellar medium.

RECONSTRUCTION FROM SOLAR SYSTEM DATA

1. Principles of approach

In order to reduce speculation to a minimum, the following principles are applied (see also Alfvén and Arrhenius, 1976).

(a) To avoid more or less arbitrary assumptions about the early states of the solar system the analysis should start with a detailed study of the present state. From this we reconstruct consecutive earlier states by relying preferentially on experimentally verified processes. Space research has in the 1970s yielded a great many new observations which provide a base for this approach.

(b) Because of the similarity between the planetary system and the satellite systems, a satisfactory set of theories covering the different aspects of the formation of the solar system should be applicable to the three regular satellite systems as well as to the planetary system. Such a general framework (we call it "hetegonic") for the formation of secondary bodies around their primaries includes all theories that satisfy the observations in all four systems. It excludes theories that limit themselves to explaining the properties of the planetary system alone or only selected aspects of it.

(c) The advances in experimental plasma physics, largely centered in thermonuclear research and exploration of the earth's magnetospheric plasmas with space probes, have brought about a sobering realization: it is almost impossible to predict the actual behavior of a plasma purely on the basis of theoretical calculations starting from fundamental princples. The properties of plasmas, in the laboratory or in space, can be understood only by highly sophisticated, diagnostic *in situ* measurements. Such measurements are now possible and have become standard procedure in the laboratory, in the planetary magnetospheres and in the heliosphere. However, neither for phenomena in distant regions of space, which are inaccessible to spacecraft, nor for events in the distant past can *in situ* diagnostics be applied. Under such conditions the most reliable approach should be to compare observable large-scale phenomena in non-accessible regions, and effects more or less ambiguously imprinted on the fossil record in meteorites, with similar phenomena known in realistic detail from the magnetosphere or from the laboratory. The less reliably observed processes can thus be treated as

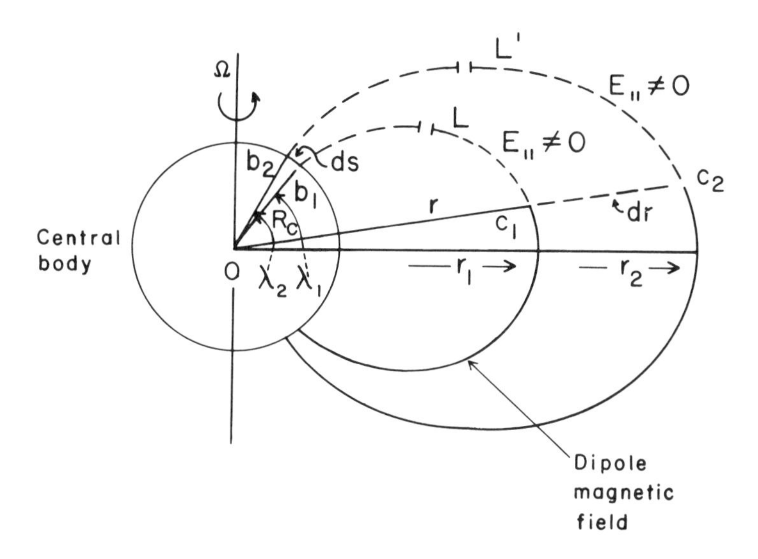

Fig. 1. In the absence of Ferraro isorotation, the angular velocity ω in the outer regions of the magnetosphere is different from the angular velocity Ω of the central body. This results in a current flow in the loop $b_1b_2c_2c_1b_1$ (shown by broken lines) which may result in the electrostatic double layers L and L'. Along part of the paths b_1c_1 and b_2c_2, the electric field has non-zero parallel components resulting in a decoupling of the plasma from the magnetic field lines (Fig. 16.3.1 of Alfvén and Arrhenius, 1976).

extrapolations of the well-established phenomena. For such extrapolations to be valid, they must be carried out in accordance with theoretically understandable and experimentally verified scaling laws. A review of these can be found in Alfvén and Fälthammar (1963).

2. Transfer of angular momentum

The transfer of angular momentum from the primary bodies in the solar system to the material now forming the secondary bodies in the system of planets and regular satellites constitutes a classical problem. This is solved, without introducing any new hypotheses, by direct application of the mechanism of partial corotation. Space experiments, particularly in the earth's magnetosphere, suggest that this phenomenon is of importance in cosmic plasma physics. Partial corotation is due to components of the electric field parallel to the magnetic field, resulting in the production of electrostatic double layers. As a consequence the angular velocity of a plasma element brought into partial corotation around a magnetized, spinning central body, is less than the driving spin angular velocity of the central body. This is in contrast to the case of Ferraro isorotation where the magnetic field lines are assumed to be frozen in, and the central body would impart its full spin angular velocity to the surrounding plasma.

A review of the theoretical and observational information on this mechanism is given by Block (1972); the resulting current systems have been studied particularly by Zmuda and Armstrong (1974). A graphic illustration of the angular momentum transfer mechanism is given in Fig. 1. The process is further discussed by Alfvén and Arrhenius (1976, Ch. 16-17; see also Alfvén *et al.* 1977).

From comparison of the expression for the tangential velocity, characteristic of the state of partial corotation, with the equation for circular Kepler motion, a general theorem follows: if in the magnetic dipole field of a rotating central body a plasma element is in a state of partial corotation, its kinetic energy is 2/3 of the kinetic energy of the circular Kepler motion at the same radial distance. The factor 2/3 derives from the geometry of the dipole field and enters because the centrifugal force makes a smaller angle with a magnetic field line than does the gravitational force. A rigorous derivation of the two-thirds law is given by Alfvén and Arrhenius (1976, Sec. 17.2 – 17.3).

In those regions of the solar system where matter still remains in a dispersed state, the 2/3 relationship remains preserved. The regions in question are Saturn's rings (Fig. 2) and the asteroid belt (Fig. 3).

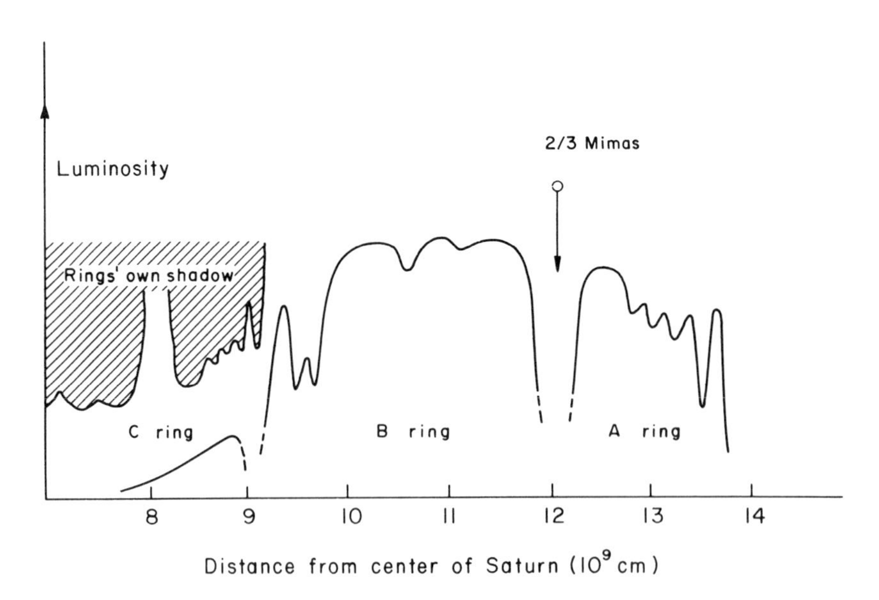

Fig. 2. Hetegonic effects in the Saturn ring system. Dollfus' (1961) photometric profile compared with Mimas' orbital distance reduced by a factory 2/3. Cassini's Division may be the "hetegonic shadow" of Mimas. In the left corner the photometric profile is turned upside down and reduced by the factor 2/3 (the ring's own shadow). The rapid drop in intensity between the *B* Ring and *C* Ring coincides with the beginning of this shadow.

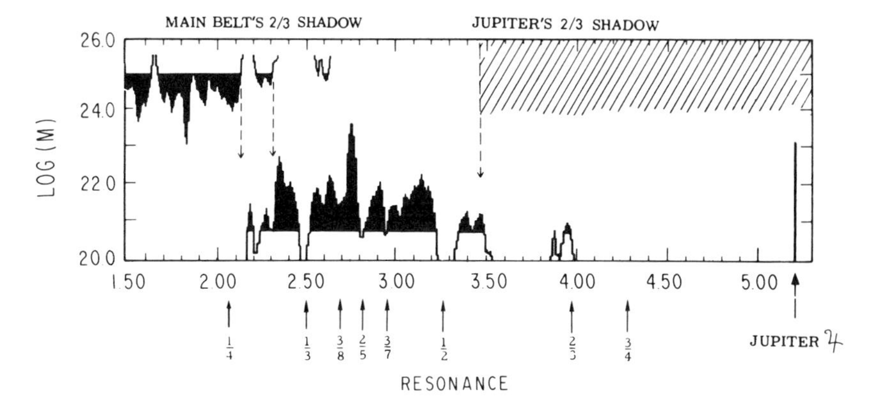

Fig. 3 Mass distribution in the asteroid belt, illustrating the 2/3 effect. The mass (Alfvén and Arrhenius, 1976, Eq. 4.3.3) in grams per radial distance interval of 0.01 AU is shown on a logarithmic scale; practically all mass in the asteroid belt is in the black regions. In analogy with Fig. 2, the distribution, diminished by a factor 2/3 and turned upside down, is shown in the upper part of the diagram; it demonstrates the "hetegonic shadow" effect which produces the inner cutoff in the asteroid belt. Similarly Jupiter's shadow, which generates the outer cutoff, is shown as well as the position of Jupiter. The Jovian resonances, associated with the Kirkwood gaps, are shown below the diagram (From Alfvén and Arrhenius, 1974).

3. Band structure of the solar system

The distributed density of matter in the planetary system and in the three regular satellite systems are shown in Fig. 4. The density distributions in these systems are intercompared in Fig. 5, where the mass of the central body (relative to the mass of the sun) is plotted along the horizontal axis. In order to compare these four systems, having widely different geometrical dimensions, the potential energy is used as an argument on the vertical axis instead of the distance from the central body. This combined diagram illustrates the band structure of the solar system with three main bands of specific gravitational energy. Whenever a band is located far enough above the surface of the central body, formation of secondary bodies is observed in this band. The generality of the rule is illustrated by the prediction of condensed matter in the innermost region of the Uranus system (see Fig. 5). Since this matter falls within the Roche limit, it should be distributed in the form of rings (De, 1978; De had six years earlier submitted this paper "On the possibility of existence of a ring of Uranus" to *Icarus*, but the paper was rejected). The existence of such rings has recently been verified observationally (Elliot *et al.*, 1977; Millis *et al.*, 1977; Bhattacharyya and Kuppuswamy, 1977; Hubbard *et al.*, 1977).

Since, as pointed out above, an acceptable theory for the formation of the solar system should give an explanation not only for the structure and properties of the planetary system, but also for the regular satellite systems, it

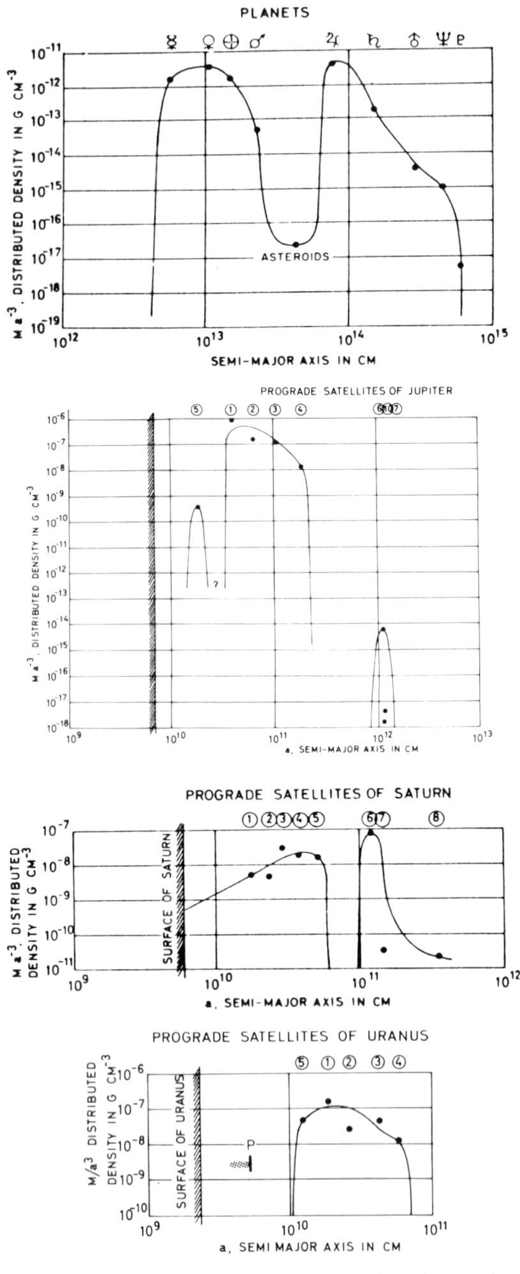

Fig. 4. Distributed densities in the planetary system and in the regular satellite systems. The symbol at P in the Uranus system marks the Roche limit and the occurrence of matter inside this limit. This was predicted from the critical velocity effect and the band structure (c.f., Fig. 5) and confirmed by the recent observations of the Uranus rings.

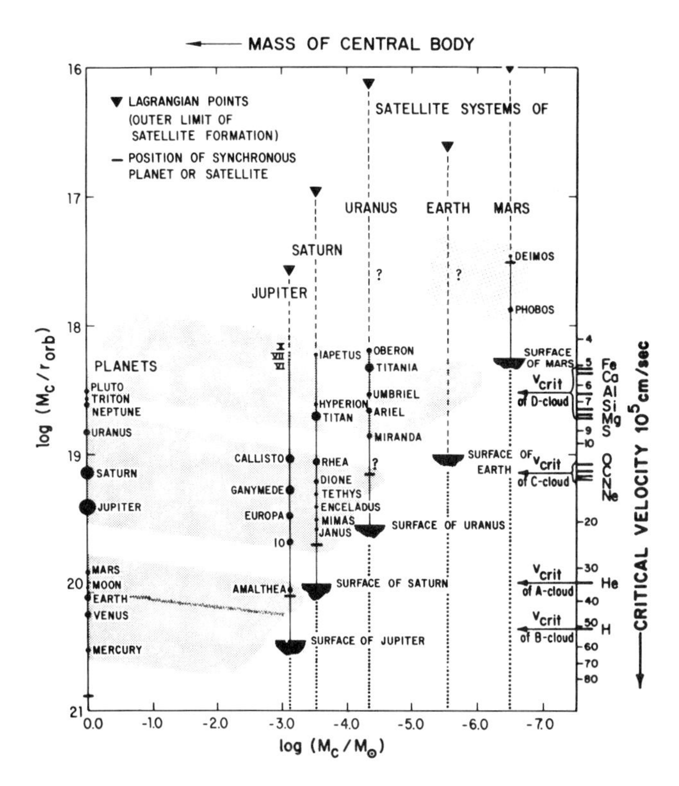

Fig. 5. Structure of the planetary and regular satellite systems. The individual systems are spaced along the horizontal axis with mass of the central body decreasing to the right; the earth, which may have had a regular satellite system, and the rudimentary Martian system, are included for comparison. The vertical scale on the left indicates the potential energy of the secondary bodies (M_c denotes the mass of the central body; r_{orb} the orbital radius of the secondary). The critical velocities of the major elements are shown on the right. The band structure, proposed to result from the critical velocity zoning of infalling neutral gas, is indicated by shading. The details of the band structure including the slope of the bands and the distribution of mass within them, and also the absences of regular satellite systems around Mercury, Venus, Earth, and Neptune, are derived by Alfvén and Arrhenius (1974, 1976). A specific prediction of the Uranus rings in the inner region marked "?" was made by De (1978; see text). (From Alfvén and Arrhenius, 1974).

seems reasonable that it should also include the collective band structure. The only theory that has thus far been advanced to explain all of these features proposes that the band structure arises as an effect of the critical velocity phenomenon. This is one of the important plasma instabilities, that has been extensively explored in the laboratory, and that we begin to understand theoretically. The empirical basis for the phenomenon is that neutral atoms and molecules, when they are accelerated in the presence of a thin collisionless plasma, become spontaneously ionized at a velocity V_{crit}, where

$$V_{crit} = \left(\frac{2\,eV_i}{m}\right)^{\frac{1}{2}}$$

with e the electron charge, V_i the ionization potential, and m the mass of the atom or molecule concerned.

The elements that make up the major part of the interstellar cloud medium have critical velocities that fall into four discrete groups. The critical velocity values are plotted on the right-hand side in Fig. 5, demonstrating the coincidence with the band locations. It is suggested that the band structure results from this selective ionization of specific element groups in specific gravitational potential energy regions around the primary bodies, and from the stopping of the infall of the ions by the magnetic field of the central body in the cases where these are magnetized. The hydromagnetic process discussed in Sec. 2 is assumed to have brought the plasma clouds into partial corotation with the central bodies. Grains that have condensed from the corotating clouds or have been electromagnetically captured in the clouds ultimately assume orbits determined by the 2/3 law (Sec. 2). The role of non-condensable plasma components in determining the localization of condensing inpurities in the plasma as well as the role of captured dust particles from the source cloud is illustrated (Fig. 5) by the A and B clouds; the source material for the terrestrial planets is condensed or captured by the plasma regions which contain mainly hydrogen and helium. Large-scale element separation into the four critical velocity groups has been observed in solar plasma emitted from reservoir regions in the sun, presumably fractionated by the same critical velocity process (Kremigis, 1977).

4. The Earliest State of the Solar System

A reconstruction of the earliest state of the solar system, which results from the above analysis is shown in Fig. 6. The protosun, formed by accretion first of dust and later also of gas, is surrounded by an extended region of magnetized plasma of essentially the same density as that of the present solar corona. Neutral gas and dust fall in from the remnants of the dark interstellar source cloud, from which the sun is assumed to be forming, and in which chemical differentiation by the critical velocity effect, condensation, and infalling dust capture take place.

Condensation from the partially corotating circumsolar plasma, and electromagnetic grain capture by the plasma is followed by orbital evolution of the individual grains which leads to the formation of populations of planetesimals in asteroid-like orbits, distributed essentially in an inner and an outer torus (A-B clouds and C-D clouds in Fig. 5). The dynamics of this

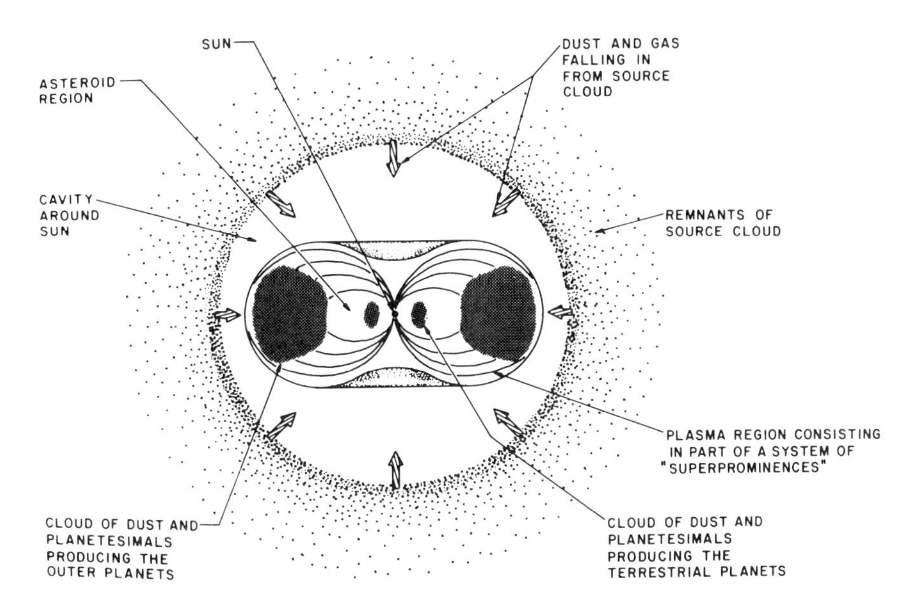

Fig. 6. Concept of the solar system in state of formation, as inferred from interstellar cloud evolution and from the present structure of planet and satellite systems.

orbital evolution and accretion process has been discussed in detail by Alfvén and Arrhenius (1976, Ch. 11-14).

RECONSTRUCTION FROM OBSERVATIONS ON THE STELLAR CLOUD MEDIUM

It is commonly assumed on the basis of astronomical observation that stars of the same type as our sun are formed in densified regions of dark interstellar clouds. If this is correct, it is also likely that the origin of our solar system is closely connected with the internal development of such clouds. Due to equipartition of kinetic and hydromagnetic energy, the interstellar cloud medium would be expected to be magnetized. Observation of other magnetized plasmas in the cosmos and in the laboratory further suggest that the distributions of fields and associated currents are highly inhomogeneous and lead to a characteristic filamentary structure of such plasmas. The filamentary structure observed in interstellar clouds should thus be treated as an indication of the state of magnetization, which is also suggested by Zeeman splitting (see the chapter by Chaisson and Vrba in this book).

Depending on the prevalence of toroidal or poloidal magnetic fields, the clouds may contract (by pinching) or be stabilized by the electromagnetic effects. It can be speculated that magnetic compression may be the main mechanism for forming interstellar clouds and keeping them together (Alfvén and Carlquist, 1978).

In the past a commonly suggested cause of star formation has been the Jeans instability. Other instabilities, discovered since may, however, be more important in this respect. One such instability is the sedimentation of dust and solid bodies toward the gravitational center, leading to a comparatively rapid gravitational accretion. This instability is analogous to the *planetesimal* accretion, generally considered to be responsible for the formation of planets and satellites. Such presumed *stellesimal* accretion of a protostar in a densified region of a cloud would be expected to lead to the formation around the protostar of a low-density cavity where neutral gas and uncharged dust particles would be depleted relatively rapidly by gravitation. A fraction of the gas falling in from the surrounding source cloud toward the central protostar would be expected to be stopped by a stellar dipole field when ionization takes place at the critical velocity limit. The resulting plasma, arrested in the discrete potential energy bands defined by the critical velocities of the major element groups, would constitute the raw material for the formation of planets and satellites.

Analysis of the evolution of a dark interstellar source cloud thus leads to a scenario for planet formation, which is reconcilable with the picture obtained from studies based on observations of the present-day solar system and on extrapolation backwards in time of successively older stages. Figure 6 represents the state of the solar system in the formative era, inferred from each of these two information sources – the present structure of planet and satellite systems, and interstellar cloud evolution.

SUMMARY

We have attempted to analyze the mechanism for protostar formation in dark interstellar clouds (Alfvén and Carlquist, 1978) and to combine the results with those of a reconstruction of the early solar system based on the present-day state of the regular systems of secondary bodies around the primaries (Alfvén and Arrhenius, 1976). These combined results define an experimentally and observationally based framework for theories describing the evolutionary chain starting from the formation of a cloud and ending with planets and satellites with their present properties. The proposed sequence of processes is the following.

1. Interstellar Medium in the Galaxy. The inferred morphology of the galactic magnetic field, the observed spectra of molecular ions, and the inferred ion molecule reactions in interstellar clouds show that the behavior of the interstellar medium is controlled by plasma effects. The existence of a non-curl-free magnetic field in the galaxy hence necessarily requires the existence of a general galactic current system, which can be assumed to be analogous to the heliospheric current system (Alfvén, 1977, 1978).

2. Formation of Interstellar Clouds. This process may be dominated by hydromagnetic effects (pinch effect) but gravitation and shock effects may also be important. With increasing density the gas temperature may drop in the densified regions, in analogy with conditions in prominences in the solar corona. The high electron temperatures inferred from microwave spectroscopic observations suggest that hydromagnetic effects continue to control the dynamics of the development, although the associated kinetic temperatures may be low. Growth of increasingly larger molecules into grains becomes an important exit channel for condensable species as the density increases (Alfvén and Carlquist, 1978; Arrhenius, 1976).

3. Star Formation – Collapse of Clouds. Densification leading to star formation may be induced by currents, some of which may be associated with shock fronts. Sedimentation of solid particles is probably an important incipient process, which means that the formation of protostars, at least in the formation of embryonic protostars, may be a non-hydromagnetic process (runaway gravitational accretion initially of solid grains and, subsequently also of neutral gas). However, the source cloud and cavity surrounding the protostar would remain dominated by the interactions between plasma and neutral gas.

4. Emplacement of Gas and Dust in Orbit Around the Protosun. The formation of the band structure and transfer of angular momentum are explained as plasma phenomena. The magnetic field of the star is not necessarily a frozen-in field from the source cloud, but may be produced by dynamo effects in its interior.

5. Accretion of Grains to Planetesimals and their Growth into Planets and Satellites. This is essentially a non-hydromagnetic process. It must be reconcilable with the evolution during the preceding and subsequent states. There is no indication of massive protoplanets which later explode or evaporate. The interplanetary medium continues to be dominated by plasma effects (Alfvén and Arrhenius, 1976, ch. 4, 11-13).

6. Formation of Regular Satellite Systems. The same process which generates planets in the observed configuration is also assumed to generate satellites. Controlling parameters in the case of formation both of the planet and of the satellite systems are the mass and the spin of the central bodies (Alfvén and Arrhenius, 1976, ch. 23).

7. Capture of Irregular Satellites. The retrograde satellites of Jupiter and Saturn are generally considered to be due to the capture, probably during the accretional phase, of planetesimals, originally in solar orbit. Triton and the moon are suggested originally to have been planets (McCord, 1966; Alfvén and Arrhenius, 1976, ch. 24) or streams of planetesimals which were captured by similar processes.

REFERENCES

Alfvén, H. 1977. *Rev. Geophys. Space Phys.* 15: 271.

____. 1978. *Astrophys. Space Sci.* In press.

Alfvén, H., and Arrhenius, G. 1974. *Astrophys. Space Sci.* 29: 63.

____. 1976. *Evolution of the Solar System.* Washington, D.C.: U.S. Government Printing Office.

Alfvén, H.; Arrhenius, G.; and Mendis, D.A. 1977. In *Comets, Asteroids, Meteorites* (A.H. Delsemme, ed.), p. 561. Toledo, Ohio: Univ. Toledo Press.

Alfvén, H., and Carlquist, P. 1978. *Astrophys. Space Sci.* In press.

Alfvén, H., and Fälthammar, C.G. 1963. *Cosmical Electrodynamics, Fundamental Principles.* 2nd ed. London: Oxford University Press.

Arrhenius, G. 1976. In *Origin of the Solar System* (S.F. Dermott, ed.). p. 522. New York: Wiley.

Bhattacharyya, J.C., and Kuppuswamy, K., 1977. *Nature* 267: 331.

Block, L. 1972. *Cosmical Electrodynamics* 3: 349. New York: Plenum.

De, B. 1978. *Astrophys. Space Sci.* In press.

Dollfus, A. 1961. In *The Solar System, Vol. III, Planets and Satellites* (G.P. Kuiper and B.M. Middlehurst, eds.), p. 568. Chicago, Ill.: Univ. Chicago Press.

Elliott, J.L.; Dunham, E.; and Mink, D. 1977. *Nature* 267: 328.

Hubbard, W.B.; Coyne, G.V.; Gehrels, T.; Smith, B.A.; and Zellner, B.H. 1977. *Nature* 268: 33.

Krimigis, S.M., and Rollof, E.C. 1977. In preparation.

McCord, T.B. 1966. *Astron. J.* 71: 585.

Millis, R.L.; Wasserman, L.H.; and Birch, P.V. 1977. *Nature* 267: 330.

Zmuda, A.J., and Armstrong, J.C. 1974. *J. Geophys. Res.* 79: 4611.

ON ANGULAR MOMENTUM TRANSFER AND ACCUMULATION OF SOLID BODIES IN THE SOLAR NEBULA

V. S. SAFRONOV and T. V. RUZMAIKINA
Институт Физики Земли АН СССР им.О.Ю.Шмидта

The average angular momentum of a fragment of interstellar medium is too large for it to contract into a star without loss of angular momentum. Numerical calculations show that the collapse of a cloud with large angular momentum ($\sim 10^{54}$ g cm^2 sec^{-1}) leads to the formation of a double or multiple system rather than to an individual star.

The solar system could have been formed from a cloud with significantly smaller angular momentum ($< 10^{53}$ g cm^2 sec^{-1}). During collapse, such a cloud develops an opaque central core before a Larson ring can form. Quasi-adiabatic heating of the core causes the Jeans length to increase and impedes the formation of a ring. Subsequent evolution of the nebula can lead to the formation of a star surrounded by a low-density envelope. We briefly discuss the following mechanisms for momentum transfer in this stage of evolution: turbulent friction, magnetic braking of rotation, and interaction between gas and dust. We note that meridional circulation cannot effectively redistribute angular momentum in a non-convective cloud.

We consider general characteristics of the process of accumulation of solid bodies in a protoplanetary cloud. Gravitational instability in the region of the inner planets requires a disk which is too flattened to be realistic. We present arguments in favor of a long accumulation time scale ($\sim 10^8$ yr) for the terrestrial planets.

An important problem of modern cosmogony is the understanding of the conditions of formation of stars with planetary systems. The solar system is the single reliable example for such a system. One of the basic characteristics of the solar system is its distribution of angular momentum. The sun contains about 99.9% of the mass and only 2% of the angular momentum. This peculiarity has led to hypotheses about the capture of planetary material by the already-formed sun. However, modern concepts of star formation and modern isotopic data support the hypothesis of the formation of the solar system from a single rotating cloud.

We begin this chapter with estimates of the mass and angular momentum of the solar nebula whose evolution leads to the formation of the solar system. We also consider the collapse of a rotating cloud without redistribution or loss of angular momentum (Sec. I). The formation of the sun with its protoplanetary cloud requires the transport of angular momentum from the interior part of the solar nebula to its exterior. An attempt to understand by which mechanism this transport occurred, and when, is given in Sec. II.

The final sections of this chapter consider the accumulation of protoplanetary bodies. Section III discusses the growth of small particles during the collapse process and after the collapse of the solar nebula. We find that during this stage, angular momentum transfer from particles to the gas is insignificant. More appreciable transfer could be possible for the motion of large bodies in a turbulent gas. Section IV discusses the conditions for the development of a gravitational instability in the dust disk, which is formed in the equatorial plane of the solar nebula, and a possible role of dust condensations. The last section deals with the evolution of a swarm of protoplanetary bodies (planetesimals), the growth of bodies and the increase of their relative velocities, the runaway growth of the largest bodies in the formation of planetary embryos, and the sweeping up of bodies by growing embryos.

I. INITIAL ANGULAR MOMENTUM OF THE SOLAR NEBULA

The present angular momentum of the planetary system is 3×10^{50} g cm^2 sec^{-1}. If we increase the masses of the planets, diluting their material with volatiles according to solar composition, the angular momentum of the solar system increases to 3×10^{51} g cm^2 sec^{-1}. Evidently, this is the minimum value of the angular momentum of the solar nebula, and approximately this value was adopted by Hoyle (1960). On the other hand, the protosun could also effectively lose its angular momentum due to mass loss (Schatzman, 1962). A rapid mass loss of $\dot{M} = 10^{-6} - 10^{-8}\ M_\odot$/yr occurs during the T Tauri phase (Rydgren *et al.*, 1976; see Rydgren's chapter). It is natural to identify the start of this stage with the onset of complete mixing in the protosun, which begins at a radius of about 50 $R_\odot$ (Hayashi *et al.*, 1962). If there is Keplerian rotation at the equator at this point, the angular momentum of the protosun is 2×10^{52} g cm^2 sec^{-1} (Okamoto, 1969). Part of the angular momentum that is lost by the protosun could be transferred to the protoplanetary disk (Schatzman, 1967; Okamoto, 1969).

Solid material could lose angular momentum by interacting with the gas component and the solar wind (Radzievskyi, 1949; Whipple, 1972; Weidenschilling, 1977). However, this cannot significantly change the angular momentum of the protoplanetary cloud (see Sec. III).

Mass loss could cause a significant change in the angular momentum of the protoplanetary cloud. By increasing the initial mass of the protoplanetary cloud to 0.1 $M_\odot$ without changing the relative distribution of surface density, we find that the angular momentum is equal to 2.4×10^{52} g cm^2 sec^{-1}. We may expect that the angular momentum of the solar nebula with a mass not exceeding 1.1 $M_\odot$ would be lower than the sum of the last two values of the angular momentum of the protosun and the protoplanetary cloud respectively, i.e., less than 4.4×10^{52} g cm^2 sec^{-1}.

A similar conclusion is reached from a numerical analysis of the collapse of rotating clouds without transfer of angular momentum. (We note that the transfer of angular momentum during the free-fall time is insignificant.) Using the two-dimensional collapse calculations of Bodenheimer and Black (1976; see their chapter) we show that the collapse of a cloud of solar mass with an angular momentum less than 10^{53} g cm^2 sec^{-1} leads to the formation of an opaque core, and does not form a ring during the transparent phase. If the angular momentum of the cloud is significantly greater (as e.g., in the model of Cameron and Pine, 1973), collapse leads to formation of a ring which will fragment into a multiple system (Black and Bodenheimer, 1976). Note that because of the rotation of the galaxy, a spherical volume of interstellar matter with a mass equal to that of the sun has an angular momentum on the order of 10^{54} g cm^2 sec^{-1}. Cameron's models in the decade following 1962 have such a value of the angular momentum for the solar nebula. However, because there is no correlation between the directions of the rotation axes of stars and the direction of the rotation axis of the galaxy, there must be a significant change in the angular momenta of individual fragments during the process of fragmentation.

Black and Bodenheimer assert that a ring forms as a result of gravitational instability, indicating that it is not possible to form a star with a planetary system in this case. Extrapolation of Black and Bodenheimer's results to lower angular momenta allows one to estimate those values of the momentum for which a ring structure does not appear at all during the stage of isothermal collapse. After an opaque core has formed, the Jeans length increases and hinders the development of a gravitational instability and formation of a ring. As is evident from numerical calculations, in the early stages of collapse of a rotating cloud, its central regions develop an oblate core of constant density and an envelope in which the density depends on radius according to $\rho \sim R^{-2}$. The radius of the core decreases as the density increases, and at a certain moment a ring-like structure develops on its periphery. At the time of ring formation, the radius of the core depends on β

(the initial ratio of the rotational energy to the gravitational energy) and M (the mass). This dependence can be approximated by

$$r_c \sim 10^{16}\ \beta M/M_\odot \text{ cm} \quad . \tag{1}$$

Noting the inverse squared dependence of the density on radius in the envelope, we find that the density of the core is proportional to β^{-2} at the moment of ring formation. If β is quite small, the core becomes opaque ($\rho_{op} \sim 10^{-12}$ g cm^{-3}) sooner than a ring is able to form on its periphery. This occurs ($\rho_c \gtrsim \rho_{op}$) for

$$\beta \lesssim 4 \times 10^{-6} \left(\frac{M}{M_\odot}\right)^{-\frac{2}{3}} \rho_s^{\frac{1}{6}}\ \rho_{op}^{-\frac{1}{2}} \tag{2}$$

where ρ_s is the initial density of the cloud. For $M \sim 1\ M_\odot$, the opaque core forms at an initial value $\beta < 4 \times 10^{-3}$, i.e., an angular momentum of the cloud $J < 10^{53}$ g cm^2 sec^{-1}. Although this estimate is valid only to order of magnitude, one may nevertheless conclude that in clouds with $J \lesssim 5 \times 10^{52}$ g cm^2 sec^{-1} and $M = 1 - 1.5\ M_\odot$, i.e., $\beta < 2 \times 10^{-5}$, ring structure does not appear during the stage of isothermal collapse. Further compression of the core takes place quasi-adiabatically.

It is more difficult to form a ring during adiabatic collapse than during isothermal collapse. Takahara *et al.* (1977) have shown that the ring forms only in rapidly rotating clouds at low temperature. At the moment of ring formation, the local value of β in the central part of the cloud is greater than 0.2. For isothermal collapse, according to the calculations of Black and Bodenheimer, the value of β for the core at the moment of ring formation is much smaller. Thus we assert, in contrast to Black and Bodenheimer, that the nature of the instability leading to ring formation is different in the cases of isothermal collapse and the adiabatic collapse of an opaque core. Note that a rotating spheroid is dynamically unstable to ring-like perturbations for $\beta >$ 0.46 (Ostriker, 1972). For the solar nebula with $J \lesssim 5 \times 10^{52}$ g cm^2 sec^{-1}, β reaches such a value for any density greater than 10^{-6} g cm^{-3}. If transfer of angular momentum is inefficient the subsequent evolution probably leads to formation of a close binary system (Whipple, 1972). To form a system like the solar system, it is necessary to have effective transfer of momentum from the central portions to the periphery. Mechanisms for such transfer are considered in the next section.

II. TRANSFER OF ANGULAR MOMENTUM

One can identify two basic questions in the problem of the transfer of angular momentum in the solar nebula. The first is related to the transfer of angular momentum which is necessary in order to form the sun and the protoplanetary disk. The second is to reproduce the current distribution of angular momentum in the solar system, i.e., to slow the sun's rotation to its current value.

In a pioneer paper, Hoyle (1960) unified these two processes by assuming that the sun slowed its angular velocity to the current value by transferring angular momentum to the protoplanetary disk via a frozen-in magnetic field. Hoyle's work stimulated the study of the role of a magnetic field in the transfer of angular momentum. However, certain aspects of Hoyle's scheme were subjected to criticism (Cameron, 1963; Safronov, 1966).

Schatzman (1962) suggested a mechanism for loss of angular momentum of a star by interaction of material ejected from its surface with its magnetic field. The ejected plasma is forced to corotate with the star to the distance where the velocity of the solar wind is equal to the Alfvén speed. This distance is much greater than the radius of the star. Thus a small amount of mass loss can cause a significant momentum loss. Mestel (1968*a,b*), and Mestel and Selley (1970) have given a detailed theory for the reduction of stellar rotation due to mass loss. We note that both the solar wind and the magnetic field are generated by convection in the outer layer of the star. Durney and Latour (1977) assumed that stars of spectral class later than F6 possess convective zones which are sufficiently extensive to generate magnetic fields. Comparison of angular rotation of F and G stars in clusters of varying ages shows that the rotation of stars decreases with a characteristic time of $(0.4 - 1) \times 10^9$ yr (Kraft, 1967; Bernacca and Perinotto, 1974). The current value for slowing the sun's rotation is $(4 - 6) \times 10^9$ yr (Modisette, 1967; Weber and Davis, 1967). Observations of T Tauri stars (Rydgren *et al.*, 1976) indicate that in earlier epochs of evolution, the solar wind was more intensive and the rotation decreased more rapidly. When a star first flares up, it is still surrounded by a gas-dust envelope. Particles of solar wind being trapped by the envelope can have a braking action on the envelope and can transfer to it an angular momentum which is greater than its initial one. Okamoto (1969) assumed that the protosun could effectively lose its momentum by Schatzman's mechanism during the stage of complete mixing. In order for the sun to lose most of its angular momentum during this stage, it would have to generate a magnetic field with an intensity of about 600 gauss at its surface. If the initial angular momentum of the solar nebula was less than 2×10^{52} g cm^2 sec^{-1} the protosun could contract without preceding loss of angular momentum to the radius $\sim 50\ R_\odot$ at which, according to Hayashi *et al.* (1977), complete mixing occurs.

We shall now consider mechanisms for the transfer of angular momentum during the stage preceding intensive mass loss from the protosun due to solar wind. Basic mechanisms for slowing the rotation by the magnetic field during this stage are reduction of rotation by generation of hydromagnetic waves (Kulsrud, 1971; Gillis *et al.*, 1974, 1977; Fleck and Hunter, 1976) or loss of mass due to rotational instabilities in the presence of a magnetic field (Vainstein and Zeldovich, 1975). During the early stages of evolution of the solar nebula, there could only be weak interstellar fields, weakened by ambipolar diffusion. They had no effect on collapse dynamics and probably cannot effectively transfer angular momentum (Mestel, 1977). Generation of the magnetic field by the dynamo mechanism in the contracting protosun can begin when the temperature in its central region becomes sufficiently high for thermal ionization of the alkali metals and other elements with low ionization potentials. The stage approximately coincides with the start of dissociation of molecular hydrogen. According to Larson's adiabat, this temperature is exceeded for a density $\rho \sim 10^{-8}$ g cm^{-3}. During the collapse process, shock waves are generated by the accretion of envelope matter onto the opaque core and the accretion of outer core matter onto the inner core, causing dissociation and ionization of hydrogen. One may therefore assume that the core will be turbulent and that the turbulence will have helicity due to the presence of rotation and a density gradient. Thus, the core has all of the ingredients for generation of a magnetic field (see e.g., Parker, 1975; Vainstein and Zeldovich, 1975). Stress of the stationary magnetic field generated by the dynamo can be estimated by equating the magnetic and Coriolis forces: $\frac{B_\phi B_r}{8\pi r} \sim \rho v_t \Omega$. The characteristic time for the redistribution of angular momentum in the core is equal to

$$\tau \sim \frac{r}{v_t} \tag{3}$$

where v_t is the turbulent velocity, and r is the characteristic radius of the core. Unfortunately it is difficult to estimate the dimensionless coefficient in this expression. The efficiency of transport of angular momentum by the magnetic field is probably not less than that produced by turbulent friction. Transport of angular momentum due to viscosity in a disk in which the gravitational forces are on the order of the centrifugal forces, is considered by Lynden-Bell and Pringle (1974). In the case of turbulent viscosity they take the dimensionless coefficient in the characteristic time of transfer of angular momentum to be 10^3. Kippenhahn and Tscharnuter (1975) have shown that the characteristic time of transport of angular momentum by turbulent diffusion exceeds the free-fall time. However, it can be comparable to the time of evolution of an opaque core which increases due to increasing core mass. The evolution time is determined by the rate of accretion and by thermal radiation from the surface of the core. Because of increasing density, the turbulent energy increases more rapidly than the gravitational energy, so that turbulence contributes to slowing the compression. However, it cannot stop contraction altogether, and it cannot lead to a cloud in a state of uniform rotation (Schatzman, 1971). The growth of velocities in turbulent motion at all scales must evidently be limited by the sound speed, since supersonic turbulence is quickly dissipated by excitation of shock waves. Thus, we believe the supersonic turbulence assumed by Prentice (1976, 1977) to be unrealistic.

Finally, we would like to point out some other mechanisms for transport of angular momentum which need to be studied for application to the solar nebula:

1. Large scale flow generated by turbulence in a rotating medium (see the review by Durney, 1976). This flow has a certain resemblance to meridional circulation which Cameron and Pine (1973) used in their model as a mechanism for transport of angular momentum. However, as Ruzmaikina and Safronov (1976) have shown, this mechanism is not effective in a non-turbulent medium.
2. Variation of the angular velocity along the axis of rotation as a result of non-uniform collapse. This can lead to the spin-up instability of Goldreich and Schubert (Benton and Clark, 1974; Goldreich and Schubert 1967), which can transfer angular momentum.
3. Hydrodynamic waves (Rossby waves and spiral density waves) generated in a rotating medium which can also transport angular momentum outward (Greenspan, 1968; Roberts and Shu, 1972; Kaplan and Pikelner, 1974).

III. ACCUMULATION OF PROTOPLANETARY BODIES

Growth of Small Particles

In a collapsing nebula without turbulent motions, particles grow very slowly because of low density during the lengthy initial phase of collapse and due to very small relative Brownian velocities of particles in the gas (Kusaka *et al.*, 1970). In a nebula with turbulent motions the velocities of particles are much greater. But the increase of velocities as the particles grow can prevent them from coalescing during collisions. According to Coradini *et al.* (1977) spherical non-magnetic particles that are elastically colliding can grow up to $r \sim 10^{-3}$ cm through the action of van der Waals and electrostatic forces. Inclusion of inelastic collisions, the irregular shape of particles, and especially the magnetic properties of metallic particles should significantly increase this limit.

The absence of a quantitative model for the solar nebula at the end of its collapse [$\rho(R)$ and $T(R)$] prevents a real estimate of the sizes of particles at this time. However, one can obtain the order of magnitude by considering the growth of particles in a uniform non-rotating collapsing sphere, assuming that they have some constant relative velocity, v. Then at the conclusion of collapse

$$r \approx \frac{\xi \nu}{4\delta} \frac{\rho_p}{\rho_g} (\rho / 6\pi G)^{\frac{1}{2}} \tag{4}$$

where ξ is the probability of coalescence at a single collision. The duration of collapse is determined by the initial gas density of the nebula ($\propto \rho_{go}^{-\frac{1}{2}}$) while the size of particles is determined by the final density ($\propto \rho_{go}^{-\frac{1}{2}}$). For $\delta = 1$ g cm^{-3}, $\rho_g = 10^{-10}$ g cm^{-3}, and $\nu = 5$ m sec^{-1}, we find $r \approx 0.1$ mm. In the hot inner part of the nebula, particles could evaporate and would resume condensation only after the nebula cools.

The collapse terminates when the rotation of the nebula becomes Keplerian, at which point it is highly flattened. The turbulence dies away and dust particles settle to the central plane $z = 0$ (Safronov and Ruskol, 1957). Larger particles settle more quickly and sweep up other particles with which they collide, increasing in size up to $r \sim 1$ cm at $z \sim 0$. Furthermore, the larger particles have a larger angular velocity (the velocities of the smaller particles are closer to the gas velocity, which is less than the Keplerian velocity because of the pressure gradient [Whipple, 1972]) and these undergo more collisions with small particles, thus growing and settling to $z = 0$ even faster.

Goldreich and Ward (1973) considered the interaction of an opaque dust disk with a surrounding gas moving with a different angular velocity and in a state of laminar or turbulent flow. The disk transfers its angular momentum to the gas and the disk material spirals toward the sun (or protosun). But as the particles settle to $z = 0$, their positions shift toward the sun only by $\Delta R \sim 10^{-3} R$. This is a reasonable picture of the final stage of disk flattening, when $\rho_p \gg \rho_g$. Weidenschilling (1977) considered in detail the interaction of individual particles with the gas as they move around the sun (the gas flows laminarly with the velocity less than the Keplerian velocity). Small particles moving with almost the same velocity as the gas have a radial component of velocity ν_r considerably higher than the tangential velocity relative to the gas. This velocity ν_r determines the frequency of collisions of particles of various sizes. Weidenschilling's calculations satisfactorily describe the initial phase of evolution of the dust layer, when $\rho_p \ll \rho_g$.

We may compare the radial velocity ν_R with the velocity at which the particles settle to the equatorial plane, ν_z, by using equations given in Safronov's book (1969). We find

$$\frac{\nu_R}{\nu_z} = \frac{4n\nu_g h_g}{3\pi V_k z} \sim \frac{h_g}{15z} \quad , \tag{5}$$

where V_k is the Keplerian velocity, ν_g is the thermal velocity of molecules,

$2h_g = H_g$ is the thickness of the gas layer, and $n \approx 1$–3. At the first stage of settling of the dust layer, $z \sim h_g$ and $v_r \ll v_z$ for most particles. The growth of particles is then determined by the difference of their settling velocities v_z. The deviation of the gas from Keplerian rotation, ΔV, begins to significantly accelerate the growth of particles only when the thickness of the dust layer becomes at least an order of magnitude smaller than the thickness of the gas layer.

Expression (5) gives the decrease of the distance of a particle from the sun as a function of the decrease of its distance from the central plane

$$\mathrm{d}R/\mathrm{d}z = v_R/v_z$$

and

$$2\frac{k_0-k}{k_0} \approx \frac{R_0-R}{R_0} \approx \frac{4n\, v_g h_g}{3\pi\, R_0 V k} \ln\frac{z_0}{z} < 10^{-2} \tag{6}$$

where $k = V_k R$ is the specific angular momentum of a particle with respect to the sun. Thus when particles settle to the central plane of gas which is rotating about the sun in a laminar fashion, they transfer to the gas only a small fraction of their orbital momentum (less than 1%). As the mass of particles is only a small portion of the gas mass, the change in the specific angular momentum of the gas in this case is even two orders of magnitude smaller.

Further evolution of the dust layer depends on whether or not there are perturbations which are capable of preventing contraction of the layer to the critical density for the onset of gravitational instability. If such perturbations are absent, the layer breaks up into many dust condensations as described by Safronov (1969), Goldreich and Ward (1973) and Genkin and Safronov (1975). On the other hand, the critical density corresponds to a high degree of flattening of the layer, and if this is not achieved, the layer does not break up and instead the particles continue to grow by coalescence during collisions. The thickness of the dust layer is proportional to the relative velocities of particles, v_p. In the earth's zone, velocities $v_p > 11$ cm sec^{-1} are sufficient to prevent attainment of the critical density in the layer.

The rate of radial motion of a particle under the influence of the gas drag force, F_D can be expressed in terms of the particle growth velocity

$$\frac{\mathrm{d}R}{\mathrm{d}t} = v_R = -2F_D/m\omega$$

$$\frac{\mathrm{d}r}{\mathrm{d}t} = \frac{\xi \rho_p v_p}{4\delta} \approx \frac{\xi v_p \sigma_p \omega}{2\pi\delta v_z} \tag{7}$$

where σ_p is the surface density of particles and v_z is their random velocity in the z direction. The drag force on particles in a gas is a complicated function of their dimensions (Whipple, 1972). For the most lengthy stage of evolution, which corresponds to large Reynolds numbers (> 800), i.e., for rather large bodies ($r > 10$ m),

$$F_D = 0.22\pi r^2 \rho_g v^2 \approx 0.22\pi r^2 \rho_g (\Delta V)^2 \tag{8}$$

and

$$\frac{\mathrm{d}R}{\mathrm{d}r} \approx -\frac{\sigma_g v_z (\Delta V)^2}{\xi \sigma_p v_p v_g \omega r} \tag{9}$$

where $\Delta V = nRT/2\mu V_k \sim 10^4$ cm sec^{-1} is the deviation of gas rotation from Keplerian velocity. The random velocities of bodies in the z direction are altogether unknown, since they are determined by turbulent motions in the gas, whose character is quite unclear. If we exclude so far unknown external sources, the turbulence may for example tend to couple to the motion of the gas ΔV and to the motion of small bodies which are under its influence, as compared with large bodies. Taking $v_z \sim v_p \sim \Delta V$ in Eq. (9), and also $T \propto R^{-\frac{1}{2}}$ in the expression for ΔV, and integrating, we find

$$\frac{\mathrm{d}R}{\mathrm{d}r} = -C\frac{R^{\frac{7}{4}}}{r} , \left(\frac{R_0}{R}\right)^{\frac{3}{4}} - 1 \approx \frac{\sigma_g}{\xi \sigma_p} \left[\frac{(\Delta V)^2}{v_g V_k}\right]_0 \ln \frac{r}{v_0} . \tag{10}$$

For $\sigma_g/\sigma_p = 500$, $\Delta V = 10^4$, $\xi = 0.5$, $R_0 = 1$ AU, $r_0 = 10$ m, $r = 50$ km, we find $R/R_0 = 0.53$. During this time the angular momentum decreases by less than 30%. With greater turbulence and smaller ξ, the variation of R will be greater. But we may expect that in reality the velocities in the z direction of bodies of similar dimensions would be smaller, and the density ρ_p and the corresponding rate of growth of bodies in Eq. (7) would be greater. Then the change in distance from the sun found above would correspond to a smaller value ξ. In the giant planet region, σ_g/σ_p is almost an order of magnitude smaller, and correspondingly the displacement of bodies in R is much less than in the region of the terrestrial planets. In summary, it is evident that the angular momentum transferred from bodies to the gas during the accumulation process was definitely less than the present angular momentum of the planetary system.

Gravitational Instability of the Disk

A solution of a number of important questions in cosmogony is lacking because of the absence of an analytic solution of the equations which describe the process of concern here (for example, the three-body problem, and accretion or collapse in a rotating medium, etc.). The possibilities of numerical methods are quite limited. From this point of view the problem of gravitational instability in the dust disk has a quite definite and rather simple solution, but it is not yet solved in the context of cosmogony. According to estimates, the gravitational instability due to axisymmetric perturbations sets in only for extreme flattening of the disk. In the earth zone, one needs $H_p/R \approx 2 \times 10^{-6}$, i.e., the thickness of the disk $H_p \approx 3 \times 10^7$ cm, which is only 1/20 of the earth's radius. In Jupiter's zone, one needs $v_{cr} \approx 270$ cm sec^{-1} and $H_p/R \approx 10^{-4}$, i.e., H_p is on the order of the radius of Jupiter. The closer to the sun, the more extreme are the conditions for the instability. In Mercury's zone one requires $v_{cr} \sim 0.4$ cm sec^{-1} and $H_p/R \sim 4 \times 10^{-8}$, i.e., H_p on the order of 7 km. But close to the sun, which is the basic source of various perturbations of nebula, such small particle velocities and such an extreme degree of flattening of the disk are quite impossible. Thus, we may assert that in the zone of the inner planets, gravitational instability in the disk did not occur; the particles instead grew by coalescing during collisions. Such coalescence was favored by small velocities of the particles and significant differences in their dimensions.

In the region of the giant planets, the conditions for gravitational instability were more favorable. As a result of instabilities in Jupiter's zone, dust condensations should have formed with initial masses of the order of 10^{22} g and dimensions of the order of 10^{10} cm (Safronov, 1960, 1969). Goldreich and Ward (1973) obtained essentially the same result, although they reached the further conclusion that the condensations themselves could be non-uniform, being composed of a large number of smaller condensations or bodies.

When this same instability condition is applied to the gaseous component of the cloud, it shows that even Jupiter and Saturn could not form as gaseous condensations of solar composition. In order to produce instability, one requires a considerably greater gas mass – greater than 60 Jupiter masses in each zone. In this case it would be impossible to understand why only 1% of the mass of the gas went into these planets while 99% was lost from the solar system. The masses of the condensations would have to exceed one tenth of the mass of Jupiter, and neither thermal dissipation nor the influence of the solar wind could destroy such condensations or remove them. It seems even less possible to form Jupiter or Saturn directly from a single condensation. For this reason we prefer models of gas accretion onto a planetary embryo which is assembled from solid material (Ruskol and Safronov, 1977), rather than various models of formation of these planets which invoke contraction or collapse of a single initially dispersed gas condensation. Unfortunately, the gas accretion models are not mathematically straightforward.

The question of whether or not there was a gravitational instability in the dust disk is somewhat beside the point since the stage of dust condensations was an intermediate one and in the terrestrial planet region was quite brief. Coalescence of condensations during collisions was accompanied by their contraction, and after their mass increased by about two orders of magnitude, they became the usual planetesimals (Safronov, 1969). But in the giant planet region the initial masses of condensations would be considerably greater ($\propto \sigma^3 R^6$) and the stage of dust condensations could be considerably prolonged, thus accelerating the growth of Uranus and Neptune which is otherwise too slow (Safronov, 1975).

The problem of transformation of dust condensations into solid bodies is closely connected with the problem of formation of circumplanetary satellite swarms around growing planets and the evolution of these swarms (Harris and Kaula, 1975; Safronov and Ruskol, 1977), and also with the problem of the origin of planetary rotation. Analytic estimates (Harris, 1977) as well as numerical calculations (Kiladze, 1977) show that the angular momentum which is acquired by a planet from bodies which impact on it in the process of accumulation from the planet's zone, is insufficient to explain the observed rotation of the planets. Thus, one may expect that most of the angular momentum acquired by the planet is the result of impact of material from the satellite swarm which encircles the planet. The explanation of rotation of planets by a two-stage acquisition of matter from the planetary zone via satellite swarms was suggested by Artemjev and Radzievskii (1965).

Evolution of a Swarm of Preplanetary Bodies (Planetesimals)

The basic characteristics of a protoplanetary swarm of bodies are the distribution of the bodies by mass and their relative velocities. These characteristics are closely related, and strictly speaking, it is only possible to determine them simultaneously. The velocities of bodies depend on their mass distribution, and vice versa; the character of the fragmentation of bodies depends on their velocities and affects their mass distribution. But, because of the complexity of such a coupled problem, the solution has been carried out in two parts; for a given mass distribution, the relative velocities of bodies has been estimated, and for given velocities the mass spectrum has been determined (Safronov, 1969). A rigorous analytic solution of the coagulation equation without fragmentation has been found for a coagulation coefficient which is proportional to the sum of the masses of the colliding bodies (Safronov, 1962). A qualitative investigation of this equation has been made for more general assumptions about the coagulation coefficient (Zvjagina and Safronov, 1971). It was shown that except for a rather narrow interval involving the most massive bodies, the mass distribution tends to an inverse power law $n(m) = cm^{-q}$ with index $q \approx 1.55 \pm 0.15$. The form of the asymptotic solution does not depend on the form of the initial mass distribution. The equation has also been studied with fragmentation processes included (Zvjagina *et al.*, 1973). For a smooth variation of parameters describing fragmentation, the solution may be approximated by an inverse power law with the index $q \approx 1.8$ everywhere except for the interval of the largest bodies. As a special case for this solution, when there is no accumulation and bodies fragment upon collision, one obtains the asymptotic solution for the asteroid mass distribution which was found by Dohnanyi (1969), Hellier (1970) and Bandermann (1972). Note that the solution of these authors with index $q = 11/6$ is only obtained when it is assumed that the nature of the fragmentation of bodies does not depend on their masses. Numerical calculations have been carried out (Pechernikova *et al.*, 1976) which describe the evolution of the mass distribution under these same assumptions.

In a system which is differentially rotating, gravitational interactions between bodies during encounters tend to increase the relative random velocities, while inelastic collisions decrease them. As velocities increase, gravitational perturbations become less effective, the relative importance of collisions increases, and velocities thus tend to decrease. As velocities decrease, the opposite occurs, so that a system sets up a quasi-equilibrium velocity distribution which is determined by the parameters of the system, the masses in particular. By balancing the energies of random motion which are gained at encounters and lost at collisions, one finds an expression for the average velocity of bodies relative to circular Keplerian motion for large mean free paths (Safronov, 1969),

$$v = \sqrt{Gm/\theta r} \tag{11}$$

where θ is a dimensionless parameter which depends on the properties of the system, and m and r are the mass and radius of the largest body in the zone under consideration. In a system of bodies of equal mass, which coalesce on collision, $\theta \approx 1$. For a power law distribution of masses, $n(m) = cm^{-q}$ at q < 2, one finds $\theta \approx 3$ to 5. For $q > 5/3$ the smaller bodies have somewhat smaller velocities. When gas drag acts on the motion of bodies, θ can become several tens. The increase in mass of bodies causes their velocities to increase, and the thickness of the swarm increases. From this we conclude that it is impossible to achieve a high degree of flattening for a swarm of massive bodies from which planets can succeed in growing in 10^4 yrs. We also conclude that the concept of "jet streams", in which collisions of bodies lead to $v \to 0$, is groundless (Safronov, 1972*b*). In the latter reference it is noted that if the largest body in the zone begins to "run away" in mass from the other bodies, its gravitational perturbations become less effective in increasing the velocities of bodies, and Eq. (11) for v should instead use m and r for the second largest body.

Öpik (1961) proposed a different mechanism for increasing the relative velocities of bodies as a consequence of their repeated encounters with a planet that moves in a slightly eccentric orbit. This mechanism is more effective than the one described above when the bodies gravitationally interact with a single large body and their velocities are small. However, the mechanism is less effective when there are two or more such large bodies in the zone. Considering the earth's zone, with the present values of the mass of the earth and the eccentricity of its orbit, the energy of random motions which is acquired by bodies due to Öpik's mechanism is about 0.02θ of the energy which they acquire as a result of their interaction with many bodies. But in Mercury's zone the situation would be quite different; at the final stage of planetary growth, Öpik's mechanism would be about 50θ times more effective, and rules out values of θ greater than 5 or 6.

The gravitational cross section for collisions of the largest bodies is proportional to the 4th power of their radii. Thus, the ratio of the mass of the largest body to the mass of the second largest body in this zone would increase with time. For θ = const., the limiting mass ratio would take the value $(2\theta)^3$. Thus, a characteristic feature of the accumulation process would be the formation of the largest bodies, which are planetary "embryos," whose mass is much larger than the masses of other bodies in their vicinity. They move in slightly eccentric orbits, and the half width of their feeding zone, ΔR, is determined mainly from the eccentricity of the orbit of bodies by $e \approx v/K_K$ and equals

$$\Delta R \approx Rv/v_k \quad . \tag{12}$$

Such zones would overlap considerably for neighboring embryos, and the general number of embryos with average mass m in the zone of a planet with mass m_p would exceed $n_e \approx (m_p/m)^{\frac{1}{3}}$, especially in the early stage when there may be several embryos along a single orbit. Consequently, the masses of the embryos would be small ($m \ll m_p$) and the total number of planetary embryos would be much greater than the present number of planets. The total mass of the embryos would be much less than the mass of all the bodies in the planetary zone. The embryos grow, consuming these bodies, their zones overlap even more, the smaller ones fall behind the growth of the largest and under its gravitational influence they depart from a circular orbit and eventually impact on the largest embryo. This rarefaction process would continue until all solid material between embryos is swept up, and until the separation between the embryos is such as to stabilize the orbits of the resulting bodies against mutual gravitational perturbations over an extended period of time. This process of competition between embryos has led to the presently observed law of planetary distances.

Using our results on the relative velocities of bodies and on a runaway of embryos, Levin (1978) has concluded that during most of the accumulation period the relative velocities had to be very small (effective $\theta \sim 10^2 - 10^3$), being determined by the second largest body and not by the first one (the embryo). In this case the accumulation of giant planets is accelerated about 10^2 times; this obviates his previous hypothesis (Levin, 1972) of a very massive protoplanetary cloud. However, this idea rasies serious objections.

1. A runaway of the largest bodies is impossible at the early stage of growth when there are several bodies along one orbit almost at the same distance from the sun. The differential rotation slowly changes the neighborhood of the first largest body. A "second" largest body which begins to lag behind in mass gradually leaves its zone and is replaced by another large body which grew earlier outside this zone without lagging behind etc.

   At the next stage when the bodies are large enough ($r \gtrsim 200$ km) and there is only one embryo along the orbit, the feeding zones of neighboring embryos overlap considerably because ΔR is several times greater than the half width of the gravitational zone of influence of the embryo. In the overlapping region the relative velocities of bodies should be sufficiently large, i.e., θ should be small. According to Wetherill (1976; see also his chapter in this book), at such conditions $\theta \approx 1 - 2$. The common region widens due to the growth of embryos. The second embryo grows more slowly than the first, but the ratio of their masses increases only a few times when the second one begins to undergo strong gravitational perturbations from the first embryo, leaves its nearly circular orbit and loses its role of being an embryo. The next farther embryo, of about the same mass as the first one, becomes included into the enlarged feeding zone of the first embryo and the process is repeated.
2. A faster sweeping out of bodies with small velocities leads to a relative increase of a portion of bodies with bigger velocities.
3. At the final stage of accumulation only relatively large velocities (corresponding to $\theta \sim 2$) permit the terrestrial planets to sweep out all bodies in their zones and accordingly make it possible to understand the law of planetary distances.
4. The suggestion of small relative velocities of bodies does not solve the accumulation problem. It leads only to the formation of bodies of lunar size due to very narrow feeding zones of the embryos. The important and longest lasting next stage of accumulation of the bodies remains beyond the scope of Levin's discussion.

Consideration of the inclination of the rotation axes of planets has placed an important limitation on the masses of bodies which have impacted on them. The largest bodies which impacted the earth could not exceed 10^{-3} to 10^{-2} Earth masses (depending on the mass distribution of these bodies). Neighboring embryos would lag behind the earth and thus would ultimately disappear in the competition. They would have a considerably larger probability of a near miss rather than a direct impact, and thus would break up into smaller bodies within the Roche limit. This break-up would insure preservation of the mass limit for impacting bodies mentioned above.

The parameter θ has a simple physical meaning; twice its value, $2\theta = \nu_e^2/\nu^2$ is equal to the ratio of the potential energy at the surface of the largest body, $\nu_e^2/2$, to the kinetic energy of relative motion of bodies, per unit mass. For $2\theta \sim 10$, collision of the largest body (the planetary embryo) with any other body leads to their coalescence, since significantly more energy goes into heat and mechanical work during collision than the 10–15% of the kinetic energy which exceeds the potential energy, and which must be lost in order to unite the bodies. According to Hartmann's experiments (1977), even low-velocity elastic collisions with no fracturing lead to the loss of about 1/4 of the impact energy during the collision. Collisions of bodies of smaller mass can also lead to their coagulation if their mass difference is large. But collision of small bodies of comparable dimensions leads to their destruction and dispersal into individual fragments.

Thus the growth of planetary embryos must proceed under the most favorable conditions, namely with a coefficient of coagulation equal to unity. The growth of such a body is described by a simple formula which was derived by Safronov in 1954,

$$\frac{dm}{dt} = 4\pi\nu^2(1+\nu_e^2/\nu^2)\sigma/P = 4\pi\nu^2\ (1+2\theta)\sigma/P \tag{13}$$

where σ is the surface density of solid material in the planetary zone which has not yet been swept up (it is a function of m and r) and P is its period of revolution. The derivation assumes that the bodies are uniformly distributed throughout the zone. For $\theta = 3$, the formula allows 98% of the earth's mass to be collected over 10^8 yrs, which corresponds to a characteristic e-folding time for the sweeping up of bodies in its zone of $\tau_0 \sim 2 \times 10^7$ yr. Subsequently some geochemists and geophysicists adopted, without any dynamical considerations, a much shorter growth time for the earth (about 10^5 yr and less), in order to obtain a high initial temperature for the earth. This time scale has also served as the basis for a hypothesis about inhomogeneous accumulation of the earth. According to the earliest of such hypotheses (Turekian and Clark, 1969) refractory iron particles were condensed from the gas first and succeeded in uniting into a single large body, the iron core, before the silicate particles began to condense. But, when it turned out that the initial condensate as the nebula cools is not iron but certain silicates, the hypothesis was modified (Anderson and Hanks, 1972) by postulating a different initial distribution of chemical elements within the earth.

Some cosmogonical models were also proposed which implied more rapid planetary accumulation (Öpik 1971; Cameron, 1973; Hallam and Marcus, 1974), but all of them had arbitrary assumptions which predetermined the results (Safronov, 1975). The most serious of them is the assumption of small relative velocities of bodies which leads to a large value of θ in Eq. (13) and thus accelerates the growth of the planets. But decreasing the velocities correspondingly shrinks the width of the planetary feeding zone and thus severely limits the planet's mass. In order to allow the planet to grow to current dimensions it is necessary to broaden the feeding zone by increasing the velocities of bodies. Thus θ is reduced to approximately 3 and the accumulation time of the planet correspondingly increases to approximately 10^8 yr. This question was carefully investigated by Weidenschilling (1976), who confirmed the time scale of 10^8 yr. Moreover, numerical calculations have shown that the velocities of bodies in the earth zone would be significantly increased as a result of their interactions with bodies ejected from Jupiter's zone (Kaula and Bigeleisen, 1975; Wetherill, 1976), as well as interactions with Venus. Wetherill finds velocities of bodies of approximately 8 km $\sec^{-1}$, which corresponds to $\theta \sim 1$.

An interesting numerical model of the accumulation process has been studied by Greenberg *et al.* (1977; see their chapter), in which a simultaneous consideration of the change of the dimensions of bodies and the change of their relative velocities is carried out. The authors considered fracturing of bodies in correspondence with the results of Hartmann (1977). They found that a few bodies grow quickly (over $\sim 10^4$ yrs) from kilometer dimensions to several hundreds of kilometers, while the remainder essentially do not grow at all. The relative velocities of bodies do not increase significantly despite the considerable growth of the largest bodies. So far, it is difficult to compare these results with those discussed above. It is probable that they are mainly the result of different assumptions about the nature of fragmentation of bodies, as well as different expressions for the relative velocities adopted by the authors. But we are convinced that the short time scale of this stage of growth does not indicate a shorter time scale for planetary growth as a whole. The final stage of planetary growth, which was not considered in this work, must be more prolonged.

Extension of the accumulation theory into the region of the giant planets reveals some important features of the process at the periphery of the solar system. The time of accumulation, which is proportional to σ/P, increases considerably with distance from the sun, and for the same value of θ as for the terrestrial planet zone, the accumulation time in Uranus and Neptune's zone becomes unacceptably large: $\sim 10^{11}$ yrs. As is evident from Eq. (13), the time may be shortened by increasing the product $\sigma\theta$. It is interesting that there exists a realistic mechanism which causes simultaneous increase of both σ and θ. As the planetary mass increases, the relative velocities of bodies in the zone increase and for a mass m_e on the order of 1/10 of the present planetary mass, the velocities reach the velocity $\sim v_k/2$ for which the bodies escape the solar system (primarily in the direction of the planet's orbital motion). Because of this effect, the velocities of bodies do not increase with further growth of the planet and $\theta = v_e{}^2/2v^2$ begins to grow.

On the other hand, ejection of bodies from the system indicates a large initial mass of material in the zone and thus a large surface density, σ. The ratio of the mass ejected by the planet to the mass which impacts on it is proportional to v_e^2/v_k^2 (Safronov, 1972*a*) and at the final stage of Jupiter's growth would be ~ 10, and for other planets somewhat smaller. Such a rate of ejection of material from the system would correspond to an initial protoplanetary cloud mass $\sim 0.1\ M_\odot$, and the total mass of ejected material would be on the order of the mass of all planets. About 1% of all the bodies would be gravitationally perturbed by stars and remain on the periphery of the solar system, forming a cometary cloud. The bodies primarily ejected in the direction of rotation would have to carry off positive angular momentum, whose order of magnitude must be on the same order of the present angular momentum of the planets. Öpik's estimates (1966) lead to a somewhat smaller quantity of ejected material.

IV. CONCLUSION

Calculations of the evolution of rotating clouds as well as estimates of the loss of angular momentum for certain processes in the solar nebula show that the initial angular momentum of the solar nebula probably totalled 2 to 5×10^{52} g cm^2 sec^{-1}. The absence of a correlation between the directions of the axes of rotation of stars and the direction of the galactic rotation axis indicates that the angular momentum produced by random motions in interstellar aggregates exceeded the angular momentum connected with the regular motion of galactic rotation.

Consideration of mechanisms for the transfer of momentum in the protoplanetary cloud involving the interaction of gas and dust as well as ejection of bodies from the solar system by growing giant planets shows that these mechanisms could lead to a loss of momentum only on the order of the present angular momentum of the solar system.

Individual stars could form from fragments with relatively small angular momentum. In this case, the central part of the collapsing cloud would achieve a sufficiently high temperature to ionize alkali metals, the generation of magnetic fields would begin and permit the protosun to lose angular momentum. Prior to this stage, the loss would have to occur through hydrodynamic motions. Further research is necessary in order to estimate the efficiency of these mechanisms.

Acknowledgment. We thank W. B. Hubbard for translating this paper into English.

REFERENCES

Anderson, D. L., and Hanks, T. C. 1972. *Nature* 237: 387.
Artemjev, A. V., and Radzievsky, V. V. 1965. *Astron. Zh.* 42: 124.

Bandermann, L. W. 1972. *Mon. Not. Roy. Astr. Soc.* 160: 321.
Benton, E. R., and Clark, A., Jr. 1974. *Ann. Rev. Fluid Mech.* 6: 257.
Bernacca, P. L., and Perinotto, M. 1974. *Astron. Astrophys.* 33: 443.
Black, D. C., and Bodenheimer, P. 1976. *Astrophys. J.* 206: 138.
Cameron, A. G. W. 1963. *Icarus* 1: 13.
____. 1973. *Icarus* 18: 407.
Cameron, A. G. W., and Pine, M. R. 1973. *Icarus* 18: 377.
Coradini, A.; Magni, C.; and Federico, C. 1977. *Astrophys. Space Sci.* 48: 29.
Dohnanyi, J.S. 1969. *J. Geophys. Res.* 74: 2531.
Durney, B. R., and Latour, J. 1977. Preprint.
Fleck, R. C., and Hunter, J. 1976. *Mon. Not. Roy. Astr. Soc.* 175: 335.
Genkin, I. L., and Safronov, V. S. 1975. *Astron. Zh.* 52: 306.
Gillis, J.; Mestel, L.; and Paris, R. P. 1974. *Astrophys. Space Sci.* 27: 167.
____. 1977. In preparation.
Goldreich, P., and Schubert, G. 1967. *Astrophys. J.* 150: 571.
Goldreich, P., and Ward, W. R. 1973. *Astrophys. J.* 183: 1051.
Greenberg, R.; Wacker, J. F.; Hartmann, W. K.; and Chapman, C. R. 1978. *Icarus*. In press.
Greenspan, H. P. 1968. *The Theory of Rotating Fluids.* Cambridge: The University Press.
Hallam, M., and Marcus, A. H. 1974. *Icarus.* 21: 66.
Harris, A. W. 1977. *Icarus* 31: 168.
Harris, A. W., and Kaula, W. M. 1975. *Icarus* 24: 516.
Hartmann, W. K. 1977. *Icarus* 33: 50.
Hayashi, C.; Hoshi, R.; and Sugimoto, D. 1962. *Prog. Theor. Phys. Suppl.* 22: 1.
Hayashi, C.; Nakazawa, K.; and Adachi, I. 1977. *Publ. Astron. Soc. Japan* 29: 163.
Hellier, B. 1970. *Mon. Not. Roy. Astr. Soc.* 148: 383.
Hoyle, F. 1960. *Quart. J. Roy. Astron. Soc.* 1: 28.
Kaplan, S. A., and Pikelner, S. B. 1974. *Ann. Rev. Astron. Astrophys.* 12: 113.
Kaula, W. M., and Bigeleisen, P. E. 1975. *Icarus* 25: 18.
Kiladze, R.I. 1977. *Bull. Abastumani Obs.* 48: 191
Kippenhahn, R., and Tscharnuter, W. 1975. *Lecture Notes in Physics* (T.L. Wilson and D. Downes, eds.), 42: 79. New York: Springer-Verlag.
Kraft, R. P. 1967. *Astrophys. J.* 150: 551.
Kulsrud, R. M. 1971. *Astrophys. J.* 163: 567.
Kusaka, T.; Nakano, T.; and Hayashi, C. 1970. *Prog. Theor. Phys.* 44: 1580.
Levin, B. J. 1972. *On the Origin of the Solar System* (H. Reeves, ed.), p. 341. Paris: C.N.R.S.
____. 1978. *Astron. Zh. Lett.* 4: 102. In Russian
Lynden-Bell, D., and Pringle, J. E. 1974. *Mon. Not. Roy. Astr. Soc.* 168: 603.
Mestel, L. 1968*a*. *Mon. Not. Roy. Astr. Soc.* 138: 359.
____. 1968*b*. *Mon. Not. Roy. Astr. Soc.* 140: 177.
Mestel, L., and Selley, C. S. 1970. *Mon. Not. Roy. Astr. Soc.* 149: 197.
____. 1977. *Star Formation.* (T. de Jong and A. Maeder, eds.), p. 213. Boston: Reidel.
Modisette, J. L. 1967. *J. Geophys. Res.* 72: 1521.
Okamoto, J. 1969. *Publ. Astron. Soc. Japan* 21: 350.
Öpik, E. J. 1966. *Mém. Soc. Roy. Sci. Liège, Ser. 5* 12: 523.
____. 1971. *Adv. Astron. Astrophys.* 8: 107.
Ostriker, J. P. 1972. *On the Origin of the Solar System* (H. Reeves, ed.), p. 154. Paris: C.N.R.S.
Parker, E. N. 1975. *Astrophys. J.* 198: 205.
Peebles, P. J. E. 1969. *Astrophys. J.* 155: 393.
Pechernikova, G. V.; Safronov, V. S.; and Zvjagina, E. V. 1976. *Astron. Zh.* 53: 612.
Prentice, A. J. R. 1976. *Astron. Astrophys.* 50: 59.
____. 1977. Submitted to *Icarus.*
Radzievski, V. V. 1949. *DAN – SSR* 67: 807.
Roberts, W. W., and Shu, F. H. 1972. *Astrophys. Lett.* 12: 49.
Ruskol, E. L., and Safronov, V. S. 1977. In *Proc. Lunar Sci. Conf.* 8th. p. 820. New York: Pergamon.

Ruzmaikina, T. V., and Safronov, V. S. 1976. *Astron. Zh.* 53: 860.
Rydgren, S. E.; Strom, F. H.; and Strom, K. M. 1976. *Astrophys. J. Suppl.* 30: 307.
Safronov, V. S. 1962. *DAN – SSR* 147: 64.
____. 1966. *Astron. Zh.* 43: 817.
____. 1969. *Evolution of the Protoplanetary Cloud and Formation of the Earth and the Planets.* Moscow: Nauka; 1972. Transl. from Russian. Jerusalem: Israel Program for Scientific Translation.
____. 1972*a*. *The Motion, Evolution of Orbits, and Origin of Comets* (G.A. Chebotarev and E.I. Kazimirchak-Polonskaya, eds.), p. 251. Dordrecht: Reidel.
____. 1972*b*. *On the Origin of the Solar System* (H. Reeves, ed.), p. 89. Paris. C.N.R.S.
____. 1972*c*. *On the Origin of the Solar System* (H. Reeves, ed.), p. 361. Paris: C.N.R.S.
____. 1975. In *Proc. Soviet American Conf. on Cosmochemistry of the Moon and Planets* (A. P. Vinogradov, ed.), p. 624. Moscow: Nauka.
Safronov, V. S., and Ruskol, E. L. 1957. *Voprosy Cosmogonii* 5: 22.
____. 1977. *Planetary Satellites* (J. Burns, ed.), p. 501. Tucson, Ariz.: Univ. Arizona Press.
Schatzman, E. 1962. *Ann. d'Astrophys.* 25: 18.
____. 1967. *Ann. d'Astrophys.* 30 (6): 963.
____. 1971. *Physics of the Solar System.* Goddard Inst. for Space Studies, X630-71-380, 413.
Turekian, K. K., and Clark, S. P. 1969. *Earth Planet. Sci. Lett.* 6: 346.
Takahara, M.; Nakazawa, K.; Narita, S.; and Hayashi, Ch. 1977. *Prog. Theor. Phys.* 58: 536.
Tscharnuter, W. 1975. *Astron. Astrophys.* 39: 207.
Vainstein, S. I., and Zeldovich, J. B. 1975. *Uspechy. Phys. Nauk. SSSR.* 106: 431.
Vitjazev, A. V.; Pechernikova, G. V.; and Safronov, V. S. 1978. *Astron. Zh.* In press.
Weber, E. J., and Davis, L. 1967. *Astrophys. J.* 148: 217.
Weidenschilling, S. J. 1976. *Icarus* 27: 161.
____. 1977. *Mon. Not. Roy. Astr. Soc.* 180: 57.
Wetherill, G. W. 1976. *Proc. Lunar Sci. Conf.* 7th. p. 3245. New York: Pergamon.
Whipple, F. L. 1972. In *From Plasma to Planets* (A. Elvius, ed.), p. 211. New York: Wiley.
Zvjagina, E. V., and Safronov, V. S. 1971. *Astron. Zh.* 48: 1023.
Zvjagina, E. V.; Pechernikova, G. V.; and Safronov, V. S. 1973. *Astron. Zh.* 50: 1261.

From
Protostars & Planets
Gehrels, ed, 1978

ACCUMULATION OF THE TERRESTRIAL PLANETS

GEORGE W. WETHERILL
Dept. of Terrestrial Magnetism
Carnegie Institution of Washington

The current status of theoretical and numerical research on the accumulation of the terrestrial planets is reviewed. At the present time it is unclear whether planets form on a short time scale ($\sim 10^4$ yr) as a result of massive gravitational instabilities, or by a sweeping-up process whereby larger bodies collide with smaller ones and accumulate into planets on a 10^7-10^8 yr time scale. Gravitational instability requires a massive solar nebula and its attendant problems, and also requires a quite special series of events if terrestrial planets, rather than gas-rich planets are to form. Consequently it is of considerable importance to learn the circumstances under which a gas-free sweeping-up growth is dynamically permissible. A major question is whether or not mutual gravitational perturbations are adequate to overcome the tendency of collisional damping and angular momentum conservation to circularize orbits, resulting in many small terrestrial planets rather than four large ones. Recent numerical studies of three consecutive stages of accumulation are discussed. In all cases, the above problem is present, and it may turn out that external perturbing bodies (e.g., Jupiter) may be required to maintain adequate orbital eccentricities in the terrestrial planet region. If this is the case, the problem of gas-free terrestrial planet formation cannot be separated from the formation of the giant planets in a gaseous medium, possibly involving gravitational instabilities.

Theories of the origin of the solar system have been primarily concerned with explaining how a portion of the interstellar medium can evolve into a star like the sun, surrounded by a distribution of dust and gas exhibiting some tendency to concentrate near the observed positions of the planets. Relatively little attention has been given to the question of the detailed way in which this matter actually coalesces to form planets exhibiting the physical and chemical characteristics of the actual planets in the solar system.

An understanding of these final stages of solar system evolution would be essential if its only purpose were to satisfactorily complete our knowledge of planetary origin. However, it is also needed for other reasons. At least at present, our knowledge of the preceding stages of stellar and planetary evolution are only poorly constrained by observation. It is likely that some "scenarios" of the course of this earlier evolution will fail the test of leading to planets with the characteristics of those observed. For this reason, quantitative studies of planetary growth are needed in order to distinguish those theories which can pass this test from those which cannot. Also, a number of important questions concerning the subsequent evolution of the planets require a rather detailed understanding of the way in which they formed. These include such things as the initial temperature profiles of the planets, their post-accumulation bombardment history, the origin of the residual "small bodies" – the comets, asteroids and planet-crossing bodies such as Pluto and "Kowal's object" (1977 UB), the inclination of planetary rotational axes, and the very difficult problem of planetary rotation periods. Also, the time scale for planet formation is required if we are to understand which stars may be expected to have terrestial-like planets and which may not. If, as seems likely, planetary growth requires times of $\sim 10^8$ yr, massive stars will be too short-lived to permit this to occur during their main-sequence lifetime.

It is probably not necessary to inform the reader that these details are not understood very well at present. In fact, perhaps a greater danger arises from the prevalent belief that no worthwhile work at all has been done on questions such as the time scale for accumulation of the planets, or the size and velocity distributions of the bodies from which the planets are formed. This belief has sometimes led to authors assuming that any opinion on these questions was as good as any other, and that these relevant quantities can be independently chosen as free parameters at one's pleasure. One purpose of this review is to introduce readers to the considerable and growing body of work which has been done on these questions. While there is still more to be learned than has been discovered, this is true of most similarly fundamental scientific areas of investigation. Nevertheless, scientific progress depends on a sympathetic but critical understanding of previous contributions, and upon development of "ground rules" defining which evolutionary trajectories are consistent with each other and which are not. Otherwise, the ladder of scientific progress will be replaced by a treadmill on which investigators will be doomed to endlessly repeat previous calculations and discussions, as an ironic consequence of their goals being yet distant.

This chapter will primarily consider a portion of the problem of planetary accumulation – that of the terrestrial planets, which contain less than 0.5% of the mass of the sun's planets. In fact, primary attention will be given to the even more specialized problem of the accretion of the terrestrial planets in a gas-free environment. The reasons for these limitations include the fact that most of the detailed work available for review has been carried out under this restriction, as well as the author's limited understanding of the more complex case of accumulation in a gaseous region. Some attention has been given to accretion in the presence of gas (Hayashi, 1977; Whipple, 1973; Weidenschilling, 1977*a, b*). It may turn out that detailed investigation of the gas-free problem will show that gas-free accumulation simply will not work, and the gaseous problem could dominate similar future discussions.

In order to place this "minor" part of the problem in a more general context, a qualitative discussion of the more complete problem will be given in the next section. This discussion must be in large part conjectural because there are so many aspects of this general problem which are poorly understood at present.

I. THE GROWTH OF PLANETS

A. Subjects on Which There is Some Agreement

Almost all proposed descriptions of the formation of the solar system include a stage in which at least as much material as is required to form the planets is present in the form of a rotating disk, of more or less the dimensions of the present solar system. At this stage there may or may not be a marked concentration of matter at the center of the disk – the protosun. The matter in the disk consists of a mixture of gas and dust, which may include fossil interstellar grains, as well as gaseous and solid species resulting from vaporization and condensation in the disk. The disk is commonly referred to as the "solar nebula," or sometimes simply "the nebula," even though different workers mean quite different things by these terms, which has led to avoidable misunderstandings.

There seems to be general agreement, in spite of other divergent views, that energy considerations preclude the disk being highly turbulent or at high temperature for more than 100 to 1000 yr (ter Haar, 1950). Insofar as such physical characteristics are required to transfer angular momentum and excess mass within and from the disk (Lynden-Bell and Pringle, 1974; Cameron, 1977), these processes must occur on this short time scale. A significant quantity of excess mass is generally thought to be necessary insofar as planet-scale gravitational instabilities within the disk are held to be an important process in planetary growth (Kuiper, 1951*a, b;* Urey, 1951, 1952, 1959; McCrea, 1960; Cameron, 1977; also see Cameron's chapter). The time scale for the development of these instabilities is short (i.e., a few rotation periods).

B. The Road to Rapid Accretion and Giant Protoplanets

The association of the processes and properties listed in the previous paragraph with one another, together with a desire for consistency, has resulted in a "family" of scenarios in which planetary formation is thought to take place on a short (i.e., $\lesssim 10^4$ yr) time scale in a massive, turbulent, high-temperature, "nebula" sometimes involving massive gravitational instabilities or "protoplanets" (see references in paragraph above; also Cameron, 1973; Cameron and Pine, 1973). Theories of this kind have heavily influenced recent cosmochemical thought regarding the composition of the planets (Lewis, 1972; Turekian and Clark, 1969); of the meteorites (Larimer, 1967; Grossman, 1972), and the moon (Ringwood, 1970), as well as discussions of the thermal histories of the moon and planets (e.g., Hanks and Anderson, 1969; Toksöz *et al.,* 1972). Although these theories have had a very widespread following, the extremely complex calculations necessary to show that they are dynamically acceptable have not yet been carried out, and it is unclear whether the necessary chain of events required to bring it all off on schedule can actually take place. This is particularly true of the formation of the terrestrial planets (McCrea, 1972; Williams and Crampin, 1971), for which quite exacting timing of events is required: (1) formation of giant gaseous protoplanets in a disk which is unstable owing to absence of a sufficiently massive central condensation; (2) completion of the raining out of a volatile-poor "proto-terrestrial planet" as a core within the giant protoplanet; (3) growth of the central condensation (the sun) to the point where the giant protoplanet is unstable with respect to tidal disruption by the sun, causing loss of the gas rather than the gas being gravitationally captured by the terrestrial planetary cores. The latter event would result in the formation of a major planet rather than a terrestrial planet.

C. The Slow Road to Planetary Formation by Continual Sweeping up of Small Bodies by Larger Ones

In contrast to the approach described above, other workers have followed a different path, leading to a quite different view of the way in which planets formed. Again starting from a flattened, rotating "nebula" containing a mixture of dust and gas, larger bodies are envisaged as growing by collision and coalescence of a swarm of smaller bodies, with gravitational instability playing, at most, a minor role. The initial stages of this growth could occur as a consequence of physical or chemical sticking of the dust grains (Urey, 1952) at the low velocities prevailing after dissipation of the turbulence in the disk after ~1000 yr (Safronov and Ruskol, 1956, 1957; Safronov, 1958). Many workers have regarded the necessity of assuming the existence of such sticking processes to be unsatisfactory. Consequently there was a general feeling of relief following the publication of the outlines of a theory by Goldreich and Ward (1973) in which small-scale gravitational instabilities in a thin central dust layer of the disk led to the formation of small (~2 km) planetesimals on a short (10-100 yr) time scale without the necessity of invoking sticking mechanisms. Calculations leading to this same result had been presented many times previously (Edgeworth, 1949; Gurevich and Lebedinskii, 1950; Safronov,

1960; Lyttleton, 1961) but for the most part they were ignored. In spite of the current popularity of starting planetary accumulation with dust-layer instabilities, as Weidenschilling (1977*a, b*) has emphasized, a very low probability of sticking need occur in order for planetesimals to grow to ~2 km before enough dust has settled to the central plane of the disk in the very thin layer required to initiate the gravitational instability. Therefore, it seems quite likely that the "sticking" mode of growth might preempt gravitational instability.

In the absence of further gravitational instabilities, the accumulation of these small planetesimals into planets must take place as a consequence of collisions between individual bodies, rather than through collective motion in the averaged field of the entire swarm of planetesimals.

Growth of planets from a planetesimal swarm of this kind has been discussed qualitatively by Chamberlin (1904), Barrell (1918), and semi-quantitatively by Shmidt (1957). Many of the concepts which figure prominently in more recent quantitative formulations of this mode of planetary accumulation were introduced in these earlier discussions. For example, the idea that a large number of asteroid-sized "planetoids" would be formed prior to the final accumulation of the terrestrial planets was seen by Barrell (1918) as a way of obtaining high initial temperatures for the earth.

> ". . . evidence favors the hypothesis that the scattered matter which was added to the nucleus to form the earth was largely of such size that the individual planetoids would have plowed through a primordial atmosphere and ocean, if such existed, and have penetrated beneath the surface of the liquid or solid body below. The energy of impact from dust-like material would be absorbed at the surface, and, as heat, quickly radiated into space. The accretion of dust would favor the growth of an earth solid throughout. Larger masses would, on the other hand, carry the energy of impact into the earth. They would not strike with the high velocities of meteors which collide with the earth since the different planetoids were traveling in the same general direction, but nevertheless a state of incandescence and liquidity would be likely to result from the sizes of the masses involved. If in addition the infall of masses was sufficiently rapid to bury the heat of previous infalls before it could be dissipated by conduction to the surface, a general heating and liquefaction of the earth would tend to take place, both from the increased compression of the deeper nucleus and the effects of impact at higher levels. . . ."

This mode of planetary growth leads to long (10^7-10^8 yr) time scale for the formation of the terrestrial planets, "second largest" bodies in the accumulation zones of the terrestrial planets in the range of 10^{24}-10^{26} g (Ceres to lunar-size), and mean impact velocities as large as 8 km sec^{-1} on the moon during its final stages of growth. The bases for these results are reviewed in the following sections of this chapter.

This approach to the growth of the terrestrial planets avoids the necessity of the precise timing of the formation of the sun, of giant protoplanets and their proto-terrestrial planetary cores, and the tidal disruption of the giant planets discussed previously. It is also possible that even if this proto-planetary chain of events occurred as required, that several planetary cores would be formed in similar orbits, which would then be expected to make close encounters to one another. This could lead to tidal disruption of the smaller bodies, and gravitational perturbation of the fragments by the surviving bodies. Thus it is possible that the gravitational instability route would also lead to a planetary swarm in the terrestrial planet region similar to that produced by the "sweeping-up" process at a later stage of its evolution. The sweeping up process also has the practical advantage of being more amenable to calculation, and therefore its predictions can be more readily compared with observations. Of course, this convenience does not bear on its validity, but should make it possible to understand the circumstances under which it could be valid.

The formation of the giant planets has also been studied using this sweeping-up approach (Safronov, 1969; Hayashi *et al.,* 1977). Detailed discussion of formation of the giant planets is beyond the scope of this chapter. However, it will be seen from the next sections that to a significant extent the course of events during the formation of the terrestrial planets is influenced by the mass and position of the giant planets, and therefore some discussion of the formation of these planets is required. Unfortunately, in this case there are a number of additional complicating problems which have yet to be resolved. These include the fact that gases (principally hydrogen and helium) must have been present in large quantities during the formation of at least Jupiter and Saturn, but were nearly absent during the final stages of the growth of Uranus and Neptune. Also, overlapping of accumulation "feeding zones" must have occurred at least for these last two planets, and ejection of mass in this region must have occurred as a consequence of strong gravitational perturbations by the large embryos in a region of the solar system with low escape velocity. Accumulation times of the order of 10^8 yr have been calculated for Jupiter and Saturn by the authors cited above, but are quite uncertain because of the difficulties just mentioned. Even greater problems exist for Uranus and Neptune, as their accumulation times are found to be greater than the age of the solar system.

Even a relatively short 10^8 yr time scale for the formation of Jupiter introduces some problems for accumulation of the terrestrial planets. Calculations by the author, of the sort described in Section IV, show that a very rapid fall-off in the initial mass of the disk with distance from the sun is required beyond the orbit of the earth to account for the small mass of Mars. This "hole" in the disk is commonly attributed to heavy bombardment of this region of the solar system by residual planetesimals from Jupiter's zone (Safronov, 1969; Weidenschilling, 1975). To be effective in preventing the "overgrowing" of Mars, the clearing of this region must occur on a time scale of $\sim 10^7$ yr, significantly earlier than that predicted by accumulation calculations for Jupiter.

Because of these difficulties with time scales for the growth of the major planets, efforts have been made by various authors to modify the calculation so at least to permit the formation of Uranus and Neptune to occur in the past rather than the future (Levin, 1972, 1978; Safronov, 1975). Something of this kind might prove correct, but it seems likely, at least to the author, that a nearly pure sweeping-up theory is very likely to always give time scale problems of various kinds, including conflicts with the observed cratering history of the moon and terrestrial planets. For this reason it seems worthwhile to keep our minds open to other ways to produce giant planets on a short time scale. For example, there may be a possibility that ring instabilities could occur in the outer solar system even for a low-mass disk ($\sim 0.1\ M_O$), as stated by Cameron (1977). In the case of Uranus and Neptune there still would remain the problem of explaining marked differences between the chemical composition of these bodies (primarily H, C, N, O compounds) and Jupiter and Saturn (primarily H, He). Although such chemical differences could arise from the ionization processes discussed by Alfvén (1954), the very large mass of C, N, O required to produce a gravitational instability is inconsistent with the low-density disk associated with this theory. If somehow all these problems could be overcome then it could turn out that the natural way to accumulate most of the mass of the volatile-rich planets would be through gravitational instability or some phenomenon essentially dependent on the presence of gas, leading to fairly rapid formation of at least the cores of the giant protoplanets. On the other hand the natural way to accumulate the terrestrial planets could still be to sweep up this smaller quantity of mass in the form of planetesimals. Of course, these ideas concerning the giant planets are highly speculative at this stage of understanding, and the author regards it as fortunate that they lie beyond the scope of the present review.

II. ACCUMULATION OF THE TERRESTRIAL PLANETS BY THE SWEEPING UP OF PLANETESIMALS: SIMPLE CONSIDERATIONS

The complete problem of the evolution of a swarm of planetesimals into terrestrial planets is very complex and at present we are a long way from even a good approximation to a definitive solution. Nevertheless it is possible to gain some insight into accumulation of planetesimals in a gas-free region by use of simple calculations using methods similar to those of elementary kinetic theory of gases. Results found in this way will be found to be useful in making starting approximations in more advanced calculations.

A. Time Scale of Accumulation

From this elementary point of view the embryo which is to become the planet can be envisaged as a body of radius R_e moving through a swarm of planetesimals at relative velocity $\bar{v}$ (Fig. 1). The rate of growth of the embryo's

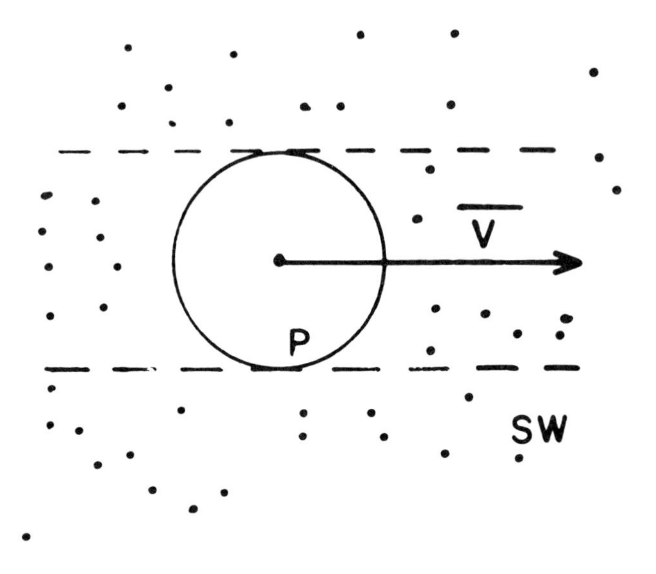

Fig. 1. Schematic representation of an embryo P moving at a mean relative velocity, $\overline{V}$ through a gas-free swarm SW.

radius will be roughly

$$\frac{dR}{dt} = \frac{\rho_{sw}\overline{v}}{4\rho_p}\left(1 + \frac{v^2{}_e}{\overline{v}^2}\right) \tag{1}$$

where ρ_{sw} is the density of the swarm, ρ_p is the density of the planetesimals being accumulated ($\sim$4g/cm^3) and v_e is the escape velocity of the embryo. The expression in parentheses represents the enhancement of the effective capture radius of the embryo as a consequence of its gravitational field.

The density of the planetesimal swarm at the midpoint of growth will be given by the ratio of its mass to its volume. At this stage mutual perturbations by members of the swarm will be small, relative to perturbations by the embryo. Referring to Fig. 2, the volume of Earth's swarm will be taken to be a ring with rectangular cross-section centered at 1 AU, and of width 2δ extending midway to the present semi-major axes of Venus and Mars. The heighth of this volume will be $2a \sin i$ where i is the mean inclination of the bodies in the swarm. Assuming energy equipartition between the three components of velocity

$$a \sin i \approx \frac{a\overline{v}}{\sqrt{3}v_c} \tag{2}$$

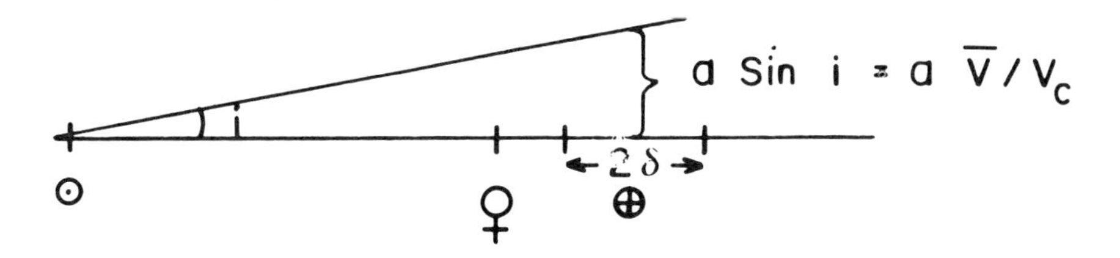

Fig. 2. Illustration of the volume of Earth's swarm, used in simple calculation. The swarm is considered to be bounded within a circular ring of thickness 2δ centered at Earth's radius and of height $2a$ and sin i.

where ν_c is the circular Kepler velocity at 1 AU (29.8 km $\sec^{-1}$). Then

$$\rho sw \approx \frac{M_P}{2} \frac{1}{\text{Volume}_{\text{swarm}}} = \frac{M_P}{2} \frac{1}{(2\pi a)(2a\bar{v})/\sqrt{3}V_c\,(2\delta)}$$

$$= \frac{\sqrt{3}M_p V_c}{16\pi a^2 \delta \bar{v}} = 3.06 \times 10^{-12} \text{ g cm}^{-3} \qquad (3)$$

where M_p is the final mass of the planet, neglecting escape from the swarm during accumulation (see next section).

Substituting (3) into (1)

$$\frac{dR}{dt} = \frac{\sqrt{3}}{64\pi} \frac{M_p V_c}{a^2 \delta \rho_p} \left(1 + \frac{V_e^{\,2}}{\bar{v}^2}\right) \qquad (4)$$

Expression (4) is independent of $\bar{v}$, except insofar as it enters into the gravitational cross-section term in parentheses. In order for accumulation to proceed at this stage it is necessary that $\bar{v}$ be large enough for bodies with semi-major axes at $(a+\delta/2)$ to have perihelia at least as small as a. The velocity of a low inclination (i.e., 5°) body in such an orbit with respect to the embryo at 1 AU will be 3.1 km $\sec^{-1}$. This is a reasonable value to use for $\bar{v}$, and is the same as the value found by use of the expression of Safronov (1969).

$$\theta \simeq 4 = \frac{V_e^{\,2}}{2\bar{v}^2} \quad . \qquad (5)$$

If lower values of $\bar{v}$ are inserted in Eq. (4) larger accumulation rates would be found. However in this case the use of Eq. (4) would not be valid, as the bodies in the swarm would not be "orbitally linked" to the embryo, which dominates the growth at this stage. This term will be used to describe the situation wherein the orbits of the bodies of the swarm are sufficiently strongly coupled by their mutual gravitational perturbations to continue to undergo major changes in their elements on a time scale short compared to 10^8 years. For small bodies this requires that there be extensive crossing of orbits. Large enough bodies can be linked even when actual intersection of orbits does not take place. The situation opposite to orbital linkage occurs when the system becomes "dead", i.e., the bodies remain sufficiently distant from one another to lead to nearly stable orbits on this time scale.

It is certainly conceivable that breakdown of linkage may occur, which would lead to a situation in which the rate of accumulation would be controlled not by Eq. (4), but by the rate at which bodies would "diffuse" by relatively weak mutual or long-range perturbations into the vicinity of the embryo, in some ways analogous to accumulation of planetesimals in a gaseous medium (Hayashi et al., 1977). This question of maintaining orbital linkage is an important matter which has not yet been thoroughly addressed in even more sophisticated treatments of planetary accumulation.

Substitution of these numerical values into Eq. (4) gives

$$\left.\frac{\mathrm{d}R}{\mathrm{d}t}\right._{\mathrm{midpoint}} = 16.7 \text{ cm/yr.} \tag{6}$$

This accumulation rate is seen to be quite low, and if it were concentrated entirely in small particles (which it is not) it could easily be swept away every month or so from one's sidewalk with a broom. At this average rate, the time required for the earth to complete its growth will be $T_{\mathrm{growth}} = 38 \times 10^6$ yr. Similar accumulation time scales of 10^7 to 10^8 yr also are found in more complex treatments of this problem. The reason is that the physics is essentially the same – including the assumption of orbital linkage.

B. Distribution of the Initial Swarm

It is obviously necessary that mass, energy, and angular momentum be conserved during the process of accumulating the planetesimal swarm into the planets. If the swarm could be treated entirely as a closed system, the initial values of these quantities would be defined by the masses and orbits of the present planets. However the system is not entirely closed; energy is lost in the form of heat when the planetesimals collide with one another and with the planetary embryos as they emerge during the later stages of accumulation. Mass, together with associated energy and angular momentum, is lost if members of the swarm are accelerated by perturbations by the embryos into orbits with perihelia beyond Mars, including those which escape the solar system in hyperbolic orbits. Angular momentum transfer occurs between the major planets and the terrestrial planets as a consequence of long-range gravitational perturbations. Although these deviations from closed system behavior cannot be known *a priori* with a high degree of certainty, their values can be estimated. This permits placing some limits on the distribution of mass in thc swarm, based on these very general principles.

Expressions for conservation of mass, angular momentum and energy can be written in the following form

$$M_{\text{initial}} = \sum_{i=1}^{4} M_i + M_{\text{escaped}} \tag{7}$$

$$L_{\text{initial}} = \sum_{i=1}^{4} M_i \, (GM_\odot)^{\frac{1}{2}} \, a_i^{\frac{1}{2}} \, (1-e_1{}^2)^{\frac{1}{2}} + L_{\text{escaped}} + \Delta L_G \tag{8}$$

$$U_{\text{initial}} = -\sum_{i=1}^{4} \frac{GM_\odot}{2a_i} + U_{\text{escaped}} + U_{DE} + U_{DP} \quad . \tag{9}$$

The sums i = 1,4 are taken over the present values for the four terrestrial planets, ΔL_G is the angular momentum exchanged between the swarm and the giant planets, U_{DE} is the energy lost as heat in planetesimal-embryo collisions, U_{DP} is the energy lost in planetesimal-planetesimal collisions.

If a fraction f_E of the final mass of the planet escaped, then

$$M_{\text{initial}} = (1+f_E)\, M_T \tag{10}$$

where M_T is the present total mass of the terrestrial planets (including the moon with the earth).

In the same way

$$L_{\text{escape}} = f_E \, L_T \tag{11}$$

$$U_{\text{escape}} = f_E \, U_T \tag{12}$$

assuming that the initial angular momentum and energies of the bodies which escaped were the same as the rest of the swarm.

The other quantities on the right hand side of Eqs. (7), (8), and (9) can be estimated. These estimates will of course be model dependent. However, this is equally true of much more complex numerical approaches to the problem of the later stages of accumulation. For a given choice of model parameters (e.g., size distribution and fraction of kinetic energy dissipated in collisions), these simple calculations permit useful estimates of the differences between the initial and final swarms which will be found by numerical calculations employing the same model.

All of the planetesimals must strike the embryo, and lose their kinetic energy of relative motion in the process. Using ∇ = 3.1 km sec^{-1} at the midpoint of accumulation, as estimated in the previous section, then

$$U_{DE} \approx \frac{\overline{v}^2}{2} \, M_T = 5.35 \times 10^{38} \text{ ergs} \tag{13}$$

i.e., ~1% of the energy of the present planets.

The energy lost in planetesimal-planetesimal collisions is more model dependent, because most of the energy will be lost in collisions between the smaller members of the swarm, and the number of smaller members is dependent on the size distribution of the swarm. One possible model used in numerical studies (Sec. IV. C.) is to consider the swarm to consist of large bodies $\sim 10^{24}$ g in mass (440 km radius), each of the bodies being accompanied by a collisional steady-state "retinue" of smaller bodies obeying the power law

$$\frac{dn}{dr} = Cr^{-3.5} \quad . \tag{14}$$

Using the "kinetic theory of gases" approach of the previous section, this size distribution leads to a rate of energy loss per second in the earth's zone (Fig. 2) at the midpoint of accumulation

$$\frac{dU}{dt} = \frac{3}{128} \frac{\overline{v}^3 \; \alpha M^2{}_e \; (1+f_E)^2}{\rho_p \, \Lambda \, r_{max}} \ln \frac{r_{max}}{r_{min}} \tag{15}$$

where M_e is the mass of the earth, Λ is the volume of Earth zone (Fig. 2), α is the fraction of the energy lost in each collision (e.g., 0.5). The result is not sensitive to the value of r_{min}. Substituting the numerical values defined above and by the earlier calculations, with r_{min} = 1 meter, this gives an energy loss rate of

$$\frac{dU}{dt} = 2.87 \times 10^{37} \; (1+f_E)^2 \text{ ergs}/10^6 \text{ yr.} \tag{16}$$

In the 38×10^6 yr accumulation time calculated as compatible with this choice of parameters, the total energy loss is

$$U_{DP} = 1.09 \times 10^{39} \; (1+f_E)^2 \text{ ergs} \tag{17}$$

in earth's zone; i.e., since f_E will turn out to $\ll 1$, it represents about a 4% correction to the total energy of the swarm, assuming a similar loss in Venus' zone. If a larger relative velocity or a size distribution in which more of the mass is concentrated in small bodies had been chosen, higher energy losses would be calculated. It will be seen that even these modest energy losses lead to significant restrictions on the distribution of the bodies in the original swarm unless some mechanism exists which permits the swarm to lose angular momentum as well. The difficulties arising therefrom will be only aggravated by larger energy losses.

The remaining quantity to be approximated is ΔL_G, the angular momentum transferred from the terrestrial planet region to the giant planets as a result of secular perturbations. This is difficult to approximate with any degree of certainty. The principal reason for this is that it is not even known when the giant planets formed relative to the terrestrial planets, as discussed in Sec. I. In addition, the positions of all the planets were probably changing during the time interval of formation of the terrestrial planets, changing both the frequencies and the amplitude of the secular perturbation forcing terms. However, if major condensations existed which gave rise to oscillations forced by secular perturbation similar to those at present, some estimate of this angular momentum transfer can be given.

As a consequence of the collisional energy dissipation which resulted in the energy loss terms U_{DE} and U_{DP}, the energies and therefore the semi-major axis of both the planetesimals and the embryos will tend to decrease with time. Since angular momentum will be conserved in these collisions, and since

$$L = (GM_\odot)^{\frac{1}{2}} \sqrt{a(1-e^2)} \tag{18}$$

a decrease in a requires a decrease in e, i.e., the orbits will tend to become more circular. In the case of the planetesimals this circularization will be prevented by perturbations of the planetesimals by the embryos, assuming orbital linkage. However, there are no comparable perturbers for the embryos themselves within the system of terrestrial planets, and one must look to the giant planets if the orbits of the embryos are to remain non-circular. In the present solar system the root-mean-square eccentricity of the earth resulting from these perturbations is $\sim$0.03, and it varies above and below this value on a time scale of $\sim 10^5$ yr. At the midpoint of accumulation of the earth, the rate at which the earth's embryo loses energy as a consequence of U_{DE} and U_{DP} will be $\sim (U_{DE} + U_{DP})/2T_{\text{growth}}$, assuming that the average semi-major axis of the swarm remains equal to that of the earth. In other words, some of the energy lost by planetesimal-planetesimal collisions must be supplied by the earth embryo in the process of maintaining the steady-state geocentric velocity of the swarm. The relationship between the energy loss of Earth's embryo, and the change in its semi-major axis will be

$$da_E = \frac{-2a^2{}_E\,(1+f_E)^2\; dU_E}{GM_\odot M_E} \simeq 0.0013(1+f_E)^2 \text{ AU}/10^6 \text{ yr.} \tag{19}$$

Conservation of angular momentum would result in reduction of the eccentricity of Earth's embryo from 0.03 to zero for a change in semi-major axis of 0.0009 AU, and according to Eq. (19) this would take place in $\sim 7 \times 10^5$ yr, in the absence of external perturbations. This is longer than the reciprocal angular velocity of even the longer period terms in the present forced secular perturbations of the earth (Brouwer and van Woerkom, 1950), and consequently this rate of dissipation will not damp the amplitude of the forced oscillations very much. Maintenance of the forced eccentricity at the value it would have had in the absence of dissipation will require transfer of angular momentum from the earth to the giant planets at a rate sufficient to offset the decrease in eccentricity caused by the decrease in semi-major axis. It follows from Eq. (18) that at constant eccentricity

$$\frac{1}{L}\frac{\mathrm{d}L}{\mathrm{d}t} = \frac{1}{2a}\frac{\mathrm{d}a}{\mathrm{d}t} \quad . \tag{20}$$

When combined with the rate of change in semi-major axis given in Eq. (19) it is estimated that 2.5% of the earth's angular momentum will be transferred to the giant planets during the course of its accumulation. The actual value will probably be somewhat less, as the amplitude of the forced oscillations will be decreased by damping during the early more rapid stages of accumulation.

The observed mass, final angular momentum and energy of the terrestrial planets, together with the preceding estimates of energy and angular momentum loss during accumulation, can be combined to obtain the initial energy and angular momentum per unit mass of the swarm by use of Eqs. (7), (8) and (9) and then dividing by M_{initial} giving

$$L_{\mathrm{initial}} = \frac{4.971 \times 10^{47} \times (1.025)}{1.188 \times 10^{28}} = 4.29 \times 10^{19}\,\mathrm{erg\text{-}sec/g} \tag{21}$$

$$U_{\mathrm{initial}} = \frac{-6.219 \times 10^{40} \times (0.95)}{1.188 \times 10^{28}} = -4.97 \times 10^{12}\,\mathrm{ergs/g} \quad . \qquad 22)$$

When formulated in this way, the distribution of angular momentum and energy in the initial swarm is expressed in a manner which is independent of the fraction of the initial mass which is lost by acceleration out of the terrestrial planet region into Jupiter-crossing. This fraction f_E will not be large. The minimum geocentric velocity for a body crossing the orbit of both Jupiter and Earth (in their present orbits) is 8.6 km sec^{-1}. Therefore, bodies with geocentric velocity of 3 km sec^{-1}, as assumed in the preceding calculations, cannot be lost to Jupiter. Only those bodies with geocentric velocities well above the average will have sufficient velocity to become Jupiter-crossing, and only a fraction of such high-velocity bodies will be perturbed into Jupiter-crossing before they impact Earth or Venus. Previously published calculations (Wetherill 1975*a,b*) show that in the complete absence of dissipation, about 18% of a swarm of objects initially in orbits with low (~1 km sec^{-1}) geocentric velocity will be accelerated by Earth and Venus perturbations (at their final masses) into Jupiter-crossing. However, because of the absence of dissipation, this swarm was accelerated to a mean geocentric velocity of $\sim$12 km sec^{-1}, much higher than in the case being discussed here. Furthermore, the smaller embryo masses lead to less acceleration. The results of calculations of accumulation including dissipation discussed in Sec. IV of this chapter lead to escape fractions of 0.3 to 2%.

The initial values of energy and angular momentum per unit mass given by Eqs. (21) and (22) will now be compared with hypothetical initial swarms in order to see what constraints conservation of energy and angular momentum place upon the initial distribution.

For this purpose, initial swarms with semi-major axis distributions

$$\frac{dN}{da} = \rho_s a^n \tag{23}$$

will be considered, in which N is the initial number of bodies per unit area with semi-major axis between a and a+da, and the exponential parameter n distinguishes between swarms with different radial density distributions. The total mass, angular momentum, and energy of the swarm will be given by

$$M_t = 2\pi\rho_s \int_{a_1}^{a_2} a^{n+1}\, da = \frac{\rho_s}{(n+2)} (a_2{}^{n+2} - a_1{}^{n+2}); (n \neq -2) \tag{24}$$

$$L_t = 2\pi\rho_s (GM_\odot)^{\frac{1}{2}} \overline{(1-e^2)^{\frac{1}{2}}} \int_{a_1}^{a_2} a^{n+3/2}\, da \tag{25}$$

$$= \frac{2\pi(GM_\odot)^{\frac{1}{2}}}{(n+5/2)} \rho_s (1-e^2)^{\frac{1}{2}} (a_2{}^{n+5/2} - a_1{}^{n+5/2}); (n \neq -5/2)$$

$$U_t = -\pi GM_\odot \rho_s \int_{a_1}^{a_2} a^n\, da = -\pi\rho_s \frac{GM_\odot}{(n+1)} (a_2{}^{n+1} - a_1{}^{n+1}); (n \neq -1) \tag{26}$$

where a_2 and a_1 are the bounds of the swarm. When n has the values excluded in Eqs. (24), (25), or (26), the integral in question is given by the substitution: $a^\circ \rightarrow \ln a$ in these equations. Averaging of the initial value of $(1-e^2)^{\frac{1}{2}}$ is permissible since the initial eccentricities must be low.

Equations (24), (25) and (26) have been numerically evaluated for different swarms, as given in Table I. The "nominal" case, corresponding to the estimates of U_{DE}, U_{DP} and ΔL_G given in this section requires the values of L and U given in the fourth entry in Table IC. Comparison with tables IA and IB shows that values of n less than $-1/2$ are excluded, i.e., the initial mass density per unit area must be nearly constant, rather than falling off approximately as $a^{-\frac{3}{2}}$, which is the case for the present solar system taken as

TABLE I.

Energy and Momentum per Unit Mass of Planetesimal Swarms

A. $a_1 = 0.4, a_2 = 1.4$

n	L (10^{19} erg sec/g)	U (10^{12} ergs/g)
-3/2	4.01	-5.91
-1	4.14	-5.54
-1/2	4.26	-5.21
0	4.36	-4.93
+1/2	4.45	-4.67
+1	4.54	-4.46

B. $a_1 = 0.6, a_2 = 1.2$

n	L (10^{19} erg sec/g)	U (10^{12} ergs/g)
-3/2	4.15	-5.23
-1	4.17	-5.12
-1/2	4.22	-5.01
0	4.27	-4.93
+1/2	4.30	-4.83
+1	4.33	-4.74

C. Required values

L (10^{19} erg sec/g)	U (10^{12} ergs/g)	**Energy Dissipation** %	**Ang. Mom Loss** %	**Remarks**
4.18	-5.23	0	0	no orbital linkage
4.18	-4.97	5	0	"no Jupiter"
4.27	-4.92	6	2	good fit to n=0 (swarm B)
4.29	-4.97	5	2.5	"nominal" case
4.43	-4.61	12	6	

a whole (Weidenschilling, 1977*c*). Some decrease in surface density with radius ($n \sim -1$) is possible if it is assumed that there is no energy dissipation (first entry in Table C). This would imply very low relative velocities, which, as discussed before, is not compatible with maintenance of orbital linkage between the embryo and the swarm.

If for some reason the secular perturbations are ineffectual in removing angular momentum (e.g., the giant planets have not yet formed) then the required values are given in the second entry of Table IC. For all the swarms calculated, the angular momentum is too high at the required energy. In fact, the extreme case of all the mass concentrated at a single radius (0.89 AU) corresponding to a minimum angular momentum for the given mass, is required to fit both the energy and angular momentum. Therefore, velocities as high as that assumed here in the nominal case, ~3 km sec^{-1}, require that there be some external mechanism for removing angular momentum from the terrestrial planets during their accumulation, unless the rate of energy dissipation is quite a bit lower than that in the model chosen as an example.

If the dissipation is slightly different from the nominal case (entry 3), a good fit is given by the narrow flat distribution (n = o) of Table IB. The differences between the energies and angular momenta of the swarms rejected and the values required may not seem great. The significance of these differences becomes more clear if their consequences are calculated. For example, consider the case where the swarm corresponding to $n = -1/2$ in Table IA is the initial swarm, and the nominal losses of energy and angular momentum occurred. If Mercury and Mars turned out to be at their proper mass and orbits, conservation of energy and angular momentum would require that Venus be at 0.63 AU (with its present mass) and the earth at 1.06 AU. Thus differences of a few percent in the initial energy and angular momentum will have a large effect on the final positions of the planets.

It is also seen that acceptable swarms do not span the entire range from Mercury's perihelion to Mars' aphelion but must be truncated at the lower end at ~0.4 AU and the upper end at ~1.4 AU. Wider swarms will have too large negative values of energy at the required angular momentum. The form of the terrestrial planet swarm thus is more a band than a broad zone, as discussed by Alfvén and Arrhenius (1976). Another characteristic of swarms with velocities in the range corresponding to maintenance of orbital linkage, and size distribution containing a steady-state population of small bodies, is that the initial mass of Earth's swarm ($a > 0.85$) is about twice that of Venus' swarm ($a < 0.85$). In order for a swarm of this kind to evolve into the present terrestrial planets, it is then necessary that material originally in Earth's swarm be accumulated by Venus and Mars. This effect will tend to reduce density differences between the terrestrial planets arising from possible initial variation in the chemical composition of the swarm as a function of heliocentric distance.

C. Conclusions

Subject to the validity of the assumptions made in the foregoing calculations, it is concluded, based on these simple considererations:

1. If velocities are to be high enough (~3 km sec^{-1}) to maintain orbital linkage during the principal stages of the embryos' growth, a small but significant (e.g., 5%) loss of energy by planetesimal-embryo and planetesimal-embryo collision will occur for size distributions of the sort considered.

2. Under these circumstances, some mechanism (i.e., Jupiter perturbations) must exist to transfer angular momentum (in this case 2.5%) out of the terrestrial planet region during the accumulation of these planets if circularization of the embryo's orbits is to be avoided. If circularization occurs, acceleration of the planetesimals by the radial component of the embryos' velocity will be ineffective, relative velocities will drop, and multi-embryo crossing may become unusual, causing further loss of velocity.
3. Initial swarms spanning the entire range of the present terrestrial planets, and with surface density decreasing with distance from the sun, are not compatible with these energy and angular momentum calculations. Rather, the swarm must initially occupy a more narrow central band, which spreads during the process of accumulation.
4. A corollary of (3) above is that significant exchange of material must occur between the "feeding zones" of the various terrestrial planets during their accumulation.

In addition to the conclusions listed above, earlier studies of the post-accumulation evolution of Earth's residual swarm (Wetherill, 1977*a*) shows that a significant fraction (~10%) of the swarm is transferred to Mars-crossing, and most of this material will "leak" back to Earth on a time scale of hundreds of millions of years. This same phenomenon should occur during this preceding accumulation stage as well. Its effect will be to impose a long-lived "tail" on the growth of the earth which, if sufficiently large, will be in conflict with the low crater densities observed on the lunar maria. This is likely to be a major constraint on such questions as the initial position of Mars' embryo, and the collisional lifetime of bodies in the region between Earth and Mars.

III. ANALYTICAL CONTRIBUTIONS OF SAFRONOV AND OTHERS

Until the mid 1970's, more detailed treatments of the sweeping-up of the planetesimal swarm have been fundamentally based on Eq. (1) (Gurevich and Lebedinskii, 1950; Safronov, 1960, 1969, also see the chapter of Safronov; Marcus, 1965, 1969; Weidenschilling, 1974, 1976; Hallam and Marcus, 1974; Hartmann and Davis, 1975). In some of these calculations the mass distribution and relative velocities of the planetesimals were treated as independent free parameters. As a consequence such theories are primarily kinetic, rather than dynamic. It is implicit in the use of Eq. (1) that account is not properly taken of the differences between the heliocentric Keplerian orbits of either the planetesimals or the embryos. This leads to such theories being of limited applicability to growth of embryos in non-circular orbits and to circumstances in which the feeding zones of several embryos merge. Nevertheless, a number of important concepts and results emerged from these studies which will prove to be very important in future work.

The most extensive dynamical treatment of the problem using this limited but fruitful approach has come from the efforts of followers of O. Yu. Shmidt to quantify his more qualitative results. The resulting theories are dynamic, rather than simply kinematic, in that the relative velocities of the planetesimals are calculated, rather than assumed. These relative velocities arise from the balance between the mutual gravitational accelerations and the loss of energy by collisional dissipation. An important result of this work is that both the acceleration and damping processes are proportional to the encounter probability and hence the total cross-section (including its enhancement by gravity) of the colliding bodies. Dependence on the size of the bodies arising from encounter probability therefore cancels. However the deflection angle caused by a gravitational perturbation at a given number of planetesimal radii is proportional to $GMp/R_p V^2$, in which M_p and R_p are the mass and radius of the perturbing planetesimal, and V is their relative velocity. The increments in velocity are proportional to this deflection angle and therefore scale with the dimensionless quantity (Safronov number)

$$\theta = \frac{GM_p}{R_p V^2} = \frac{1}{2} \frac{V^2{}_{\text{escape}}}{V^2} \quad . \tag{27}$$

Thus two swarms differing in total mass and size of bodies, but composed of bodies of uniform size, and with the same value of θ, can both be at the steady state for which increments in velocity caused by gravitational perturbations will be balanced by collisional decrements in velocity. Furthermore, two swarms composed of bodies unequal in size will have the same velocity distribution provided that the sets of $\theta_1, \theta_2, \theta_3, \ldots$ of the individual bodies are the same in both swarms. In this case it is convenient to specify M_p, $V^2{}_{\text{escape}}$ and a single θ in Eq. (27) as pertaining to the largest body of the swarm. Doing this does not mean that only the largest body is effective as a perturber. It simply represents the fact that if the size distribution is specified, the individual values of θ are determined once θ for the largest body is specified.

The value of θ corresponding to this steady state will depend on the efficacy of collisions in reducing the relative velocity of the colliding bodies, i.e., it will depend on the physical parameters of the bodies. If these physical properties and the size distribution of the swarm remain constant as the planetesimals grow by sweeping up smaller members of the swarm, then θ will remain constant during this growth. Under these circumstances the relative velocities of the planetesimals will increase as they grow, and will be a fixed fraction of the escape velocity of the larger body. Using plausible values of the energy dissipation in collisions, Safronov (1969) shows that $\theta \approx 3$ to 5, and the mean relative velocity of the bodies will remain well below $\sim 1/3$ the escape velocity of the largest body. Experiments by Gault et al. (1963) and calculations by O'Keefe and Ahrens (1977) show that planetesimals will grow, rather than become smaller by fragmentation, if the impact velocity is lower than several times the escape velocity. Therefore, the size distributions in which smaller bodies exist which are within a factor of $\sim 10^3$ to 10^5 the mass of the largest body, these smaller bodies will grow along with the largest body. Very small bodies will fragment. If the size below which fragmentation occurs is proportional to the escape velocity of the body with this critical size, introduction of fragmentation will not invalidate the above conclusions, again insofar as there is no significant change in the mass distribution or the physical properties of the swarm as growth and fragmentation proceed.

As discussed above, the mean relative velocities of planetesimals in the swarm will be determined by their size distribution. However, inspection of Eq. (1) shows that the rate of growth of the bodies, and therefore the evolution of their size distribution, is dependent on their relative velocities. In the case of fragmentation, Dohnanyi (1969) has shown that under a broad range of assumptions, bodies with the same fragmentation properties will achieve a steady-state size distribution with a differential radius dependence

$$\frac{dN}{dr} \propto r^{-3.5} \quad . \tag{28}$$

Zvyagina et al. (1973) have presented calculations leading to approximately steady-state power laws with similar exponents for combined accumulation and fragmentation, as will occur during the growth of planets from a swarm. However, it is not clear if and when such steady states will be realized, and as noted above, the validity of the arguments regarding the constancy of θ and the growth of the relative velocities "in step" with the escape velocities are dependent upon the size distribution being at least in some approximate sense constant over significant increments of growth.

It can easily be shown that a constant size distribution is not valid at *all* stages of growth. For example, at the large end of the distribution, expressions for the ratio of the radii of the largest and second largest bodies, R, can be given (Wetherill, 1976)

$$R = \frac{2\theta \left(\frac{R_0 - 1}{R_0 - 2\theta}\right) e^{(2\theta - 1/2\theta + 1)\lambda t^*} - 1}{\left(\frac{R_0 - 1}{R_0 - 2\theta}\right) e^{(2\theta - 1/2\theta + 1)\lambda t^*} - 1} \tag{20}$$

where R_0 is the value of this ratio at an earlier time $t = 0$, and θ is the Safronov number (Eq. 27). The time-like variable $t^* = \frac{1}{\rho_{c0}} \int_0^t c(t')\, dt'$, where ρ_{c0} and $\rho_c(t')$ are the masses and densities of the swarm at $t = 0$ and at later times t'. This expression for R is time dependent, and approaches a constant value only asymptotically (Safronov, 1969)

$$\frac{R_1}{R_2} \rightarrow (2\theta)^3 \quad . \tag{30}$$

Numerical examples of such "runaway accretion" are given by Hartmann and Davis (1975), and this phenomenon was used by Safronov (1969) to limit the size of the second largest body in Earth's zone to $\sim 10^{-3}$ the mass of the earth. On the other hand, as explained elsewhere (Wetherill, 1975*a*), if θ can be maintained at a low value, i.e., ~ 1, the effects of runaway accretion will be greatly reduced.

Also, there is no reason to believe that the initial size distribution of the planetesimals, whether established by "sticking" or local gravitational instabilities will be near the steady-state distribution given by Eq. (28). In fact, any plausible initial size distribution resulting from these processes will be characterized by an approximate upper limit at which dN/dr drops rapidly, and below which this quantity varies more slowly and smoothly. Therefore the steady-state size distribution must be generated during the process of accumulation rather than be present at the outset. Insufficient work has been done to discuss further the circumstances under which an initial distribution will actually evolve into a stable steady-state power law with an approximately constant exponent, as opposed to one with a more complex size and time dependence. Numerical work by Greenberg et al. (1978; see their chapter in this book), discussed in the next section, suggests a relationship between maintenance of orbital linkage and establishment of a steady-state power-law size dependence. If this proves to be the case, it could turn out that variants of the theory which "work", i.e., lead to planets with masses and orbits similar to those observed, are characterized by a power-law size distribution and approximately constant θ, whereas others are not.

Insofar as these basic assumptions, made by Safronov, are correct, to a remarkable degree the entire history of planetary accumulation is determined by the extent to which energy is dissipated in the form of heat during collisions. Thus the time scale for accumulation of the earth ($\sim 10^8$ yr), the size of the second largest body ($\sim 10^{25}$ g), the final relative velocities (~ 3 km sec^{-1}), and the initial temperature profiles of the planets cease to be independent parameters, but are linked through the physical properties of the accumulating material.

This result is obviously too simple to be entirely correct. Insofar as it is at least to some extent a description of the processes by which a swarm can evolve into planets, it will provide a conceptual basis for thinking about more complex formulations of the problem. The identification of various approximate steady states should prove particularly valuable in providing "constants of the motion" amid otherwise overwhelmingly complex phenomena. Future progress will to a large extent consist of examination of the domain of applicability of these fundamental ideas.

IV. RECENT NUMERICAL STUDIES OF TERRESTRIAL PLANET ACCUMULATION

A. The Need for Such Studies

The analytical approach, described in the previous section, leads to great complexity when one attempts to extend it to include the fact that the planetesimals and embryos are moving in different heliocentric Keplerian orbits. In their present form, calculations using this approach do not include effects such as the eccentricity and inclination of the embryos' orbits or the consequence of merging of "feeding zones", i.e., the crossing of more than one embryo by the planetesimal orbits. Also, in the analytical work it was necessary to treat the coupled problems of the evolution of the size and velocity distributions of the planetesimals as separate problems, and work back and forth between these problems in order to find a consistent simultaneous solution. This made it very difficult to consider the problem of whether or not this consistent solution was the one into which an actual swarm will evolve.

In order to overcome these difficulties, during the past few years there has been increasing interest in constructing computer simulations of various evolutional stages in the accumulation of a planetesimal swarm. The early results of such simulations show great promise at least providing "answers" to problems difficult to treat analytically. Even more important, the results of the calculations can be expected to provide new fundamental insights into processes occurring during accumulation rather than being limited to simply "what the computer tells me."

B. Investigations Directed Toward an Understanding of the Titius-Bode Law of Planetary Distances

An early attempt at simulation of this kind was that of Hills (1970). The motivation of Hills' work was to understand the origin of the Titius-Bode "Law" of planetary distances (Titius von Wittenberg, 1766; Bode, 1972). In its most complete form, as given by Blagg (1913) and Richardson (1944, 1945), the semi-major axes of the planets are given by

$$A_n = A(1.728)^n \; [B + f(\alpha + n\beta)] \tag{31}$$

where n is the number of the planet (or satellite) (including the asteroids as $n = 5$), A, B, α, β are constants, and f is a periodic function that corrects the expression for deviations from the exact geometric progression (see discussion by Nieto, 1972).

In Hills' work the dynamic evolution of a system consisting of a small number of massive bodies was followed by two-dimensional integration of the equations of motion. Although neither accumulation nor fragmentation were included in the calculations, this investigation can be considered a precursor of a more recent treatment of a similar problem by Cox (1977). The usefulness of Hills' study was limited inasmuch as the numerical integrations were not continued long enough to achieve dynamical stability. Nevertheless, the bodies did seem to evolve toward more stable widely spaced configurations, rather than continue to accelerate one another towards higher relative velocities while remaining strongly coupled to one another. Thus, even when planetary masses are exaggerated and collisional dissipation of energy is ignored, orbital linkage is not necessarily maintained.

Dole (1970) carried out a numerical simulation, which has been extended by Isaacman and Sagan (1977) that is in a way complementary to that of Hills, in that accumulation was included, but dynamical perturbations were ignored. In these studies, again a spacing of the semi-major axes of the planets was found which was similar to those in the actual solar system, and therefore to that extent they reproduced the Titius-Bode Law. However, in this case it seems likely that this result is primarily an artifact of the calculation. Dole, as well as Isaacman and Sagan, adopted an initial swarm of dust and gas containing bodies of uniform eccentricity. Embryos were injected into this swarm and gravitationally accumulated dust (and gas if they grew large enough). Embryos in crossing orbits coalesced. The size of the region swept out by an embryo is principally limited by exhaustion of dust in orbits crossing that of its final embryo. The n'th planet will therefore sweep out a feeding zones bounded on the outer edge by dust with perihelia at a_n and on the inner edge dust with aphelia at a_n. The semi-major axis a_b of the boundary of the feeding zone between the planets n and $(n+1)$ will be given by the relationships

$$a_b(1-e) = a_n \tag{32}$$

$$a_b(1+e) = a_{n+1} \quad .$$

The ratio of the semi-major axes of the planets will be

$$\frac{a_{n+1}}{a_n} = \frac{(1+e)}{(1-e)} \quad . \tag{33}$$

Because the eccentricity of the swarm was taken to be independent of distance from the sun, the right hand side of Eq. (33) is a constant. This is the relationship required to reproduce the Titius-Bode Law (31), ignoring the correction term in brackets. In Dole's calculations, the constant value of e was taken as $e = 0.25$, leading to a "Law"

$$a_n = A(1.667)^n \tag{34}$$

not very different from that given by Eq. (31). The various values of e used by Isaacman and Sagan gave rise to different laws of the form

$$a_n = Ab^n \tag{35}$$

where the geometric progression ratio b appears to be primarily dependent on e, as given by Eq. (33). In these calculations the random manner of injection of the "seed nuclei" will lead to deviations from Eqs. (34) and (35), and such deviations are seen in the results reported. However, it seems likely that the trend to large spacings for greater semi-major axes is imposed by the assumption of constant eccentricity. Furthermore, as pointed out by Nieto (1972), Dole's result that only the more distant planets were capable of gravitationally accreting gas, is simply a result of their having large feeding zones, in accordance with Eq. (33), and therefore being able to accumulate sufficiently massive cores.

The calculations of Dole, and Isaacman and Sagan cannot be considered realistic simulations of the solar nebula, if only for the reason that the high eccentricities of the dust orbits assumed would be very quickly eliminated by gas drag caused by the large quantities of gas in the model. The "residual truth" behind these rather unrealistic models is that in the absence of gravitational perturbations or damping, the spacing of the planetary distances, i.e., the point at which orbital linkage breaks down, is determined by the initial eccentricities assumed. Actual spacings are found only with unreasonably high initial eccentricities. The work of Hills suggests that excessively large dynamic perturbations may be needed to break out of this difficulty and produce spacings as large as those found in the present solar system. Although these results fall far short of explaining the law of planetary distances, they suggest the possibility of difficulty in maintaining adequate orbital linkage in more realistic models. Achievement of a good quantitative understanding of the circumstances by which an accumulating system will remain "live" is an important short-range goal for research in this field.

C. More Realistic Simulations of More Limited Problems

Significant progress has been made by foregoing (at least at the outset) the hope of solving the major problems of planetary growth with a few hours of computer time, but rather by concentrating on more realistic simulations of limited stages in the accumulation process. Three such studies have been presented. These are the treatment of the earliest stage of accumulation of a swarm of $\sim$1 km planetesimals (Greenberg et al., 1977, see also their chapter in this book), a two-dimensional simulation of close encounters and coagulation of a swarm of $\sim 10^{26}$ g planetesimals (Cox, 1977; Cox, Lewis and Lecar, 1977), and simulation of the final 2/3 of the growth of terrestrial planet embryos from a swarm of $\sim 3 \times 10^{24}$ g bodies (Wetherill, 1977*b,c,d*).

The initial stages of accumulation. In the work of Greenberg et al. the initial swarm was modeled as consisting of 10^{12} bodies 1 km in diameter located in a band 0.053 AU in width, thus approximately representing the surface density in the present region of the terrestrial planets. The initial orbits were nearly circular and coplanar ($e \sim i = 5 \times 10^{-4}$). The bodies are considered as being divided among "bins" according to size, with an average value of eccentricity and inclination assigned to each bin. The evolution is followed in a series of time steps. In each time step the redistribution of bodies between the bins as a consequence of collision (including both coagulation and fragmentation) is calculated by a simple probabilistic expression similar to Eq. (1). The values of e and i are re-averaged for each bin.

A result of this calculation is shown in the chapter by Greenberg *et al.* in this book. At the low mass end, the deficiency of bodies <1 km in diameter is being eliminated by fragmentation of larger bodies. Bodies >1 km diameter are formed by coalescence which takes place because the initial velocities are comparable to the escape velocity (O'Keefe and Ahrens, 1977). With the passage of time larger and larger bodies are formed. This may represent an approach to a steady-state size of the sort described by Zvyagina *et al.* (1973). During this time the eccentricities and relative velocities remain low (<1 m $\sec^{-1}$) because the bodies are too small to accelerate one another appreciably.

However, in spite of the fact that almost all the mass is in the small (<10 km) bodies, a few bodies ~300 km diameter are produced at this stage. They are insufficiently numerous to produce much acceleration of smaller objects. Because of their greatly enhanced gravitational cross-section at the prevailing <1 km $\sec^{-1}$ velocities, they grow rapidly relative to the smaller bodies. Thus, large bodies are produced on a very short time scale. The calculations cannot be carried beyond ~300 km, as at this stage several assumptions involved in the calculation break down.

These results have an important bearing on the matter of achievement of Safronov's steady-state size and velocity distributions, as discussed in Sec. III. In the small mass range the beginning of an approach to a steady-state size distribution appears as well underway as would be expected after only $\sim 2 \times 10^4$ yr. The only unexpected development is the emergence of potentially "runaway" ~300 km bodies at this stage. If these large bodies are able to dominate subsequent accumulation, they could prevent the establishment of a steady state in the intermediate mass range. This in turn could lead to a premature breakdown of orbital linkage, following coalescence of adjacent large objects, leading to wide and stable spacing between $\sim 10^{26}$ g bodies.

The early emergence of these large bodies is strongly dependent on their remaining at extremely low velocities. Their rapid growth would be greatly reduced if they were accelerated to even as little as 10 m $\sec^{-1}$. Accelerations of this magnitude could be produced by giant planets if they are present at this stage. Greenberg et al. have calculated such effects caused by commensurability resonances with Jupiter (see their chapter). Otherwise, there seems to be a real possibility that accumulation will "bog down" into a system of lunar-size bodies moving in nearly concentric orbits, unless the mutual perturbations of these large bodies themselves are sufficient to maintain orbital linkage. The results of Cox, discussed below, suggest that this may not be the case.

Late stages prior to the emergence of the final embryos. The work of Cox (1977), and Cox, Lewis, and Lecar (1977) starts at a stage beyond that at which Greenberg et al. left off, and assumes a distribution of 100 bodies of 1.2×10^{26}g between 0.5 and 1.5 AU with initial eccentricities distributed randomly between zero and 0.15. The two-dimensional evolution of this swarm is calculated by numerical integration of the equations of motion, or by use of an approximation to this developed by these authors. Bodies which approach within the sum of their geometric radii are assumed to coalesce. Fragmentation is not included.

The integrations are carried out until orbital linkage breaks down, i.e., the system becomes gravitationally "dead". At this point a system of planets with relative masses and spacings are found which are not dissimilar to those of the present terrestrial planets.

However, when a maximum initial eccentricity of 0.05 is used instead of 0.15, a larger number of small planets, more closely spaced, are produced. This is reminiscent of the calculations of Isaacman and Sagan. The mutual perturbations of the bodies do not appear to be sufficient to keep them in crossing or nearly crossing orbits and thereby maintain orbital linkage as long as is necessary to reproduce the present planetary system. It is possible that this is an artifact of the two-dimensional calculation. In two dimensions at a given number of planetary radii, the ratio of the number of close strongly perturbing encounters to the number of impacts increases linearly with the encounter distance. In three dimensions this ratio increases with the square of the encounter distance. Therefore, in three dimensions it may be expected that mutual perturbations will be more effective before coalescence occurs. On the other hand, these results may again suggest that there is a real problem with the efficacy with which the mutual planetesimal perturbations can maintain orbital linkage over the full range of planetary growth.

Late stages subsequent to the emergence of the final embryos. During 1977 the present author has carried out numerical studies of the final 2/3 of the accretion of the terrestrial planets. The motivation behind the work was to gain insight into the modifications required in Safronov's theory, as a consequence of the real planetary embryos being in non-circular inclined orbits. In Safronov's results the value of θ increases, and the relative velocity of the swarm decreases, at the stage in planetary growth when the gravitational perturbations by the embryo become dominant relative to the mutual perturbations of the planetesimals. This is a consequence of the constant value of the Jacobi parameter in the restricted 3-body problem, which precludes increase in relative velocity solely through perturbations by the embryo.

The restricted 3-body approximation can break down as a consequence of a finite eccentricity of the embryos. Therefore acceleration could still occur, and could lead to planetesimals crossing the orbits of both Earth and Venus, and even more serious breakdown of the restricted 3-body approximation. Semi-quantitative calculations (Wetherill, 1976) suggested that as a consequence of these effects, under some circumstances the emergence of closely-spaced dominant embryos in the positions of Earth and Venus may have led to an *increase* in the relative velocities of the residual swarm, rather than to a decrease as calculated by Safronov.

In order to pursue this question more quantitatively, a numerical simulation of this final stage of growth was carried out. A principal tool in this work is a program based on that of Arnold (1965), which has been extended to encompass planetary growth by the simultaneous tracing of the orbital evolution of a set of initial orbits chosen to represent the planetesimal swarm.

The general operation of this program is as follows:

1. Initial masses and orbital elements of the planetary embryos are specified (usually the present orbits of the terrestrial planets).
2. Orbital elements are specified for a number (usually 100) of planetesimals in initially embryo-crossing orbits. The distribution of elements of this swarm is chosen to conform with the considerations of energy and angular momentum conservation discussed in Sec. II.
3. The Öpik (1951) collision formula is used to calculate the probability distribution of the time of an encounter of each planetesimal with each embryo, and a time and planetary embryo for the next encounter of each planetesimal is chosen by a random procedure weighted by these probabilities.
4. Starting at time zero, the planetesimal with the first encounter time is identified.
5. The details of this encounter (distance, azimuth, and position of the embryo in its orbit) are chosen by appropriately weighted random selection.
6. The change in orbital elements of both the planetesimal and the embryo caused by gravitational perturbation during the encounter are calculated. The elements specified in Steps 1 and 2 are updated in accordance with these changes.
7. If the encounter happened to be within one planetary radius (as augmented by its gravitational field), an impact is scored and the mass of the embryo is increased. In addition, encounters within a specified number of planetary radii (usually 5) are scored as both impacts and encounters in accordance with a procedure that increases the statistical

significance of the results of these calculations (Wetherill, 1975*b*). This procedure can be shown to be rigorously valid in the usual Monte Carlo calculations in which the orbital evolution of a single planet-crossing body is following. It is only approximately valid in the case of multiple evolution and embryo growth. Comparison with calculations not using this procedure demonstrate that this approximation is not a bad one, but requires further study to estimate its validity quantitatively. The net effect of the procedure is to increase the mass of the planets in steps of 1/2500 of the total initial mass of the swarm.

8. The changes in orbital elements of the planetesimal resulting from mutual planetesimal collision since its last planetary encounter are calculated. The "drag" parameter assumed here is the principal free parameter of the calculation. In the calculations carried out thus far, this drag is assumed to be proportional to the time elapsed since the last encounter of this particular planetesimal with an embryo, to the number of planetesimals in the swarm with semi-major axis between the perihelion and aphelion of the chosen planetesimal, and to the relative velocity of the planetesimal with respect to a body in circular orbit at the semi-major axis of the planetesimal.
9. The changes in the orbital elements of the embryo resulting from impact are calculated, making use of conservation of energy (including heat) and angular momentum.
10. The time and planet for the next encounter of this particular planetesimal are selected as in Step 3. This is added to the list generated by Step 3.
11. As in Step 4, the next nearest encounter time and its associated planet and planetesimal are identified.
12. The energy and angular momentum of the planetesimal are not conserved in the "damping" calculation of Step 8, as the reaction of the swarm must be included in the total balancing of these quantities. The effect of this is approximated by calculating the total change in energy (including heat) and angular momentum following each "pass" through the program. These quantities are then stored in energy and angular momentum "banks", the contents of which are aliquoted to each planetesimal the next time it is selected. (In some cases, a procedure for permitting secular perturbations to maintain the eccentricity of the embryos in the face of their tendency to be more circular was introduced; non-conservation of angular momentum arising in this way was not "banked" as it was transferred outside of the system.)
13. Steps 5 through 10 are repeated for this planetesimal and planet.
14. This procedure is continued until every planetesimal has exhausted its allowed number of impacts (usually 25) in accordance with the procedure mentioned in Step 7, has been ejected from the solar system in a hyperbolic orbit, or has been transferred to a long-lived orbit in the

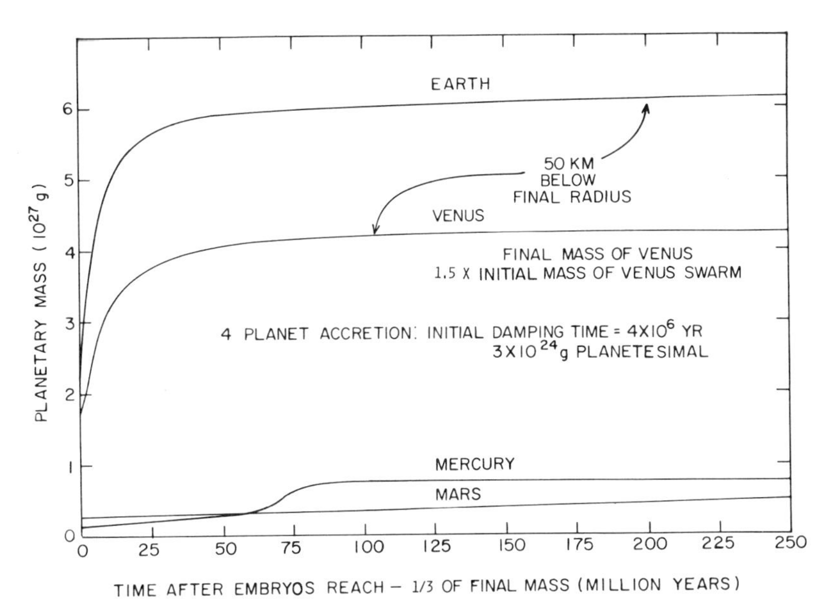

Fig. 3. Results of a typical simulation of the final 2/3 of terrestrial planet growth. Accumulation of planets is nearly complete after $\sim 10^8$ yr, but a long-lived "tail" is imposed on the growth of Earth and Mars by transfer of Earth planetesimals to long-lived Mars-crossing orbits. As a result of stochastic fluctuations in the extent of "swarm-stealing", the calculated final planet masses do not agree with actual values.

asteroid belt. Ejection or asteroid-belt orbit occurs only rarely (~1% for reasonable choices of initial orbit and dissipation parameter.

15. The final mass and orbits of the planets can be compared with those observed at present. The size of each planetary embryo as a function of time is plotted, as well as the geocentric velocity distribution of the swarm as a function of time. The calculation produces data which can be used to investigate such questions as the extent to which planetesimals with orbits initially in the vicinity of a particular embryo contribute to the growth of other embryos, the extent of migration of the semi-major axes of the planets during growth, and the relative roles of mutual collision and planetary impacts as dissipative mechanisms.

The results of a typical calculation of this kind are shown in Figs. 4, 5, and 6. The time scale for growth is $\sim 10^8$ yr, with most of the growth occurring during the 25×10^6 yr after the starting time, i.e., when the embryo's are 1/3 grown. The effect of Mars imposing a long-lived tail on the growth of the earth as discussed at the end of Sec. II is observed. Phenomena which might be invoked to eliminate this tail can be suggested (Wetherill, 1977*a*) but sound too specialized at this stage of understanding, and this difficulty should be kept in mind.

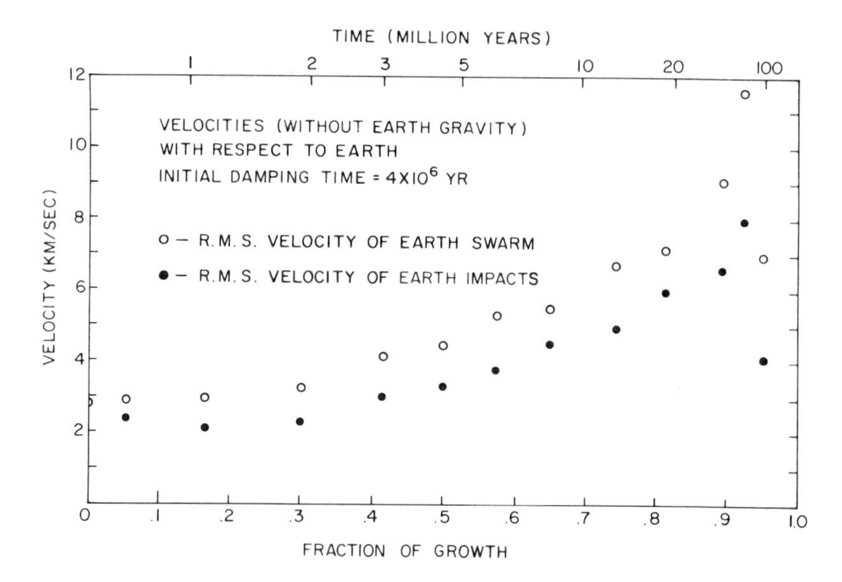

Fig. 4. Increase of root-mean-square velocities Earth swarm and Earth impacts during the growth of Earth. The erratic fluctuations during the final 10% represent small-number statistics after Earth's swarm is depleted by capture and replenished erratically by Mars-crossing bodies.

In these calculations the dissipation parameter is treated as a free coefficient in the above Step 8. The particular value chosen for the example shown is about the same as that estimated in Sec. II. Accumulation times are to some extent dependent on the choice of this coefficient. However, within the range of larger values permitted in order to maintain adequate orbital linkage (about a factor of 10), the time scale remains $\sim 10^8$ yr. Mid-point velocities (Fig. 5) remain near their initial value, but rise to ~ 6 km sec^{-1} during the final stages of accumulation. Larger values of dissipation result in lower velocities; just how low is dependent on whether or not secular perturbations are permitted to maintain the eccentricities of the embryos at ~ 0.01. If these eccentricities fall too low, the crossing of two planets becomes unusual and relative velocities fall to ~ 1 km sec^{-1}.

The collisional energy lost as heat for the final 2/3 of growth was found to be 5.9%, of which 73% was lost in planetesimal-planetesimal collisions. In this case only 1.3% of the initial angular momentum was lost, a lower value than estimated. This appears to be a consequence of less need for secular perturbations to supply angular momentum, probably as a consequence of a greater share of the necessary loss being provided by the swarm, rather than by the embryos. One percent of the mass was found to escape to Jupiter and beyond.

The final mass found for the earth is somewhat high, that for Venus

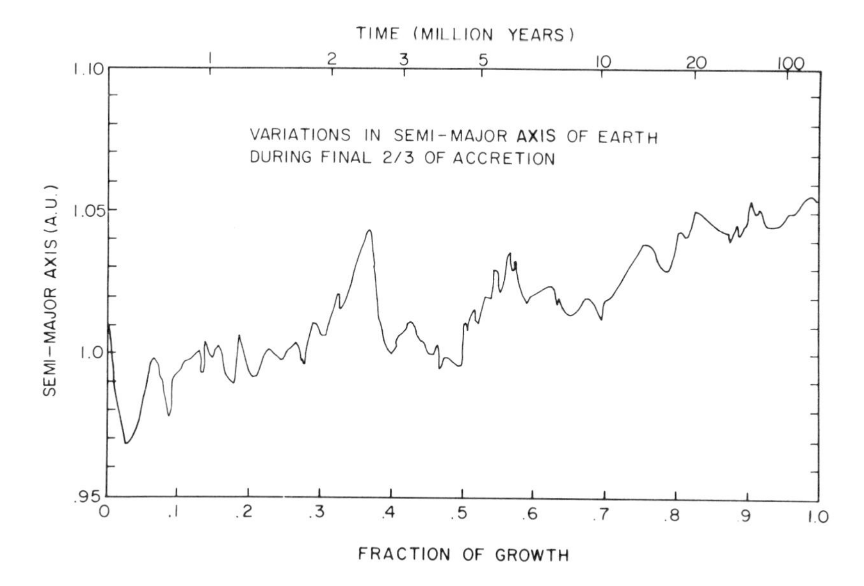

Fig. 5. Variation in semi-major axis of Earth resulting from impacts and reaction to perturbations of planetesimals. The final semi-major axis (1.05 AU) differs from the actual value as a consequence of fluctuations in the planetary masses and angular momentum distribution.

rather low. Stochastic fluctuations in these values, as well as for the final semi-major axes of these planets (Fig. 6), are found from one calculation to another. These arise from instabilities wherein one of the embryos can grow faster than the others by capture of planetesimals into its zone from the zone of the other body. Venus normally does this more than Earth, and in the case shown here the final mass of Venus was 50% greater than the initial mass of Venus' swarm (including its embryo). Even so, Venus did not grow enough. In other cases Venus "overgrew". These instabilities never led to gross "runaways" but cause ~20% fluctuations in final masses. It seems likely that similar instabilities would have occurred in the real solar system if the planets accumulated in this way.

Much larger fluctuations are found for Mars and Mercury. Indeed, for the case shown the relative mass of these two planets is inverted. The reverse situation, where Mars overgrows relative to Mercury, is also found, as well as more nearly correct values. These phenomena arise from the small number of bodies required to make a major change in the masses of these planets, exaggerating the instabilities discussed for Earth and Venus. Its importance in the real solar system will depend on the actual mass distribution of the planetesimals.

An important result found in this work is that when the effect of the bodies being in Keplerian orbits is included, as is the case here, conservation of angular momentum plays a role comparable to collisional damping in driving orbits toward circularity. This effect will be reduced by external secular perturbations, provided that adequate perturbers (e.g., giant outer planets) exist at this stage.

This work is in a preliminary state. At present the validity of the simple manner in which damping (Step 8) and angular momentum and energy conservation (Step 15) are handled is being investigated. In any case, it does seem likely that if the terrestrial planetary system can make it to the stage in which the final embryos lose their orbital linkage, the residual swarm can then complete its growth. However, this could require the presence of something equivalent to Jupiter, unless the swarm is more deficient in small bodies (less dissipative) than assumed.

V. SUMMARY AND PROGNOSIS

Although it is premature to be at all certain about these matters, the theoretical and numerical investigations carried out to date permit some conjectures concerning the circumstances under which the terrestrial planets could grow by sweeping up of planetesimals from a gas-free swarm in the manner discussed in this chapter.

It seems likely that the key question will be whether or not dominant, non-linked, embryos can avoid emerging prematurely. Factors favoring such prematurity include high-collisional damping, strict conservation of angular momentum, size distributions favoring small bodies, and the largest body becoming markedly larger than the remaining bodies at an early stage. It seems likely that plausible values of the relevant parameters will exist which will avoid such prematurity, but possibly only if some sort of "Jupiter" is formed early. Perhaps the most fruitful approach at present will be to press forward our understanding to the point where it is possible to make a better estimate of the probability of sweeping-up growth turning out to be correct without the existence of Jupiter. If it does seem probable that an evolution of this kind *can* occur, then one can justifiably interpret imperfectly definitive observational and experimental evidence as increasing the plausibility that it *did* occur. If otherwise, it would be worthwhile to try and find natural ways to form Jupiter on a short ($\lesssim 10^7$ yr) time scale. At the same time, it will be important for all workers in areas for which these considerations are relevant, to consider seriously the possibility that this type of accumulation occurred, and include the consequences thereof in their thinking. It must also be remembered that no alternative mode of accumulation has yet been subjected to this degree of scrutiny, and that for these alternative "scenarios" even greater problems than those alluded to above are likely to arise when this is actually done.

REFERENCES

Alfvén, H. 1954. *On the Origin of the Solar System.* London: Oxford University Press.
Alfvén, H., and Arrhenius, G. 1976. *Evolution of the Solar System.* NASA SP-345. Washington, D.C.: U.S. Government Printing Office.
Arnold, J. R. 1965. *Astrophys. J.* 141: 1536.
Barrell, J. 1918. *The Evolution of the Earth and its Inhabitants.* New York: Yale University Press.
Blagg, M.A. 1913. *Mon. Not. Roy. Astr. Soc.* 73: 414.
Bode, J. E. 1772. *Anleitung zur Kenntniss des gestirnten Himmels.* (2nd ed.). p. 462. Leipzig.
Brouwer, D. and van Woerkom, A.J.J. 1950. *Astron. Papers U. S. Naval Obs. Naut. Almanac Office* 13: 85.
Cameron, A. G. W. 1973. *Icarus 18: 407.*
____. 1977. Center for Astrophysics Preprint No. 841. Submitted to *Moon and Planets.*
Cameron, A. G. W., and Pine, M. R. 1973. *Icarus* 18: 377.
Chamberlin, T. C. 1904. *Carnegie Inst. Wash. Year Book* 3: 195.
Cox, L. P. 1977. *Bull. Am. Astron. Soc.* 9: 544.
Cox, L. P., Lewis, J. S.; and Lecar, M. 1977. *Icarus.* In press.
Dohnanyi, J. W. 1969. *J. Geophys. Res.* 74: 2531.
Dole, S. H. 1970. *Icarus* 13: 494.
Edgeworth, K.E. 1949. *Mon. Not. Roy. Astr. Soc.* 109: 600.
Gault, D. E.; Shoemaker, E. M.; and Moore, H. J. 1963. NASA TN-1767.
Goldreich, P., and Ward, W. R. 1973. *Astrophys. J.* 183: 1051.
Greenberg, R.; Wacker, J.; Chapman, C.R.; and Hartmann, W.K. 1978. *Icarus.* In press.
Grossman, L. 1972. *Geochim. Cosmochim. Acta* 36: 597.
Gurevich, L.E., and Lebedinskii, A.I. 1950. *Seriva Fizich.* 14: 765.
Hallam, M., and Marcus, A. H. 1974. *Icarus* 21: 66.
Hanks, T.C., and Anderson, D.L. 1969. *Phys. Earth Planet. Int.* 2: 19.
Hartmann, W. K., and Davis, D. R. 1975. *Icarus* 24: 504.
Hayashi, C.; Nakazawa, K.; and Adachi, I. 1977. *Publ. Astron. Soc. Japan* 29: 163.
Hills, J. G. 1970. *Nature* 225: 840.
Isaacman, R., and Sagan, C. 1977. *Icarus* 31: 510.
Kuiper, G. P. 1951*a*, *Proc. Nat. Acad. Sci.* 37: 1.
____. 1951*b*. In *Astrophysics* (J.A. Hynek, ed.), Ch. 8. New York: McGraw Hill.
Larimer, J. W. 1967. *Geochim. Cosmochim. Acta* 31: 1215.
Levin, B.J. 1972. *On the Origin of the Solar System* (H. Reeves, ed.), p. 2. Paris: C.N.R.S.
____. 1978. *Moon and Planets* (special Protostars and Planets issue). In press.
Lewis, J. S. 1972. *Earth Planet. Sci. Lett.* 15: 286.
Lynden-Bell. D., and Pringle, J.E. 1974. *Mon. Not. Roy. Aster. Soc.* 168: 603.
Lyttleton, R.A. 1961. *Mon. Not. Roy. Astron. Soc.* 122: 399.
Marcus, A.H. 1965. *Icarus* 4: 267.
____. 1969. *Icarus* 11: 76.
McCrea, W.H. 1960. *Proc. Roy. Sci. London* 256A: 245.
____. 1972. *On the Origin of the Solar System.* (H. Reeves, ed.) Paris: C.N.R.S.
Nieto, M.M. 1972. *The Titius-Bode Law of Planetary Distances: Its History and Theory.* Oxford: Pergamon.
O'Keefe, J.D., and Ahrens, T.J. 1977. *Science* 198: 1249.
Öpik, E.J. 1951. *Proc. Roy. Irish Acad.* 54A: 165.
Richardson, D.E. 1944. *Pop. Astron.* 52: 497.
____. 1945. *Pop. Astron.* 53: 14.
Ringwood, A.E. 1970. *Earth Planet. Sci. Lett.* 8: 131.
Safronov, V.S. 1958. *Rev. Mod. Phys.* 30: 1023.
____. 1960. *Voprosy Kosmogonii* 7: 59.
____. 1969. *Evolution of the Protoplanetary Cloud and Formation of the Earth and Planets.* Moscow: Nauka Press; NASA TT F-677, 1972.
____ 1975. *Proceedings of the Soviet-American Conference on the Cosmochemistry of*

the Moon and Planets. Moscow: Nauak Press; (transl. in English) NASA SP-370. p. 797, 1977.
Safronov, V.S., and Ruskol, E.L. 1956. *Dokl* AN SSSR 108 (3): 413.
——. 1957. *Voprosy Kosmogonii* 5: 22.
Shmidt, O. Yu. 1957. *Four Lectures on the Theory of the Earth's Origin* (3rd ed.). Iz datel'stvo AN SSSR.
ter Haar, D. 1950. *Astrophys. J.* 111: 179.
Titius von Wittenburg, J.D. 1766. *Bertrachtung über die Natur,* vom Herrn Karl Bonnet, p. 7. Leipzig. (Translation into German from French.)
Toksöz, M.N., Solomon, S.C.; Minear, J.W.; and Johnson, D.H. 1972. *The Moon* 4: 190-213; 5: 249.
Turekian, K.K., and Clark, S.P., Jr. 1969. *Earth Planet. Sci. Lett.* 6: 346.
Urey, H.C. 1951. *Geochim. Cosmochim. Acta* 1: 209.
——. 1952. *The Planets: Their Origin and Development.* New Haven: Yale University Press.
——. 1959. *J. Geophys. Res.* 64: 1721.
Weidenschilling, S.J. 1974. *Icarus* 22: 426.
——. 1975. *Icarus* 26: 361.
——. 1976. *Icarus* 27: 161.
——. 1977*a*. *Comets, Asteroids, Meteorites* (A.H. Delsemme ed.), p. 541. Toledo: University of Toledo Press.
——. 1977*b*. *Mon. Not. Roy. Astr. Soc.* 180: 57.
——. 1977*c*. *Astrophys. Space Sci.* 51: 153.
Wetherill, G.W. 1975*a*. *Proc. Lunar Sci. Conf.* 6: 1539. Pergamon: New York.
——. 1975*b*. *Proc. Soviet-American Conf. on Cosmochem. of the Moon and Planets* p. 411. Moscow: Nauka Press. (NASA SP-370 p. 553, 1977.)
——. 1976. *Proc. Lunar Sci. Conf.* 7: 3245. New York: Pergamon.
——. 1977*a*. *Proc. Lunar Sci. Conf.* 8: 1. New York: Pergamon.
——. 1977*b*. *EOS* 58: 428.
——. 1977*c*. *Bull. Am. Astron. Soc.* 9: 545.
——. 1977*d*. *Carnegie Inst. Wash. Year Book* 76. p. 761.
Whipple, F.L. 1973. In *Evolution and Physical Properties of Meteorites,* p. 355. NASA SP-319. Washington, D.C.: U.S. Government Printing Office.
Williams, I.P., and Crampin, D.J. 1971. *Mon. Not. Roy. Astr. Soc.* 152: 261.
Zvyagina, Y.V.; Pechernikova, G.V.; Safronov, V.S. 1973. *Astron. Zh.* 50: 1261.

THE ACCRETION OF PLANETS FROM PLANETESIMALS

RICHARD GREENBERG, WILLIAM K. HARTMANN,
CLARK R. CHAPMAN, AND JOHN F. WACKER
Planetary Science Institute

The planetesimal hypothesis for aggregating nebular condensates into full-size planets has received wider acceptance than the protoplanet model because it seems viable and seems to fit meteoritic and planetary evidence. We have devised a numerical simulation of collisional interaction and gravitational aggregation that readily grows a few discrete 1000-km planets from a swarm of kilometer-scale planetesimals. But uncertainties remain in understanding how the discrete small planets so produced eventually aggregate into full-size planets, and also how the initial condensates grow into kilometer-scale planetesimals. At the end of our simulation, the small planets become isolated in circular orbits so that further accretion is impeded. We discuss several mechanisms which may permit later-stage growth, including the gravitational effects of a first-formed planet elsewhere in the nebula. Orbital resonances would impede growth of small planetesimals, but could enhance the later-stage growth of larger bodies. The earliest stage of evolution, from condensate grains to planetesimals, requires aggregation mechanisms different from those applicable at kilometer scales. Gas effects, including Brownian motion and drag, inhibit grain aggregation unless grains are exceptionally "sticky." Perhaps local gravitational instabilities are required to produce first-generation planetesimals. Accretion of planetesimals and planets may proceed in a similar fashion in other single-star systems, but maintenance of high velocities, in a binary system for instance, would yield comminution. Comminuting planetesimals would produce exceptional quantities of dust that might be injected into interstellar space.

From a host of observational and theoretical considerations, developed since the speculations of Kant and Laplace, it is now believed that planetary systems form from a swirling disk of dust and gas that accompanies the formation of appropriate stars. There is some direct evidence about the properties of nebular disks with solid grains that may be forming planetary systems at the present time. The solar system, the only case where we know that some such planet-forming process once occurred, gives more evidence, but to reconstruct the process we must reason backwards more than four and a half billion years from the present physical and chemical state of the solar system to the early dusty state. In this chapter, we discuss attempts to model the physical processes in the evolution of planetary material from initial nebular models that are in accord with known astrophysical data and processes.

Modeling of the subsequent evolution of nebular disks has tended to fall into two rather separate categories – the "protoplanet model," in which the nebula fragments directly into a few giant blobs or rings that subsequently condense into planets, and the "planetesimal model," in which condensed grains aggregate into larger grains and gradually accrete through stages of progressively larger bodies all the way to full-scale planets. The actual planet-forming process may include elements of both approaches; for instance, the influential treatment of Safronov (1972; also see Safronov and Ruzmaikina chapter in this book) invokes local gravitational instabilities to produce planetesimals of kilometer scale, but then relies on direct accretion through collisional interaction to produce full-scale planets.

The protoplanet model is applicable if the nebula is subject to very large scale gravitational instabilities on a time scale more rapid than direct accretion of grains or planetesimals. Earlier work on the protoplanet model has been reviewed by Williams (1975); more recently Cameron (see his chapter in this book) has described a model of the solar accretion disk that exhibits large-scale instabilities.

The planetesimal approach, however, has received considerable attention in the 1970's. It has seemed especially attractive to planetary scientists because many physical and chemical properties of meteorites (e.g., the absence of noble gases, cooling time data, evidence for collisional shock) can be most readily understood in terms of small parent bodies. Further, the cratered surfaces of planets demonstrate that at least the final stages of accretion involved planetesimal-sized bodies, and Wetherill (1977, see also his chapter in this book) has shown that the dynamical survival time scales of such bodies agree with the observed decay rates for the early intense lunar cratering. Safronov (1972) has shown that even planetary obliquities can be understood as effects of the last collisions with large planetesimals. (See also the comments by Wasson and Kaula in support of the planetesimal hypothesis at the end of this chapter.) Although some arguments in favor of the planetesimal hypothesis assume that meteorites, asteroids, and other small bodies present in the solar system today are remnants of the planetesimals that formed the larger planets, it is conceivable that meteoritic material might be a natural outcome of a protoplanet evolutionary history.

Without attempting a judgement on the relative likelihood of the protoplanet or planetesimal scenarios, we will hereafter concentrate on a discussion of the latter approach. As we shall show, not only is this approach more tractable but it seems viable, at least for aggregating planets from bodies of roughly kilometer scale. Still problematical is the *early stage* of accretion in which condensate grains accumulate into km-scale planetesimals. Some workers (e.g., Safronov, 1972; Goldreich and Ward, 1973) believe that the most likely process of early-stage accretion is direct formation due to local gravitational instabilities in a thin dust layer formed at the midplane of the nebular disk. We shall address the question whether a grain-by-grain process of accretion can occur instead. We shall show that the force by gravity is far too weak to overcome effects of the gas which necessarily accompanies the earliest stages of nebular evolution; thus if grain-by-grain accretion is to compete successfully with local gravitational collapse, grains must be exceptionally "sticky".

We define the *intermediate stage* of accretion as covering the transition from an essentially continuous medium of small planetesimals to a system of widely separated, discrete, small planets. This stage extends from about 1-km diameter – the size of Goldreich-Ward planetesimals – to about 1000 km, the size at which our own simulation of collisional interaction and gravitational aggregation (to be described in detail below) breaks down due to the isolation of the largest bodies in nearly circular orbits. Although we find some important differences with the results of Safronov for this regime, the planetesimals readily and rapidly form the required planetary embryos.

The *final stage* of accretion involves the sweeping up of interplanetary debris by the discrete embryos. This stage has been modeled by several workers, e.g., Weidenschilling (1974), Hartmann and Davis (1975), and Hayashi *et al.* (1977). In this chapter we address several problems concerning the interface between the intermediate and final stages. For instance, the emergence of a few 1000-km embryos in the intermediate stage may be insufficient to guarantee further accretion. We describe how the gravitational influence of a first-formed planet (e.g., Jupiter) located elsewhere in the system may assist this transition.

The whole planetesimal approach to planetary formation involves a transition from an early, essentially *astrophysical* phase involving opaque dusty nebulae around infrared stars, grain formation, and gravitational instabilities in a nearly continuous medium to a late, essentially *planetary* phase which may be studied from the impact record of the last accreted planetesimals onto a few discrete planets and from the properties of remaining interplanetary debris. The intermediate stage forms a natural chronological interface between the astrophysicists' and the planetary scientists' points of view, and hence is central to the topic of this book.

For the analytical treatment of the intermediate phase, Safronov decoupled the evolution of the size distribution of planetesimals from the evolution of their random relative velocities. He used a size distribution with an assumed power law and the exponent selected such that most of the mass was in the largest bodies, in order to show that relative random velocities among planetesimals approach the order of the escape velocities of the largest size bodies. This equilibrium solution balances the damping effects of energy loss due to collisions and gas drag against the gravitational stirring by close encounters. The stirring is ultimately driven by the Keplerian shear (differential Keplerian angular velocities) of the system. Safronov in turn used this equilibrium solution to study the evolution of the size distribution.

In order to obtain a coupled treatment of the size and velocity distributions, we have developed a computer simulation of collisional evolution. This numerical approach permits us to incorporate an algorithm for the results of collisions that is not restricted by artificial simplifications required for analytic solution. Instead, we can incorporate as detailed a model of collisional outcomes as is commensurate with our knowledge of the behavior of materials upon collisional impact. Moreover, the algorithm can be updated as new experimental data become available.

In this chapter we briefly outline our numerical model, including a description of the algorithm for collisional outcomes. More details are discussed by Greenberg *et al.* (1978) and Hartmann (1978). We delineate some of the assumptions of our model to clarify the conditions under which our results would represent the actual behavior of growing planetesimals, and we point out areas in which further experimental and analytical work would help us to make the simulation more realistic. We shall discuss whether or not collisional growth can compete with gravitational instability as a means of rapid aggregation of grains during the early stage of accretion. Other experiments show some conditions under which comminution and generation of interstellar dust grains become possible. We will show that intermediate-stage planet growth is a very rapid phenomenon for a wide range of plausible initial conditions and material properties. The results indicate that evolution proceeds along lines very different from those that Safronov envisioned, but the system does make the transition from an essentially continuous medium to a system containing about one discrete planet $\sim$1000 km in diameter per AU^2 of the planetesimal disk. Such planets might serve as seeds for final stage accretional growth of full-size planets.

THE MODEL

In our simulation, the particle size distribution is represented by the number of particles per logarithmic increment in diameter. Typically, we consider about 20 increments spanning the size range of a few meters to a few thousand kilometers. Associated with each size increment (or "bin") are an orbital inclination, i, and eccentricity, e, which represent respectively the out-of-plane and in-plane components of the particles' random velocities with respect to purely circular orbits. We consider the events in a series of small time steps. In each step, the probable number of collisions between particles in each pair of size bins is computed using a "particle-in-a-box" estimate. The "box" may be thought of as moving in a circular Keplerian orbit at the particular distance from the sun under consideration and it covers some narrow increment of semi-major axes. The number of collisions for a given pair of size bins is proportional to the number of particles per area of the disk in each size bin, the mean relative velocity, the collision cross-section (enhanced by the gravitational cross-section) and the length of time; it is inversely proportional to the thickness of the disk which depends on the inclinations.

The relative motion of particles may be thought of as having two components. The first component is random motion, analogous to thermal motion in gas dynamics. The random motion is represented by orbital eccentricities and inclinations. A particle with $e = i = 0$ would have no random motion. The second component of relative motion is a systematic one; across the box under consideration in the direction radial to the sun, there is on the average a uniform variation of velocity parallel to the box's orbital motion. This systematic component is due to the increase in orbital period with semi-major axis as described by Kepler's third law. We call this relative motion "Keplerian shear." The shearing behavior would be most plainly seen in the limit where all particles have $e = i = 0$, so that the only relative motion of particles in the box would be due to this systematic component. But, even when e and i are significant, Keplerian shear is superimposed on the random motion. The gravitational deflection of particles passing one another due to Keplerian shear may be important, as we discuss below, for stirring up random velocities. However, for computing collision probabilities and collision velocities, we must assume that random motions are sufficiently large that they dominate over Keplerian shear. This assumption is necessary because the classical particle-in-a-box analysis is not amenable to inclusion of systematic motion in determination of collision probabilities and velocities. Our numerical results show that this assumption holds true throughout intermediate-stage growth, and breaks down only at the end of that stage when ~1000-km bodies in nearly circular orbits are formed.

Given the masses and velocities involved in each collision, we determine the outcome of the collision in terms of resulting size distribution of ejecta, debris, or fragments and resulting relative velocities, using all applicable experimental data. The population distribution is then adjusted according to the number and outcome of collisions of each of several types discussed below. The values of e and i are modified by averaging in the new relative velocities of those particles which come out of collisional events.

The e and i values are further modified by computing the average gravitational deflection per time step as each size particle moves through the field of particles of every other size. In computing this gravitational stirring effect, Keplerian shear velocities must be included, as well as random relative velocities, because Keplerian shear is ultimately the only energy source capable of balancing the velocity damping effects of collisions. This conversion of systematic to random motion is computed according to a method suggested by Ward (1976), and described in more detail by Greenberg *et al.* (1978). By enhancing random motion, Keplerian shear can contribute indirectly to determination of collision probabilities and impact velocities in subsequent time steps, although it does not contribute directly to the calculation of relative velocities, as discussed above.

An implicit assumption in our model is that a single characteristic e and i for each size bin provides adequate representation of the random velocity distribution. Ideally, we would like to have velocity distribution for each size bin, but that would add an additional dimension to the problem and is beyond our resources for numerical study. Nevertheless, the present model is much more realistic than previous ones in which a single random velocity was used to characterize all sizes.

To date, we have assumed that there is no gaseous medium during the intermediate stage of accretion. While this assumption is plausible *a priori,* we shall show that our results give such rapid aggregation that gas present at the end of the early stage may not have had time to be removed before intermediate-stage growth begins.

Our algorithm for the outcome of collisions is based on the available body of theoretical and experimental work with rocky materials. Data are sparse compared with the wide range of masses and velocities required (e.g., Gault *et al.,* 1963; Stöffler *et al.,* 1975). In order to help correct this situation, we have performed experiments (Hartmann, 1978) which extend to the previously neglected low-velocity regime (1-1000 m sec^{-1}). Nevertheless our model still requires considerable extrapolation, which must be reconsidered as further relevant data become available.

The details of the collision-outcome algorithm are discussed elsewhere (Greenberg *et al.,* 1978), but a general summary is appropriate here. The results of impact are divided into four categories.

1. Outcome 1 is pure rebound without chipping or cratering on either surface. This occurs for impact velocities less than some critical value, 0-40 m sec^{-1}, depending on the material properties. The coefficient of restitution must be specified and might range from 0.8 for solid rock down to 0.01 for a regolith.
2. For higher impact velocities, the surfaces of both impacting bodies are locally damaged, e.g., chipped or cratered (Outcome 2). Parameters must be estimated for the size of craters (taken proportional to impact energy), ejecta size, and velocity distribution (taken on the basis of experiments to follow power laws), and the coefficient of restitution of the colliding bodies (somewhat larger than for Outcome 1).
3. For still higher velocities, or for colliding mass ratios far from unity, the

smaller body receives energy per volume that exceeds its "impact strength" and catastrophic fragmentation therefore occurs (Outcome 3). Typical experimental values of impact strength range from 3×10^7 ergs cm^{-3} for solid rock down to $\sim 10^4$ ergs cm^{-3} for weak material similar to carbonaceous chondrites. We must apply velocity and size distributions for the fragmental debris as well as for the crater ejecta from the larger body.

4. Finally, for Outcome 4, the collision is so energetic that the impact strength of even the largest body of the pair is exceeded. Both bodies fragment catastrophically and the fragments are redistributed in size and velocities.

In all collisions the actual impact velocity consists of the mean relative approach velocity computed from the values of e and i, plus the colliding bodies' mutual gravitational acceleration. The separation velocities of fragmental debris, crater ejecta, and rebounding bodies are all compared with the escape velocity of the colliding system to determine how much material escapes and how much is gravitationally bound. This gravitational binding is the only means of aggregation in our model; no *ad hoc* sticking mechanisms are assumed.

INTERMEDIATE-STAGE GROWTH

For initial conditions at the end of the early stage of aggregation, we take 10^{12} bodies of diameter 1 km, as suggested by the gravitational instability calculations of Goldreich and Ward (1973). This corresponds to a surface density for the particulate disk of ~ 8 gm cm^{-2}, the value obtained by smearing the mass of the terrestrial planets over their portion of the solar system. The central body is given the sun's present mass while the evolution is considered at 2.7 AU, but similar results should apply at 1 AU as well. For initial relative velocities, we have selected values a few times escape velocities of the 1-km bodies. Gravitational stirring would prohibit much lower velocities, and higher velocities would yield comminution rather than growth, as we discuss in a later section. (Our initial velocities are somewhat higher than an equilibrium value, but we will show that our system relaxes toward typical escape velocities in most cases.) The initial values of e and i were taken to be 5×10^{-4}, which correspond to random velocities of about 7 m sec^{-1}, ten times the escape velocity.

The first numerical experiment was performed with material parameters intermediate between the extremes of solid basalt and loosely bonded regolith. The resulting evolution is shown in Figs. 1 and 2. For the first 14,700 yr, the evolution consists of collisions which crater the 1-km bodies, damping their random velocities, eroding them slightly, and generating a small amount of ejecta. Once the relative velocities of the 1-km bodies become low enough, about five times their escape velocities, they begin to accrete one another because their low rebound velocities permit gravitational binding. In the next $\sim$10,000 yr, aggregation of a few 500-km

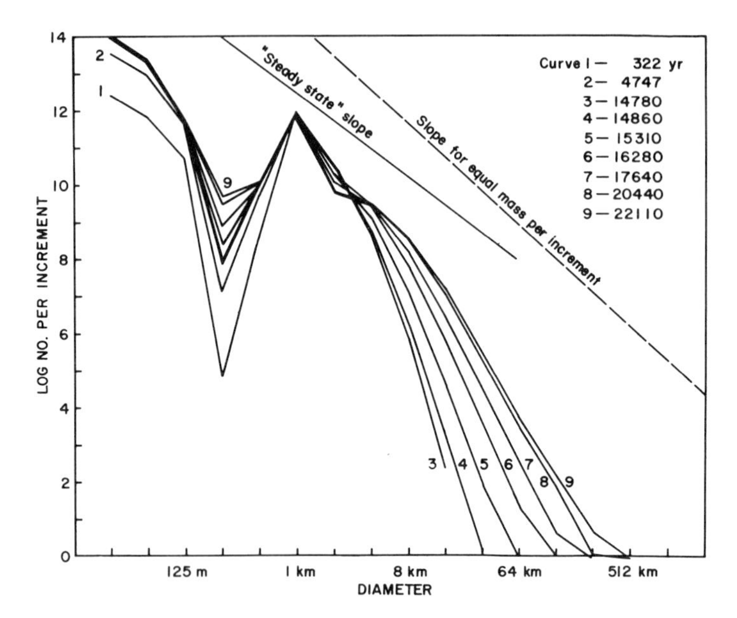

Fig. 1. Particle size distribution as a function of time for a material intermediate between loosely bonded regolith and solid rock. The number of particles is approximately per AU^2 of the disk. Each increment is a factor of 2 in diameter. Comparison with the slope that corresponds to equal mass per log increment in diameter shows that the bulk of the mass remains in bodies ~ 1 km throughout this phase of evolution. The theoretical steady-state size distribution, $dN/dr \propto r^{-3.5}$ (c.f., Wetherill's chapter) is not reached during this phase.

bodies occurs. Thus, intermediate-stage growth seems to have occurred remarkably fast, resulting in creation of a few discrete small planets from the initially nearly continuous system. The growth through this stage seems to be an orderly process of gradual growth, in which material is continually fed through the intermediate-size ranges (10-100 km), rather than "runaway growth" of the largest body. At the end of this stage, most of the mass of the system still resides in km-scale planetesimals. Smaller debris and ejecta have negligible mass, and the slope of the size distribution from 1-500 km is steeper than the "equal mass" slope. At least during this stage of planetary growth, our result contradicts Safronov's assumption that most mass would be contained in the largest bodies. Accordingly, we find that all random velocities remain near the escape velocities of the 1-km bodies, and not near that of the largest bodies as Safronov found based on his assumptions. Of course, such lower velocities enhance the prospects for further growth, rather than fragmentation. Levin (1978) has also suggested on analytical grounds that random velocities might be considerably less than the escape velocity of the largest bodies. It must be emphasized that we (and Levin) confirm Safronov's result that relative velocities are on the order of escape velocities of those bodies which dominate the population. We and Levin differ from

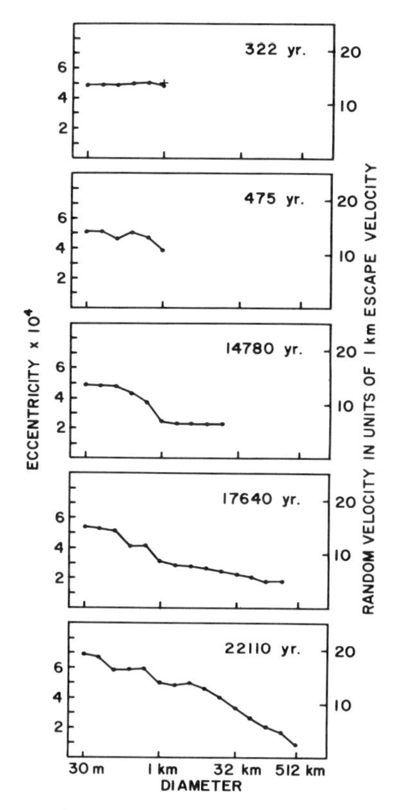

Fig. 2. Eccentricity distribution as a function of time and particle size corresponding to the size distribution shown in Fig. 1. Inclinations are similar. Initial value, e = 5×10^{-4}, is shown by +. Corresponding random velocities are shown on right hand scales in terms of 1 km escape velocity, v_e ~65 cm sec^{-1}.

Safronov in finding that, in at least some stages of planet growth, the largest bodies do not dominate.

The discontinuity in the size distribution between one and two kilometers is due to the slight increase in relative velocities as larger bodies are produced. This increase prohibits further mutual accretion of 1-km objects.

The largest bodies have low random velocities due to the drag imposed by the massive medium of smaller particles, and the lack of objects capable of gravitationally stirring them. The velocities are so low, in fact, that for bodies larger than about 500-1000 km, our assumption that random velocities dominate over Keplerian shear in controlling relative motion is violated. Therefore, our simulation in its present form is not appropriate for following further evolution beyond the intermediate stage.

We have also run numerical simulations for material properties with the extreme parameters for loosely bonded regolith and for solid rock. For weak material, the evolution is similar to that of the intermediate material, as long as initial random velocities are only a few times the escape velocity of the initial bodies. For rock, the high coefficient of restitution prohibits accretion because the initial bodies reach an equilibrium random velocity that is too high. However, such solid rock is an implausible material in the early solar system. If the coefficient of restitution of the rock is reduced to ~0.1, in modeling an energy-damping surface layer (Hartmann, 1978), accretion can occur and the evolution is similar to the other cases. Changing the initial number of bodies changes the time scale roughly in inverse proportion.

Over a wide range of parameters, the basic conclusions are the same. The intermediate stage starting from kilometer-size planestesimals yields rapid growth of a few 500-km planets (~$1/AU^2$), in a swarm of planetesimals which contain most of the system's mass. Alternately, before seeing our results one might have imagined the intermediate stages to have been characterized by such rapid growth of the smaller bodies that the largest ones grew fast in numbers compared to their growth in size. If that were the case, the bulk of the mass would have resided in the larger bodies and the later stages of planet growth would have involved their mutual accretion, rather than accretion of planetesimals by a few seed planets hundreds of kilometers in diameter. (A size distribution with most of its mass in the larger bodies is observed today in the asteroid belt, but this is a product of comminution rather than accretional evolution.)

In contrast, the situation we obtain suggests that subsequent evolution might involve the largest bodies acting as seeds for agglomeration of the remaining mass of planetesimals. However, the same problem that invalidates following further evolution with our simulation, namely the nearly circular, coplanar orbits of the largest bodies, introduces serious questions as to how such subsequent growth might have occurred. (See the chapter by Wetherill in this book and his comment at the end of our chapter.) The large bodies will only be able to encounter and accrete those remaining planetesimals having orbital semi-major axes close enough that Keplerian shear can bring them within the collision cross-section.

There may be several ways in which subsequent evolution might proceed, however. Perhaps as the zone of nearby planetesimals is swept clear by accretion, the capture cross-section of a growing seed planet would grow so as to dominate an increasingly large annulus of planetesimals. If this fails to occur, the isolation of the largest bodies might still be only temporary. Diffusion by interactions of small-scale planetesimals from adjacent zones might tend to feed material into the large body's accretion zone. If that mechanism were slow or ineffective, continued collisional evolution among planetesimals in other zones would grow other 500-km scale bodies by the same process which led to the isolated first generation of large bodies. Eventually, there might be enough of these large bodies that they would begin to perturb one another towards more eccentric orbits thereby providing access to one another and to any remaining planetesimals. Since the relative velocities due to stirring would be of the order of the escape velocity of the large bodies, collisions thus promoted would probably result in accretion (Hartmann, 1978).

INFLUENCE OF THE FIRST-FORMED FULL SIZE PLANET

The isolation of the 500-km size bodies might alternately be broken by the effects of an earlier-formed full-size planet elsewhere in the system. Perhaps such a planet formed directly due to gravitational collapse in a locally unstable region or due to collisional accretion in a region with especially sticky material. Weidenschilling (1974) hypothesized that Jupiter grew earlier than the terrestrial planets because low temperatures would have led to early condensation of more, and possibly stickier, solid matter. However an earlier planet formed, it could modify the collisional evolution elsewhere in several ways: (a) gravitational scattering of planetesimals from its own neighborhood; (b) secular perturbations on the collisionally evolving population; and (c) forced resonances near orbital commensurabilities.

The first influence, gravitational scattering of planetesimals from the first-formed planet's own neighborhood, provides a high-velocity flux of material to other regions. For example, Weidenschilling (1975) and Kaula and Bigeleisen (1975) have proposed models in which planetesimals scattered by close encounters with Jupiter have different effects on each of the early terrestrial bodies and account for important differences in observed physical properties.

An earlier-formed planet would also perturb the orbits of bodies that lie sufficiently distant such that they are not directly scattered. As Wetherill has suggested (personal communication; see also his chapter in this book), "secular" perturbations might tend to drive eccentricities and inclinations up to significant values depending on the mass and the orbital values of e and i for the perturbing planet. In classical perturbation theory the forced in-plane and out-of-plane oscillations would have phases and amplitudes dependent on the semi-major axis of the perturbed body. Hence, for all the perturbed bodies near a given semi-major axis, the forced components of e and i would be expected to add nothing to their relative motion. Relative velocities would still be governed by the free components of e and i, or by Keplerian shear, just as in the unperturbed case.

However, if the secular perturbations are superimposed on a collisionally evolving system, we find that the oscillations induced by secular perturbations may depend on the size of the perturbed body. They may differ in phase and amplitude according to size so that relative motion may be enhanced. To demonstrate this phenomenon, let us consider the secular effects on eccentricity. We define $h \equiv e \cos \tilde{\omega}$ and $k \equiv e \sin \tilde{\omega}$, where $\tilde{\omega}$ is the longitude of a perturbed body's perihelion. Thus h and k specify the amplitude and phase of the in-plane oscillation with respect to circular motion. We shall use primes to denote corresponding orbital elements of the perturbing planet. From standard secular theory (c.f., Greenberg, 1975), we have

$$dh/dt = -Ak + Bk' \quad \text{and} \quad dk/dt = Ah - Bh' \tag{1}$$

where A and B are constants and are both of the order of the mean motion, n times the perturber's mass, m' expressed in units of the sun's mass. The orbit of the perturber precesses at a rate N governed by the mass of the planetesimal disk so that

$$h' = e' \cos Nt \quad \text{and} \quad k' = e' \sin Nt. \tag{2}$$

Objects in the system undergo additional orbital perturbations due to collisions. For example, depending on the size, eccentricity may be damped by interactions with smaller particles. Expressing this damping in the form $de/dt = -Ce$ and adding it to the secular theory, we obtain

$$dh/dt = -Ak + Be' \sin Nt - Ch$$

and

$$dk/dt = Ah - Be' \cos Nt - Ck\,. \tag{3}$$

The solution is

$$h = Ke^{-Ct} \cos(At + \delta) - \frac{Be'}{[(N-A)^2 + C^2]^{\frac{1}{2}}} \cos(Nt + \Delta)$$

$$k = Ke^{-Ct} \sin(At + \delta) - \frac{Be'}{[(N-A)^2 + C^2]^{\frac{1}{2}}} \sin(Nt + \Delta) \tag{4}$$

where $\tan \Delta = C/(N-A)$, $0 \leqslant \Delta \leqslant \pi$, and K and δ are integration constants. The first terms in both expressions represent the free component of the eccentricity. Note that it is damped on a time scale $1/C$. The second terms represent the forced component. Evidently, both phase Δ and the amplitude of the forced component depend on the value of the damping coefficient, C. Therefore any differential damping of e according to size in the evolving swarm, when coupled with secular perturbations, could contribute to relative velocities between bodies of different sizes.

To get some idea of a plausible magnitude of this effect, consider relevant values for Jupiter today: $A \sim B \sim 10^{-3}$/yr $\gg N$ and $e' = 0.05$. With $C < A$, the amplitude of the forced eccentricity would be ~0.05, 100 times the typical eccentricities of our numerical experiments (e.g., Fig. 2). This value suggests that secular perturbations are a powerful way to enhance orbital eccentricities. However, enhancement of relative velocities depends upon the differential response of various sized bodies. The value 0.05 represents only an upper limit to the relative motion that might be induced and there are arguments to show that differential response would actually be much less. From our numerical experiments we find that $C \lesssim 10^{-4}$/yr. Hence, according to Eqs. (4), secular perturbations may generate differences between forced eccentricities of different sized planetesimals only as great as ~0.005. Such a value would correspond to a substantial fraction of the spacing between seed planets and would certainly be adequate to break the isolation of the largest bodies.

On the other hand, we find that typical values of C towards the end of the evolution phase that we have simulated are less than $\sim 10^{-5}$/yr for the large seed bodies, due to their low area/mass ratios, and for the 1-km bodies, due to the low total mass of subkilometer bodies. Hence, the difference between forced eccentricities of the seed bodies and the portion of the population that contains the bulk of the population's mass is $\lesssim 0.0005$, a value which yields no significant enhancement of relative velocities above the random velocities. The indication is that, while secular perturbations may have played an important role in the growth of planets, it is by no means obvious that they could have induced the relative velocities necessary to deliver planetesimals to the seed planets.

Furthermore, any such forced relative motion is seen to depend directly on the value of the perturber's eccentricity, e'. Conceivably an early-formed planet would have an orbital eccentricity less than Jupiter's present value. Secular effects might have been important if the eccentricity of the perturber were considerably larger than 0.05 or if its precession rate, N, matched that of the perturbed bodies, A. In the latter case, called "secular resonance," the forced eccentricity would be very large and very sensitive to C, according to Eqs. 4. In that case, at certain locations, the secular effects of an early-formed planet might be important. Secular resonances do exist in parts of the present asteroid belt.

Another type of resonance that can operate over long distances, but which does not require a significant e or i for the perturber, occurs when a particle's orbital period is near a small whole number commensurability with the period of the perturbing planet. Consequent repetitive mutual configurations induce a forced eccentricity in the particle's orbit, the magnitude of which increases with decreasing distance from the exact commensurability (c.f., Greenberg, 1977). From here on, we shall use the word "resonance" only in this context, as distinct from secular resonances.

The possible importance of such resonances in the planet-building process was stressed by Safronov (1972, p. 89) and by Kuiper (1974). Several properties of the present planetary distribution suggest that an accretional model governed by resonances may be relevant. The asteroid belt spans orbital radii which correspond to the important low-order commensurabilities with Jupiter's orbital period; planetesimals in the belt never grew to diameters much greater than 1000 km. (Had they ever exceeded 1000 km they would still survive according to Chapman and Davis [1975].) The density distribution within the belt appears to be governed by resonances, with either gaps or concentrations at commensurable distances. In the outer solar system there are striking near-commensurabilities between adjacent planets (Wilkins and Sinclair, 1974); satellite systems contain a statistically significant excess of resonances (Goldreich, 1965); and the structure of Saturn's rings appears governed by resonances with other satellites (Franklin and Colombo, 1970). The terrestrial planets do not exhibit such striking mutual commensurabilities, but this might be explained by the shift in resonance positions which would have occurred in the presence of the early inner disk of material (more dense than in the outer solar system), just as resonances may be shifted in Saturn's rings according to the theory of Franklin and Colombo (1970).

Just as for secular perturbations, the forced response of collisionally evolving particles to resonant perturbations depends on the degree of eccentricity damping, which in turn is a function of particle size. The nature of the response for any particular size depends on the population's size and velocity distribution, for it is that distribution which determines the magnitude of the e damping term. Without making unwarranted assumptions about the population the problem is analytically intractable. Thus we have modified our numerical model to incorporate the synergistic effect of resonance into the collisional evolution. The modifications are based on the analytic study of resonance in a damping medium by Greenberg (1978). The study showed that the free component of the eccentricity is damped exponentially just as in the secular case. In contrast to the secular case, at any given semi-major axis the phase of the forced component is independent of the damping coefficient of C. However the amplitude of the forced eccentricity is $\sim n(m'/M)/C$, strongly dependent on C. Thus substantial resonance-induced increases in relative velocity amongst particles are possible if C differs for various sized objects, especially if C is small.

The resonance version of our numerical simulation includes for each size bin, in addition to a free e and i, an eccentricity component in phase with the forced eccentricity. The mean relative velocities between particles of different sizes now depend not only on the phase-mixed values of e and i, but on differences in amplitude between the in-phase components. As before, in each short time step the collision probabilities and outcomes are computed for each pair of size bins. Gravitational stirring due to close encounters is also taken into account. Again, the corresponding changes in population are computed, carefully separating the random phase components of any resulting velocities from the components in phase with the forced eccentricity. All changes are then made, except for those in eccentricities due to smaller particles. The latter changes in the total eccentricity for each size bin are used to evaluate C. Interactions with larger particles would represent large discontinuous changes which cannot properly be part of the perturbation theory used in the resonance analysis. Then the value of C is used with the results of the resonance analysis to compute the additional changes in the free eccentricity and in the in phase component of the eccentricity as the system damps toward the forced eccentricity.

Two important restrictions to this procedure must be monitored. One is that the analysis is valid only for eccentricities $\lesssim 0.1$, so numerical experiments in which this limit is exceeded must be disregarded. In addition, the phase of the forced eccentricity is such that secular variations in semi-major axis are introduced (Greenberg, 1978). If these variations are large, mass will tend to be concentrated on one side of the resonance region with possibly important implications for collisional history, but our program cannot presently follow evolution in this additional dimension and the results are therefore valid only for cases in which such variations are small.

For our first numerical experiment with the resonance version, we considered a planet of fixed mass 10^{28} gm (~1 earth mass) at the present position of Jupiter. The particles were taken to be at the semi-major axis which corresponds to the 2:1 resonance with the perturber. Initially the e component in phase with the forced eccentricity was taken to be zero for all particles. All other initial conditions and parameters were taken as identical to the case of non-resonant intermediate material described earlier (Figs. 1 and 2).

In this case, early evolution is similar to the non-resonant case. However, by the time accretion has produced some 16-km bodies (~15,000 yr from start), the in-phase eccentricity has reached values ~0.08, with differences $\sim 10^{-2}$ over a few size bins. These differences are much greater than the free components of e and i ($\lesssim 10^{-3}$) so relative velocities indeed are substantially greater than in the non-resonant case. However, rather than helping promote growth of the larger bodies by breaking their isolation, the higher velocities tend to promote comminution processes (erosion and catastrophic fragmentation). By t = 24,000 yr the bulk of the mass resides in bodies a few meters in diameter with growth of bodies larger than 16 km suppressed.

Lowering the mass of the perturber by a factor of 3 yields similar results although some particles ~ 32 km are produced before resonance-induced comminution begins to dominate. We conclude that resonance is a growth-inhibiting effect in the intermediate stage of planetary aggregation. This result supports the contention that resonances with Jupiter may have prevented growth in the asteroid belt.

In order to promote growth we need to enhance relative velocities toward the end of the intermediate stage, but not before. At this point the isolated seed-planets require agitation to get them out of isolation, but they are sufficiently large so as to resist comminution by impacts at enhanced velocities. For example, we have run one resonant case in which we start near the final conditions from the case of non-resonant intermediate material after a few 250-km bodies have been produced. The perturber's mass was taken as 5×10^{28}gm. For about 4000 yr growth continues although at a somewhat slower rate than in the non-resonant case. The largest body barely reaches 500 km, but the numbers of bodies of sizes >10 km increase substantially. More important, the in-phase eccentricities of the largest bodies differ from those of smaller bodies by an amount $\sim 10^{-2}$ which introduces much greater relative motion (~150 m $\sec^{-1}$) than in the non-resonant case (~10m $\sec^{-1}$). Further evolution cannot be followed because the absolute values of in-phase eccentricities begin to exceed ~0.1, the upper limit for applicability of the analysis. However, the results do demonstrate that the isolation of the largest bodies can be broken by resonance effects.

In this section we have addressed two possible effects of a first-formed planet on collisional interaction and accretion occurring elsewhere in the nebula. Secular perturbations have the potential for major effects on the collisional interactions, especially if the eccentricity of the perturbing planet is very large or in the vicinity of secular resonances. Otherwise, eccentricity damping must be significantly different for the 500-km seed planets than for the kilometer-scale bodies which contain the bulk of the mass, in order for secular perturbations to promote further accretion. Also of importance for planetesimal collisional evolution are the effects of a first-formed planet on populations near orbital resonances. We have shown that velocities may be enhanced so much that 1-km particles will fragment and comminute each other rather than accrete. But, if such resonances become important relatively late in the intermediate stage of accretion, they may assist temporarily isolated seed-planets to accrete the remaining planetesimals. If the accretion of the terrestrial planets were assisted by Jupiter resonances, it might be required that Jupiter formed roughly contemporaneously with the terrestrial planets and not by direct gravitational collapse at an early stage.

COLLISIONAL EVOLUTION IN THE EARLY STAGE OF AGGREGATION

The success of our model in producing rapid growth from kilometer to thousand kilometer dimensions by collisional evolution raises the question of whether collisional accretion is so efficient that it may have competed with gravitational instability as the dominant mechanism for early-stage aggregation. To put the question another way, is it possible that the initial dust condensates could have grown all the way to planetary dimensions solely by collisional accretion, without any gravitational instabilities being involved at all? A finding that collisional accretion operated on a shorter time scale than gravitational collapse would obviate the need for such Goldreich-Ward processes for forming planetesimals.

First we need to ask whether the invoked processes, such as the collisional mechanics observed at centimeter scale in our laboratory experiments, can be applied all the way down to the micron dimensions of condensed grains. While our collisional-mechanical analyses appear applicable to particles sensibly smaller than a kilometer, they are not applicable without modification to the initial microscopic particles, if for no other reason than that they ignore effects of the gaseous medium in which the earliest condensed particles surely existed.

In fact, these gas effects will prevent collisional accretion of the smaller early planetesimals in size ranges less than some meters. Collisional accretion requires that any initial velocities of the particles must be able to relax to lower values until relative velocities are comparable to the escape velocities of the largest particles in the swarm, in the case of clean rock particles, or within perhaps two orders of magnitude in the case of weak aggregates (Hartmann, 1978). Thus, any process that maintains velocity above this limit will prevent aggregation because most particles will rebound at more than escape velocity.

The nebular gas has just this effect of maintaining high velocities for at least two reasons, namely Brownian motion and gas drag. Brownian velocities of microscopic particles imbedded in gas far exceed their own escape velocities. According to classical Brownian theory, the grain velocity is simply the thermal velocity characteristic of a macro-molecule of grain size

$$V_{\text{thermal}} = 3kT/\sqrt{m} = 4.5\ (10^{-7})/\sqrt{m} \qquad (5)$$

for a gas at temperature 500 K. (The results derived below are insensitive to nebular temperature.) For sufficiently small grains, therefore, the thermal motions keep the grains moving at speeds much greater than their own escape velocities. Figure 3 shows these two velocities as a function of particle mass and radius for a wide range of particle sizes (assuming $\rho = 3$ g cm^{-3}).

Gas drag also increases the relative velocities of particles (Weidenschilling, 1977). Particles of certain sizes drift at velocities as high as 10^4 cm sec^{-1} with respect to larger and smaller particles, as shown by the dashed curves in Fig. 3. The figure shows that both the thermal Brownian effects and the Weidenschilling drag effects (somewhat more model dependent) maintain grain-grain encounter velocities as much as seven orders of magnitude greater than the escape velocity. Also shown in Fig. 3, as a speckled band, is the velocity range at which rocky particles of comparable size begin to fragment one another. Above this region comminution is enhanced and accretion does not occur.

In vacuum chamber tests with powder targets that simulate the lunar regolith at gas densities of the order 10^{-6} gm cm^{-3}, we have observed most of the impact energy being absorbed by the powder, with rebound or ejection velocities probably less than 10^{-2} times the impact velocity. In other words, projectiles could be accreted by dust-ball targets even if they struck at speeds on the order of 100 times the escape velocity or perhaps even more. Therefore, Weidenschilling's velocity curve (on the right in Fig. 3), calculated for a gas density of 10^{-7} gm cm^{-3}, may not rule out accretion among loosely aggregated planetesimals from sizes of some meters upward.

Nonetheless, the Brownian velocities alone make it very difficult to see how collisional accretion could produce larger planetesimals from an initial swarm of micron-sized condensates. Figure 4 shows some results from our numerical modeling program, modified to take into account thermal motions. The encounter velocities of grains are allowed to evolve as before, except that they are constrained not to drop below the Brownian velocities shown in Fig.

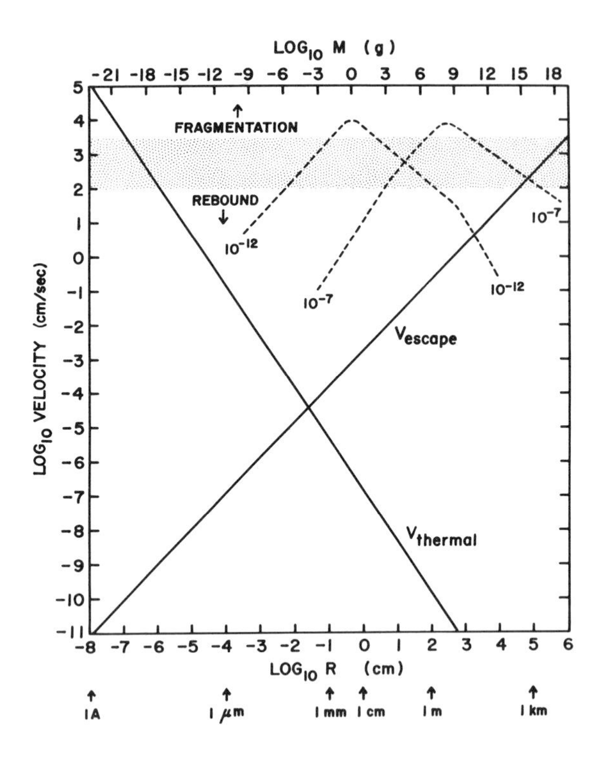

Fig. 3. Comparison of thermal Brownian velocities and escape velocities (solid curves) for particles of various sizes. For particles smaller than $\sim 10^{-2}$ cm thermal velocities imparted by the gaseous medium maintain dust at speeds higher than escape velocity for the same particles, preventing accretion in a uniform-sized swarm. Dashed lines show additional velocity imparted by effects of gas drag, as computed by Weidenschilling (1977) for solar nebulae of two specified densities (g cm^{-3}). The speckled band shows velocity range where fragmentation sets in (Hartmann, 1978).

3. Evolution in three particle swarms is shown, each swarm consisting of the particle diameter shown at the top. Two time steps, *A* and *B*, are shown. As judged from Fig. 3, the thermal gas effects prohibit growth of the particles smaller than a few centimeters and no evolution is seen (the collision velocities being too slow to cause fragmentation and production of a fragment tail at small sizes in the size distributions). However, at sizes above a few centimeters, growth actually occurs, and the diagram demonstrates collisional aggregation at $\gtrsim$ 10-cm sizes. These results consider only the Brownian effects of the gas; inclusion of the drag effect, at least in nebular gases less dense than about 10^{-7} gm cm^{-3}, would be expected to inhibit growth even at larger sizes up to some meters, in keeping with the earlier discussion.

We conclude that planets cannot grow from initial, uniform, micron-sized

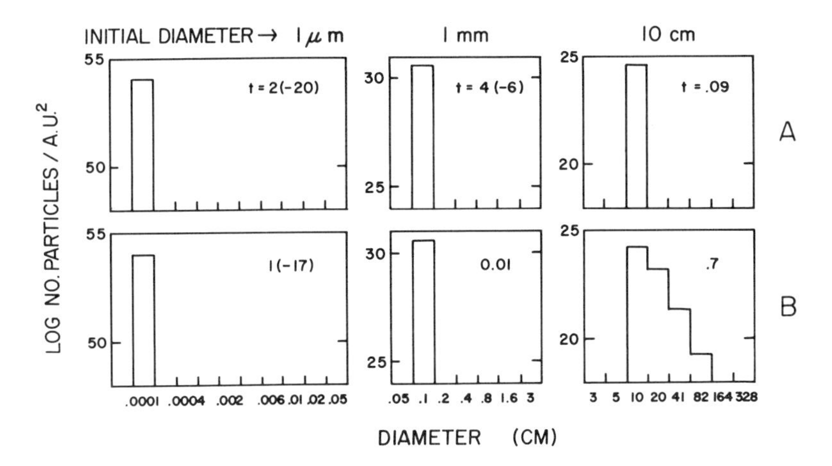

Fig. 4. Size distributions at two time steps, A and B, illustrating limiting effects of thermal Brownian motion on aggregations of dust particles in four swarms. The swarms initially contained uniform-sized particles, specified at top. Values of t in boxes give time in years. Dust particles in first swarm (1 μm, left column) were assumed to have mechanical properties of strong, hard, clean silicate mineral grains; larger particles in other swarms were assumed to have mechanical properties of aggregates of such grains (e.g., loosely bonded regolith in Greenberg *et al.,* 1978. Collision velocities of co-orbiting particles were not allowed to relax below thermal Brownian velocities for each given size. Growth occurs among the 10-cm particles with the time scales indicated, but Brownian motions (and possibly other effects, see text) prevent submillimeter particles from accreting.

condensates by collisional accretion in which binding is due to gravity alone. Perhaps some less understood process (e.g., electrostatic binding, cold-welding, or other *ad hoc* "stickiness" with large sticking coefficient) may be more efficient in aggregating the smallest particles up to sizes where collisional accretion based on gravity takes over. But in the absence of such a process, it appears that local, patch instabilities such as described by Goldreich and Ward (1973) are necessary to bridge the size range of microns to at least several meters. While so far we have provided a quantitative treatment only for Brownian motion, other gas effects including the drag effects of Weidenschilling and even more speculative effects of turbulence must be addressed to calculate precisely the size at which gas no longer affects particle orbits.

It would be desirable some day to model more precisely the evolution through the micron to kilometer size range since this is the phase of evolution that might be most clearly manifested in observations of newly forming planetary systems. Once larger planetesimals have formed, the surface area of planetary materials and concomitant opacity will have dropped below detectability.

THE QUESTION OF EXTRA-SOLAR PLANETARY SYSTEMS

Many astrophysicists have stated or implied that planet formation is a common process in the universe, given the dusty clouds observed around proposed newly-formed stars. Yet we have only one firm example of a planetary system and as long as obstacles have remained in modeling the complete accretion of dust into full-sized planets a possibility has remained that the sun's planetary system is the result of unusual circumstances and that other solar systems are rare. Thus, our success in attaining rapid accretion through the difficult intermediate size ranges from 1 to 1000 km, as well as tracing a plausible planet-forming sequence from 1 μm to 1000 km, increases our expectation that planetary systems formed around some other stars.

The best candidate stars would be single. In binary and multiple systems, orbits would be more irregular, maintaining collision velocities at higher values, and thus making accretion less likely.

While our numerical results to date have concentrated on plausible early solar system models, we have made preliminary studies of other possible initial conditions (e.g., high velocities) that might limit planet formation in the general stellar case. One example is a case with extremely weak material comparable to the loosely bonded regolith discussed earlier, except that the initial random velocities were higher, about 10 v_e. Figure 5 permits comparison of growth stages of the population in this experiment (first column) with the intermediate material (second column) and the solid rock case (third column). The evolution is shown in four arbitrarily chosen stages, whose ages (in years) are shown in the upper right. In all three cases initial bodies are 1 km across (Stage A), perhaps the product of Goldreich-Ward contraction, with random velocities exceeding their escape velocities. Evolution to Stage B primarily involves damping of random velocities. In each case, except for solid rock, crater ejecta produce a small particle "tail" in the distribution at Stage B. Once relative velocities are sufficiently damped, the systems in the right and middle columns undergo accretion. By the time bodies a few tens of km in diameter are produced (Stage C), the solid rock still has not produced any small particle debris while the intermediate material continues to augment the debris even as accretion occurs. These two systems show remarkably similar size distributions when 500- to 1000-km bodies are produced (Stage D). By this time, impact velocities are great enough to produce a small particle tail even for the case of solid rock.

The system with higher initial random velocities (left column) behaves markedly differently from either of the other cases after Stage B. The velocity never gets low enough to permit accretion. In the computer program as set up for each of these three runs, debris smaller than 30-m diamater are ignored (removed from the system, in effect). In physical reality, small debris would be removed from the system by Poynting-Robertson effects, radiation pressure, the Yarkovsky effect (see Glossary for definitions), etc., but

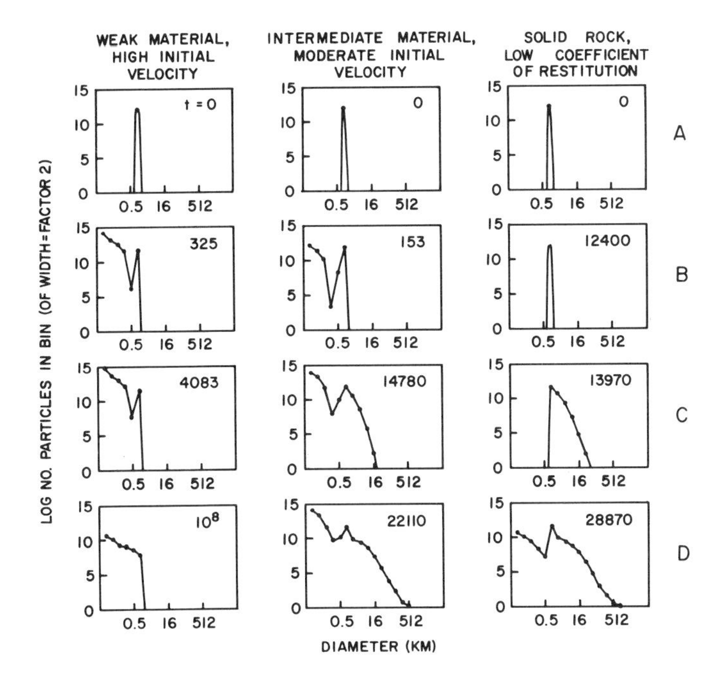

Fig. 5. A comparison of growth stages for three different planetesimal systems. Each chart shows a size distribution; ages in years are given in upper right. On the left, initial velocities are chosen to be high enough so that much mass is ground into small particles and lost before planets can accrete. In contrast, for the case with stronger material and lower initial velocities (middle column), growth begins after a period of small particle production. For the solid rock case (on the right), growth begins even before production of any small particles.

predominantly at much smaller diameters. Nonetheless, we believe the mass loss occurring here is conceptually significant. What happens in this system is that before the velocity can relax to permit efficient accretion, sufficient mass is lost that the system is worn down (c). No growth occurs even after 10^8 yr (D). This system thus demonstrates conceptually how a sufficiently high-velocity regime may completely inhibit growth of a planetary system and produce only a swarm of asteroid-like bodies.

PRODUCTION OF INTERSTELLAR DUST DURING PLANETESIMAL COLLISIONS

Some of our unpublished experiments suggest that rock fragmentation will produce debris extending down to the size of the homogeneous grains in the shattered material. These grains have greater strength than the parent rock which they compose, and thus they are more likely to be left intact. The best-known representatives of primitive solar system material are carbonaceous chondrites, whose little-altered matrices contain crystals in the size range of 0.3 to 5 μm (Green, Radcliffe, and Heuer, 1971). Our work thus suggests that the collisional evolution of planetesimals would produce abundant μm-scale rocky or icy particles. These would have been produced when the solar nebula was clear to radiation, most of the early obscuring dust having been already concentrated in the growing planetesimals. Soter, Burns, and Lamy (1976) show that particles of metallic, icy, or silicate composition in the size range 0.05 to 1 μm would be driven out of the solar system by radiation pressure. Thus planetesimal systems would be sources of observed interstellar μm-scale grains of silicate and perhaps icy and metallic composition, as earlier suggested by Herbig (1970) and Hartmann (1970). Further modeling of the evolution of the grains is required in order to estimate the rate of grain supply from different protoplanetary systems.

CONCLUSION

Collisional accretion appears to be a viable, and seemingly unavoidable, mechanism for intermediate-stage growth from a swarm of planetesimals into a system containing a few discrete seed planets. Some other mechanism must be invoked to explain growth of condensate grains up to at least tens of meters. Gravitational instability in the particulate disk seems a plausible means of achieving that early-stage growth up to kilometer-scale bodies. The late stage of growth in which the seed planets accrete the remaining material now presents difficulties due to the isolated, circular orbits generated from the intermediate collisional stage. But once the first full-sized planet forms, orbital resonance might break the isolation of those seed planets with commensurable periods. Subsequent growth would culminate in the production of the surface impact structures and the residual orbiting debris that remain as the only observational record of planetary aggregation.

Acknowledgements. We thank G. W. Wetherill for his stimulating discussion and criticism of this work. This research is supported by the National Aeronautics and Space Administration.

COMMENTS

Kaula: The starting conditions for the model of Greenberg *et al.* are most logically those created by a "Gureyev-Lebedinskii" or "Goldreich-Ward" or "patch" instability: the gravitational instability of the relatively thin layer of dust at the midplane, separate from the gas. The surface density required for this instability depends on the degree the layer is stirred up by turbulence. For a minimum nebula, patch instabilities result in ~ 1-km planetesimals. For a ~ 1 $M_\odot$ nebula, the planetesimals are ~100 km in size which leaves the problem of how to get rid of excess mass.

There also are global instabilities, but if the nebula is viscous enough for outward transfer of angular momentum, it is hard to understand why they should result in giant protoplanets rather than in some sort of convective motion.

Wasson: The chondrites are sedimentary rocks, each grain of which formed in an environment different from that of the neighboring grains. The simplest explanation of this structure is that the grains formed in a variety of (nebular) environments, were mixed by various dynamic processes, and ultimately agglomerated, perhaps by the Safronov-Goldreich-Ward mechanism. Other mechanisms, specifically those involving protoplanets, cannot easily explain these grain-to-grain differences.

Greenberg: These comments lend further strong support to the planetesimal hypothesis. Hopefully the protoplanet scenario will be studied, just as the planetesimal scenario is undergoing study by ourselves and others, to a level of detail such that it may be possible to determine definitively which hypothesis best explains the meteorite record and most plausibly follows from a reasonable solar nebula.

Kaula: What is the effective partitioning of impact energy to heat energy and kinetic energy of fragments that you use? Do you have a positive correlation of the portion to heat with impact velocity (due to phase transitions), as indicated by the work of O'Keefe and Ahrens?

The aforestated partitioning should apply to catastrophic collisions (your category No. 4) as well as lesser cratering events, since it depends on the early-stage thermodynamics and hydrodynamics associated with the initial shock front, etc.

Greenberg: The calculations of O'Keefe and Ahrens (1977) give 10% of the energy going into ejecta kinetic energy for impact of gabbroic anorthosite. In our algorithm for the solid rock extreme, this figure is 15%; for the loosely bonded regolith it is 0.3%; and for the "intermediate material" it is 3%. Our values are based on experimental determination of ejecta mass and velocity and agree well with the energy partitioning of O'Keefe and Ahrens. (See Greenberg *et al.*, 1978 for details.) Catastrophic fragmentation is so different in nature from local cratering that it is hard to believe that the result of O'Keefe and Ahrens applies. In the examples discussed here, we conservatively give the debris 50% of the impact energy. If we lower this parameter, it serves to further promote accretion on impact.

Wetherill: As one works out details as you are doing, problems are certain to arise for planetary formation. One fairly easy way out of this is to invoke Jupiter perturbations. But then some way must be found to make Jupiter on a short time scale, perhaps by gravitational instability.

Greenberg: Perturbations by a first-formed planet may indeed be a key feature of the growth of other planets. We have discussed in this chapter the possible roles such perturbations may have played, but more work is required to model these effects in adequate detail.

REFERENCES

Chapman, C.R., and Davis, D.R. 1975. *Science* 190: 553.

Franklin, F., and Colombo, G. 1970. *Icarus* 12: 338.

Gault, D.E.; Shoemaker, E.M.; and Moore, H.J. 1963. *NASA Tech. Note D-1767.*

Goldreich, P. 1965. *Mon. Not. Roy. Astr. Soc.* 130: 159.

Goldreich, P., and Ward, W.R. 1973. *Astrophys. J.* 183: 1051.

Green, H.W.; Radcliffe, S.V.; and Heuer, A.H. 1971. *Science* 172: 936.

Greenberg, R. 1975. *Mon. Not. Roy. Astr. Soc.* 170: 295.

_____. 1977. *Vistas in Astronomy* 21: 209. (A. Beer, ed.) Oxford: Pergamon.

_____. 1978. *Icarus* 33: 62.

Greenberg, R.; Wacker, J.F.; Hartmann, W.K.; and Chapman, C.R. 1978. *Icarus.* In press.

Hartmann, W.K. 1970. *Icarus* 10: 201.

_____. 1978. *Icarus* 33: 50.

Hartmann, W.K., and Davis, D.R. 1975. *Icarus* 24: 505.

Hayashi, C.; Nakazawa, K.; and Adachi, I. 1977. Preprint KUNS 379. Kyoto University.

Herbig, G.H. 1970. *Mém. Roy. Soc. Sci. Liège, Ser. 5,* 19: 13.

Kaula, W.M., and Bigeleisen, P.E. 1975. *Icarus* 25: 18.

Kuiper, G.P. 1974. *Cel. Mech.* 9: 321.

Levin, B.J. 1978. *Moon and Planets* (Special Protostars and Planets issue). In press.

O'Keefe, J.D., and Ahrens, T.J. 1977. In *Proc. Lunar Sci. Conf.* 8th p. 3357. New York: Pergamon.

Safronov, V.S. 1972. *NASA TT F-677.*

Soter, S.; Burns, J.A.; and Lamy, P.L. 1976. In *Comets, Asteroids, Meteorites – Interrelations, Evolution and Origins* (A.H. Delsemme, ed.), p. 121. Toledo, Ohio: University of Toledo.

Stöffler, D.; Gault, D.E.; Wedekind, J.; and Polkowski, G. 1975. *J. Geophys. Res.* 80: 4062.

Ward, W.R. 1976. In *Frontiers of Astrophysics* (E.H. Avrett, ed.), p. 1. Cambridge, Mass: Harvard University Press.

Weidenschilling, S.J. 1974. *Icarus* 22: 426.

_____. 1975. *Icarus* 26: 361.

_____. 1977. *Mon. Not. Roy. Astr. Soc.* 180: 57.

Wetherill, G.W. 1977. *Proc. Lunar Sci. Conf.* 8th, p. 1. New York: Pergamon.

Wilkins, G.A. and Sinclair, A.T. 1974. *Proc. Roy. Soc. London A* 336: 85.

Williams, I.P. 1975. *The Origin of the Planets.* London: Adam Hilger Ltd.

PART VI

Cores and Stellar Winds

THE FU ORIONIS PHENOMENON

GUNNAR WELIN
Astronomiska Observatoriet Uppsala

The FU Orionis phenomenon consists of a (quasi?) permanent brightness increase of several magnitudes observed in a small number of young stars, one of which was known to be a T Tauri type before the flare-up. Post-outburst spectra are of F-G giant/supergiant type.

The causes of these flare-ups are as yet unknown. Different ideas are discussed from statistical and phenomenological points of view. The simplest hypothesis, viz. that the outbursts are due to the dissipation of dust envelopes around relatively massive stars, appears to be at least as tenable as any other explanation.

In a discussion of possible effects on circumstellar matter it is suggested that the occurrence of the phenomenon may result in nearly total loss of protoplanetary matter. If the FU Orionis stars are more massive than about 1.5 solar mass, this may be responsible for the seeming lack of planets around such stars.

In 1936-37 a previously unknown star in the dark nebula B35 rose from about 16 to 10 mag over a period of half a year. The star has remained nearly as bright since then. Its increase in brightness was discovered some time later by Wachmann (1939). Subsequent investigations revealed that the star had been slightly variable before the outburst. It was given the variable star designation FU Orionis.

Later spectral studies have shown FU Orionis to be a peculiar F type giant or supergiant surrounded by a "shell" of lower temperature. A review of the optical observations has been given by Herbig (1966). The star has also been observed to be quite bright in the infrared (e.g., Mendoza, 1968).

The pre-main sequence nature of FU Orionis was recognized by Herbig (1966). He suggested an interpretation of the flare-up as the free-fall collapse to the top of the Hayashi track of a star of about one solar mass. This interpretation was, however, ruled out when in 1969-70 the T Tauri star V1057 Cygni (LkHα 190) underwent a similar brightening (Welin, 1971), thus establishing that stars already formed are the progenitors of FU Orionis stars. After the flare-up this star has changed its spectral type from (possibly B to) A to F giant, along with a decrease in luminosity of about two magnitudes. The spectrum is now composite of F II shortwards of about 5000 Å, and G Ib longwards of this wavelength. Later, the star V1515 Cygni has also been included among the FU Orionis stars, after an increase in brightness that extended over some twenty years. The observations of these two Cygnus stars have been summarized by Herbig (1977).

Subsequently a possible fourth member of the class has been discovered to increase in brightness by Cohen, Kuhi, and Harlan (1977 *a, b*). According to an objective-prism plate taken in September 1977 with the Uppsala-Kvistaberg Schmidt telescope, the previously quite strong Hα emission in this object has decreased considerably. This is similar to the development of Hα emission in V1057 Cygni during its flare-up.

Additional stars that may belong to the FU Orionis class (although then their flare-ups must have occurred before deep photography of the sky began) are, e.g., SU Aurigae (suggested by Herbig, 1977) and UHα 34 (Welin, 1975). From low-resolution spectograms of UHα 34 obtained in 1976 at the Haute-Provence Observatory, it appears that at least Hβ to Hδ are produced in a rising shell, as is the case in V1057 Cygni. Better spectroscopic observations, e.g. at Hα of this rather faint and reddened star would be valuable in determining whether it really belongs to this class.

PROPOSED INTERPRETATIONS

So far no generally accepted interpretation of the FU Orionis phenomenon has been presented. There are two ways in which the different hypotheses could be separated: (1) according to which stars may be involved, and (2) according to what mechanisms are at work.

Which Stars Become FU Orionis Stars?

In a statistical analysis Herbig (1977) arrived at the conclusion that the FU Orionis phenomenon is recurrent, and develops in most or all T Tauri stars at intervals of roughly 10^4 yr. This was based on the number of relatively bright T Tauri stars, the observed frequency of FU Orionis outbursts, and the probable mean duration (5×10^6 yr) of the T Tauri stage. He rejected the idea that the phenomenon occurs only once in a given star, and only in relatively massive stars, a notion forwarded by, e.g., Grasdalen (1973) and Welin (1976).

Two objections may be raised against Herbig's arguments. The first is that nothing is known about the time spent by a massive star in the T Tauri stage, except that it is likely to be appreciably shorter than for less massive stars. Secondly, all the observed FU Orionis stars seem to belong to rather evolved stellar aggregates containing early-type stars on or near the main sequence, whereas in Herbig's scheme they should occur in T associations of all conceivable ages. Thus the outbursts may well have occurred at the termination of the T Tauri phase of these stars.

Another statistical approach is to relate the number of FU Orionis stars to the birthrate of stars throughout the galaxy. Putting the observed frequency of stars/year as F, the number of flares in each star as n, and the quotient of stars that flare up at all as Q, we obtain $F/nQ = CN$, where C is the degree of completeness (i.e., the number of detected outbursts divided by the total number of such events in the galaxy during the same time) and N is the number of stars born each year in the whole galaxy.

If Herbig's conclusion is used in a straightforward manner, $F = 3/80$, $n \simeq 500$, and $Q = 1$. This leads to $CN \simeq 7.5 \times 10^{-5}$. N is usually estimated to be about two stars per year. C consists of two factors, the ratio of practically surveyable volume of space (with an estimated radius of about 1 kpc) to the total volume in which star formation is taking place, and the ratio of detected to undetected events in the surveyed volume. If stars are formed at a smooth rate all over the galactic plane the first factor of C would be about 4×10^{-3}. This implies that we have missed something like 99 out of 100 FU Orionis outbursts inside of 1 kpc during the last 80 years, which seems rather improbable. Otherwise the rate of star formation must be considerably smaller than commonly believed.

If the FU Orionis stars are assumed to be more massive than $1.5\,M_\odot$, Q can be estimated from the standard birthrate function to be 1/15. Also, n is taken to be 1. With four observed outbursts in 80 years we get $CN = 0.75$. This suggests that a burst of star formation has occurred in the "local" region within the last few million years. Here one may point to the presence of Gould's Belt, where the concentration of nebulae offers the potentialities of such a burst. All the known FU Orionis stars lie fairly close to this belt – the Cygnus stars also lie in the local spiral arm.

The value of CN may be lowered if we permit the stars to be subject to a small number of consecutive outbursts, separated by a relatively short time. Here the three flare-ups of P Cygni between 1600 and 1715 may be mentioned, although it is not ascertained whether P Cygni should be included among the FU Orionis stars.

As the available timescale of observation is so short and the number of observed events so small, these statistical considerations should not be given undue weight. A smooth *mean* rate over a long period is likely to show events sometimes clustered together, and sometimes widely separated, during any shorter interval. Thus the three latest maxima of FU Orionis-like outbursts have been observed during a period less than ten years; all four are contained within slightly more than 40 years; and none is known during the 40 years preceding the outburst of FU Orionis, whether due to failure to detect them, or to a real absence. Hence it is as yet impossible to judge how representative the observed frequency may be.

Circumstantial evidence for the assumption that not all T Tauri stars become FU Orionis stars could be found in the pre-outburst spectrum of V1057

Cygni (reproduced by Herbig, 1977), that belonged to a rather special subgroup of the T Tauri class, lacking stellar absorption lines and with a steep decrement of Balmer line emission. A similar spectrum is found in V1331 Cygni (LkHα 120), that was estimated by Kuhi (1964) to have a mass of about 4 $M_{\odot}$. The spectral characteristics of FU Orionis and V1515 Cygni at minimum are unfortunately unknown.

It may also be noted that preliminary results presented by Kohoutek and Meinhardt (1975) suggest that the spectral type of V1057 Cygni as reflected in the colors of the star did not change appreciably during the flare-up (i.e., that it was already an early-type star before the outburst). Likewise, the new Cohen-Kuhi-Harlan star has been assigned an early A type (Cohen, *et al.*, 1977*a*), indicative of a stellar mass of more than 1.5 $M_{\odot}$.

Another argument used by Herbig (1977) to refute the idea of only one flare-up per star is the lack of great numbers of F-G giants/supergiants, or faint near-main sequence stars in the young stellar aggregates. However, this is also based on the assumption that virtually every star undergoes an FU Orionis-like outburst. If only the more massive stars do this, the number of post-outburst objects will be considerably smaller. Furthermore, these are likely to become early-type stars, which are often found in young clusters. The transient stage of a star as an F-G giant/supergiant is of unknown duration, but may be short enough to account for the scarcity of such stars. A lower limit might be deduced from SU Aurigae, which has remained bright at least since about 1850. If the flare-up of P Cygni in 1600 represents the same basic phenomenon, then a high upper limit to the time necessary for such a massive star to pass from its quasi-supergiant stage to a Be star lies around 250-300 yr. The appropriate scaling of this limit to less massive stars remains, however, an open question.

What Causes the Outbursts?

Regarding the ideas about mechanisms at work in an FU Orionis type flare-up, the main dividing line goes between changes (more or less drastic) in stellar structure, and the removal of a dense circumstellar dust envelope. Other ideas have been reviewed, and found unsatisfactory by Herbig (1977). A hitherto unexplored possibility lies in interactions in a newly formed binary system.

Among the proponents of structural change, Herbig has argued that the spectral changes observed in V1057 Cygni necessitate the assumption that the outburst was not caused by the simple removal of a dust screen. He also calculates changes in the radius of the star during and after the outburst. Since, however, these calculations are made under the assumption of intrinsic changes, the result of increasing radius during the flare-up is self-evident, and cannot be considered as proof of these intrinsic changes – the outburst may just as well have been due to the dissipation of a dust shell as to a radius increase. Furthermore, the mechanism responsible for temporarily increasing the radius of the star, and at the same time heating its atmosphere (as supposed by Herbig), is entirely unknown, and apparently quite energy consuming.

It has been suggested by Grasdalen (1973) that the increase in brightness may be identified with the increase in luminosity at the transition from hydrodynamic to radiative-equilibrium contraction calculated by Larson (1972) to occur in relatively massive stars. However, in the calculations this increase falls short of the observed by several magnitudes.

The notion that the flare-up of V1057 Cygni might be due to the disruption of a dust envelope was first forwarded by Cohen and Woolf (1971), following their discovery of strong infrared radiation presumably due to dust close to the star. The idea was supported by theoretical calculations made by Larson (1972). In view of the small or nonexisting color changes during the outburst of V1057 Cygni referred to above, the dust would have to be almost neutrally absorbing. Such dust has been observed around other stars in corresponding evolutionary stages (Strom *et al.*, 1972; c.f. below).

It is true that the post-outburst development of the spectrum of V1057 Cygni shows that the structure of the atmosphere of the star has changed. These changes may, however, as well be the consequences of the removal of a dust shell, as of any other cause of the flare-up. One possibility is that a former "chromosphere," responsible for the T Tauri type emission-line spectrum, has expanded and cooled to form an opaque shell or shells around the star giving rise to the observed F- and G-type spectra, thus hiding the underlying, hotter star. The decrease in luminosity, interpreted by Herbig (1977) as a shrinking (and cooling) of the star, may follow the slow regaining of part of the circumstellar dust envelope. If so, another flare-up might be expected within a relatively short time – c.f., the succession of three outbursts in P Cygni. That this decrease is more prominent in V1057 Cygni than in FU Orionis may be due to a difference in stellar mass. Considering the probably greater distance of V1057 Cygni, and its position within a local dark cloud with an R-value likely to be larger than 3 (from the colors of neighboring LkHα 192, according to Haro, 1972), this star attained a higher maximum brightness, and may be considered the more massive of the two. A minor part of the stellar radiation may be converted to mechanical energy, used in maintaining a strong stellar wind, observed as shortward displacement of hydrogen and calcium absorption lines and in heavy turbulence in the shells, seen as appreciable broadening of other spectral lines. It seems likely that V1057 Cygni is seen almost pole-on (c.f., Herbig, 1977), and hence this broadening would not be caused by axial rotation.

The dust-dissipation hypothesis may also be able to explain the difference in rise-time between FU Orionis and V1057 Cygni on the one hand, and V1515 Cygni on the other. A rotating dust shell would be thicker nearer the equator. Thus a star seen nearly pole-on (as apparently is the case for V1057

Cygni) may brighten considerably faster than a star seen from lower latitudes. The rotation of a star associated with a cometary nebula may be supposed to be coaxial with the axis of the nebula. If so, the position of V1515 Cygni at the edge of its elliptical reflection-nebulosity ring suggests that the axis of this star may indeed be inclined to the line of sight.

EFFECTS ON CIRCUMSTELLAR MATTER

An attempt to estimate the influence of an FU Orionis-like outburst on the primitive solar nebula was made by Herbig (1977). He found that large dust grains were likely to be melted out to a distance of about 0.5 AU, and that the strong solar wind would effectively clear the inner region of the finer dust.

It appears, however, that he underestimated the effects. The solar wind was considered only as a few, rapidly moving shells of ejecta. In the case of FU Orionis the "shell" has persisted essentially unchanged for about 30 years. Thus the mass ejection would take place continuously, and not in discrete shells. Also, the influence of radiation pressure on the dust grains was neglected. At maximum brightness grains of a radius less than several microns would be driven outwards by radiation pressure alone. Furthermore, the "melting distance" would be moved outwards if the effective temperature of the star at maximum is taken higher than the seemingly low estimate of 6500 K used by Herbig. Both radiation pressure and the stellar wind would exert a force also on circumstellar gas. According to, e.g., Gehrz and Woolf (1971), the gas would be rather strongly coupled to the dust and hence driven outwards along with it. The efficiency of this process is dependent both on the distribution of dust-grain sizes, and on the relative amounts of gas and dust.

If, as Herbig supposes, solar-type stars are subject to multiple FU Orionis-like outbursts, it seems the only way that enough protoplanetary matter at adequate distances could be preserved is by the formation of quite large grains at a very early stage in the evolution of the primeval nebula. Moreover, these grains would have to attain the proper amount of angular momentum at formation, since otherwise they would rapidly fall into the protostar. Maybe even the formation of protoplanets is required at this early stage; there may not remain large enough masses of nebular gas to form giant planets, even though the gas would be constantly replenished by the stellar wind. This issue should be more closely investigated.

If the FU Orionis stars are recruited from only the more massive stars the picture seems somewhat clearer. Such stars are likely to begin ejecting matter at a high rate in their T Tauri stage. Thus V1331 Cygni appears to lose about $6 \times 10^{-7} M_{\odot}$ per year (Kuhi, 1964), a much higher value than found in normal T Tauri stars. The presence of an appreciable stellar wind in pre-outburst V1057 Cygni is suggested by the suppression of H-line emission, presumably by shortward-shifted Hϵ absorption (Herbig, 1977). Hence the fine-grained dust is probably removed from the inner parts of the circumstellar nebulae, leaving particles that are just large enough to balance gravity against radiation and stellar wind pressure. The size of these particles would be such as to produce an almost grey absorption of starlight.

If these particles are destroyed or dispersed during an FU Orionis-like outburst – either causing the flare-up, or as a consequence of it – the star may find itself entirely devoid of material for planet formation. Thus the idea that the FU Orionis phenomenon is restricted to relatively massive stars may be related to the possible lack of planetary systems around these stars, as is reflected in their rapid rotation.

Acknowledgments. I am indebted to B. Gustafsson for many critical discussions.

REFERENCES

Cohen, M.; Kuhi, L. V.; and Harlan, E. A. 1977*a*. *Astrophys. J.* 215: L127.
______. 1977*b*. *I. A. U. Circ.* No. 3095.
Cohen, M., and Woolf, N. J. 1971. *Astrophys. J.* 169: 543.
Gehrz, R. D., and Woolf, N. J. 1971. *Astrophys. J.* 165: 285.
Grasdalen, G. L. 1973. *Astrophys. J.* 182: 781.
Haro, G. 1972. *Inf. Bull. Var. Stars* No. 714.
Herbig, G. H. 1966. *Vistas in Astronomy* Vol. 8 (A. Beer, ed.) p. 109. Oxford: Pergamon.
______. 1977. *Astrophys. J.* 217: 693.
Kohoutek, L., and Meinhardt, R. 1975. *Astron. Ges. Mitt.* 36: 101.
Kuhi, L. V. 1964. *Astrophys. J.* 140: 1409.
Larson, R. B. 1972. *Mon. Not. Roy. Astron. Soc.* 157: 121.
Mendoza V, E. E. 1968. *Astrophys. J.* 151: 977.
Strom, S. E.; Strom, K. M.; Brooke, A. L.; Bregman, J.; and Yost, J. 1972. *Astrophys. J.* 171: 267.
Wachmann, A. A. 1939. *Beob. Zirk.* 21: 12.
Welin, G. 1971. *Astron. Astrophys.* 12: 312.
______. 1975. *Astrofizika* 11: 261.
______. 1976. *Astron. Astrophys.* 49: 145.

HERBIG-HARO OBJECTS AND THEIR INTERPRETATION

K. H. BÖHM
University of Washington

After a brief general discussion of the relation between Herbig-Haro objects and the problem of star formation, the basic observational results are summarized. Discussions of the (geometric) structure, variability, continuum and line spectra, observed velocity fields and polarization data are presented. Information is presented regarding the basic physical parameters including electron densities and temperatures. We emphasize the very small filling factors of individual condensations ("nuclei") and the resulting low masses of 1 to 10 earth masses per nucleus. A study of the chemical composition shows no depletion of Ca (in comparison to normal Population I abundances) due to dust formation. An unexpectedly low flux in the C I λ 9849 line seems to indicate a low abundance of atomic carbon.

We discuss the arguments for identifying the cooling regions of shock waves as the line-forming regions of Herbig-Haro objects. A number of other properties of these objects including time scales and sizes of nuclei in objects like HH 1 and HH 2 can be understood in terms of the shock wave theory. The possibilities of identifying the source of the shock waves observationally and theoretically and relating them to the process of star formation are discussed. There is fairly strong evidence that the shock formation is a consequence of the presence of strong stellar winds in T Tauri stars and related young objects. One of the remaining fundamental problems is the relation between objects that show reflection effects (like HH 24) and the ones that do not (HH 1 and HH 2).

A number of different arguments show that Herbig-Haro (H-H) objects (Herbig, 1951; Haro, 1952; Ambarzumian, 1954) are definitely related to the processes of star formation and/or early stellar evolution. The two most convincing of these arguments are:

1. H-H objects occur only in moderately to heavily obscured regions in which also T Tauri stars and other young objects are present (Herbig, 1962*a*).
2. The small nebulae surrounding T Tauri (Burnham's Nebula) has all the spectroscopic properties of an H-H object (Herbig, 1950, 1951; Schwartz, 1974). A similar H-H like nebula surrounds HL Tauri.

Additional evidence for the fact that H-H objects are related to the processes of very early stellar evolution comes from their connection with (probably young) infrared objects (Strom, Strom and Grasdalen, 1974; Strom, Grasdalen and Strom, 1974). Strom *et al.* have argued that H-H objects cannot be older than about 10^5 yr.

Though it would be presumptuous to claim that we can show a proven relation between Herbig-Haro objects and processes of planet formation, it is intriguing to note that the individual condensations (knots) have (visible) masses of 1 to 10 earth masses (Böhm, Siegmund and Schwartz, 1976). This coincidence may be accidental especially since it refers to the gaseous mass only which consists mostly of volatile elements. On the other hand, it is puzzling that objects of such a small mass seem to behave as independent entities (e.g., with regard to their variability, see Herbig, 1969, and their spectra, see Böhm and Schwartz, 1973).

Recently it has become clear that the formation of emission lines in H-H objects occurs in cooling regions of shock waves (c.f., Schwartz, 1975; Böhm, Siegmund and Schwartz, 1976; Münch, 1977; Raymond, 1977; Dopita, 1977*b;* Schwartz, 1977; Böhm 1975). The exact role of these observed shock waves in the processes of star formation is not yet entirely clear. However, a connection with a strong stellar wind generated in the very early phases of stellar evolution seems to be one promising suggestion (Schwartz, 1975; Strom, Strom and Grasdalen, 1975; Schwartz, 1978). This idea would also explain why Herbig-Haro objects occur only in regions of recent star formation.

In this chapter we shall try to summarize the present empirical information on Herbig-Haro objects and the reasonably obvious deductions about the physical properties of these objects which can be made from the observations. In this way we can get information about the thermodynamic parameters, like T_e, N_e and their inhomogeneities as well as about the structure, velocity fields, masses, chemical composition and possibly the dust content of H-H objects. We shall put special emphasis on the implications of the shock wave theory for the observable properties of H-H objects and on some aspects of the abundances of elements which are relevant to the problems of dust formation and destruction in H-H objects.

OBSERVATION OF H-H OBJECTS AND DIRECT INTERPRETATION

Geometric Structure

Important observational results about H-H objects were already given in the discovery papers by Herbig (1951) and Haro (1952). Originally it seemed that the geometric structure of H-H objects always shows small condensations (knots) imbedded in a somewhat more diffuse medium (c.f., the photographs in Herbig's 1974 catalog). However, if the typical "H-H" emission-line spectrum is used as a criterion for classification also, relatively extended H-H objects without a condensation structure are found (Strom, Strom and Grasdalen, 1974; Strom, Grasdalen and Strom, 1974; Schwartz, 1977). In the classical Herbig-Haro objects, e.g., HH 1 and HH 2, the condensation structure is pronounced and the individual condensations seem to form separate entities varying independently (Herbig, 1969, 1973) and showing quantitatively different spectra (Böhm and Schwartz, 1973; Böhm, Siegmund and Schwartz, 1976; Schwartz, 1976). The condensations seem to have radii of about 250 to 500 AU.

It is a strange fact that the condensation structure seems to be so pronounced in the classical H-H objects (as given in Herbig's 1974 catalog) and that the individual condensations seem to be independent of each other in these objects whereas this is not the case in some of the more recently discovered H-H objects (c.f., Strom, Grasdalen and Strom, 1974).

Spectroscopic evidence shows that at least the classical H-H objects (like HH 1 and HH 2) are even more inhomogeneous than the direct observations indicate. There are not only large density differences between the condensations and the surrounding regions, but even within a condensation less than 1% of the volume is filled with matter sufficiently dense for the formation of the largest part of the emission-line spectrum. This fact leads to the very low masses of the visible parts of condensations (~1-10 earth masses, see above).

Variability

At least a considerable fraction of Herbig-Haro objects (if not all) is variable (c.f., Herbig, 1957, 1969, 1973; Strom, Grasdalen and Strom, 1974). The variability has been studied in detail and over a sufficiently long interval only for HH 1 (Herbig, 1973) and HH 2 (Herbig, 1957, 1969). It is important to note in this context that HH 1 and HH 2 are definitely not reflection nebulae as has been pointed out by Schmidt and Vrba (1975; see also Böhm, Siegmund and Schwartz, 1976). We are therefore certain that the observed variations refer directly to the line emitting regions.

The fundamental results of these studies are:

1. The brightness of individual condensations seems to vary independently of each other.
2. The time scale of the variations seems to be of the order of ten years.

It is not yet known to what extent these conclusions can be applied to other H-H objects.

How are the observed brightness variations related to changes in physical parameters? The information about this problem is even much more limited than that of the brightness variations themselves. It obviously requires spectrophotometric observations at various times and their interpretation in terms of, say, T_e and N_e. Very preliminary results are available for HH 1 and HH 2H, seen in Fig. 1. These are based on the results by Böhm, 1956; Böhm, Perry and Schwartz, 1973; Böhm, Siegmund and Schwartz, 1976; Böhm and Brugel, 1978.

Figure 1 shows the path described by HH 1 and HH 2H in a T_e-N_e diagram as a function of time. In the case of HH 1 we have also indicated the size of the crossing regions (c.f., Seaton, 1954; Böhm, Siegmund and Schwartz, 1976) in the T_e-N_e diagram for most of the forbidden-line ratios (except [O II] 3726/3729 and [S II] 6717/6731 which indicate lower density regions). It should be noted that the size of the crossing region usually appears considerably larger than the T_e-N_e uncertainty, due to errors in the collision strengths and transition probabilities, and it depends as well on the line ratios used. Correspondingly the overlap of the crossing regions for HH 1 in 1969 and 1973 probably does *not* imply that there was any real change in the physical parameters. However, as emphasized earlier (Böhm *et al.*, 1976) the interpretation of the mean-

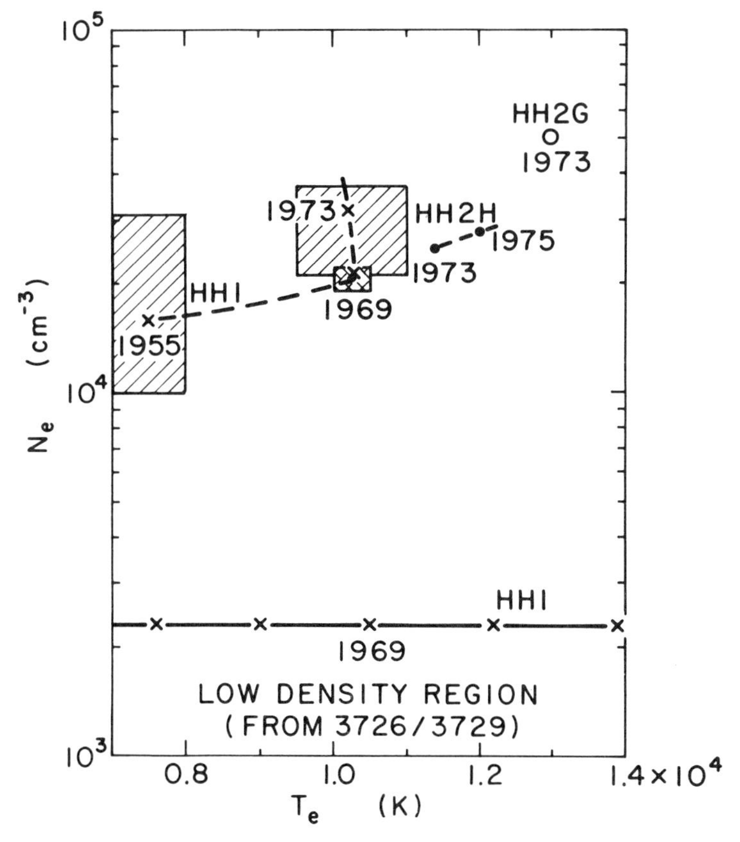

Fig. 1. The variation of the electron temperature T_e and the electron density N_e between 1955 and 1973 in HH 1 (crosses) and between 1973 and 1975 in HH 2H (dots). The size of the "crossing regions" (see text) in the T_e-N_e diagram for HH 1 is indicated by the hatched areas. The position for HH 2G in 1973 is also plotted. We also show the electron density for the low density component as derived from the [O II] 3726/3729 ratio. No time variation is known for this component.

ing of the changes in HH 1 is difficult because the main region of line emission has moved from the southeast part to the northwest part since 1954 (see Herbig, 1973).

The positions of HH 2H in 1973 and 1975 in the T_e-N_e plane are also shown in Fig. 1. In this case the time interval is too short to determine whether the apparent slight changes in T_e and N are real.

For comparison purposes we have also plotted the density in HH 1 determined from the [O II] 3726/3729 ratio which gives information about the low-density component (Böhm *et al.*, 1976). No information is yet available about time variations of this component (if there is any).

Basic Physical Parameters and Inhomogeneities in H-H Objects

Herbig-Haro objects show a very faint continuous (or quasi-continuous) spectrum (Herbig, 1951; Böhm, Schwartz and Siegmund, 1974) and an emission line spectrum consisting of a large number of forbidden lines (including [O I], [O II], [O III], [N I], [N II], [C I], [Ne III], [Ca II], [Fe II], [Fe III], [Cr II]) and permitted lines of H I, He I and Ca II. Since the forbidden lines are excited by electron collisions (c.f. Böhm, 1956, 1975; Osterbrock, 1958), we can use standard methods (Seaton, 1954) to determine T_e and N_e from the forbidden-line ratios. As is well known every observed line ratio gives a curve in the T_e-N_e plane. The center of the crossing region is then assumed to define characteristic (mean) values of T_e and N_e. The size of the crossing region is determined by (1) errors of the spectrophotometric observations, (2) errors in the transition probabilities and collisions strengths, and (3) inhomogeneities in N_e and T_e.

In many instances the third case will be the most important effect and we can derive some information about the (spatial) variation of T_e and N_e within the object from the T_e-N_e diagram (Böhm, Perry and Schwartz, 1973; Böhm *et al.*, 1976). It has been found that individual condensations of H-H objects in general show a well defined and relatively small crossing region except for the ratios [S II] 6717/6731 and [O II] 3726/3729. The latter ratio typically leads to a value of N_e which is smaller by a factor of $\geqslant 10$ than that indicated by the general crossing region. These effects together with the small emission measure of individual condensations lead us to a heuristic model in which even within a condensation (having a radius of about 200-500 AU) only a few tenths of a percent of the total volume is filled with matter with electron density of $3 - 6 \times 10^4$ cm^{-3}, whereas the largest part of the condensation contains matter with only $N_e \sim 10^3$ cm^{-3}. The latter is identical to the electron density in the immediate surroundings of the H-H object. The situation is illustrated in Fig. 2 (T_e-N_e diagram). A very schematic description of the resulting empirical model is shown in Fig. 3.

The typical electron temperatures cover the range from 7.5×10^3 K (SE part of HH 1 in 1955, see Böhm, 1956; Böhm *et al.*, 1976) to 13×10^4 K (HH

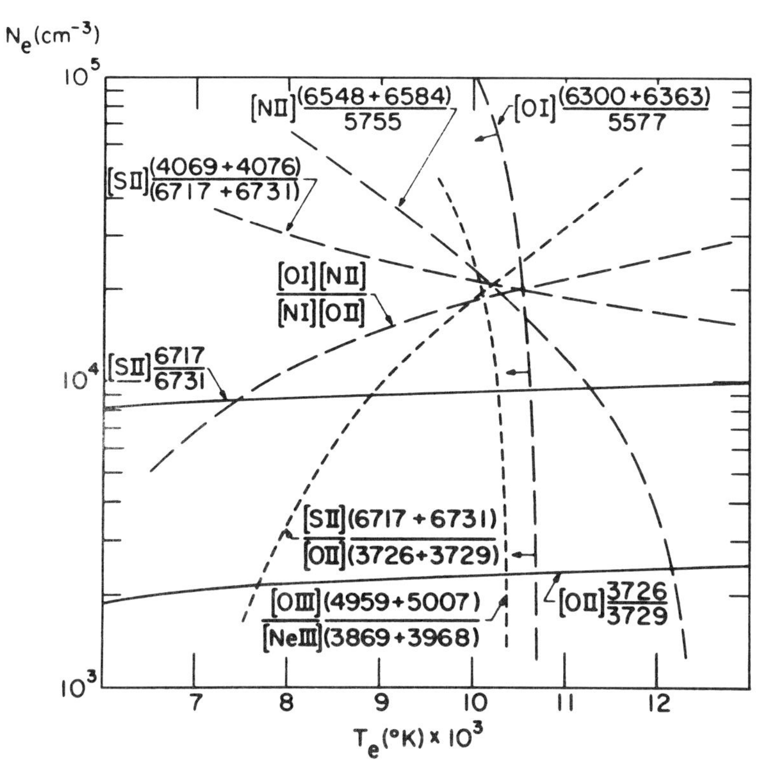

Fig. 2. T_e-N_e diagram for HH 1 in 1969. (See text for detailed explanations.) The diagram shows clearly the well defined "main crossing region" from which we derive the physical conditions in the high-density component. We also see the influence of the low-density component in the lower N_e values derived from the [S II] 6767/6731 and the [O II] 3726/3729 ratio (from *Astrophysical Journal* 179: 149, 1973).

2G in 1973, Böhm *et al.*, 1976). The electron densities in the high-density components lie in the range between 1.6 x 10^4 (HH 1 in 1955) and 5 x 10^4 cm^{-3} (HH 2G in 1973). Since the hydrogen is on the average only 50% ionized the total particle density will be greater by, very approximately, a factor 3 than these values.

Chemical Composition

A study of the abundances of chemical elements in Herbig-Haro objects is of great interest since it permits us to find out whether there is any depletion of the refractory elements due to dust formation. The question is especially whether or not Fe and Ca are depleted.

Preliminary abundance analyses are available for HH 1, HH 2H and, with much larger uncertainties, HH 2G (Böhm *et al.*, 1973; Böhm *et al.*, 1976). The

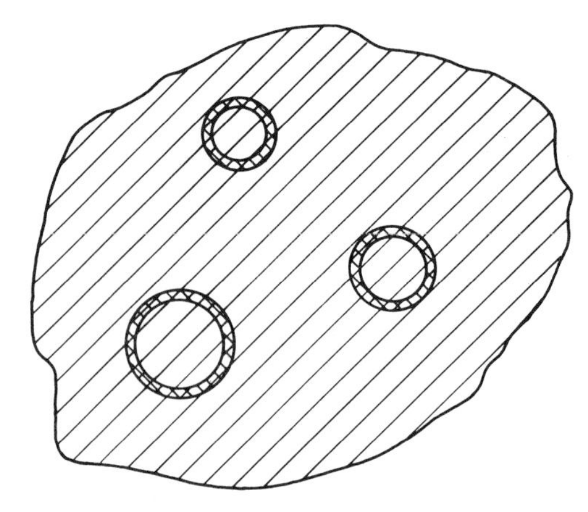

Fig. 3. Schematic picture of the empirical model of a "classical" H-H object (like HH 1, HH 2) showing the condensation structure and the small filling factor within a condensation. Information obtained from spectrophotometric data and from direct photographs has been used in the derivation of this model.

results for HH 1 and HH 2H are shown in Fig. 4. The abundances A are given relative to the cosmic abundances A_c as listed by Allen (1973). Obviously we can determine the element abundances only by summing over the *observable* ions of an element. Since we often observe only one ion per element it is not surprising that in these cases A/A_c is somewhat smaller than one. With this fact in mind we do not see any indication of a sizable metal depletion in Fig. 4. This is especially significant in the case of Ca which in normal H I regions shows the highest degree of depletion; values of less than 10^{-3} for the ratio of interstellar to cosmic abundance have been found in some cases (c.f., Spitzer and Jenkins, 1975). Taking into account the fact that Ca II has an ionization energy of only 11.87 eV, the Ca II abundance – of about 25% of the total Ca abundance for Population I – is surprisingly high. The normal (cosmic) abundance of Ca in H-H objects is independently confirmed by the shock wave calculations of Raymond (1977, see below). Using cosmic abundances he predicts a Ca II K emission-line flux which agrees with the observations. Certainly the Ca abundance in H-H objects does not differ by more than a factor of 2 from its cosmic abundance. Figure 4 does show one discrepancy which we do not yet understand. The C I abundance, determined from the 9849 line, is unexpectedly low – ~4% of the cosmic C abundance. This is surprising since the ionization energies of C I and Ca II differ very little. In fact, Raymond's calculations predict a 9849 line which is about 8 to 20 times stronger than the observed line (depending on the shock wave parameters used, see below). One possible explanation is that most of the carbon is still bound in graphite particles.

In view of the fact that Ca shows no depletion it would be important to check the Fe abundance. Fe is observed in the form of [Fe II] and very few

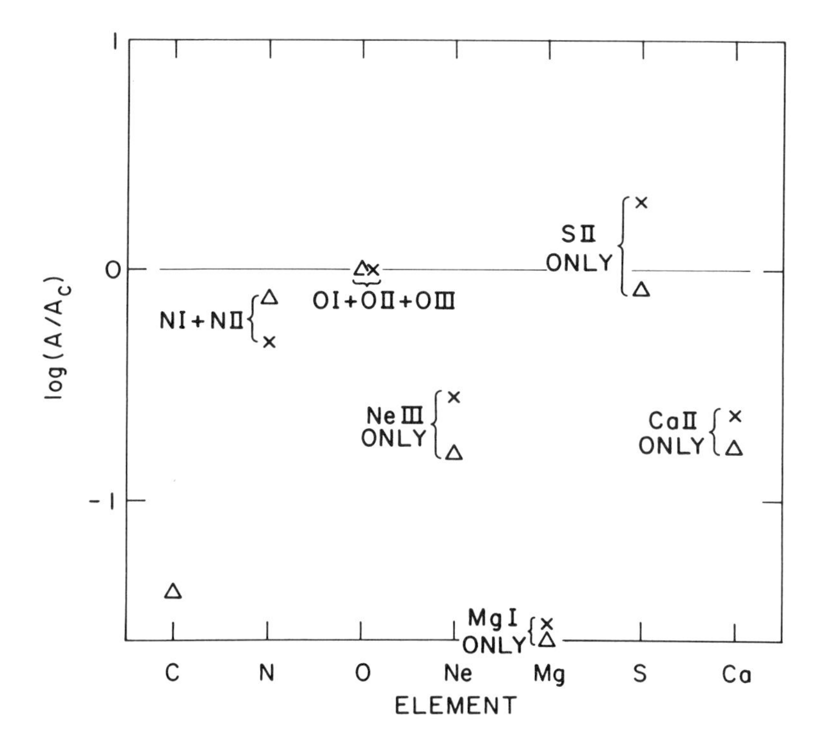

Fig. 4. The logarithm of the ratio of the observed abundance A to the "cosmic" abundances A_c (see text). Crosses refer to HH 1, triangles to HH 2H. The abundances shown here have been determined only from the ions shown in the diagram.

[Fe III] lines. Though transition probabilities for the observed lines are available from the work of Garstang (1962, 1967) we do not have reliable information about collision strengths for Fe II. In a very preliminary study using only few [Fe III] lines Böhm and Brugel (1978) found no apparent depletion of Fe.

Consequently we can state that there is no indication of metal depletion due to dust formation in H-H objects. This is surprising since we would expect that dust formation is important in regions of star formation. One possible explanation would be that the dust has been destroyed by the shock waves which seem to be present in H-H objects (see below).

Velocity Fields in H-H Objects

In connection with the shock wave interpretation of H-H objects the observed velocity field is of great interest. Early investigations showed that the average radial velocity of most Herbig-Haro objects seemed to be negative with respect to the surrounding interstellar gas. Radial velocities in the range −35 to -240 km sec^{-1} with an average value near -60 km sec^{-1} were observed (Herbig, 1962*b;* Strom, Grasdalen and Strom, 1974; Böhm, Perry and Schwartz, 1973; Münch, 1977). If interpreted as an expansion of the individual H-H object – or even the individual condensation – it leads to the intriguing question why we cannot see the backside of the object, which should exhibit a redshift. More recent studies by Herbig (1977) and Schwartz (1978) have indicated that the situation is considerably more complicated than we had assumed. Herbig found that in HH 32A one of the velocity components coming from the extension (outside the main nucleus of the H-H object) shows a very broad line centered at +280 km sec^{-1}! Recently Schwartz (1977) studied the radial velocities in a number of condensations ("nuclei") of HH 2. The results are rather complex and are certainly not restricted to blueshifts for all condensations. On the other hand, the list by Strom, Grasdalen and Strom (1974) shows that 14 out of 15 observed H-H objects have, in part large, negative radial velocities with respect to the surrounding gas. Consequently it is hard to avoid the conclusion that most H-H objects do really show negative radial velocities. However, it follows from the work by Herbig (1977) and Schwartz (1978) that there are also some rather drastic exceptions.

In most H-H objects the line widths do not seem to be very large. Böhm, Perry and Schwartz (1973) found line widths $\lesssim 1$ Å for HH 1. The much more precise measurements by Schwartz (1978) show line widths between 1 and 1.8 Å for HH 1. According to the same reference, line widths in the range 0.7 to 1.4 Å are observed in HH 2. Strom *et al.* (1974) find velocity dispersions of more than 165 km sec^{-1} in at least two H-H objects.

Polarization and the Presence of Infrared Sources

Strom, Strom and Grasdalen (1974), Strom, Grasdalen and Strom (1974), Strom, Strom and Kinman (1974) and Vrba, Strom and Strom (1975) found moderate to strong linear polarization in a number of H-H objects. Strom and his collaborators have considered this as strong evidence for their reflection hypothesis which is based on the assumption that the light that we receive from a Herbig-Haro object is reflected light coming originally from a strongly obscured "pre-T Tauri" star. This star may be visible as an infrared source. The most convincing case has been presented for HH 24 (Strom, Strom and Kinman, 1974). In this case the electric vector maximum in 2 condensations (*A* and *E*) is almost exactly perpendicular to the direction towards a nearby infrared object.

On the other hand, in the case of the classical brightest H-H objects HH 1 and HH 2 the observational evidence clearly shows that they are not reflection nebulae (Schmidt and Vrba, 1975; see also Schwartz, 1976; Böhm and Schwartz, 1973; Böhm, Siegmund and Schwartz, 1976).

It is still an open question whether or not these results imply that there are two different classes of Herbig-Haro objects. Some authors find that this conclusion must be drawn (c.f., Schmidt and Vrba, 1975; Gyul'budagyan, 1975) whereas others believe that it can be avoided by interpreting the observational results somewhat differently (c.f., Haro, 1976; Münch, 1977).

THEORETICAL INTERPRETATION OF H-H OBJECTS

Some General Remarks

During the last few years considerable evidence has accumulated which seems to show that the emission-line spectrum is formed in cooling regions of shock waves (Osterbrock and Schwartz, 1974, personal communication; Schwartz, 1975; Böhm *et al.*, 1976; Münch, 1977; Dopita, 1977*a, b;* Schwartz, 1978). This fact, of course, does not permit us to determine immediately the evolutionary role of H-H objects. At present we do not even know how the shock waves are generated. Nevertheless the description goes considerably beyond the purely heuristic explanation using the T_e-N_e diagram (see above) and the use of empirical filling factors. It permits a direct explanation of the observed degrees of ionization and it gives a more convincing description of the physical conditions in the regions of line formation. In order to keep our discussion simple we shall describe everything as if the line formation occurs always at the place where we observe it. (This is of course true for objects like HH 1, HH 2). We are aware of the fact that at least in some cases the reflection effect will have to be taken into account for a complete description. Other hypotheses for the explanation of the line formation have been suggested, e.g., ionization by moderately fast protons (Osterbrock, 1958; Magnan and Schatzman, 1965; Gyul'budagyan, 1975) or by transition radiation (Gurzadyan, 1974, 1975). However, we feel that at the present time the shock wave theory is considerably more successful in making detailed predictions of the observed spectra as well as in meeting objections than other theories of line formation. Consequently we shall restrict the following discussion to the shock wave theory.

The Shock Wave Interpretation of the Spectra of H-H Objects

Osterbrock and Schwartz (1974, personal communication) drew attention to the fact that there is a great quantitative similarity between the typical spectra of H-H objects and those of supernova remnants like e.g., N 49 whose spectrum has been studied by Osterbrock and Dufour 11973). These authors had shown that the spectrum of N 49 can be approximately explained by the theoretical predictions of shock wave spectra made by Cox (1972*a*). These facts immediately lead to the conclusion that very probably the spectra of H-H objects are also formed in cooling regions of shock waves. In order to appreciate the need for this conclusion we must remember that earlier there had been no suggestion to permit even a qualitative explanation of the peculiar features of H-H spectra especially the great strength of the neutral lines and the moderate strength of the Ca II K line. The first step in the direction of a really quantitative explanation was made by Raymond (1977) who predicted emission-line spectra under Herbig-Haro conditions.

If our knowledge of HH 1 and HH 2 can be used as a guideline then H-H objects are typically imbedded in rather high-density interstellar gas of $N_e \sim 10^3$ cm^{-3} or slightly less (Böhm, Perry and Schwartz, 1973). Consequently the calculations of Cox (1972*a*) are not directly applicable since they assume a shock wave moving into a medium with $N_e \sim 1$. Raymond (1977) has investigated three models of shock waves relevant to the problem of H-H objects having the following basic parameters

(1) $n_o = 300$ cm^{-3}, $V_s = 60$ km sec^{-1}, $H = 10^{-7}$G

(2) $n_o = 350$ cm^{-3}, $V_s = 60$ km sec^{-1}, $H = 10^{-7}$G

(3) $n_o = 300$ cm^{-3}, $V_s = 70.7$ km sec^{-1}, $H = 10^{-7}$G

with n_0 the pre-shock particle density, V_s the shock velocity and H the pre-shock magnetic field strength (G in gauss).

For models (2) and (3) the cosmic abundances as listed by Allen (1973) have been used. The abundance set for model (1) also agrees with the cosmic abundances except that C and Fe have been reduced by a factor 5, Mg has been reduced by a factor 40, while Ne has been increased by a factor 2.

A comparison of the theoretically predicted line fluxes for the three models with the observed ones for HH 2H is made in Fig. 5. In order to appreciate the results we must remember that it would of course be sufficient if one shock

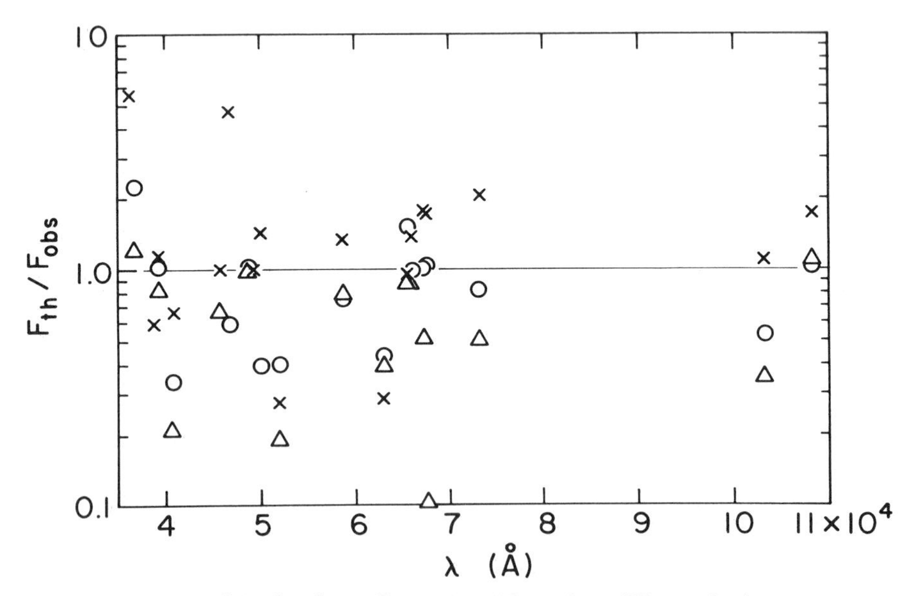

Fig. 5. The ratio of the line fluxes F_{th} predicted from three different shock wave models (Raymond, 1977) to the observed line fluxes F_{obs} as a function of the wavelength. Circles correspond to the theoretical Model (1), crosses to Model (2) and triangles to Model (3) (see text). In judging the scatter in this diagram we have to keep in mind that the measured line fluxes cover a range of about two orders of magnitude.

wave model would agree with the observations. Secondly we should bear in mind that no special effort has been made to adjust the shock wave parameters in such a way that optimal fitting was achieved. Finally we must keep in mind that the observed line fluxes (used in the diagram) cover a range of almost two powers of ten.

It is therefore quite encouraging that the large majority of points does not deviate from the line $F_{th}/F_{obs} = 1$ by more than a factor of ~ 2. Most of those few lines which do show larger deviations from $F_{th}/F_{obs} = 1$ are relatively faint (e.g., [N I] 5200 Å) and may contain errors in their spectrophotometry. One more serious discrepancy occurs in the line [O I] 6300 Å. Although this line is strong and there are no doubts about the spectrophotometry, the theoretical prediction is lower by a factor 2.3 than the observed value. However, even this cannot be an argument against the shock wave theory in general but only against certain details since the same type of discrepancy is also observed in some supernova remnants (see Danziger and Dennefeld, 1974). Essentially the same result is obtained if we carry out this comparison for other HH objects (e.g., for HH 1, see Böhm *et al.*, 1976) for which detailed spectrophotometric data are available (see Dopita, 1977*b*).

Our general conclusion is that for those H-H objects for which detailed spectrophotometry is available the emission-line spectrum seems to be formed in the cooling regions of shock waves. No explanation can yet be given for the very faint continuum (c.f., Böhm, Schwartz and Siegmund, 1974; Böhm, 1975; Schwartz, 1976) which shows a rather strange energy distribution in visible light and an increase towards the ultraviolet which may be related to the ultraviolet excess in T Tauri stars.

The important remaining questions are:

1. What other properties of H-H objects can be explained in terms of the shock wave hypothesis?
2. Where do the shock waves originate?
3. What role do they play in the process of star formation and early stellar evolution?

These questions will be discussed in the following two sections.

Possible Explanation of Other Properties of H-H Objects by the Shock Wave Hypothesis

While the philosophy of the shock wave interpretation of the *spectra* is fairly straightforward we now have to face the question as to which other properties of H-H objects need to be explained and which properties should be considered as accidental. At present the observational material is much too limited to give a definite answer to this type of question.

Should the size, the typical time scale and the filling factor of the individual condensations (nuclei) be explained by the theory? Is the fact that many (but not all) H-H objects consist of a number of individual condensations (Herbig, 1969, 1973, 1974) significant? I personally tend to answer yes to all these questions but no general consensus has been reached concerning this point.

It is interesting to note that the visible sizes, the time scales, the filling factors and the expansion velocities of individual condensations can be predicted if we assume that the individual condensations are approximately spherical blast waves which expand into a medium with $N_e \sim 10^3$ cm^{-3} (Böhm, 1978*a*), where we have taken this number from the observations (see above). The theory is based on the work of Cox (1972*b*). The required energy input into a "typical" condensation is approximately 7 x 10^{41} ergs. For these parameters we find (Böhm, 1978*a*) that the shock should be visible for about 14 years, that during this time the radius of the condensation should expand from about 600 to 900 AU and that the (matter) expansion velocity near the end of the visible phase is ~ 59 km sec^{-1} ($V_s \sim 79$ km sec^{-1}). The filling factor is ~ 7 x 10^{-3}.

All these numbers are in reasonable agreement with observations of HH 2 and HH 1. Further observational and theoretical studies are needed in order to find out whether or not this agreement is fortuitous.

A very different approach to the problem has been used by Schwartz (1978). He does not try to explain sizes, time scales, etc. of condensations but considers them more or less accidental. Schwartz assumes that the visible condensations are due to bow shocks which are formed when a (quasi-stationary) stellar wind hits very small clouds of high-density material that are already present. This approach is somewhat analogous to that of McKee and Cowie (1975) and of Sgro (1975) who consider the bow shocks of cloudlets created by a supernova blast wave. The great advantage of Schwartz' hypothesis is that it connects the Herbig-Haro objects with a well-known phenomenon, namely the strong stellar wind in the early phases of stellar evolution (see Kuhi, 1964; Strom *et al.*, 1975). However, I tend to consider it as a drawback that the sizes, time scales of variability and their variation independent of each other – at least in HH 2 (see Herbig, 1969) – appear as accidental in this theory.

Further detailed theoretical studies based on more comprehensive information will be necessary before we can be certain about the details of the shock wave mechanism in H-H objects.

The Problem of the Origin of the Shock Waves and their Relation to Star Formation and Early Stellar Evolution

If we accept the conclusion that the emission lines must be formed in cooling regions of shock waves we are immediately led to the problem of the origin of these shocks. In the hypothesis of Schwartz (1978) the problem is reduced to the origin of the strong stellar wind. In our own hypothesis (Böhm, 1978*a*, see above) we only know that we need a total energy input of 10^{41} to 10^{42} erg for every H-H condensation.

One fundamental difficulty in every theory that has been suggested so far (except for the reflection hypothesis which, however, is not relevant to objects like HH 1 and HH 2) is the following: A starlike (or protostar-like) object is needed very close to the H-H objects as the energy source of the shock wave or the stellar wind. No such object has yet been observed in cases like HH 1 or HH 2, but there is some hope that eventually we may discover central objects in at least HH 1 and HH 2. Schmidt and Vrba (1975) have found infrared sources within these objects (in contradistinction to the infrared sources studied in connection with other H-H objects found; e.g., by Strom, Strom and Grasdalen, 1974). Schmidt and Vrba (1975) find K magnitudes of 12.30 (HH 1) and 12.62 (HH 2). They point out that the infrared luminosity could be attributed to free-free transitions. However, this requires a filling factor which is about 20 times larger (see Schwartz, 1978) than that determined from the spectral lines in the visible part of the spectrum (Böhm, Siegmund and Schwartz, 1976). Consequently it is still possible that there are real infrared sources in HH 1 and HH 2. An even stronger argument in favor of such sources is the large K-L color index in HH 2 ($\sim$2.2 mag). In this connection it is necessary to remember (see Böhm, 1978*b*) that e.g., HH 1, HH 2 and HH 3 cannot obtain their energy from a stellar wind coming from the obvious source, the nearby young object V380 Orionis. This would require an energy supply rate of at least 2.5×10^{38} erg sec^{-1} ($\sim 0.6 \times 10^5 L_\odot$) in the form of a stellar wind alone which we consider, to say the least, as very improbable (Böhm, 1978*b*).

Another way of looking at our problem in very general terms would be the following. The total energy output of the northwest part of HH 1 in lines-plus-continuum in the spectral range 3300 Å $\leqslant \lambda \leqslant$ 10,900 Å is $\sim 2 \times 10^{32}$ erg sec^{-1}, $\sim 0.05 L_\odot$ (Böhm, 1975). The total visible mass in this region is only 7 earth masses. This makes it probable that, regardless of the mechanism, some, as yet undiscovered, object is needed to supply the energy. As we saw above this object has to be very close to, if not inside, the H-H object.

What do the shock waves whose effects are seen in H-H objects have to do with the processes of star formation or early stellar evolution? It seems probable that we shall be able to answer this question once we know what type of objects supply the energy to the shock waves (or, in Schwartz' theory, supply the stellar wind).

If the variability of the condensations is really due to the shock waves being non-stationary (as assumed by Böhm, 1978*a*) it would be necessary for some sort of instability to be present in the process of star formation which leads to the occasional generation of shock waves. If Schwartz' (1978) approach is correct the main problem would be the generation of a strong stellar wind in the very early phases of stellar evolution. We know from observation that at least some young stellar objects show a very strong stellar wind (Kuhi, 1964; Schwartz, 1975; Kuhi and Ulrich, personal communication, 1978). In Schwartz' theory we may consider the H-H objects as probes which can supply us with information about the strong stellar winds in regions of star formation.

Acknowledgments. I am very much indebted to G. H. Herbig, J. Raymond and R. D. Schwartz for making their results available in advance of publication. This study has been supported by a grant from the National Science Foundation.

REFERENCES

Allen, C. W. 1973. *Astrophysical Quantitites* (3rd ed.) London: Athlone Press.
Ambarzumian, V. A. 1954. *Comm. Burakan Obs.* No. 13.
Böhm, K. H. 1956. *Astrophys. J.* 123: 379.
______. 1975. In *Problems in Stellar Atmospheres and Envelopes* (B. Baschek, W. H. Kegel and G. Traving, eds.) p. 205. Heidelberg-New York: Springer-Verlag.
______. 1978*a*. *Astron. Astrophys.* 64: 115.
______. 1978*b*. In *Proc. I.A.U. Colloq. No. 42* (W. Strohmeier, ed.) Bamberg Ser. Var. Star Coll. p. 3.
Böhm, K. H., and Brugel. 1978. In preparation.
Böhm, K. H.; Perry, J. F.; and Schwartz, R. D. 1973. *Astrophys. J.* 179: 149.
Böhm, K. H., and Schwartz, R. D. 1973. *Bull. Amer. Astron. Soc.* 5: 437.
Böhm, K. H.; Schwartz, R. D.; and Siegmund, W. A. 1974. *Astrophys. J.* 193: 353.
Böhm, K. H.; Siegmund, W. A.; and Schwartz, R. D. 1976. *Astrophys. J.* 203: 399.
Cox, D. P. 1972*a*. *Astrophys. J.* 178; 143.
______. 1972*b*. *Astrophys. J.* 178; 159.
Danziger, I. J., and Dennefeld, M. 1974. *Astron. Astrophys.* 36: 149.
Dopita, M. A. 1977*a*. *Astrophys. J. Suppl.* 33: 437.
______. 1977*b*. Paper presented at I.A.U. Symp. No. 76 (Y. Terzian, ed.) Dordrecht: Reidel.
Garstang, R. H. 1962. *Mon. Not. Roy. Astron. Soc.* 124: 321.
______. 1967. In *I.A.U. Symp. No. 34* (D. E. Osterbrock and C. R. O'Dell, eds.) p. 134. Dordrecht: Reidel.
Gurzadyan, G. A. 1974. *Astron. Astrophys.* 33: 307.
______. 1975. *Astrofizika* 11: 531.
Gyul'budagyan, A. L. 1975. *Astrofizika* 11: 511.
Haro, G. 1952. *Astrophys. J.* 115: 572.
______. 1976. *Bol. del Instituto de Tonantzintla* 2: 3.
Herbig, G. H. 1950. *Astrophys. J.* 111: 11.
______. 1951. *Astrophys. J.* 113: 697.
______. 1957. In *I.A.U. Symp. No. 3* (G. H. Herbig, ed.), p. 3. Cambridge: University Press.
______. 1962*a*. *Adv. Astron. Astrophys.* 1: 75.
______. 1962*b*. *Astron. J.* 67: 645.
______. 1969. In *Nonperiodic Phenomena in Variable Stars* (L. Detre, ed.), p. 75. Dordrecht: Reidel.
______. 1973. Comm. 27. *I.A.U. Inf. Bull. Var. Stars* No. 832.
______. 1974. *Lick Obs. Bull.* No. 658.
Kuhi, L. V. 1964. *Astrophys. J.* 140: 1409.
Magnan, C., and Schatzman, E. 1965. *C. R. Acad. Sci. Paris* 260: 6289.
McKee, C. F., and Cowie, L. L. 1975. *Astrophys. J.* 195: 715.
Münch, G. 1977. *Astrophys. J.* 212: L77.
Osterbrock, D. E. 1958. *Publ. Astron. Soc. Pacific* 70: 399.
Osterbrock, D. E., and Dufour, R. J. 1973. *Astrophys. J.* 185: 441.
Raymond, J. 1977. Ph.D. Dissertation, University of Wisconsin.
Schmidt, G. D., and Vrba, F. J. 1975. *Astrophys. J.* 201: L33.
Schwartz, R. D. 1974. *Astrophys. J.* 191: 419.
______. 1975. *Astrophys. J.* 195: 631.
______. 1976. *Publ. Astron. Soc. Pacific* 88: 159.
______. 1977. *Astrophys. J.* 212: L25.
______. 1978. Submitted to *Astrophys. J.*

Seaton, M. J. 1954. *Mon. Not. Roy. Astron. Soc.* 114: 154.
Sgro, A. G. 1975. *Astrophys. J.* 197: 621.
Spitzer, L., and Jenkins, E. G. 1975. *Ann. Rev. Astr. Astrophys.* 13: 133.
Strom, K. M.; Strom, S. E.; and Grasdalen, G. L. 1974. *Astrophys. J.* 187: 83.
Strom, K. M.; Strom, S. E.; and Kinman, T. D. 1974. *Astrophys. J.* 191: L93.
Strom, S. E.; Grasdalen, G. L.; and Strom, K. M. 1974. *Astrophys. J.* 191: 111.
Strom, S. E.; Strom, K. M.; and Grasdalen, G. L. 1975. *Ann. Rev. Astr. Astrophys.* 13: 187.
Vrba, F. J.; Strom, S. E.; and Strom, K. M. 1975. *Publ. Astron. Soc. Pacific* 87: 337.

THE SPECTRAL APPEARANCE OF SOLAR-TYPE COLLAPSING PROTOSTELLAR CLOUDS

CLAUDE BERTOUT
Landessternwarte Königstuhl
and
HAROLD W. YORKE
Max-Planck-Institut für Physik und Astrophysik, München

We review the spectral properties of collapsing protostellar clouds, based on radiative transfer computations in hydrodynamic protostar models. The basic results of protostar evolution computations in spherically symmetric and axially symmetric geometries, as they pertain to the appearance of protostars, are briefly reviewed. We discuss the continuum appearance of spherically symmetric protostars with various masses. We also present recent results for the appearance of the continuum for an axially symmetric protostellar cloud. We describe the line formation problem and preliminary results for an OH molecule in an axially symmetric collapsing cloud. Theoretical and observational results are obtained for the last evolutionary phase of protostars, known as the YY Orionis phase, when the stellar core first becomes visible in the optical range.

Rotating collapsing clouds are in general less luminous and cooler than corresponding non-rotating clouds, due to the longer evolutionary time scale. Nevertheless, high-resolution ($\lesssim 1$ arsec) studies at infrared wavelengths ($\lambda \sim 10$ μm) are sufficient to reveal some details of possible disk-like condensations. Extreme caution must be exercised when interpreting line widths in order to derive maximum cloud infall velocities.

I. THE EVOLUTION OF PROTOSTELLAR CLOUDS

Protostellar clouds are the smallest fragments of a larger molecular cloud which are gravitationally unstable; that is, the internal gravitational forces of attraction overcome the combined forces of internal gas pressure, magnetic fields, turbulence and rotation. It is from this material that stars are formed either individually or as members of a multiple system. It is from this material also that planets are formed.

An important aspect of the study of the formation of protostars and planets is the qualitative and quantitative theoretical prediction of the physical structure and appearance of a protostellar cloud undergoing collapse and evolving towards the final stage of a main-sequence star surrounded by a system of planets, planetoids and comets. The basic features of the evolution of such a collapsing cloud have been discussed extensively in the literature (Larson, 1973, 1977; Kippenhahn and Tscharnuter, 1975; chapter by Bodenheimer and Black in this book). In spite of the large differences in the initial conditions and assumptions adopted by various authors, the results of different numerical collapse calculations are at least qualitatively quite similar. In all cases the collapse proceeds non-homologously; the central regions of the protostellar cloud collapse most rapidly and a small condensation in hydrostatic equilibrium is formed.

During the subsequent evolution the central condensation becomes more dense, while at the same time it accretes material from the surrounding infalling envelope. During this phase of mass accretion the details of the numerical calculations and their assumed initial conditions become important. Thus, the initial mass, density and temperature of the protostellar cloud, its angular momentum, the optical properties of the dust and angular momentum transfer rates are all important parameters governing the evolution during the accretion phase.

The accretion phase ends when no new material falls into the center (or onto the disk which forms when considering rotating clouds). The reason for the end of the accretion phase could be attributed to possibly four effects.

1. All the material in the protostellar cloud has fallen onto the core.
2. Part of the cloud is expelled due to radiation pressure on the dust, the occurrence of stellar winds, or ionization of the envelope surrounding the accretion core.
3. The overproduction of energy in a non-thermally adjusted core leads to overheating and rapid expansion of the outer layers of the core when nuclear burning starts, similar to the "helium flash" in more evolved stars.
4. The transfer of angular momentum towards the outer regions results in the stopping of the collapse of material with high angular momentum.

All of these mechanisms should have important observational consequences. When discussing the continuum appearance of protostellar clouds during the various phases of collapse, the density and temperature distribution of the absorbing material as well as its optical properties are needed. For the line formation problem, the velocity field and population distribution of the atom or molecule considered must also be known. In general, there are many free parameters which can be adjusted to fit the observations of very young stars or suspected protostars. In the following, we shall restrict ourselves to the results of known hydrodynamic collapse calculations in the spherically symmetric and axially symmetric cases. Examples of calculations of such continuum radiation transfer are presented by Yorke (1977*a*) for the high-mass (50 $M_\odot$ and 150 $M_\odot$) spherically symmetric, collapsing protostars computed by Yorke and Krügel (1977), and by Bertout (1976) for the collapsing 1 $M_\odot$ protostar computed hydrodynamically by Appenzeller and Tscharnuter (1975). The resulting theoretical spectra show good agreement with infrared observations of protostellar candidates and very young stars. Line profile formation in spherically symmetric low-mass protostars prior to the onset of hydrogen burning has been investigated by Bertout (1977*a*) and Wolf *et al.* (1977), who compare computed line profiles with observed high-dispersion spectra of YY Ori stars.

In spite of the success of these spherically symmetric investigations in explaining many qualitative features of young pre-main-sequence objects, it has become apparent that more sophisticated models are needed to obtain better qualitative agreement with observations. Furthermore, the existence of multiple stellar systems and planetary systems (at least one) clearly indicate that rotation must play an important role during the formation process. Thus, we shall include in the following the theoretical results obtained for axially symmetric, rotating protostellar clouds. But first let us summarize the hydrodynamic evolution of spherically symmetric, and then the axially symmetric, protostellar clouds, emphasizing the effect that the hydrodynamic evolution should have on the appearance of the spectrum. For a more complete review the reader is referred to the chapter by Bodenheimer and Black.

A. Spherically Symmetric, Hydrodynamic Calculations.

Just as in the case of main-sequence stars, it is often useful to distinguish between high-mass ($M \gtrsim 3M_\odot$) and low-mass ($M \lesssim 3\, M_\odot$) protostellar clouds. There are several reasons for this, which may be summarized as follows:

1. For a high-mass protostar hydrogen burning will start before the accretion of material has stopped. The low-mass protostar, however, evolves towards the main sequence slowly, contracting on a Kelvin-Helmholtz time scale, even after the envelope is exhausted. Thus, the low-mass protostar should first become optically visible on the lower end of its Hayashi track, whereas stars of higher mass will spend part of their main-sequence lifetime completely obscured optically.

2. As can be seen from Fig. 1, constructed from the results of a number of authors, the protostars of high mass will not accrete all of their available protostellar material, for a variety of reasons. According to Appenzeller and Tscharnuter (1974) the rapid compression of the high-mass protostellar core will, if the core cannot thermally adjust fast enough, lead to an

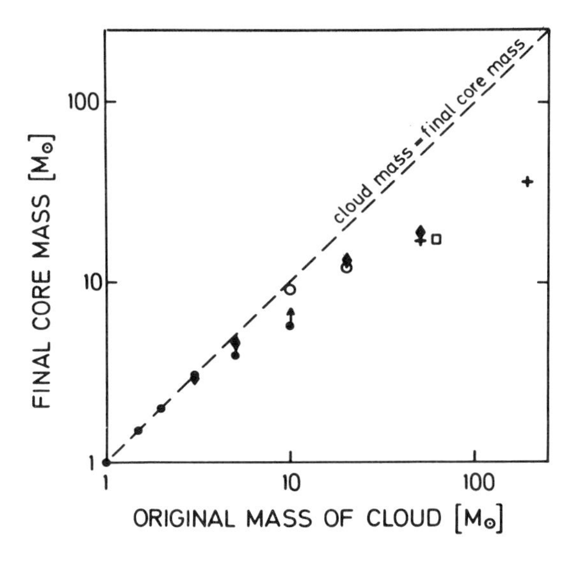

Fig. 1. The results of calculations of the spherical collapse of protostellar clouds is displayed together on a plot of the "initial cloud mass" versus "final mass of the core." The results of Larson (1972) are denoted by filled circles (●), Westbrook and Tarter (1975) by filled diamonds (♦), Appenzeller and Tscharnuter (1974) by an open square (□), Yorke and Krügel (1977) by crosses (+), and Yorke (1978) by open circles (○). Note that the formation of massive protostars is less efficient and the corresponding points therefore fall far below the dashed line for which the initial cloud mass is equal to the final mass of the resulting star.

overproduction of energy and can thus cause overheating and rapid expansion of the outer layers of the core. In the case of a 60 $M_\odot$ protostellar cloud only 16 $M_\odot$ remained in the core. Kahn (1974) has shown that radiative acceleration should become important in the envelopes of accreting high-mass protostars. In a detailed numerical calculation of the evolution of protostellar envelopes, Yorke and Krügel (1977) have found that an increasing fraction of the envelope is blown off by radiative acceleration when one considers clouds of higher mass. In these calculations, a 50 $M_\odot$ protostellar cloud produced a 16 $M_\odot$ star (32% efficiency) whereas a 150 $M_\odot$ cloud was necessary for producing a 36 $M_\odot$ star (24%). The 5 $M_\odot$ and 10 $M_\odot$ cases presented by Larson (1972*b*) are only lower limits to the final mass of the core, because these numerical calculations were stopped once hydrogen burning started.

For the low-mass protostar gravity is always larger than the radiative acceleration on the protostellar material, provided that the dust is "frozen" into the gas. During late evolutionary stages when the density in the envelope has decreased to the extent that dust and gas are no longer effectively coupled, radiative acceleration could rid the envelope of its dust, but certainly not of its gas.

3. Another distinction between high- and low-mass protostellar clouds is

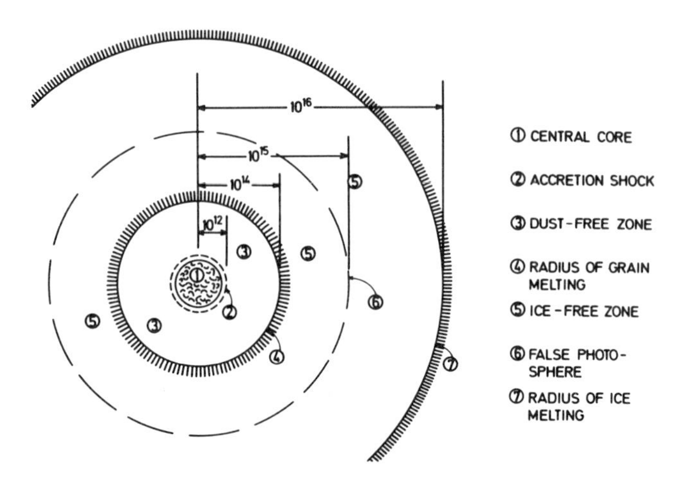

Fig. 2. Schematic representation of the structure of a collapsing protostellar cloud after the formation of a central (accreting) core (see text).

the types of objects one can identify with each. Due to their higher luminosity and larger fraction of ionizing radiation, the protostellar clouds of higher mass should be observable as "hot spots" of enhanced molecular radiation, masers, luminous infrared sources and compact H II regions. The low-mass protostars, on the other hand, are to be associated with weaker infrared sources during the phases in which they are obscured optically, and with (pre-main-sequence) T Tauri and YY Orionis stars when they first become optically visible.

In spite of these basic differences, there are many similarities between spherically symmetric, accreting protostars during the accretion phase. In Fig. 2 we display schematically the structure of the protostellar cloud during the accretion phase. The approximate sizes of the various regions are given in cm for a $20\,M_\odot$ cloud after about one half of the mass has accreted onto the core (Yorke, 1978).

For all cases calculated to date the core is always surrounded by an accretion shock front, where the kinetic energy of the (almost) freely falling material is converted into heat and then radiated away. During most of the accretion phase nearly all of the luminosity is produced in this shock front. This can be qualitatively explained by the following crude argument. The ratio of the accretion time scale (defined by $t_{acc} = M/\dot{M}$, where M is the core mass, $\dot{M}$ the mass accretion rate) to the Kelvin-Helmholtz time scale of the core (defined by $t_{KH} = \dfrac{GM^2}{R_*}L$, where R_* is the core radius, L the intrinsic core luminosity without the contribution from the shock) is equal to the ratio of L to the shock luminosity $L_{shock} = \dfrac{GM\dot{M}}{R_*}$. Written as an equation

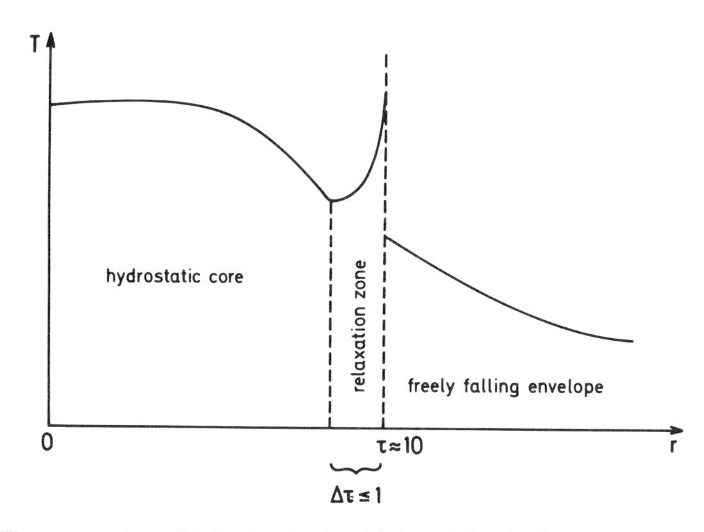

Fig. 3. The temperature distribution in the vicinity of the shock front surrounding an accreting protostar is displayed schematically. Note the pre-heating of material in the freely falling envelope, the overshoot at the discontinuous shock and the relaxation zone, where the heat is radiated away (from Tscharnuter, 1976).

this becomes $t_{acc}/t_{KH} = L/L_{shock}$. On the one hand, this equation tells us why the shock luminosity is so much greater than the intrinsic core luminosity during the accretion phase when $t_{acc} \ll t_{KH}$. (Without accretion the core can evolve only as fast as it can radiate away its internal energy, which is slow compared to how fast it accretes material when the core mass is small.) On the other hand, one can also qualitatively understand why the more massive stars reach the main sequence before the accretion stops. Considering similar envelope structure, L increases faster with the core mass than the shock luminosity; when the energy is transported by radiation, for instance, one expects $L{\sim}M^3$, whereas it can be shown that $L_{shock}{\sim} M^{1.5}\ \rho_s r_s^{0.5}$, where ρ_s, r_s are the upstream density and radius of the shock, respectively. Therefore $t_{KH}{>}t_{acc}$ for a sufficiently high core mass.

In Fig. 3 we display schematically the temperature distribution in the immediate vicinity of the accretion shock front, as has been calculated by Winkler (1977), for a 1 $M_\odot$ protostellar cloud. Because he did not rely on "artificial viscosity" to smear out the shock front, Winkler was able to resolve the shock structure. Between the mechanical shock, where the kinetic gas temperature changes discontinuously, and the hydrostatic core of the protostar there is a relaxation zone where the heat is radiated away. In front of the shock the freely falling material in the envelope is preheated by the sum of the shock luminosity and intrinsic core luminosity. Typical values of the total luminosity for the solar mass case, given similar initial conditions, vary from several solar luminosities to a maximum of 200 $L_\odot$. Because of the relatively high luminosities, the radiation field in the immediate vicinity of the core and accretion shock is much too intense for dust grains to survive and out to a radius r_m (defined as the radius at which grains are destroyed) the gas is dust-free and essentially opacity-free. Beyond this "sublimation" radius, r_m, the radiation field has become sufficiently diluted geometrically so that the more refractory grains, graphites and silicates can survive. A typical value for r_m is about 10 to 100 times the core radius.

Even though the outer regions of the protostellar envelope may be shielded from the optical and ultraviolet radiation from the core, the radiation field is still too intense for the more volatile grain materials consisting of H_2O-ices, NH_3-ices, etc. Only in the far outside regions of the cloud, typically of the order of 100 times the graphite or silicate sublimation radius, is it possible to have ice-coated grains. We shall see in the next section how this can affect the infrared continuum appearance of such protostars.

Whether or not the core of the protostar is at all visible optically depends on the amount of dusty material in the envelope. In the schematic diagram of Fig. 2 all of the optical and ultraviolet radiation from the core is completely absorbed and reradiated in the infrared in a thin ($\Delta r \approx 10^{12}$ cm) transition region just outside the dust sublimation radius. Thus, the infalling dusty envelope acts as a false photosphere. The object appears to be an infrared source with a temperature $\sim 10^2$ K to 10^3 K.

In Sec. II we estimate the total optical depth of the envelope by use of a crude analytic model and thus predict when the protostar should be optically visible.

B. Axially Symmetric, Hydrodynamic Calculations.

It has become increasingly apparent that rotation plays a major role in the evolution of protostars and the formation of planets (see the chapter by Bodenheimer and Black). We have restricted ourselves in our solutions of the radiative transfer equation in axially symmetric, rotating protostellar clouds to numerical hydrodynamic calculations by Tscharnuter (unpublished). It should be noted that Völk *et al.* (1978) have also used these hydrodynamic results in calculating the dynamics of the dust and formation of a thin dust disk. These models have a distinct advantage over earlier calculations because the evolution of the optically thick core could be followed to a more advanced stage.

The early collapse proceeds in a non-homologous manner similar to that encountered in the spherically symmetric calculations. Even if the initial angular velocity of the cloud material is so large that centrifugal forces are greater than or equal to gravity in some regions, a central condensation will form due to the infall of material of relatively low angular momentum close to the axis of rotation. The exact structure of this central condensation varies greatly from author to author and depends sensitively on both the initial conditions and the numerical method used. If the initial rotational velocity is low, a somewhat flattened optically thick core can form before rotational effects become important, and the subsequent evolution should proceed in a manner similar to the spherically symmetric case. The hydrostatic core is surrounded by an accretion shock which is stronger at the poles than at the equator. For higher values of initial angular velocity, the collapse at the equator can be stopped and even reversed. The result is a flattened accretion disk; shock fronts form at the poles only. In some cases calculated, a torus or ring will form in the equatorial plane, in other cases it will not. This matter is discussed by Bodenheimer and Black in this book.

At this point the axially symmetric calculations deviate greatly from the corresponding spherically symmetric results. First of all, the relevant time scales governing the further evolution in the rotating case are, in addition to the Kelvin-Helmholtz and accretion time scales, the time scale for the transport of angular momentum. This has important consequences for the luminosity and spectral appearance of the protostellar cloud.

As shown by these hydrodynamic calculations, the relatively slow evolution of the rotating and flattened central condensations – slow in comparison with corresponding non-rotating protostars – means that the energy output per unit time (the luminosity) is also lower. Whereas the Kelvin-Helmholtz contraction and the conversion of the kinetic energy of the infalling material into heat in an accretion shock front are the main sources of luminosity for the spherically symmetric protostar (both sources come from the reservoir of gravitational potential energy); the central condensation of high angular velocity can no longer evolve as fast as it can radiate away its internal energy. Depending on how fast angular momentum can be transfered out of this core, the available gravitational energy is released over a longer period of time. The combined results of calculations for hydrodynamic and continuum radiation transfer show also that the temperatures at the center of rotating and collapsing protostellar clouds are thus much lower than in the corresponding non-rotating cases; this also reflects the fact that the heat produced by compression can be radiated away at a greater rate from the surface of a flattened disk-like core than from the surface of a spherical one. The rate of heat production by the viscous turbulent friction assumed for the calculations was not high enough to affect the continuum appearance significantly during the early evolutionary stages covered by the calculations. Thus, the collapsing protostellar cloud of high initial angular momentum should appear as a much fainter and cooler infrared object than such a cloud with little or no angular momentum. This argument holds as long as the core does not fragment.

In Figs. 4*a* and *b* we show the temperature and density contours obtained by Tscharnuter at two evolutionary stages at which 0.02 $M_\odot$ (0.12 $M_\odot$) accreted onto the flattened ring-shaped core. The total luminosity is L = 0.095 $L_\odot$ (0.166 $L_\odot$), the evolutionary age after the collapsing cloud becomes optically thick is 580 yr (4160 yr) and the ratio of total rotational energy, E to the gravitational potential energy, W of the core is E/W = 0.08 (0.36). (We

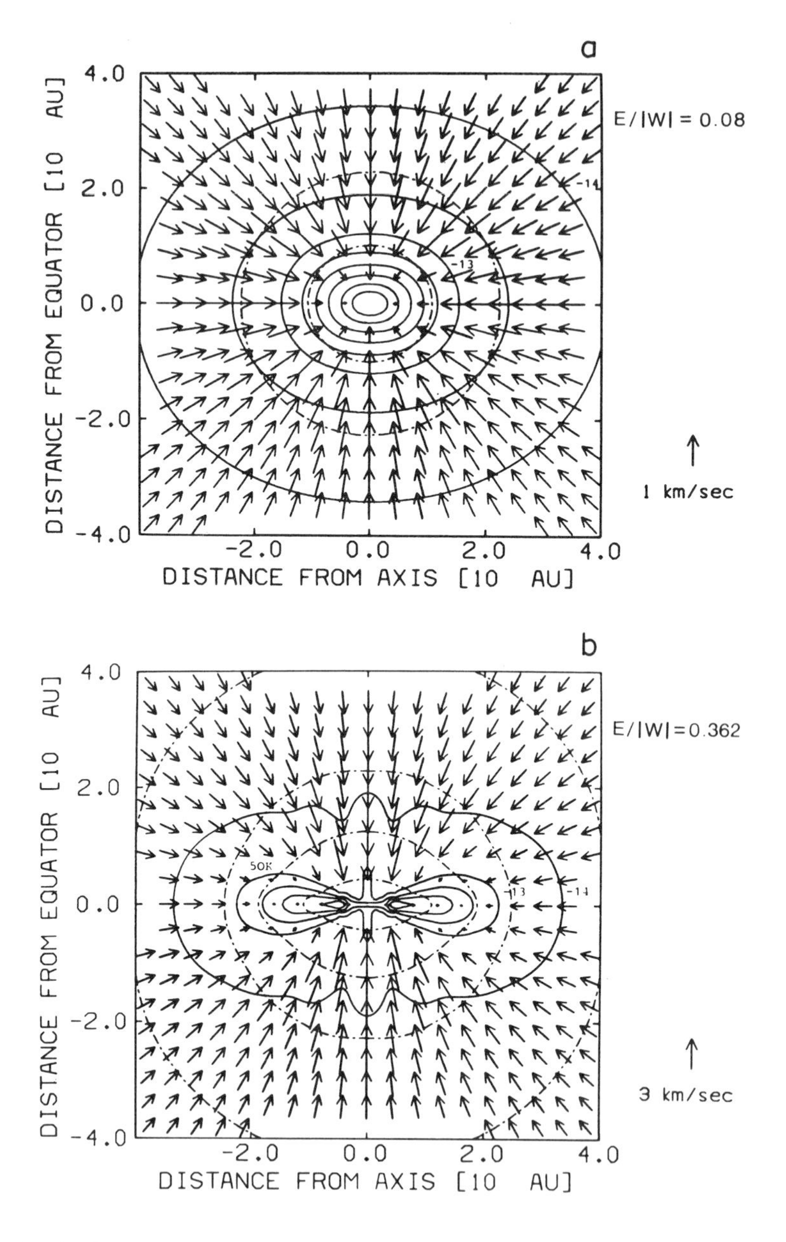

Fig. 4. (*a*) The temperature (dashed-dotted contour line), density (solid lines), and velocity structure (arrows) for Tscharnuter's Case B of a rotating protostellar cloud after $0.02 M_{\odot}$ has accreted onto the central core. Each contour line represents a change of 0.5 orders of magnitude. Values for log ρ along several contour lines are indicated. The outermost temperature contour is 30 K. (*b*) Same as (*a*) except now $0.12\, M_{\odot}$ are in the core. Note the abrupt change in velocity from several km sec^{-1} to almost zero at the accretion shock fronts, where density contour lines are bunched.

give relevant parameters for Fig. 4*b* in parentheses.) Since a disk or ring is dynamically unstable for $E/W > 0.24$, one would expect the ring displayed in Fig. 4*b* to have fragmented into a binary or multiple system before this stage. The central temperature T = 304 K (150 K) and the central density $\rho_c = 1.2 \times 10^{-10}$ g cm^{-3} (5×10^{-12} g cm^3) are lower than the maximum values $(T_c)_{max}$ = 800 K and $(\rho_c)_{max} = 1.2 \times 10^{-9}$ g cm^{-3} attained at an intermediate evolutionary stage. Thus, for the case displayed in Fig. 4*b* the maximum density has shifted outwards from the center to a ring.

C. The equation of radiative transfer

We shall describe the results of numerical calculations and solutions of the time-independent, non-relativistic equation of radiation transfer as

$$\frac{dI_\nu}{ds} = - \sigma_\nu^{ext} (I_\nu - S_\nu) \tag{1}$$

where I_ν is the radiation intensity at the frequency ν, σ_ν^{ext} is the extinction coefficient, S_ν is the monochromatic source function, and s is the path length along a given line of sight. For the time-dependent equation of radiative transfer one must include $1/c(\partial I_\nu/\partial t)$, where c is the speed of light.

In a spherically symmetric, protostellar cloud the intensity $I_\nu = I_\nu(r,\theta,t)$ is a function of four variables: the frequency ν, the distance from the cloud's center r, the angle θ between the photon direction and the radial direction, and time t. In an axially symmetric (i.e., rotating) cloud the intensity I_ν = $I_\nu(r,\mu,\theta,\phi,t)$ is a function of six independent variables: frequency ν, position (r,μ) where r is the radial distance from the center of the cloud and μ is the cosine of the angle between the position vector and the symmetry axis, photon direction (θ,ϕ), and time t.

It should be noted that many approximate methods for solving Eq. (1) such as the diffusion approximation and the Eddington approximation must be used with extreme care when calculating the radiation field in a collapsing protostellar cloud simply because the mean free path of the photons λ_{ph} = $1/\sigma_\nu^{ext}$ in the frequency range in which most of the radiative energy is being transferred is of the order of the dimensions of the object considered. Thus, one should mistrust methods which depend on the cloud being optically thick or optically thin.

For the solution of the line transfer problem and the continuum radiation problem it should be further noted that in order to calculate the extinction coefficient and the source function of Eq. (1) one must first know something about the radiation field. The kinetic temperature of the gas and dust grains and the population levels of the atoms and molecules considered, for example, have to be calculated simultaneously with the radiation field.

Thus, a complicated iterative procedure must be used in which the radiation field is first calculated by approximate means. From this the extinction and emission properties of the absorbing material can be calculated. Then the radiation field can be recalculated, and so on. For many of the present calculations for axially symmetric radiative transfer that is frequency-dependent, the iterative procedure was stopped at this point before the above "and so on." An estimate of the error encountered can be obtained by comparing similar test calculations for the spherically symmetric case, all of which were iterated until a self-consistent solution had been obtained.

II. THE APPEARANCE OF THE CONTINUUM OF COLLAPSING PROTOSTELLAR CLOUDS.

A. Grain Properties

Before one can solve the equation of radiative transfer in the envelopes around protostars, it is necessary to specify the properties of the absorbing material. Although a great deal of information on the extinction properties of the interstellar medium exists in the form of detailed extinction curves for a number of stars, ranging from infrared to ultraviolet frequencies, it has proved difficult to extract specific information about the nature of the dust grains responsible for this extinction. It seems likely, however, that the interstellar dust is not homogeneous in composition or size; more than one type of grain contributes to the extinction (Aannestad and Purcell, 1973). The 2200 Å bump is commonly attributed to small graphite particles (sublimation temperature $T_s \sim 2000$ K) and the 10μ absorption features to silicate grains ($T_s \lesssim 1500$ K). However, it is not possible to explain all of the extinction properties of the interstellar medium with only a mixture of graphites and silicates (Bedijn, 1977). The most common explanation is the presence of volatile materials ($T_s \lesssim 150$ K) which accrete onto the surfaces of graphite and silicate grains. The chemical composition of this material is not clear. In the dense clouds associated with star formation, the condensation of more than one type of molecule onto the grains seems likely. Here the grains are for the most part well shielded from the various destructive influences such as cosmic rays and hard radiation (optical and ultraviolet starlight). In the following we shall refer to the hypothesized mantle material as "ice", even though frozen H_2O may not be the main constituent.

In Figs. 5*a* and *b* we display the frequency dependent optical properties for the refractory grains and for the core-mantle grains (*b*) which were used in all our calculations for continuum radiative transfer. For the extinction cross-section and albedo of the core-mantle grains we adopted the results of Leung (1975) as obtained by using Mie theory assuming a given size distribution of ice-coated silicates. All resonances except the feature at 3.1μ have been smoothed. The refractory grains have been modeled by including

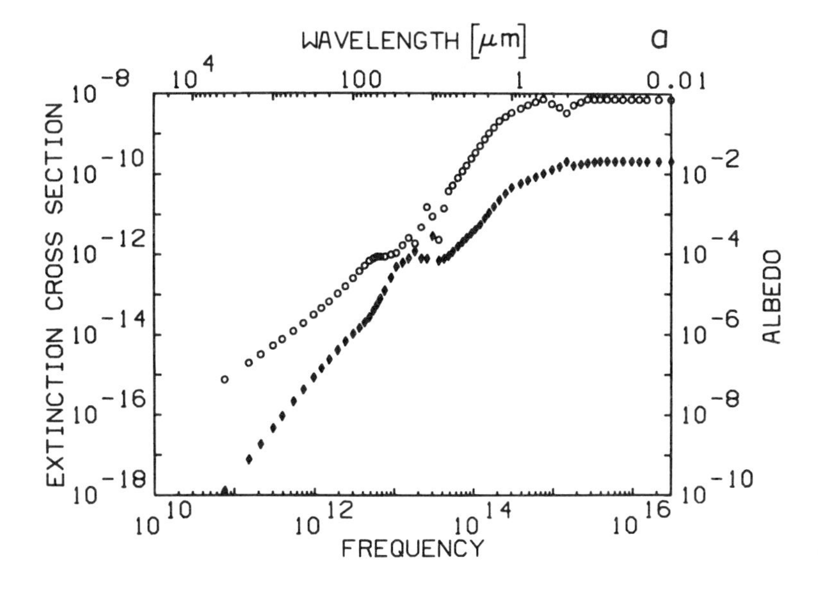

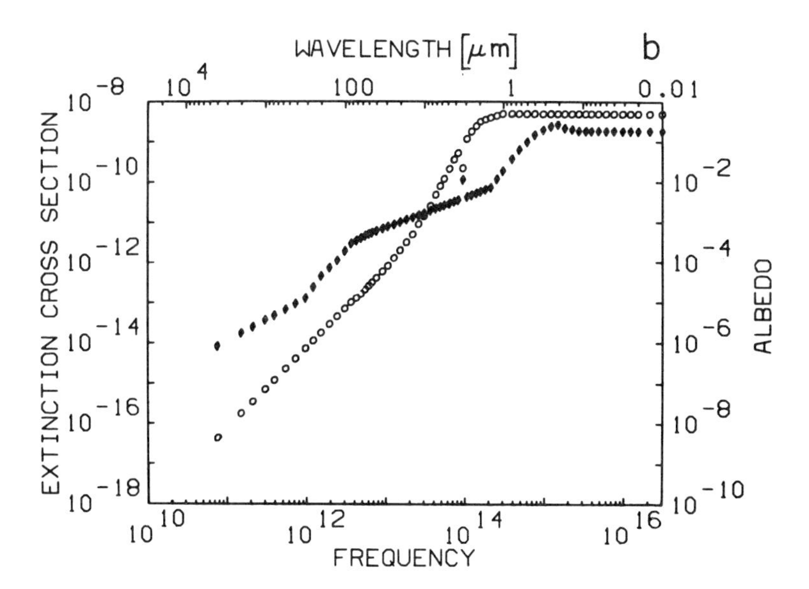

Fig. 5. (*a*) Optical properties of refractory grains as a function of frequency (wavelength) as assumed for continuum calculations. The extinction cross-section (in cm^2) is represented by filled diamonds (♦) and the albedo by open circles (○). (*b*) Assumed optical properties of volatile "ices" as a function of frequency (wavelength).

the 10μ feature and broad ~16μ feature attributed to silicate grains and the 2200 Å absorption feature normally attributed to small graphite particles. Here, optical and ultraviolet data for interstellar dust grains measured by Lillie and Witt (1976) and the infrared properties of the lunar silicate sample No. 14321 were incorporated. Sublimation temperatures ranging from 1500 K to 1700 K were assumed for the refractory particles, whereas for the mantles a sublimation temperature of 150 K was used.

B. An Estimate of the total optical depth of the dusty envelope

As mentioned above, the continuum appearance of protostellar clouds depends on the total optical depth of the envelope surrounding the central core. In the following we describe a simple but useful procedure for estimating the total optical depth at a given frequency and thus predict when the protostar should become visible optically. We assume first that the density in the envelope is distributed according to

$$\rho = ar^{-\alpha} \tag{2}$$

where ρ is the gas density and a is a constant to be determined. The parameter $\alpha = -\mathrm{d}\ln\rho/\mathrm{d}\ln r$ is equal to 3/2 in a steady-state and freely falling envelope surrounding a gravitating core. With the distribution given by Eq. (2) the mass of the envelope is given by

$$M_e = \frac{4\pi a}{3-\alpha} R^{3-\alpha} \tag{3}$$

where R is the radius of the cloud. If we now set R equal to the Jeans radius for gravitational instability of a cloud of initial mass M_0 and initial temperature T_0 we find

$$R = (1.5 \times 10^{17}\ \mathrm{cm}) \left(\frac{M_0}{1M_\odot}\right)\left(\frac{T_0}{10\ \mathrm{K}}\right)^{-1} \tag{4}$$

where we have assumed the hydrogen to be molecular.

Because the principal contribution to the opacity comes from dust in the envelope, we calculate the optical depth of the envelope by

$$\Delta\tau = \int_{r_m}^{R} \rho\kappa\, \mathrm{d}r \tag{5}$$

where κ is the opacity at the wavelength considered of the dusty gas outside the sublimation radius r_m. Thus, we find

$$\Delta\tau = A \left(\frac{\kappa}{200\ \mathrm{cm^2 g^{-1}}}\right) \left(\frac{M_e}{1\ M_\odot}\right) \left(\frac{M_0}{1 M_\odot}\right)^{\alpha-3} \left(\frac{T_0}{10\ \mathrm{K}}\right)^{3-\alpha} \left(\frac{r_m}{1\ R_\odot}\right)^{1-\alpha}$$

$$A(\alpha) = 2 \times 10^3\ (2 \times 10^6)^{\alpha-\frac{3}{2}} \left(\frac{3-\alpha}{\alpha-1}\right) \quad (6)$$

where we have assumed $R \gg r_m$ and $1 < \alpha < 3$. Considering the evolution of a protostellar cloud of one solar mass with an initial temperature of 10 K and a dust opacity of 200 $\mathrm{cm^2 g^{-1}}$ in the optical range we find $r_m < 100\ R_\odot$ (the exact value is not too critical because it enters into the estimate only to the $1-\alpha$ power) for silicate grains and estimate that the envelope of the 1 $M_\odot$ protostar is optically thin only when much less than 0.1% of the original mass is still in the envelope. Considering now a 10 $M_\odot$ protostellar cloud and assuming the same values for κ and T_0, we find $r_m \approx 1000\ R_\odot$ and estimate for $\alpha = 3/2$ that $\tau = 0.6$ when $M_e/M_0 = 0.01$. We shall see in the following that these are indeed typical values when compared to detailed numerical computations of 3 $M_\odot$ and 10 $M_\odot$ protostellar clouds.

C. Results of spherically symmetric calculations

The spectral appearance of the continuum of solar mass collapsing protostellar clouds has been presented by Larson (1969) and Bertout (1976). Bertout has assumed dust absorption opacities in the visual region of the electromagnetic spectrum which are much lower than the κ = 200 $\mathrm{cm^2 g^{-1}}$ (the true value may be even higher) used in Sec. II.B. Thus, Bertout predicts relatively high fluxes of optical radiation when several percent of the original cloud mass is still in the envelope, whereas our present estimates show that the spherically accreting protostar should be obscured optically until much less than 1% of the original mass is in the envelope. However, observations of YY Orionis stars seem to indicate that the core of the protostar is already visible when about 2% of the original cloud mass is still in the envelope (Appenzeller, personal communication, 1978).

Two different paths can be followed to interpret the discrepancy between theoretical estimates and observations.

1. If the assumption of spherical symmetry in the collapsing cloud is to be retained, the number of absorbing grains in the envelope has to be reduced. This could result either from the growth of the dust particles with time (Morfill *et al.*, 1978) or by a separation of dust and gas during the accretion of the final few percent of material. (When the dust is blown outwards, the total optical depth decreases.)
2. However, the assumption of spherical symmetry is probably unrealistic. Observations of YY Orionis stars first suggested that the calculations for

spherically symmetric collapse were in fact a good approximation of the physical reality (Appenzeller and Wolf, 1976). Recent detailed calculations indicate that typical YY Orionis line profiles are also produced in rotating axially symmetric envelopes (see Sec. III.B). If the assumption of spherical symmetry is abandoned, the stellar core might be seen at earlier evolutionary stages if a dust accretion disk were formed around the star. In fact, Kuhi (see his chapter in this book) suggested that an accretion disk could probably account for the observed properties of most T Tauri line profiles. Another possible explanation is the formation of holes in the dust distributions around the star, for which a possible mechanism has been proposed by Wolf *et al.* (1977). This process requires no angular momentum.

Indeed, it is likely that all possibilities discussed here might be at work at the same time in the envelope of a real protostar.

The spectra of protostellar clouds with $M \gg 1\ M_\odot$ have been calculated by Yorke (1977*a*). The principal feature of the more massive protostellar clouds is a double-peaked spectrum, the near infrared peak arising from the inner "false photosphere" ($T \sim 1000$ K), and far infrared radiation from the cool material ($T \sim 100$ K) which had been blown outwards by radiative acceleration on the dust. Yorke (1977*a*) compared these calculated spectra with near- and far-infrared measurements of several protostellar candidates and shows that there is qualitative agreement.

We present here the infrared appearance of 10 $M_\odot$ and 3 $M_\odot$ collapsing protostellar clouds, computed (Yorke, 1978) by solving the coupled equations of (a) hydrodynamics of gas, (b) hydrodynamics of the dust, and (c) frequency and spherical radiation transfer that is angle-dependent in the envelope. Because the core was not included in these calculations, the radius of the core had to be specified as an inner boundary condition. This was done in a way consistent with previous hydrodynamic results; details of these calculations will be presented elsewhere.

In Table I and Fig. 6 we summarize some of the results obtained for the 10 $M_\odot$ case, also discussed by Yorke (1977*b*). For these calculations an initial uniform density, $\rho_0 = 10^{-19}$ g cm^{-3} and zero velocity were assumed. The age, Δt of the central core after its initial formation, the mass of the core, M_{core}, the total luminosity, L, the mass flux, $\dot{M}$ onto the core and the visual extinction, A_v are given in Table I at several evolutionary times in a calculated sequence. Each evolutionary age displayed has been labeled A, B, etc.; the spectra of the corresponding models are shown in Fig. 6. Model D (not displayed in Fig. 6) was a very compact H II region of radius $R \approx 5 \times 10^{16}$ cm expanding at a rate of 10 km sec^{-1} and a powerful far-infrared source. Approximately 0.5 $M_\odot$ of the original cloud material was not able to fall onto the core due to radiation acceleration on the grains.

In Table II and Fig. 7 we show some of the results from a numerical calculation of the collapse of a 3 $M_\odot$ protostellar cloud with an initial density

TABLE I

Core and Envelope Parameters of a Collapsing 10 $M_\odot$ Protostellar Cloud at Various Evolutionary Times.

Model	Δt (10^3 yr)	M_{core} ($M_\odot$)	L ($L_\odot$)	$\dot{M}$ (10^{-6} $M_\odot$/yr)	A_v (mag)
A	7.6	1.5	850	242	5700
B	26.6	5.3	3630	141	1520
max *L*	45	7.1	7710	72	650
C	106	9.1	5530	14	133
D	169	9.5	4960	1	3

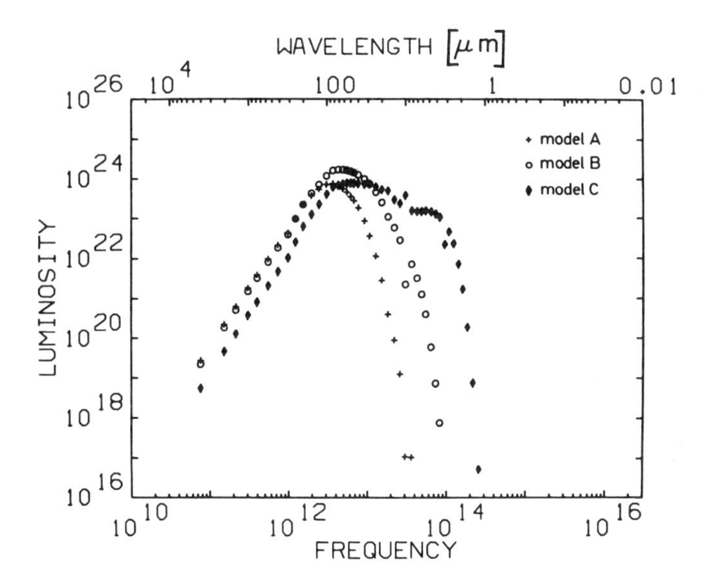

Fig. 6. The spectra from a collapsing $10M_\odot$ protostellar cloud during various evolutionary phases as described in Table I and in the text.

of $\rho_0 \simeq 2 \times 10^{-20}$ g cm^{-3}. After reaching a maximum total luminosity of over 10^2 $L_\odot$ when a little more than 70% of the cloud material has fallen onto the hydrostatic core, the luminosity from the accretion shock drops off as the mass of the envelope decreases. The spectra corresponding to the evolutionary models listed in Table II are displayed in Fig. 7. The spectra show that for the early phases, dust absorption features at 10μm and 3.1μm are seen in absorption only. Note that for Model *F* the 10μm feature can be seen in emission, the 3.1μm feature not at all, and the 2200 Å feature in absorption. At an intermediate evolutionary time to that displayed by models

TABLE II

Core and Envelope Parameters of a Collapsing Spherically Symmetric 3 $M_\odot$ Protostellar Cloud at Various Evolutionary Times after the Formation of a Central Core.

Model	Δt (10^3 yr)	M_{core} ($M_\odot$)	L ($L_\odot$)	$\dot{M}$ (10^{-6} $M_\odot$/yr)	A_v (mag)
A	18	0.46	8	15.6	4712
B	67	0.97	15	8.2	1183
C	156	1.52	24	5.0	580
D	332	2.12	111	2.3	147
E	598	2.50	90	0.94	66
F	1170	2.85	97	0.33	17
G	1345	2.90	96	0.20	0.66

F and *G*, the density of the envelope decreased to the point that gas and dust were no longer effectively coupled. (The Coulomb interaction of charged grains and the partially ionized gas was included but turned out to be ineffective at this stage.) The dust was then blown outwards due to radiative acceleration and the optical depth of the envelope was thus decreased. At the same time, the infrared flux at wavelengths between $\lambda = 1\mu m$ and $10\mu m$, principally due to the hot dust close to the central source, decreased. The resulting spectrum given for Model *G* illustrates this phenomenon.

D. Results of axially symmetric calculations

In order to calculate theoretically the appearance of the continuum of a rotating protostellar cloud, it is necessary to solve simultaneously the equations of radiative transfer and hydrodynamics. In this section we present a summary of recent (unpublished) results obtained by solving the continuum radiative transfer problem in rotating protostellar clouds, using the hydrodynamic models of Tscharnuter (1978) discussed previously. More details of these calculations will be presented elsewhere. Unlike the numerical method used for the spherically symmetric protostellar clouds considered in Sec. II.B, the frequency-dependent and direction-dependent equations of radiative transfer were not solved simultaneously with the hydrodynamic equations of motion. Instead, Tscharnuter solved the "moment" equations of radiative transfer, assuming the Eddington approximation. Gray opacities obtained from the Planck and Rosseland means of the frequency-dependent dust opacities discussed in Sec. II.A were used.

The temperature and densities thus obtained from Tscharnuter's calculations were used to solve the direction- and frequency-dependent

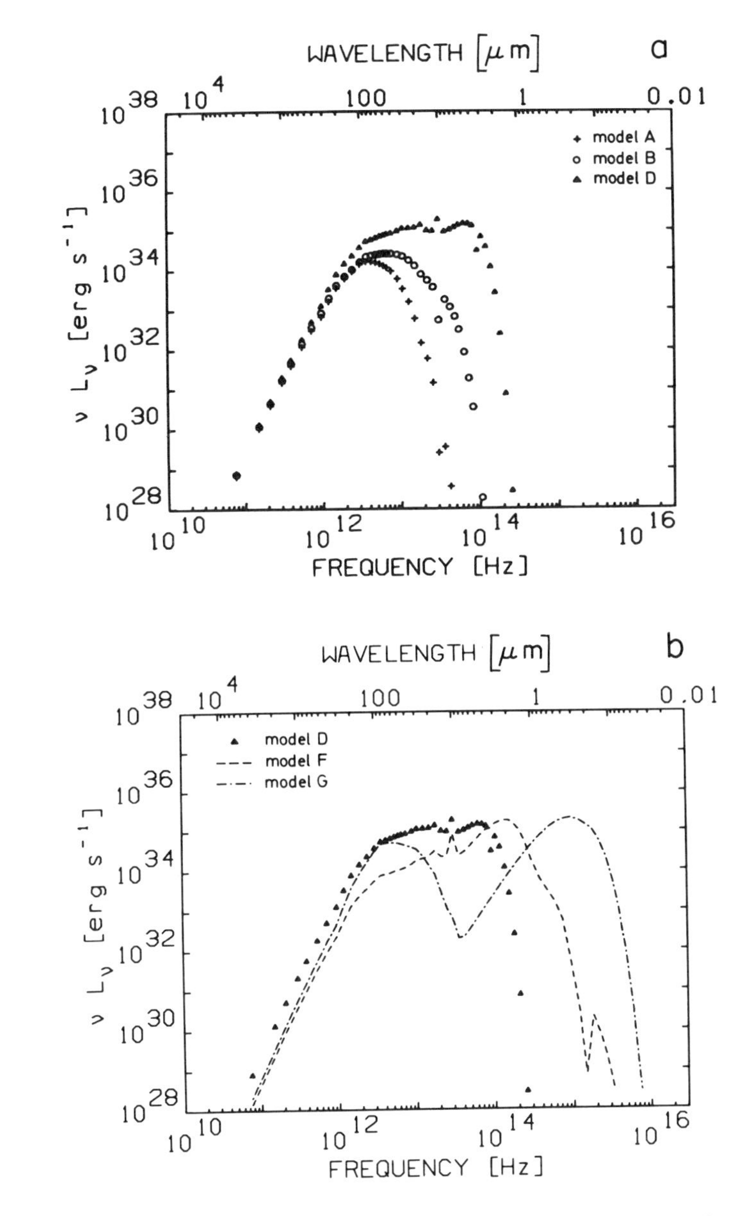

Fig. 7. Two plots of spectra of a collapsing $3M_\odot$ protostellar cloud at various evolutionary times as described in Table II and in the text.

equation of radiative transfer. Because these dust temperatures may be either too low or too high in optically thin regions (the Eddington approximation assumes an isotropic radiation field), absolute magnitudes of the radiative flux are probably inaccurate. The relative spatial contrast at a given frequency

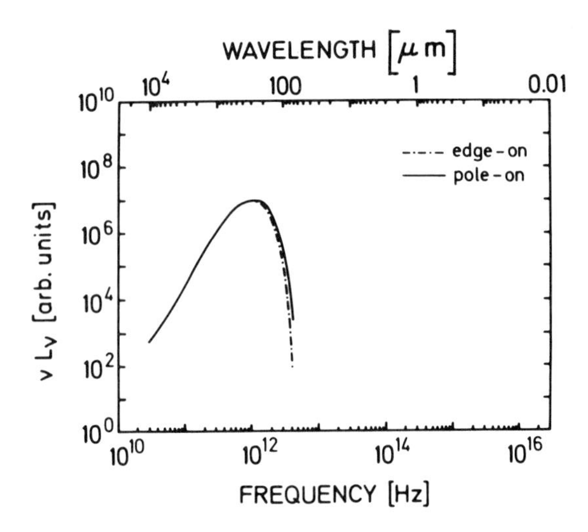

Fig. 8. The spectrum of the 3 $M_\odot$ rotating protostellar cloud after 0.12 $M_\odot$ has accreted onto the stellar core (see text and Fig. 4*b*). Two possible orientations are shown. Note that there is little change in the total spectrum as it appears pole-on (0°) compared to the edge-on (90°) view.

and the spectral distribution of the radiation in a given direction are not too sensitive to the exact values of temperature in optically thin regions, however, and we are therefore justified in displaying in Fig. 8 and Fig. 9 theoretical spectra and isophote maps of the protostellar cloud for a given early, evolutionary stage. Absolute magnitudes of the flux densities are not given.

In Fig. 8 we display the total spectral distribution of the radiative flux νL_ν as seen from two directions, edge-on (90°) and pole-on (0°). In both cases, the resulting spectrum is that of a cool ($T \sim 15$ K) source, which only appears to be slightly warmer when viewed pole-on. The warmer central regions with $T_c = 150$ K are thus completely obscured and only the outer regions are seen at this stage. This was also the case at an earlier evolutionary time when the central temperatures were in excess of 800 K. In Fig. 9 theoretical isophote maps of the protostellar cloud at 100μm and as viewed assuming different tilts of the axis of rotation towards the observer, are shown. (The projection of the symmetry axis would be a vertical line at the center of each diagram.) Here, structure on a scale somewhat smaller than 100 AU is noticeable at the larger angles of inclination. The edge-on view is characterized by a "double-source" appearance, caused by the disk's absorption in the equatorial plane and the preferential heating of dust close to the rotational axis.

Similar isophote maps at 10μm were not produced for this early evolutionary stage because of the low radiative flux expected at this wavelength. In analogy to the 3 $M_\odot$ spherically symmetric calculations, however, the protostellar cloud is expected to become brighter at 10 μm during more advanced evolutionary stages. The maps displayed in Fig. 9 can

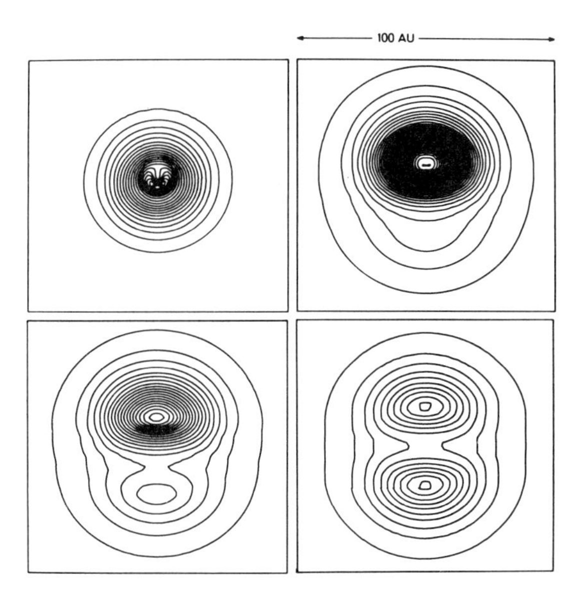

Fig. 9. Theoretical isophote maps at 100μm of the central (100 AU) regions of a 3 $M_\odot$ rotating protostellar cloud (see text and Fig. 4*b*) after 0.12 $M_\odot$ has accreted onto the core. The four orientations shown are from left to right and top to bottom 15°, 60°, 75°, and 90°, measured from the axis of rotation. (The projection of the axis of rotation onto each map would be a vertical line through the center). The central region appears brighter when viewed pole-on. In arbitrary units the outmost contour levels for 60°, 75°, 90° correspond to an intensity level of 1 increasing linearly by one unit per contour. For the 15° view, however, the outermost contour corresponds to an intensity level of 5; the contour interval is 5 units.

be considered illustrative of the continuum appearance of the central regions.

From the above discussion it should be apparent that continuum observations of rotating protostellar clouds would be difficult to make and difficult to interpret during these early evolutionary stages. We therefore focus our attention on the formation of molecular lines and the interpretation of their profiles.

III. LINE FORMATION IN COLLAPSING SOLAR-TYPE PROTOSTARS

Spectral lines emitted by a protostar are found in different regions of the spectrum, depending on the evolutionary stage of the observed stellar object. Several processes are known to favor the formation of complicated molecules in the low-temperature dusty environment of infalling protostellar envelopes as discussed in the previous sections. Cosmic rays and photons allow for the dissociation or ionization of the different molecular species, whereas chemical reactions can take place in the gas or on the surfaces of the dust grains. While the envelope is optically thick in the visible region of the spectrum, we expect to observe molecular lines in the radio and infrared spectral ranges. Later in the history of the protostar, the infalling cloud becomes transparent to the radiation emitted in the accretion shock. At this stage, the molecules could be destroyed and the optical and ultraviolet line spectrum emitted in the shock front and in the denser parts of the infalling envelope as well as in the photospheric late-type stellar spectrum would become observable.

We shall discuss in some detail the two evolutionary phases sketched above. In Sec. III-A, work done on molecular compositions and molecular line formation of solar-type collapsing clouds is briefly reviewed, as are line transfer calculations, not in local thermodynamic equilibrium, for a multi-level OH molecule in the axisymmetrical protostar models of Tscharnuter (1978). In Sec. III-B we review work done on line transfer problems in the last phases of the hydrodynamical protostellar evolution and compare theoretical results to observations of YY Orionis stars.

Most of the authors who have worked on the problem of line transfer in collapsing clouds and protostars have used the escape probability formalism to solve the radiative transfer problem. We therefore review briefly the basic assumptions of the escape probability theory, due to Sobolev (1960) and further developed by Castor (1970), Lucy (1971), Grachev and Grinin (1975), Marti and Noerdlinger (1977), Bertout (1977*a*), and others. Surdej (1977) has summarized the mathematical formalism in a very readable way. The basic assumption of this theory is that the macroscopic flow velocity along a line of sight must change by an amount equal to the microscopic thermal velocity of the emitting atoms or molecules over a distance small compared with the dimension of the envelope. Then each frequency in the line profile is formed in thin shells where the radial velocity with respect to the observer is constant. The equation of a constant radial velocity surface (hereafter CRVS) is given by

$$\frac{\nu - \nu_0}{\nu_0} = \frac{\mathrm{v}_R}{c} \tag{7}$$

where ν is the frequency considered, ν_0 the line center frequency, v_R the radial velocity, and c the light velocity. The shape of the CRVS for two different *radial* velocity fields is shown in Fig. 10. The set of CRVS in Fig. 10*a* illustrates the cases of radial velocity fields accelerating outwardly or decelerating inwardly; Fig. 10*b* shows the cases of decelerating outflow and accelerating inflow. In the case of Fig. 10*a*, a line of sight intersects a given CRVS only once, so that the radiative transfer problem is local. In the case of Fig. 10*b*, a line of sight generally intersects a CRVS twice. In principle, an interlayer coupling term must therefore be taken into account in the derivation of the source function. However, Marti and Noerdlinger (1977) have shown that neglecting the interlayer coupling resulted in an error of typically only 15% in the source function. Hummer and Rybicki (1978)

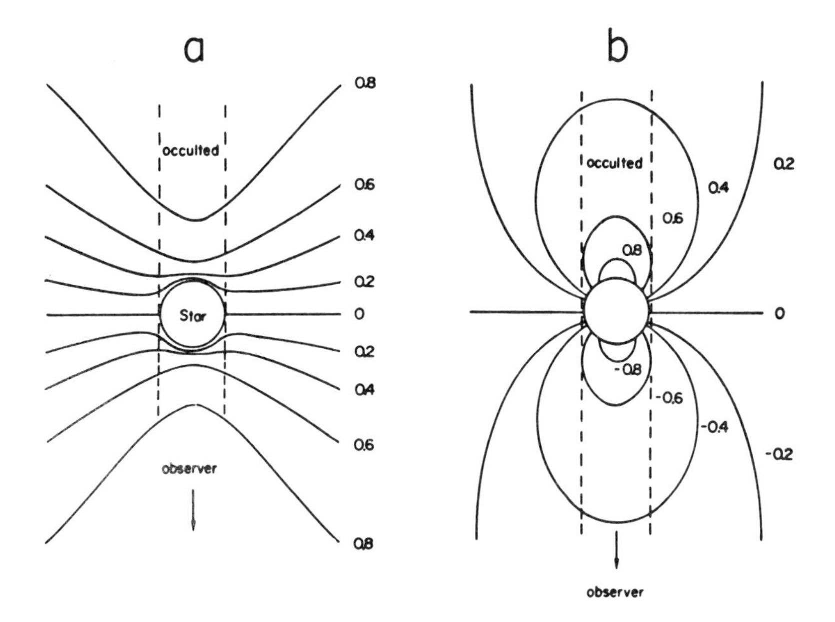

Fig. 10. (*a*) Constant radial velocity surfaces for an outflowing accelerating or an infalling decelerating atmosphere. The velocity law is $v(r) = v_\infty (1 - r_c/r)^{\frac{1}{2}}$ and the numbers indicate the ratio v_R/v_∞ for each surface. (*b*) Constant radial velocity surfaces for an outflowing decelerating or an infalling accelerating atmosphere. The velocity law is $v(r) = v_c(r_c/r)^{\frac{1}{2}}$ and the numbers indicate the ratio v_R/v_c for each surface. (From Kuan and Kuhi, 1975.)

found that errors up to 50% in the line intensity were introduced if the interlayer coupling was neglected. However, they pointed out that the source function was not qualitatively affected by the non-local contribution due to the interlayer radiative coupling.

As a first approximation, one can therefore consider that the radiative transfer problem is essentially local for both cases illustrated above. Due to this simplification, it is possible to consider complicated geometries and velocity fields, as well as multi-level atoms and molecules. Most authors have restricted themselves to spherical symmetry and power law velocities. The first axisymmetrical calculations in the two-level case have been reported by Bertout (1977*b*). More recently, a computer code allowing for the treatment of multi-level atoms and molecules in axially symmetric envelopes and velocity fields has been developed (Bertout, 1978).

Rybicki (1970) and Castor (1970) have derived quantitative criteria for defining the validity domain of the Sobolev method. However, this point seems to need more detailed investigation according to Rybicki. Comparative studies, using different numerical codes, are highly desirable in order to better understand the limits of the Sobolev approximation. However, protostellar envelopes probably are nearly ideal candidates for the Sobolev method:

1. Free-fall conditions in the envelope govern the process of accretion over most of the hydrodynamic evolutionary time, at least for low-mass solar-type protostars (as long as radiative pressure effects can be neglected). Therefore, we expect strong velocity gradients over the envelope.
2. The gas temperature in the envelope is low, so that the thermal line width remains small. To give a numerical example, a high estimate for the thermal width of a hydrogen line at a late evolutionary stage can be gained by using the spherical model calculations by Winkler (1978). The temperature just above the shock is $T \approx 6000$ K, corresponding to a hydrogen thermal width of $\Delta \nu \cong 10$ km s^{-1}. Now the flow velocity changes by $\Delta \nu$ over a distance $\Delta r \cong 8 \times 10^9$ cm, which is indeed negligible compared to the overall radius of the envelope. At earlier stages of the collapse, similar estimates can be made for molecules. We therefore have some confidence that the Sobolev formalism gives a fairly good solution of the radiative transfer inside collapsing clouds.

A. The Molecular Stage

A satisfactory solution of the molecular line transfer problem in solar-type protostellar envelopes can be given only after the problem of molecule formation has been solved in the protostellar cloud; this is because the density distribution of the different molecules over the cloud is then known as a function of time. One could hope to solve the problem consistently with the hydrodynamical evolution and the complete radiative transfer, but this will remain a dream for some years to come. However, first steps in this direction are being made. For example, Iglesias (1977) recently computed the chemical evolution of low-density molecular clouds as a function of time; Gerola and Glassgold (1977) calculated the chemical, thermal and hydrodynamic evolution of a collapsing massive cloud ($M = 2.7 \times 10^4$ $M_\odot$); and Hertel (1976) computed the steady-state molecular composition of a low-mass spherical protostar. Below we shall briefly review Hertel's results.

Chemistry of Molecules in Low-Mass Protostellar Envelopes. To calculate the steady-state molecular population of the protostellar cloud, Hertel sets up a chemical reaction network consisting of 125 gas-phase reactions for 49 molecules, ions and atoms formed from H, He, C, N, and O. Furthermore, Hertel includes in his system of equations a reaction describing the formation of molecular hydrogen from atomic hydrogen on the surface of dust grains at low temperature, and three reactions of molecular hydrogen and helium with cosmic protons. Not included in the equation system are three-body collisions, reactions triggered by ultraviolet photons, and reactions taking place on the surface of grains except for the one mentioned above. He

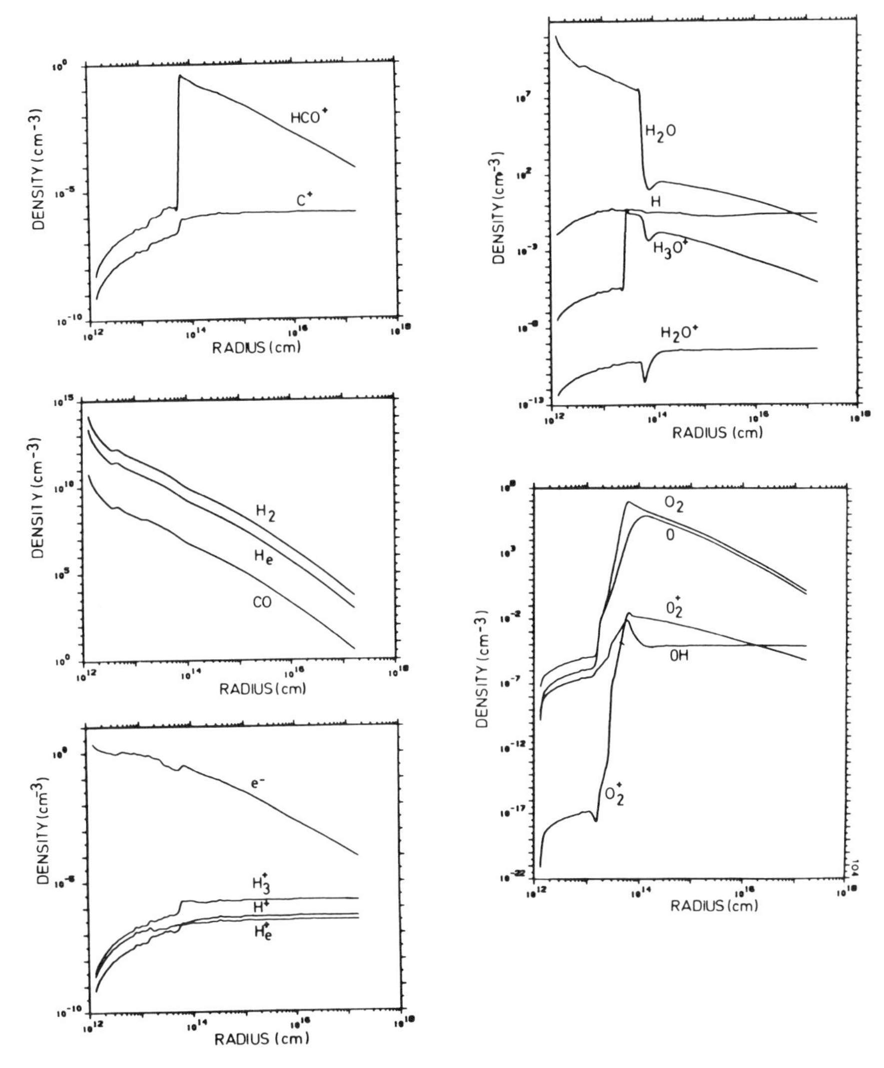

Fig. 11. Variations of the most abundant species in a low-mass spherical protostellar envelope at an early stage of its evolution. (From Hertel, 1976.)

included this particular equation because it has been studied in detail by Hollenbach and Salpeter (1970) and because it is the only reaction which builds H_2 from H in the low-temperature range. Invoking the present lack of detailed knowledge of the processes taking place on grain surfaces, Hertel does not include further reactions of this type in his network, although they probably play an important role for the formation of heavy neutral particles (Iglesias, 1977). Neglecting the reactions due to ultraviolet stellar photons is a good approximation during early stages of protostellar evolution, and three-body collisions may be neglected as long as the gas densities are $\lesssim 10^{11}$ cm^{-3} (Herbst and Klemperer, 1973; see the chapters of E. Herbst and Watson). Using the density and temperature distributions of a spherical solar-mass protostar model calculated by Bertout (1976), Hertel solves the system of equations iteratively. Figure 11 shows his results for the most abundant species in the spherical protostellar envelope. While He and CO, for example, closely follow the nearly free-fall density distribution present in the overall envelope, some species such as OH are seen to have a constant density over a large N_{H_2} and T interval. The variation of H_2O is seen to be extremely sensitive to the density of N_H once a critical temperature has been reached. Although these calculations have been made only for one particular evolutionary time when 4% of the initial mass has accreted onto the core, they hint at the qualitative variation of molecular abundance in a low-mass protostellar envelope. Using calculations of this nature, one can significantly reduce the number of free parameters usually available in the radiative transfer calculations for molecular clouds.

Line Formation in Molecular Clouds. Work on line transfer in large collapsing molecular clouds has been done by a number of authors, including Goldreich and Kwan (1974), Scoville and Solomon (1974), de Jong *et al.* (1975), Snell and Loren (1977), and Lucas (1976). Most of these authors have investigated clouds of uniform density with a velocity field given by

$$v(r) = r\sqrt{\frac{2GM}{R^3}} \tag{8}$$

where M is the mass of the cloud and R its outer radius. Equation (8) is *not* a free-fall velocity law as usually understood in gravitational collapse, although it is often referred to as the "free-fall law." In fact, Eq. (8) represents the velocity of a homologously collapsing sphere under the influence of self-gravity. Since all numerical protostar calculations indicate that the collapse proceeds in a non-homologous way, with formation of one, or several accretion bodies at the center of the cloud early in the collapse history, the relevance of Eq. (8) in the case of collapse is indeed questionable. The free-fall velocity field relevant to the protostellar formations in spherical geometry is given by

$$v(r) = \sqrt{\frac{2GM_c}{r}} \tag{9}$$

where M_c is the mass of the central accreting stellar object. An important difference between Eqs. (8) and (9) is that Eq. (8) leads to a set of open

CRVS's (Fig. 10*a*) while Eq. (9) leads to closed CRVS (Fig. 10*b*). This has important consequences for the radiative transfer problem, which will be further discussed at the end of this section. Of the above authors only Snell and Loren use a true free-fall velocity field, while de Jong *et al.* consider the case of an isothermally collapsing sphere, which is known to correspond to the early stage of the protostellar collapse (Larson, 1969). All of these authors, however, have considered the case of spherical low-density clouds, which might bear little resemblance to the solar-type nebulae, due to the very different density scales.

Hertel's (1976) thesis is the only work we are aware of which deals with low-mass protostars in the molecular stage. Hertel does not use the escape probability formalism but rather a sophisticated numerical code solving the transfer equation at the line frequencies. Hertel applied his method to the case of OH molecules in a protostar envelope, using the calculations described in the preceding paragraph to derive the OH density at each radial point. He found that the infalling envelope was optically thin in all calculated transitions, and obtained a population inversion in the 18-cm transition in a spherical shell of about 10^{15} cm diameter. However, he did not include the 2.8 K background radiation in his calculation, although it is the major source of continuum at 18 cm and did not derive integrated line profiles, but rather line profiles over a line of sight, which makes it difficult to compare with observational work. In the following paragraph, we present a preliminary report on OH line formation in the low-mass rotating protostars of Tscharnuter, in an attempt to further sophisticate the theoretical models.

OH Line Formation in Axisymmetrical Low-Mass Protostars. A major reason for choosing OH as a test molecule in low-mass protostars is that it has been observed near reflection nebulae containing T Tauri variables, although rarely in the exact direction of T Tauri stars (see e.g., Johansson *et al.*, 1974). Clearly OH emission is associated with regions of stellar formation, and might be present in stages of the stellar evolution preceding the T Tauri phase, i.e., in low-mass solar-type clouds. A review of the observational situation has been given by Turner (1973).

In this report we use the simplified OH molecule model derived by Hertel. The six rotation levels taken into account in the calculations are shown in Fig. 12. The hyperfine structure is neglected. Hertel gives an upper limit of 100 K for the validity of this model. Five lines have been calculated, including the 18-cm transition between the Λ-components of the ground state. The transition probabilities are also shown in Fig. 12. Three pumping mechanisms have been discussed for the OH molecule:

1. Infrared pumping (Litvak, 1969) at 2.8 μm can produce Λ-doublet inversions if the temperature of the OH gas exceeds 1500 K. Such a high temperature is not found in the models of Tscharnuter.
2. Ultraviolet pumping (see e.g., Turner, 1970) cannot be used here because

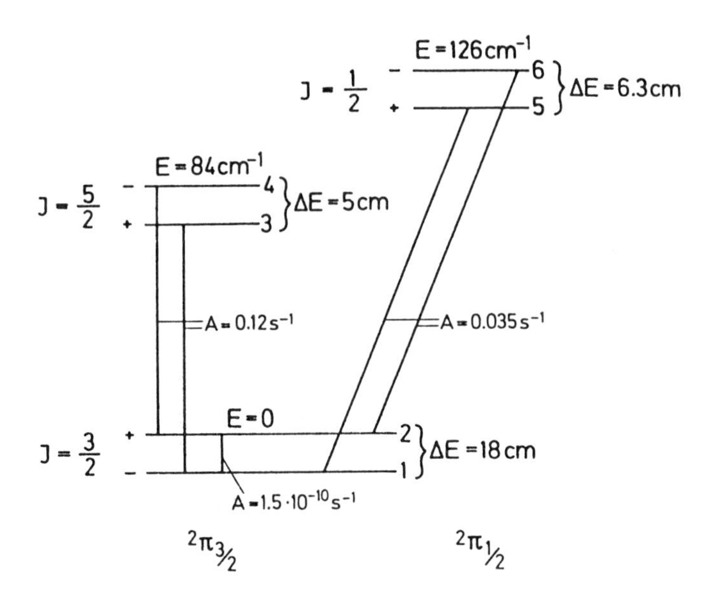

Fig. 12. Simplified six-level OH molecule used in the calculations. The five lines calculated are indicated with their transition probabilities.

of the lack of ultraviolet sources in the protostellar cloud.

3. Collisional pumping with H_2 and H due to Gwinn *et al.* (1973) can produce actual population inversions, provided that the kinetic temperature of the cloud is greater than 25 K and $n_{H2} \gtrsim 10^3$ cm^{-3} (Turner, 1973). In our calculations we used the collisional rates calculated by Gwinn *et al.*

The details of the numerical method used for calculating the line transfer in Tscharnuter's model will be published elsewhere. To understand the results it is sufficient to know that the optical depth in the line is given by

$$\tau_{ij} = \frac{k_{ij}}{|\mathrm{d}v_r/\mathrm{d}s|} \tag{10}$$

where k_{ij} is the absorption coefficient, and v_r is the radial velocity in the s direction. The mean intensity integrated over the line profile is calculated according to

$$\bar{J}_{ij} = (1-\beta_{ij})S_{ij} + W\beta_{ij}\, I_{ij}{}^{c} + (1-W)\beta_{ij}\, I_{ij}^{BB} \tag{11}$$

where S_{ij} is the source function, β_{ij} the escape probability in the transition i–j, and BB indicates black body radiation. W is a spherical dilution factor, calculated according to

$$W = \frac{1}{2}\left[1 - \left(\frac{1-r_c{}^2}{r^2}\right)^{\frac{1}{2}}\right] . \tag{12}$$

The term

$$J_c = W \beta_{ij} I_{ij}^c \tag{13}$$

represents the contribution from the continuum radiation of the core to the line intensity field, and

$$J_{BB} = (1-W) \beta_{ij} I_{ij}^{BB} \tag{14}$$

is the contribution due to the isotropic background radiation at 2.8 K. In the numerical code, an iterative procedure was used to calculate consistently the statistical equilibrium populations and the intensity field.

At this stage of the calculations, we are more interested in a qualitative understanding of the radiative transfer phenomena inside a low-mass protostar than in quantitative comparison with observational work. Several simplifying approximations were therefore made in the calculations. First, we neglected the interlayer coupling in finding the source function. The error introduced by this approximation has been discussed in Sec. I. We have a prototype code which takes into account the coupling term, but at this stage it still requires too much computer time to be used for such a complicated geometry. Test calculations however seem to indicate that the results presented here are not qualitatively affected by this approximation. We also assumed the central optically thick core to be spherically symmetric, so that the dilution factor, W in Eq. (11) is a radial function. Since the core radius used was 4×10^{13} cm in an overall cloud of 5×10^{17} cm, this approximation should not be critical. Another approximation is the use of W itself for determining the continuum contribution in the line. Use of W assumes that the escape probability is an isotropic function, which is not the case in the velocity fields used. Both these approximations can be removed by calculating the true escape probability that an emitted photon hits the core. However, it demands another time-consuming two-dimensional numerical integration. Test calculations using both solutions indicate that the results will not be qualitatively affected by the use of W in Eq. (11). We therefore adopted these approximations at this stage of the calculation.

Calculations were carried out for the models discussed in Sec. I.B of this chapter and at different angles of sight. In all calculations an OH density of 10^{-2} cm^{-3} was assumed, corresponding to the highest value found by Hertel (1976) in his calculations. Furthermore, the OH density was assumed to be constant over the densities and temperatures found in the cloud, in agreement with Hertel's results.

Fig. 13. Intersections of the constant radial velocity surfaces with an equal density surface ($\rho = 10^{-15}$ g cm^{-3}). Velocity and density distributions are taken from the hydrodynamic axisymmetrical calculations of Tscharnuter (c.f., Fig. 4*a*). Four different angles of inclination shown are from left to right and top to bottom 1°, 30°, 60°, 90°.

Figure 13 is an attempt to illustrate the geometry of the CRVS for the complicated velocity field present in the protostellar envelope. It represents the intersections of a CRVS set with an equal-density surface in the envelope. In the first quadrant of the figure, the protostar is seen practically pole-on; the CRVS are of the form illustrated in Fig. 10*a*, so that the intersections with the equal-density surface are almost circular. The innermost circle corresponds to the highest radial velocity. In quadrants 2 and 3, the angle of sight is increased, and in the fourth quadrant the protostar is seen edge-on. Due to the rotational velocity component, the maximum radial velocity is displaced to the right of the figure's center.

Figure 14 shows for the hydrodynamic model of Fig. 4*a* the variation of optical depth, source function and mean integrated intensity at a given frequency as a function of radius along the curve of intersection of the corresponding CRVS and the plane defined by the line of sight from observer to cloud center and its perpendicular in the equatorial plane. The angle of sight is 75° and the line frequency corresponds to $(\nu - \nu_0)/\nu_0 = 8 \times 10^{-7}$ at $\lambda_0 =$ 18 cm. The optical depth increases rapidly with radius. This is due to the velocity gradient, which becomes smaller in the outer parts of the envelope, thus reducing the probability of photon escape. The two peaks in the optical

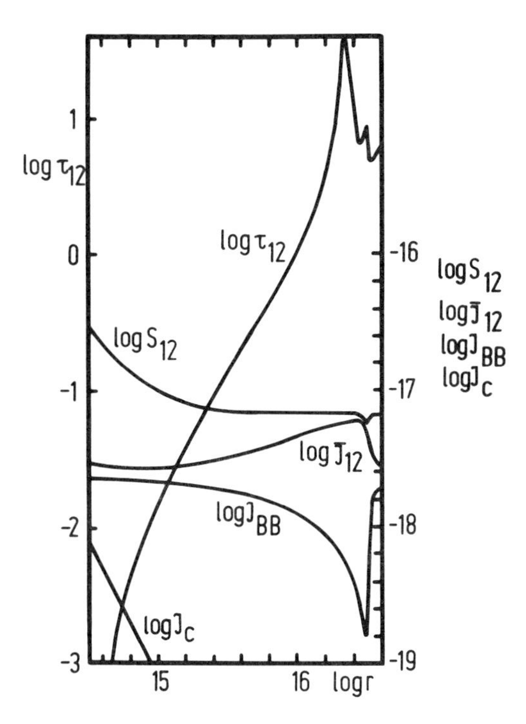

Fig. 14. Variation of the optical depth, source function and integrated mean intensity for the OH 18-cm line along the intersection of a constant radial velocity surface (CRVS) with a plane. See text for explanations and details.

depth correspond to the directions where the velocity gradient changes sign. A description of this effect in the spherical case has been given by Bertout (1977*c*). The quantities J_C and J_{BB} are defined in Eqs. (13) and (14). In the inner parts of the cloud, most of the contribution to the intensity field is given by the 2.8 K continuum radiation. With increasing optical depth the source term becomes more important in Eq. (11) and the line approaches local thermal equilibrium. However, the large radial velocity gradient near the cloud boundary (c.f., Fig. 5) significantly increases the escape probability in this region, thus dethermalizing the transition. Qualitatively, the overall features of the 18-cm transition indicated in Fig. 14 were found for the other line frequencies as well as at the different evolutionary stages which we investigated.

In the calculations, population inversion occurred in the denser parts of the cloud. However, due to the small number-density of OH used in the calculations and due to the large velocity gradients present in the cloud, the optical depth in the 18-cm transition never became more negative than about -10^{-2}. The radius at which population inversion of the 18-cm Λ-doublet disappeared corresponded to a temperature of about 40 K. Some models with arbitrarily increased OH density were computed in order to study the maser phenomenon. Due to high optical depth in the outer layers of the envelope, maser radiation occurring in the inner shells is absorbed there. Consequently, emission is barely seen on the "red" side of the line. On the "blue" side, however, maser emission can be seen due to the shape of the CRVS's (c.f., Fig. 10*a*). A double peaked structure of the maser line was found in the calculated examples. The secondary peak occurred at the frequency corresponding to the most extended CRVS completely enclosed in the amplification region. Calculations taking the non-local coupling into account must be undertaken in order to quantitatively study this effect. 18-cm lines showing double structure have been observed in several clouds (Turner, 1973).

Before closing this section, we should like to point out a consequence of the closed CRVS associated with the free-fall collapse, namely that in radioastronomy the beam-filling factor depends on the dimensions of the CRVS rather than on the dimension of the observed cloud. In other words, the beam-filling factor is frequency dependent (for the free-fall case in spherical geometry, for example, a factor 2 change in frequency corresponds roughly to a factor 16 change in the projected area of the CRVS). In the case of a continuum source distributed over all the beam area (for example, the isotropic 2.8 K background continuum), the amount of radiation due to the continuum at a given line frequency will therefore depend on the frequency considered. This implies that the width of the observed line does not give much information about the velocity field present in the envelope. Only the smallest radial velocities corresponding to the CRVS of relatively large projected areas comparable to the beam size contribute significantly to the line width. The high-velocity contributions are "washed-out" in the continuum flux unless the mean intensities at these velocities are extremely large. To illustrate this point, we consider the 18-cm transition and the 120-μm line (1-3 transition in Fig. 12). Both lines are formed in the same region, so that one should expect that they would give the same information on the cloud velocity field. However, the continuum sources are different for the two lines. For the 18-cm line, the 2.8 K background radiation dominates, and fills up the beam outside the CRVS. Since dust is not taken into account in our calculations, core radiation is the dominant continuum source at 120 μm. As a consequence of these different continuum sources, the widths of the two lines were found to be different, ranging from 1.2 km sec^{-1} in the case of the 18-cm line to 2.4 km sec^{-1} for the 120-μm line, which also shows the typical inverse P Cygni profile due to geometrical absorption of the core continuum radiation in a collapsing envelope. Since the geometry of the CRVS is independent of the overall cloud size, the width of the radio lines in collapsing clouds should not be used to derive the velocity distribution of these clouds, even if the cloud should fill the beam area. A more detailed description of this effect will be published elsewhere.

B. Late Protostellar Stages – the Optically Visible Core

Observational data of low-mass protostars are available principally in the near infrared, optical and near ultraviolet parts of the spectrum and therefore pertain to the relatively advanced stages of protostellar evolution when the central core becomes optically visible. There is little doubt now that the YY Orionis stars, first observed by Walker (1972), are indeed low-mass protostars at the end of their hydrodynamical evolution. Such an interpretation of the properties observed in YY Orionis stars was suggested earlier by several authors, including Larson (1972), Walker (1972), and Appenzeller and Tscharnuter (1975), and later confirmed by detailed comparison of protostar models with observations. From the calculations of protostar models, it is expected that the last phase of the accretion process should last long enough to be observable.

Observations of YY Orionis stars. The YY Orionis stars are those T Tauri stars which sometimes show a displaced absorption component at the red edge of certain emission lines, in particular at the edges of Balmer lines (with the exception of Hα, for which redward displaced absorption has never been observed in YY Orionis stars; Herbig, 1977). Most of the YY Orionis stars exhibit a strong ultraviolet excess (defined by U–B $\lesssim$ O) with a record value of U–B = –1.02 for WZ Cha. In their other properties, YY Orionis stars resemble T Tauri stars. The spectrum is often "veiled" by continuous emission in the blue spectral range, making it difficult to discern the underlying late-type absorption spectrum. Large variations of the Balmer lines are recorded on a time scale of days, and variations of the continuum level on the same time scale is indicated by the photometric measurements in the U and B filter ranges. Also, most YY Orionis stars exhibit strong infrared excess.

A list of the 24 YY Orionis stars known in September 1977 has been given by Appenzeller (1977). This list is shown in Table III. Most of these stars are faint, as were the nine YY Ori stars first identified by Walker (1972); but in the last few years two luminous stars have been found. They are S CrA (m_v = 11.49) and CoD –35°10525 (m_v = 11.62). For these two stars, coudé spectrograms of high dispersion have been reported (Wolf *et al.*, 1977; Appenzeller *et al.*, 1978; Rydgren, 1977). The spectra of S CrA and CoD –35°10525, like in other YY Orionis stars, is dominated in the blue spectral region by strong emission lines of Ca II (H and K) and hydrogen (Balmer lines). However, the emission-line spectrum of S CrA is much richer than the spectrum of CoD –35°10525. Figure 15 shows low-dispersion spectrograms of S CrA (from Appenzeller and Wolf, 1976), and includes identification of the strongest lines. The rapid variations of the line spectrum is noticeable in this series of spectrograms. The spectrum of CoD –35°10525 does not show the Fe II and Ti II emission lines found in S CrA. However, more He I and He II emission lines are found in CoD –35°10525 than in S CrA.

A common feature of the high-dispersion spectrograms of YY Orionis

TABLE III

The Presently Known YY Orionis Stars

Star	*V*	*U-B*	**Location**
UZ Tau f	12.88	-0.18	Tau-Aur
DL Tau	13.42	-0.39	Tau-Aur
CI Tau	13.05	-0.06	Tau-Aur
DM Tau	13.93	-0.34	Tau-Aur
DR Tau	13.38	0.00	Tau-Aur
SU Ori	14.50	—	Orion Neb.
XX Ori	14.87	-0.38	Orion Neb.
YY Ori	13.55	-0.39	Orion Neb.
SY Ori	13.70	—	Orion Neb.
BO Ori	14.02	+0.70	Orion Neb.
CE Ori	14.60	—	Orion Neb.
NS Ori	14.20	—	Orion Neb.
LX Mon	14.96	-0.61	NGC 2264
MM Mon	14.14	-0.01	NGC 2264
MO Mon	13.54	-0.16	NGC 2264
TW Cha	13.43	-0.19	Cha T Ass.
VW Cha	12.64	-0.13	Cha T Ass.
VZ Cha	13.23	-0.48	Cha T Ass.
WZ Cha	15.48	-1.02	Cha T Ass.
LkH 450-6	13.68	-0.42	Lup T Ass.
CD-35°10525	11.62	-0.06	Lup T Ass.
EX Lup	13.63	0.00	Lup T Ass.
Haro 1-1	13.34	-0.06	B 42
S CrA	11.49	-0.16	CrA T Ass.

stars is the highly complex structure of the Balmer lines. Figure 16 shows the variations of the S CrA Balmer line profiles observed on two consecutive nights (from Wolf *et al.*, 1977), and Fig. 17 illustrates the structure of the Balmer lines in CoD -35°10525. A typical Balmer line profile consists of:

1. A blue-shifted emission component, the strength of which seems to stay approximately constant with time;
2. an absorption component located approximately at the line center (slightly blue displaced in the case of S CrA, slightly red shifted for CoD -35°10525);
3. a red-shifted emission component separated from the blue-shifted emission by about -200 km s^{-1}, this component varying quickly in the case of S CrA;
4. a red-shifted absorption at about -300 to -400 km s^{-1};
5. a third emission component at the red edge of the profiles, in the case of

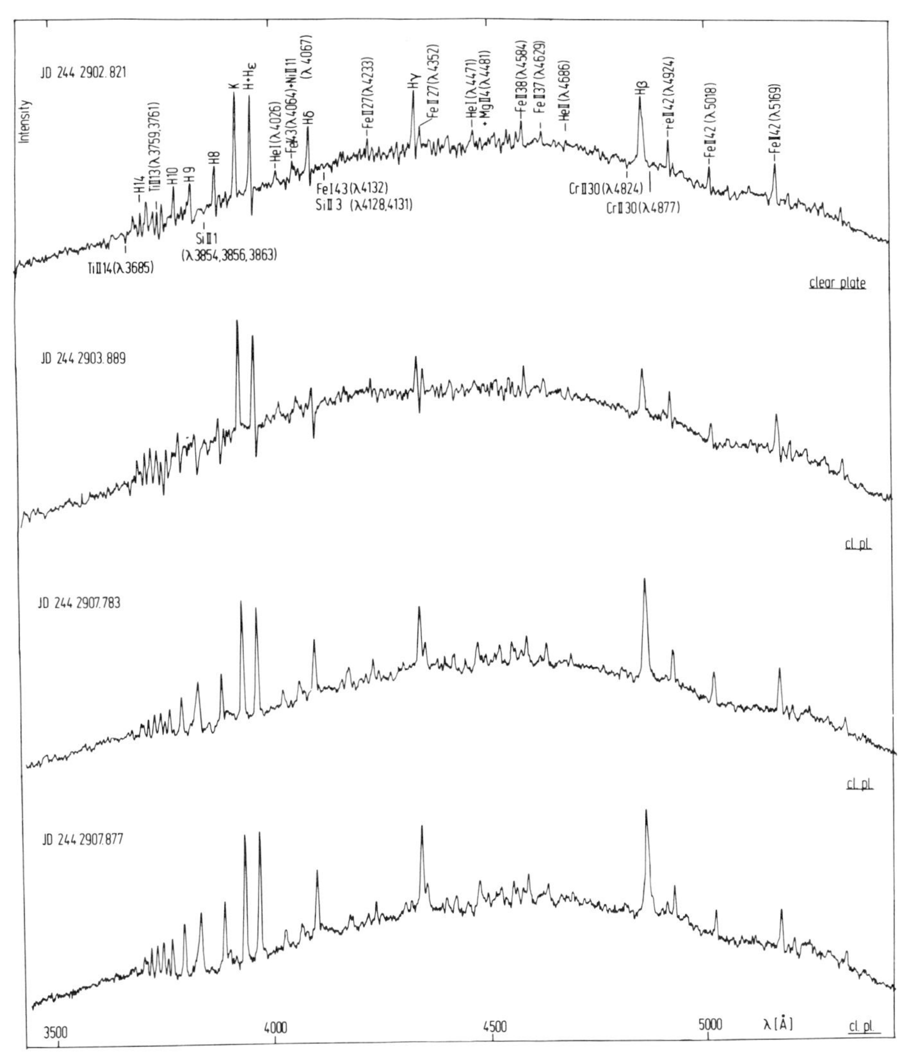

Fig. 15. Low dispersion spectrograms of S CrA. The most prominent lines are identified. Note the short time scale variations of the Balmer lines. (From Appenzeller and Wolf, 1976.)

CoD −35°10525 and possibly also present in S CrA although less conspicuously.

Simultaneous moderate dispersion spectroscopy and photometry of YY Orionis stars have been reported by Walker (1977), Bertout *et al.* (1977), and Mundt and Wolf (1977). For YY Orionis itself, Walker found no correlation of the hydrogen Balmer line intensities with the luminosity, and no strong correlation of the Ca II line intensities with the H intensities. Furthermore, Walker did not find any correlation between the strength of the hydrogen absorption components and the luminosities. A correlation analysis based on 20 spectrograms with dispersions of 20 Å/mm to 40 Å/mm and simultaneous UBV photometry was made for S CrA by Mundt (personal communication). He found that the intensities of the Balmer emission lines were not correlated

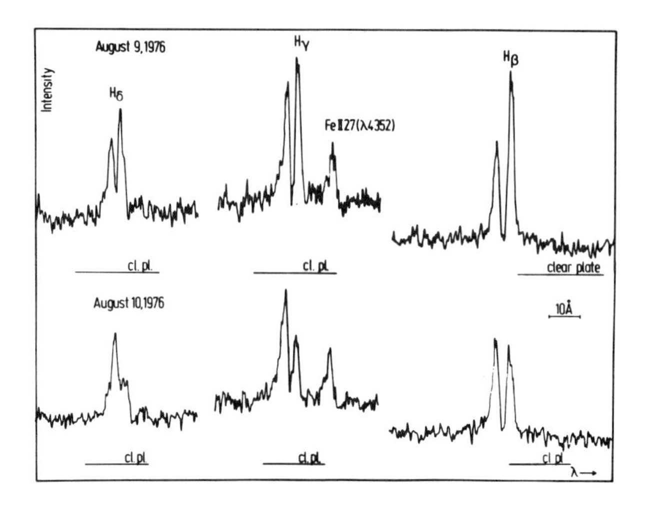

Fig. 16. Balmer line profiles of S CrA as observed on high-dispersion spectrograms (from Wolf *et al.*, 1977). A detailed description of the profiles is given in the text.

with the luminosity. However, the Ca II lines are correlated with H emission, whereas the intensities of the different metal lines are correlated among themselves but not with the hydrogen intensities. Mundt found also that for S CrA the equivalent width of the red displaced absorption component of the Balmer lines seems to be correlated with the strength of the ultraviolet excess, in the sense that the equivalent width increases when U–B decreases. Also an increase of the equivalent width with increasing blue luminosity was found.

Between 1975 and 1977 the number of known YY Orionis stars increased from nine to twenty-four. The percentage of these stars among T Tauri stars with strong ultraviolet excess has been estimated to be about 75% (Appenzeller, 1977). This would correspond roughly to 40% of all T Tauri stars. Because of the high discovery rate of YY Orionis characteristics when a systematic search for such properties is made, one may suspect that most of the T Tauri stars undergo YY Orionis phases during their evolution.

Line Formation in YY Orionis Envelopes. Assuming that a typical YY Orionis star is a spherically symmetric protostar at the end of its hydrodynamic evolution, Bertout (1977*c*) calculated the formation of line profiles in a dust-free collapsing envelope, while assuming a two-level atom model. Most of the free parameters were eliminated by using hydrodynamical models of a spherical protostar of solar mass at the end of the collapse phase (Appenzeller and Tscharnuter, 1975). The results showed good agreement with the available low-dispersion line profiles of YY Orionis stars. However, this simple collapse model could not reproduce the line profiles seen on high-dispersion spectrograms. A detailed model of the YY Orionis phenomenon, based on the interpretation of the complex Balmer line profiles

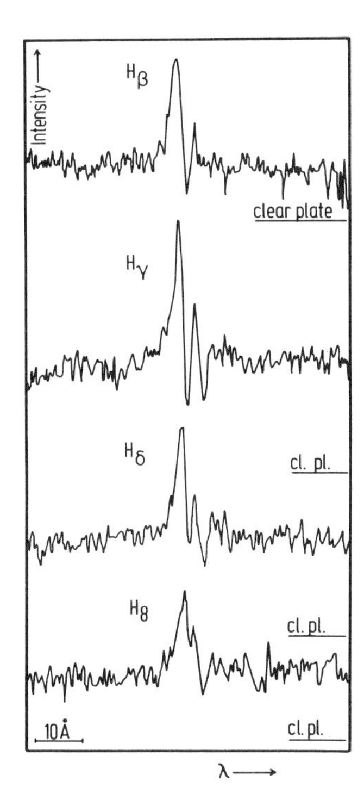

Fig. 17. High dispersion observations of the Balmer line profiles of CoD $-35°10525$ (from Appenzeller *et al.*, 1977).

discussed above, has been proposed by Wolf *et al.* (1977). This model is consistent with the overall properties of the protostellar collapse found in hydrodynamical calculations; but it indicates that such effects as radiative pressure in the lines, which have been ignored in the evolution calculations, can play a role in the collapse dynamics at late evolutionary stages.

Basically, Wolf *et al.* show that due to the Doppler shift, the Lyman line radiation emitted from the shock cooling region and relaxation zone (see Fig. 3) which surrounds the stellar core can be absorbed only in the outer layers of the protostellar envelope where the velocity of the infalling material is of the same order of magnitude as the velocity in the post-shock region. Because of the high-absorption cross-section of neutral hydrogen for Lyman line radiation, the absorption will occur in a thin shell where the velocity gradient is negligible. Therefore, a Lyman α photon will be trapped in the shell until it undergoes a rare collisional de-excitation. As a result of this resonance scattering process, many hydrogen atoms will be in the $n = 2$ quantum state and are therefore potential Balmer line absorbers. Since this effect occurs in regions of low velocities, additional Balmer line absorption will occur near the line center as observed in the Balmer line profiles of YY Orionis stars.

Although the exact calculation of radiative pressure on the layers absorbing Lyman line radiation is difficult, it can easily be shown that the maximum radiative force which acts on the envelope is much smaller than the gravitational force, i.e., the collapse of the *whole* envelope can neither be reversed nor halted by radiative pressure effects in the lines. However, the absorption will occur in thin shells in which the velocity gradient is negligible. For S CrA, only 10^{-7} of the total flux in $L\alpha$ is sufficient to decelerate or stop the collapse in such a thin shell, resulting in a non-stationary gas flow in parts of the envelope. Due again to the Doppler shift, the radiative force increases as long as the velocity is directed inward, and decreases when outward. Furthermore, a shell supported against gravity by radiation pressure from below is Rayleigh-Taylor unstable. Therefore, outward motions as well as non-radial motions might be expected in the envelope.

The optically thin, shock cooling region emits not only Lyman line radiation, but also Lyman continuum photons, which in contrast to the line radiation can be absorbed by the free falling material just above the shock. Therefore a shock precursor region of highly ionized matter can be formed. This region will contribute recombination radiation to the Balmer lines, mostly on the red side of the line due to occultation effects by the stellar core. The recombination emission rate is strongly dependent on the density in the shock precursor region, and the size of the precursor region is expected to vary with the energy output of the shock cooling region, so that rapid variations of the Balmer line component due to recombination is expected. A detailed model of these variations would require a solution of the radiative transfer problem across the shock. This is a highly complicated problem which demands that the coupling of hydrodynamics and radiative transfer be taken into account. In the absence of computed solutions, only guesses at the structure of this region can be made.

The line profile calculations indicate that the strength of the recombination components strongly depend on the radius of the recombination region. A radius of about 1% of the stellar core is sufficiently large for reproducing the line strengths observed in S CrA. If recombination is neglected, but partial ionization assumed in the precursor region, a largely red-displaced emission is still found at the edge of the profile (as observed in CoD −35°10525). The size of this component is strongly dependent on the assumed ionization degree. Using the model sketched above, Wolf *et al.* were able to reproduce the line profiles observed in S CrA. As already noted the profiles of CoD −35°10525 can be reproduced by a similar model without recombination emission in the precursor shock. In fact, many profiles of normal T Tauri stars can be qualitatively reproduced by a model combining infall and rotation (Bertout, 1977*b*). In Fig. 18, we show for purpose of comparison an Hδ line profile observed by Rydgren (1977) and a profile calculated according to the above model and using the S CrA parameter derived from the observations by Wolf *et al*, except for the maximum infall velocity which was assumed to be 600 km s^{-1} in this calculation. Some

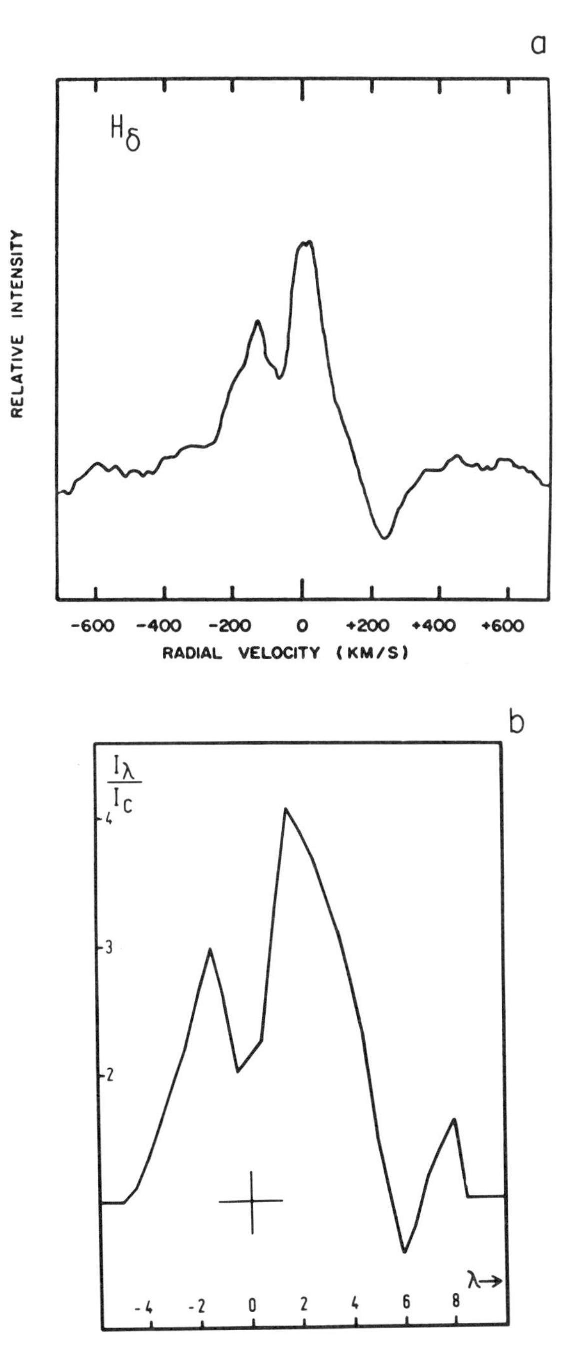

Fig. 18. (*a*) Hδ line profile of S CrA observed by Rydgren (1977) and (*b*) the computed profile using the model of Wolf *et al.* (1977). See text for explanations.

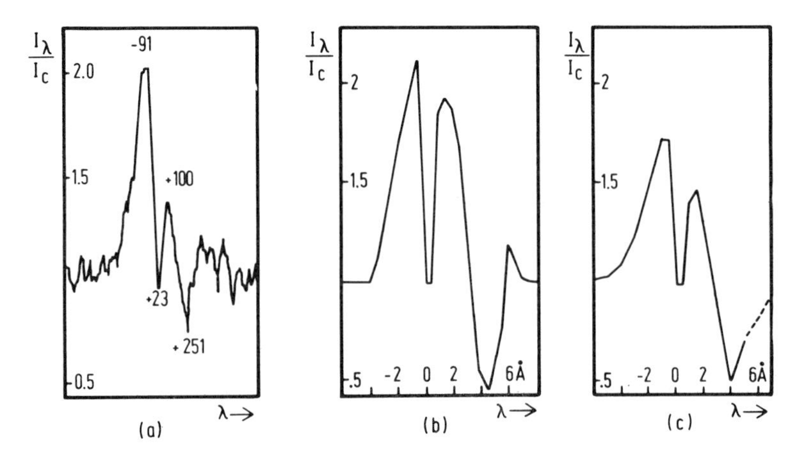

Fig. 19. (*a*) Hδ line profile of CoD−35°10525. Relative velocities of maxima and minima are as indicated. (*b*) Profile computed assuming a radial infall. (*c*) Profile computer assuming infall and rotation (from Bertout, 1977*b*). Further explanations are given in the text.

quantitative differences with the profiles of Wolf *et al.* resulted from the use of a different code. The simultaneous infall and outflow in the envelope is treated in the framework of the escape probability theory (Bertout, 1977*a*) which is possible because the same set of CRVS can represent both an accelerating infall and/or a decelerating outflow.

It has been suggested by Appenzeller *et al.* (1978) that CoD −35°10525 might be in a later evolutionary stage than S CrA. They based their suggestion on the comparison of the two stars' properties with the hydrodynamical collapse calculations and it is consistent with the fact that no recombination radiation is necessary to reproduce the Balmer lines of CoD −35°10525. Recombination is known to occur only when the shock precursor region is almost completely ionized (Bertout, 1977*c*). A lower rate of mass infall from the envelope would result in a lower shock-cooling radiation output and a lower photoionization rate in the precursor region.

In Fig. 19, an Hδ line profile of CoD −35°10525 (Plot *a*) is compared with two calculated profiles (Plots *b* and *c*). Plot *b* profile was calculated using the model of Wolf *et al.* but neglecting recombination radiation, and assuming a radial collapse, while Plot *c* is a profile calculated in a test of the numerical code described in Sec. II. In Plot *c* a two-level atom was used, and collapse and rotation were assumed, using the velocity field described by Ulrich (1976). The chosen parameters roughly correspond to a protostar of solar mass, and the angular momentum value was chosen so that all the matter in the envelope can hit the stellar core. This example shows that rotation does not change the qualitative properties of the YY Orionis line profiles and that a better agreement with some observed profiles is reached by introducing rotation in the model. Restricting the YY Orionis stars to pole-on or slow-rotating stars, as had been proposed by Appenzeller and Wolf for explaining the observed large infall velocities and the close agreement with spherically symmetric collapse calculations, therefore does not seem necessary. However, the small width of the helium lines in the best studied YY Orionis stars indicate that the rotational velocity should not exceed 50 to 100 km sec^{-1}.

The question of the appearance and disappearance of the red-displaced absorption components in the spectra of YY Orionis stars deserves further investigation. Calculations using the simplified two-level atom model indicate that the strength of the absorption component strongly depends on the overall radius of the emitting envelope and on the collisional de-excitation rate (Bertout, 1977*b*). This suggests that the Balmer Hα line is not seen in absorption because it is produced in a region of smaller size than the other members of the Balmer series. A multi-level line-transfer calculation (non LTE) of a hydrogen atom in the collapsing envelope should further clarify this point.

The main uncertainties in the line profile calculations in YY Orionis envelopes are introduced by our lack of knowledge of the radiative transfer phenomena across the shock front. A detailed solution to this problem is highly desirable. Another problem requiring detailed investigation is the mechanism proposed by Wolf *et al.* to stop the collapse in the outer layers of protostars. Although it is essentially a three-dimensional problem (due to the expected Rayleigh-Taylor instability of the layers where collapse is halted), one-dimensional hydrodynamical calculations might give valuable hints on the coupling between shock luminosity and density perturbations in the envelope. Quasi periods with time scales of several days, which have been observed in several T Tauri stars (Mauder and Schulz, 1977), might be reproduced by such a model. As pointed out by Wolf *et al.* (1977) and Appenzeller (1977), simultaneous mass accretion from the inner envelope layers and mass outflow from the outer envelopes layers seem to be natural consequences of detailed collapse models of protostars. Therefore, it would not be surprising if the properties of normal T Tauri stars and of YY Orionis stars could be explained by the combination of both effects.

Acknowledgments. We thank W. Tscharnuter for the use of his models and for many stimulating discussions. A major part of this work was done while C. B. was a guest at the Max-Planck-Institut für Physik and Astrophysik in Munich. C. B. thanks R. Kippenhahn for the invitation, and K. H. Winkler for introducing him to the IBM 360/91 at Garching. Additional calculations were done on the IBM 370/168 of the University Computing Center in Heidelberg. This work was supported in part by the Deutsche Forschungsgemeinschaft.

REFERENCES

Aanestad, P.A., and Purcell, E. M. 1973. *Ann. Rev. Astron. Astrophys.* 11: 309.
Appenzeller, I. 1977. In *Proc. IAU Colloq. No. 42.* (R. Kippenhahn, J. Rahe and W. Strohmeier, eds.). In press.
Appenzeller, I.; Mundt, R.; and Wolf, B. 1978. *Astron. Astrophys.* 63: 289.
Appenzeller, I., and Tscharnuter, W. 1974. *Astron. Astrophys.* 30: 423.
____. 1975. *Astron. Astrophys.* 40: 397.
Appenzeller, I., and Wolf, B. 1976. *Astron. Astrophys.* 54: 713.
Bedijn, P. J. 1977. Doctoral Dissertation, University of Leiden.
Bertout, C. 1976. *Astron. Astrophys.* 51: 101.
____. 1977*a*. *Mitt. Astron. Ges.* In press.
____. 1977*b*. In Proc. *IAU Colloq. No. 42* (R. Kippenhahn, J. Rahe and W. Strohmeier, eds.). In press.
____. 1977 *c*. *Astron. Astrophys.* 58: 153.
____. 1978. In preparation.
Bertout, C.; Krautter, J.; Moellenhoff, C.; and Wolf, B. 1977. *Astron. Astrophys.* 61: 737.
Castor, J. I. 1970. *Mon. Not. Roy. Astr. Soc.* 149: 111.
de Jong, T.; Chu, S.-I.; and Dalgarno, A. 1975. *Astrophys. J.* 199: 69.
Gerola, H., and Glassgold, A. E. 1977. Preprint.
Goldreich, P., and Kwan, J. 1974. *Astrophys. J.* 189: 441.
Grachev, S. I., and Grinin, V. P. 1975. *Astrophysics* 11: 20.
Gwinn, W. D.; Turner, B. E.; Goss, W. M.; and Blackman, G. 1973. *Astrophys. J.* 179: 789.
Herbig, G. H. 1977. *Astrophys. J.* 214: 747.
Herbst, E., and Klemperer, W. 1973. *Astrophys. J.* 185: 505.
Hertel, G. 1976. Ph.D. Dissertation. MPI-PAE/Astro 107, München.
Hollenbach, D., and Salpeter, E. E. 1970. *J. Chem. Phys.* 53: 79.
Hummer, D. G., and Rybicki, G. B. 1978. *Astrophys. J.* 219: 654.
Iglesias, E. 1977. *Astrophys. J.* 218: 697.
Johansson, L. E. B.; Hoeglund, B.; Winnberg, A.; Nguyen-Q-Rieu; and Goss, W. M. 1974. *Astrophys. J.* 189: 455.
Kahn, F. D. 1974. *Astron. Astrophys.* 37: 149.
Kippenhahn, R., and Tscharnuter, W. 1975. *Lecture Notes in Phys.* 42: 79. New York: Springer-Verlag.
Kuan, P., and Kuhi, L. V. 1975. *Astrophys J.* 199: 148.
Larson, R. B. 1969. *Mon. Not. Roy. Astr. Soc.* 145: 271.
____. 1972. *Mon. Not. Roy. Astr. Soc.* 157: 121.
____. 1973. *Ann. Rev. Astron. Astrophys.* 11: 219.
____. 1977. *Star Formation* (T. de Jong and A. Maeder, eds.), p. 249. Dordrecht: Reidel.
Leung, C. M. 1975. *Astrophys. J.* 199: 340.
Lillie, C. F., and Witt, A. N. 1976. *Astrophys. J.* 208: 64.
Litvak, M. M. 1969. *Astrophys. J.* 156: 471.
Lucas, R. 1976. *Astron. Astrophys.* 46: 473.
Lucy, L. B. 1971. *Astrophys. J.* 163: 95.
Marti, F., and Noerdlinger, P. D. 1977. *Astrophys. J.* 215: 247.
Mauder, H., and Schulz, E. 1977. In *Proc. IAU Colloq. No. 42* (R. Kippenhahn, J. Rahe and W. Strohmeier, eds.). In press.
Morfill, G.; Röser, S.; Tscharnuter, W.; and Völk, H. 1978. *Moon and Planets* (special Protostars and Planets issue). In press.
Mundt, R., and Wolf, B. 1977. In *Proc. IAU Colloq. No. 42* (R. Kippenhahn, J. Rahe and W. Strohmeier, eds.). In press.
Rybicki, G. 1970. *Nat. Bur. Standards Special Publ.* 332: 87.
Rydgren, A. E. 1977. Preprint.
Scoville, N. Z., and Solomon, P. M. 1974. *Astrophys. J.* 187: L67.
Snell, R. L., and Loren, R. B. 1977. *Astrophys. J.* 211: 122.

Sobolev, V. V. 1960. *Moving Envelopes of Stars.* Cambridge, Mass.: Harvard University Press.
Surdej, J. 1977. *Astron. Astrophys.* 60: 303.
Tscharnuter, W. 1976. *Computer Phys. Comm.* 12: 1.
_____. 1978. *Moon and Planets* (special Protostars and Planets issue). In press.
Turner, B. E. 1970. *J. Roy. Astr. Soc. Canada* 64: 282.
_____. 1973. *Astrophys. J.* 186: 357.
Ulrich, R. K. 1976. *Astrophys. J.* 210: 377.
Völk, H. J.; Jones, F. C.; Morfill, G.; and Röser, S. 1978. *Moon and Planets* (special Protostars and Planets issue). In press.
Walker, M. F. 1972. *Astrophys. J.* 175: 89.
_____. 1977. In *Proc. IAU Colloq. No. 42* (R. Kippenhahn, J. Rahe and W. Strohmeier, eds.) In press.
Westbrook, C. K., and Tarter, C. B. 1975. *Astrophys. J.* 200: 48.
Winkler, K.-H. 1977. Ph.D. Dissertation, University of Göttingen.
Wolf, B.; Appenzeller, I.; and Bertout, C. 1977. *Astron. Astrophys.* 58: 163.
Yorke, H. W. 1977*a*. *Astron. Astrophys.* 58: 423.
_____. 1977*b*. In *Proc. IAU Colloq. No. 42* (R. Kippenhahn, J. Rahe and W. Strohmeier, eds.). In press.
_____. 1978. In preparation.
Yorke, H. W., and Kruegel, E. 1977. *Astron. Astrophys.* 54: 183.

INTERPRETING INFRARED OBSERVATIONS OF T TAURI STARS

A. E. RYDGREN
Rensselaer Polytechnic Institute

Infrared observations of T Tauri stars may yield useful information on the early stages of planetary systems. The interpretation of these observations is unfortunately hampered by uncertainties in the correction for interstellar extinction. Nevertheless, the observed spectral energy distribution of the typical T Tauri star does not favor the optically-thick dust shell model for the infrared excess. Additional work is needed to determine if the present observations in the infrared actually contain evidence of planetary material.

Our solar system apparently originated in a dusty preplanetary disk associated with the early sun. Infrared (IR) observations of stars currently in such an evolutionary state should provide important information on the formation of planetary systems. The T Tauri stars may be relevant to this question, since lower mass stars ($M \lesssim 2\ M_{\odot}$) are first clearly seen in this evolutionary phase and T Tauri stars are known to have IR excesses. However, one must understand the visible plus infrared spectral energy distributions (hereafter abbreviated SED) of these stars before one can determine if the IR excesses actually contain information about preplanetary material. In this chapter, we briefly examine the question of interpreting the SEDs of T Tauri stars, with emphasis on the IR wavelength region; other aspects are reviewed in the chapters of Imhoff, Kuhi and Ulrich.

HISTORICAL REVIEW

The first evidence that T Tauri stars are unusual in the infrared wavelength region was presented by Mendoza (1966, 1968). He obtained optical and IR photometry for a sample of brighter T Tauri stars and showed that the IR fluxes were greater than expected for the spectral types and estimated interstellar reddening. Mendoza suggested that these IR excesses might arise from circumstellar dust or from the interaction of cosmic rays with the mass outflow from these stars.

The subsequent interpretation of the IR excesses was strongly influenced by Larson's (1969, 1972) hydrodynamical models for spherically-symmetric protostellar collapse. These calculations showed that a central core forms and accretes the surrounding remnant material over an extended ($\sim 10^6$ yr) period. Thus the IR excesses seen by Mendoza found a ready explanation in the optically-thick shells of remnant gas and dust predicted by Larson's models. Walker's (1972) spectroscopic evidence for mass infall in some T Tauri stars added credence to this view. The combined optical and IR photometry of Low *et al.* (1970) showed that the observed SED of T Tauri does indeed have the double-peaked structure expected for a star surrounded by a circumstellar dust shell. Observations of R Mon (Low and Smith, 1966) have provided further evidence for circumstellar dust shells around very young stellar objects.

Extensive observations of T Tauri stars in the 2 to 20 μm wavelength region have been reported by Cohen (1973*a*). These data show the IR SEDs of many T Tauri stars to be remarkably flat. The lack of concurrent optical photometry has prevented a detailed analysis, but Cohen (1973*b*) has suggested that a circumstellar dust shell model might explain the observations if scattering and non-gray opacity were properly taken into account.

Rydgren, Strom and Strom (1976) obtained spectra and optical plus IR photometry for a large sample of T Tauri stars in the Taurus and Rho Oph dark clouds. They suggested that the veiling in the optical region in T Tauri stars is due to hydrogen continuous emission from a hot gaseous envelope and that the IR counterpart of this emission contributes to the IR excess. They also showed that the SEDs of many T Tauri stars can be understood as the sum of the appropriate late-type photosphere and the emission from a hot gaseous envelope, provided the interstellar extinction in the Taurus and Rho Oph dark clouds is anomalous ($R > 4$).

A detailed examination often shows that reality is more complex than our first theories suppose. Thus, in the case of the T Tauri stars, research over the past several years has brought a number of problems more clearly into focus but has not provided many satisfactory answers. In the following sections, we consider the problem of correcting for interstellar extinction and what can be learned from the observed SEDs of T Tauri stars, and then conclude with a discussion of unresolved questions.

PROBLEMS IN CORRECTING FOR INTERSTELLAR EXTINCTION

The observed SED of a T Tauri star must be corrected for interstellar extinction before it can be compared with theoretical models and dissected into its component parts. Unfortunately, the intimate relationship between T Tauri stars and dense interstellar dust clouds means that members of even the nearest T associations may suffer significant interstellar extinction. One cannot generally rely on a mean color excess for a group of T Tauri stars, but must obtain an individual reddening estimate for each star. Several serious sources of uncertainty are encountered in correcting the observed SED of a T Tauri star for interstellar extinction:

1. Emission lines and spectral veiling can affect the broadband colors of T Tauri stars from the ultraviolet at least into the near IR. The effect in K- and M-type stars is toward smaller (bluer) B-V and V-R colors, so that the true color excess may be underestimated. Thus, even when the photospheric spectral type is known, the intrinsic colors and corresponding color excesses may be quite uncertain. In the most extreme cases (e.g., RW Aur, DG Tau, S CrA), the line emission and spectral veiling preclude assignment of a spectral type.
2. Circumstellar dust shells may be present around T Tauri stars. When the intrinsic SED of the star must be considered an unknown, it is very difficult to separate the effects of circumstellar absorption from interstellar extinction.
3. There is growing evidence that the interstellar extinction in dense interstellar dust clouds is anomalous (e.g., Carrasco, Strom, and Strom, 1973; Whittet and van Breda, 1975; Herbst, 1975, 1977; FitzGerald, Stephens and Witt, 1976). Since many T Tauri stars are embedded in such clouds, the adoption of a "normal" extinction law could lead to a serious overestimate of the true IR excess. Moreover, it is not clear that a single extinction law will necessarily apply to all parts of a particular dark cloud.
4. Although T Tauri stars are known irregular variables at all wavelengths, the photometry at visible and IR wavelengths used to construct the observed SED is rarely simultaneous. This procedural problem can be overcome through properly coordinated observations.

The problems noted above make it difficult to obtain the intrinsic (dereddened) SED of the typical T Tauri star* and thus hinder our attempts to discern the exact nature of the IR excess. Perhaps the best way to avoid these sources of uncertainty is to concentrate observational work on T Tauri stars for which the interstellar extinction appears minimal. Warner, Strom and Strom (1977) point out that the members of NGC 2264 should meet this condition.

INFERENCES FROM THE OBSERVED SPECTRAL ENERGY DISTRIBUTIONS

The observed SEDs of T Tauri stars, even if uncorrected for interstellar extinction, fortunately contain some information on the physical nature of

---

*By "typical" T Tauri star, I have in mind a star with a late K or early M spectral type, a luminosity of no more than five solar luminosities, and an inferred mass of about one solar mass or less. It is important to distinguish such stars from the brightest, best-studied T Tauri stars which are characterized by somewhat earlier spectral types, higher luminosities, and larger inferred masses.

these objects. In preparation for this writing, the observed SEDs of some 65 T Tauri and related stars were plotted from broadband photometry available to the author. The discussion which follows is based largely on this body of data.

An optically thick circumstellar dust shell around a T Tauri star should manifest itself as broad Planck-like emission with an effective temperature $\lesssim 10^3$ K. If the dust shell is sufficiently important, the SED should show one maximum from the reddened stellar photosphere and a second maximum farther in the IR due to the dust re-emission. To the author's knowledge, T Tau itself (Fig. 1) is the only T Tauri star which clearly shows such structure (Low *et al.*, 1970; Rydgren, Strom and Strom, 1976). The early-type stars HD 97048 (in Cha T1) and AB Aur also show IR excesses of the form expected from circumstellar dust shells (see Fig. 1). Both stars are near the ZAMS at spectral type B9 and probably have masses ($M \approx$ 2-3 $M_\odot$) similar to that inferred for T Tau. These observations suggest that circumstellar dust shells may not be rare for young stars in this mass range.

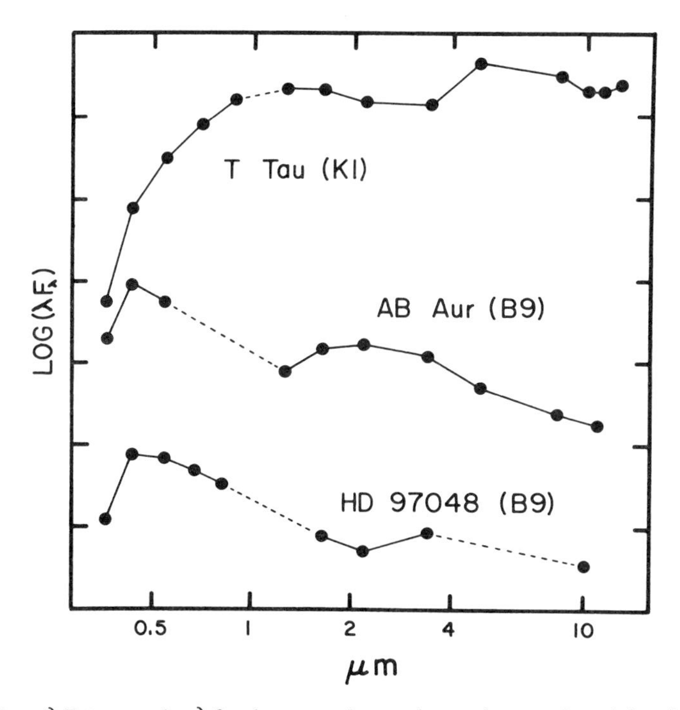

Fig. 1. Log (λF_λ) versus log λ for three stars (spectral types in parentheses) showing probable circumstellar dust shells. Data are taken from Mendoza, 1968 (T Tau); Rydgren, Strom and Strom, 1976 (T Tau); Rydgren unpublished (AB Aur, HD 97048); Strom *et al.*, 1972 (AB Aur); Gillett and Stein, 1971 (AB Aur); Grasdalen *et al.*, 1975 (HD 97048). The tick marks on the vertical scale indicate intervals of 0.5 in log (λF_λ) (same for all four figures).

Most of the T Tauri stars in our sample have lower luminosities and later spectral types than T Tau, and inferred masses of less than 2 $M_{\odot}$. The SEDs of these stars normally show a single maximum, usually in the wavelength interval $1 < \lambda < 2$ μm (see Fig. 2 for three typical examples). This seems best understood as the photospheric maximum shifted to longer wavelength by interstellar reddening, since it is difficult to explain a large IR excess shortward of 2 μm by circumstellar dust emission. The observed peak at the H filter (1.6 μm) relative to the adjacent J and K filters is a characteristic of K- and M-type photospheres (Gingerich and Kumar, 1964; Woolf, Schwarzschild and Rose, 1964) and supports this interpretation.

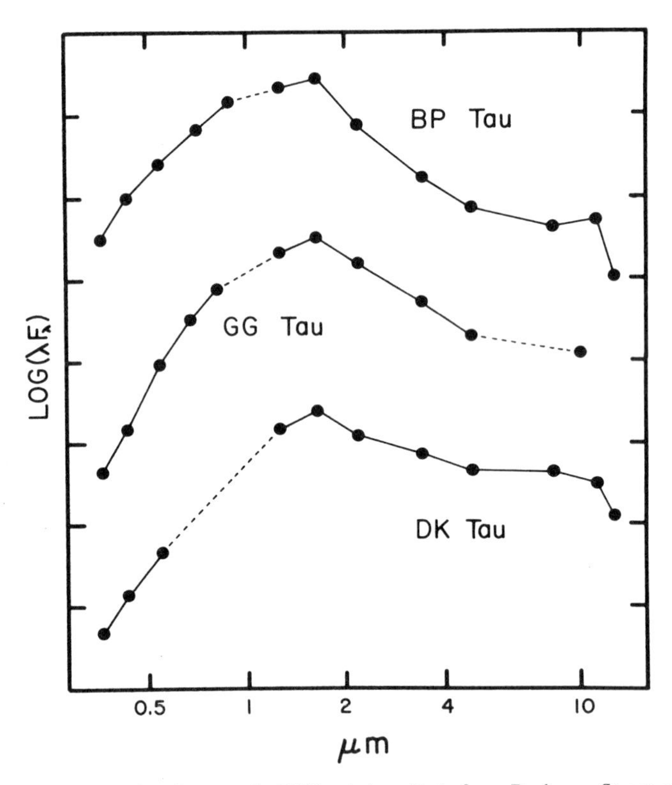

Fig. 2. As for Fig. 1, for three typical T Tauri stars. Data from Rydgren, Strom and Strom, 1976 (all 3 stars); Rydgren, unpublished (GG Tau); Mendoza, 1968 (BP Tau).

Even in those T Tauri stars not showing direct evidence of circumstellar dust shells, the observed SEDs do not fall off longward of the photospheric maximum as rapidly as expected for a late-type photosphere. Among the possible explanations for such excesses are (1) hydrogen continuous emission (the IR counterpart of the visible region veiling emission), (2) re-emission from modest circumstellar dust shells (sufficiently weak that a second maximum is not apparent in the broadband IR photometry), and (3) disk continuum emission of the form suggested by Lynden-Bell and Pringle (1974). In any case, the SED of the typical T Tauri star is clearly not dominated by a circumstellar dust shell, and any reradiation from circumstellar dust is much less prominent than in the class prototype. The remarkable flatness of the observed SEDs of extreme T Tauri stars such as RW Aur and S CrA (Fig. 3) seems consistent with the view that these particular stars are dominated by emission from hot circumstellar gas.

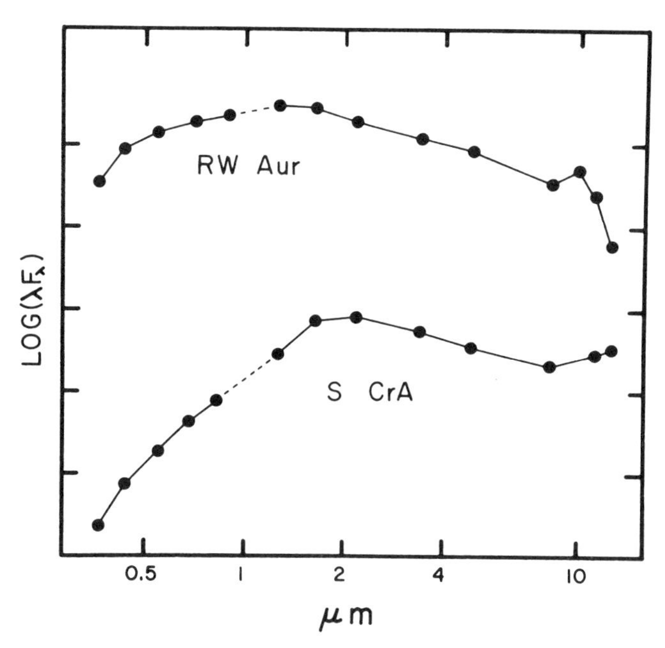

Fig. 3. As for Fig. 1, but for two extreme T Tauri stars. Data from Mendoza, 1968 (RW Aur); Rydgren, Strom and Strom, 1976 (RW Aur); Rydgren, 1977 (S CrA); Glass and Penston, 1975 (S CrA); Vrba, Strom and Strom, 1976 (S CrA).

If some circumstellar dust is present around a T Tauri star, one might expect to see emission in the 10 μm and 20 μm silicate features. Observations of some of the brighter T Tauri stars by the Stroms (see Rydgren, Strom and Strom, 1976) reveal that 10 μm silicate emission features are relatively common, even in stars which do not show direct evidence of optically thick circumstellar dust shells (Fig. 4). Substantial flux is also observed at 20 μm; the flux in the combined silicate features typically represents about five percent of the total stellar luminosity. A crude analysis of the 10 μm to 20 μm flux ratio suggests a silicate temperature on the order of 200 K. The observations need to be extended to less luminous T Tauri stars to determine if silicate emission is present in these stars as well.

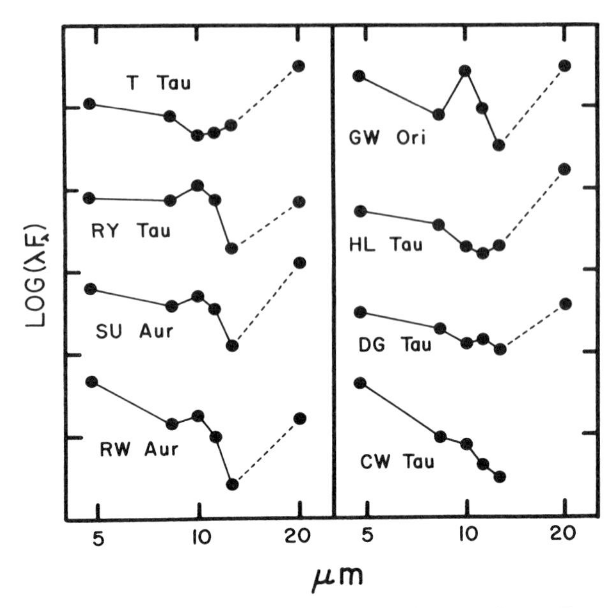

Fig. 4. The region of the silicate features in eight bright T Tauri stars. Data are taken from Rydgren, Strom and Strom, 1976, except for GW Ori which is from S. E. and K. M. Strom (unpublished).

The observed SEDs of a few T Tauri stars rise steeply from the visible into the IR, suggesting extreme interstellar reddening or circumstellar extinction (e.g., HL Tau, XZ Tau, VV CrA). The implied visual extinctions are sufficiently large that one might expect to see the 3.1 μm ice feature or the 10 μm silicate feature in absorption. Observations of HL Tau (Cohen, 1975; Rydgren, Strom and Strom, 1976) show that both features are present in absorption. It remains to be seen if these features arise from intracloud extinction or from circumstellar material associated with HL Tau. Absorption features at 3.1 μm are not seen in a number of other T Tauri stars (Cohen, 1975), most of which should have smaller visual extinction than HL Tau.

DISCUSSION

The view that the typical T Tauri star is enshrouded in an optically-thick circumstellar dust shell now seems doubtful. The observed SEDs provide no compelling evidence for such a model, and the apparent circumstellar dust shell in T Tau should probably not be considered optically thick. Calculations of protostar collapse which assume spherical symmetry appear to be an oversimplification, since angular momentum considerations clearly favor collapse to a disk structure. Unfortunately, a disk which is sufficiently thin will intercept relatively little stellar radiation and might be difficult to detect in the IR. The possibility of self-luminous viscous disks around very young stars (Lynden-Bell and Pringle, 1974) is an intriguing idea which may be relevant to the T Tauri stars. In either case, theory and observation suggest that remnant material not accreted into the central star should be sought in a disk rather than a spherical shell structure.

The relationship between circumstellar dust around a T Tauri star and a possible strong mass outflow is not entirely clear. One would expect such a wind to tend to disperse the more tenuous circumstellar material (e.g., near the stellar poles). However, the possible formation of dust in such an outflow should also be examined (see Elmegreen, 1978).

It seems likely that there is some remnant material (probably in the form of a disk) around many T Tauri stars. However, the unambiguous detection of such preplanetary material may not be easy. The silicate emission features at 10 and 20 μm can provide direct evidence of warm, finely divided silicate material. The results for the brightest T Tauri stars are encouraging, but the observations must be extended to less luminous stars to determine if silicate features are characteristic of the class as a whole. Continuum emission from dust particles could also be contributing to the observed excess longward of the photospheric maximum in the typical T Tauri star. Simultaneous monitoring of the brightness variations of T Tauri stars at visible and IR wavelengths has been suggested as a way to separate the emission due to circumstellar dust from hydrogen continuous emission. Anticorrelated visible and IR variations could indicate dust (absorbed visible flux re-emitted at longer wavelengths), while correlated variations are expected from a hot gaseous envelope. Preliminary efforts (Cohen, 1973*a*; Cohen and Schwartz, 1976) reveal a bewildering variety of behavior; additional monitoring of selected stars is clearly needed. The effect of deviations from spherical and axial symmetry in both cases must also be considered.

The present observational evidence (e.g., Rydgren, Strom and Strom, 1976) points to a time scale on the order of 10^6 yr for the T Tauri phenomenon (ages according to the evolutionary tracks of Iben and Talbot, 1966). Thus the collapse of any disk of remnant material to a relatively thin configuration (i.e., most of the stellar flux escapes unimpeded) apparently must occur on a time scale shorter than this.

As the physical processes in T Tauri stars become better understood and the question of circumstellar dust and dark-cloud extinction laws are resolved, it should be possible to begin ordering T Tauri stars by age and to sketch the earliest evolution of stars like the sun. Together with the important work on the early history of our solar system, this should lead to a better understanding of the process of planetary formation.

COMMENTS

Kuhi: What are the error bars on the silicate features? Does not the presence of these features argue for the existence of circumstellar dust?

Rydgren: I do not know what the formal errors are, but a comparison of results from different nights for a given star supports the reality of the features shown. On your second question, yes, the silicate features do imply the presence of circumstellar dust in some form.

Greenberg: In answer to Kuhi's question about the paradox of "silicate" emission features with no optical obscuration of the star: if the silicate particles are small (say $\leqslant 0.05$ μm in radius or even a bit larger) they would show up in emission but their wavelength dependence of extinction is such that they would be relatively unimportant in the infrared and visual part of the spectrum.

REFERENCES

Carrasco, L.; Strom, S. E.; and Strom, K. M. 1973. *Astrophys. J.* 182: 95.
Cohen, M. 1973*a*. *Mon. Not. Roy. Astron. Soc.* 161: 97.
______. 1973*b*. *Mon. Not. Roy. Astron. Soc.* 164: 395.
______. 1975. *Mon. Not. Roy. Astron. Soc.* 173: 279.
Cohen, M., and Schwartz, R. D. 1976. *Mon. Not. Roy. Astron. Soc.* 174: 137.
Elmegreen, B. G. 1978. *Moon and Planets* (special Protostars and Planets issue). In press.
FitzGerald, M. P.; Stephens, T. C.; and Witt, A. N. 1976. *Astrophys. J.* 208: 709.
Gillett, F. C., and Stein, W. A. 1971. *Astrophys. J.* 164: 77.
Gingerich, O., and Kumar, S. S. 1964. *Astron. J.* 69: 139.
Glass, I. S., and Penston, M. V. 1975. *Mon. Not. Roy. Astron. Soc.* 172: 227.
Grasdalen, G.; Joyce, R.; Knacke, R. F.; Strom, S. E.; and Strom, K. M. 1975. *Astron. J.* 80: 117.
Herbst, W. 1975. *Astron. J.* 80: 498.
______. 1977. *Astron. J.* 82: 902.
Iben, I., and Talbot, R. J. 1966. *Astrophys. J.* 144: 968.
Larson, R. B. 1969. *Mon. Not. Roy. Astron. Soc.* 145: 271.
______. 1972. *Mon. Not. Roy. Astron. Soc.* 157: 121.
Low, F. J.; Johnson, H. L.; Kleinmann, D. E.; Latham, A. S.; and Geisel, S. L. 1970. *Astrophys. J.* 160: 531.
Low, F. J., and Smith, B. J. 1966. *Nature* 212: 675.
Lynden-Bell, D., and Pringle, J. E. 1974. *Mon. Not. Roy. Astron. Soc.* 168: 603.
Mendoza V., E. E. 1966. *Astrophys. J.* 143: 1010.
______. 1968. *Astrophys. J.* 151: 977.
Rydgren, A. E. 1977. *Publ. Astron. Soc. Pacific* 89: 557.
Rydgren, A. E., Strom, S. E.; and Strom, K. M. 1976. *Astrophys. J. Suppl.* 30: 307.
Strom, S. E.; Strom, K. M.; Yost, J.; Carrasco, L.; and Grasdalen, G. 1972. *Astrophys. J.* 173: 353.
Vrba, F. J.; Strom, S. E.; and Strom, K. M. 1976 *Astron. J.* 81: 317.
Walker, M. F. 1972. *Astrophys. J.* 175: 89.
Warner, J. W.; Strom, S. E., and Strom, K. M. 1977. *Astrophys. J.* 213: 427.
Whittet, D. C. B., and van Breda, I. G. 1975. *Astrophys. Space Sci.* 38: L3.
Woolf, N. J.; Schwarzschild, M.; and Rose, W. K. 1964. *Astrophys. J.* 140: 833.

T TAURI STAR EVOLUTION AND EVIDENCE FOR PLANETARY FORMATION

CATHERINE L. IMHOFF
University of Arizona

The T Tauri stars have often been referred to as pre-main-sequence objects which might resemble our young sun and its solar nebula. They appear to be roughly solar mass stars about a million years old, evolving along their quasi-static equilibrium tracks. The occurrence of mass loss, rapid rotation, and infrared radiation from a disk of dust and gas in the T Tauri stars indicates similar conditions in the early life of the sun. However, the observational evidence for these phenomena is conclusive only for rapid rotation. The interpretation of the mass flow of the material near the T Tauri stars is still open to question. The debate over the source of the infrared excess may be irrelevant to the problems of planetary formation, since the region giving rise to the radiation is fairly close to the star. In any case, planetary formation has probably occurred before the T Tauri phase.

In trying to understand the early evolution of the sun and its planetary system, it is logical to look at other stars presently going through their earliest stages of life. One such group of stars is the T Tauri stars. These variable stars are thought to be roughly equal to the sun in mass but have an age of about a million years. They are blessed with a bewildering variety of peculiarities, such as complex emission lines, absorption-line veiling, ultraviolet and infrared excesses, and spectral and photometric variability. A review of the T Tauri stars and other pre-main-sequence objects has been given by Strom *et al.* (1975); also see the chapters by Abt, Ulrich, Kuhi, Rydgren, Böhm and Welin. An historical perspective on the stars has recently been published by Imhoff (1978*a*).

Several of the processes thought to go on in the T Tauri stars may have a bearing on the early evolution of the sun. The solar wind may be a feeble remnant of an early, energetic outflow of mass, such as is seen in some T Tauri stars. Likewise, one may surmise that the sun's rotation has slowed greatly from its pre-main-sequence value. Finally, the infrared excesses of the T Tauri stars may arise from a thick disk of dust and gas in which planetary formation could occur. How conclusive a case can one build for protoplanets among the protostars? What kind of evidence do the T Tauri stars really give about pre-main-sequence evolution and planetary system formation?

THE EVOLUTIONARY STAGE OF THE T TAURI STARS

The argument that the T Tauris are pre-main-sequence stars rests largely on their elevation above the main sequence in the Hertzsprung-Russell (HR) diagram, as well as their association with O and B stars. Estimates of the masses, ages, and evolutionary states of these stars have been made by comparing their observed luminosities and spectral types with theoretical evolutionary tracks (Mendoza, 1968; Imhoff and Mendoza, 1974; Cohen and Kuhi, 1976; Rydgren *et al.,* 1976; Cohen, 1978). Using "classical" evolutionary tracks, such as those calculated by Iben (1965), masses from 0.3 to 3 $M_\odot$, ages* from 10^5 to 5×10^6 yr, and radii from 1 to 6 $R_\odot$ have typically been found. Most of the well-known T Tauri stars appear to fall on the quasi-static equilibrium (radiative) portions of their evolutionary tracks (Imhoff and Mendoza, 1974; Strom *et al.,* 1975; Rydgren *et al.,* 1976), but others lie on the hydrodynamic (convective) portions of their tracks (Walker, 1972; Cohen and Kuhi, 1976; Cohen, 1978). The T Tauri phase seems to first appear in stars with relatively high luminosities and ages of about 10^5 yr. The stars rapidly dim as they approach quasi-static equilibrium, then evolve more slowly to the main sequence, gradually losing their T Tauri peculiarities.

The comparison of the observed characteristics of the stars with the theoretical evolutionary tracks is itself not a trivial matter. The effective temperatures are derived from the spectral types. However, the spectra are not normal dwarf spectra but are affected by line emission, continuous emission, and rotational broadening (Herbig, 1962). Many stars exhibit no absorption spectrum at all, making classification impossible. Herbig and Rao (1972) list the most reliable spectral types, although revisions have been given for a few stars by Herbig (1977*a*), and Joy and Abt (1974). An added complication is the variability of the spectra.

The calibration of effective temperature with spectral type is largely a matter of faith. The calibration for normal dwarfs, given by Johnson (1966), is generally used, although its relevance for pre-main-sequence stars is in doubt, even if it is modified to account for the higher luminosities of the stars (Imhoff and Mendoza, 1974). In addition, the transference of the stars' temperatures to the theoretical HR diagram requires that the theoretical models themselves treat the outer envelopes and atmospheres of the model stars accurately.

The observed luminosities of the stars are fairly easy to obtain by integrating their fluxes, from broadband photometry, over a wide wavelength range. In the absence of complicating factors, the observed luminosity is a lower limit to the true luminosity of the star (Imhoff and Mendoza, 1974). However, since perhaps 2/3 of the stellar flux is re-emitted in the infrared (Imhoff, 1973), an understanding of the emitting region is required. Specifically one

*These are ages from $t = 0$ in the stellar models, a subtle distinction which probably does not significantly affect their age estimates.

needs to know the geometry of the shell, whether it is optically thick or thin, and whether the excess is due to gas or dust or both. Unfortunately the observational evidence is not yet conclusive and these factors are largely unknown.

The largest correction to the luminosity is the amount of extinction affecting the stars (the problem is reviewed in Rydgren's chapter). It may be both interstellar and circumstellar. There have been many suggestions of abnormal extinction laws, with $R = 6$, and of significant amounts of neutral extinction (Strom, 1972; Imhoff and Mendoza, 1974; Rydgren *et al.*, 1976; and others). The radiative transfer in the circumstellar shell needs to be considered.

The combined uncertainties in the effective temperatures and luminosities of the T Tauri stars make their positions in the theoretical HR diagram rather imprecise. However, it is hoped that the consideration of a large enough sample of stars will ameliorate these difficulties. For instance, one may argue from Iben's (1965) tracks that any star with log $T_e \leqslant 3.65$ is in the convective portion of its evolutionary track; according to Johnson's (1966) calibration this would correspond to a star of type K5 or later. A quick check of Herbig and Rao's (1972) catalog shows that out of 257 T Tauri or possible T Tauri stars, 180 have continuous or unclassified spectra, 47 have G and K spectral types, and only 30 have K5 or M spectral types. So the "convective" T Tauri stars appear to be a significant minority of all T Tauri stars.

More recent theoretical work, employing non-homologous hydrodynamic models such as those by Larson (1969; see the chapter by Bodenheimer and Black) indicates a different scenario for the newly formed stars from that previously discussed. The core of a protostellar cloud is expected to collapse first and begin to evolve, but it will be concealed by the surrounding cloud of gas and dust. A star may evolve nearly to the main sequence before its obscuring cloud is dispersed and the star revealed.

However, observations appear to disagree with this view (Rydgren *et al.*, 1976; Cohen and Kuhi, 1976; Cohen, 1978). Many stars are found to have relatively high luminosities and low temperatures – stars which are predicted by "classical" evolutionary theory but which should not be visible according to non-homologous models. In addition, the T Tauri stars show photospheric spectra and only moderate infrared excesses, indicating that they are not heavily obscured. The most likely explanation for this disagreement with Larson's (1969) model is that, while the star is still quite young (10^5 yr), it undergoes a high rate of mass loss. The mass loss prevents further infall of matter and clears out much of the surrounding cloud (Rydgren *et al.*, 1976).

An interesting sidelight to the consideration of the evolutionary state of the T Tauri stars is the "FU Orionis phenomenon." At least three stars (FU Orionis, V 1057 Cygni, and V 1515 Cygni) have experienced sudden dramatic increases of brightness. The phenomenon's importance to this discussion lies in the fact that V 1057 Cygni was apparently an ordinary T Tauri star before its outburst. The source of this phenomenon has been attributed to a wide variety of reasons: (1) free-fall collapse to the top of the Hayashi track of a star (Herbig, 1966); (2) an unpredicted hydrodynamic form of evolution in a massive star (Grasdalen, 1973); (3) the disruption of a dust envelope (Cohen and Woolf, 1971; Welin, 1978); and (4) a normal recurrent surface disturbance in T Tauri stars (Herbig, 1977*b*). Recent discussions, and opposing views, of this problem have been given by Herbig (1977*b*) and Welin (see his chapter in this book).

A very fundamental point should be noted. It is commonly assumed that all stars of a certain mass range go through a "T Tauri phase" in their early evolution. Herbig (1970*b*) has argued that the mean duration of such a phase is around 5 x 10^6 yr. Certainly the stars exhibiting T Tauri characteristics are seen over a large portion of the HR diagram. It is not inconceivable, however, that there is something peculiar about these particular stars that sets them apart from "normal" stars – strong magnetic fields, rapid rotation, interactions with surrounding gas and dust clouds, or undetected close binaries. Better statistical information on regions of star formation, including those of faint stars, and an improved understanding of the T Tauri peculiarities are required to settle this question. It may be noted that Kuhi's (1978) results for rotation in NGC 2264 indicate no difference betweem Hα emission (including T Tauri stars) and non-emission stars. In addition, Herbig's (1977*a*) work on the radial velocities of T Tauri stars is consistent with a normal proportion of spectroscopic binaries.

One may conclude that the T Tauri stars are in a normal phase of evolution for young stars of approximately solar mass. The T Tauri characteristics are exhibited over a range of ages and masses; most are evolving in quasi-static equilibrium along their radiative tracks but a significant portion is still in hydrodynamic evolution. Their peculiarities result from mass flow and circumstellar gas and dust, which doubtless has an influence on their immediate environments and thus may affect planetary formation.

THE T TAURI ENVIRONMENT AND PLANETARY FORMATION

There are three T Tauri peculiarities which may be related to planetary formation and which will be examined here. They are the occurrence of mass loss, the presence of infrared-emitting circumstellar shells, and rapid rotation.

Mass Loss

The idea of mass loss in T Tauri stars is derived from the occurrence of asymmetric emission-line profiles in their spectra. Several T Tauri stars have strong emission lines with blue-shifted absorption components, analogous to those seen in the hotter P Cygni stars. The first models were constructed by Kuhi (1964), who interpreted the line profiles to mean that the stars are losing mass in extended shells of hot gas. The mass loss rates he found ranged up to

10^{-7} $M_{\odot}$/yr; over the evolution of the star to the main sequence this could prove to be a significant fraction of the star's original mass.

Herbig (1970*a*) suggested that the emission lines and ultraviolet excess in the T Tauri stars are chromospheric in origin, rather than arising in an extended shell. However, a velocity field, and thus some mass loss, is still required to fit the line profiles, according to the calculations of Dumont *et al.* (1973).

A subset of the T Tauri stars, known as the YY Orionis stars, have *inverse* P Cygni line profiles (Walker, 1966). Walker (1972) suggested that these stars are the younger T Tauri stars in which mass infall is still occurring and which could be described by Larson's (1969) hydrodynamical calculations. The emission lines and ultraviolet excess could be produced by the impact of the infalling material. Another possibility is that the YY Orionis stars are the less massive T Tauri stars, in which mass loss (by whatever mechanism) never becomes important. The estimates of the numbers of YY Orionis stars are around 35% (Walker, 1972) to 40-50% (Appenzeller, 1977), but Herbig (1977*a*) found none in the sample of stars he studied, including YY Orionis itself. However, the definition of "YY Orionis star" is biased in favor of including the star in the subclass. The line profiles are variable and may change from P Cygni to inverse P Cygni, but few spectra are taken of any one star. A star will be put into the YY Orionis subclass if it shows an inverse P Cygni line profile only once; thus a star could have P Cygni line profiles 95% of the time and still be put into the subclass. A thorough study of the variations of the line profiles in a number of stars is needed to clarify the real differences between the YY Orionis subclass and the other T Tauri stars.

Ulrich (1976) has reinterpreted the "normal" P Cygni line profiles originally examined by Kuhi in terms of mass loss. Ulrich showed that the line profiles may signify mass *infall,* as a circumstellar cloud of gas falls in a spiral path inward to the star's surface. This ambiguity in the interpretation of the line profiles is quite unsettling. It is clear that more attention must be paid to the details of the line profiles and their variability before the mass flow near the T Tauri stars is thoroughly understood. The chapter by Ulrich in this book addresses the problems of interpreting the T Tauri emission-line profiles.

Infrared-Emitting Circumstellar Shells

The T Tauri stars are among the first stars to be discovered to have infrared excesses (Mendoza, 1966). The traditional interpretation is that the infrared radiation arises from warm (about 700 K) dust in a shell or disk around the star (Low and Smith, 1966; Mendoza, 1968). Such a description might lead one to envision a "stellar nebula" of gas and dust around the stars from which protoplanets might condense. However, a debate has arisen over the source of the infrared excess. Rydgren *et al.* (1976) have argued that the infrared radiation arises not from the dust but from bound-free and free-free emission from the hot gas surrounding the stars. Rydgren discusses this question further in this book.

The author (Imhoff, 1978*b*) has begun to calculate stellar energy distributions for stars surrounded by dust shells. Particular attention is being paid to the energy distribution expected from an optically thin dust shell, in which the temperature of the dust grains is set essentially by the distance from the star. Some preliminary results may be given here.

The energy distribution due to an optically thin dust shell is determined by the temperature of the star, the evaporation temperature of the grains, the density distribution of the dust, and the optical properties of the grains. For this discussion the grains are assumed to be large and non-volatile, as might be expected for grains close to a star. Then the grains act as black bodies and cause neutral extinction. The evaporation temperature may be assumed to be 1500 or 2000 K; the density distributions used are $\rho \sim r^{-2}$ and $\rho \sim r^{-1}$.

The resultant energy distributions due to optically thin dust shells look like black body distributions with $T = T_{evap}$, with a broadening of the distribution to longer wavelengths. In other words, the greatest contribution to the energy distribution (for $\lambda \leqslant 5\mu m$) is due to the hottest grains, while the cool grains contribute less. This is understandable from the radiation laws of black bodies and from the decrease in density with distance from the star (where the cooler grains lie).

When a relatively cool (K2) star is surrounded by such an optically thin dust shell, the resultant energy distribution has a single, very broad maximum. It has been asserted by Rydgren *et al.* (1976; see also Rydgren's chapter) that optically thick dust emission necessarily manifests itself by a second peak in the energy distribution. However, this will not be so if (1) the dust is relatively hot, (2) the peak flux from the dust is only a fraction of that of the star, or (3) the dust is optically thin.

Thus far it has been possible to fit the energy distribution observed for RY Tauri very well with the optically thin dust shell model, but more work must be done before any firm conclusions can be drawn. In any case, the result will show that *both* the gas emission and the dust emission models can reproduce the observed energy distributions of the T Tauri stars.

The debate over the source of the infrared excess in T Tauri stars may be of little consequence, however, to the question of planetary formation. The hot gas required to produce free-free and free-bound infrared emission must be quite close to the star; a gas temperature of about 10^4 K is required. In the optically thin dust shell models, virtually all the infrared flux is emitted within 30 stellar radii of the star, corresponding roughly to the orbit of Mercury in the solar system. So the dust shell models do not give much information on a solar system scale.

There is, in fact, a good argument that, if the T Tauri stars are associated with planetary systems, they have already formed. The clouds of dust and gas around the stars are thin enough to allow stellar radiation to come through, implying that the condensation (or dispersal by mass loss) of most of the material in the "stellar nebula" has already occurred. Any protoplanets themselves would not be detectable. As pointed out by Kuhi (1978), this allows one to set a limit on the time when planetary formation must occur. If the planets are largely formed before the T Tauri phase, then they must do so within the first 10^5 or so years of their lifetimes.

Rapid Rotation

One of the basic questions concerning the solar system is the reason for the distribution of angular momentum between the sun and the planets. The sun currently rotates at about 2 km sec^{-1}, but the T Tauri stars indicate much larger pre-main-sequence values. Herbig (1962) originally explained their broad absorption lines as due to rotation, with values of $v \sin i$ from 20 to 65 km sec^{-1}. Willson (1975) has shown that rotation is important in the shells of gas close to the stars. Kuhi (1978) found that observations of the stars in NGC 2264, a young cluster including several T Tauri stars, are consistent with velocities of up to 150 km sec^{-1}. It is clear that rapid rotation is the norm in young stars.

One would like to know how quickly the rotation slows with the age of the star. Skumanich (1972) has plotted the change in rotational velocities with age for G stars in several clusters and the sun. Herbig's (1962) measurements of $v \sin i$ for four G- and K-type T Tauri stars average 39 km sec^{-1}, which when corrected statistically for $\sin i$ gives a mean rotational velocity of 50 km sec^{-1} (one should keep in mind the very small sample involved). Kuhi (1978) found a maximum rotational velocity of 150 km sec^{-1} for the F, G, and K stars in NGC 2264. All these values are graphed in Fig. 1.

The new data agree well with Skumanich's (1972) relation. He found that the rotational velocity decreases approximately as $t^{-0.5}$. This appears to hold true for stars of ages from less than 10^6 yr to the sun at 4.5×10^9 yr. The results indicate that the slowing of a star's rotation must be a gradual braking process, associated with the mass loss and magnetic fields of the star. The mass loss may be responsible for carrying off much of the star's angular momentum. This line of reasoning indicates that the slow rotation of the sun may have little to do with the large amount of angular momentum carried by the planets. If planetary systems have already formed around the T Tauri stars, as discussed earlier, then it seems unlikely that the angular momentum of the rotating star is somehow transferred to the planets, as was once thought.

It is interesting to note, however, that if the sun were rotating at 150 km sec^{-1} its angular momentum would roughly equal that of the planets. One could conclude that the solar nebula, well *before* the T Tauri phase, may have participated in a general rotation. As the planets formed and the nebula thinned, the revolution of the planets and the rotation of the sun became decoupled.

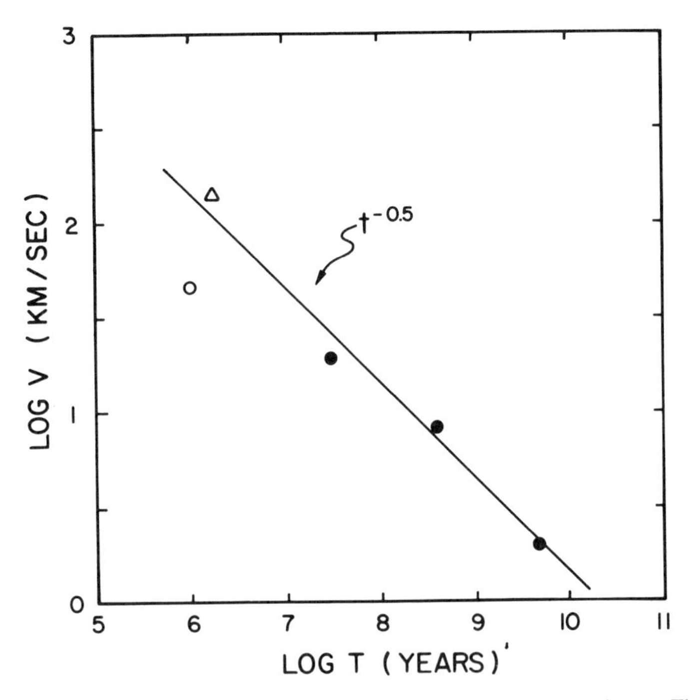

Fig. 1. The change in rotational velocity with age for G stars and T Tauri stars. The three filled circles indicate data for the Pleiades, Hyades, and the sun from Skumanich (1972). The open circle represents the mean of the values determined by Herbig (1962) for four T Tauri stars, corrected for sin i. The open triangle is the maximum velocity in NGC 2264 found by Kuhi (1978). The line represents Skumanich's relation of $t^{-0.5}$.

The sun itself then slowed down as it lost mass, during its T Tauri phase and afterwards, reaching its present slow rotation only after billions of years.

CONCLUSION

The T Tauri stars are stars of intermediate mass with ages of about a million years. Most of them are evolving along their quasi-static equilibrium tracks while some are still undergoing hydrodynamic evolution. They are surrounded by moderate shells of gas and dust which largely account for their peculiarities. It seems likely that planetary formation, if it is to occur, has already happened before a star reaches the T Tauri phase. If so, then one may set some limits to the behavior of the young solar system. The star is rotating rapidly and may be undergoing significant mass loss. Its nebula has largely dissipated except for an infrared emitting region close to the star.

Acknowledgments. I am indebted to various colleagues for discussions on some of the topics presented in this chapter, especially L. Kuhi, M. Cohen, L. A. Willson and K. Day, and to two referees for their comments.

REFERENCES

Appenzeller, I. 1977. *The Interaction of Variable Stars with their Environment* (W. Strohmeier, ed.). Bamberg Ser. Var. Star Coll.

Cohen, M. 1978. *Moon and Planets* (special Protostars and Planets issue). In press.

Cohen, M., and Kuhi, L. V. 1976. *Astrophys. J.* 210: 365.

Cohen, M., and Woolf, N. J. 1971. *Astrophys. J.* 169: 543.

Dumont, S.; Heidmann, N.; Kuhi, L. V.; and Thomas, R. N. 1973. *Astron. Astrophys.* 29: 199.

Grasdalen, G. L. 1973. *Astrophys. J.* 182: 781.

Herbig, G. H. 1962. *Adv. Astron. Astrophys.* 1: 47.

______. 1966. *Vistas in Astron.* 8: 109. (A. Beer and K. A. Strand, Eds.) Oxford: Pergamon Press.

______. 1970*a*. *Évolution Stellaire avant la Sequence Principale* (Colloq. Int. d'Astrophys. Liège). p. 13.

______. 1970*b*. *Spectroscopic Astrophys.* (G. H. Herbig, ed.). p. 237. Berkeley, California: University of California Press.

______. 1977*a*. *Astrophys. J.* 214: 747.

______. 1977*b*. *Astrophys. J.* 217: 693.

Herbig, G. H., and Rao, N. K. 1972. *Astrophys. J.* 174: 401.

Iben, I. 1965. *Astrophys. J.* 141: 993.

Imhoff, C. L. 1973. Master's Thesis, The Ohio State University, Columbus, Ohio.

______. 1978*a*. *Astronomy Quarterly* 1: 213.

______. 1978*b*. In preparation.

Imhoff, C. L., and Mendoza V., E. E. 1974. *Rev. Mex. Astron. Astrof.* 1: 25.

Johnson, H. L. 1966. *Ann. Rev. Astron. Astrophys.* 4: 193.

Joy, A. H., and Abt, H. A. 1974. *Astrophys. J. Suppl.* 28: 1.

Kuhi, L. V. 1964. *Astrophys. J.* 140: 1409.

______. 1978. *Moon and Planets* (special Protostars and Planets issue). In press.

Larson, R. B. 1969. *Mon. Not. Roy. Astron. Soc.* 145: 271, 297.

Low, F. J., and Smith, B. J. 1966. *Nature* 212: 675.

Mendoza V., E. E. 1966. *Astrophys. J.* 143: 1010.

______. 1968. *Astrophys. J.* 151. 977.

Rydgren, A. E.; Strom, S. E.; and Strom, K. M. 1976. *Astrophys. J. Suppl.* 30: 307.

Skumanich, A. 1972. *Astrophys. J.* 171: 565.

Strom, S. E. 1972. *Publ. Astron. Soc. Pacific* 84: 745.

Strom, S. E.; Grasdalen, G. L.; and Strom, K. M. 1975. *Ann Rev. Astron. Astrophys.* 13: 187.

Ulrich, R. 1976. *Astrophys. J.* 210: 377.

Walker, M. F. 1966. *Stellar Evolution* (R. F. Stein and A. G. W. Cameron, eds.). p. 405. New York: Plenum Press.

______. 1972. *Astrophys. J.* 175: 89.

Willson, L. A. 1975. *Astrophys J.* 197: 365.

SPECTRAL CHARACTERISTICS OF T TAURI STARS

LEONARD V. KUHI
University of California at Berkeley

The basic observational facts concerning T Tauri spectra are briefly described. Their characteristics are discussed, on the basis of scanner observations, with regard to general appearance, variability, frequency of the occurrence of Fe II, He I and Na I emission, and underlying photospheric absorption. Statistics are presented on the type of H α emission line profiles: 10% are single and symmetrical, 60% are double-peaked with a blue-shifted absorption component, 5% are classical P Cygni, another 5% are inverse P Cygni, and ~ 20% are inadequately observed. Variability of profiles occurs on very short time scales. Some implications for modeling the atmospheres of T Tauri stars are mentioned along with their lack of importance for the formation of planets which seems to have occurred before the T Tauri phase.

T Tauri stars are found in dark clouds such as the Taurus-Auriga complex and clusters such as Orion and NGC 2264. Radial velocity studies in Taurus (Herbig, 1977*a*) show no significant difference between the T Tauri stars as a group and the velocities determined from the microwave measurements of various molecules in the cloud. Thus one can conclude that the T Tauri stars are intimately associated with the cloud and hence must be intrinsically very young. Their close association with early-type stars in young clusters leads to the same conclusion. In addition, color-magnitude diagrams of young clusters (e.g., NGC 2264; Walker, 1956) show that T Tauri stars are located above the zero-age main sequence and hence that they must still be gravitationally contracting towards the main sequence. Recent attempts at determining their location on a luminosity-effective temperature diagram (Cohen and Kuhi, 1976) indicate the most T Tauri stars lie in the region of the Hayashi convective tracks for ~0.5 to 1.5 $M_{\odot}$ stars. Hence, they are among the youngest objects (in an evolutionary sense) that can be considered as bona fide stars. Consequently, they have the potential of being extremely useful in the study of the early history of the solar system provided that their somewhat bizarre spectral characteristics can be correctly understood.

BASIC CHARACTERISTICS

The basic defining characteristics were described in some detail by Herbig (1962) and may be summarized briefly as follows. The underlying absorption spectrum is usually G, K or M and occasionally is completely obliterated by what appears to be an overlying continuum. The emission-line spectrum displays an enormous range in intensity: the weakest cases have only H α and the Ca II H and K lines in emission; the moderate stars have stronger hydrogen lines, and various lines of Fe I and Fe II have made their appearance; the strongest cases have very strong hydrogen lines, a very well-developed Fe II spectrum, lines of He I, and a number of other metals. Forbidden lines such as O I are often present without regard to the overall general strength of the emission-line spectrum and occasionally include S II and O II. The forbidden lines come from the low-density nebulous regions that are intimately associated with many T Tauri stars. An ultraviolet excess is often present in the strong-line stars and modest infrared excesses are observed in most T Tauri stars. The ultraviolet excess is most likely due to bound-free emission of hydrogen; the infrared excess has two possible explanations: thermal radiation from dust (Cohen, 1973) and free-free emission from hydrogen or even H^- (Rydgren, Strom and Strom, 1976). Rydgren (see his chapter in this book) has discussed the difficulties involved in making a clear-cut distinction between the two models. Large values of R (ratio of total to selective extinction) are required for the free-free model, but a distinct blackbody infrared energy distribution is seen only for a very small number of T Tauri stars.

Figures 1 and 2 illustrate a series of typical well-developed T Tauri spectra as observed with the Wampler cassegrain scanner of the 3-meter reflector at Lick Observatory. The resolution is only 7 Å but is sufficient to show the diversity and similarity of T Tauri spectra. These spectra are part of a massive project undertaken by Cohen and Kuhi to ascertain the spectral properties of some 300 stars in the catalogue published by Herbig and Rao (1972) as well as of H α emission objects discovered in the original surveys but not included in the catalogue because of the total lack of information about them. In Fig. 1, YY Ori shows rather mild T Tauri characteristics, namely H α, β and γ in emission, very weak Fe II emission, He I 5876 in

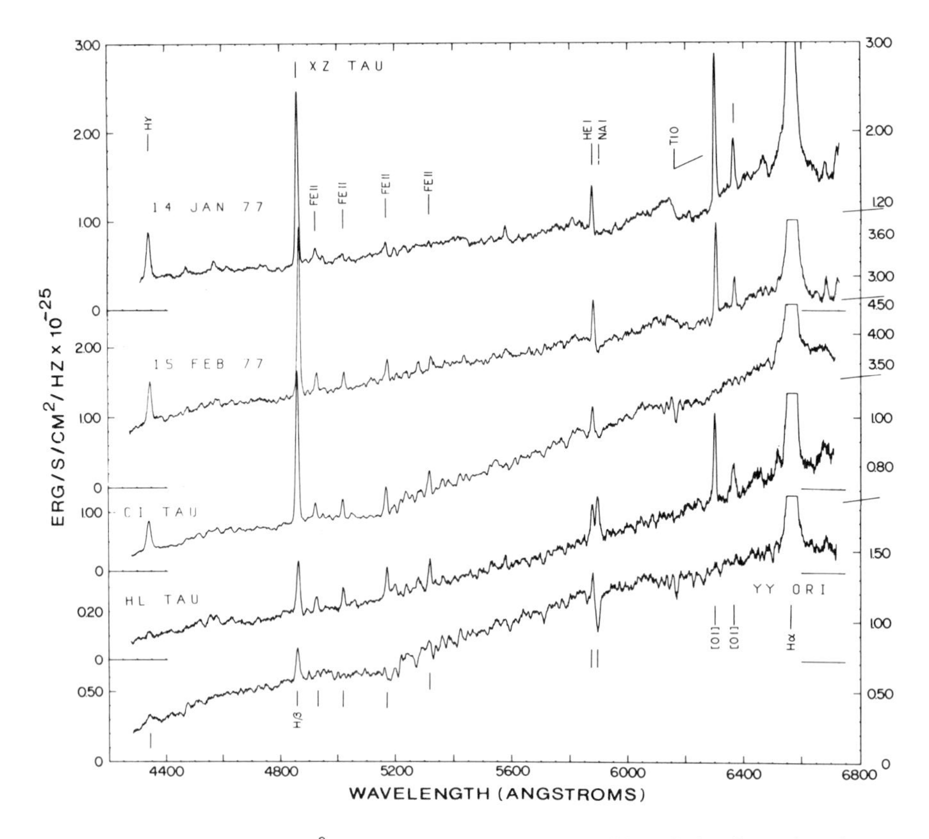

Fig. 1. Scanner spectra (7 Å resolution) of T Tauri stars with emission lines of moderate strength. The horizontal lines on the right correspond to the zero-levels indicated on the left. The two spectra of XZ Tau were obtained one month apart and show the effect of the overlying continuum emission on the Ti O bands.

emission and the underlying absorptions and continuum shape of a K 5 star. CI Tau shows stronger Fe II and hydrogen emission but the underlying K star is still clearly discernible. In HL Tau very few, if any, absorption features remain whereas in RW Aur and DG Tau (Fig. 2) the emission spectrum is so strong that no underlying photospheric spectrum is visible. The forbidden oxygen lines are also very strong. The appearance of the Na I D lines ranges from strong absorption to strong emission and their presence does not seem to be correlated with any other feature.

Figures 1 and 2 also show the extreme variability of the spectra of T Tauri stars. The spectrum of XZ Tau on 14 January 1977 shows moderate line emission with quite deep Ti O absorption bands indicating an underlying spectral type of ~M1. Yet one month later the line spectrum is a little stronger but the Ti O band strength has decreased considerably so that one might classify it as ~M0. This change can be due to either an increase in the strength of overlying continuum emission (the so-called "blue" continuum

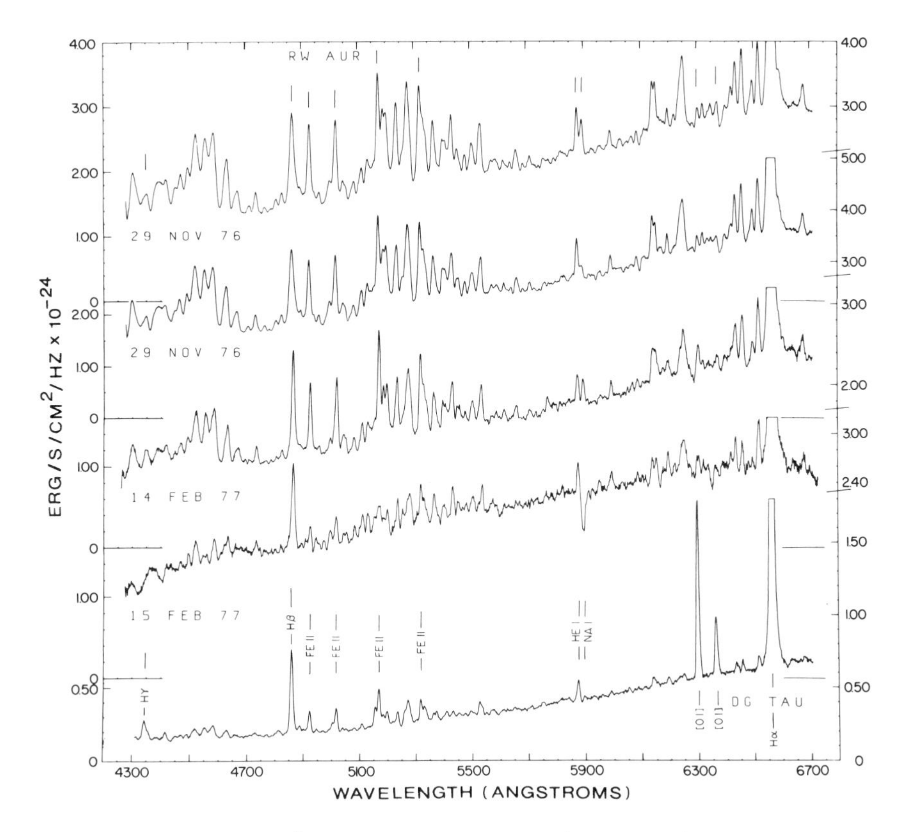

Fig. 2. Scanner spectra (7 Å resolution) of DG Tau and RW Aur which have strong to very strong emission lines. The horizontal lines on the right correspond to the zero-levels indicated on the left. The two spectra of RW Aur labeled 29 November 76 were obtained two hours apart; the other two were obtained one day apart. Together these spectra show the extreme variability occurring on very short time scales.

which extends to the red in extreme cases) or an increase in the photospheric temperature so that the degree of dissociation of Ti O has increased. The "blue" continuum emission has often been associated with a strong emission-line spectrum but this is not the case for XZ Tau. The spectrum of RW Aur is remarkable in the strength of its emission lines and in its variability. Figure 2 shows two sets of spectra taken two hours apart (29 November 76) and one day apart (14 and 15 February 77). The Na I D emission has decreased by a factor of ~2 in only two hours while the Fe II and hydrogen have changed very little. This rapid variability must put severe constraints on the extent of the region responsible for the Na I D emission. In 24 hours the Fe II emission strength has dropped dramatically, Na I has changed from emission to absorption while the He I remains almost unchanged and the hydrogen lines have decreased only slightly. Spectra of RW Aur obtained in the fall of the same year show even weaker emission lines and reveal the appearance of an underlying K-type spectrum. Similar variations have been seen in most of the twenty strong-line stars observed by Cohen and Kuhi over a period of three years. The conclusions are as follows:

1. The Na I D lines are the most unstable and significant changes can occur in a few hours.
2. The Fe II lines can vary significantly in 24 hours.
3. The hydrogen lines and the "blue" continuum can vary significantly on a time scale of weeks to months.
4. The forbidden lines exhibit very little change on time scales of up to a few years.
5. The He I lines can vary significantly on a time scale of days.
6. The variability itself is extremely irregular: some stars (e.g. RW Aur) show spectral changes with every observation whereas others (e.g. CI Tau) may be essentially constant in spectral appearance with only an occasional burst of extreme variability.

The ~ 300 stars from the Herbig and Rao (1972) catalogue observed by Cohen and Kuhi can also provide some statistics (although woefully incomplete) on the relative frequency of the occurrence of various emission characteristics. For example, He I emission occurs in 40% of the T Tauri stars, Fe II in 18% and Na I in 7% (as seen as 7 Å resolution). Also approximately 10% of them have a strong ultraviolet excess. However, since the stars have all been selected on the basis of detectable H α emission, the sample is necessarily biased and no statistical discussion can be made concerning the relative frequency of various T Tauri characteristics among the entire pre-main-sequence sample of contracting stars. In addition, most stars have been observed only once and the sample as a whole therefore represents an instantaneous picture, obtained at different times, of stars whose intrinsic spectral variability can be quite extreme.

SPECTRAL LINE PROFILES

When an underlying photospheric spectrum is visible, the absorption lines are considerably broadened compared to those of normal G, K or M dwarfs. This broadening was measured by Herbig (1957) for four T Tauri stars and, if attributed to rotation, implies values of $v \sin i$ as large as 65 km sec^{-1}. These values have been confirmed by Kuhi (unpublished) using a power spectrum technique to improve the signal-to-noise ratio. In NGC 2264 Kuhi finds values of $v \sin i$ up to ~100 km sec^{-1} from which the presence of a strong stellar wind is inferred in order to remove enough angular momentum to slow down the rotation rate so that it agrees with that of the solar type main-sequence stars. However, this wind can be as much as 10^5 times larger than the present solar wind and yet be completely undetectable in the stellar spectrum.

A much stronger stellar wind was suggested by Kuhi (1964) who used a very crude model for the production of the emission lines of hydrogen and Ca

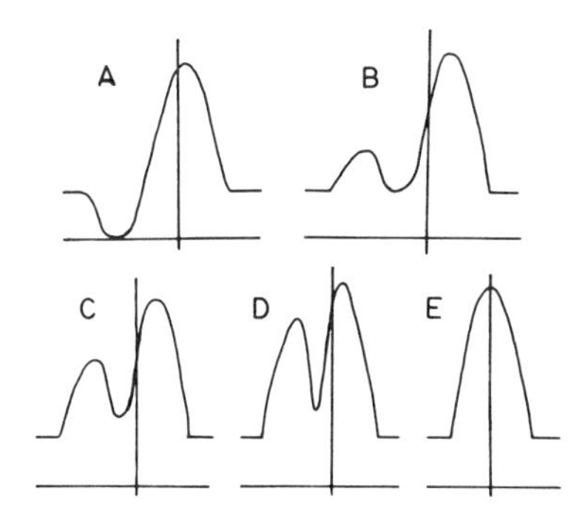

Fig. 3. Schematic profiles of typical H α emission lines encountered in T Tauri spectra. Wavelength increases to the right. Profiles showing evidence for infalling material are similar but have the direction of increasing wavelength reversed.

II and obtained mass-loss rates ranging from 10^{-8} to 5×10^{-7} $M_\odot$/yr for eight stars. Since then, the presence of a very strong wind has been used by many solar system theorists to provide a massive cleansing action on the early solar nebula, effectively removing all the debris left over from the formation of planets and blowing it out into interstellar space. Little additional work was done on the emission-line profiles for approximately ten years because of the difficulty in obtaining spectra of adequate resolution of such faint stars. However, the widespread use of image tubes has considerably increased the number of stars studied and more reliable statements about the line profiles can now by made.

Figure 3 illustrates schematically the basic types of profiles that are found. They range from the classical P Cygni type profile (Fig. 3*A*), with strong shortward displaced absorption going well below the continuum, to symmetrical profiles with no hint of any absorption. The double-peaked profile is a very common one for hydrogen and the absorption feature is nearly always displaced towards shorter wavelengths. The shortward displaced absorption component has traditionally been used to infer the presence of mass outflow (Kuhi, 1964). Profiles showing the same general features but with the direction of increasing wavelengths reversed would then indicate the presence of infalling matter. Such profiles having strong longward displaced absorption components are referred to as inverse P Cygni profiles and have also been observed in some T Tauri stars (Walker, 1972).

The largest sample of T Tauri spectra is that used by Herbig (1977*a*) in his radial velocity study which covers a rather limited wavelength range but does include H α. Additional spectra have been obtained by Kuhi for his rotational velocity study. This material can be combined to provide some statistics (again very biased) on the frequency of occurrence of various types of line profiles among the ~75 stars represented by ~150 spectra. By far the most common H α profile is double-peaked with a shortward displaced absorption; ~50% of the well-exposed spectra show this profile (Fig. 3*B,C,D*) with the blue peak having velocities of −100 to −200 km sec^{-1}; the red +50 to +150 km sec^{-1} and the absorption feature −50 to −150 km sec^{-1}. Another 40% appear to have single peaks often because of overexposure of H α. Twelve such "single-peak" stars have spectra obtained with proper exposure and higher resolution; about half of their profiles also show a shortward-displaced absorption component at −50 to −100 km sec^{-1}; the other half show only a symmetrical line profile with no velocity displacement (Fig. 3*E*). A classical P Cygni profile (Fig. 3*A*) is seen in only three stars: V 1331 Cygni (= LKH α 120), AS 353A and DR Tau. Four stars show redward-displaced absorption components of some kind: DE Tau, UX Tau, DK Tau and DN Tau. The corresponding velocities range up to +60 km sec^{-1}. No classical inverse P Cygni profile (Fig.*A*) but with λ-direction reversed) is seen. One can conclude from this sample that ~60% of the H α profiles are double-peaked with a shortward-displaced absorption feature, ~10% are single, ~5% are classical P Cygni profiles, ~5% show some longward-displaced absorption feature, and ~20% are still inadequately observed.

The other significant sample of T Tauri stars was that studied by Walker (1972) who concentrated on a small subset having a strong ultraviolet excess. Of 25 stars, nine show redward-displaced absorption components in the hydrogen lines but only rarely in the Ca II K line. The profiles are often classical inverse P Cygni (Fig. 3*A* reversed) and the corresponding velocities are +150 to +400 km sec^{-1}. Walker interpreted these profiles as showing clear evidence of infalling matter. The appearance of the profile changes with time; for example, the prototype YY Ori was observed on 14 nights and did not show redward-displaced absorption on six of them. Other stars observed more than once showed the red component 30 to 50% of the time. Walker also searched for inverse P Cygni profiles in another 50 T Tauri stars in Orion and found only one further example. The ultraviolet excess stars constitute ~11% of the Herbig and Rao (1972) catalogue so that ~4% of the 300 T Tauri stars can be expected to show inverse P Cygni profiles if Walker's results can be extrapolated to the entire catalogue.

However, it appears that not only ultraviolet excess stars can show inverse P Cygni profiles in their hydrogen lines. Rydgren, Strom, and Strom (1976) found such a profile in DM Tau; Appenzeller and Wolf (1977), Wolf, Appenzeller and Bertout (1977) and Rydgren (1977) have observed it in S Cr A; Boesgaard and Herbig (personal communication) and Kuhi in AA Tau; and Kuhi in DK Tau. In both S Cr A and AA Tau, the presence of redward-displaced absorption components is extremely variable: the hydrogen lines show inverse P Cygni profiles on one night and a completely symmetric profile with no absorption at all on the next. The inverse P Cygni structure also shows up more clearly in Balmer lines of high *n* (principal quantum number). Curiously, in S Cr A the Ca II K line does not show the inverse P Cygni profile seen in the hydrogen lines. Also Rydgren's hydrogen-line profiles show the common double-peaked structure but with an added redward-displaced component as well! The inclusion of these objects would raise the fraction of stars showing inverse P Cygni profiles to ~5%. Appenzeller (1977) claims to find a much higher percentage (~50%) from a survey of 18 southern emission-line objects; but the detection of any absorption feature in half his cases is quite marginal and not at all convincing. Even so, these objects are all strong ultraviolet excess stars and hence the bona fide detection rate showing inverse P Cygni profiles is very similar to Walker's results.

Inverse P Cygni profiles have also been noted by Kuhi (1975) for the Na I D lines in BP Tau. The emission is extremely variable with time and the hydrogen lines show only symmetrical profiles with no absorption components of any kind. Ulrich (chapter in this book) has noted that in AS 205 the Na I D lines show inverse P Cygni profiles whereas the hydrogen lines have the common double-peaked profile (Fig. 3*b*) with a blue absorption component. Thus it would seem that both infall and outflow are present in the same star at the same time if these profiles are interpreted in the traditional manner. The extreme variability of the Na I D emission lines (on a time scale of hours) imposes a severe restriction on any uniform flow model (infall or outflow). The Na I D lines may be produced in local flare and prominence activity which is totally unrelated to the overall mass flow.

DISCUSSION

The theoretical interpretation of the spectra of T Tauri stars is not in a very satisfactory state. The models fall into three general categories:

1. Stellar wind with or without rotation (e.g., Kuhi, 1964; Willson, 1975),
2. Chromospheric (e.g., Dumont *et al.*, 1973) and
3. Accretion (e.g., Walker, 1972; Ulrich, 1976; Bertout, 1977).

I do not propose to discuss any models here (see Ulrich's chapter) other than to indicate what constraints the above observations seem to place upon them. The distribution of line profiles suggests that unambiguous cases of strong outflow (i.e. classical P Cygni profiles) are rare (~5%). V 1331 Cygni is a good case and may be an example of a longer time scale version of the eruptive phenomena which Herbig (1977*b*) suggests every T Tauri star may undergo. The most common profile is the double-peaked one with a blue absorption feature (70%) which is followed by the symmetrical single-peak profile at 20%. This is most simply interpreted qualitatively via the Sobolev approximation[a] for a stellar wind model as a combination of outflow and rotation (Doazan, 1965). The theoretician's thin rotating disk would produce a sharply defined double-peaked emission feature with a central absorption. Adding spherical outflow would increase the total emission considerably by

[a]The Sobolev approximation allows one to calculate an emission-line profile without solving the spherical equation of transfer because of the large velocity gradient in the line of sight.

making the emitting region more like a flattened sphere than a thin disk and would move the absorption component to shorter wavelengths. Such rotationally flattened envelopes would be seen from different inclinations and hence should exhibit a significant change in emission line widths. In particular, single-peaked lines should be much narrower than double-peaked ones because the former would be produced when the flattened configuration is seen face-on. The observed range in H α line widths is approximately a factor of two in the expected direction and suggests that this type of model is worth further consideration. Willson (1975) reached the same conclusion from her study of the Fe II emission lines. By the same Sobolev-type considerations, infall seems to be the only viable way of explaining that the remaining 5% of the stars show inverse P Cygni profiles. However, the great problem here is their variability: such absorption components appear on one night and are absent on the next. This implies a very variable accretion rate and large density fluctuations in the infalling material. The rapid spectral changes are a problem for any model but could be most easily explained on the chromospheric model with rapid changes (e.g., due to flares) in the mechanical energy input into the line-forming region. The behavior of the Na I D lines could then be due to cooler infalling material from such flare or prominence activity in which the matter involved did not reach escape velocity.

It is also likely that the interpretation of the line profiles is completely dominated by radiative transfer effects because of the low velocity gradients in the line of sight. These have not been handled properly in any of the models available in the literature. Nor have models been constructed to explain with the same basic set of parameters all, or even a large fraction, of the line profiles of H, Ca II, Na I, Fe II, and He I observed at the same time in one star. There are clearly many paths for the theoretician to explore. However, the observers still have an enormous task to accomplish before the final word can be said about the flow of matter near T Tauri stars. Systematic studies of the variability of line profiles in individual stars and of their statistics in general are urgently needed; the former are especially critical to any flow model.

Finally, something should be said about the evolutionary state of T Tauri stars and their connection, if any, to problems of solar system formation. Their location on the HR diagram suggests that most of them are still highly convective; their photospheric line widths suggest a high rotational velocity and their emission spectra indicate extreme instability. It thus seems likely that they are in the transition phase between convective and radiative equilibrium and also between a more or less steady stellar wind and the earlier accretion or collapse phase. One might view the emission spectrum as the result of whatever adjustment is necessary in the atmosphere, envelope or interior to produce the stable state of a non-emission pre-main-sequence star. The stars showing hydrogen inverse P Cygni profiles could represent the last vestiges of the accretion phase; those with P Cygni profiles the last major outflow before a normal stellar wind takes over; and the other 90% of the stars some combination of outflow and rotation as well as large-scale prominence and flare-type activity.

Do they have anything to do with the formation of planets? This seems unlikely because the dimensions of all the models (crude as they are) are on the order of a few solar radii to a few 100 $R_\odot$ at most for the line-emitting region. This is considerably smaller than any protoplanetary disk discussed by solar system theoreticians. Nor is there much evidence for very dusty T Tauri envelopes nor for any infrared difference between the presumably younger YY Orionis stars and the regular T Tauri stars (Cohen, personal communication). This all suggests that the planet-forming stage has already passed and that the T Tauri phase has very little to do with it. The T Tauri phase occurs too late to be of much interest in this connection, although the wind may still play a significant role in post-planet-formation cleanup of the interplanetary medium.

Acknowledgments. The scanner observations were carried out with the support of a grant from the National Science Foundation.

REFERENCES

Appenzeller, I. 1977. *Astron. Astrophys.* 61: 21.
Appenzeller, I., and Wolf, B. 1977. *Astron. Astrophys.* 54: 153.
Bertout, C. 1977. *Astron. Astrophys.* 58: 153.
Cohen, M. 1973. *Mon. Not. Roy. Astr. Soc.* 164: 395.
Cohen, M., and Kuhi, L.V. 1976. *Astrophys. J.* 210: 365.
Doazan, V. 1965. *Ann. d'Astrophys.* 28: 1.
Dumont, S.; Heidman, N.; Kuhi, L.V.; and Thomas, R.N. 1973. *Astron. Astrophys.* 29: 199.
Herbig, G.H. 1957. *Astrophys. J.* 125: 612.
____. 1962. *Adv. Astron. Astrophys.* 1: 47.
____. 1977*a*. *Astrophys. J.* 214: 747.
____. 1977*b*. *Astrophys. J.* 217: 693.
Herbig, G.H., and Rao, K.N. 1972. *Astrophys. J.* 174: 401.
Kuhi, L.V. 1964. *Astrophys. J.* 140: 1409.
____. 1975. *Publ. Astron. Soc. Pac.* 87: 502.
Rydgren, A.E. 1977. *Publ. Astron. Soc. Pac.* 89: 557.
Rydgren, A.E.; Strom, S.E.; and Strom, K.M. 1976. *Astrophys. J. Suppl.* 30: 307.
Ulrich, R.K. 1976. *Astrophys. J.* 210: 377.
Walker, M.F. 1956. *Astrophys. J. Suppl.* 2: 365.
____. 1972. *Astrophys. J.* 175: 89.
Willson, L.A. 1975. *Astrophys. J.* 197: 365.
Wolf, B.; Appenzeller, I.; and Bertout, C. 1977. *Astron. Astrophys.* 58: 163.

THE STATUS OF T TAURI MODELS

ROGER K. ULRICH
University of California, Los Angeles

Possible models for the T Tauri phenomenon are reviewed in the light of recent observational evidence. It is proposed that the Na D lines are a sensitive diagnostic for the gas flow in the envelope surrounding the stars. This diagnostic indicates that either inflow or outflow of gas can occur in conjunction with the spectral characteristics of T Tauri stars. It is pointed out that the kinetic energy density of the gas flow if thermalized would be characterized by a much higher temperature than is inferred from the spectra. Unless something with macroscopic order like a magnetic field serves as an intermediate agent, such a hot gas must be found between the stationary star and the flowing envelope. Two models without such an agent are discussed – accretion and continuous flaring.

T Tauri stars are defined according to Herbig (1962) as those stars which show irregular variability and have emission at Hα and Fe I and Fe II. They are almost invariably associated with nebulosity. The resonance pumped lines of Fe I at 4063 and 4076 Å are considered definitive characteristics of the class. A number of other properties are typical but not uniformly found, such as ultraviolet excess emission and infrared excess emission. The [O I] forbidden lines are also frequently seen in emission. The presence of these lines indicates that some of the gas near the T Tauri stars is in a low state of ionization. The Ca H and K lines are frequently in emission and do not obey the Wilson Bappu relation (Wilson and Bappu, 1957; Kuhi, 1965).

The Balmer emission is usually non-thermally broadened by an amount which indicates the presence of velocities on the order of 300 km sec^{-1}. The Balmer lines often show structure which has been interpreted as resulting from absorption. These features are most often seen blue-shifted so that the T Tauri stars are generally believed to involve strong stellar winds. Walker

(1972) discussed a subclass of the T Tauri stars typified by YY Orionis which showed red-shifted absorption features in the higher Balmer lines. He interpreted these stars as being in a matter accreting stage of evolution. In all other respects these stars satisfy the defining criteria of the T Tauri class.

The emission characteristics defining T Tauri stars require a gas of low ionization. The Fe lines are probably collisionally excited except for the two Fe I resonance pumped lines. The Fe I and Fe II lines suggest a gas which is gently heated. On the other hand, the continuum radiation, especially the ultraviolet, indicates a somewhat hotter gas. Rydgren, Strom and Strom (1976) have suggested a model involving a 20,000 K gas to explain the continuum radiation. Rydgren discusses these data regarding the continuum in detail in this book. They have found that a moderately high density (Ne $\approx$ 10^{12}) and optical depth effects were required to explain the observed Balmer line to continuum ratio. In contrast, the [O I] and occasionally [Fe I] emission indicates a much lower density. Undoubtedly, different regions contribute to the spectrum.

There are three related classes of objects which may have some bearing on the T Tauri models. These objects are the Herbig (1970) Ae and Be stars, the Herbig-Haro objects (see the chapter by Böhm in this book) and the FU Ori objects (see Welin's chapter in this book). These objects all seem to be somewhat more luminous than the typical T Tauri star. The spectra of the latter two types seem to clearly require a strong stellar wind. Moderate resolution spectra ($\lambda/\Delta\lambda = 10^4$) obtained at Lick Observatory, Mt. Wilson and Palomar Mountain by Ulrich in collaboration with Knapp, Shectman and Wright of two of the FU Orionis objects (FU Ori and the Cohen, Kuhi and Harland 1977 object) and V1331 Cygni, a possibly related T Tauri star, all show strong P Cygni line profiles in the Balmer lines and violet displaced absorption in the Na D lines. It may be that the flare-up of the FU Orionis stars is related to the establishment of a strong stellar wind. The Herbig-Haro objects may simply be such FU Orionis objects buried in thick nebulosity.

There are two related questions which must be addressed by a model for the T Tauri phenomenon:

1. What are the energetics of the gas which lead to the observed emission properties?
2. What are the kinematics of the gas flow?

The first question may have an ambiguous answer. The emission depends on the density, temperature and source of excitation and ionization. Different physical processes can lead to similar conditions and thus be equally satisfactory or unsatisfactory in providing an answer to Question (1). For example, the physical conditions in the optically radiating layer of the accretion model proposed by Ulrich (1976) are very similar to chromospheric conditions. Several authors (Dumont *et al.*, 1973; Herbig, 1970) have suggested that the same mechanism producing the solar chromosphere is also operating actively in the T Tauri stars. Implicitly such a model must involve physical conditions similar to those in the solar chromosphere. Clearly the accretion model and the chromospheric models will have a similar degree of success in reproducing the T Tauri emission properties. This question of uniqueness must be considered in the evaluation of models for the energetics of T Tauri stars.

One possible source of the excitation is X radiation produced by the cooling of a gas 10^6 to 10^7 K. One may be tempted to reject such a hypothesis on the grounds that there is no evidence for such a hot gas. Unfortunately, unless the far-ultraviolet or X-ray spectral regions can be observed, a hot gas is extremely difficult to detect. In fact, many of the optical peculiarities are a natural consequence of X-ray heating. They could also be a consequence of other heating mechanisms as well. Even He II 4686 does not come from the primary cooling of the hot gas so its presence or absence is not diagnostic. Only direct observations of a line like O VI 1034 or soft X rays in the range 15-40 Å would be definitive. Present techniques and instruments do not permit the study of objects as faint as T Tauri stars in these spectral ranges.

The second question is more difficult to answer because it involves a vector field and thus has many more degrees of freedom. For the same reason an answer to this question will lead to a more definitive understanding of the T Tauri phenomenon. Because the kinematics involve the velocity of the gas flow, the crucial evidence is to be found in the line profiles of the various atomic transitions observed from the T Tauri stars. Profiles have been published by Kuhi (1964), Dibaj and Esipo (1969); Salmanov (1974); Kolotilov and Zaitseva (1974); Appenzeller and Wolf (1977); Wolf, Appenzeller and Bertout (1977) and Rydgren (1977). The last three papers have been confined to the star S CrA, and only the first two papers have presented data for more than one or two stars. For the related Herbig Ae/Be stars, Garrison and Anderson (1977) have presented Hα profiles for 14 stars. Recently, Ulrich in collaboration with Knapp, Wright and Shectman has undertaken a program of moderate resolution observations of T Tauri Hα, Hβ and Na D line profiles. Dispersions of 7 to 34 Å/mm have been used at lick Observatory, Mt. Wilson and Palomar Mountain. Some results from this program are discussed below.

I. OBSERVATIONS OF T TAURI LINE PROFILES

Walker (1972) has presented evidence that stars in a class typified by YY Ori are accreting matter. The higher Balmer lines and the Ca II K line showed an inverse P Cygni profile while the lower Balmer lines showed a more complex behavior. Theoretical work by Ulrich (1976) indicated that non-spherically symmetric accretion can produce blue-shifted absorption-like features. For reasons discussed in Sec. II, the presence of an absorption-like feature at a particular velocity indicates that matter is moving at that velocity

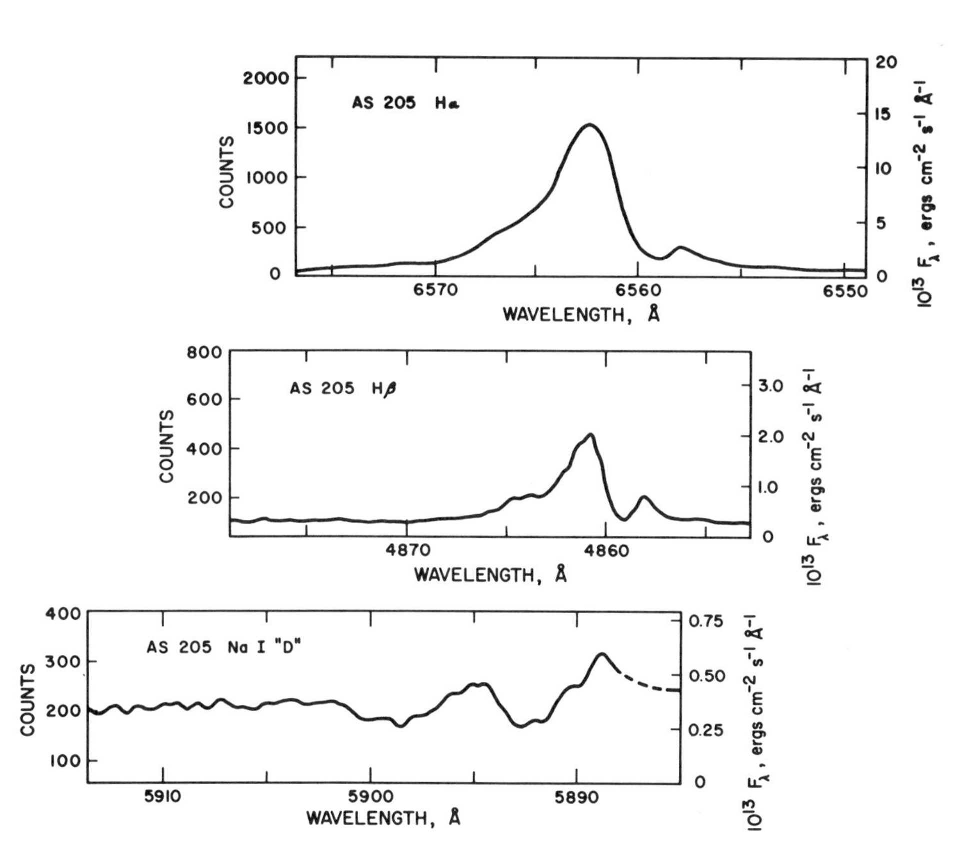

Fig. 1. Line profiles for Hα, Hβ and the Na D lines observed on HJD 2443297.764 from T Tauri star AS 205. The wavelengths shown are in the observatory velocity frame. The wavelength scale is reversed from the normal presentation. The Balmer lines are P Cygni Type III, the Na D lines are inverse P Cygni.

only for resonance lines. Unfortunately, the Ca II lines in the blue are studied in T Tauri stars only with great difficulty. Rydgren (1977) has also found that for the star RS CrA the higher Balmer lines have ambiguous structures similar to YY Ori stars. However, the Ca II K line did not show the inverse P Cygni profile. Recent observational work by Ulrich and Knapp at Mt. Wilson using a coudé scanner like the one described by Shectman and Hiltner (1976) has demonstrated clearly that the Balmer lines are not reliable as indicators of the direction of gas flow. Figures 1 and 2 show Hα, Hβ and Na D line profiles for the stars AS 205 and AS 353. The resolution was 0.7 Å and the flux scales were derived by comparison to 3 standard stars (Stone, 1974) and are accurate to 7%. Although the Balmer lines are superficially similar, the Na D lines have oppositely shifted absorption. The detailed difference between the Balmer profiles is that Hα and Hβ go below the continuum level

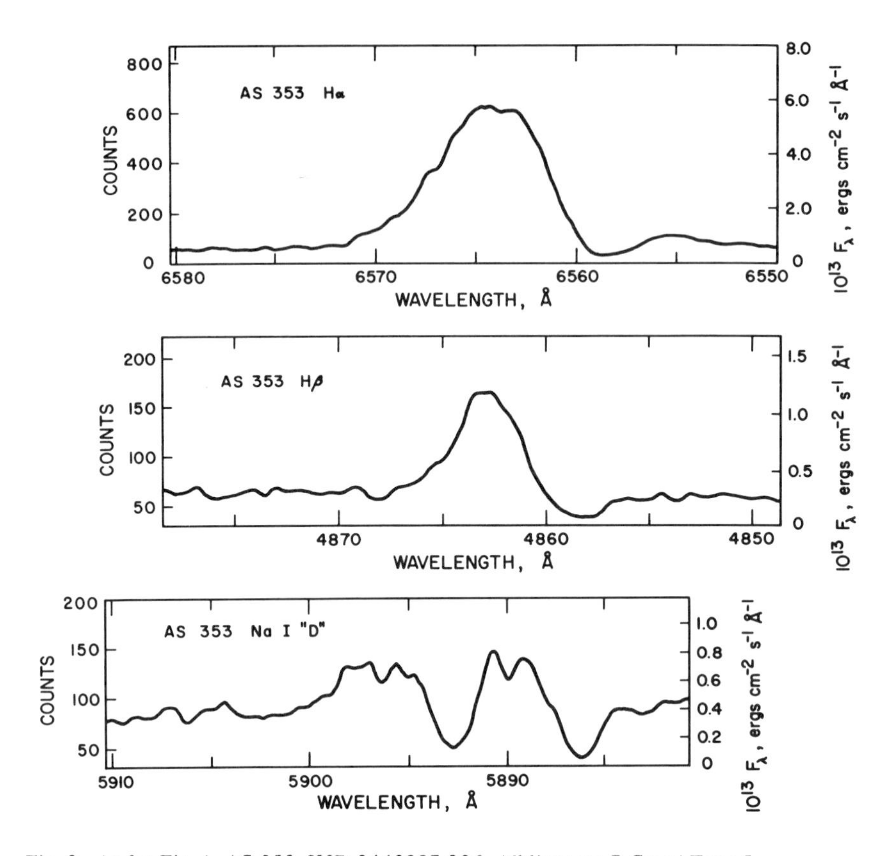

Fig. 2. As for Fig. 1. AS 353, HJD 2443297.926. All lines are P Cygni Type I.

on the blue side by the same fraction in AS 353 while the absorption feature from AS 205 does not go below continuum in Hα but does go below continuum in Hβ.

The survey of the Balmer and Na D lines was extended by Ulrich and Knapp in November 1977 to include 16 additional T Tauri stars. The results of this survey are summarized in Table I. The entries followed by a question mark showed probable red-displaced absorption but the spectra have photon statistics which are too poor to make a certain identification possible. In every case with a red displaced absorption component, except AS 205, no emission at the Na D lines is seen. Consequently there is no possible doubt that these absorption components result from matter moving towards the star at a location between us and the star. The Na atoms are definitely being accreted. The case of AS 205 is not unique in having the absorption-like feature in the Balmer lines blue-shifted while the D lines are red-shifted. We are then faced with assuming that the hydrogen and sodium are moving in opposite directions or with assuming that the hydrogen feature represents a

TABLE I

Na D Line Profile Summary[a]

Star	Blue Absorption	Red Absorption	Symmetric Wide	Absorption Narrow	Emission
Lk Hα 264				✓	✓
Lk Hα 272		✓?		✓	
DH Tau			✓		
DI Tau				✓	
HL Tau	✓			✓	✓
XZ Tau					
UY Aur		✓		✓	
GM Aur			✓		
HI Ori	✓				
EZ Ori				✓	
YY Ori	✓				
LL Ori			✓		✓
MM Mon		✓		✓	
MO Mon		✓			
AS 205		✓			✓
SR 9		✓?			
AS 353	✓				✓
V 536 Aql		✓?			
Lk Hα 120	✓				
DI Cep				✓	✓

[a] See Appendix at the end of this chapter for notes to this table.

velocity which is deficient in emitting gas as in Ulrich's (1976) model. Because of the shortness of the dynamical time interval appropriate to regions where velocities of 300 km sec^{-1} occur, and the fact that the absorption depth indicates a substantial fraction of the line of sight is covered by both the hydrogen and sodium absorbers, the assumption that these two species are moving in opposite directions cannot remain valid long. The only alternative interpretation of these line profiles is that one of the apparent absorption features is due to lack of emission. Since the sodium lines involve little or no emission, it is clear that the abosrption-like feature in the Balmer lines must be due to lack of emission. For this to be a viable interpretation, we must be able to explain the lack of a true absorption feature in the Balmer lines at the same velocity as the sodium feature. The following section presents a discussion which shows that the sodium absorption feature can be present without an Hα absorption feature for a range of accretion rates. Although this discussion is simplified in a number of respects, it is based on much more elaborate calculations which include all collisional processes and variable optical depth effects in a 20 level model hydrogen atom. These calculations are being carried out by S. Drake at U.C.L.A.

II. THE RELATIVE OPTICAL DEPTH IN THE Na D LINES AND THE BALMER LINES

The profiles of the Balmer lines are important in constraining the kinematics of the gas; however, there are theoretical reasons why the interpretation of these lines is not straightforward. Because the Balmer lines are not resonant lines, neutral hydrogen may be present between us and the star moving with high velocity and still not produce a noticeable absorption line.

I present here a simple estimate for the optical depth in the Balmer lines and the Na D lines. An attempt has been made to formulate this estimate in such a way that is is applicable to a range of both outflow and inflow models. The basic assumption is that the gas goes through a phase where its temperature is high enough for the thermal velocity to be comparable with the escape velocity. This hot gas produces ultraviolet radiation and soft X rays which ionize and heat the surrounding layers which in turn are responsible for the observed emission. Under these conditions the ionization is primarily radiative and the ground state of hydrogen is highly overpopulated relative to excited states. Because of this last point we can neglect ionizations from excited states and stimulated emissions.

A basic difference between the Balmer line optical depths and the Na D optical depths comes about because absorbers for the Balmer lines are produced by the ionizing photons. The column depth of hydrogen in the n=2 state thus depends on the square of the accretion rate – once through the flux of ionizing photons and once through the column depth of hydrogen in the n=1 state. The situation for Na D is more normal and the column depth depends only on the first power of the accretion rate. The number of hydrogen atoms in the n=2 state is in part determined by the rate of recombination and the escape probability of Lyman α photons. If the emitting region is ionized throughout by ultraviolet radiation or soft X rays, the optical depth in Lyman α is 10^5 or less. Escape through the Lyman α wings occurs once in every 10^5 scatterings. Thus the population of hydrogen in the n=2 state is

$$N_2 \cong \frac{\alpha_B N_{\mathrm{H}^+} N_e}{\beta_{21} A_{21}} \tag{1}$$

where α_B is the case B recombination coefficient, β_{21} is the escape probability for Lyman α, while the other quantities have their standard meanings. The electron and ion density are related to the ionization rate by

$$\alpha_B N_{H+} N_e = R_{bf} N_{H^0} \quad (2)$$

where $R_{bf} = \int \frac{4\pi J_\nu}{h\nu} d\nu$ is the photo ionization rate. Thus,

$$\frac{N_2}{N_{H^0}} = \frac{R_{bf}}{\beta_{21} A_{21}} . \quad (3)$$

We must now estimate the ionization rate. I assume that the observed line width of 200-400 km sec^{-1} is due to moving matter with the appropriate velocity near the star. For accretion, the velocity represents the terminal free-fall velocity. For a wind, this represents the maximum ejection velocity. In either case, gas with a temperature characterized by

$$T \approx \frac{1}{2} \frac{M_H v^2}{k} = \frac{\bar{E}_\nu}{k} \quad (4)$$

is likely to be present on the stellar surface where the moving gas interacts with the hydrostatic protostellar core. I further assume that the process generating the outflow velocity or the shock terminating the infall flow produces X rays at roughly the same rate as the flow of kinetic energy into or out of the hydrostatic core. This assumption gives a flux of energy, F which is

$$F = \frac{1}{8} \frac{m v^2}{\pi r^2} . \quad (5)$$

The flux of hard photons is characterized by an energy $\bar{E} = kT$ where T is given by Eq. (4). The rate of ionization produced by this flux is either

$$R_{bf} = \frac{\overline{\sigma F}}{\bar{E}\nu} \quad (6a)$$

or

$$R_{bf} = \frac{\sigma F}{23 \text{ eV}} \quad (6b)$$

depending upon whether the gas is mostly ionized or neutral respectively. The cross-section σ is roughly $1.5 \times 10^{-23} (\bar{E}_\nu)^{-3}$ where $\bar{E}_\nu$ is in keV according to Brown and Gould (1970).

The optical depth in either Hα or Na D at line center is

$$\tau_0 = \frac{\pi^{\frac{1}{2}} e^2}{m c \mathrm{v}_D} f_{ij} N_i \ell \tag{7}$$

where ℓ is a path length, v_D is a Doppler line broadening parameter which includes the effect of a macroscopic velocity gradient, and all other quantities have their standard meanings. The column depth $N_i \ell$ is related to the mass flow rate by

$$N_{\mathrm{H}} \ell = \frac{\dot{M}}{4\pi \mathrm{v} r} .$$

A dimensionless constant of order unity arising from the r dependence of v has been omitted from Eq. (8). The difference between Na D and Hα comes about because of the different behavior of N_i/N_{H} for these two species. For the D lines I take $N_{\mathrm{Na}}/N_{\mathrm{H}} = 1.9 \times 10^{-6}$. For H$\alpha$ we must use Eqs. (3), (4), (5) and (6*b*) to get N_2/N_{H}. The resulting equations for τ_0 in Hα (9*a*) and Na D (9*b*) are

$$\tau_0 = 5.6 \times 10^{17} \frac{\dot{M}^2}{\beta_{21} r^5 \mathrm{v}^5 \mathrm{v}_D} \tag{9a}$$

$$\tau_0 = 9.4 \times 10^{11} \frac{\dot{M}}{\mathrm{v}\, \mathrm{v}_D\, r} \tag{9b}$$

where the units of $\dot{M}$, v, v_D and r are respectively $M_\odot$ yr^{-1}, 100 km sec^{-1}, 100 km sec^{-1} and $R_\odot$. As a numerical example I take $r = 3$, $\mathrm{v} = 3$, $\mathrm{v}_D = 1$, and $\beta_{21} = 3 \times 10^{-4}$. For this example we have for Hα (10*a*) and Na D (10*b*)

$$\tau_0 = 3.2 \times 10^{16} \dot{M}^2 \tag{10a}$$

$$\tau_0 = 1.0 \times 10^{11} \dot{M} \tag{10b}$$

so that for $10^{-11} < \dot{M} < 7 \times 10^{-9}$, τ_0 (Hα) < 1 and τ_0 (Na D) > 1. Thus I conclude that the Na D lines should show evidence for absorption in the surrounding cloud at lower mass flow rates than will Hα. Furthermore, as long as a significant fraction of the energy flux associated with the mass motion is converted into Hα radiation, we expect the emission to be detectable at 20% of the continuum flux in a 4000 K star as long as $\dot{M} > 5 \times 10^{-11}$. Thus there is a substantial range in M for which the Na D lines can provide a definitive means of determining the direction of the mass flow.

Collisional ionization of the Na could easily reduce the absorption optical depth given above. The importance of this effect must be evaluated on a case by case basis. For cases where the Na is Na II, the inverse line P Cygni profile should occur in the Ca II H and K lines.

III. STATUS OF THE MODELS

On the basis of the observed Na D line profiles, it now appears that the gas flow may be in either direction. We are unable to discern any pattern which would indicate the factor which leads to the reversal of the flow from accretion to a stellar wind. We regard the answer to this question as a very important objective of the study of young stars. This reversal of the flow undoubtedly is a key event in the life of a star. For now, we can examine the details of the theoretical models for either direction of flow.

Stellar Wind Models

The concept of a stellar wind has appeal because we can relate this concept to the known solar wind. Thus it seems natural to discuss the early stages of the evolution of a star like the sun in terms of an enhanced stellar wind. This plan meets with a fundamental difficulty which is not as widely recognized as it should be; namely, the solar wind is driven by the thermal expansion of a gas hot enough that the sound velocity is a substantial fraction of the escape velocity. For the sun the temperature required to achieve the appropriate sound velocity is $1 - 3 \times 10^6$ K. A gas at this temperature has very little neutral hydrogen and little recombination. Thus the solar wind does not radiate effectively in the Balmer lines. Balmer radiation is important only for temperatures in the range 8000 to 25,000 K. Such a low temperature could not thermally drive a stellar wind from any T Tauri star. Thus, the concept of an enhanced solar wind is not compatible with strong Balmer radiation. In fact, low temperatures are always used in modeling T Tauri envelopes and the question of the cause for the motion is ignored. A cause that is effective for P Cygni stars, namely radiation pressure, cannot be appealed to for T Tauri stars because of the low luminosity.

In spite of the poor state of theoretical development of the stellar wind models, they form the conceptual basis for most discussions of T Tauri stars. A proper model of this sort should start with a well-defined set of physical assumptions and then proceed to derive all observable properties. Instead, due to the complexity of the observed properties, individual properties have been studied by partial models. Models of this type are those used by: (1) Kuhi (1964) to reproduce line profiles on a line-by-line basis; (2) Kuan (1975,

1976) to reproduce the relative line strengths; and (3) Rydgren, Strom, and Strom (1976) to reproduce the continuum energy-flux distribution. To a degree these partial models are complementary in that the adoption of one partial model does not necessarily exclude another partial model. The principal defect in this approach is the absence of a unifying theory.

The specific questions left unanswered in the above partial theories are:

1. What is the source of momentum which leads to the stellar wind?
2. What is the source of energy which leads to ionization and the emission of radiation from the T Tauri stars?

The similarity between T Tauri stars and flare stars noted by Haro may provide a basis for answering these questions and lead to a unifying theory. The similarity in the photometric behavior of these two classes of objects has also been noted by Kuan (1976). In fact, a few stars are members of both classes. A model based on this idea might be termed a continuous flare model.

The most natural answers to questions (1) and (2) in the context of a stellar wind model are somewhat incompatible. Expansion of a hot gas to a velocity on the order of the speed of sound is the most straightforward source of the momentum for the wind. This is what happens in the solar wind. Unfortunately a high enough temperature to provide the observed velocity is too high to permit the observed emission of Balmer radiation. The observed radiation is most readily explained as a result of ionization and heating by high-energy photons or particles. Unfortunately, high-energy photons or particles as a source of momentum require excessively high rates of energy input. Thus, these two aspects of T Tauri stars require separate but hopefully related explanations. The essential problem is that the expansion temperature characteristic of the wind velocity is much higher than the material temperature required by the emission characteristics. This problem can be resolved if the matter cools from the high temperature needed for the expansion to the lower temperature needed for the radiation. For spherically symmetric expansion this cooling occurs over a distance on the order of R, the radius of the star. By the time the radiation temperature is reached, the density is too low to yield the observed line intensities. However, flares provide a concentrated source of energy for the gas. The cooling then occurs as the gas expands from an initially small volume. The temperature of the gas drops as the gas volume increases. Since the initial volume may be small in this type of model, the temperature drop may be rapid. Thus, X rays and the high velocity can be produced in the high-density flare regions and the Balmer radiation can be produced by secondary radiation resulting from the absorption of the X rays by the surrounding, expanding gas. This type of model needs to be explored in detail before we can evaluate its validity.

The existing wind models are not entirely satisfactory even within the context of their intended application. The line profile models by Kuhi (1964) used a very approximate radiative transfer treatment; he assumed that the hydrogen radiates as if it were part of a static H II region. Furthermore, he needed a separate layer of cool atoms exterior to the emission producing region in order to explain the absorption line features. The column density of absorbers needed for different Balmer lines depended on the line being modeled even though the lines were observed simultaneously. This approach is unsatisfactory.

The calculations by Kuan (1976) involve two serious defects which render the reliability of the quantitative results unreliable. First, he treated cases with decelerating envelopes but neglected the effects of cross-talk between layers along a line of sight having the same radial velocity. The magnitude of the cross-talk terms can be as large as 30 to 40% of the local terms. Second, Kuan used collision rates not specific to hydrogen. These rates differ from the more appropriate values given by Johnson (1972) by as much as a factor of two. In spite of these two serious quantitative difficulties, the qualitative character of Kuan's results is undoubtedly correct and his work serves as a useful guide to the interpretation of the Balmer line intensities in the T Tauri stars.

Accretion Models

The observed radiation from a star undergoing accretion comes from four regions:

(1) the infalling envelope,
(2) a disk in the equatorial plane,
(3) a hot region just interior to the accretion shock, and
(4) the interior of the static core of the star.

Various theoretical treatments have emphasized only one or at most two of the above regions. The work by Bertout (1977, see the chapter by Bertout and Yorke) has emphasized Region (1) and shown how it can produce an inverse P Cygni profile. The work by Ulrich (1976) emphasized Region (3) and showed how this region can produce a P Cygni profile of Type III (erroneously referred to as Type II by Ulrich, 1976). These lines generally show an absorption feature which is blue-shifted and superimposed on an emission line. The minimum intensity in these features generally does not go below the continuum level (Beals, 1950). Lynden-Bell and Pringle (1974) primarily discussed Region (2) although they did not calculate line profiles. Region (4) is probably responsible for the nearly normal stellar spectrum occasionally observed in T Tauri stars.

No complete treatment taking into account all four regions is yet available. In addition, the treatment of the radiative transfer in the accretion calculations to date is far from satisfactory. The best is the work by Bertout in which a two level atom is treated with an escape probability theory. Such a model cannot realistically represent the Balmer line transfer problem for which most data is available. In particular, the argument presented earlier in this chapter indicates that for the case considered by Bertout τ (Hγ) $\cong 10^5$ to

10^6 rather than 1 to 10 as in Bertout's calculations. A similar conclusion is reached by noting that Bertout's pressure of 15 to 30 dynes cm^{-2} corresponds to the pressure of the lower chromosphere of the sun. The Balmer line absorption cores are formed in this layer so that $\tau = 1$. However the scale height in the sun is 10^8 cm whereas the region of line formation in Bertout's model is on the order of 10^{11} cm. Thus the optical depth should be at least 10^3.

The treatment by Ulrich (1976) is very much simpler than that by Bertout and no real attempt was made to determine the rate of accretion relevant to the model. The primary emphasis was on the kinematic structure of non-spherically symmetric accretion. Ulrich's neglect of the precursor region and the disk cannot be justified and this neglect constitutes a serious defect in the model as presented. Very low accretion rates may yield models for which these assumptions are justified.

The work by Lynden-Bell and Pringle did not address the line profile question and thus cannot be considered in the same category as the two preceding treatments. The line profile produced by a disk is well defined and does not match the observed profiles except in a few cases. I am aware of three T Tauri stars which have shown line profiles that are compatible with emission from a disk. These stars are AS 209, DI Cep and possibly VV Ser. Further observations can test this suggestion. If a disk is truly responsible for the observed line profiles, then the profiles should remain compatible with this interpretation for future observations. If a disk is not responsible for the emission then the variation in the line profiles normally found in these stars should result in profiles which are not compatible with the disk model after only a few additional observations. The Lynden-Bell and Pringle type of disk may occur prior to the T Tauri phase of evolution so that the observations do not rule out their model as an early stage in the process of star formation.

IV. SUMMARY

No models are yet completely satisfactory. The observations now indicate that no unique model can account for all T Tauri stars. In particular both mass outflow and inflow are observed in systems fitting the definition of the class. For stars like AS 205, the lack-of-emission hypothesis proposed by Ulrich (1976) to account for absorption-like features in the Balmer lines appears to be required. There are now three such cases. Thus, although Ulrich's specific model involves theoretical difficulties, a modified form of this model may be successful. Similarly, the theoretical work by Bertout (1977) is not yet in a form to make detailed comparisons to the observations meaningful in a quantitative sense; nonetheless, the part of the problem he treats is essential and must be included in a complete model. The models most in need of further work are the wind models. The only line profiles calculations with such a model are those by Kuhi (1964). Substantially improved radiative transfer methods are now available and should be used for the line profile calculations for the wind model.

Acknowledgments. As a guest investigator at the Hale Observatories I am especially grateful to G. Knapp for permission to use some of the data from our joint observing program prior to its regular publication. I would also like to thank M. Jura for providing several helpful comments to this manuscript. The assistance of S. Shectman was indispensable in the acquisition of the data. The data reduction programs were kindly made available by K. Clarety. This research is supported in part at the University of California, Los Angeles by a grant from the National Science Foundation.

Notes Added in Proof

Correspondence with Bertout has clarified the source of the factor of 10^5 in $\tau(H\gamma)$ by which our calculations differ. Bertout calculated the population of the $n = 2$ state LTE formulae at a temperature of 2000 K and finds the hydrogen is mostly ionized. My estimate is based on non-LTE calculations with Drake's code at temperatures between 5000 and 15,000 K. These calculations show that the gas is mostly neutral. At 20,000 K we find that the ratio of ionized to neutral hydrogen is in the range of 200 – 500 to 1 for 10^9 cm$^{-3} < N_e < 10^{13}$ cm^{-3}. At higher temperatures we can recover a very low value of N_2/N_H such as is required by Bertout's model. It is difficult to estimate the temperature in the circumstellar envelope so no clear choice of $\tau(H\gamma)$ can be made at present.

Observations of AS 209 show that it does not have line profiles compatible with disk emission.

R. Thompson at Steward Observatory has pointed out to me that geometries may exist in which a Lynden-Bell and Pringle type disk intercepts little of the ionized radiation illuminating the circumstellar region. Thus the disk may not contribute any Balmer line emission. If such a model can be found, the line profile consideration discussed here will be irrelevant. Scattered Lyman continuum photons must be shielded as well as photons from the star itself.

APPENDIX

Notes to Table I[a]

Lk Hα 264: Emission intensity 50% above continuum; FWHM = 350 km s^{-1}; absorption core at zero velocity is unresolved.

Lk Hα 272: Sharp absorption FWHM 60 km s^{-1}; continuum depressed 10% redward to +300 km s^{-1}.

DH Tau:	Broad shallow absorption band, the two lines blended together; possibly rotation-broadened with v sin i = 300 km s^{-1}; no visible core absorption.
DI Tau:	Absorption profile v-shaped with FWHM = 150 km s^{-1}.
HL Tau:	Classic P Cygni profile with a zero velocity absorption superimposed; blue-shifted absorption at −130 km s^{-1}.
XZ Tau:	No clear feature; there may be a sharp emission core superimposed on a wider absorption.
UY Aur:	Weak absorption feature at zero velocity; broad absorption (FWHM = 225 km s^{-1}) centered at +130 km s^{-1}.
GM Aur:	Absorption feature virtually identical to DI Tau.
HI Ori:	Absorption centered at −45 km s^{-1}; an absorption feature at +225 km s^{-1} suggested the line profile may be complex.
EZ Ori:	Absorption FWHM = 90 km s^{-1}; zero velocity.
YY Ori:	Absorption FWHM = 120 km s^{-1} centered at −90 km s^{-1} relative to the LSR; the Ca I feature is also at −90 km s^{-1} so YY Ori may be a binary.
LL Ori:	Emission superimposed on absorption as in XZ Tau.
MM Mon:	Sharp core (FWHM = 90 km s^{-1}) at zero velocity; absorption starting at zero velocity and 40% below continuum intensity increasing to continuum intensity at v = +420 km s^{-1}.
MO Mon:	Poor photon statistics; absorption centered at ν - +120 km s^{-1} - relative to LSR.
AS 205:	Profile given in Fig. 1. The dashed line resulted from data lost in an image tube defect.
SR 9:	Marginal redward absorption at 10% below continuum extending to +200 km s^{-1}.
AS 353:	Profile given in Fig. 2.
V 536 Aql:	Poor photon statistics; possible broad absorption at +150 km s^{-1}
Lk Hα 120:	Absorption at −300 km s^{-1}.
DI Cep:	Emission at a peak intensity of 1.6 times continuum; FWHM = 180 km s^{-1} for emission; sharp, unresolved absorption core at zero velocity.

[a] Most velocities are referred to the Ca I 5857.5 feature which appears to be photospheric.

REFERENCES

Appenzeller, I., and Wolf, B. 1977. *Astron. Astrophys.* 54: 713.

Beals, C. S. 1950. *Publ. Dom. Astrophys. Obs.* 9: 1.

Bertout, C. 1977. *Astron. Astrophys.* 58: 153.

Brown, R.L., and Gould, R. J. 1970. *Phys. Rev. D.* 1: 2252.

Dibaj, E. A., and Esipov, W. F. 1969. In *Non-Periodic Phenomena in Variable* Stars (L. Dètre, ed.), p. 107. Dordrecht: Reidel.

Dumont, S.; Heidmann, N.; Kuhi, L. V.; and Thomas, R. N. 1973. *Astron. Astrophys.* 29: 199.

Garrison, L. M., and Anderson, C. M. 1977. *Astrophys. J.* 218: 438.

Herbig, G. H. 1960. *Astrophys. J. Suppl.* 4: 337.

____. 1962. *Adv. Astron. Astrophys.* 1: 47.

____. 1970. In *Evolution Stellaire avant la Sequence Principale. Mém. Roy. Soc. Sci. Liège Ser.* 5,9: 13.

Johnson, L. C. 1972. *Astrophys. J.* 174: 227.

Kolotilov, E. A., and Zaitsev, G. V. 1974. *Astrophys.* 10: 228.

Kuan, P. 1975. *Astrophys. J.* 202: 425.

____. 1976. *Astrophys. J.* 210: 129.

Kuhi, L. V. 1964. *Astrophys. J.* 140: 1409.

____. 1965. *Publ. Astron. Soc. Pacific* 77: 253.

Lynden-Bell, D., and Pringle, J. E. 1974. *Mon. Not. Roy. Astr. Soc.* 168: 603.

Rydgren, A. E. 1977. *Publ. Astron. Soc. Pacific* 89: 557.

Rydgren, A. E.; Strom, S. E.; and Strom, K. M. 1976. *Astrophys. J. Suppl.* 30: 307.

Salmanov, I. R. 1974. *Astrophys.* 10: 181.

Shectman, S. A., and Hiltner, W. A. 1976. *Publ. Astron. Soc. Pacific* 88: 960.

Stone, R. P. S. 1974. *Astrophys. J.* 193: 135.

Ulrich, R. K. 1976. *Astrophys. J.* 210: 377.

Walker, M. F. 1972. *Astrophys. J.* 175: 89.

Wolf, B.; Appenzeller, I.; and Bertout, C. 1977. *Astron. Astrophys.* 58: 163.

Wilson, O. C., and Bappu, M. K. V. 1957. *Astrophys. J.* 125: 661.

Glossary, Acknowledgments, and Index

GLOSSARY[a]

a	semi-major axis of an orbit
Å	angstrom = 10^{-8} cm
A_V	interstellar extinction of starlight in visible light; A_V is for the V filter of the UBV photometric system
α	right ascension
albedo	geometric albedo: ratio of planet brightness at zero phase angle to the brightness of a perfectly diffusing disk with the same position and apparent size as the planet; Bond albedo: fraction of the total incident light reflected by a spherical body
Alfvén speed	the speed at which hydromagnetic waves are propagated in a magnetic field: $v_A = B/(4\pi\rho)^{\frac{1}{2}}$ where B is the strength of the magnetic field and ρ is the space density
Alfvén waves	waves moving perpendicularly through a magnetic field. They are caused by the oscillation of magnetic lines of force by the motions of the fluid element around its equilibrium position, which in turn is caused by the interactions between density fluctuations and magnetic variations

[a]In addition to *Astrophysical Quantities* by C.W. Allen (London: Athlone Press, 1973) we used *Glossary of Astronomy and Astrophysics* by J. Hopkins (Chicago: University of Chicago Press, 1976).

Allende	a meteorite often discussed in this book. It is a large, carbonaceous chondrite which fell near Allende, Mexico in 1969. For other meteorites, see the list in the Index under meteorites, and for a discussion of Allende see Boynton's chapter
ambipolar diffusion	refers to the relative motion of the charged components of a gas and the neutral components under the action of electromagnetic forces. Since the charged particles (electrons or ions) are subject to direct interaction with the electromagnetic field, they can move relative to the neutrals, which are then dragged along by the collisions with the ions or electrons. The separation of the charged and neutral componnents by this mechanism is given the general name "ambipolar diffusion" (Spitzer, L., Jr. 1968. *Diffuse Matter in Space,* p. 238. New York: Interscience Press)
arcmin	one minute of arc = 1′
arsec	one second of arc = 1″
AU	the mean distance between Earth and Sun (1.496×10^{13} cm $\simeq$ 500 light seconds)
Balmer lines	the spectral lines of hydrogen corresponding to electron energy transitions involving the second energy level
bifurcation	(1) division into two branches, parts, or aspects; (2) the point at which the division occurs
blocking temperature	the temperature below which permanent magnetization is locked into ferromagnetic materials for long times
B meteoroids	small interplanetary particles whose orbits are significantly altered by light pressure at the instant they are released from a larger parent body; *B* is the term used for the ratio of radiation pressure force to gravitational force
Bonner-Ebert condition	for a gas to satisfy this condition, $M_{BE} \gtrsim 1.2\, C^4/(G^3 P_0)^{\frac{1}{2}}$, where C is the isothermal speed of sound in the gas, G is the constant of gravitation, and P_0 is the surrounding pressure (see Mouschovias' chapter)
breccias	rocks characterized by a broken, fragmented internal texture
c	speed of light in a vacuum = 3×10^{10} cm sec^{-1}
C3V	a combination of the two methods of classification of mete-

oritic types C3 and CV as can be seen in Table I of the chapter by Herndon and Wilkening

CAI — Ca-, Al-rich inclusions

CCF — carbonaceous chondrite fission

chondrite — a stony meteorite usually characterized by the presence of chondrules. Type I carbonaceous chondrites contain no chondrules

chondrules — small spherical grains varying from microscopic size to the size of a pea, usually composed of iron, aluminum, or magnesium silicates. They occur in abundance in primitive stony meteorites

CIS — circumionization shell

cm — centimeter

comminution — the breaking up or grinding into small fragments

Copernicus — Orbital Astronomical Observatory (OAO-3), launched August 21, 1972 (a = 7123 km, e = 0.00083, i = 35.°0) equipped with an ultraviolet telescope, a steerable X-ray telescope, and gamma-ray detectors

Coriolis effect — the acceleration which a body in motion experiences when observed in a rotating frame. This force acts at right angles to the direction of the angular velocity

CPM stars — stars exhibiting common proper motion

CRVS — constant radial velocity surface

1D, 2D, 3D — one-, two-, three-dimensional; used in this book for calculations of star formation

δ — declination

δ — magnitude of the isotopic anomaly = $|R_n - R_a|/R_n$; see "isotopic anomalies" below

Eddington approximation — an approximation used in the study of radiative transfer. It is the assumption that the ratio of the second moment of the radiation field to the mean intensity is everywhere equal to 1/3, the value of this ratio for an isotropic field

e-folding time — the time within which the amplitude of an oscillation increases or decreases by a factor e (e = 2.718 . . .)

euhedral crystals — crystals in igneous rocks with well-developed faces

eV	electron volt = 1.60×10^{-12} ergs
Faraday rotation	rotation of the plane of polarization of linearly polarized radiation when the radiation passes through a plasma containing a magnetic field that has a component in the direction of propagation. Depolarization can be caused by this effect on different interstellar clouds along the line of sight
G	gauss = 1 oersted = 79.58 amp-turn/m
G	universal gravitational constant = 6.67×10^{-8} dyn $cm^2 g^{-2}$
GeV	giga electron volt = 10^9 eV
globules	also called Bok globules. Well-defined, roundish dark nebulae with angular diameters mostly between 3 and 20 . Their radii are in the range 0.15 to 1.2 pc (3.0×10^4 to 2.5×10^5 AU)
Gould Belt	the local system of stars and gas within about 300 pc of the sun. It is a belt inclined about 10-20° to the galactic plane in which the greatest concentration of naked-eye O and B stars occurs
Gyr	giga-year = 10^9 yr
Hayashi track	a nearly vertical track, in the Hertzsprung-Russell diagram, of stellar evolution toward the main sequence during phases when the star is largely or completely in convective equilibrium. The luminosity, originally very high, decreases rapidly with contraction, but the surface temperature remains almost constant
Henry's solubility law	the pressure dependence of solubility $C/P = K$ where C is the concentration of the gas in solution, P is the partial pressure and K is the distribution coefficient or Henry's constant
Herbig-Haro object	an object with many of the characteristics of a T Tauri star (e.g., its spectrum shows a weak continuum with strong emission lines), believed to be a star in the very early stages of evolution. All known Herbig-Haro objects have been found within the boundaries of dark clouds. They are strong infrared sources and are characterized by mass loss
HPBW	half power beam width
HR diagram	Hertzsprung-Russell diagram: in present usage, a plot of bolometric magnitude against effective temperature for a group of stars. Related plots are the color-magnitude plot

(absolute or apparent visual magnitude against color index) and the spectrum-magnitude plot (visual magnitude versus spectral type, shown in the figure below), the original form of the HR diagram

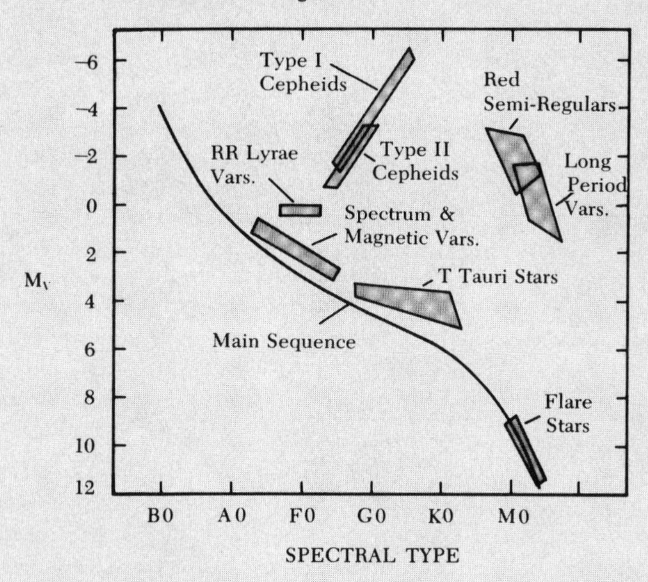

IAU	International Astronomical Union
IMF	initial mass function
IR	infrared
ISM	interstellar medium
isochron	the term denotes a straight line containing all the points representing the same time or age. In geochronology it usually refers to a line constructed either by plotting the ratio $^{87}Sr/^{86}Sr$ as a function of the ratio $^{87}Rb/^{86}Sr$ or by plotting the ratio $^{207}Pb/^{204}Pb$ or $^{208}Pb/^{204}Pb$ as a function of the ratio $^{206}Pb/^{204}Pb$ for minerals or rocks of the same age
isopedion contours	contours of equal magnetic field strength
isotopic anomalies	let R_n be the normal ratio of two nuclides and R_a the anomalous ratio; then $\delta = \lvert R_n - R_a \rvert / R_n$ is the magnitude of the anomaly. The isotopic anomalies in meteorites are reviewed in Reeves' chapter
JD	Julian Day number

Jeans mass — (or Jeans limit, Jeans criterion) the mass necessary for protostellar collapse $M_J \geqslant 1.795 \left(\frac{RT}{G}\right)^{\frac{3}{2}} \rho^{-\frac{1}{2}}$, where R is the gas constant, T the cloud temperature, G the gravitational constant, and ρ the cloud density (see introductory chapter by Clayton)

Jy — Jansky; 1 Jy = 10^{-26} W m^{-2} Hz^{-1}

k — Boltzmann constant = 1.38×10^{-16} erg deg^{-1}; alternately, = 8.62×10^{-5} eV deg^{-1}

kcal — kilo-calorie = 4.185×10^{10} erg

Kelvin contraction time — (or Kelvin-Helmholtz time scale) the time it takes a star to contract gravitationally from infinite radius down to its present radius by radiating its thermal energy (for the sun, $\sim 2 - 3 \times 10^7$yr). The Kelvin time is roughly equal to (gravitational binding energy)/luminosity

keV — kilo-electron volt = 10^3 eV

km — kilometer = 10^5 cm

kpc — kilo-parsec = 10^3 pc

kyr — kilo-year = 10^3 yr

$L_\odot$ — solar luminosity = 3.826×10^{33} erg sec^{-1}

ℓ — galactic longitude (see Fig. 3 of the chapter by W. Herbst and Assousa)

Lagrangian point — see "Trojans"

Landé factor — the constant of proportionality relating the separations of lines of successive pairs of adjacent components of the levels of a spectral multiplet to the larger of the two J-values for the respective pairs. The interval between two successive components J and $J + 1$ is proportional to $J + 1$

lightyear — the distance that light travels in one year in a vacuum = 9.46×10^{17} cm

line strength — the intensity of a spectrum line

Lorentz force — the force on a charged particle moving in electric and magnetic fields, equal to the particle's charge times the sum of the electric field and the cross product of the particle's velocity with the magnetic flux density

LSR	local standard of rest. A frame of reference in which the mean motion of stars in the immediate neighborhood is zero. In such a reference system, the motions of stars in the solar neighborhood (a volume of space ~ 100 pc in diameter) average out to zero. It is a coordinate system in which the origin is a point in the galactic plane moving in a circular orbit around the galactic center with a period equal to that of the sun
LTE	local thermodynamic equilibrium
μ	molecular weight
μG	microgauss; 1 μG = 10^{-6} gauss
μm	1 μm = 1 micrometer = 1 micron = 10^{-4} cm
M_{bol}	bolometric absolute magnitude; measure of the total amount of energy radiated by a star at all wavelengths
M_J	see "Jeans mass"
$M_\odot$	solar mass = 1.989×10^{33} g; occasionally, the symbol $m_\odot$ is used
mag	astronomical magnitude proportional to $-2.5 \log_{10} I$, where I is the intensity
magmatic differentiation	(1) the process by which the different types of igneous rock are derived from a single parent magma; (2) the process by which ores are formed by solidification from magma; also known as "magmatic segregation"
main sequence	the principal sequence of stars in the HR diagram, containing more than 90% of the stars we observe, that runs diagonally from the upper left (high temperature, high luminosity) to the lower right (low temperature, low luminosity). A star appears on the main sequence after it has started to burn hydrogen in its core, and is estimated to stay on the main sequence until it has used up about 12% of its hydrogen (for a 1 $M_\odot$ star, about 10^{10} yr). See figure under "HR diagram"
MeV	million electron volts = 10^6 eV
mG	milligauss = 10^{-3} gauss
micrometeorite	a very small meteorite or meteoritic particle with a diameter generally less than a millimeter

Mie theory	a theory of light scattering by small spherical particles
millibarn	a unit of area used in measuring cross sections; 1 millibarn $= 10^{-3}$ barn $= 10^{-27}$ cm^{-2}
mol	1 mol = 1 mole = 6.023×10^{23} atoms (or molecules)
Myr	mega-year = 10^6 yr
Ne-E	a component of noble gases in meteorites characterized by $^{20}Ne/^{22}Ne < 1.4$. Such a composition is not formed by any known process (radioactive decay, cosmic-ray interactions, etc.) or source (sun, terrestrial atmosphere). Hence, it is ascribed to a pre-solar system source (Black, D. 1972. *Geochim. Cosmochim.* 36: 377)
nuclide	a species of atomic nucleus, analogous to the word "isotope" for a species of atom. The word is also used to distinguish between atomic nuclei that are in different energy states
OB association	associations of stars of spectral type O – B2. About 20 are known
Ockham's razor	*Entia non sunt multiplicanda* ("entities are not to be multiplied"). A doctrine formulated by William of Ockham in the fourteenth century. Any hypothesis should be shorn of all unnecessary assumptions; if two hypotheses fit equally well, the one that makes the fewest assumptions should be chosen
Oort limit	the density limit of all matter in the solar neighborhood = $0.13\,M_\odot$ pc^{-3} $= 8.8 \times 10^{-24}$ g cm^{-3} $\simeq 4.4$ atoms cm^{-3}
pc	parsec; 1 parsec = the distance where 1 AU subtends 1 arcsec = 206,265 AU = 3.26 light year $= 3.086 \times 10^{18}$ cm
P Cygni star	the spectrum of P Cygni shows strong emission lines with blue-shifted absorption components presumably due to mass loss by an expanding shell of low-density matter
Periodic Table	see figure on next page
phase Q	that part of the HF/HC1 resistant meteorite residue which is soluble under oxidizing acid treatment. Although only a few hundredths of a percent of the meteorite by weight, it contains up to 99% of the heavy noble gases of the meteorites. It is a mixture of Fe, Ni- and Fe, Cr sulfides
PMS	pre-main sequence (see "main sequence")

Period	IA	IIA	IIIB	IVB	VB	VIB	VIIB	VIII	VIII	VIII	IB	IIB	IIIA	IVA	VA	VIA	VIIA	O
1	1 H Hydrogen 1.00797																	2 He Helium 4.0026
2	3 Li Lithium 6.939	4 Be Beryllium 9.0122											5 B Boron 10.811	6 C Carbon 12.01115	7 N Nitrogen 14.0067	8 O Oxygen 15.9994	9 F Fluorine 18.9984	10 Ne Neon 20.183
3	11 Na Sodium 22.9898	12 Mg Magnesium 24.312											13 Al Aluminum 26.9815	14 Si Silicon 28.086	15 P Phosphorus 30.9738	16 S Sulfur 32.064	17 Cl Chlorine 35.453	18 Ar Argon 39.948
4	19 K Potassium 39.102	20 Ca Calcium 40.08	21 Sc Scandium 44.956	22 Ti Titanium 47.90	23 V Vanadium 50.942	24 Cr Chromium 51.996	25 Mn Manganese 54.938	26 Fe Iron 55.847	27 Co Cobalt 58.933	28 Ni Nickel 58.71	29 Cu Copper 63.54	30 Zn Zinc 65.37	31 Ga Gallium 69.72	32 Ge Germanium 72.59	33 As Arsenic 74.922	34 Se Selenium 78.96	35 Br Bromine 79.909	36 Kr Krypton 83.80
5	37 Rb Rubidium 85.47	38 Sr Strontium 87.62	39 Y Yttrium 88.905	40 Zr Zirconium 91.22	41 Nb Niobium 92.906	42 Mo Molybdenum 95.94	43 Tc Technetium (98)	44 Ru Ruthenium 101.07	45 Rh Rhodium 102.905	46 Pd Palladium 106.4	47 Ag Silver 107.870	48 Cd Cadmium 112.40	49 In Indium 114.82	50 Sn Tin 118.69	51 Sb Antimony 121.75	52 Te Tellurium 127.60	53 I Iodine 126.904	54 Xe Xenon 131.30
6	55 Cs Cesium 132.905	56 Ba Barium 137.34	57 La ★ Lanthanum 138.91	72 Hf Hafnium 178.49	73 Ta Tantalum 180.948	74 W Wolfram 183.85	75 Re Rhenium 186.2	76 Os Osmium 190.2	77 Ir Iridium 192.2	78 Pt Platinum 195.09	79 Au Gold 196.967	80 Hg Mercury 200.59	81 Tl Thallium 204.37	82 Pb Lead 207.19	83 Bi Bismuth 208.980	84 Po Polonium (210)	85 At Astatine (210)	86 Rn Radon (222)
7	87 Fr Francium (223)	88 Ra Radium (226)	89 Ac ★★ Actinium (227)															

★ 6	58 Ce Cerium 140.12	59 Pr Praseodymium 140.907	60 Nd Neodymium 144.24	61 Pm Promethium (147)	62 Sm Samarium 150.35	63 Eu Europium 151.96	64 Gd Gadolinium 157.25	65 Tb Terbium 158.924	66 Dy Dysprosium 162.50	67 Ho Holmium 164.930	68 Er Erbium 167.26	69 Tm Thulium 168.934	70 Yb Ytterbium 173.04	71 Lu Lutetium 174.97
★★ 7	90 Th Thorium 232.038	91 Pa Protactinium (231)	92 U Uranium 238.03	93 Np Neptunium (237)	94 Pu Plutonium (242)	95 Am Americium (243)	96 Cm Curium (245)	97 Bk Berkelium (249)	98 Cf Californium (251)	99 Es Einsteinium (254)	100 Fm Fermium (255)	101 Md Mendelevium (256)	102 No Nobelium (254)	103 Lw Lawrencium (257)

(Courtesy Hubbard Scientific Co., Northbrook, IL)

Population I stars	(sometimes called "arm population") young stars with relatively high abundances of metals. They are usually found in the disk of a galaxy, especially the spiral arms, in dense regions of interstellar gas
Population II stars	(sometimes called "halo population") older stars with relatively low abundances of metals. They are usually found in the nucleus of a galaxy or in globular clusters
Poynting-Robertson effect	a light pressure effect on small grains. The time it takes for particles to move into the sun $t = 7.0 \times 10^6\ r p a q$ yr, (radius r in cm, density p in g cm^{-3}, a and q in AU) where a and q are the semi-major axis and perihelion distance of the initial particle orbit.
p-process	the name of the hypothetical nucleosynthetic process thought to be responsible for the synthesis of the rare heavy proton-rich nuclei which are bypassed by the *r*- and *s*-processes. It is manifestly less efficient (and therefore rarer) than the *s*- or *r*-process since the protons must overcome the Coulomb barrier, and may in fact work as a secondary process on the *r*- and *s*-process nuclei. It seems to involve primarily (p, γ) reactions above cerium (where neutron separation energies are low). The *p*-process is assumed to occur in supernova envelopes at a temperature $\gtrsim 10^9$ K and at densities $\lesssim 10^4$ g cm^{-3}
pulsar	an object which has the mass of a star and a radius no larger than that of Earth and which emits radio pulses with a very high degree of regularity (periods range from 0.03 sec for the youngest to more than 3 sec for the oldest). Pulsars are believed to be rotating, magnetic (surface magnetic fields of 10^{10} to 10^{14} gauss are estimated) neutron stars which are the end products of supernovae
quasar	an object with a dominate starlike component, with an emission line spectrum showing a large redshift. The light of most if not all quasars is variable over time intervals between a few days and several years, so their diameters must not be much larger than the diameter of the solar system; yet they are the intrinsically brightest objects known. The basic problem of quasars is that they emit too much radiation in too short a time from too small an area
R	the ratio of total to selective extinction in interstellar space; $R = \dfrac{A_V}{E_{B-V}}$

$R_\odot$	solar radius = 6.96×10^{10} cm
R association	group of reflection nebulae
Rayleigh-Taylor instability	an instability which develops in the case of a light fluid (or a magnetic field) supporting a heavy fluid against gravity (or an effective gravity); protrusions of the heavy fluid "sink" into the light one
REE	rare earth elements = lanthanide series in the Periodic Table
refractory	a material of high melting point, or the property of resisting heat
regolith	the layer of fragmentary debris produced by meteoritic impact on the surface of the moon or a minor planet
Reynolds number	a dimensionless number ($R = L\mathrm{v}/\nu$, where L is a typical dimension of the system, v is a measure of the velocities that prevail, and ν is the kinematic viscosity) that governs the conditions for the occurrence of turbulence in fluids
Roche limit	the minimum distance at which a satellite under the influence of its own gravitation and that of a central mass, about which it is describing a Keplerian orbit, can be in equilibrium. For a satellite of negligible mass, zero tensile strength, and the same mean density as its primary, in a circular orbit around its primary, this critical distance, at which the satellite will break up, is 2.44 times the radius of the primary
Rossby waves	cyclonic convection waves in a rotating fluid
r-process	the capture of neutrons on a very rapid time scale (i.e., one in which a nucleus can absorb neutrons in rapid succession, so that regions of great nuclear instability are bridged), a theory advanced to account for the existence of *all* elements heavier than bismuth as well as the neutron-rich isotopes heavier than iron. The essential feature of the *r*-process is the release of great numbers of neutrons in a very short time (less than 100 seconds). The presumed source for such a large flux of neutrons is a supernova, at the boundary between the neutron star and the ejected material
SED	spectral energy distribution
Sobolev approximation	turbulent velocities are small compared to systematic collapse velocities
solar wind	a radial outflow of energetic charged particles from the

solar corona, carrying mass and angular momentum away from the sun. Mean number density of solar wind (1971), 5 per cm^3; mean velocity at Earth 400 km sec^{-1}; mean magnetic field 5×10^{-5} gauss; mean electron temperature 20,000 K; mean ion temperature 10,000 K. The sun ejects $\sim 10^{-13}\ M_\odot$ per year via the solar wind

spallation — a nuclear reaction in which the energy of each incident particle is so high that more than two or three particles are ejected from the target nucleus and both its mass number and atomic number are changed

spectral types B, A, etc. — see "HR diagram"

***s*-process** — (slow neutron capture) a process in which heavy, stable, neutron-rich nuclei are synthesized from iron-peak elements by successive captures of free neutrons in a weak neutron flux, so there is time for β-decay before another neutron is captured (c.f., *r*-process). This is a slow but sure process of nucleosynthesis which is assumed to take place in the intershell regions during the red-giant phase of evolution, at densities up to 10^5 g cm^{-3} and temperatures of about 3×10^8 K (neutron densities assumed are 10^{10} cm^{-3})

star identification — bright variable stars have letters, for example AZ Orionis, ZZ Ori. Most other stars have catalog numbers. For instance, CoD $-35°$ 10525 is star 10525 in the $-35°$ declination zone of the Cordoba Durchmusterung, or Lk Hα indicates a star in the Hα survey of the Lick Observatory

Stokes parameters — four parameters to describe fully a beam of polarized light. They involve the maximum intensity, the ellipticity, and the direction of polarization

supernovae Type I — supernovae which have a spectrum characterized by detailed structure without a well-defined continuum, and appear to be deficient in hydrogen; they occur in the halo and in old disk populations (Type II population, ironically) in galaxies

supernovae Type II — supernovae which have a spectrum characterized by a continuum with superimposed hydrogen lines and some metal lines; they occur in the arms of spiral galaxies, implying that their progenitors are massive stars (Type I stars)

T association — group of T Tauri stars

T_e	effective temperature, in flux $F \propto T_e{}^4$
Trojans	Trojan asteroids occur in two of the *Langrangian Points,* (namely the ones preceding and following Jupiter in its orbit) which may be defined as points in the orbital plane of two massive bodies in circular orbits around a common center of gravity where a third body of negligible mass can remain in equilibrium
T Tauri stars	eruptive variable subgiant stars associated with interstellar matter and believed to be in the process of gravitational contraction
UBV	the photometric system developed by H. L. Johnson, with ultraviolet, blue and visual filters and colors expressed in *B-V* and *U-B* magnitudes. Filters at other wavelengths are also used and indicated with letters *R, I, H, J, K, L, M,* etc., and colors may be expressed as *K-L,* for instance
UV	ultraviolet
v_A	Alfvén speed (see above)
van Rhijn distribution	the relative numbers of stars of different luminosities or masses in a given volume of space
Van't Hoff's solubility law	Van't Hoff's equation applied to equilibrium solubility of gases in solids is $\log k = -\Delta H/2.303\ RT + C$, where k is distribution coefficient, ΔH is the heat of solution, T the absolute temperature, R the gas constant, and C a constant of integration
virial theorem	for a bound gravitational system the long-term average of the kinetic energy is one half of the potential energy
VLBI	very long baseline interferometry
$v \sin i$	rotation velocity as seen at the apparent inclination of the rotation axis of the star
Wilson Bappu relation	a linear relation between the width of the K_2 emission core in the resonance line of Ca II at 3933 Å, detectable for late-type stars, and the absolute magnitudes of the stars
X	relative abundance of hydrogen
Y	relative abundance of helium
Yarkovsky effect	perturbation of a small body's orbit by re-radiation of solar energy in a direction asymmetric with respect to the sun due to the body's rotation

YY Ori	an extremely young star (younger than T Tau, although YY Ori stars possibly are a subclass of T Tau stars) in the Orion Nebula. YY Orionis stars are low-mass stars in the gravitationally contracting stage in which the star is still accreting matter from the protostellar cloud
Z	relative abundance of heavy elements
ZAMS	zero-age main sequence (see "main sequence")
Zeeman Effect	spectral line broadening due to the influence of magnetic fields. A multiplet of lines is produced, with distinct polarization characteristics. The Zeeman Effect is measured by measuring the difference between right-hand and left-hand polarization across a spectral line

ACKNOWLEDGMENTS

The following people helped to make this book possible, in organizing, writing, refereeing or otherwise. The members of the Organizing Committee and Chairpersons are indicated with an asterisk (*).

*H. Abt, Kitt Peak Nat. Obs., Tucson, Ariz.
S. Aiello, Cattedra di Fisica dello Spazio, Univ. di Firenze, Firenze, Italy.
L. Alaerts, Enrico Fermi Inst., Univ. of Chicago, Chicago, Ill.
H. Alfvén, Dept. of Appl. Phys. and Inform. Sci., Univ. of Calif., La Jolla, Calif.
C. C. Allen, Lunar and Planetary Lab., Univ. of Ariz., Tucson, Ariz.
I. Appenzeller, Landessternwarte, Königstuhl, Heidelberg, W. Germany.
*J.R. Arnold, Dept. of Chemistry, Univ. of Calif., La Jolla, Calif.
*G. Arrhenius, Dept. of Appl. Phys. and Inform. Sci., Univ. of Calif., La Jolla, Calif.
G. E. Assousa, Dept. of Terrestrial Magnetism, Carnegie Inst. of Wash., Washington, D.C.
L. H. Auer, High Altitude Obs., Boulder, Colo.
W. A. Baum, Lowell Obs., Flagstaff, Ariz.
R. Beer, Jet Propulsion Lab., Pasadena, Calif.
C. Bertout, Landessternwarte, Königstuhl, Heidelberg, W. Germany.
*D. C. Black, Ames Research Center, Moffett Field, Calif.
L. Blitz, Dept. of Astronomy, Columbia Univ., New York, N.Y.
*P. Bodenheimer, Lick Obs., Univ. of Calif., Santa Cruz, Calif.
K. H. Böhm, Dept. of Astronomy, Univ. of Washington, Seattle, Wash.
*B. J. Bok, Steward Obs., Univ. of Ariz., Tucson, Ariz.
A. P. Boss, Dept. of Physics, Univ. of Calif., Santa Barbara, Calif.
W. V. Boynton, Lunar and Planetary Lab., Univ. of Ariz., Tucson, Ariz.
W. K. Brown, Los Alamos Scientific Lab., Los Alamos, New Mex.
D. E. Brownlee, Dept. of Astronomy, Univ. of Washington, Seattle, Wash.
J. A. Burns, College of Engineering, Cornell Univ., Ithaca, N.Y.
R. F. Butler, Dept. of Geosciences, Univ. of Ariz., Tucson, Ariz.
*A. G. W. Cameron, Center for Astrophysics, Harvard College Obs., Cambridge, Mass.
H. Campins, Dept. of Space Physics and Astronomy, Rice Univ., Houston, Tex.
E. J. Chaisson, Center for Astrophysics, Harvard College Obs., Cambridge, Mass.
*C. R. Chapman, Planetary Science Inst., Tucson, Ariz.
R. Chevalier, Kitt Peak Nat. Obs., Tucson, Ariz.
D. D. Clayton, Dept. of Space Physics and Astronomy, Rice Univ., Houston, Tex.
M. Cohen, Dept. of Astronomy, Univ. of Calif., Berkeley, Calif.
G. J. Consolmagno, Lunar and Planetary Lab., Univ. of Ariz., Tucson, Ariz.
*G. V. Coyne, Lunar and Planetary Lab., Univ. of Ariz., Tucson, Ariz.
G. Crozaz, Dept. of Earth and Planetary Sciences, Washington Univ., St. Louis, Mo.

D. D. Cudaback, Radio Astronomy Lab., Univ. of Calif., Berkeley, Calif.
J. Davies, Dept. of Geophysics and Astronomy, Univ. of British Columbia, Vancouver, B.C., Canada.
K. L. Day, Lunar and Planetary Lab., Univ. of Ariz., Tucson, Ariz.
J. Degewij, Lunar and Planetary Lab., Univ. of Ariz., Tucson, Ariz.
P. Demarque, Yale Univ. Obs., New Haven, Conn.
S. F. Dermott, Center for Radiophysics and Space Research, Cornell Univ., Ithaca, N.Y.
B. Donn, Goddard Space Flight Center, Greenbelt, Md.
M. J. Drake, Lunar and Planetary Lab., Univ. of Ariz., Tucson, Ariz.
E. Dwek, Dept. of Physics and Astronomy, Univ. College, London, England.
W. Dziembowski, N. Copernicus Astronomical Center, Warszawa, Poland.
P. Eberhardt, Physikalisches Inst., Univ. Bern, Bern, Switzerland.
B. G. Elmegreen, Center for Astrophysics, Harvard College Obs., Cambridge, Mass.
D. M. Elmegreen, Center for Astrophysics, Harvard College Obs., Cambridge, Mass.
N. J. Evans II, Dept. of Astronomy, Univ. of Texas, Austin, Tex.
S. Fair, Alltype, Tucson, Ariz.
S. W. Falk, Enrico Fermi Inst., Univ. of Chicago, Chicago, Ill.
S. Federman, Dept. of Physics, New York Univ., New York, N.Y.
*G. B. Field, Center for Astrophysics, Harvard College Obs., Cambridge, Mass.
B. Fitton, European Space Agency, Noordwijk, the Netherlands.
R. C. Fleck, Dept. of Physics and Astronomy, Univ. of Florida, Gainesville, Fla.
T. N. Gautier, Lunar and Planetary Lab., Univ. of Ariz., Tucson, Ariz.
N. Gehrels, Dept. of Physics, Calif. Inst. of Technology, Pasadena, Calif.
*T. Gehrels, Lunar and Planetary Lab., Univ. of Ariz., Tucson, Ariz.
F. C. Gillett, Kitt Peak Nat. Obs., Tucson, Ariz.
R. T. Giuli, Johnson Space Center, Houston, Tex.
P. Goldreich, Geology Div., Calif. Inst. of Technology, Pasadena, Calif.
J. Gradie, Lunar and Planetary Lab., Univ. of Ariz., Tucson, Ariz.
J. Graham, Obs. Interamericano de Cerro Tololo, La Serena, Chile.
*J. M. Greenberg, Sterrewacht, Leiden, the Netherlands.
R. Greenberg, Planetary Science Inst., Tucson, Ariz.
A. S. Grossman, Dept. of Physics, Iowa State Univ., Ames, Iowa.
L. Grossman, Dept. of Geophysical Sciences, Univ. of Chicago, Chicago, Ill.
B. Hapke, Dept. of Earth and Planetary Sciences, Univ. of Pittsburgh, Pittsburgh, Penn.
A. Harris, Jet Propulsion Lab., Pasadena, Calif.
*W. K. Hartmann, Planetary Science Inst., Tucson, Ariz.
G. H. Herbig, Lick Obs., Univ. of Calif., Santa Cruz, Calif.
E. Herbst, Dept. of Chemistry, College of William and Mary, Williamsburg, Va.
W. Herbst, Dept. of Terrestrial Magnetism, Carnegie Inst. of Wash., Washington, D.C.
J. M. Herndon, Dept. of Chemistry, Univ. of Calif., La Jolla, Calif.
R. A. Hesler, Univ. of Ariz. Press, Tucson, Ariz.
S. Holbrook, Tucson, Ariz.
D. Hollenbach, Radio Astronomy Lab., Univ. of Calif., Berkeley, Calif.
J. Horner, Northern Ariz. Univ., Flagstaff, Ariz.
V. Horner, Lunar and Planetary Lab., Univ. of Ariz., Tucson, Ariz.
R. Howell, Lunar and Planetary Lab., Univ. of Ariz., Tucson, Ariz.
W. B. Hubbard, Lunar and Planetary Lab., Univ. of Ariz., Tucson, Ariz.
D. R. Huffman, Dept. of Physics, Univ. of Ariz., Tucson, Ariz.
*D. M. Hunten, Lunar and Planetary Lab., Univ. of Ariz., Tucson, Ariz.
J. H. Hunter, Dept. of Astronomy, Univ. of South Florida, Tampa, Fla.
C. L. Imhoff, Steward Obs., Univ. of Ariz., Tucson, Ariz.
S. Isobe, Tokyo Astronomical Obs., Univ. of Tokyo, Mitaka, Tokyo, Japan.
J. R. Jokipii, Lunar and Planetary Lab., Univ. of Ariz., Tucson, Ariz.
F. C. Jones, Goddard Space Flight Center, Greenbelt, Md.
M. Jura, Dept. of Astronomy, Univ. of Calif., Los Angeles, Calif.
A. R. Kassander, Vice Pres. for Research, Univ. of Ariz., Tucson, Ariz.
W. M. Kaula, Dept. of Earth and Space Sciences, Univ. of Calif., Los Angeles, Calif.
J. F. Kerridge, Inst. of Geophys. and Planetary Phys., Univ. of Calif., Los Angeles, Calif.

R. I. Klein, Kitt Peak Nat. Obs., Tucson, Ariz.
R. F. Knacke, Dept. of Earth and Space Sci., State Univ. of New York, Stonybrook, N.Y.
M. Kobrick, Jet Propulsion Lab., Pasadena, Calif.
M. Kondo, Dept. of Earth Science and Astronomy, Univ. of Tokyo, Tokyo, Japan.
*Z. Kopal, Dept. of Astronomy, The University, Manchester, England.
B. K. Kothari, Dept. of Chemistry, Univ. of Calif., La Jolla, Calif.
L. V. Kuhi, Dept. of Astronomy, Univ. of Calif., Berkeley, Calif.
C. Lada, Center for Astrophysics, Harvard College Obs., Cambridge, Mass.
K. R. Lang, Dept. of Physics, Tufts University, Medford, Mass.
J. W. Larimer, Dept. of Geology and Center for Meteorite Studies, Ariz. State Univ., Tempe, Ariz.
R. B. Larson, Yale Univ. Obs., New Haven, Conn.
J. Lattimer, Dept. of Astronomy, Univ. of Illinois, Urbana, Ill.
M. J. Lebofsky, Lunar and Planetary Lab., Univ. of Ariz., Tucson, Ariz.
T. Lee, Enrico Fermi Inst., Univ. of Chicago, Chicago, Ill.
C. M. Leung, Nat. Radio Astronomy Obs., Charlottesville, Va.
*B. J. Levin, Astronomical Council, Academy of Sciences, Moscow, USSR.
E. H. Levy, Lunar and Planetary Lab., Univ. of Ariz., Tucson, Ariz.
R. S. Lewis, Dept. of Chemistry, Univ. of Chicago, Chicago, Ill.
F. J. Low, Lunar and Planetary Lab., Univ. of Ariz., Tucson, Ariz.
O. K. Manuel, Dept. of Chemistry, Univ. of Missouri, Rolla, Mo.
S. H. Margolis, Enrico Fermi Inst., Univ. of Chicago, Ill.
S. Marinus, Lunar and Planetary Lab., Univ. of Ariz., Tucson, Ariz.
M. A. Matthews, Tucson, Ariz.
M. S. Matthews, Lunar and Planetary Lab., Univ. of Ariz., Tucson, Ariz.
L. McCrumb, Max-Planck-Inst. für Chemie, Mainz, W. Germany.
R. J. McGonegal, Dept. of Astronomy, Univ. of Toronto, Toronto, Canada.
S. C. McMillan, Univ. of Ariz. Foundation, Tucson, Ariz.
W. H. Mendell, Johnson Space Center, Houston, Tex.
*D. A. Mendis, Dept. of Appl. Phys. and Inform. Sci., Univ. of Calif., La Jolla, Calif.
*L. Mestel, Astronomy Center, Univ. of Sussex, Brighton, Sussex, England.
G. E. Morfill, Max-Planck-Inst. für Kernphysik, Heidelberg, W. Germany.
D. Morrison, NASA Headquarters, Washington, D.C.
T. Ch. Mouschovias, Dept. of Astronomy, Univ. of Illinois, Urbana, Ill.
M. L. Norman, Lawrence Livermore Lab., Livermore, Calif.
T. Owen, Dept. of Earth and Space Sci., State Univ. of New York, Stonybrook, N.Y.
D. A. Papanastassiou, Div. of Geological and Planetary Sciences, Calif. Inst. of Technology, Pasadena, Calif.
E. N. Parker, Inst. for Nuclear Studies, Univ. of Chicago, Chicago, Ill.
S. J. Peale, Dept. of Physics, Univ. of Calif., Santa Barbara, Calif.
D. Phinney, Lockheed Electronics Co., Houston, Tex.
S. S. Prasad, Jet Propulsion Lab., Pasadena, Calif.
R. Pratap, Physical Research Lab., Ahmedabad, India.
J. Rahe, Remeis Obs. Bamberg, Univ. Erlangen, Nüremberg, W. Germany.
G. A. Ransford, Jet Propulsion Lab., Pasadena, Calif.
S. Rasool, NASA Headquarters, Washington, D.C.
*H. Reeves, Section d'Astrophysique, Centre d'Etudes Nucleaires de Saclay, Gif-sur-Yvette, France.
J.H. Reynolds, Dept. of Physics, Univ. of Calif., Berkeley, Calif.
*G. H. Rieke, Lunar and Planetary Lab., Univ. of Ariz., Tucson, Ariz.
N. G. Roman, NASA Headquarters, Washington, D.C.
S. Röser, Max-Planck-Inst. für Kernphysik, Heidelberg, W. Germany.
T. V. Ruzmaikina, O. Schmidt Inst. of Physics of the Earth, Moscow, USSR.
E. Rydgren, Dept. of Physics, Rensselaer Polytechnic Inst., Troy, N.Y;
*V. S. Safronov, O. Schmidt Inst. of Physics of the Earth, Moscow, USSR.
J. M. Scalo, Dept. of Astronomy, Univ. of Texas, Austin, Tex.
D. N. Schramm, Enrico Fermi Inst., Univ. of Chicago, Chicago, Ill.
R. D. Schwartz, Dept. of Physics, Univ. of Missouri, St. Louis, Mo.

M. Schwarzschild, Princeton Univ. Obs., Princeton, N.J.
E. H. Scott, Ames Research Center, Moffett Field, Calif.
R. C. Shirkey, Dept. of Physics, New Mexico State Univ., Las Cruces, New Mex.
J. Silk, Dept. of Astronomy, Univ. of Calif., Berkeley, Calif.
H. Smith, Lunar and Planetary Lab., Univ. of Ariz., Tucson, Ariz.
*C. P. Sonett, Lunar and Planetary Lab., Univ. of Ariz., Tucson, Ariz.
*L. Spitzer, Jr., Princeton Univ. Obs., Princeton, N.J.
G. Sprock, Alltype, Tucson, Ariz.
R. Stein, Dept. of Astronomy, Michigan State Univ., East Lansing, Mich.
J. R. Stephens, Dept. of Chemistry, Univ. of Calif., La Jolla, Calif.
N. Stevens, Alltype, Tucson, Ariz.
H. E. Suess, Dept. of Chemistry, Univ. of Calif., La Jolla, Calif.
S. Sutton, Alltype, Tucson, Ariz.
P. A. Strittmatter, Steward Obs., Univ. of Ariz., Tucson, Ariz.
E. F. Tedesco, Lunar and Planetary Lab., Univ. of Ariz., Tucson, Ariz.
P. Thaddeus, Goddard Space Flight Center, Inst. for Space Studies, New York, N.Y.
R. I. Thompson, Steward Obs., Univ. of Ariz., Tucson, Ariz.
A. G. G. M. Tielens, Sterrewacht, Leiden, the Netherlands.
J. E. Tohline, Lick Obs., Univ. of Calif., Santa Cruz, Calif.
M. Townsend, Univ. of Ariz. Press, Tucson, Ariz.
B. M. Trivedi, Dept. of Geophysics and Planetary Physics, Univ. of Calif., Los Angeles, Calif.
W. Tscharnuter, Max-Planck-Inst. für Physik and Astrophysik, Münche, W. Germany.
R. K. Ulrich, Dept. of Astronomy, Univ. of Calif., Los Angeles, Calif.
V. Vanýsek, Dept. of Astronomy and Astrophysics, Charles Univ., Prague, Czechoslovakia.
C. D. Vesely, Lunar and Planetary Lab., Univ. of Ariz., Tucson, Ariz.
K. Villere, Ames Research Center, Moffett Field, Calif.
H. Völk, Max-Planck-Inst. für Kernphysik, Heidelberg, W. Germany.
J. Young, Univ. of Ariz. Press, Tucson, Ariz.
*F. J. Vrba, Naval Obs., Flagstaff, Ariz.
J. F. Wacker, Lunar and Planetary Lab., Univ. of Ariz., Tucson, Ariz.
M. Walker, Lick Obs., Univ. of Calif., Santa Cruz, Calif.
R. M. Walker, McDonnell Center for the Sapce Sci., Washington Univ., St. Louis, Mo. Mo.
G. J. Wasserburg, Div. of Geological and Planetary Sciences, Calif. Inst. of Technology, Pasadena, Calif.
J. T. Wasson, Inst. of Geophysics, Univ. of Calif., Los Angeles, Calif.
W. D. Watson, Dept. of Physics and Astronomy, Univ. of Illinois, Urbana, Ill.
A. Waxman, Dept. of Astronomy and Astrophysics, Univ. of Chicago, Chicago, Ill.
A. B. Weaver, Exec. Vice President, Univ. of Ariz., Tucson, Ariz.
W. B. Weaver, Monterey Inst. for Research in Astronomy, Carmel Valley, Calif.
S. J. Weidenschilling, Dept. of Terrestrial Magnetism, Carnegie Inst. of Wash., Washington, D.C.
G. Welin, Astronomiska Obs., Uppsala, Sweden.
M. W. Werner, Dept. of Physics, Calif. Inst. of Technology, Pasadena, Calif.
G. W. Wetherill, Dept. of Terrestrial Magnetism, Carnegie Inst. of Wash., Washington, D.C.
*F. L. Whipple, Center for Astrophysics, Harvard College Obs., Cambridge, Mass.
M. P. Wijesinghe, Dept. of Physics, Univ. of Sri Lanka, Vidyodaya Campus, Sri Lanka.
*L. L. Wilkening, Lunar and Planetary Lab., Univ. of Ariz., Tucson, Ariz.
B. A. Wilking, Lunar and Planetary Lab., Univ. of Ariz., Tucson, Ariz.
*I. P. Williams, Dept. of Applied Mathematics, Queen Mary College, London, England.
L. A. Willson, Dept. of Physics, Iowa State Univ., Ames, Iowa.
K.-H. Winkler, Max-Planck-Inst. für Physik und Astrophysik, München, W. Germany.
H. A. Wootten, Dept. of Astronomy, Univ. of Texas, Austin, Tex.
H. W. Yorke, Max-Planck-Inst. für Physik und Astrophysik,München, W. Germany.

INDEX